SYSTEMATIC GEOGRAPHY

BRIAN KNAPP
Leighton Park School, Reading

Illustrated by Brian Knapp and Duncan McCrae

London
ALLEN & UNWIN
BOSTON SYDNEY

Allen & Unwin (Publishers) Ltd,
40 Museum Street, London WC1A 1LU, UK

Allen & Unwin (Publishers) Ltd,
Park Lane, Hemel Hempstead, Herts HP2 4TE, UK

Allen & Unwin Inc.,
8 Winchester Place, Mass. 01890, USA

George Allen & Unwin (Australia) Ltd,
8 Napier Street, North Sydney, NSW 2060, Australia

First published in 1986

British Library Cataloguing in Publication Data

Knapp, Brian
Systematic geography.
1. Geography—Text-books—1945-
I. Title
910 G128
ISBN 0-04-910080-7

Set in 9 on 10 point Palatino by
Mathematical Composition Setters Ltd, Salisbury, UK
and printed in Great Britain by
The Bath Press, Avon

PREFACE

Geographers have always been concerned to describe and explain the main physical features and the patterns of human activity of the world. They have tried to show how people influence, and are influenced by, the natural environment. But, whereas in the 19th century geographers could assume that many of the activities of people were largely in the hands of the environment, their counterparts today must take account of the environment's diminishing role. Indeed, people are now often a major landscaping force on a short timescale.

There are many disciplines studying the interaction of people and the world. Some, such as engineering, dealing with the stability of hillslopes or the firmness of natural materials for building foundations, are primarily concerned with the natural world and have helped to identify the basic rules by which nature operates. Similarly, social scientists have intensively studied the growth and structure of the world's population. But they have not primarily been concerned with the distribution of people in relation to environment. Yet the world does not function in independent compartments: each aspect interacts constantly with every other. The value of geographical studies is to focus on this interaction. And such a perspective will be ever more valuable in a world under increasing pressure for space and resources.

This book aims to go some way to providing an outline overview of people and their environment, first by looking at the main natural forces that have shaped the world, then examining the global pattern of human activity. Throughout the book there is a particular emphasis both on important areas in which people are often in conflict with the environment and on some of the possible solutions to their problems.

This book, painted with a broad brush, attempts to summarise some of the work of those who, with finely detailed and precise efforts, have built up a store of knowledge on which our present understanding rests. In translating detail into generalisation I hope I have not failed them too badly.

BJK Sonning Common, Oxon, May 1985

CONTENTS

LIST OF TABLES

Part A

Earth

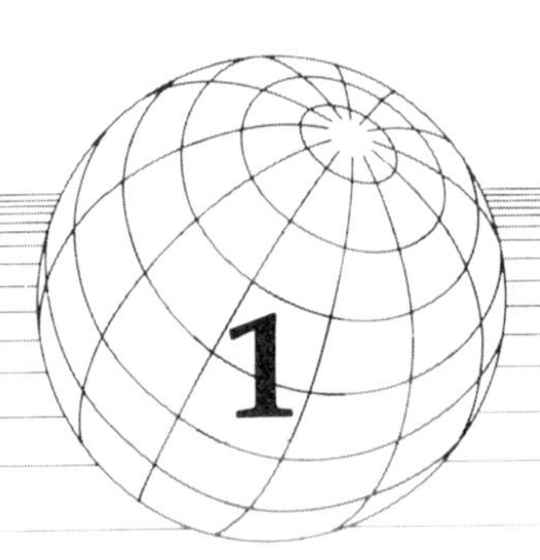

The geological foundation

1.1 Perspectives on the Earth

The Earth is simply too large to understand as a whole.

Our everyday geological experience is on a much smaller scale. We notice the rocks beneath our feet when we walk in mountains, we see rocks in cliffs while on holiday by the coast; we watch reports of spectacular volcanic eruptions or the devastating effects of earthquakes. However, the complexities of the Earth defy perspective and unification when seen on a local scale; to achieve these we need to move from the particular to the general, to look at the structure of the Earth and to see how this has produced the crustal rocks and caused the distribution of continents and oceans. Alternatively, we can see how the constant movements within the Earth influence the distribution of continents and oceans, mountains and plains, and how this distribution not only explains the global pattern of the mineral wealth on which industry depends, but also hints at the whereabouts of yet undiscovered resources.

The activities of mankind are influenced by major movements of parts of the Earth's crust: not only do these movements determine the dimensions of the land on which we live, the location of resources, the proportions of the land beneath each climatic belt and the shape of the land surface; they also determine the pattern of earthquakes and volcanoes that play such an important rôle in so many lives. Seen in the global context, each volcanic eruption is no more than a sparkler in a huge fireworks display, a perfectly normal and unremarkable event in the shaping of the Earth's surface. But the perspective gained by seeing how the Earth's features are interconnected provides the basis for learning how to live more safely with the mobile crust on which we depend for everything.

1.2 The Earth's interior

In an attempt to understand the structure of the Earth's interior, geologists have assembled a mass of evidence. From an examination of the surface rocks they have shown that there has been frequent uplift, subsidence, deformation and reworking of the Earth's rocks; these are the result of interactions between the erosive processes induced by the atmosphere and forces acting deep within the planet. But most evidence concerning the Earth's structure has been obtained indirectly from studies of seismic (earthquake) waves, of the varying patterns of the Earth's magnetism and gravitational field, and of materials ejected by volcanoes.

Seismic waves penetrating the Earth show it to have a solid **inner core** surrounded by a liquid **outer core**. Both parts of the core are probably alloys of nickel and iron; both are hotter than 4000 °C (Fig. 1.1). As the Earth rotates, the liquid outer core moves also; these movements are believed to be the source of the planet's magnetic field.

Outwards from the core is a thick shell called the **mantle.** Like the core, the mantle appears to consist of a solid inner shell surrounded by a thin outer shell, the outermost part of which is weak. The outer mantle probably consists mainly of solid rock with but a few per cent of liquid, but this liquid is quite sufficient to ensure that the mantle deforms under stress.

In comparison with the mantle and core, the **crust** is merely a 'paper-thin' surface layer, varying in thickness between 5 km under the oceans to over 100 km under major mountain chains. The oceanic crust is made primarily of **basaltic** materials, rich in the same minerals as the mantle and having a specific

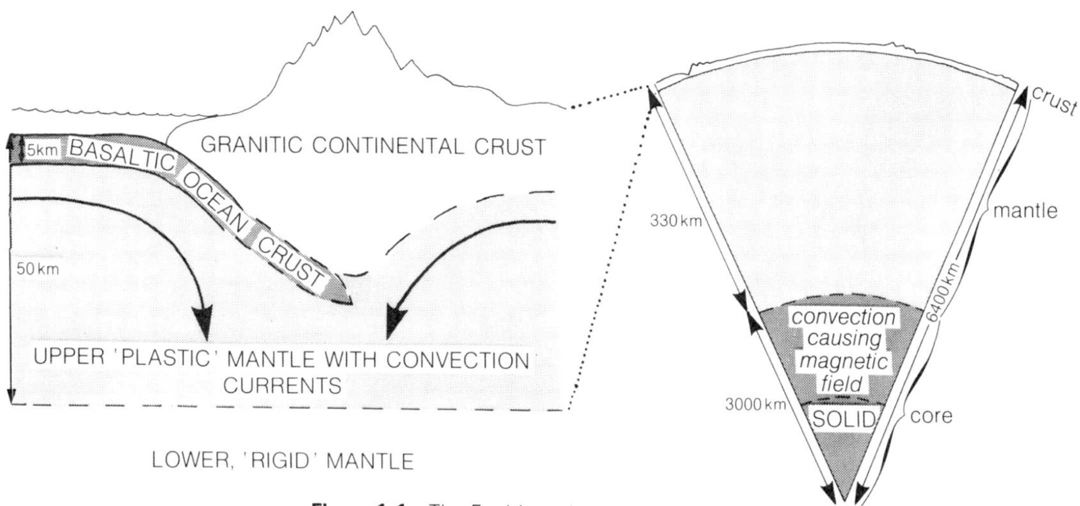

Figure 1.1 The Earth's major structural divisions.

gravity of 3.3; the continental crust is mainly composed of **granitic** materials, rich in silica and aluminium, and having a specific gravity of 2.8. Although basaltic rocks exist below the continents, continental rocks are absent from the oceans, suggesting that the granitic materials are in some way derived from the basalts.

Part of the upper mantle (from 75 km to 175 km below the surface) is weak and can deform under pressure: this helps to explain why some of the surface stands higher than the rest, giving continents and oceans. The relationship between low-density granitic crust and the underlying higher-density basaltic mantle is comparable to a block of wood floating in water. When applied to the Earth's crust, this principle of flotation or buoyancy is called **isostacy.** Thus the higher elevations of mountains are counterbalanced by greatly thicker crust pushing deep into the upper mantle (Fig. 1.2). These deep wedges of compensating granite crust are often referred to as **mountain roots**. Furthermore, the principle of isostacy can be used to show that, if the mountains are made smaller by erosion, the unloading involved – which would leave the crust out of isostatic equilibrium – is balanced by a compensating rise of the root. Thus mountain ranges are much more enduring features of the Earth's surface than would be the case if only the bulk currently visible had to be eroded.

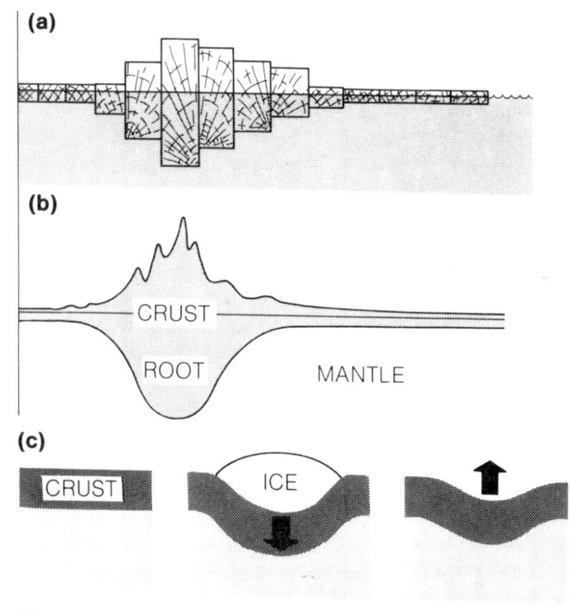

Figure 1.2 The similarity of blocks of wood floating in water (a) to the crust floating in the mantle (b) provides the reasoning for mountain roots. Mountain root materials are detected by gravity surveys. (c) Isostatic recovery operates at many scales and even the added weight of ice in the Ice Age was sufficient to depress the continental crust locally. However, isostatic recovery of a continent is not a steady process, because the crust is floating in a material that does not readily deform. Isostatic recovery occurs episodically, with a period of rapid rise being separated by a long period of stability. This is because the crust is rigid, has considerable strength and is able to withstand a substantial degree of local unloading before it adjusts. Such irregular movements are important in the formation of some river terraces and raised beaches (p. 00).

1.3 Ocean-floor spreading

Within the past forty years, careful surveying of the ocean floor has revealed a mountain system greater than anything on land (Fig. 1.3). Dominating the

Figure 1.3 The distribution of the Earth's major landforms.

oceans, the mountain chains – known as **mid-ocean ridges** – stretch some 40 000 km and link the oceans together. In the Atlantic the ridge is placed almost centrally, rising from **abyssal plains**, through a zone of rugged foothills, to an **axial rift** valley whose faulted boundaries are partly covered with the outpourings of recent underwater volcanoes. The ridge rises to the surface at Iceland, where the rift structure can be examined in greater detail (Fig. 1.4).

Iceland's landscape is dominated by chains of active volcanoes. Some are violent and produce steep-sided cones of **ash**, but the majority erupt relatively quietly from long fissures, sending sheets of highly mobile basaltic **lava** to cover the nearby rocks. Dating techniques have shown that the rocks are progressively older with increasing distance from the zone of active volcanicity. This evidence suggests that the ocean floor is **spreading** away from the ridge.

The mid-Atlantic ridge can be traced around southern Africa, where it splits into two: one arm reaches up to the Indian Ocean as far as the Red Sea; the other curves around Antarctica before heading into the Pacific Ocean. Where the Indian Ocean ridge reaches the Red Sea there is a consistent pattern of rifting, stretching from the Sea of Galilee in Israel, through the Red Sea and into the East African rift valley (Fig. 1.5). This rift is lined with such famous volcanoes as Mt Kilimanjaro, Mt Kenya and the Ruwenzoris. The African rifting pattern provides further evidence that oceanic ridges are associated with regions where the crust is moving apart. Indeed,

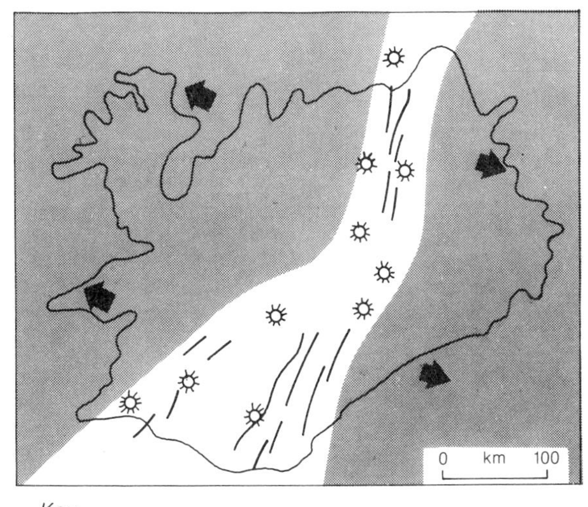

Figure 1.4 The main zones of volcanic activity in Iceland.

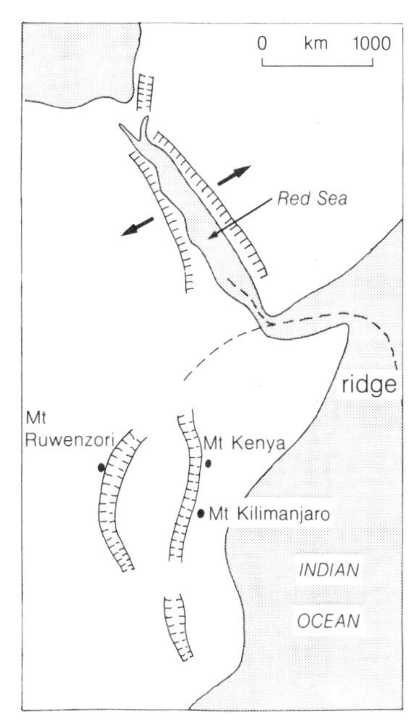

Figure 1.5 Major fault structures of East Africa and the Middle East.

ocean ridges appear to be regions where material from the mantle is rising to the Earth's surface as **dyke intrusions** and spilling out over the ocean floor as basaltic lava flows. With each new intrusion the ridge material is pushed apart and the ocean floor spreads a little further (Fig. 1.6).

Because the newly created crust is thin, it loses heat rapidly. In turn this causes molten mantle rock – **magma** – to be cooled by contact with the undersurface of the crust, which thereby begins to thicken by **underplating**, much as surface ice thickens downwards in a still pond. At first the underplating occurs quickly because heat loss through the thin crust is rapid; but as the crust thickens and the rate of heat loss decreases, underplating becomes less effective and a more or less stable thickness is attained. The thicker crust is more difficult to disrupt and it helps restrict the zone of volcanic activity to a narrow band astride the ridge.

Spreading occurs irregularly in time and unevenly along the length of the mid-ocean ridges. For example, measurements indicate that the rate of ocean-floor spreading varies from an average of 1 cm/yr to 7 cm/yr in the Pacific. Furthermore, spreading occurs from several **spreading centres** which act as foci for movement (Fig. 1.7a). Thus the ocean floor spreads in an arc, with the necessary curved motions of the ocean floor being accommodated by pieces of crust slipping past each other along **transform(tear) faults** (Fig. 1.7b). Transform faults are also major foci of

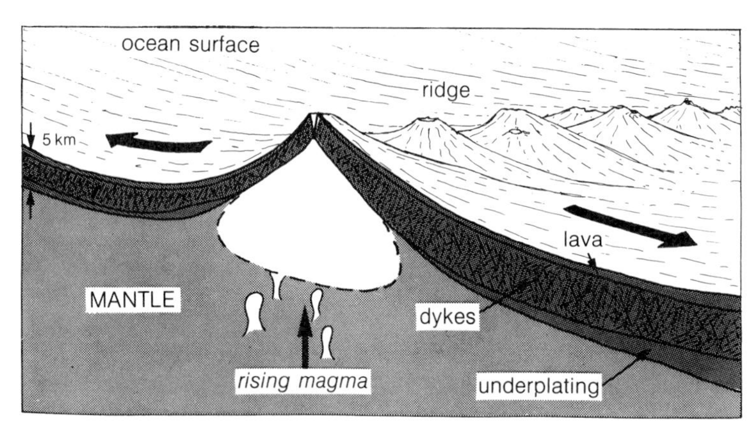

Figure 1.6 Major mid-ocean structural features.

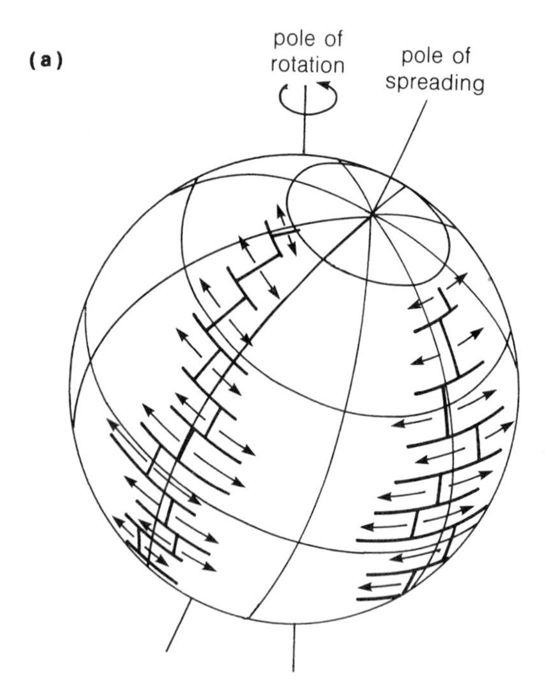

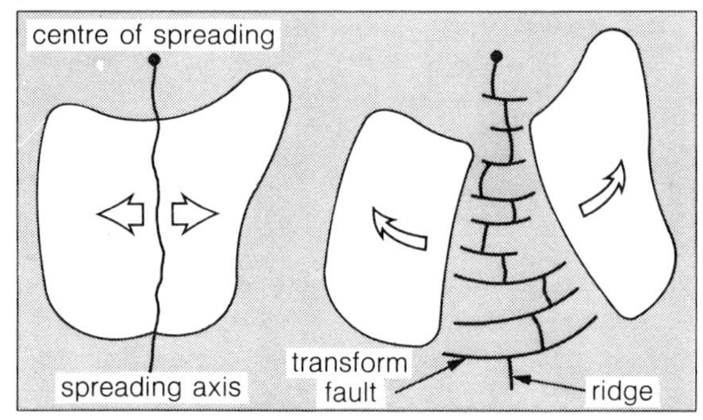

Figure 1.7 (a) A geometrical relation between ridge axis and fracture zone becomes evident if one conceives of lines of latitude and longitude drawn about a 'pole of spreading' rather than the pole of rotation. In each ocean, the fracture zones are perpendicular to the spreading axis, and the rate of spreading (arrows) varies directly with distance from the Equator. (b) If an ocean opens by spreading about a centre of rotation, the mid-ocean ridge splits into segments, each segment bounded by long curving transform faults.

earthquake activity. The global pattern of transform faults shows very clearly on Figure 1.3 and the foci of spreading can be located from the distribution of the faults.

1.4 Continental margins

The oceans contain the highest mountains and also the deepest trenches in the world. Most of the **ocean trenches** are narrow, with an arc-shaped plan, and nearly all lie parallel to strings of volcano-dominated islands (Fig. 1.8). Furthermore, most of the **island arcs** and trenches lie in a ring surrounding the Pacific Ocean, their active volcanoes earning the name of 'Pacific ring of fire'.

The pattern of earthquake foci in this region indicates that the ocean crust is actively dipping down beneath the continental margins (Fig. 1.8b). The trench indicates the position of the descending basaltic ocean crust, any low-density sediments that have accumulated on its surface being scuffed up against the edge of the continent. As the crust descends, it melts and becomes reabsorbed within the mantle. Some of this melting basaltic crust is believed to provide the material for the volcanoes that push through the crust and create the bulk of the island arcs. However, this material is no longer pure basalt, for only the more volatile part of the basaltic crust is able to migrate upwards through fractures in the crust. This 'distilled' fraction contains a lower

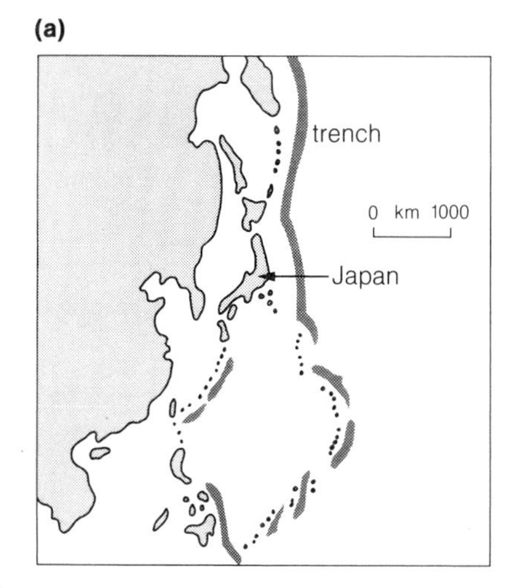

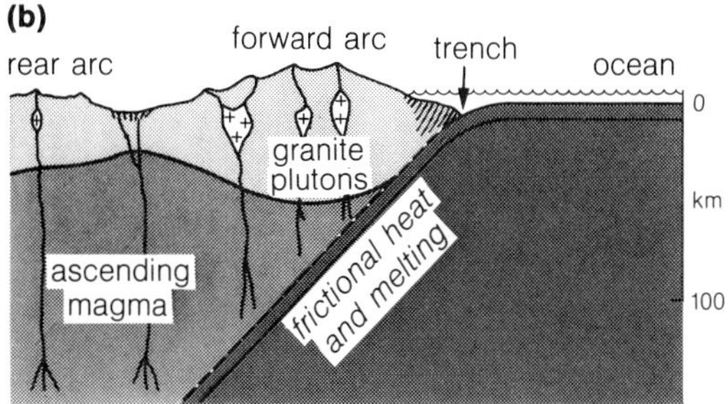

Figure 1.8 (a) Part of the western Pacific Ocean, showing the parallel arcs and trenches. (b) A model of the structure of the western Pacific Ocean emphasising the role of subduction.

proportion of iron and magnesium than the pure basalt; it forms a range of materials from **andesite**, through **dacite**, to **rhyolite**. These are the materials from which the **central vent volcanoes** are built. The movement of the ocean crust back into the mantle is called **subduction** and the region in which it occurs is the **subduction zone.**

1.5 Plate tectonics

As shown above mid-ocean ridges can be interpreted as places where new ocean crust is created and from which it spreads away; ocean marginal trenches are places where the crust is reabsorbed into the mantle. Between them lie the vast **abyssal plains**, whose apparently quiescent floors are continually moving – gigantic conveyor belts transporting material from ridge to trench.

A map of the distribution of ridges, trenches, transform faults, volcanoes and earthquakes shows that these features cluster to form lines (Fig. 1.9). Between the lines there are large parts of the Earth's surface which are quiet and stable, lacking any major crustal activity. It therefore seems probable that the Earth's crust comprises a number of rigid and stable **plates**. Most of these plates contain both continent and ocean. Their margins are formed by:

(a) spreading centres;
(b) subduction zones;
(c) transform faults.

Thus the North American plate, for example, is bounded to the east by the mid-Atlantic spreading centre; to the south by a subduction zone complex

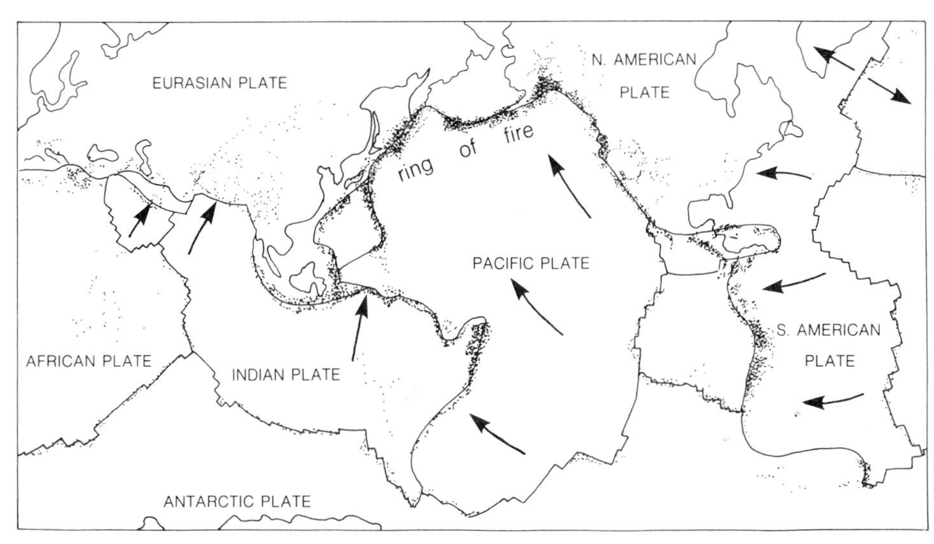

Figure 1.9 The distribution of volcanoes and earthquakes suggests that the Earth's crust is divided into a number of rigid plates.

running through the Caribbean Sea; and along the western boundary by a further subduction zone. This last feature includes a number of transform faults, the largest of which is the San Andreas fault, which crosses California.

All the crustal plates can be described in a similar way, and the study of their movements and formation constitutes the science of **plate tectonics** (from the Greek word *tekton*, meaning 'build').

1.6 Continental drift

The global 'jigsaw' of crustal plates, together with the pattern of their building and destruction, provides a scientific basis for studying the distribution of the Earth's major surface features. But it is the *movement* of the continents over the face of the Earth, rather than their construction and destruction, that has attracted most attention, ever since the first maps of the Atlantic Ocean showed a remarkable 'fit' between the opposing coasts of Africa and South America (Fig. 1.10). In this restricted sense the movement of continents is referred to as **continental drift**, but we now understand that it is the *plates*, and not simply the continents, that are the fundamental units.

A wealth of substantial evidence has now accumulated that corroborates hypothesis based on the initial observation of the 'jig-saw' fit. For example, it has been shown that geological structures on continents now thousands of kilometres apart contain matching patterns (Fig. 1.11). Futhermore, the distribution of climatically influenced rock formations (such as coal, which requires that there was once a tropical to subtropical rainforest environment) and glacial deposits shows a pattern that makes little sense until the continents are regrouped (Fig. 1.12). More recently, the discovery of fossil magnetism, locked within the iron compounds of many rocks since their formation, has revealed the earlier positions of the continents as they moved over the surface of the globe.

It was from evidence of geological matching that Alfred Wegener, in 1912, first proposed a coherent theory of continental drift. He suggested that some 200 million years ago the present continents were part of one large 'supercontinent', which he called **Pangea**. Occupying a severe indentation in this continent was a sea (which he named **Tethys**) that was the ancestral Mediterranean (Fig. 1.13).

Figure 1.10 Although there is a superficial resemblance between the coasts of South America and Africa, there is a much better fit between the true continental margins at the limits of the continental shelves.

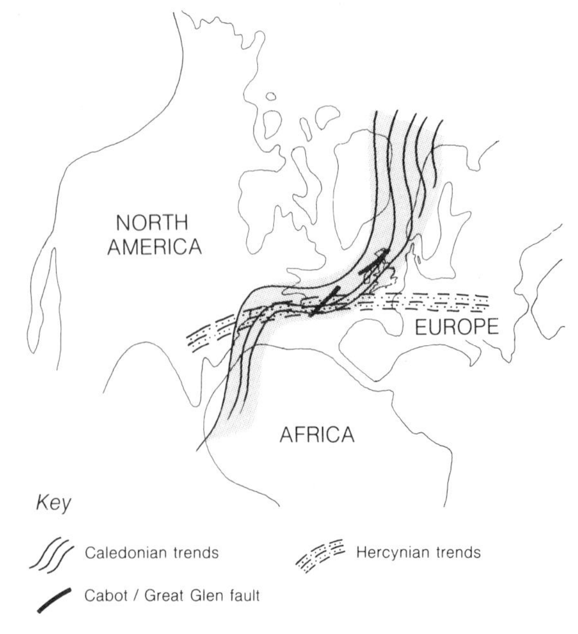

Figure 1.11 The correspondence between similar structures in North America and Europe suggests they were once joined.

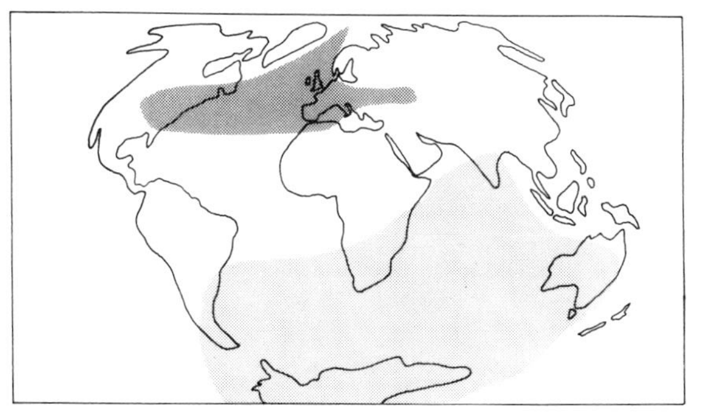

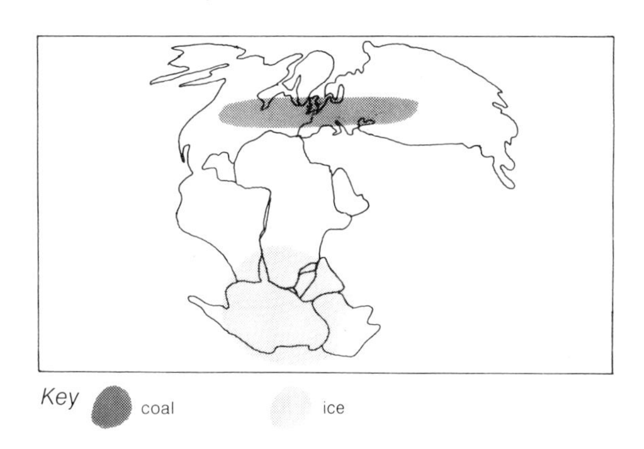

Figure 1.12 The bizarre distribution of contemporary coal or glacial deposits as seen on a map of the present land masses (left) makes sense only if the continents have both moved and were once joined together (right).

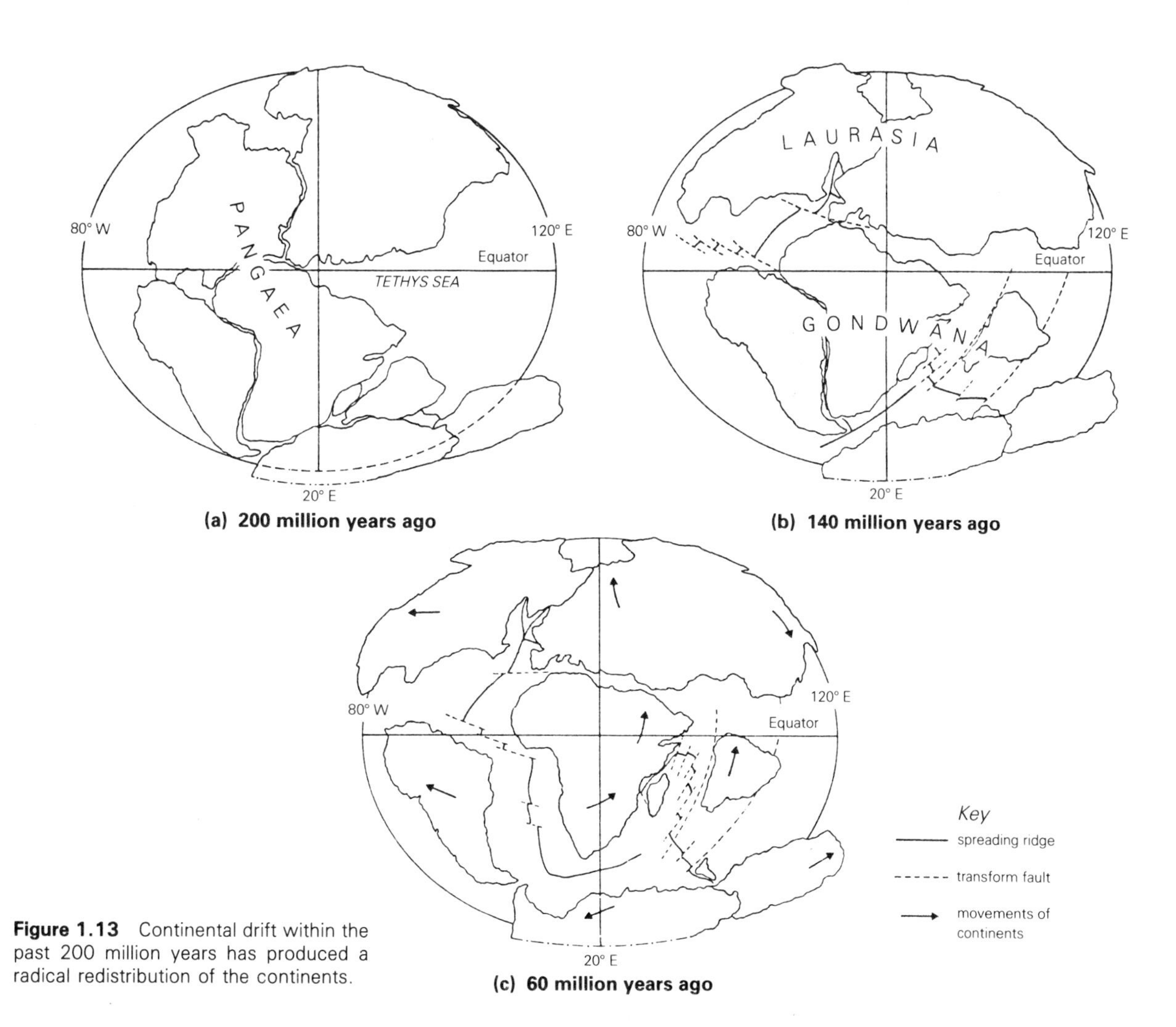

Figure 1.13 Continental drift within the past 200 million years has produced a radical redistribution of the continents.

Pangea was centred in the Southern Hemisphere, but from about 200 million years ago the Atlantic spreading-centre developed, and what is now North America began to drift away from what became North Africa and South America (while remaining attached to what became Europe). At the same time a subduction zone developed along the margins of Tethys as India drifted northwards, while a massive transform fault developed between Australia and India. Some 140 million years ago the Atlantic spreading centre was sufficiently well developed to split North America from Europe and South America from Africa. As the Americas drifted eastwards, a subduction zone developed along their Pacific margin, and mountain chains began to form. By 60 million years ago India was beginning to ride into Asia. The Tethyan Sea was gradually subducted completely with the surface sediments being crushed up into the Himalayas, Kun Lun Shan and the intervening Tibetan plateau.

The plates containing Australia and Antarctica were the last to separate, with Australia finally drifting northwards into the Pacific ocean.

1.7 Continent building

Mountain chains – elongated regions of elevated rocks having continental dimensions – contain, and are the primary sources of, all three major types of rock (Fig. 1.14). A typical mountain chain marginal to a continent would contain a central core of igneous rocks formed of large, deep-seated granite bodies called **batholiths**. Next to this core would be more mobile igneous rocks that have penetrated the overlying rock cover to form **dykes** and **sills**. Surrounding the core lies a zone of metamorphic rocks whose minerals indicate that much of the metamorphism occurred both when the rocks were at great depth, and due to proximity to the hot igneous rock core.

ROCKS OF THE EARTH'S SURFACE

There are three basic rock types:

(a) **igneous rocks**, derived directly by crystallisation of magma from the deep crust or mantle (e.g. granite, basalt);
(b) **sedimentary rocks**, derived from the erosion of igneous, metamorphic or pre-existing sedimentary rocks exposed to atmospheric agencies and formed in sheets (**strata**), mainly deposited on sea floors (e.g. sandstone, shale, limestone); and
(c) **metamorphic rocks**, that are baked and/or crushed rocks of any origin, many of which have been formed by the partial remelting of the original rock while deep below the Earth's surface – such rocks include slate, gneisss and schist.

Some of the more important rocks are:

Igneous rocks

Andesite is a fine-grained volcanic rock or lava, characteristically medium-dark in colour; it contains 54–62 per cent silica with moderate amounts of iron and magnesium.

Basalt is a fine-grained volcanic rock or lava, characteristically dark in colour; it contains 45–54 per cent silica and is rich in iron and magnesium.

Dacite is a fine-grained volcanic rock or lava, characteristically light in colour and containing 62–69 per cent silica together with moderate amounts of sodium and potassium.

Granite is a light-coloured, coarse-grained, deep-seated rock consisting of readily visible crystals of **quartz**, **feldspar** and **mica**; it contains 69 per cent or more of silica.

Rhyolite is a fine-grained volcanic rock or lava, characteristically light in colour; it contains 69 per cent or more of silica and is rich in sodium and potassium.

Sedimentary rocks

Limestone is any sedimentary rock composed chiefly of calcium or magnesium carbonate; the term includes **chalk**, a fine-grained pure white material mainly made from skeletons of plankton.

Sandstone is a sedimentary rock composed chiefly of sand grains, which are often cemented by calcium carbonate or iron oxide.

Shale is a sedimentary rock composed chiefly of clay-sized material; it frequently shows a tendency to break up into thin plates.

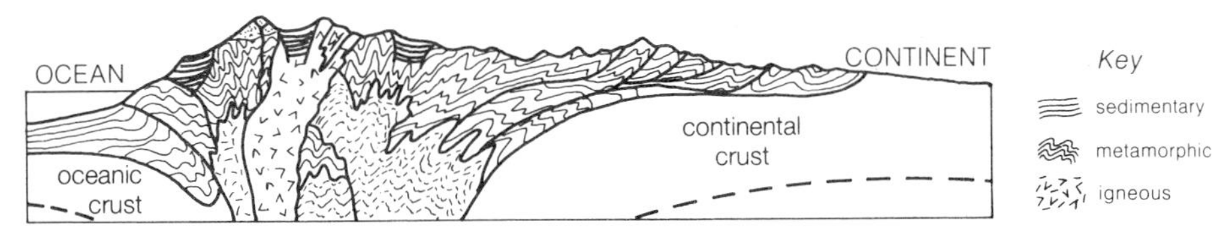

Figure 1.14 Cross section through a typical mountain chain marginal to a continent.

Above the metamorphic rocks, beds of **ash** and **lava** are common; these indicate that, during mountain building, magma reached the surface and volcanoes formed. Many ocean-margin mountain chains still contain active volcanoes. Finally, away from the centre of the mountain system lie sheets of sedimentary rocks. Some of these were formed by the erosion of former mountain chains and have been folded and faulted by mountain-building stresses; others are the erosion products of the present mountains and have been spread out in sheets of unwarped material.

A mountain chain represents tens of kilometres' thickness of material, much of which is of sedimentary origin. This means that a mountain chain must have been preceded by a long period of sedimentation in a trough-like depression. These large depressions are called **geosynclines**; their oceanward sides slope steeply down to contain a deep trench. The sedimentary rocks in a geosyncline (and thus in the mountain chain that succeeds it) are often significantly different from those deposited in shallow seas. As well as sandstones, limestones and shales, there are great thicknesses of poorly sorted material called **turbidites**. The turbidites are a mixture of sands and clays, and they show where graded shallow-water sediments have slumped off the continental shelf and into the deeper ocean-margin region of the geosyncline without having had time to become graded again. Turbidites are commonly interbedded with **mudstones**, which derive from clay that has accumulated on the seabed during the quieter conditions between successive slumpings of material from the shelf. Interbedded with both mudstones and turbidites are sheets of lava and ash, showing that as the geosyncline developed, it also stretched the crust, and thus allowed fractures to develop through which magma rose to the surface as deep volcanoes.

The transformation of a geosyncline into a mountain chain – **orogeny** – would not have been a rapid spectacular event, but a slow, progressive change involving many periods when mountains began to be uplifted, but were eroded away.

Early stages of an orogeny involve the progressive downwarping of a large section of crust (the geosyncline) and the accumulation of sediments within it (Fig. 1.15a). The sea bed off the eastern coast of North America disguises a geosyncline being filled

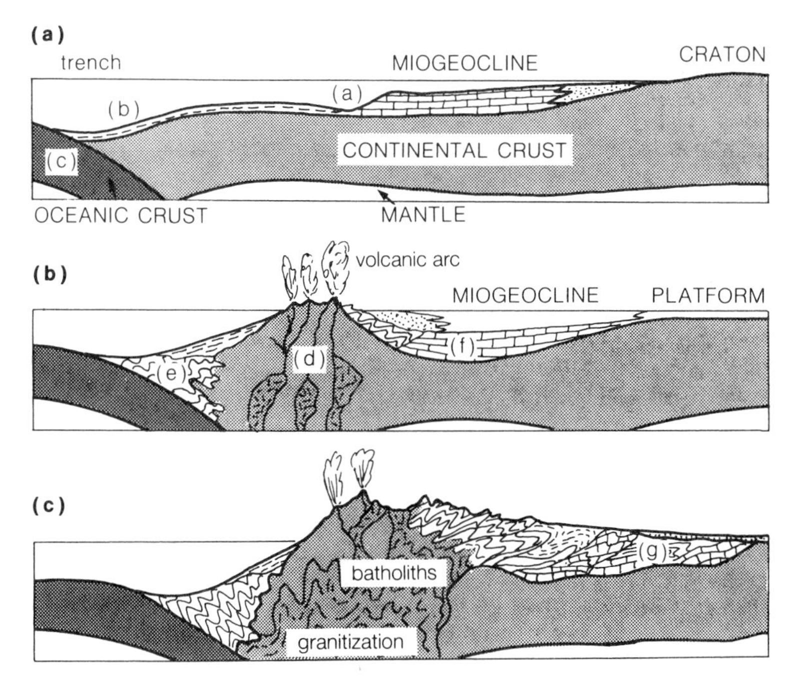

Figure 1.15 A model for the development of a mountain chain (see text for explanation).

with material eroded from the Appalachians (Fig. 1.16). However, in all cases, the formation of a geosyncline reveals active crustal downwarping such as might be produced at a subduction zone of a plate margin. Near the sagging continental margin there are mostly shallow-water sediments such as sandstone and limestone; farther out in the deeper part of the geosyncline turbidites predominate. In addition, material accumulating on the ocean side of the geosyncline is continuously scraped off and added to the geosyncline as the ocean floor subducts. Eventually subduction becomes more active: molten material rises from the region of crustal melting and an island arc forms (Fig. 1.15b). Volcanic materials are now added to the sequence of geosynclinal materials. As subduction proceeds, sediments are 'rucked up' against the arc, causing intense folding and faulting. Further melting of the ocean crust produces large quantities of magma which begin to bulk up the lower levels of the geosyncline, simultaneously causing widespread metamorphism. All the time, underplating occurs in the same manner as beneath the ocean crust, gradually thickening and stabilising the continental margin.

Eventually the descending ocean plate pushes the geosynclinal rocks onto the margin of the continent, causing landward buckling of sediments and producing a still thicker crust (Fig. 1.15c). The crust then rises into mountains thereby restoring isostatic equilibrium. Thus it is at the culmination of the orogenic cycle that the mountain chain is elevated to its full height.

Successive orogenic cycles add new mountain chains; continents grow by **marginal accretion**. The oldest parts of continents are therefore the interiors. Stable central regions of continents, originally mountainous but long since eroded (to land of low relief, with granite and metamorphic rock cores laid bare), are called **cratons** or **shields.** Such cratons now comprise the majority of the continental crust.

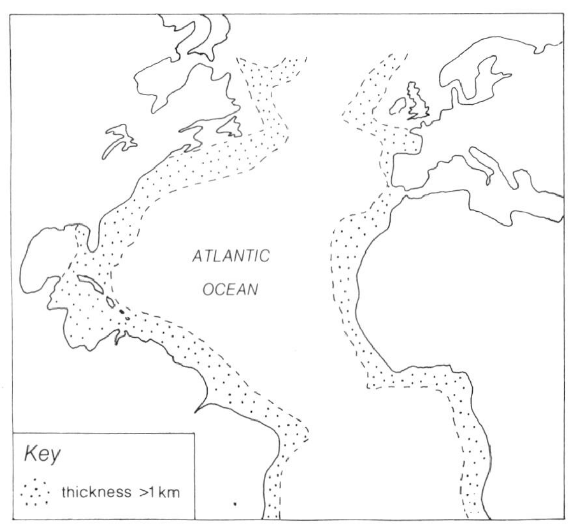

Figure 1.16 Great thicknesses of sediment are forming off the coast of North America, within the Caribbean and off the coast of Africa. These are indications of the location of geosynclines.

1.8 Convection

The mechanism for plate movements is found within the weakest zone in the upper mantle. Measurements of heat flow show that in spreading centres heat flows vigorously upwards, whereas in subduction zones poor heat flows only poorly. In effect, the plates seem to be driven by convection currents within the upper mantle. Material rises at spreading centres and sinks at subduction zones, with lateral movement between. However, the exact pattern of convection that could account for the complex movement of the continents has not yet been identified.

Tectonic processes and man

2.1 The importance of tectonics

Chapter 1 outlines the way the Earth's major surface landforms were created by the processes of plate tectonics. These processes have also been responsible for the distribution of the world's major energy and raw-material resources; as such they are of vital interest to man (see Chs. 38 & 40). In the context of the Earth's history, the tremors from a single earthquake or the eruption of a single volcano are but tiny events; yet these are the major manifestations of crustal instability and it is these that have by far the greatest direct impact on people's lives.

2.2 Earthquakes

Earthquakes occur when stresses in the Earth's crust build to such a magnitude that the rocks break apart or scrape past each other. When a break occurs, stored energy is unlocked and a train of powerful shock waves is released from the point of fracture (the **focus** or **hypocentre**; see Fig. 2.1). Waves reaching the surface move outwards in a pattern similar to that created when a stone is dopped into a pond. Because enormous quantities of energy are required to set the ground in motion, the surface waves are rapidly smoothed and damage is confined

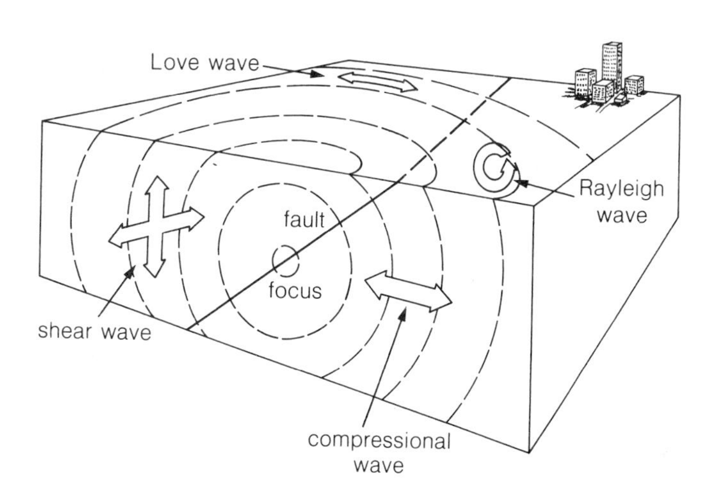

Figure 2.1 Schematic illustration of the directions of vibration caused by body and surface seismic waves generated during an earthquake. When a fault ruptures, seismic waves are propagated in all directions, causing the ground to vibrate at frequencies ranging from about 0.1 to 30 Hertz. Buildings vibrate as a consequence of the ground shaking; damage takes place if the building cannot withstand these vibrations. Compressional and shear waves cause high-frequency (greater than 1 Hertz) vibrations which are more efficient than low-frequency waves in causing low buildings to vibrate. Rayleigh and Love waves mainly cause low-frequency vibrations which are more efficient than high-frequency waves in causing tall buildings to vibrate. Because amplitudes of low-frequency vibrations decay less rapidly than high-frequency vibrations as distance from the fault increases, tall buildings located at relatively great distances (90 km) from a fault are sometimes damaged.

EARTHQUAKE MAGNITUDES

The most commonly used scale of earthquake severity was devised by C. F. Richter from measurements of the amplitude of the shock waves. On this scale, magnitude 2.0 can just be felt by people without the help of instruments. An increase of one unit (1.0) represents a tenfold increase in the wave amplitude.

Structural damage to buildings is usually caused by tremors with a severity of over 4.0. The largest magnitudes ever recorded were 9.0, the waves having energy ten million times greater than waves of magnitude 2.0. The San Francisco earthquake of 1906 had a magnitude of 8.3; the San Fernando earthquake of 1971 was of magnitude 6.5 (Fig. 2.1).

to within a few tens of kilometres of the focus.

Earthquakes mostly occur along active plate boundaries. Although they are not unknown even in Britain, which is far from a plate boundary, they are much more common in places such as California that lie on or near a boundary. The subduction zone that underlies California has thrown up ranges of mountains and has thereby profoundly influenced the landscape (Fig. 2.2). Landforms are either block-faulted mountains, such as the Coast Ranges, or basins, such as that occupied by the Salton Sea. These block movements are still continuing and, as one block periodically jerks past another, earthquakes are generated. The frequency of earthquakes makes California one of the world's most dangerous places.

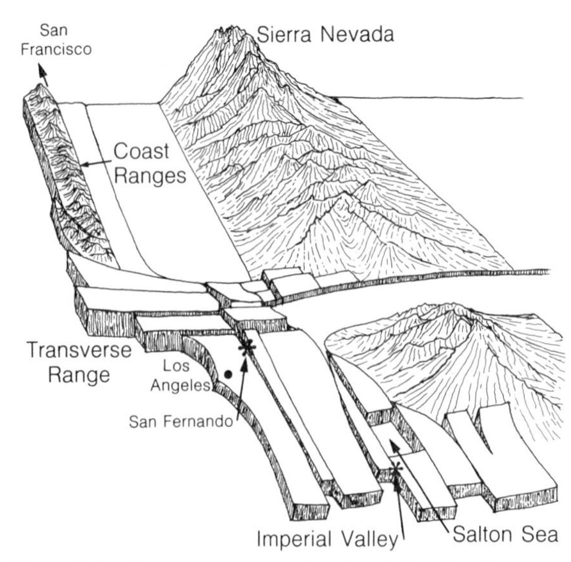

Figure 2.2 The movements of blocks of the Earth's crust have primarily been in a north–west orientation except where they encounter the deep roots of the Sierra Nevada. At this point the faults are deflected to the west giving transverse ranges. Notice how the Salton Sea is in a downfaulted block.

Although the Californian fault pattern is very complex, one particular fault stands out: the San Andreas fault system. The fault moves on average at about 6 cm/yr, although the actual pattern of movements is very irregular (Fig. 2.3) – there may be two earthquakes within a few months, then a quiescent period of many years. Each earthquake relieves stresses built up since the previous 'quake; in general, the longer the period between earthquakes, the larger the next earthquake will be.

In Southern California faulted blocks regularly slip and generate earthquakes (Fig. 2.2). For example, on 9 February 1971, a severe earthquake (Richter scale magnitude 6.6) occurred in the San Fernando valley just NE of Los Angeles. The point of original rupture was at a depth of about 13 km, but the rupture then propagated southwards and upwards, reaching the ground and causing noticeable surface movement for 15 km. On average there was a lateral slip of 2.4 m and a vertical slip of 0.9 m. Although not very severe, it was sufficient to bring down bridges over a freeway (Fig 2.4) and buckle the pillars supporting a hospital. Over the following two days, continuing readjustment of neighbouring parts of the fault caused 17 aftershocks. On 15 October 1979, a magnitude 6.5 earthquake with its focus on the US/Mexican border

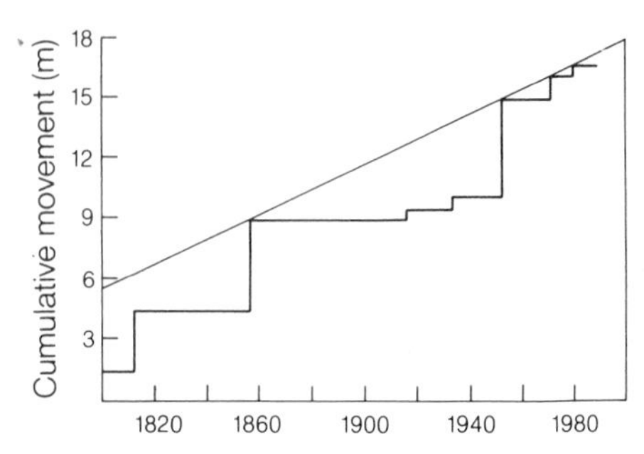

Figure 2.3 On average, the San Andreas Fault slides northwards at 6 cm/yr. However, the process is irregular, and the magnitude of earthquakes very variable.

Figure 2.4 The magnitude 6.6 earthquake in the San Fernando valley caused bridges over the main state freeway to collapse.

caused a surface rupture 31 km long, with an average lateral displacement of 0.8 m. This caused $21 million damage even though the Imperial Valley in which it occurred is an agricultural rather than an urban area. By contrast, northern California is a region where earthquakes are rare. On 18 April 1906, an earthquake of magnitude 8.3 cost 700 lives and resulted in extensive damage to the whole of San Francisco (Fig. 2.5). An earthquake of a similar magnitude today might cost 5000 lives, 700 000 injuries and $24 *billion* damage from ground shaking alone. The San Andreas fault has not moved in the San Francisco region since 1906, which suggests that when the next earthquake occurs, the effects are likely to be horrendous.

Earthquake hazard

Earthquakes often cause buildings to collapse under intense vibration, but the main damage is usually caused by their side-effects (Fig. 2.6). Water pipes are ruptured, preventing fire fighting; landslides occur; and the saturated alluvium on which buildings have been constructed is liquefied. Because these side-effects are so important, earthquake protection measures should include designing special structures that will withstand vibration and restricting construction in areas prone to liquefaction or landsliding. In San Francisco, despite the dangers, none of these precautions has so far been effectively implemented.

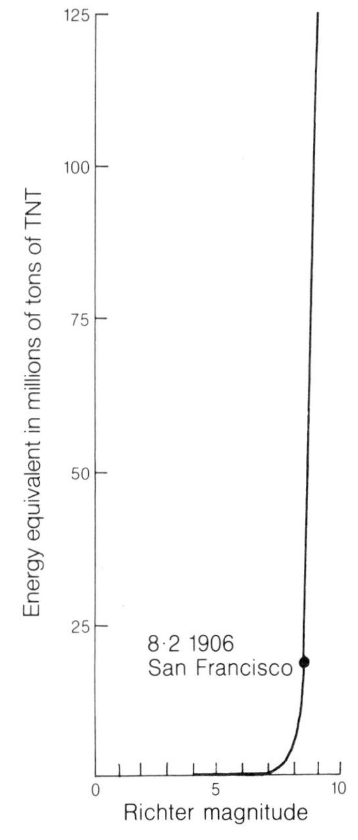

Figure 2.5

It is possible to identify the main hazards resulting from earthquakes as:

(a) direct damage to buildings, especially hospitals that would need to tend to the injured from a wide area;
(b) landslides and rockfalls triggered by the vibration;
(c) liquefaction of saturated ground, causing it to lose strength and allow buildings to founder;
(d) rupture of services, liability of fire and shortage of drinking water;
(e) damage to dams and canals with consequent risk of flooding;
(f) collapse of bridges and disruption of transport services;
(g) damage to nuclear power stations, especially reaction vessels.

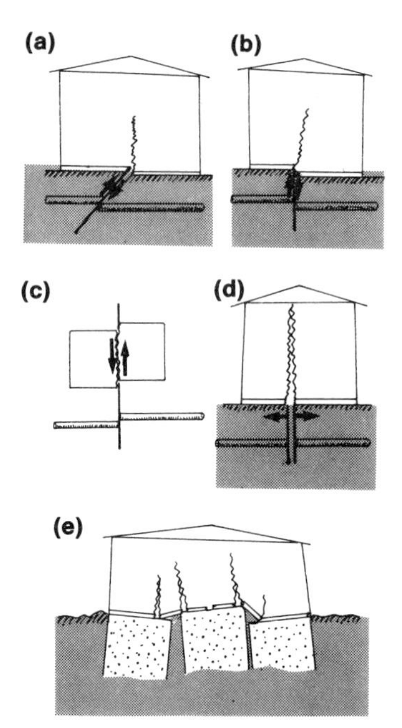

Figure 2.6 Basic types of ground ruptures and associated pipeline and building damage in the northern San Fernando Valley.

Man-made earthquakes

Earthquakes may seem totally beyond man's control, for they are mostly generated tens of kilometres below the Earth's surface. However, although they cannot be prevented, the worst of their effects can be curbed.

Man-made earthquakes have occurred when some of the world's largest reservoirs were filled, or following large-scale deep-pumping operations designed either to dispose of toxic waste or to flush oil and gas to the surface. Increased earthquake activity as a result of human activity was first observed during the filling of Lake Mead, a vast body of water held up behind the Hoover Dam on the Colorado river in the USA; further signs were later detected during the filling of other giant reservoirs such as Lake Kariba (Zambia/Zimbabwe border). However, the true seriousness of the problem was realised when earthquake tremors shook loose a piece of the mountain wall overlooking the Vajont reservoir in Italy. As the mountainside fell into the reservoir, it displaced water over the dam crest, flooding the valley below and killing 2000 people.

These problems are associated with reservoirs that are either very large (capacity more than 10^9 m^3) or simply very deep (over 100 m) and where a delicate balance of stresses in the crustal rocks was upset as the reservoir water percolated down old fault lines. Faults may have been stable for millions of years, but it should not be assumed that all original stresses causing the earthquake have now been released; on the contrary there is always a residual stress following failure. Over a long period of time stresses build up again and reach a value close to failure; the injection of water adds further pressure that allows some faults to be reactivated and so may initiate earthquakes. The most likely place for such earthquakes is where the water is deepest and exerts most pressure – just behind the reservoir dam!

Figure 2.7 Landforms influenced by igneous activity. (a) *Flood basalts* produced by highly mobile basalt welling up deep fractures, then flooding out over the landscape, solidifying as sheets. Many flows of varying size give a step or **trap** topography (examples: Colombia River Basin, USA; Deccan traps, India; Antrim Plateau, Northern Ireland). (b) *Composite volcanoes* (stratovolcanoes, central vent volcanoes) are cones built from poorly mobile andesite, dacite or rhyolite. They consist of layers of **pyroclastics** (ash, cinders, blocks and bombs) and tongue-like flows of lava. Lavas either erupt through their vent crater, or through fissures on the flanks of the cone. The flank eruptions cool to produce radiating dykes that act as ribs to strengthen the cone. Central vent materials cool to produce a pipe-like plug (see (c)). (c) Erosion of igneous rocks produces a wide variety of landforms, as shown above. Composite volcanoes may be eroded to reveal the more resistant plug and dykes; lava flows may form cap rocks or resistant strata that control the pattern of erosion in a plateau or escarpment. Subsurface intrusions may eventually be exposed to form large masses of resistant rock that stand proud as uplands (e.g. Dartmoor) or as domes (e.g. Sugar Loaf, Rio de Janeiro), or as dykes and sills that are the sites of waterfalls (e.g. High Force, Northumberland). (d) Dykes (foreground) strike out to sea. In the distance the island of Pladda (Arran, Scotland) is capped by a sill. ▷

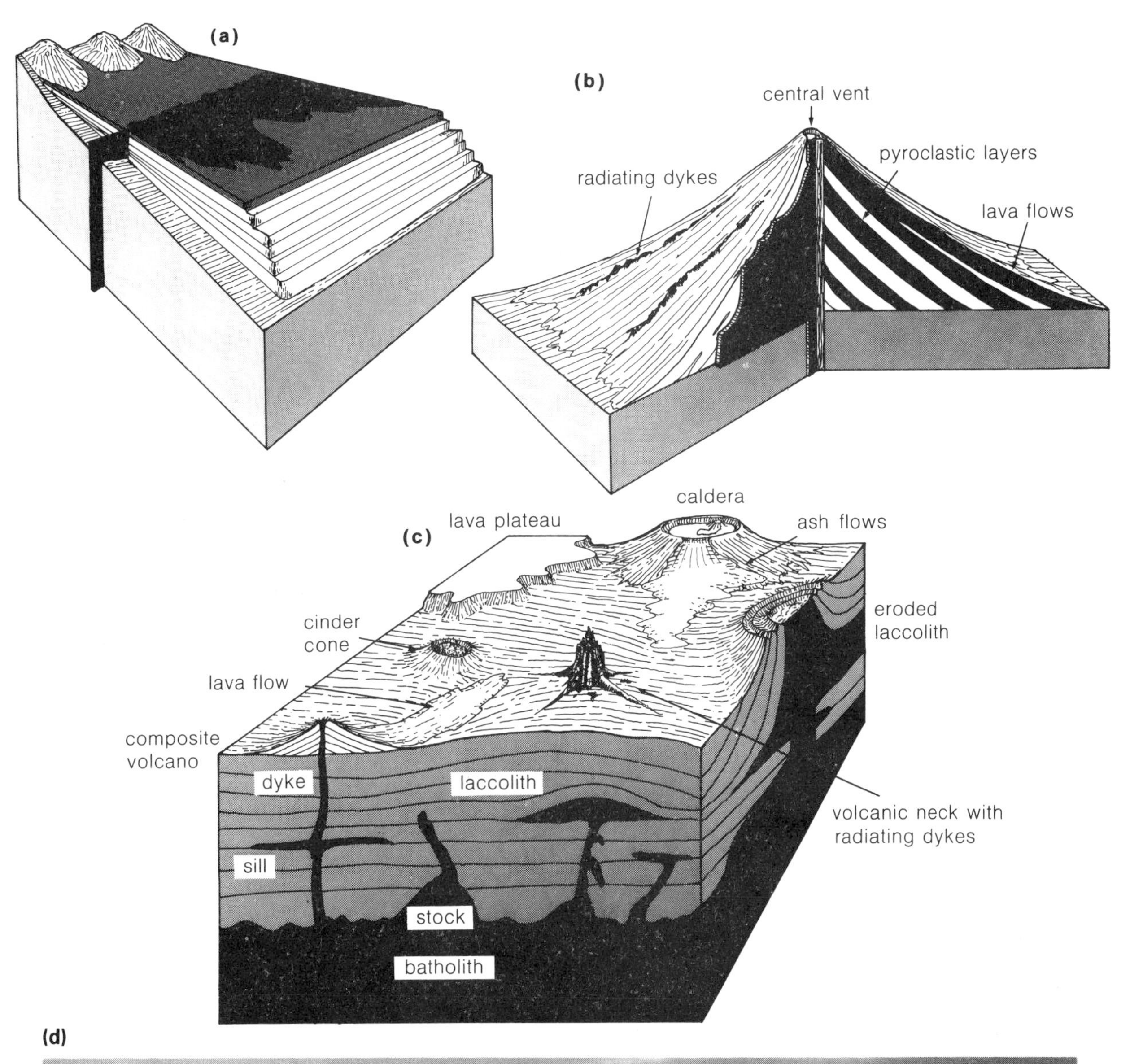
(a)
(b)
central vent
radiating dykes
pyroclastic layers
lava flows
caldera
lava plateau
ash flows
(c)
cinder cone
eroded laccolith
lava flow
composite volcano
dyke
laccolith
volcanic neck with radiating dykes
sill
stock
batholith

(d)

Most of the large man-made earthquakes have occurred when lakes were filled quickly or when deep pumping forced large quantities of water onto the ground at high pressure. By filling reservoirs or pumping liquids slowly, it is possible to allow stress that has accumulated in the rock to be relieved in small steps, so causing a large number of small (and relatively safe) tremors rather than a few large damaging ones. Indeed, some scientists believe that instead of waiting for the San Andreas fault to cause a major earthquake, it would be wise to inject water at depth along the fault and thus prevent the disaster that otherwise awaits San Francisco.

2.3 Volcanoes

Volcanoes occur where deep fractures in the crust allow magma to rise to the surface. Usually a volcano at the surface is accompanied by the development of a subterranean magma reservoir. The growth of the reservoir can often be detected by an elevation of the ground surface just before an eruption. But, as with a balloon, although the magma chamber can be observed getting larger, neither the precise place of rupture nor the exact moment of eruption can be forecast accurately.

When an eruption begins, the sequence of outpourings is similar to that of liquid from a bottle of fizzy drink that has been shaken before the cap is released: at first spray flies everywhere, then the liquid wells up and begins to froth out and run down the sides. The proportion of spray (ash and gas) to liquid (lava) varies with the chemical composition of the source magma and the length of time since the previous eruption. In general, the longer the time between eruptions, the greater will be the silica content of the magma, the proportion of ash to lava, and the violence of the eruption.

Most volcanoes have erupted a variety of materials in their long histories, from viscous rhyolite, through andesite to highly mobile basalt. In general, however, basaltic lavas are most commonly ejected from oceanic-fissure volcanoes, whereas andesite or rhyolite are most commonly associated with land-based volcanoes that build cones (Fig. 2.7).

Hazards from volcanoes

The human impact of a volcanic eruption varies considerably (Fig. 2.8). Some volcanoes are in sparsely populated areas and even when they erupt violently they have little impact on human life. Thus, although the recent eruption of Mt St Helens volcano on 18 May 1980 blasted a mixture of hot, dry rock and gas (called a **glowing avalanche** or **nuée ardente**) that felled giant trees and laid them out like matchsticks there was little loss of life (Fig. 2.9).

By contrast, island arcs (associated with subduction zones) are mostly heavily populated and volcanic activity *is* a major hazard. Most island arcs are formed from relatively alkaline materials which weather to yield well-drained, fertile soils. In such situations people are repelled by the volcanoes as an ever-present threat to life, but drawn to them as a source of the fertile soil on which their livelihood may depend. In Indonesia, Japan, the Philippines and the Caribbean, elaborate flights of terraces are built to stabilise the soil and allow cultivation on the flanks of all the volcanic mountains (Fig. 2.10). When Mt Pelée, on the Caribbean island of Martinique, erupted in 1902, the glowing avalanche that raced down the mountainside destroyed not only plantation crops but also 30 000 people, boiling them in gases at nearly 1000 °C or choking them with ash.

Some volcanoes, especially those whose dominant material is basaltic, tend to erupt more regularly and

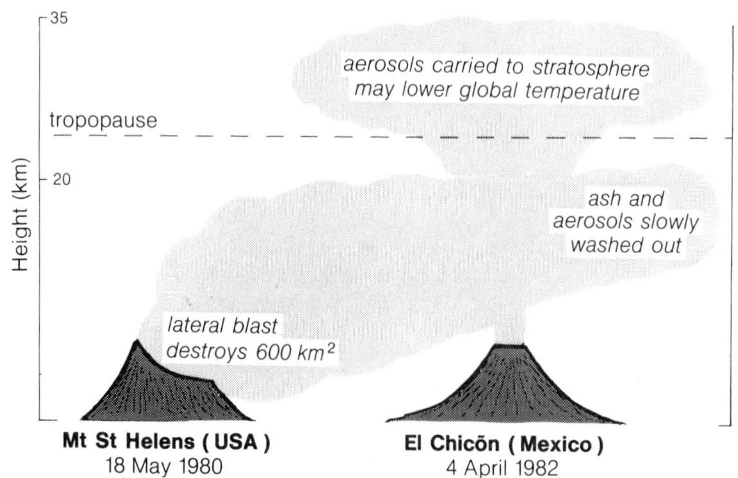

Figure 2.8 Two volcanoes of comparable size but with vastly different environmental impacts.

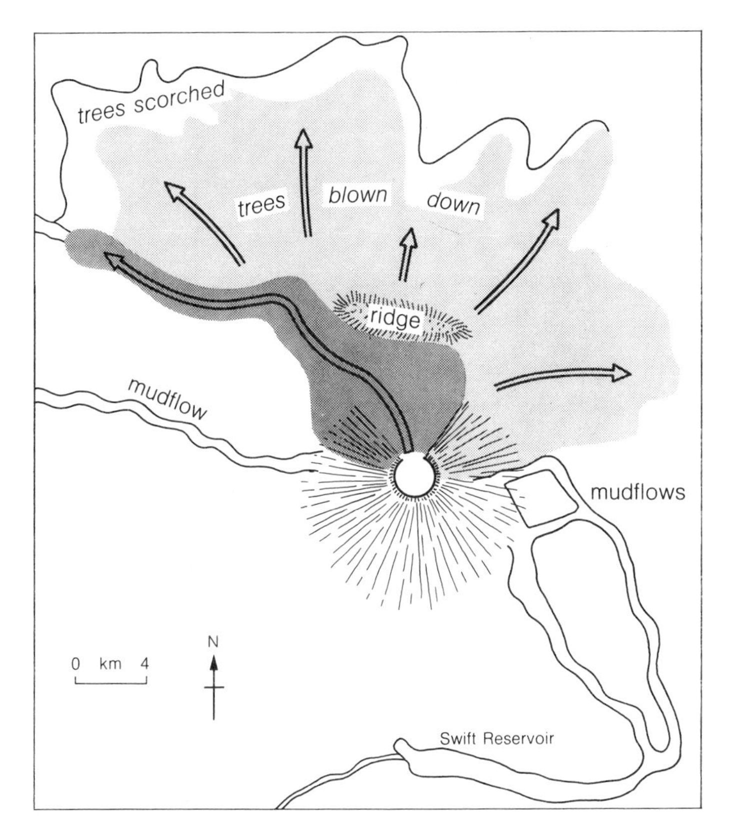

Figure 2.9 Forms of destruction and mudflows following the Mt St Helens eruption. Notice that the glowing avalanche (darker shading) was fully deflected by a ridge whereas the gas cloud (lighter shading) was only partly deflected. There was very little settlement within the blast zone.

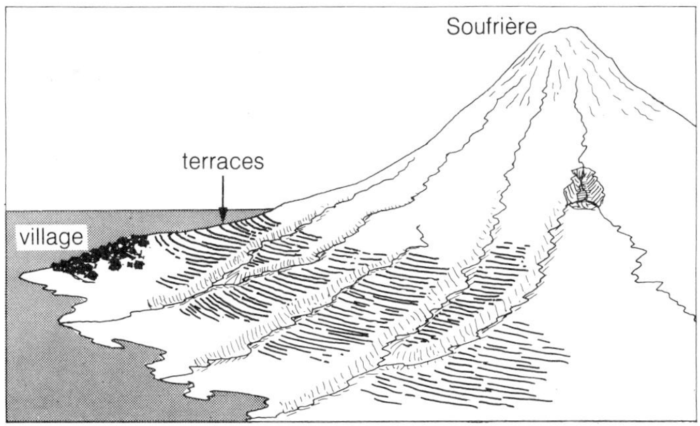

Figure 2.10 Plantations on the flanks of Soufrière volcano in the Caribbean.

less violently. Mt Etna, the world's largest continental volcano, which dominates the island of Sicily, erupts on average once every six years, sending a tongue of lava flowing down its flanks. Mt Etna is a particularly valued resource: not only do its basalts weather into fertile soils, but the highly permeable lava and ash soak up winter snowmelt and give a constant supply of spring water through the arid summer months (Fig. 2.11). Because of these agricultural and water-supply benefits, Etna is ringed with settlements, including the city of Catania. The volcano is a constant hazard: although its eruptions are never devastatingly violent, they occur unpredictably from any point on the flanks as well as from the central vent.

The hazard is restricted as new lava tends to flow around 'dams' created by accumulations of old lava; nevertheless, large sections of many settlements

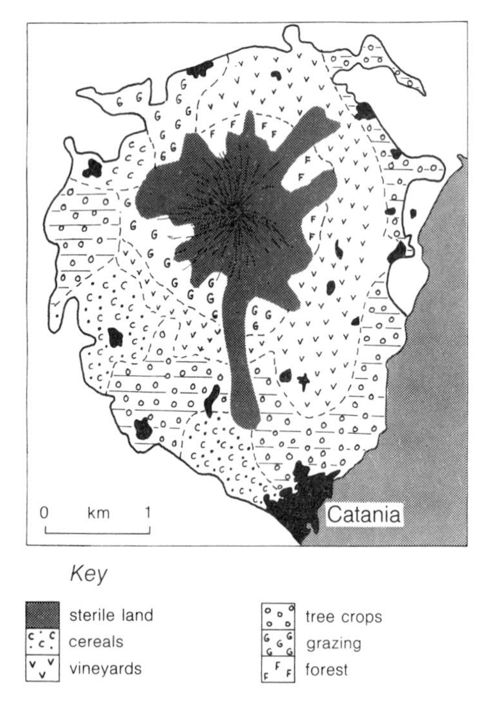

Figure 2.11 Extensive agriculture on the flanks of Mt Etna reflects the fertility of the soils. The 'sterile' land is that recently affected by lava flows or acidic materials.

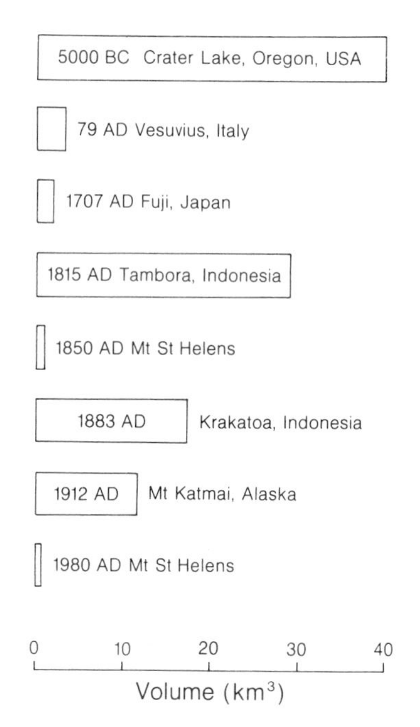

Figure 2.13 Some of the large volcanic eruptions in historic times.

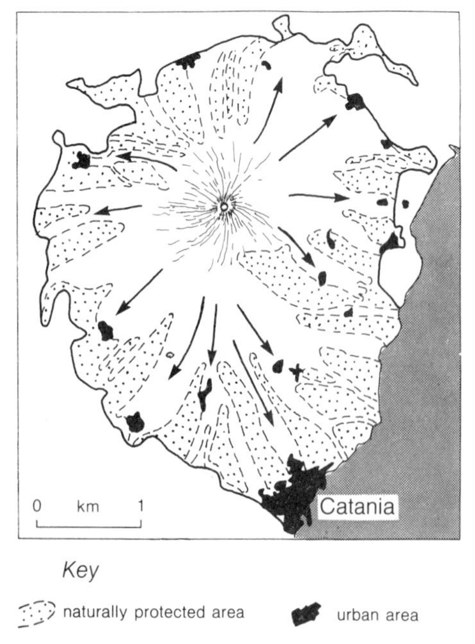

Figure 2.12 A lava hazard map prepared from a study of the topography and the likely sources of eruptions.

remain at risk. It is hoped that the preparation of a hazard map will encourage the island's planning authorities to locate new housing and industries within the safest zones (Fig. 2.12).

Although the recent eruption of Mt St Helens was spectacular, it was small in comparison with other eruptions throughout history (Fig. 2.13). The fact that further large eruptions are inevitable poses a permanent threat to those living near active volcanoes. However, the hazards involved are not confined simply to engulfment by ash or lava. Many volcanoes are capped with glaciers whose instantaneous melting would cause devastating floods and mudflows, and the torrential rain that always accompanies violent eruptions where hot gases rise through the air (see Ch. 3) will add to the flood hazard.

The major hazards are:

(a) ashfall, which smothers houses, causes roofs to collapse and buries crops;
(b) lateral blasts;
(c) glowing avalanches (nuées ardentes);
(d) lava flows;
(e) mudflows;
(f) floods;
(g) fires.

The long term consequences of the Mount St Helens eruptions are still to be fully realised. Peter J. Smith reports

Falling out with the rest of society

DURING the eruptions of Mount St Helens last summer, the media had a great time with pictures of cars buried in debris and speculation about red sunsets halfway round the world. But these were some of the immediate, and comparatively trivial, effects of that spectacular volcanic activity. The Mount St Helens eruptions were the first to take place in the north western United States since agriculture of any importance began there; and only now are the agro-economic and ecological consequences of the 1980 ash-falls becoming clear.

From the large May 18 eruptions alone, almost two cubic kilometres of ash fell on farm and forest land in Washington and surrounding states. Some areas of eastern Washington itself received up to 300 tonnes of ash per hectare, forming an uncompacted layer up to 15 centimetres thick in parts. The potential seriousness of such a deluge may be judged from the fact that in 1979 the areas to be worst affected by ash had accounted for 1.4 billion dollars of crops (mainly fruit, hay, cereals, potatoes, and legumes) and 270 million dollars of livestock products.

The full consequences of the 1980 eruptions may never be known, and where they become apparent they are often difficult to cost. But R. J. Cook and colleagues at the US Department of Agriculture and Washington State University have been trying to build up as complete a view as possible, coming up with some predictable but also some unexpected conclusions.

Fine ash can have a number of effects on plant life. The sheer weight of ash may force foliage to the ground (lodging); photosynthesis may be greatly reduced by ash acting as a light barrier; leaves may be salt-damaged by wet ash; and plant temperatures may be reduced because light grey ash reflects more of the incoming sunlight than does green foliage. All have played a part in Washington in recent months.

Alfalfa, just ready for cutting last May, proved particularly susceptible to lodging, which caused an estimated crop loss of 35 million dollars. Moreover, the alfalfa hay that farmers did manage to harvest contained up to 60 per cent ash, which greatly increased the unit cost of transport to dairy farms. By contrast, winter wheat suffered little, largely because the ash was more rapidly rainwashed from the smoother foliage. The above-normal summer rainfall helped here, producing in its own right an increased crop yield that more than offset the worst that Mount St Helens could do.

Apple leaves, being hairy (pubescent), trapped a lot of ash, which reduced photosynthesis by up to 90 per cent. The result was premature fruit drop of about 10 per cent, or 15 million dollars in financial terms. Peaches, apricots, raspberries and strawberries fared much worse, however, because the fruit itself is pubescent. It also proved resistant to washing, so that after May 18 much of the produce became unsaleable. Almost half of the strawberry crop, for example, was simply left to rot in the fields.

Cook and his colleagues have estimated crop losses so far in eastern Washington to be about 100 million dollars, or about 7 per cent of the expected crop value. But great though this is, it is far smaller than agricultural experts expected, even allowing for further loss in 1981.

As for animal life, it is quite impossible to be anything like as precise about the financial effects. Insects were particularly badly affected, not only because the weight of ash made flight difficult but also because the airborne ash particles were abrasive and thus damaged the insects' epicuticular wax layer, causing rapid desiccation and death.

Fifty to 100 per cent of house flies, cockroach nymphs and orchard mason bees died by desiccation within four - eight hours of exposure to Mount St Helens ash. Damaged yellowjackets died within three - six hours even when given water in the laboratory, although mealworm and moth larvae were hardly affected by the ash at all.

Of course, it's an ill wind . . . . Some of the insects killed were crop pests, particularly Colorado beetles and grasshoppers. Indeed in some areas the ash saved money by making the usual grasshopper spraying unnecessary. On the other hand, the insects most severely affected were beneficial species such as bees and wasps. In the worst ash zones, 80 per cent of bee colonies were destroyed or seriously damaged, resulting in reduced pollination and hence seed production.

Ironically, the abrasive properties of the ash harmed not only delicate insects but also robust farm machinery, by accelerating gear and bearing wear by a factor of three-four. This problem (not to mention fuel costs) was also amplified by increased machinery usage, for the many Washington farmers who have been developing no-till agriculture in recent years are now finding it necessary to plough the newly fallen ash into the soil.

Not all the problems resulting from the Mount St Helens eruptions have yet been fully evaluated. For example, the vast amounts of ash still lying in uncultivated regions could well affect the water run off this winter and thus change river flow and reservoir recharge. Whether or not this will seriously disrupt the hydrological characteristics that farmers have come to rely on remains to be seen. Either way, the consequences of Mount St Helens will be with the people of the north western United States for years and decades, if not centuries, to come.

Dr Peter J. Smith is Editor of Open Earth.

Part B

Atmosphere

Atmospheric processes

3.1 The influence of the atmosphere

The atmosphere may be only a thin veil of gases surrounding the Earth (in much the same proportion as the skin to the rest of an apple), but its constant movement influences everything we do. The day-to-day movement provides our weather and affects such immediate issues as what clothes we should wear; the average pattern of weather (**climate**) over many years affects more long-term issues such as what crops we grow and where we live. Moreover, variations in climate determine landscape weathering processes, soil formation and erosion. Only the major landforms of the Earth are not influenced by climate; instead, by way of continent and ocean distribution, they influence it.

3.2 Development of our understanding

Early ideas

The weather has always fascinated people because of its effect on their lives. But early meteorologists, even when standing on a hilltop, could see no farther than a few tens of square kilometres; the other 99.999 per cent of the atmosphere remained out of sight over the horizon. Their problem was akin to trying to discover how a motor car works when all you can see is a wheel nut. As a start, they studied reports from sailors on global voyages, which contained a wealth of information about winds, clouds and temperature, albeit in a rather generalized form.

Even the earliest maps of airflow revealed a fairly simple distribution, the main changes occurring parallel to lines of latitude (Fig. 3.1). Near the Equator is a region where the winds blow only fitfully; this they called the **doldrums**. By contrast, somewhat further towards the poles were regions where winds blew more reliably; because of their usefulness to European sailors seeking trade with other continents, these were called the **trade winds**. Polewards of the trades lie zones of calms (the **horse latitudes**) at about 30°; beyond these bands of **westerly winds** that carried the sailors home – blustery, variable winds often bringing storms and long gales.

Airflow maps were essential early steps to understanding the atmosphere, but they raised many questions: why should there be winds in some places and not others; why should the winds change direction; and why should some winds be more reliable than others?

Global airflow maps were soon complemented by measurements of air temperatures and rainfall. While some people set out to climb mountains to find out how the weather varied with altitude, the more intrepid made measurements from hot-air balloons. These measurements demonstrated that air temperature and pressure decreased with altitude, and early

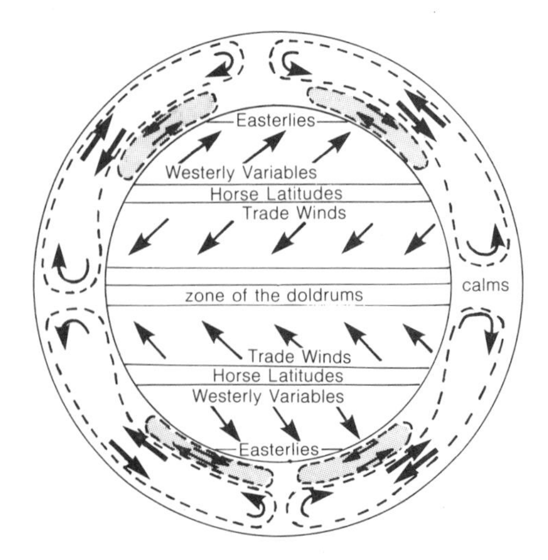

Figure 3.1 The pattern of global surface winds was one of the first aspects of the atmosphere to be mapped. This 1857 diagram shows the average airflow in only the simplest way.

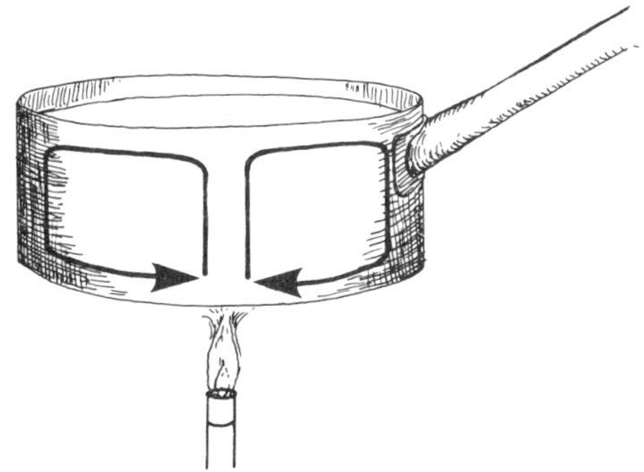

Figure 3.2 Scientists such as Edmund Halley were quick to recognise the similarity between water and air circulation. Water is heated from below, becomes less dense and rises to the surface, its place being taken by more dense cooler water, thereby establishing a circulation. In the atmospheric circulation the major heating source is the Earth's surface.

Figure 3.3 The Earth's circulation on 29 November 1978 as seen from Meteosat 1 in stationary orbit over the Equator. Note that the polar horizons are at about 60° latitude.

meteorologists quickly realised that the atmosphere is warmed from below – a very unstable situation which will tend to generate convectional movements within the atmosphere similar to those of water heated in a pan (Fig. 3.2). In 1686 Edmund Halley (of comet fame) used just such observations – of water circulating in a pan – to draw an analogy with the Earth. Halley reasoned that, because the equatorial air near the ground is much warmer than the air near the Poles, air would rise at the Equator and draw a replacement supply from the Poles. The movement of air from Pole to Equator would be felt as a surface wind. Having conceived such a circulation, he proposed also that there must be a return of air to the Poles at high level to make the system continuous. Then, because each hemisphere is a mirror image of the other, he argued there must be one giant **circulation cell** in each, having a joint source in the rising equatorial air (Fig. 3.1).

Current understanding

Today we have the advantage of a satellite's-eye view. From an orbit 35 000 km above the Equator, we can now see, picked out in patterns of cloud, air movements beyond even the wildest imaginings of Halley (Fig. 3.3). And we can also see that what appeared to the early meteorologists as a simple pattern of latitudinal belts is really very complex. For example, the belt of westerlies carries with it a series of swirling catherine wheels of air and cloud; the clear horse-latitude region is crossed by a broad band of cloud; while the cloud pattern shows that the tropical region of rising air is separated into many small cells. Today wide-ranging and sophisticated measurements have enabled scientists to develop a much more detailed pattern of the structure of the atmosphere and the forces that drive its circulation.

3.3 Energy in the atmosphere

Energy from the Sun

The Earth's atmosphere is driven by the energy received from the Sun. The amount of solar energy received (**insolation**) depends primarily on the Earth's position in the Solar System. To appreciate this, compare the Sun with a car headlamp seen at night far off across a plain. Initially it is only a small twinkle of light; it is virtually incapable of lighting or heating the place where we stand. Gradually, however, as the car comes closer, the light takes on a definite shape. Nearby shadows form as the beam becomes more intense. Eventually the car comes very close and bathes everything in light. In a similar way the radiation from the Sun is spread so widely when it reaches such far-off planets as Saturn that they remain virtually unheated and have low surface temperatures – the temperature on Saturn is −150 °C. By contrast, Mercury is very close to the Sun and has a scorching surface temperature of 400 °C. The Earth experiences moderate temperatures because throughout its elliptical orbit it is just far enough away to prevent overheating, but not so far as to experience intense cold.

At the outer limits of the atmosphere above the equator the energy of the Sun is equal to about 1.5 kW of energy over every square metre (equivalent to standing about half a metre away from an electric fire). In the long term the Earth is not getting hotter,

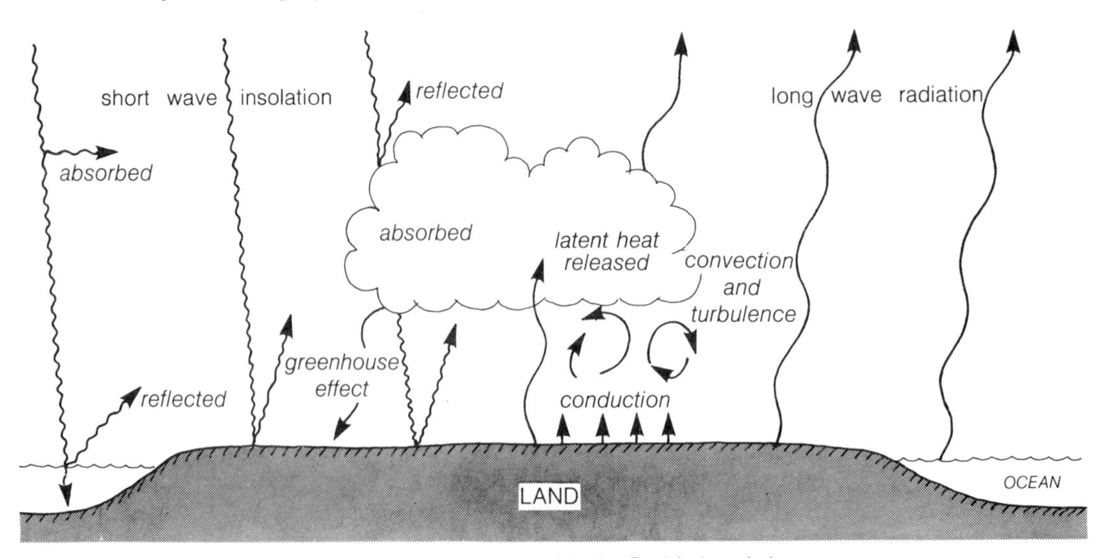

Figure 3.4 Transfers involved in the Earth's heat balance.

so the same amount of energy must be lost from the Earth: it is a matter of balancing the energy books (Fig. 3.4).

Radiation by Sun and Earth

The *short* wavelength of the radiation emitted by the Sun (solar radiation) is related to the Sun's high surface temperature. By contrast the low temperature of Earth's surface dictates that its radiation to space (**terrestrial radiation**) occurs at *long* wavelengths. Shortwave light ranges from visible to ultraviolet and beyond, whereas longwave light constitutes infra-red radiation. The Sun's radiation is more obvious because it includes visible light; it is much more difficult to appreciate the balancing radiation from the Earth. Perhaps it is useful, therefore, to think of the relationship between Sun and Earth as a central heating system : the Sun acts as the boiler; the Earth as one vast radiator which gives out as much heat as the boiler puts in, but hasn't warmed up enough to glow.

The structure of the atmosphere

The upper atmosphere is principally segregated into bands (or shells) of varying chemical composition (Fig. 3.5). The innermost shell (called the **troposphere**) includes the Earth's cloud systems and has a relatively high proportion of both water vapour and carbon dioxide. Outside this is a shell of thinner air (**the stratosphere**) which contains little water vapour or carbon dioxide and in which ozone (a form of oxygen produced by irradiation) is an important constituent. Such differences in chemistry determine the temperature structure of each shell. The stratosphere absorbs solar radiation strongly, is warmest in its outer layers and becomes cooler towards the Earth – an arrangement which keeps warm air above cold and imparts considerable stability. However, in the troposphere direct absorption of solar radiation is small compared with that of terrestrial radiation – water vapour and carbon dioxide absorb most strongly the infra-red radiation produced by the Earth. As a result the troposphere is warmed mainly from below and decreases in temperature with distance from the ground. Sometimes instability caused by cool air above warm results in widespread overturning and the formation of clouds. The great stability of the stratosphere restricts the height at which instability can occur (by confining it to the troposphere) and thus acts somewhat like 'a lid' that keeps in the boiling contents of a saucepan.

Because the troposphere is largely heated from below, it is important to be aware of the factors which influence the amount of ground surface heating. For example, a cloud is made from countless microscopic droplets of water which, like cat's-eyes in a road, are extremely good reflectors. A thick cloud layer can reflect over 80 per cent of the insolation reaching the troposphere back to space (Fig 3.6). It is equally effective in reflecting terrestrial radiation back to the ground; only a portion of the insolation reaching the ground is absorbed because the reflectivity of most surfaces is substantial. The ratio of reflected light to incident light (the **albedo**) for snow and ice ranges from 60 per cent to over 90 per cent, depending on how much dirt has blown onto their surfaces; bare rock often reflects up to 30 per cent. The shimmering waters of the oceans, whose waves sparkle so

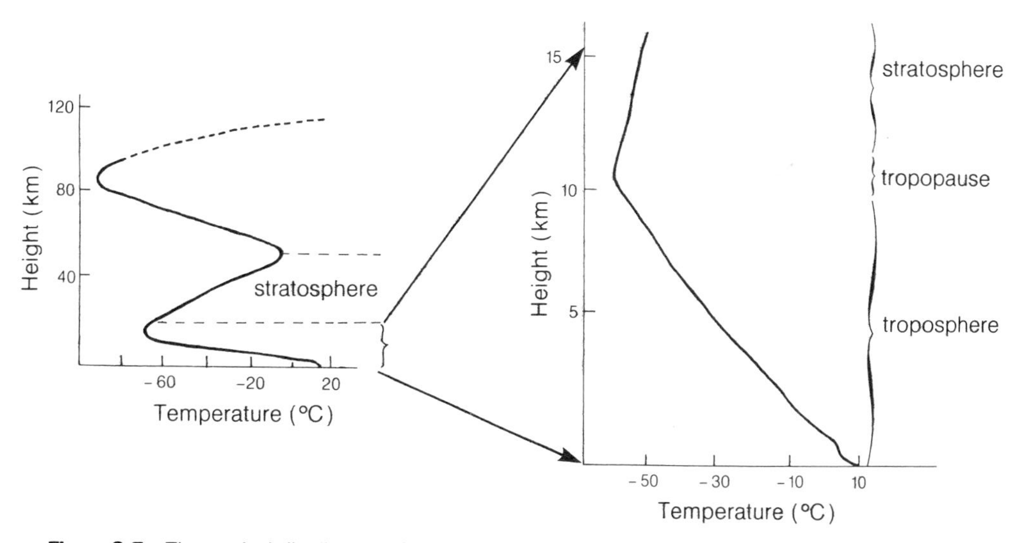

Figure 3.5 The vertical distribution of temperature in the atmosphere, revealing several thermal layers.

Figure 3.6 The Earth radiates in the long-wave infra-red part of the spectrum. This satellite image is taken in the infra-red range, with the Earth's tones a reflection of transmitted heat. The hottest regions show as dark tones, and the lighter tones indicate cold. Space is shown white on this image because it is the coldest region. White clouds show where radiation (greenhouse effect) is occuring in the lower layers of thick clouds. Thinner clouds show as grey tones because some radiation can pass through. Compare this with Figure 3.3.

delightfully, are also returning solar radiation – virtually 100 per cent of it when the Sun's rays glance over the surface at a low angle, and still over 10 per cent even when the Sun is high-angled and the sea looks dull and lifeless.

The global energy budget

The amount of energy absorbed by the Earth's surface and troposphere varies with the angle of received insolation. As with a spotlight on a stage, the energy from a high-angled Sun is concentrated on a small area and the surface heating is great, but the energy from a low-angled Sun is spread widely and surface heating is much less (Fig. 3.7). Thus each square metre of the Arctic regions receives less than a third as much energy as that reaching each square metre near the Equator. Terrestrial radiation varies much less strongly with latitude (Fig. 3.8). As a result, in areas near the Equator there is a surplus of incoming over outgoing radiation, whereas polewards of 38° the situation is reversed. This difference in heat energy between equatorial and polar regions provokes convection, the fundamental driving force for the atmospheric circulation.

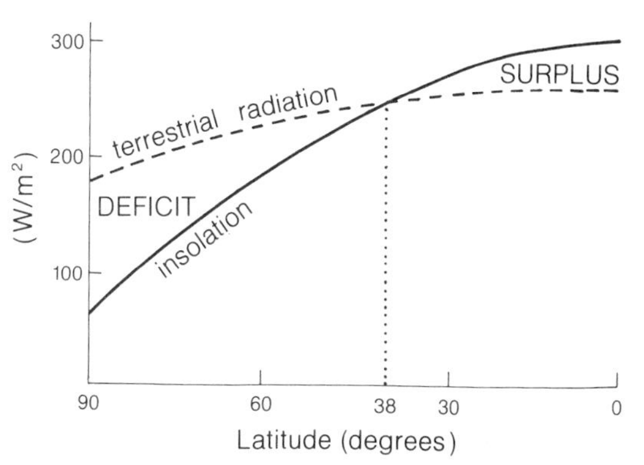

Figure 3.8 Changes in solar and terrestrial radiation over the Earth. Notice that there is surplus energy equatorwards of 38° and a deficit towards the pole.

The greenhouse effect

Water vapour and carbon dioxide gas in the troposphere absorb most strongly in the long wavelengths and therefore mostly trap terrestrial rather than solar radiation. However, there are several other important means of transferring heat to the troposphere. Most important are the processes of direct heating by convection as warm air rises, and the release of latent heat as water vapour condenses to droplets during cloud formation. Direct upward transfer of heat by conduction from air in contact with warm ground is of minor significance: air is a very good insulator.

Cloud droplets, water vapour and carbon dioxide gas are of much greater importance in influencing the temperature structure of the troposphere. Not only do water vapour and carbon dioxide absorb terrestrial radiation, and hence warm the troposphere directly; they also radiate heat partly back to the ground. Furthermore, cloud droplets reflect longwave radiation very effectively (Fig. 3.4). Thus water droplets, water vapour and carbon dioxide act in such a way as to delay the radiation from the Earth to space and temporarily retain heat in the troposphere. This action is commonly called the **greenhouse effect** by analogy with the way air becomes warmed when trapped in a greenhouse. The greenhouse effect is believed to keep the temperature of the troposphere on average some 25 °C warmer than it would otherwise be.

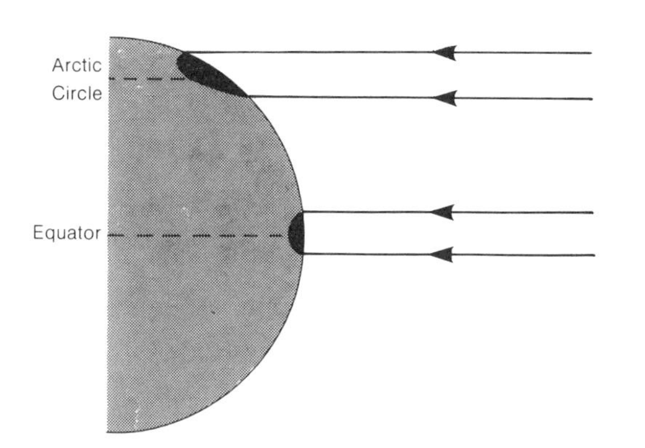

Figure 3.7 Solar radiation reaches the Earth as parallel rays. However, because of the curvature of the Earth, equivalent amounts of solar energy have to heat three times the area at the Arctic Circle compared with at the Equator.

3.4 Characteristics of the troposphere

Pressure, temperature and humidity

The most important characteristic of the troposphere is variation in **air pressure**. At sea level, the combined mass of gas molecules, drawn towards the Earth's surface by gravity, creates a downward pressure sufficient to balance a column of mercury 760 mm high. This is defined as equivalent to 1000 millibars (1000 mb) or 1 bar. With increasing altitude there is a progressively smaller mass of overlying gas so the pressure also decreases progressively (Fig. 3.9).

Changes within the troposphere are commonly measured by balloons sent aloft with monitoring equipment. The rate of decrease, or **lapse**, of temperature with height (as measured by such equipment) is called the **environmental lapse rate**, usually abbreviated to ELR. It is simply a description of the overall temperature regime of the troposphere above a particular place at a particular time.

Air has humidity: it contains a certain amount of water vapour, which it acquires mainly from evaporation of ocean waters. However, the point at which it becomes saturated is strongly dependent on its temperature: in general, the cooler the air, the less water vapour it can contain. Humidity is most usefully expressed as nearness to saturation or **relative humidity**: the ratio between – at the same temperature – the amount of vapour actually in the air to the maximum that *could* be held. Saturated air

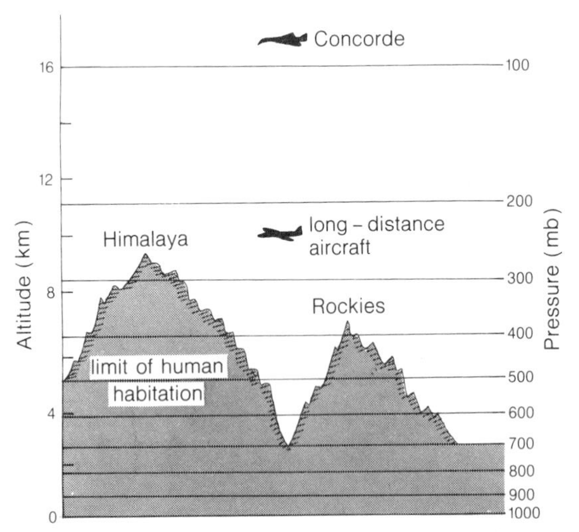

Figure 3.9 The pressure in a column of still air decreases progressively more slowly with altitude. As a result, half of the atmosphere occurs within the bottom 5 km.

has a relative humidity of 100 per cent; air containing only a quarter of the vapour it could hold has a relative humidity of 25 per cent.

The temperature at which air becomes saturated is crucial in studying air movements. **Dew point** is the temperature at which condensation occurs; as vapour condenses, latent heat energy is released into the air. Because of the great contrast between saturated and unsaturated air, the latter is often simply called 'dry', and no reference is made to its degree of saturation.

Stability and instability

THE GAS LAWS

The fact that the troposphere is heated from below, its temperature decreasing with height, does not mean that it is inherently unstable and liable to overturn. In reality, the picture is more complex and can be understood only by reference to the Gas Laws.

Scientists have discovered that gases (including air) behave in very precise ways when their temperatures or pressures are altered. Two main relationships, summarised as laws, are important.

With a constant mass of gas at a constant temperature, *pressure* (*P*) *and volume* (*V*) *are inversely related*. As pressure increases, air is compressed and reduced in volume. Because the same number of air molecules now occupy a smaller space, the air is denser. (This is what happens when a bicycle tyre is pumped up.) The relationship can be written

$$P = \frac{1}{V} \cdot K \qquad \text{(Eqn 1)}$$

Where K is a constant. This is **Boyle's Law**.

For a constant mass of gas at a constant pressure, *volume* (V) *is directly related to temperature* (T). As air is warmed, it expands; because the same number of molecules now occupy a larger space, the air is less dense. The relationship can be written

$$T = K \cdot V \qquad \text{(Eqn 2)}$$

This is **Charles' law.**

Combining the two laws,

$$PV = K \cdot T \qquad \text{(Eqn 3)}$$

thereby showing that *pressure, volume and temperature are all interdependent: any change in one will cause compensating changes in others* (provided that other conditions remain constant).

However, atmospheric volume is not easily measured. Since density is both inversely related to volume and easier to measure, Equation 3 is usually written:

$$P = R \cdot \rho \cdot T \qquad \text{(Eqn 4)}$$

Where R is a constant and ρ is the air density. The following discussion of air stability is based on the interpretation of Equation 4.

AIR STABILITY

Air is a good insulator and exchanges heat with its surroundings very slowly. Thus any change in the characteristics of one 'parcel' of air is not readily transmitted to the adjacent environment. (This is an important property without which the gas laws would not be valid.) Changes in the characteristics of a parcel of air exhibiting no exchange of heat with the surrounding environment are described as **adiabatic**.

Using the gas laws as a basis, it is possible to predict the result of a parcel of air undergoing adiabatic changes. For example, if a 'dry' air parcel rises through the troposphere, its pressure will fall and it will become cooler at a rate of 1 °C/100 m of ascent. This **dry adiabatic lapse rate** is abbreviated to **DALR**.

Because *saturated* air continuously releases latent heat and therefore cools more slowly on ascent, predicting the behaviour of a parcel of saturated air is more difficult. The amount of latent heat released depends on the absolute moisture content of the air. However, in mid-latitudes a lapse rate of 0.5 or 0.6 °C/100 m of ascent is common. The **saturated adiabatic lapse rate (SALR)** increases with continued ascent until, when there is little moisture left to condense and therefore little heat to be released, the SALR is virtually identical to the DALR.

It is common for the ELR (environmental lapse rate) to be about 0.65 °C/100 m. Thus , for example, an air parcel will continue to ascend (become unstable) only if it decreases in temperature at a rate less than that of the environment through which it is moving (i.e. SALR < ELR or DALR < ELR) (Fig. 3.10a). This is the cause of much cumulus cloud.

As a corollary, air can not ascend if its potential lapse rate is greater than the surrounding environment, for then it would be cooler at every height (i.e. SALR > ELR or DALR > ELR) (Fig. 3.10b). Thus the environment is stable and will not yield cloud unless air is *forced* aloft.

In many cases the environmental temperature gradient lies between the SALR and DALR values, so if dry air can be forced up to a level where saturation occurs, it will then rise spontaneously (Fig. 3.10c). The conditions of stability are of vital importance in the study of weather.

As an example, consider a parcel of unsaturated air at the surface which is at 18 °C. The gas laws suggest the air will cool at the DALR (1 °C/100 m), and the temperature will have dropped to only 8 °C by the time it has risen to 1 km (Fig. 3.11, bottom left). However, the air already at 1 km is 15 °C (i.e. 7 °C warmer than the air parcel at this height). If any surface air ascends to this height it will be colder, and therefore more dense, than the surrounding air. It will sink back to the surface again. This is important, because it shows why air can be warmest at the ground yet still not rise.

Stability, however, is a very fragile state. Suppose the environment has a temperature profile like that of Figure 3.11, bottom right. At 1 km surface air will have cooled to 8 °C, but the surrounding environment is only −2 °C. On this occasion the 'surface' air parcel has remained warmer (and therefore less dense) than its new environment and it can continue to rise. Of course, air must now flow in to replace the

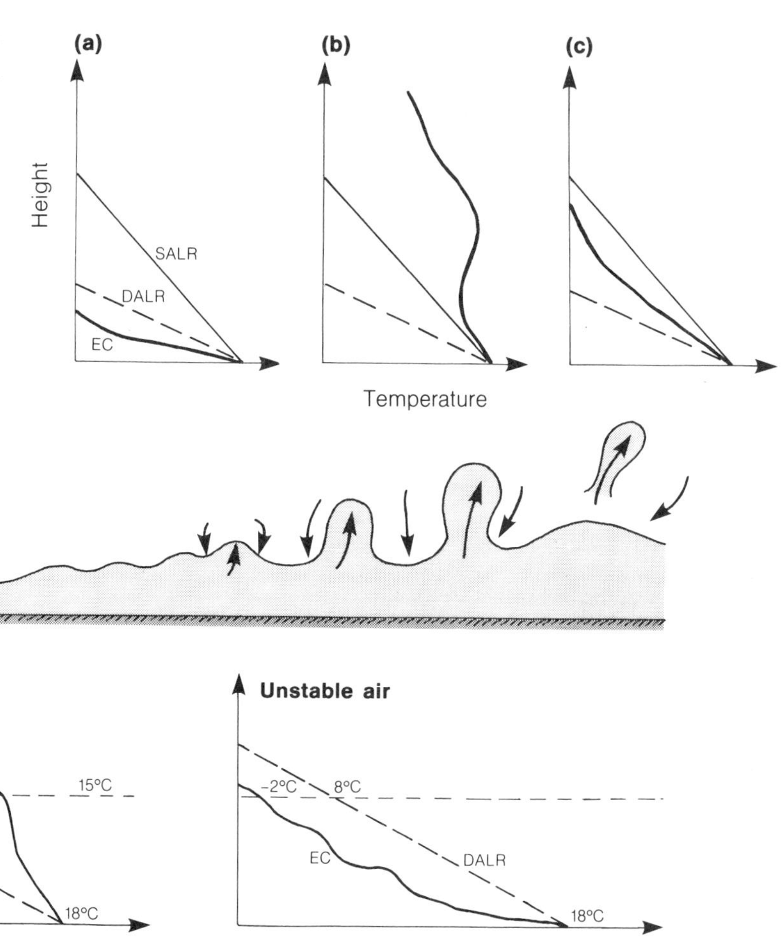

Figure 3.10 The stability of air refers to its tendency to undergo vertical motion. Thus **absolute stability** (a) is achieved if the environment does not allow either unsaturated or saturated air to rise spontaneously after an initial impetus. By contrast, **absolutely unstable** (b) conditions allow both saturated and unsaturated air to rise. A **conditionally unstable** situation (c) exists if saturated air can rise spontaneously but dry air cannot. Each of these conditions is determined by the relative lapse rates of dry, saturated and environmental air.

Figure 3.11 The stability of air, as described in the text. The upper part of the diagram shows the distribution of warm and cool air associated with stability and instability.

rising surface air, and overturning (**convection**) currents quickly become established. Even the whole Earth's *general* circulation is powered by vertical movements such as these. In practice convection occurs as 'bubbles' of warm air that break away from the warm surface air. As they rise compensatory cold air sinks nearby. As we shall see, this pattern is woven into the fabric of much of the world's weather.

Condensation

Condensation occurs when air cools below its dew point. This point is reached in a number of ways:

(a) by direct radiation cooling of air in contact with the ground, which normally produces radiation fog;
(b) by convective ascent of parcels of warm air;
(c) by forced uplift at a topographic barrier (hills, mountains, cliffs), or when warm air rises over cold air along a front;
(d) by turbulent mixing of warm and cool air.

We encounter condensation in such everyday phenomena as droplets on a cold milk bottle or on blades of grass in the cool of early morning (when the condensation is called **dew**). Condensation occurs more readily *onto* surfaces than spontaneously in the air. Within the troposphere, condensation occurs on **hygroscopic** ('water-loving') **nuclei** such as microscopic particles of salt, clay and other small pieces suspended in the air. These 'condensation nuclei' often occur in vast numbers and even build up in such concentrations as to create a haze in the sky (the famous Los Angeles smog, a special case of this, is produced mainly by hydrocarbons emitted from car exhausts). Indeed nuclei are often so plentiful – a million in every litre of air is not uncommon – that they share the condensing vapour, limiting the size of the cloud droplets that can form.

Most water droplets are between 5 and 10 millionths of a metre across (5–10 μm), and individual droplets cannot be seen with the naked eye, even in the densest fog. Very little air movement is needed to keep such tiny drops suspended and forever out of reach of the ground, so that cloud formation does not automatically result in precipitation. Indeed, the smallest raindrops to reach the ground (as 'drizzle') are about 100 μm across, and 'true' raindrops are usually over 1 mm in diameter. These large sizes compensate for the evaporation losses as droplets fall through the unsaturated air below the cloud. Even the smallest raindrops reaching the ground must be over ten times the size of the average cloud droplet, and there must be a mechanism to enable droplets to grow large enough to fall as rain or snow.

Precipitation

The temperature at which water freezes depends on air pressure. In the low pressures that prevail several kilometres above the ground water can remain liquid at temperatures as low as –30 °C. Only at very high levels (where temperatures are below –30 °C) can ice crystals rather than water droplets form. At these high levels there are many fewer nuclei around which ice crystals can develop. There are often no more than a hundred such nuclei in each litre of air, and ice crystals are correspondingly rare. As a result, tall clouds may have many small water droplets in their lower levels and rather fewer ice crystals near the top (Fig. 3.12).

Precipitation is formed mainly in the middle of these deep clouds, where ice crystals fall into a region containing very cold water droplets. Here, both ice crystals and water droplets are enveloped in air saturated with water vapour. Because ice attracts water molecules more strongly than water, vapour is

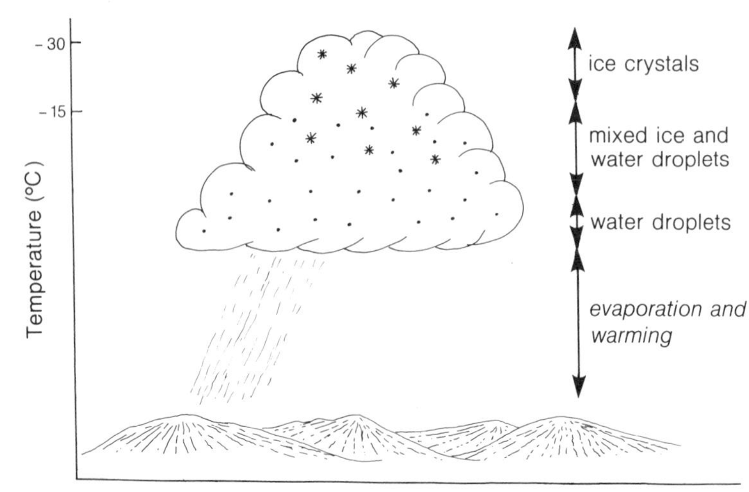

Figure 3.12 Whether rain or snow are produced and fall to the ground depends on the success of complicated growth mechanisms within the middle regions of a cloud.

attracted to the ice crystals, which then begin to grow. As the growth of ice crystals withdraws vapour, the air becomes saturated with respect to ice, but it becomes *unsaturated* with respect to water, and water droplets begin to evaporate. This added supply of vapour allows the ice crystals to grow further. Thus the ice crystals compensate for their small numbers by 'eating up' the water droplets, growing ever larger until they become heavy enough to fall against the gentle upward-moving air of the cloud. Often by this stage many crystals have aggregated into **snowflakes**. As these large flakes fall, they may melt in warmer air and reach the ground as **rain**. However, if the lower air is still cold (usually below 4 °C) the crystals melt only partially to give **sleet** or remain fully frozen as **snow**. In honour of the Norwegian meteorologist who first suggested that in temperate and cold climates snow and rain have this common origin, the snowflake-growth mechanism is called **Bergeron process**.

Temperatures below −12 °C are needed if the Bergeron process is to work well, but many tropical clouds are too warm for ice crystals to form. In the very deep, warm thunderclouds of the tropics, droplets vary greatly in size, the larger ones falling faster than the smaller. As large droplets fall through the cloud of smaller ones, they whisk these together: some **coalesce** into drops large enough to fall as rain (Fig. 3.13). Nevertheless, in these large clouds, updraughts of several metres per second occur, speeds at which raindrops of ordinary size cannot fall. Droplets falling within tropical clouds are therefore usually much larger than those in mid-latitudes. Sometimes droplets are carried in these updraughts towards the cloud top and begin to freeze, acquiring a coating of ice (called **rime**); at other times they are

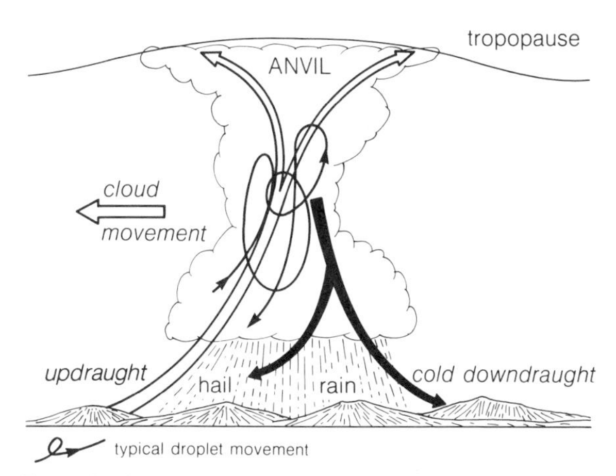

Figure 3.13 A cumulonimbus thundercloud. The pattern of rising and falling air currents helps ice crystals or raindrops to grow.

forced rapidly downwards by downdraughts whereupon they gather a coating of more small water droplets before being carried aloft once more where the surface water again freezes. After several cycles, large **hailstones** may form, some the size of golf balls or even oranges, although most small hailstones melt on their downward path, finally breaking up and reaching the ground as raindrops.

Cloud types

CUMULIFORM CLOUDS

The clouds we see in mid-latitudes are not typical of the whole world. Although in mid-latitudes **cumulus** clouds (Fig. 3.14) are associated with occasional

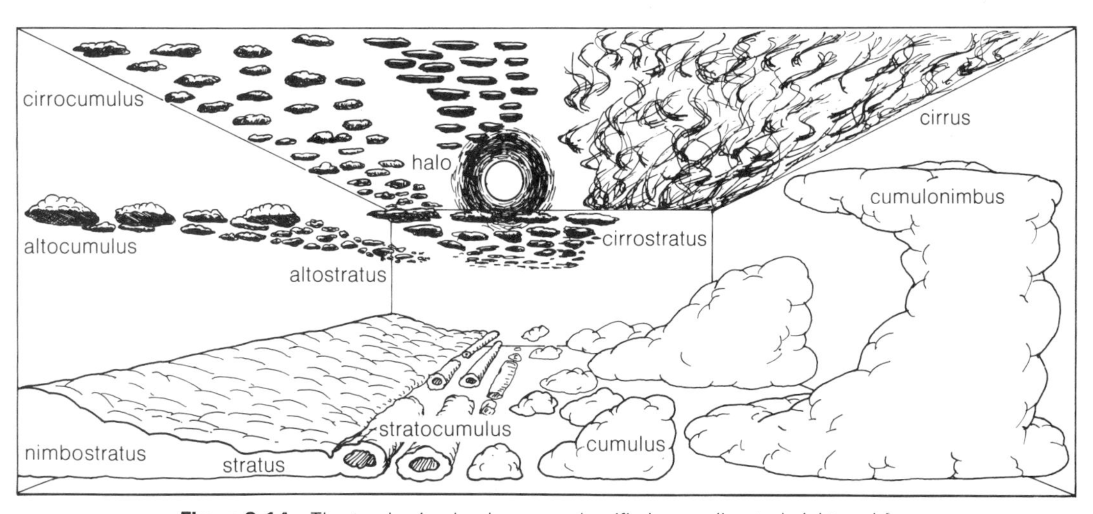

Figure 3.14 The ten basic cloud groups classified according to height and form.

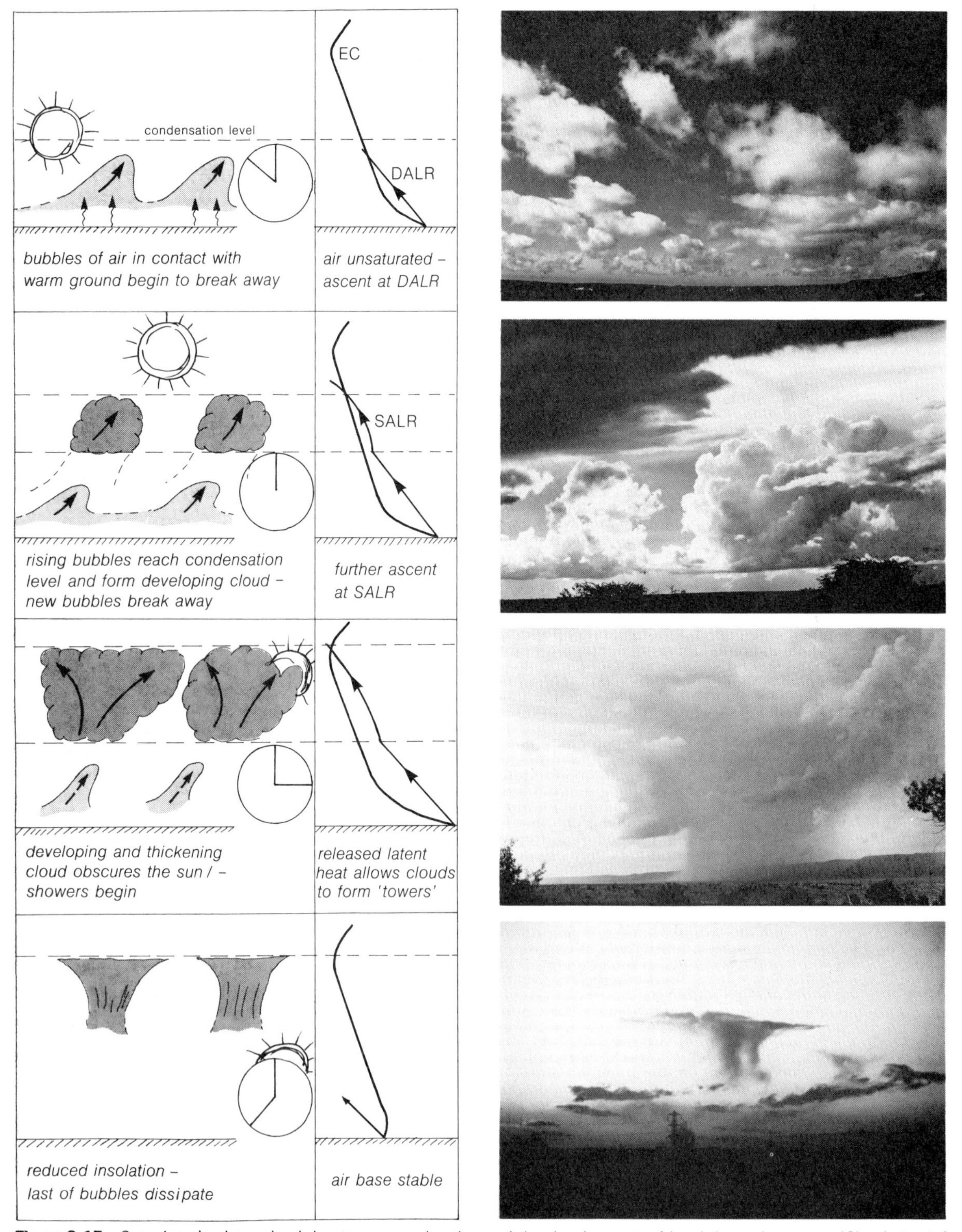

Figure 3.15 Cumulus cloud over land due to summer heating and the development of local thermal sources. 'Cloud streets' (parallel rows of cumulus clouds) are common in such circumstances.

showery interludes between the more general clouds of depressions, in most tropical and near-tropical regions they dominate the sky. These areas, where each **cumulonimbus** tower cloud (latin *nimbus*, meaning 'rain-bearing') shows the global circulation punching vast amounts of energy aloft, receive three quarters of the total world rainfall.

Because of their crucial role in the circulation of the atmosphere, the life cycle of **cumuliform** clouds is of vital interest (Fig. 3.15). They are born in the rising 'bubbles' of air called **thermals**; they grow and eventually die, within a few minutes or at most a few hours.

In the world of clouds, fat is beautiful. The fatter they are, the longer they survive, the more vigorously they behave and the taller they grow. Small clouds (which have small amounts of energy) may live only for 5–10 minutes and pump up 1000 tonnes of water vapour; a giant thunderstorm tower cloud can be active for several hours and in this time carry aloft more than 50 million tonnes of water vapour.

A rising cloud tends to drag in (**entrain**) part of the surrounding unsaturated air. Only the central region of a cloud escapes **mixing**, and the fatter the cloud, the larger the proportion of the central region that remains unaffected by such mixing. As a result, fat clouds continue growing upwards, whereas small clouds evaporate away into the surrounding unsaturated air. But even the largest clouds eventually stop growing and, with no more energy to sustain them, they dissipate and make room for the new ones to grow.

STRATIFORM AND CIRRIFORM CLOUDS

Stratiform and **cirriform** clouds are mostly layered, spreading out across the sky in varying patterns from heavy grey sheets to thin wispy veils (Fig. 3.14). They are a main component of cloud patterns in the mid-latitudes where they provide the bulk of the catherine-wheel cloud distributions that spiral into the centres of depressions, and they are also quite commonly found fringing mountain ranges.

Stratiform and cirriform clouds often form sheets, indicating that the air is not being affected by convection. These clouds therefore have no natural buoyancy. However, as air brushes over the land there is the additional disturbing effect of frictional turbulence (Fig. 3.16). This produces **stratocumulus**, or 'roll', cloud.

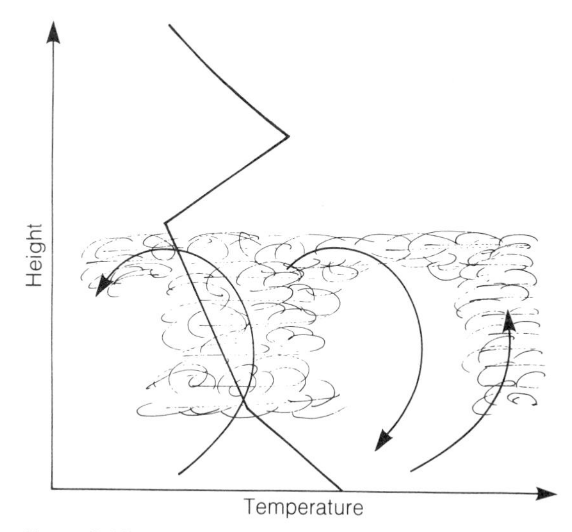

Figure 3.16 Forced ascent by turbulence brings cold, drier air down and takes moister air up in the lowest 1 km. Stratus or stratocumulus clouds form as a grey 'lumpy' sky, sometimes with a rolled appearance. Darker regions show ascent, lighter regions indicate descending (drying) air.

Stratus clouds are produced by condensation in air that remains stable, air in which the release of latent heat is insufficient to cause buoyancy. In calm conditions air often cools to its dew point by radiative cooling, producing the thin veil of stratus that is a common feature of early morning. 'Radiation' stratus is responsible too for the persistent grey sky know as 'anticyclonic gloom'. Stratus can also be produced by forced ascent in an airflow moving over hills, mountains or a range of coastal cliffs and is widespread in a depression where warm and cold air meet. Here it frequently attains a great thickness and is called **nimbostratus**, dominating the higher and thinner bands of stratiform clouds.

In a depression, stratus clouds have their own life cycle, although each cycle is much longer and not as obvious as that for cumulus clouds. As warm air is pushed ever higher at a frontal zone it must shed progressively more vapour and so the cloud thins, becoming transformed first to **altostratus** then perhaps to **cirrostratus** before finally fading away. **Cirrus** clouds are high-altitude clouds formed of ice crystals; they owe their veil-like appearance to the small number of particles at these levels that can act as nuclei during freezing.

Global circulation

4.1 Global air movements

As Halley so perceptively realised, the rise of warm air over the near equatorial regions must initiate a circulatory system. He envisaged a single global convection cell in each hemisphere, as a result of the heat imbalance over the Earth. However, there are several other factors to consider, each of which has a profound influence on the final circulation pattern. For example, the fact that polar regions occupy a very small area when compared with near-equatorial regions means that only a small proportion of the warm air from the tropics could ever be accommodated near the Poles; the majority must be recycled in lower latitudes (Fig. 4.1). Furthermore, the Earth's rotation sets up a force (the **coriolis** force) that deflects any movement in a meridional (i.e. north–south) direction (Fig. 4.2). This is because the rate of rotation varies with latitude. For example, just to stay still with respect to the Earth's surface, air at the Equator must move at 470 m/s; at 60° a speed of only 224 m/s is required; while air near the Poles hardly needs to move at all in order to maintain its position. The result of air conserving its initial west–east momentum is that in moving polewards it is always moving *faster* in a zonal (i.e. east–west) direction than is the area of land over which it arrives, while that moving equatorwards moves more slowly. When the speed of the air compared with that of the land is greater, the air movement gives a westerly wind; when it is less, it gives an easterly wind. This is true in both hemispheres. Although the effect is hardly noticeable between 0° and 10° latitude, by the mid-latitudes the rate of change is substantial and its

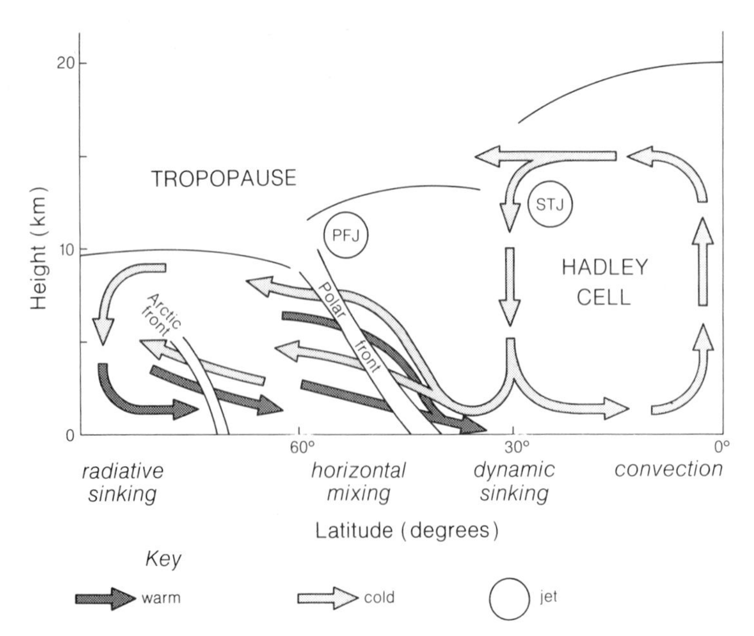

Figure 4.1 A generalised view of the global air circulation showing the change from vertical to horizontal air movement. This limits the convection (Hadley) cell to within 30° of the Equator.

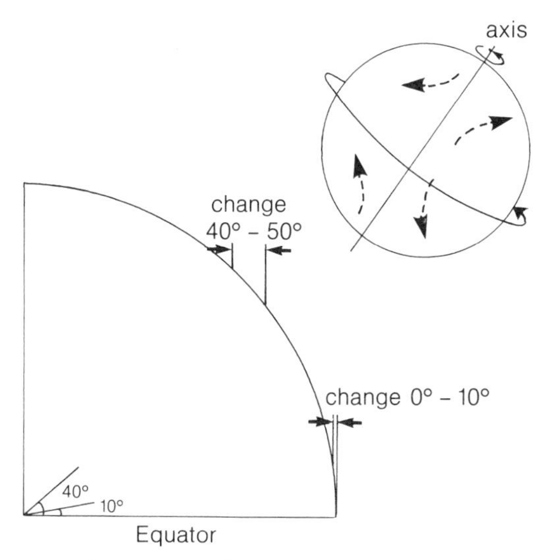

Figure 4.2 The deflecting Coriolis force (dashed arrows) varies with latitude from zero at the Equator, reaching a maximum at the pole.

effects add considerably to the complexity of the global circulation.

The most important effect of wind deflection occurs in the upper air at about 30° where an east–west 'tunnel' of fast-moving air forms, the core of which is called the **subtropical jet stream**. (Fig. 4.3). This air flow is of great importance to the climates of the sub-Tropics and especially to the pattern of monsoons.

Another major change in airflow occurs in mid-latitudes, as the troposphere becomes progressively shallower (Fig. 4.1). In this region vertical circulation becomes restricted and movement, in revolving eddies or spirals, is observed. Here *horizontal* rather than vertical motions dominate the circulation. The horizontally revolving spirals are known as **depressions** if air is being drawn towards (**converging**) on them, and **anticyclones** if it is being pushed away (**diverging**). Depressions and anticyclones are rather like the eddies created in a stream as a fast-moving thread of water drags in slower water from near the banks. In the troposphere, however, the fast-moving air lies *above* the eddies, in the form of a meander stream of upper air called the **polar jet stream**. The polar jet stream is more pronounced than the sub-tropical jet stream: it is reinforced by the large temperature contrasts created as cold air moving from the Poles meets warm air from the tropics.

4.2 The general circulation

Only near the Equator, where the coriolis deflecting force is weak and the troposphere very deep, can a thermal circulation cell be maintained. This cell dominates the circulation between the Equator and latitude 30°. It is called the **Hadley cell** in honour of George Hadley who, in 1735, was the first person to recognise the importance of the coriolis force.

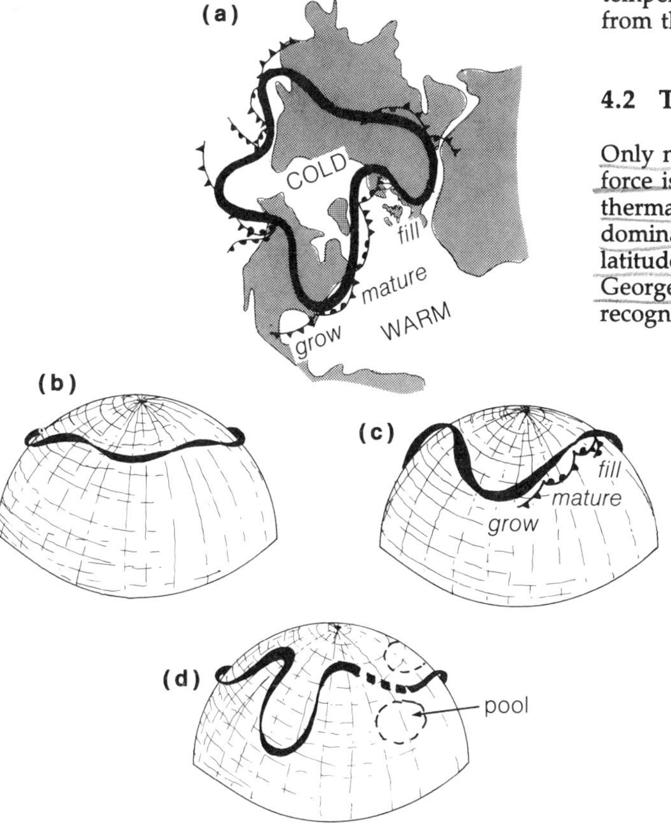

Figure 4.3 (a) Upper air waves seen on a chart centred on the North Pole. The lines are 500 mb contours. Note the relationship of upper air waves and surface depressions. Depressions are 'born' below a poleward band of the wave, travel along it, becoming first mature, then filling and dissipating. The central fast wave-like thread in the upper westerlies can meander (b) in a very gentle way (high zonal index), (c) with greater undulation (low zonal index) or (d) can cut off major loops leaving isolated pools. Beneath equatorward pools lie deep cold depressions, while poleward pools are seen as surface 'blocking' highs that may not shift position for weeks.

In the Hadley cell the main energy is derived from rising thermal plumes of cumulonimbus cloud which release large quantities of latent heat into the upper air. These are the atmospheric pistons that drive the powerhouse of the atmospheric engine. At any one time there may be up to five thousand of these giant pistons at work – growing, maturing and then fading away, to be replaced by others in their turn. At the poleward limit of the Hadley cell, the decreasing space available in the troposphere forces some of the warm upper air to return to the Equator. This is accomplished by widespread **subsidence** and an outflow of air at low levels. This subsidence forms the **subtropical high pressure belt** that encircles the Earth, centred on about 30° latitude. The dynamic subsidence involved is sufficient to overcome any surface instability and prevent air rising. On the equatorward side of the subsiding air lies the subtropical jet stream; underneath it lie the world's great deserts.

If the tropics can be regarded as the powerhouse of the atmosphere, then the mid-latitudes can be compared with gear wheels in the general circulation, their horizontal motions feeding warm air to fuel the high latitudes, and conveying cold arctic air back for warming near the Equator. The position of the gears is determined by the upper-air polar jet stream. Unlike the subtropical jet, whose position is related to coriolis deflection alone, the polar jet is 'tied' in its position by the location of major mountain ranges (such as the Rocky Mountains) and by contrasts in temperature (between continents and oceans). For example, as air flows over the Rockies it compresses into about half the thickness it had over the oceans and then expands again as it flows onto the lowland beyond (Fig. 4.4). Compression causes the airflow to diverge and slow down; expansion has the opposite effects. These contrasts set up an oscillating motion which makes the air move first polewards then equatorwards in the lee of the mountain. Expansion also occurs as air is warmed by contact with a warm ocean, and contraction (compression) as it is cooled by contact with a cold land mass. Thus a combination of topographic barriers and temperature contrasts help create and then fix the general position of the jet stream.

Just as in a river there are sometimes large meander loops and sometimes small, so the number of meanders in the upper air tends to vary between three and five. Sometimes extreme meandering can cut off an entire loop, leaving a pool of either warm or cold air isolated to one side (Fig. 4.3d). These pools of air are crucial parts of the global circulation, for they allow air to escape from the mid-latitudes back to the Equator or Poles.

Upper air movements in mid-latitudes are responsible for the pattern of 'highs' and 'lows' that travel around the globe. For example, as air meanders

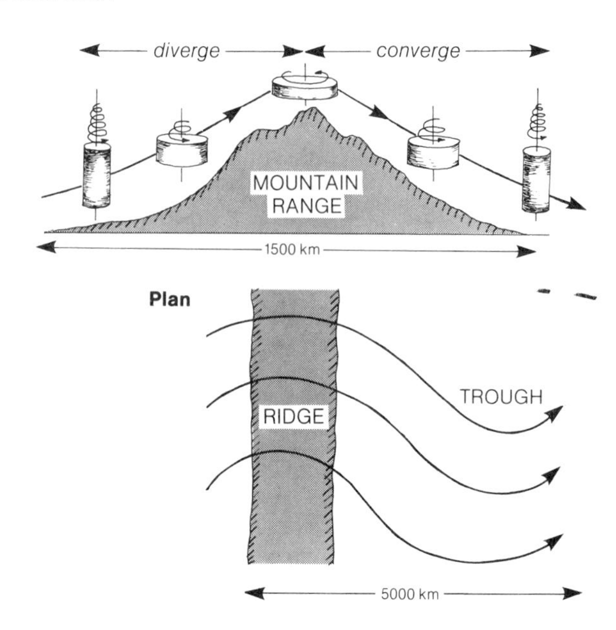

Figure 4.4 As air is forced over major mountain barriers compression and expansion cause an upper air wave to form.

polewards it accelerates, causing more air to leave the poleward limb than arrives (Fig. 4.5). Such upper divergence draws in air from the lower troposphere, leading to convergence of air near the surface and **cyclogenesis** (the formation of depression families). Conversely as air meanders equatorwards it decelerates and more air arrives than leaves. The resulting convergence aloft pushes air towards the surface, leading to divergence near the surface and the formation of an anticyclone. Britain lies just to the south of the average path traced by depression families and, consequently, its weather patterns are dominated by alternations of 'highs' and 'lows'.

4.3 Ocean currents

Ocean waters are a vast heat source for the atmosphere. Unlike land, which conducts heat poorly and has a low thermal capacity, the surface waters of the oceans are deeply penetrated by insolation, which provides direct warming as far as 100 m deep. However, solar radiation penetrates effectively only when the Sun is nearly overhead, when solar rays strike the sea at an acute angle; the majority of the radiation at other times is reflected. As a result, ocean waters in high latitudes are poorly heated.

The great contrast between heat received and stored by the oceans in low latitudes and that by oceans in high latitudes is responsible for the convectional flow of ocean waters (Fig. 4.6). The water flows towards the poles, and because warm water is less

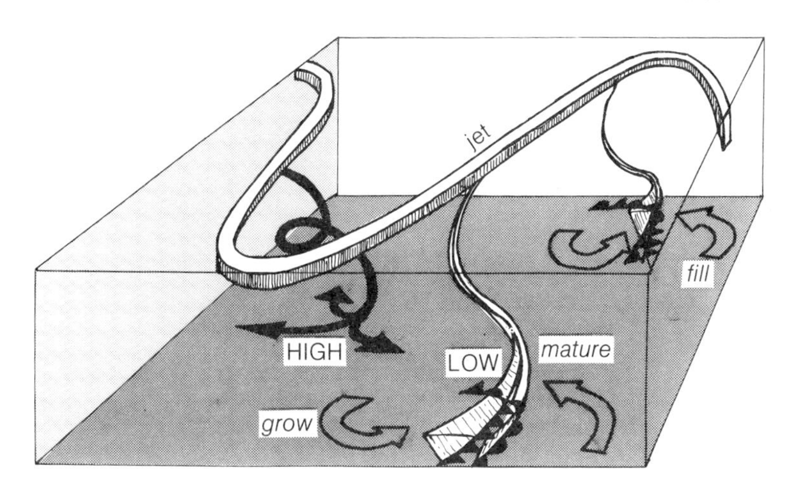

Figure 4.5 Airflow on the equatorward limb of an upper-air wave is decelerating and more air arrives than departs. This convergent flow forces air towards the ground, creating a compensating region of divergent air flow called an anticyclone. The situation is reversed on the poleward limb, with air in the high-level wave accelerating. More air leaves than arrives and the divergence draws in air from the ground, where compensating surface convergence is reflected in the formation of a depression family.

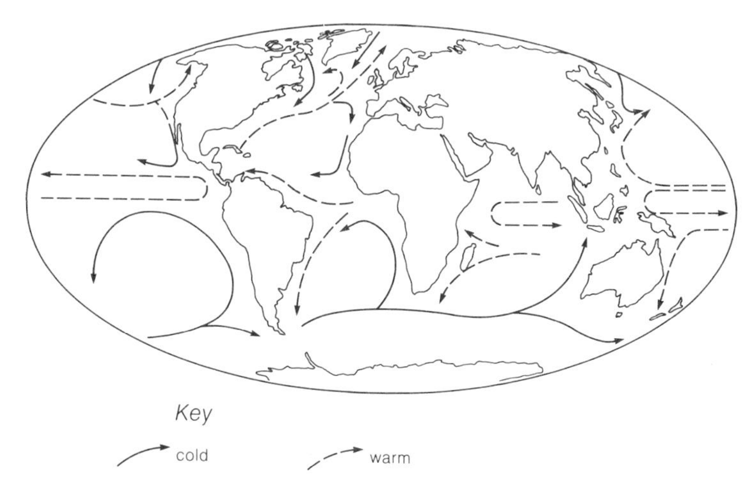

Figure 4.6 The generalised pattern of ocean currents.

dense than cold, it tends to move in the upper levels of the oceans; the cold dense water usually returns towards the Equator at depth. In direct contrast to columns of air in the atmosphere, the oceans are heated from above and therefore resist vertical overturning. However, movement of ocean water is affected by the Earth's rotation, causing water to be deflected to the right of its path.

The main north–south movements of ocean water (**ocean currents**) are driven by temperature contrasts, but the east–west movements are partly wind-driven. The constant northeasterly trade winds, for example, help transfer water from the coast of Africa, across the Atlantic where it builds up the level of water in the Gulf of Mexico.

Warm currents tend to form well defined threads within otherwise cold water; return currents of cold water are usually much more diffuse. Both warm and cold currents have an important effect on climate, because they determine the heat input to the low levels of the troposphere. Westerly winds blowing over the Gulf Stream, for example, are warmed at the base and can therefore hold increasing quantities of moisture as they travel across the Atlantic. Furthermore, the warming also increases environmental lapse rates, increasing the likelihood of convective instability. By contrast, as the trade winds blow over the cold waters of the eastern Atlantic they are cooled, the environmental lapse rate decreases and stability is increased.

4.4 The global pattern

The complex pattern of airflows that go to make up the general circulation of the atmosphere prove to be far removed from the simple one-cell model. Air does not always flow smoothly from one part of the Earth to another; in many areas it flows fitfully, accumulating an imbalance in one region, then compensating with a rapid and irregular flow of air to

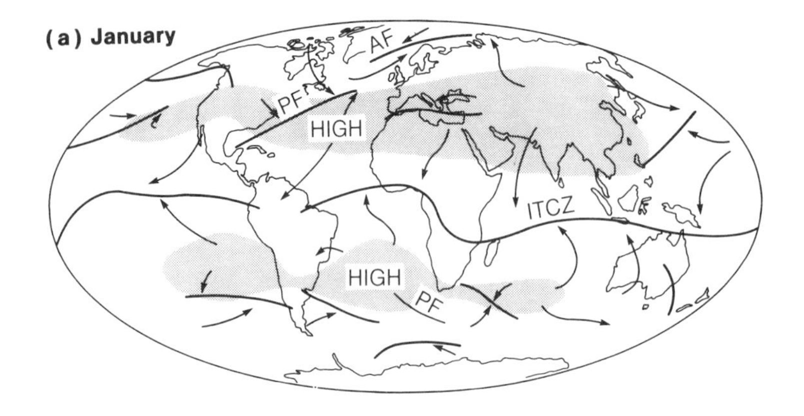

Key

PF	polar front
AF	arctic front
ITCZ	inter – tropical convergence zone

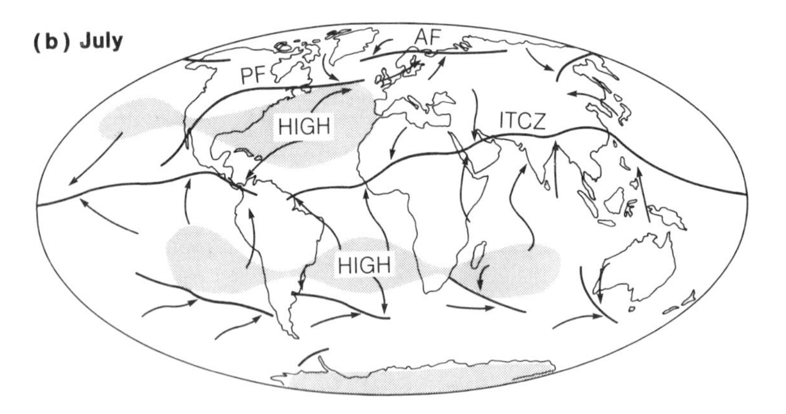

Figure 4.7 (a) Average surface circulation in January, showing mean sea-level isobars (continuous lines), mean winds (broken arrows), mean positions of fronts (heavy lines) and air masses. (b) Average surface circulation in July, showing mean sea-level isobars (continuous lines, mean winds (broken arrows), mean positions of fronts (heavy lines) and air masses.

another. It is this irregularity, accentuated by an element of random flow common to all turbulent motion, that has made it so difficult to replace Halley's model with one that matches up to observation. Thus, although the Halley model gave way to another with three interlocking cells, even this is now recognised as too great a simplification. A less 'tidy' pattern (Fig. 4.1) is probably more appropriate. Furthermore, the effects of the continents and oceans are so profound that a description of the atmosphere along a predominantly oceanic meridian may not fit the atmospheric flow over a predominantly continental meridian (Fig. 4.7). Places where convergent air meets near the Equator thus form an intermittent *zone* (the **intertropical convergence zone or ITCZ**), whereas convergent air in mid-latitudes is represented by a fragmented pattern of fronts, strong over oceans, weaker over continents. These basic elements of the circulation (Fig. 4.6) enable us to describe and explain the world's climates (Chapter 5).

Climates

5.1 Classification

The problem

Geographers have been closely involved with meteorologists in describing the pattern of the Earth's climates. Indeed, the crucial role of climate in determining Earth surface processes is underlined by the use in geography of such terms as 'desert' landforms and 'tropical humid' weathering. Climate also exerts fundamental control on the distribution of natural vegetation, on soils and on the activities of man. For all these reasons climate is central to geographical studies.

Climates reflect the long-term patterns of weather and thus rely on weather measurements: rainfall, pressure, temperature, humidity, cloudiness, wind speed, insolation. It was therefore relatively straightforward for early climatologists to attempt to characterise regions with similar weather patterns by employing long-term average figures of each type of measurement. Foremost among the measurements employed were those of temperature and rainfall.

TEMPERATURE

Temperature has been regarded as a crucial indicator of climate because of its influence over plant growth. For example, an average monthly temperature above 6 °C is required for plant growth; 25 °C represents optimal growing temperature; and many plants cannot tolerate frost ($T = 0$ °C) or very high temperatures ($T > 40$ °C). Temperature regimes also play an important role in influencing both human and animal activities. They are closely related to the energy balance of the Earth and show greatest change with latitude (Fig. 5.1). Seasonal contrasts are at a minimum near the Equator, where the Sun remains at a high elevation all year, and increase towards high latitudes as changes in the seasonal altitude of the Sun become more pronounced.

RAINFALL

Rainfall patterns are similarly fundamental in establishing climatic regimes. Terms such as 'wet', 'humid', 'sub-humid', 'semi-arid', and 'arid' are common in climatic description. Rainfall, like temperature, places strict controls on plant growth and hence, at least indirectly, on human activity.

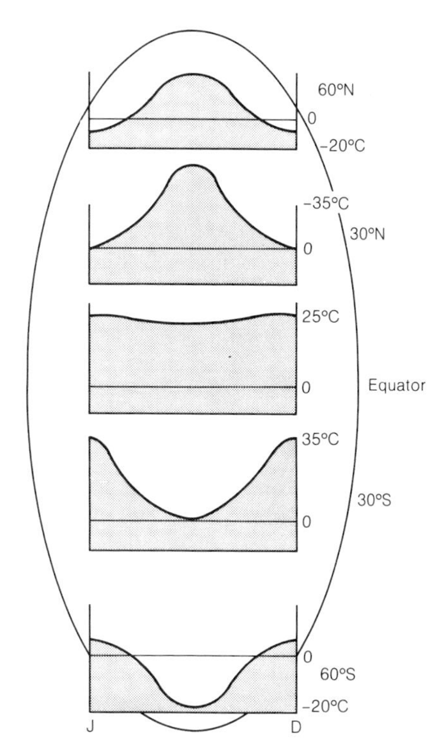

Figure 5.1 Characteristic thermal regimes on a hypothetical continent. Note that *continental* (a large seasonal temperature range) and *maritime* (small seasonal temperature range) influences have been disregarded.

Both an excess of rainfall (causing waterlogging and leaching in soils) and a deficit (causing plants to restrict their growth rate or even to wilt and die) are the most important controls. However, the effects of rainfall are complicated by temperature; rainfall is a less than satisfactory indicator of some climates. For example, in tropical regions deserts often receive less than 250 mm of rainfall a year. The same rainfall in Arctic regions would keep soils permanently saturated.

Rainfall is not related to latitude in the same way as temperature; rather it depends on the position of moisture-bearing winds (Fig. 5.2). Thus rainfall becomes greater both equatorwards and polewards of the subtropical high pressure belt and is higher on the eastern seaboards of continents in tropical regions (because of the trade winds) and on the western seaboards of the mid-latitudes (because of the westerlies). For all these reasons rainfall *patterns* tend to be more important than *totals*, especially in regions that have strongly developed wet and dry seasons. There are three important patterns:

(a) uniformly distributed (e.g. mid-latitudes, equatorial regions);
(b) seasonal, with rain heaviest in the *hottest* part of the year (e.g. tropics);
(c) seasonal with heaviest rain in the *coolest* part of the year (e.g. Mediterranean).

Alternative systems

There have been many attempts to combine both temperature and rainfall in describing areas with similar weather patterns. The system derived by Köppen in the early part of this century (Fig. 5.3) is most widely used. However, it is possible to combine observations of rainfall, temperature, insolation and evapotranspiration in a way that provides a pattern of climates and that is also of wide practical value. This was achieved by Thornthwaite who used the **soil-water budget** as the centre of his classification (Fig. 5.4; see also the water cycle).

The soil-water budget includes all or some of the following elements:

(a) a period of surplus when rainfall exceeds evapotranspiration and soil-moisture storage is full;
(b) a period when evapotranspiration exceeds rainfall and the soil may move into deficit;
(c) a period when rainfall exceeds evapotranspiration but when the soil is still not saturated, soil moisture storage is being replenished and there is no runoff.

The degree and duration of deficit are also vital information for assessing irrigation need and the demands on water supply.

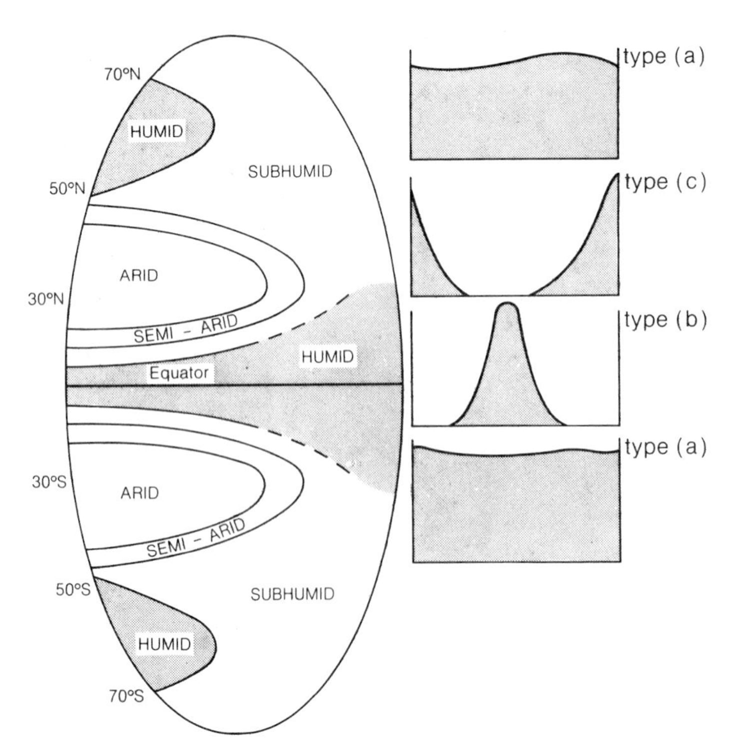

Figure 5.2 Characteristic precipitation regimes on a hypothetical continent.

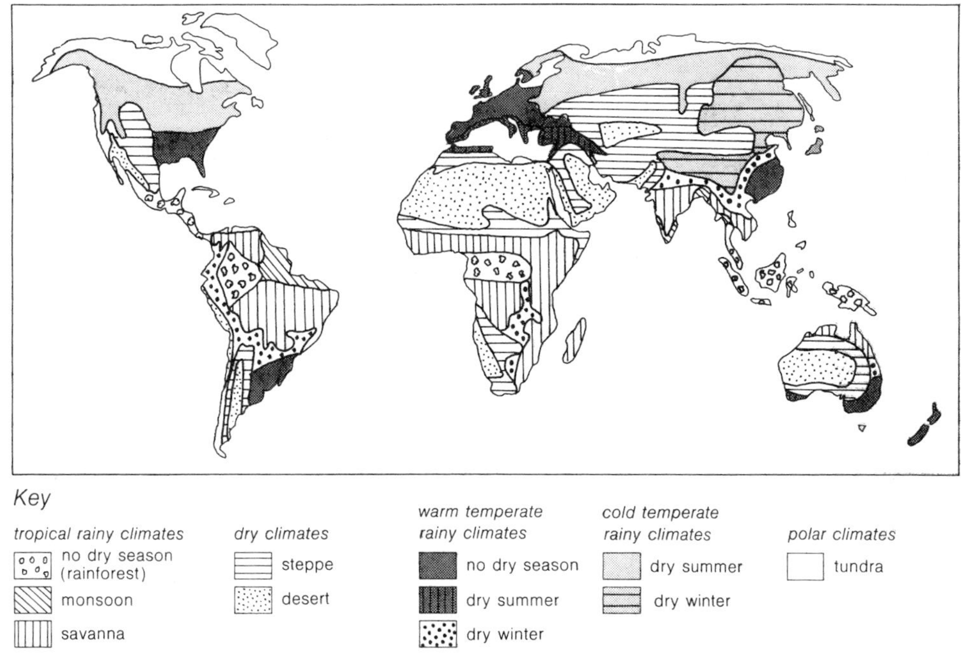

Figure 5.3 Climates of the Earth, according to Köppen.

Low-latitude climates can be defined on Thornthwaite's classification by a total annual water deficit greater than 1300 mm, **mid-latitudes** by an annual water deficit of 1300 mm to 525 mm, and **high latitudes** by a water deficit of less than 525 mm.

Moist and **dry climates** are distinguished by regarding dry climates as having a total annual soil-moisture deficit of 150 mm or larger and no monthly period of water surplus. Divisions within dry and moist climates can then be based on monthly values of soil-moisture storage. Thus Singapore, classified as equatorial by Köppen, has no monthly soil-moisture deficit and the monthly surplus available for runoff is large (Fig. 5.5a). By contrast the 'subtropical monsoon' classification of Köppen has a long period of soil-moisture deficit and a much shorter period of surplus. In the rainy season, part of the rainfall is used to replenish the deficit, but the rainfall intensity is so high there is a large surplus for runoff (e.g. Mercara, India, Fig. 5.5b). A semi-arid climate such as that of Khartoum has no month when the soil moisture is recharged, leaving a deficit throughout the year. Mid-latitude humid climates (cool temperate, western-margin climate of Köppen) such as that of Britain has a relatively short and modest period with soil-moisture deficit and moderate runoff.

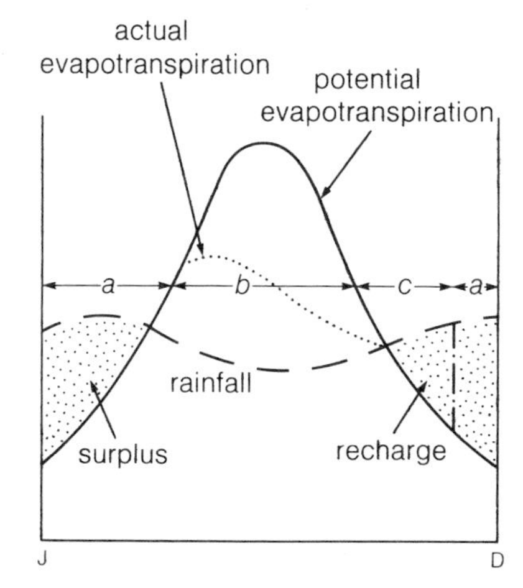

Figure 5.4 The basic elements of a water budget. The pattern shown is appropriate for Britain.

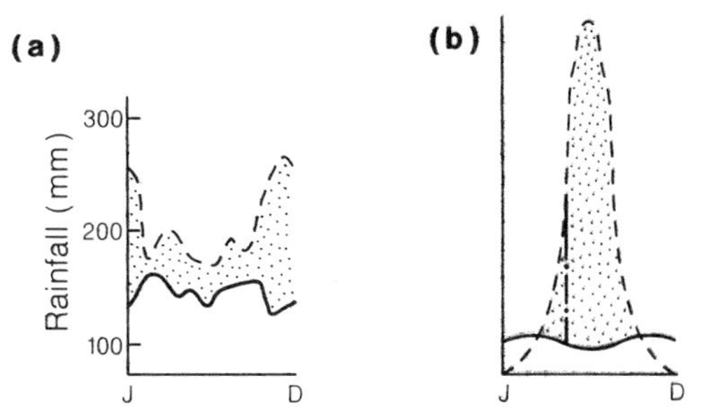

Figure 5.5 Water balance diagram for (a) Singapore, and (b) Mercara, India.

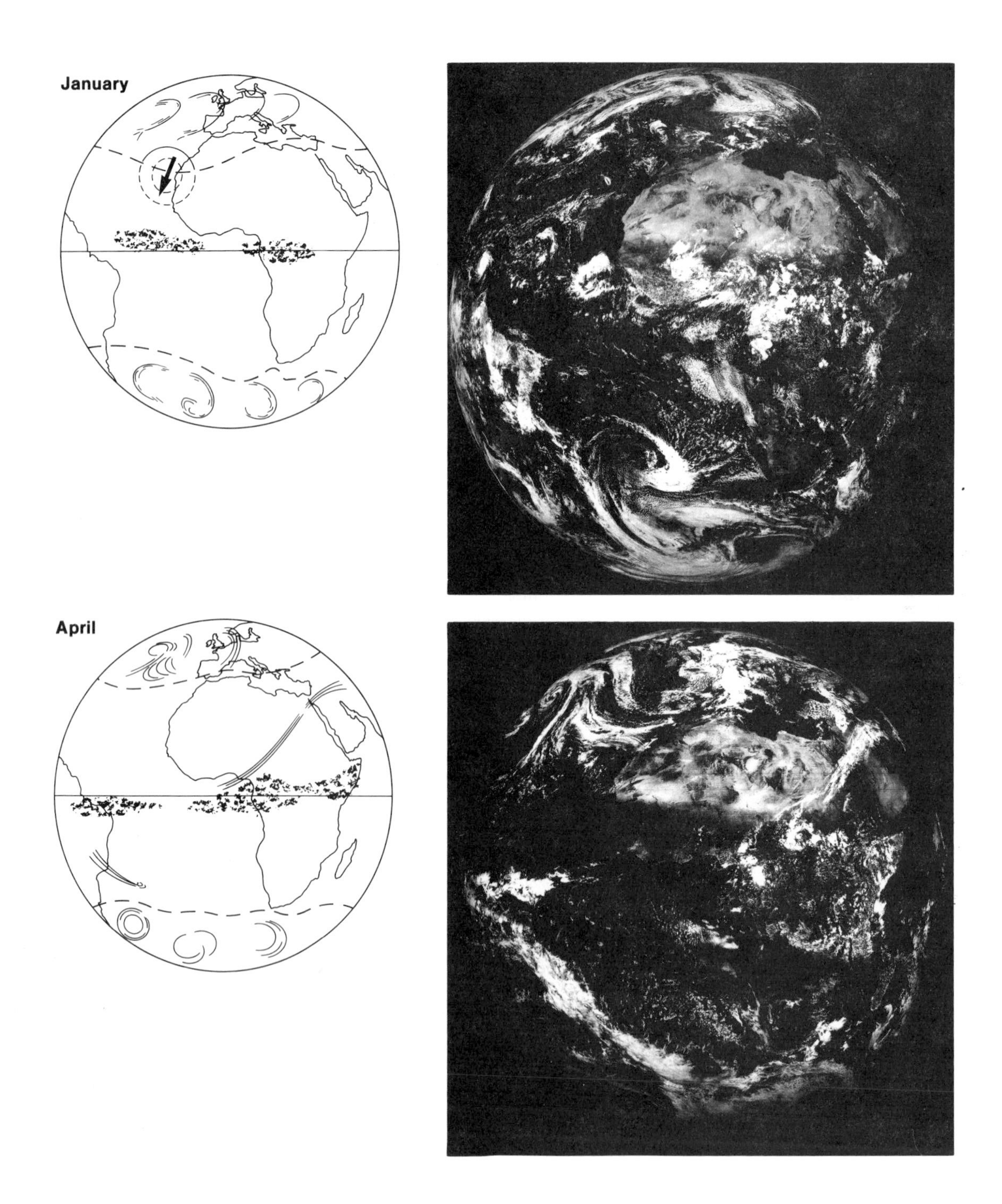

Figure 5.6 Global weather patterns.

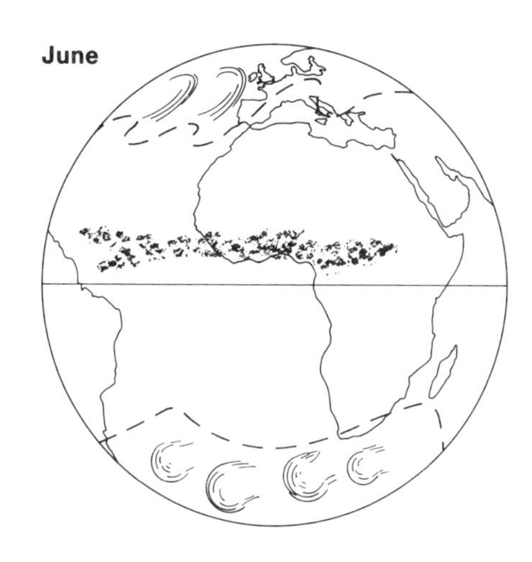
June

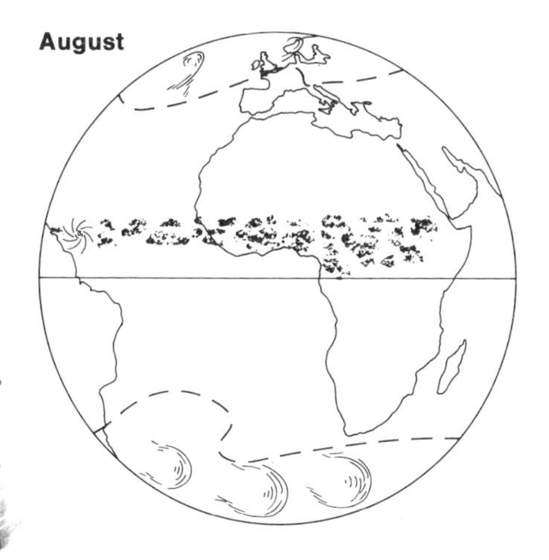
August

Key

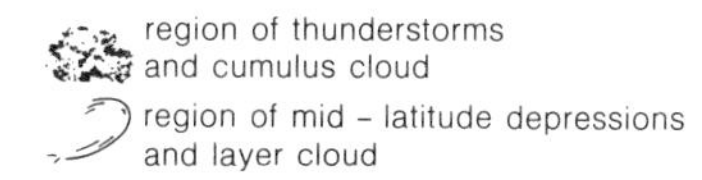
region of thunderstorms
and cumulus cloud
region of mid – latitude depressions
and layer cloud

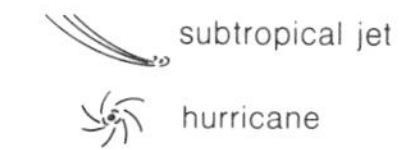
subtropical jet
hurricane

5.2 Global weather patterns

An outline

Because the global circulation is driven from the tropics, it is logical to begin to look for global weather patterns in low latitudes. Here the first task is to distinguish regions of cloud from those with clear skies. In Figure 5.6, this is done for one day in each season. The cloud patterns also provide many clues to the nature and variability of the weather. For example, the irregular belt of patchy cloud in the tropics is made from clusters of massive individual thunderstorm clouds (the pistons of the weather machine). Because each photograph shows cloud clusters in different places, separated by regions of clear sky, it is very easy to deduce that rainfall must be localised, torrential and very variable. Compare the positions of this 'central' band from January to August (Fig. 5.6) and its latitudinal shift becomes apparent. Such a shift may leave some regions permanently beneath the cloud band, but clearly others will only have seasonal cloud. In other words these photographs illustrate the contrast between **equatorial** (no dry season) and **tropical** (one wet, one dry season) climates, normally only seen by graphs of rainfall.

Using the near-equatorial cloud band as a base, areas nearer the Poles can now be examined, noting first the clear skies that occur above the world's great deserts. Look, for example, at the vast areas of clear skies that persist from one photograph to the next over northern Africa. Here the air is subsiding despite the intense ground heating that must occur with clear skies. Normally, surface heating would lead to instability and convection, but in this area the dynamic circulation of the Earth dominates over local heating. The result is the presence of permanent subtropical high-pressure cells, regions in which air subsides before spreading out at low level both towards the tropics and mid-latitudes. The dark tones on the infrared image at about 30°N and 30°S confirm that these are truly the hottest places on Earth.

The photographs also show the way cool air from mid-latitudes can return to the Equator by travelling through a region of subsidence. For this to happen the subtropical high pressure system must occasionally break down. The January photograph, for example, reveals a pool of cold air, marked by spiralling bands of cloud, moving equatorwards. It is being warmed and therefore becoming increasingly unstable by passing over the ocean off the coast of northwest Africa. In this case, though, rain is unlikely: the shallow cumulus clouds tell us that high-level subsidence is still keeping the 'lid' firmly on the air, merely allowing it to drift equatorwards at low altitudes. But it is easy to see how a less effective 'lid' would allow such unstable air to rise and create the conditions for an intense storm, of the type characteristic of desert regions.

The effects of continents

Chapter 4 outlines the importance of continents and oceans in distorting the general circulation. There is a significant impact on the atmospheric circulation because the thermal capacity of oceans is large enough to subdue seasonal changes of temperature, while the lesser thermal capacity of the continents enhances the contrasts between both summer and winter and day and night. Continental influences are reflected in summer surface heating, creating a thermal low-pressure system which may be powerful enough to override the dynamic effects of the general circulation. Similarly, in winter, surface cooling can generate a thermal high.

Differences between ocean and continent at the same latitude are often reflected in the patterns of convective cloud. The season of maximum cloud on land follows the movement of the overhead Sun more closely than does the cloud over the oceans (Fig. 5.1). Furthermore, cloud development demands a continuing supply of moisture. An airflow track that results in a long passage across a continent will remove much moisture. Thus 'continentality' is associated not only with greater seasonal and diurnal temperature regimes, but also with lower rainfall. Moreover, because the direction of moisture-bearing winds tend to be constant at each latitude, there is usually a rainfall gradient across a continent (Fig. 5.7). Such an effect shows clearly in southern Africa, where the cloudless skies over the Kalahari desert in southwest Africa are strikingly different from the

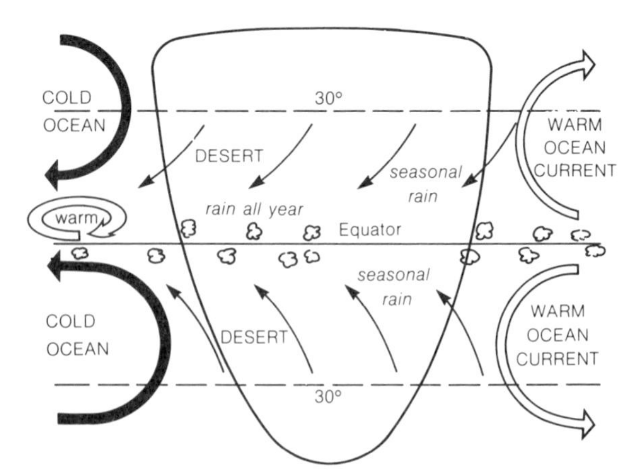

Figure 5.7 A hypothetical continent surrounded by ocean and derived in part from observations on Figure 5.6. The actual pattern of climate for any continent may vary significantly from this generalised model (compare with Figs 5.2 & 3).

cloudier coasts of the southeast. Here the southwesterly trade winds blow on shore along the east coast, bringing moisture and cloud, but they reach the west coast after a long journey across Africa and have virtually no moisture remaining.

Sometimes the contrasts imposed by the continents are reinforced by the effects of ocean currents. Ocean currents wash western seaboards with cold waters and eastern coasts with warm waters. The dry air flowing from such regions as the Kalahari or Atacama deserts is therefore cooled in its lower levels and rendered even more stable. This reinforcement creates the driest regions on Earth.

Monsoons

Monsoon (Arabic, meaning 'season') is a term used to describe some tropical and near-tropical climates characterised by a period (or season) of very intense rainfall which often begins abruptly. Monsoons are climatic *modifications* to the normal pattern of wet and dry seasons that prevails throughout much of the tropics. They are thought to reflect the interaction of continental effects with upper air waves.

Abrupt changes in the weather patterns are not common features of climatic regimes; where they occur, they indicate special local conditions. For this reason it is not possible to explain all monsoon areas in the same way. In West Africa, for example, the rainy season is associated with the poleward advance of equatorial air (Fig. 5.8). This air gathers moisture from the Atlantic Ocean and, given sufficient depth of unstable atmosphere, forms spectacular cumulonimbus tower clouds (Fig. 5.9). However, its source (over the Atlantic) and the rain falling from its clouds make this air cooler than the hot dry air that flows at low levels from the subtropical high-pressure zone over the Sahara. The Saharan outflow forms part of the northeasterly trade winds (called the **Harmattan** in West Africa) and is the surface-return airflow associated with the Hadley cell. When dry hot air and warm moist equatorial air converge at the ITCZ, the equatorial air undercuts the dry air: as it is so shallow, no clouds can form. Only beyond the zone of undercutting is the equatorial air deep enough to allow cloud development, which then occurs as a broad band or **squall line**. This comprises groups (clusters) of convective tower clouds which travel westwards, intensifying over Nigeria as they are fed with moist air, maturing and then dissipating in the cooler air over the Atlantic. It is the arrival of the squall line that signifies the onset of the monsoon.

Figure 5.9 Cumulonimbus tower clouds along a squall line in Africa.

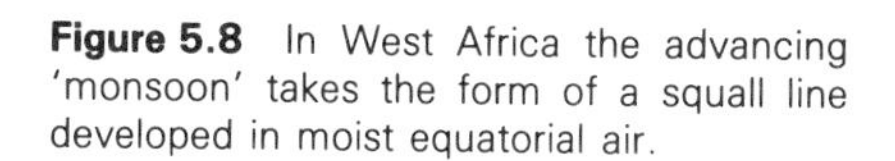

Figure 5.8 In West Africa the advancing 'monsoon' takes the form of a squall line developed in moist equatorial air.

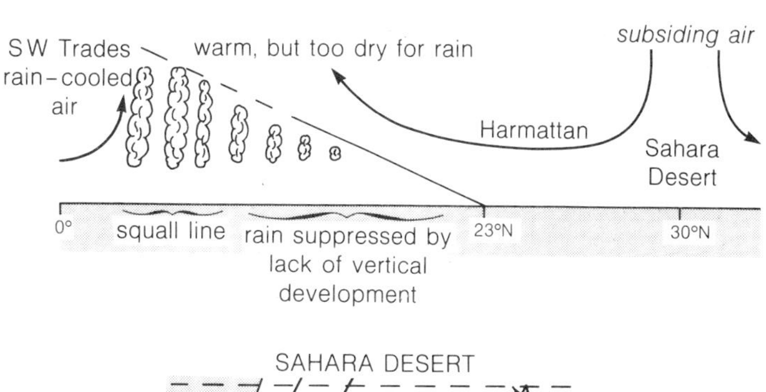

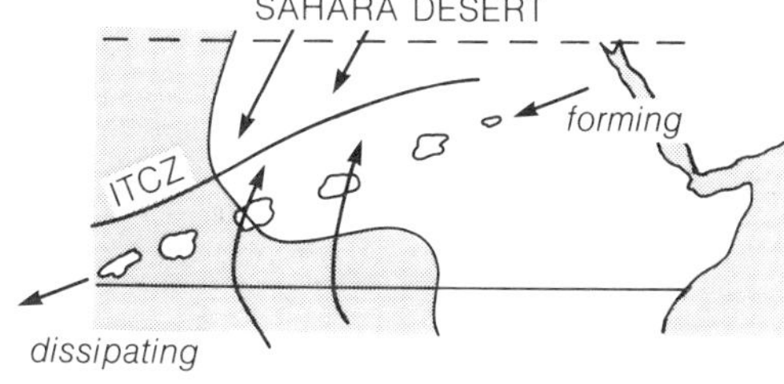

By contrast, in the well-known case of India the onset of the monsoon is believed to be caused by the blocking effect of the Himalayas on the seasonal movement of the pressure systems. At its simplest, the rainy season in India can be said to be the result of continental heating. In summer, hot air over the subcontinent creates a thermal low-pressure region which draws air in from the nearby oceans (Fig.5.10) and initiates convectional rainstorms. Towards autumn, as the continent cools down, the low weakens and the rain ceases. In winter the cool subcontinent creates a thermal high-pressure zone which inhibits rainfall. Nevertheless, the alternation of the winter high, with its subsiding dry air, and the unstable low pressure of summer occurs only gradually; it cannot be directly responsible for the onset of the monsoon. In India the monsoon begins particularly abruptly (Fig. 5.11).

The trigger mechanism for the Indian monsoon is believed to be related to the seasonal migration of the high-level westerly air flow (subtropical jet) which, in winter, is split into two parts. One of these passes to the south of the Himalayas across the Indo-Gangetic plain. As spring approaches and the poleward progression of the global circulation begins, the southerly arm of this high-level airflow is brought against the mountains where, for a time, it remains trapped. However, towards the end of May the jet abruptly diverts to the north of the Himalayas, allowing moist equatorial air to flow in over the subcontinent, steered by a high-level easterly air flow. It is the advance of the upper easterlies that produces the northwestern extension of the monsoon over India. Typically the monsoon breaks on June 1 in south-east India, but only reaches Pakistan on July 1. Following the change in flow systems, the monsoon merges into a normal rainy season of merging convectional rainstorms which slowly declines as autumn approaches, ground heating is reduced and the upper airflow again reverses.

Hurricanes

Hurricanes are another striking part of the near-tropical circulation (Fig. 5.6, August; Fig. 5.12). Like the monsoons, hurricanes are responsible for violent rainstorms, but they also bring fearsome onshore winds. Because of the terrible destruction that they wreak, they are of special interest to climatologists. Their prediction is vital if loss of life is to be minimised.

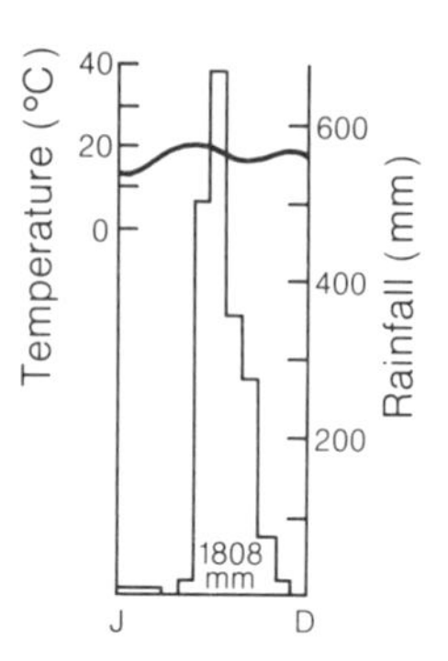

Figure 5.11 Climatic data from Bombay, India.

More air converges near the Equator than over any other part of the Earth. Yet despite the scale of convergence and the organisation of the clouds into clusters along squall lines, there is no sign of the spiralling motion so characteristic of depressions in mid-latitudes because here there is no coriolis 'twisting force'. Hurricanes are atmospheric pistons out of control – areas of rapidly converging cloud that have broken away from the equatorial band and wandered into latitudes where the Earth's rotation provides the twisting force required (Fig. 5.12). They are very seasonal in occurrence because breakaways are only possible when the intertropical convergence zone is near the limit of its poleward movement. Consequently, the Northern Hemisphere experiences hurricanes (typhoons, tropical cyclones) between July and September; the Southern Hemisphere experiences them on a more limited scale between January and March.

Although hurricanes are only half the diameter of a small mid-latitude depression (perhaps 200 km across), the winds in the hurricane are fearsome. Central pressures of below 950 mb keep winds revolving at over 33 m/sec within a mass of spiralling bands of cumulonimbus cloud towering over 12 km high. Like all thunderstorms, hurricanes derive their energy mainly from latent heat released as moisture

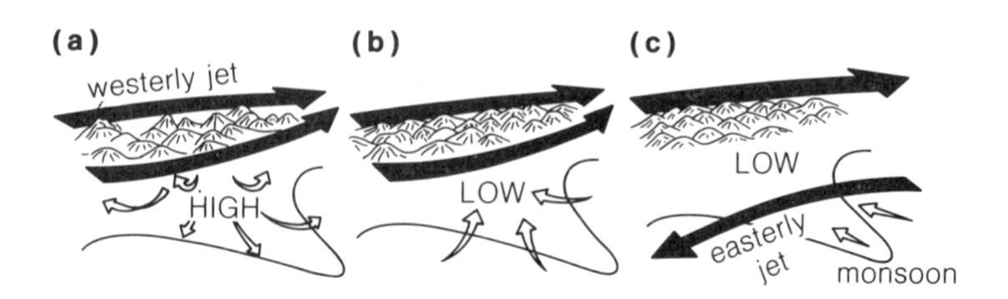

Figure 5.10 The causes of the monsoon. The rainy season is due to summer heating over India. A low pressure region forms and moist air is drawn in from the ocean. The abrupt start to the monsoon is caused by an upper movement of an easterly jet stream.

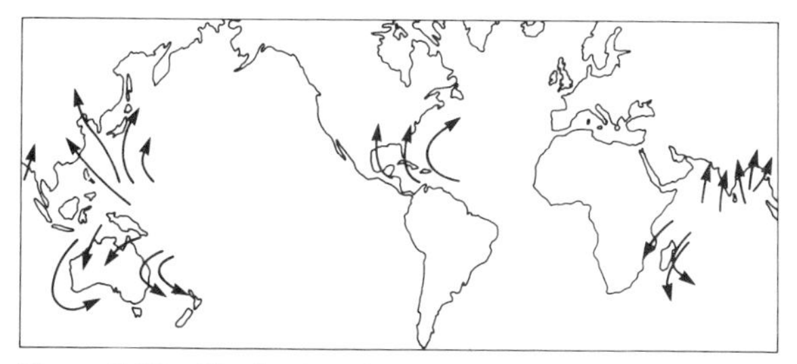

Figure 5.12 Distribution of tropical cyclones (hurricanes).

condenses within the cloud. Hurricanes are not sustained unless they remain over warm oceans where evaporation can continue to replace moisture lost as rainfall. When they finally move to colder seas or onto land, the injection of heat is so drastically reduced that they rapidly lose their energy and degenerate into forms more like mid-latitude depressions.

5.3 Mid-latitude weather patterns

Sources of air

The subtropical high-pressure belts encircling the Earth and centred at about 30° are the most important features of the global circulation. Within them horizontal pressure gradients are slight and wind speeds low. It therefore takes some weeks before the air flows out to more active regions, and in this time the surface air becomes uniform in temperature and humidity. All such large volumes of air with uniform characteristics that flow to other regions are called **air masses**. The outflow of air from the subtropical high pressure over the Atlantic (near the Azores islands) is one of the main **source areas** for the mid-latitudes (Fig. 5.13). Air flowing across the North Atlantic becomes uniformly cool and humid, and provides another source region of air; in winter, slow outward flow of cold air over Europe forms a further source region, in this case containing cold dry air. Altogether there are eight air masses that bring distinctive airflows over the British isles.

As air slowly spreads from source regions, it moves over areas where surface temperatures are different and it consequently becomes modified. Modifications are noticeable first in the lower air layers. For example, an air mass moving polewards is cooled in its lowest layers, lessening the temperature gradient within the air and increasing stability (Fig. 5.14). By contrast, air moving equatorwards is warmed in its lowest layers, especially as it flows over the warm ocean current called the North Atlantic drift, and temperature gradients steepen. This has the effect of making the air progressively unstable and more likely to produce convective cloud (Fig. 5.15b). In summer, warm land areas have the same effect, leading to

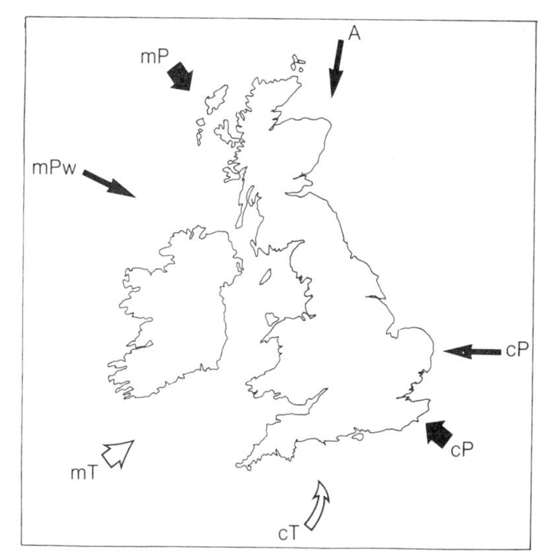

Figure 5.13 Air flows towards the British Isles from a number of contrasting source regions, including both those of cold and warm origin. The diagram illustrates the relative frequency of each air mass type at Kew.

Symbol	*Air mass name*	*Source regions*
mP	polar maritime	north Atlantic
mA	arctic	Polar ice cap
mT	tropical maritime	Azores
cP	polar continental	European mainland
mPw	warm polar maritime	Atlantic
cT	tropical continental	Sahara

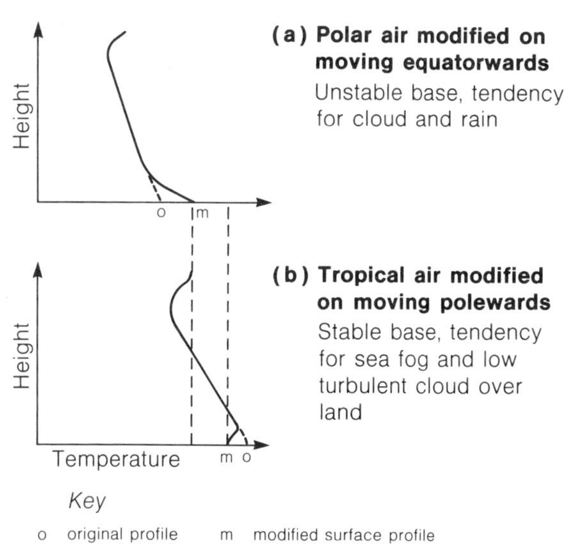

Figure 5.14 Air mass modifications on reaching 50°N.

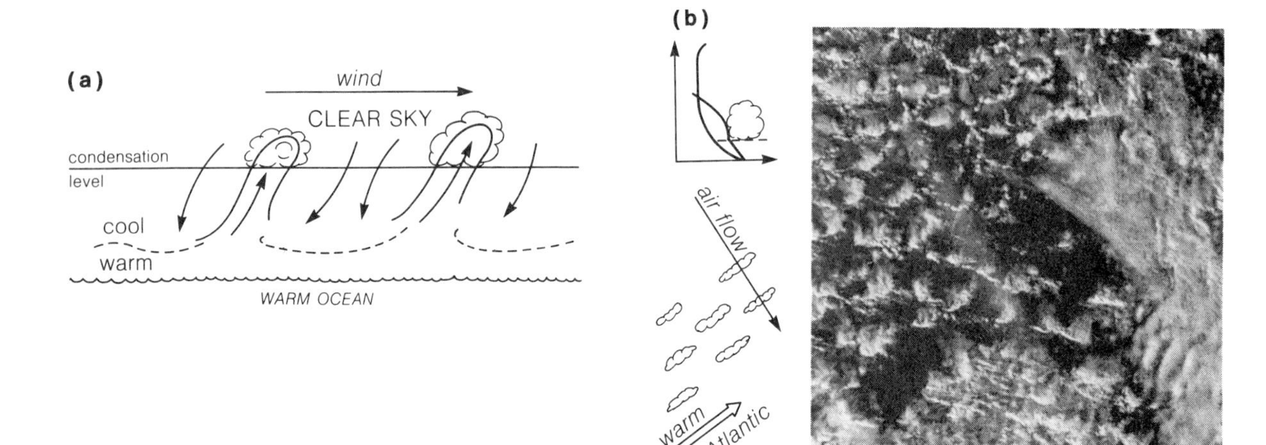

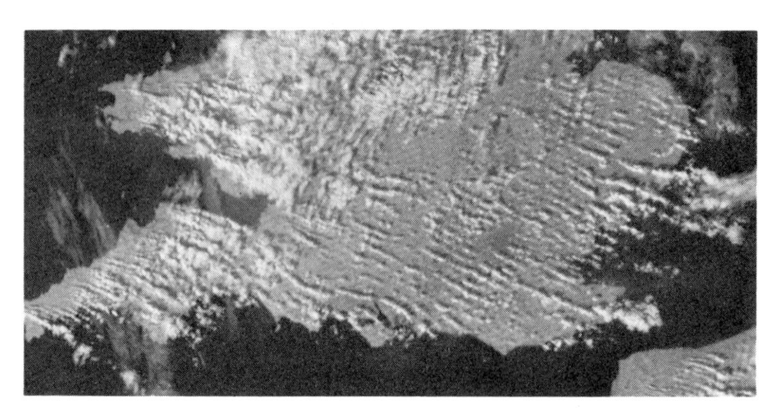

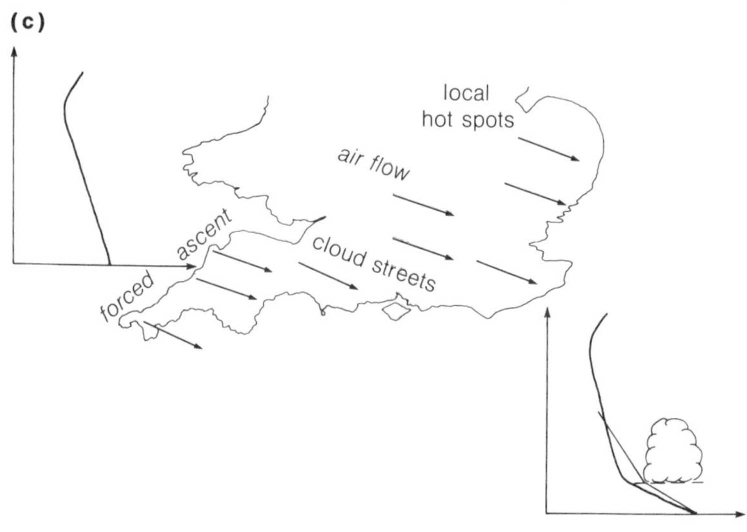

Figure 5.15 (a) The formation of cumulus cloud as cool airflows over a warm ocean, warming the base and producing a steep (unstable) air temperature profile. A pattern of cloud and clear sky is typical. (b) This satellite photo shows cumulus patterns over the ocean as cool air from the Arctic flows over warm waters of the North Atlantic Drift. The lowest layers are heated and become increasingly unstable. (c) This satellite photograph of southern Britain and northern France shows cloud street patterns (mid-afternoon in August). Forced ascent at the coast and inland local 'hot spots' combine to produce unstable air. Cloud height is limited by inversion aloft and shallow 'fair weather' cumulus prevails.

widespread instability at some distance inland from coasts (Fig. 5.15c). Each airflow can be identified from charts showing the surface air pressure and is associated with characteristic patterns or synoptic situations (Fig. 5.16).

Air pressure and winds

Air pressure is recorded on a weather map (**synoptic chart**) by means of 'contour' lines called **isobars**. Isobar values are given in millibars (mb) with 1000 mb representing average pressure at the Earth's surface. Pressure differences seen on a surface synoptic chart reflect the uneven distribution of the weight of air overlying the Earth's surface. Thus an increase of pressure from 980 mb to 1020 mb (represented by ten isobars) indicates an increase of four per cent in the weight of overlying air. Winds are produced as air flows from regions of high pressure to those of low pressure. The speed of air flow is related to the pressure gradient and is therefore indicated by the spacing of the isobars on the chart. Hence closely spaced isobars on synoptic charts are associated with strong winds. However, due to the Earth's rotation and the coriolis force, air always travels along a curved path, flowing at an angle to the line of the steepest pressure gradient. Thus air in the Northern Hemisphere spirals anticlockwise as it flows inward towards the centre of a depression and clockwise as it flows outward from an anticyclone.

The relationship between wind and pressure gradient is neatly expressed by **Buys Ballot's Law**: in the Northern Hemisphere, if you stand with your back to the wind, the low pressure is on your left-hand side.

Regions of interaction

Air moving from the subtropical and polar regions towards the mid-latitudes interacts across a broad zone called the **polar front**. Here upper air waves (highlighted by the polar jet stream) cause air to converge and diverge in such a way as to form **anticyclones** and **depressions**. The pattern of alternating anticyclones and depressions is the most characteristic element of mid-latitude weather.

Near the ground anticyclones contain downward and outward (diverging) flows of air (Fig. 5.17) and as they are steered eastwards along the path of the upper air wave, they bring calm and stable weather. Depressions, on the other hand, are regions of surface–air convergence in which air masses of contrasting properties are brought together. As the warm air ascends, drawn up by the high-level divergence, it rides over the cold air ahead (Fig. 5.18). In this flow, extensive sheets of stratiform cloud develop. As condensation proceeds, large quantities of latent heat are released which intensify the depression by leading to increased buoyancy and the localised growth of convective clouds immediately ahead of the surface warm front. This produces a belt of moderate rain in which are embedded convective cells of higher-intensity rain.

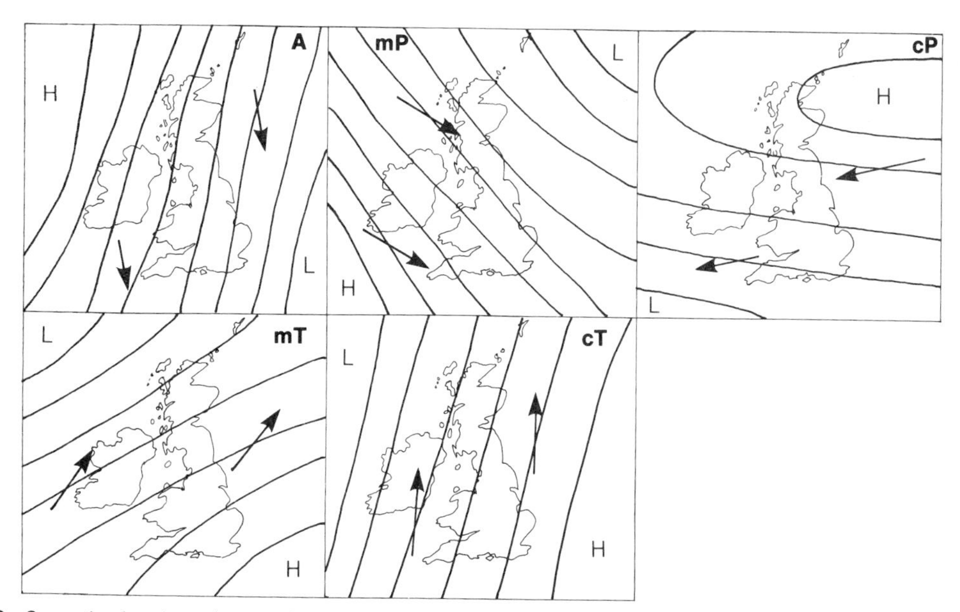

Figure 5.16 Synoptic situations that result in different air masses over the British Isles. Generalised isobars are shown, and arrows indicate surface wind directions; see also the key to Figure 5.13.

Figure 5.17 Characteristic synoptic charts for anticyclones and depressions over Britain. (a) Anticyclone, weak pressure gradient gives light and variable winds; (b) occluding depression, steep pressure gradient gives strong winds.

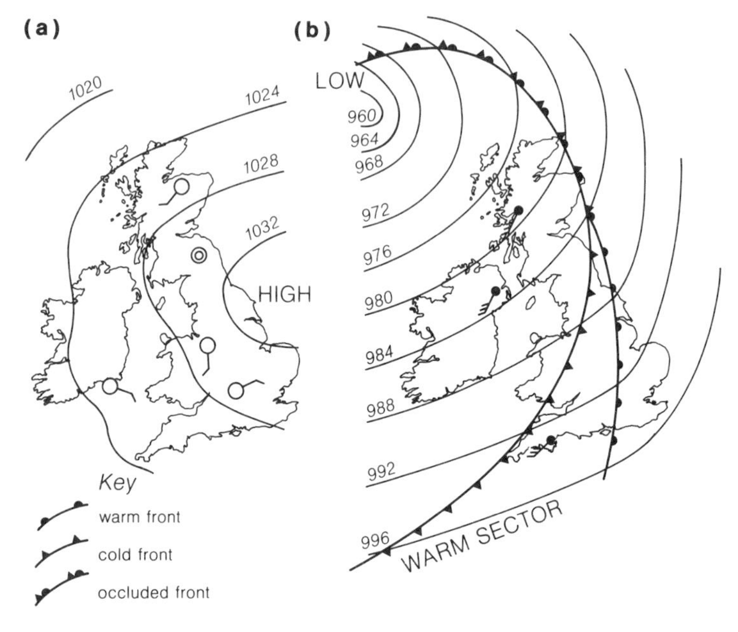

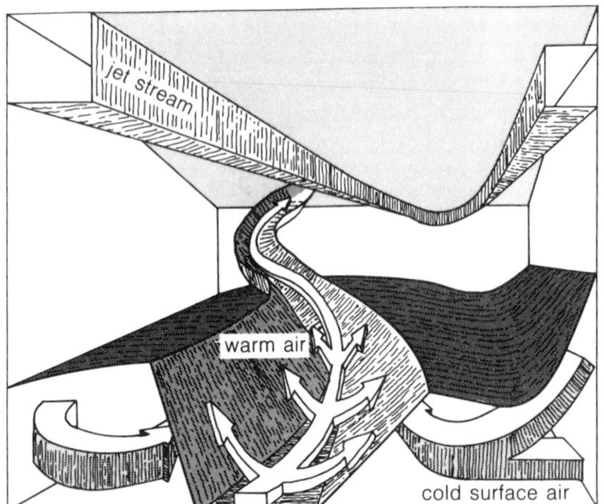

Figure 5.18 This model of a depression shows warm moist air ascending and merging with overlying cold dry air just ahead of the surface warm front. Overlying cold air causes convective instability and gives higher intensity periods of rain embedded in the general frontal rain produced by the mass rise of the warm air.

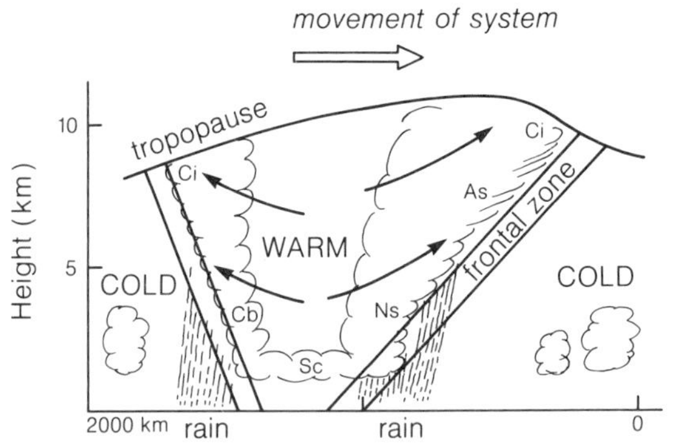

Figure 5.19 Warm air forced to rise at frontal zones produces deep cloud; by contrast, slight upward movement in the centre of the warm sector restricts cloud to shallow stratocumulus.

The warm air becomes trapped in a wedge shape between the cold air ahead and behind. This region of the depression is called the **warm sector**, each boundary of the sector being marked by uplift **frontal zones** (Fig. 5.19). However, there may be considerable contrasts between the form of uplift at the leading edge of the warm sector (the **warm front**) where the slope of the front is less than 1°, and the trailing edge (the **cold front**) where the front slopes at 2°. In general the steeper cold front produces much more active features as air moves rapidly upwards. The cold front is also usually by far the most conspicuous feature of a depression when seen on satellite photographs, marked by the readily-identifiable long curving band of cloud (Fig. 5.22).

Warm air carried aloft on the warm-sector 'conveyor' is drawn into upper air waves and thereby transfers energy to the high latitudes. On the other hand, cold air is frequently transferred towards the

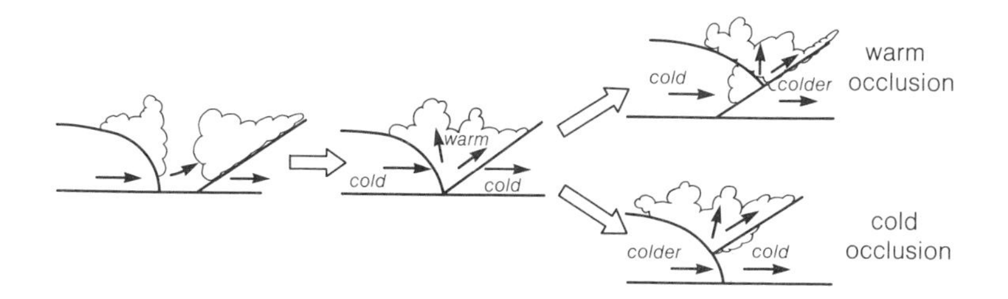

Figure 5.20 Occlusion characteristics.

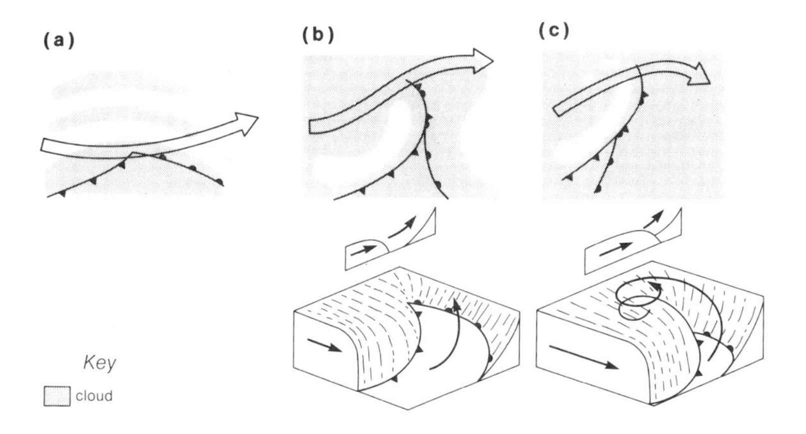

Figure 5.21 Schematic patterns of cloud cover (shaded) observed from satellites in relation to surface fronts.

Equator in an irregular manner at low levels, usually through outbursts of pools of air. Some places seem to be more favourable for such outbursts than others – one example is the Great Plains region in the lee of the Rocky mountains.

Occlusions

Occlusions develop because the cold front of a depression tends to move faster than the warm front, forcing the warm-sector air completely off the ground (Fig. 5.20). However, the cold air *ahead* of the warm sector rarely has the same temperature as the cold air *behind* it: each region has followed a different path, so each has been modified differently. Thus, when both regions of cold air meet, a new front is formed: this is called an **occluded front**. There are two possible results:

(a) the trailing air is colder and therefore undercutting, giving a **cold occlusion**;
(b) the trailing air is less cold, because it has been following a southerly Atlantic route, and yields a **warm occlusion**.

Occlusion cloud bands have many of the properties of clouds created in the warm sector, for the warm air continues to circulate aloft (Fig. 5.21).

Dynamics

The mid-latitudes form a particularly dynamic region within the general circulation and the patterns here need to be studied over time as well as in space. In this section consecutive satellite images give a 'time-lapse' effect of weather patterns during a week in the Atlantic, highlighting the drama played out by interacting air masses (Fig. 5.22).

(1) *3.8.1981* A depression north-east of Scotland has begun to fill and dissipate; a new depression begins to form south-east of Greenland
(2) *4.8.1981* The occluded depression north of Scotland has filled and moved away, while the new depression has grown and moved; it now lies south of Iceland. Cold polar air is being drawn into the depression, giving cumulus cloud bands (Fig. 5.21); the uniform white mass of cloud indicates the development of stratus cloud at the fronts.
(3) *5.8.1981* The depression has occluded rapidly and moved to a position off the Norwegian coast. A new depression is forming to the south-west of Iceland.
(4) *6.8.1981* The new depression remains stationary near Iceland while its predecessor continues to fill over Norway. A local thermal low has caused cloud over Britain and France.

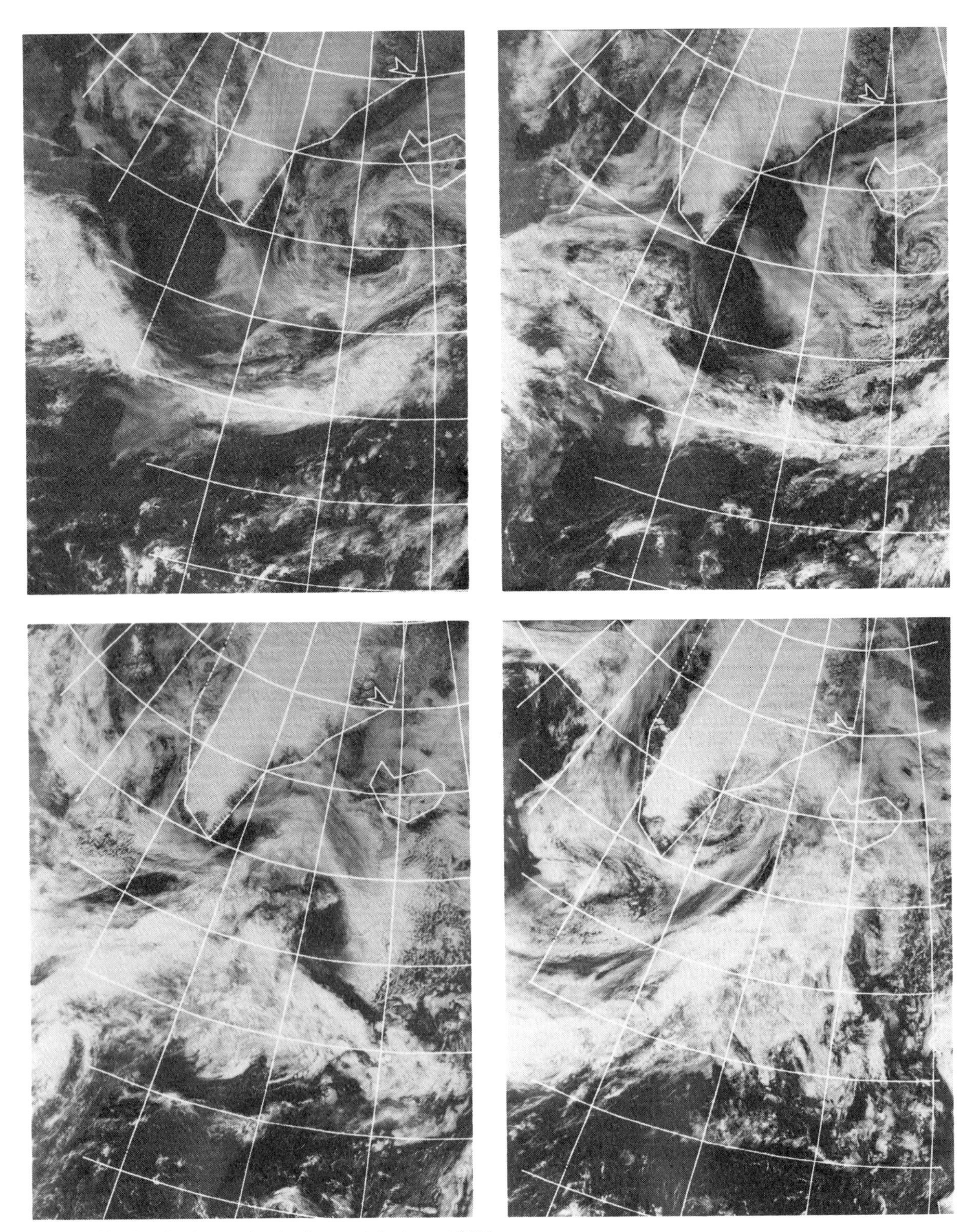

Figure 5.22 Satellite images for the first week in August 1981.

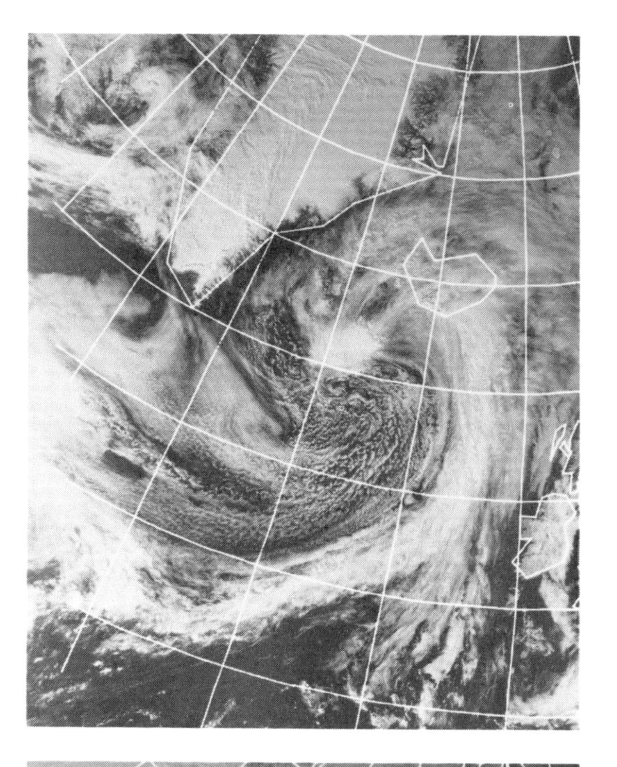

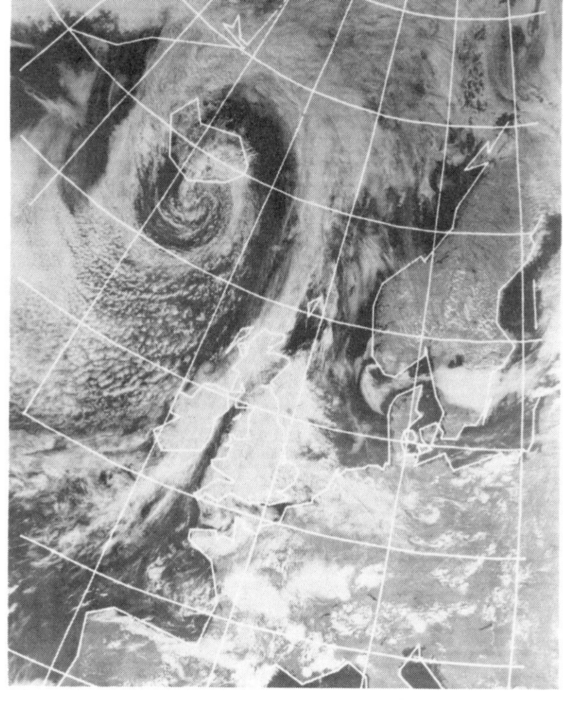

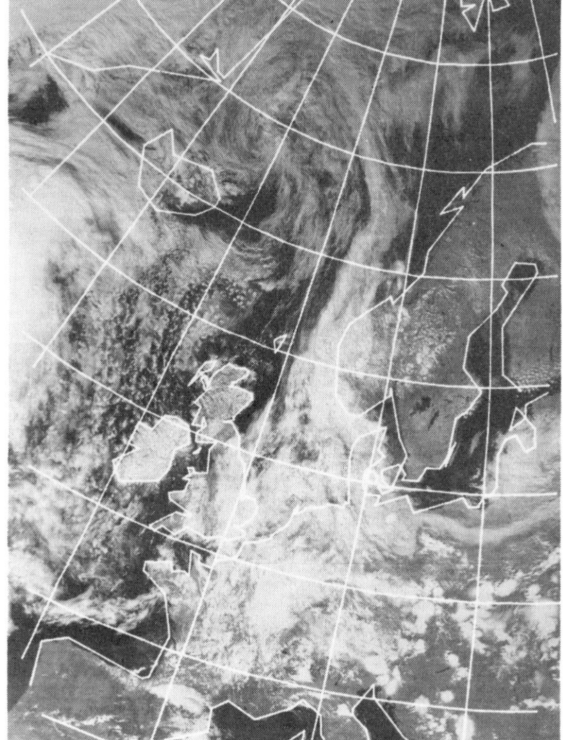

(5) *7.8.1981* the Icelandic low is now moving eastwards again with a pronounced band of cloud marking the cold front.
(6) *8.8.1981* The depression matures and moves east. The main cloud identifies the cold front, the warm front being less easy to locate. Unstable polar air is drawn south over the Atlantic to yield widespread cumulus cloud.
(7) *9.8.1981* The depression matures, moving northwards, but leaving a wide trail of frontal cloud over Britain. Meanwhile a new depression forms over the western Atlantic...

5.4 Local variations

Local-scale processes are largely created by small thermal contrasts over short distances, but they can modify the regional weather considerably, especially in quiet synoptic conditions.

Coastal climates

Distinctive coastal or lake climates develop because the thermal capacity of water is much greater than that of land; the temperature of water therefore fluctuates much less rapidly, both diurnally and seasonally. This constancy in turn reduces the temperature fluctuation in the overlying air. On the other hand the temperature of land and of the air overlying it vary much more rapidly and widely. Coastal climates develop primarily as a response to temperature differences between air overlying land and air overlying water. In all cases airflow is produced in sympathy with a *thermal* low developing over the warm surface and a *thermal* high where the surface is relatively cold. Because air flows from high- to low-pressure regions, an offshore breeze, or **land breeze**, is formed if the sea is warmer than the land (Fig. 5.23a), whereas if the land is warmer than the sea the breeze blows on shore, giving a **sea breeze** (Fig. 5.23b).

Coastal airflows influence the cloud patterns considerably because moist air flowing over an increasingly warm surface becomes progressively unstable and increases the likelihood of convective cloud formation. Thus, an offshore breeze may give rise to cloud over the sea, while an onshore breeze frequently produces cloud just inland.

Local climates can also develop in coastal zones under synoptic conditions and a prevailing wind. Thus, for example, when air is blown from sea to land, the cold land makes the air more stable and cloud forms only where high ground forces ascent (Fig. 5.23c). Where the sea off shore is warm but the coastal waters are cold due to an upwelling coastal current, onshore breezes bring warm air first over a cold sea, causing persistent fog, and then over warmer land where the fog disperses (Fig. 5.23d). The Californian coast is famous for its coastal fogs, which are produced by just such a combination of warm and cold seas. Similarly, if offshore winds take warm air over a cool sea, extensive offshore fog banks may develop. Some of the most extensive fog banks occur off the Newfoundland coast where warm air that has been in contact with the Gulf Stream drifts over the cold Labrador current that flows down from the Arctic.

Mountain and valley climates

Regions with pronounced relief develop two distinctive forms of climate.

THE FÖHN EFFECT

One kind of climate is produced when air moves over a topographic barrier high enough to produce condensation (Fig. 5.24). Because it warms at the DALR, the air sinking on the lee side of the barrier is warmer and drier than at a corresponding height on the windward side where part of the cooling takes place at the SALR. The air flow is called a **föhn** wind in the European Alps, but comparable warm mountain winds occur in other mountain ranges. For example the **Chinook** wind brings warm air down onto the Great Plains in the lee of the Rockies. If these warming winds occur in spring, they often start avalanches.

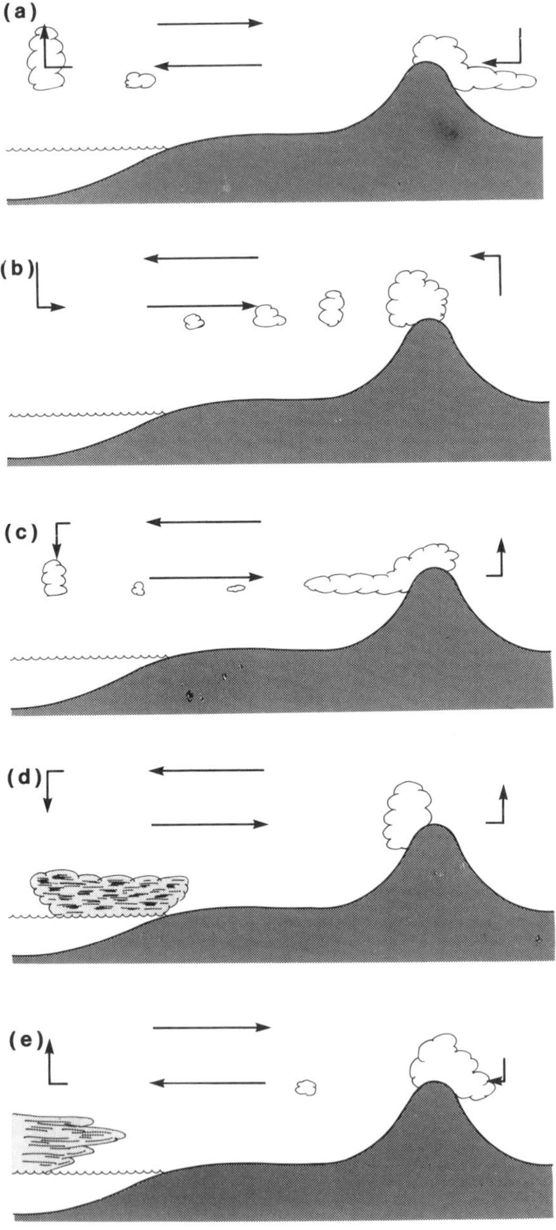

Figure 5.23 Some of the local climates produced near coasts.

VALLEY AND MOUNTAIN BREEZES

Mountains and valleys may have breezes similar to coastal breezes, being formed by contrasts in heating and cooling. In the early morning the upper, sunlit, slopes of the valley and hills warm up, while the valley floor remains shaded and cold (Fig. 5. 25). The

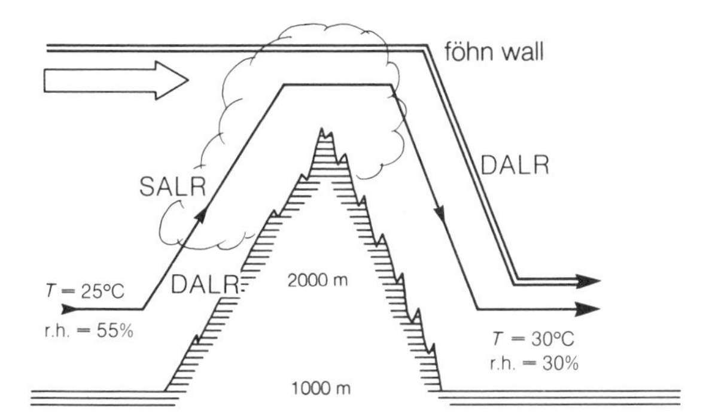

Figure 5.24 Topographic barriers. As air rises on the windward side of the mountain it first cools at the DALR. At $T = 15°C$ (the dewpoint temperature) condensation occurs and further cooling is at the SALR. Continued ascent produces cloud and releases rain, thus removing moisture from the air. On the leeward side of the mountain, the air sinks and begins to warm. Cloud evaporates and the air warms at the DALR throughout. When it reaches the plain the air is both *warmer* and *drier*. Such sinking air is called the **föhn**, and the distinctive edge to the mountain cloud is called the **föhn wall**. The double arrows indicate how even air aloft can be warmed without undergoing any ascent on the windward side of the mountain.

pressure gradient produced by the temperature contrasts causes air to rise from the valley floor and begin a **valley breeze**. By midday the valley floor may have warmed sufficiently for it to become the hottest region. Convective instability may then cause parcels of air to break away and act as a source of cumulus cloud. In the evening and during the night air flows down the valley as mountain and hillslopes lose heat by radiation; the air in contact with them cools and begins to roll down into the valley, creating a **mountain breeze**.

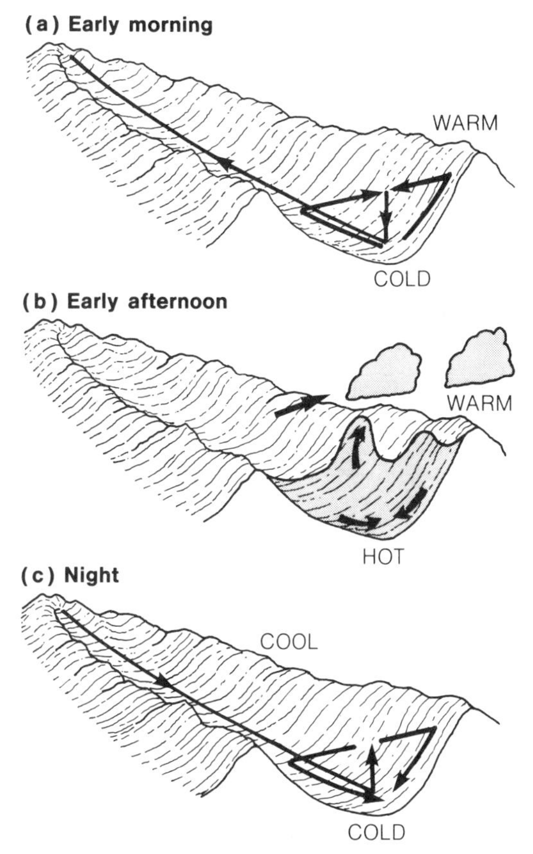

Figure 5.25 Valley and mountain breezes. (a) Early morning: upper slopes heat up and draw air from the valley to give a **valley breeze**. (b) Early afternoon: valley heats up, thermals break away and act as a source of cumulus cloud. (c) Night: upper slopes cool and air sinks to give a **mountain breeze**.

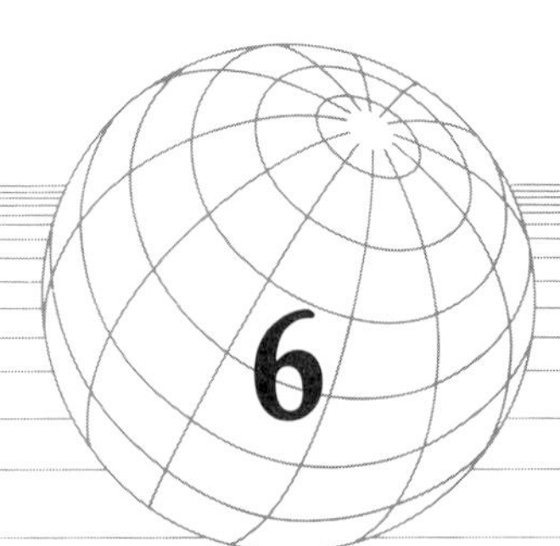

Man-made climates

6.1 The urban climate

The growth of urban areas, and especially industrial urban areas, has created an entirely new form of climate. By building in a formerly rural environment, people have altered virtually every aspect of the local climate to some degree (Table 6.1). Where once winds blew over forests and grasslands, they now sweep over an irregular surface of offices, factories and homes. Where the Sun once shone onto green vegetation with an albedo of 0.4, there is now tarmac and concrete with an albedo that may be as little as 0.1. Where plants once transpired and so both maintained a relatively high humidity and helped keep temperatures down, there are now large areas of buildings. Where once there was clear sky, there are now pollutants that reduce insolation, increase rainfall and decrease visibility.

Aspects of urban climates

TEMPERATURE

The best known effect of the urban environment is the creation of a **heat island** (Fig. 6.1). This is particularly apparent at night in cities of the middle and high latitudes. The heat island forms because:

(a) buildings store daytime heat better than does the surrounding countryside;
(b) at night stored heat is easily conducted to the surface where it can warm the air;
(c) radiation loss to space is restricted by the close proximity of buildings;
(d) less wind turbulence means that there is less mixing in cold air;
(e) there is an input of heat from fires, car exhausts etc;
(f) there is little vegetation to transpire and cool the air.

Table 6.1 Average changes in climatic elements caused by urbanisation.

Element	Comparison with rural environment
contaminants	
condensation nuclei and particulates	10 times more
gaseous admixtures	5 to 25 times more
cloudiness	
cover	5 to 10 % more
fog, winter	100 %
fog, summer	30 % more
precipitation	
totals	5 to 10 % more
days with less than 5 mm	10 % more
snowfall	5 % less
relative humidity	
winter	2 % less
summer	8 % less
radiation	
global	15 to 20 % less
ultra-violet, winter	30 % less
ultra-violet, summer	5 % less
sunshine duration	5 to 15 % less
temperature	
annual mean	0.5 to 1.0 °C more
winter minima (average)	1 to 2 °C more
heating degree days	10 % less
windspeed	
annual mean	20 to 30 % less
extreme gusts	10 to 20 % less
calms	5 to 20 % more

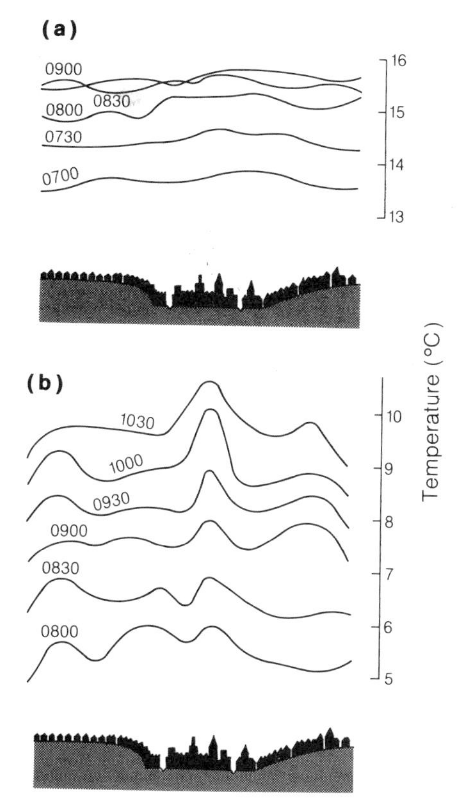

Figure 6.1 Reading's Heat Island. (a) No heat (wind speed 16 kts); (b) pronounced heat island over CBD with secondary peaks over residential suburbs (wind speed 10 kts).

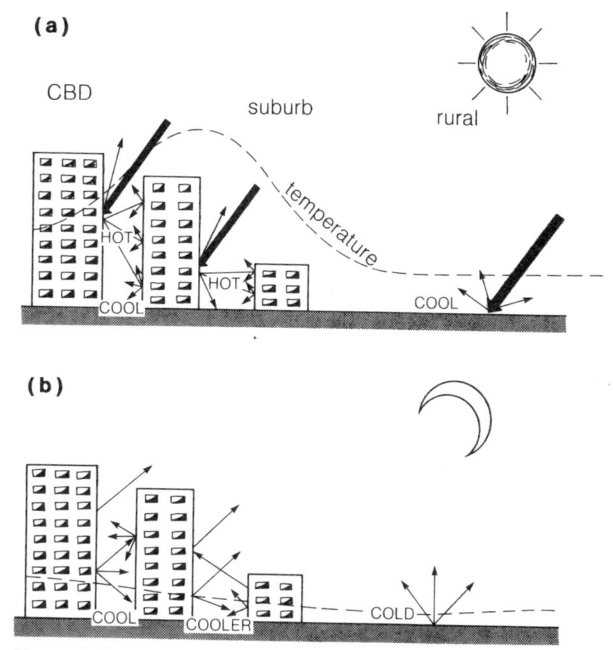

Figure 6.2 Controls on the heat island. (a) At night as heat is trapped within building space; (b) by day as heat is trapped in the upper regions of building space, creating cooler conditions at street level within a high-rise CBD.

However, distinctive urban climates are also formed during the summer daytime. In the morning, insolation begins to warm the city, but it does not penetrate into the enclosed spaces between tall buildings (Fig. 6.2). Indeed, air at street level among tall office blocks is cooler than air at street level within an urban area where buildings are shorter and so allow sunlight to penetrate. As a result the city develops three daylight (summer) temperature zones:

(a) a relatively cool centre;
(b) a warmer, encircling urban ring;
(c) a cooler rural fringe.

In middle and high latitudes the urban effect on temperatures in winter is entirely favourable: it reduces the duration of snow cover and helps in retaining heat. Fuel may therefore be conserved and people may feel more comfortable. However, in summer in mid-latitudes, and throughout the year in low-latitudes, urban structures compound heat, *add* to the energy used to run air-conditioning plants and *add* to the physiological stress on the inhabitants.

WIND

Cities usually have much rougher upper surfaces than do either grassland or forests. Wind speeds are therefore reduced, especially in areas of tall buildings (Fig. 6.3), which also restrict turbulence and prevent adequate ventilation reaching street level. As a result, car exhaust gases and other forms of pollution created close to the ground are not effectively removed and they accumulate as a health hazard. It requires a 22 kt wind to ventilate London thoroughly and remove the heat island.

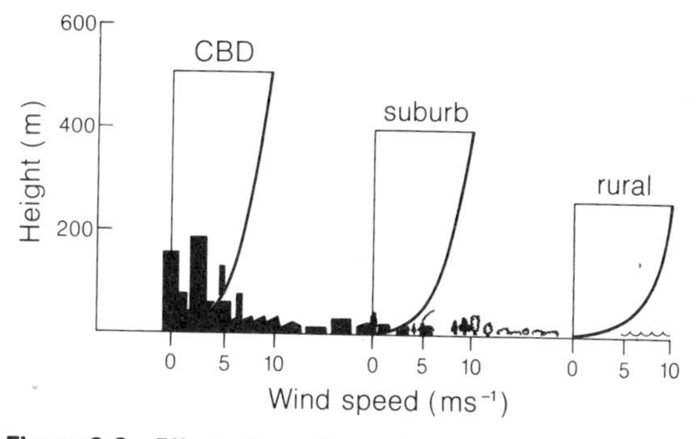

Figure 6.3 Effect of terrain roughness on the wind speed profile. With decreasing roughness, the depth of the affected layer becomes shallower and the profile steeper.

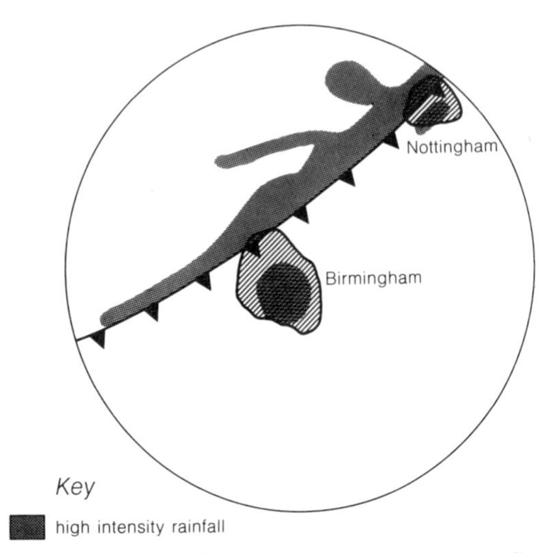

Figure 6.4 Radar shows how large urban areas influence stability in the region of a front.

RAIN

The impact of cities on rainfall is the most disputed effect of man's activities. In theory the increased emission of gases and particles (pollutants) by houses, factories and cars increases the number of condensation nuclei in the air and thus makes cloud formation more likely. A build-up of heat over the city also increases the instability of the lower air layers (the lapse rate increases) and makes convective ascent more likely. Figure 6.4, for example, shows how the large urban areas of Birmingham and Nottingham receive high-intensity rain ahead of the approaching front.

VISIBILITY

Every time you breathe, you inhale about a hundred million small liquid and solid particles (an **aerosol**) composed of such things as sulphuric acid, soil and organic compounds. The aerosols created by man are **pollutants**. Pollution from urban sources leads to changes in the composition and transparency of the air, especially on days with light winds when there is a temperature inversion that prevents convective turbulence. A dome-shaped pall of haze commonly builds up under these circumstances and, if acted on by strong sunlight, chemical reactions occur that produce the **photochemical smog** for which Los Angeles is renowned. It is quite common for the concentration of sulphur dioxide to be an order of magnitude greater over the city than over the nearby countryside, especially in winter. When heating systems are in use sulphur dioxide and dust particles act as particularly effective condensation nuclei and lead to the formation of (acid) fog – smog.

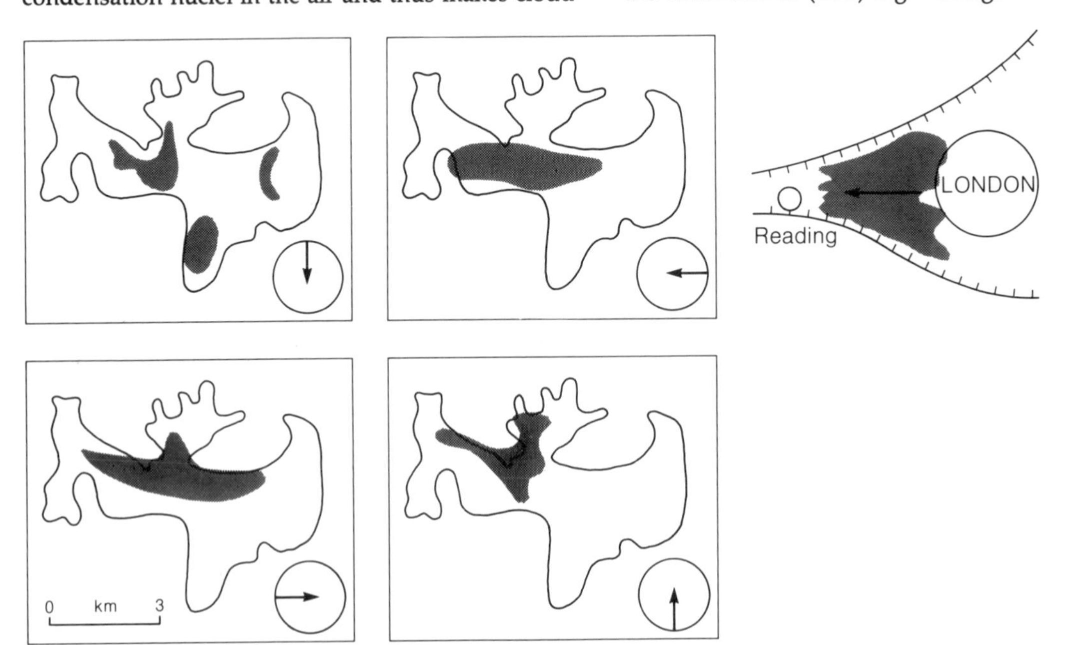

Figure 6.5 Changes in the pattern of smoke pollution over Reading, Winter 1961–2. Clearly the direction of the wind is important, but the level of pollution with an easterly wind was three time higher than with a westerly because up to 40% of Reading's pollution originates from the London area, 60 km away. Conditions are particularly severe in winter when anticyclone air gives light easterly winds that allow pollution-laden air held close to the ground by the temperature inversion to be funneled over Reading by the shape of the Thames Valley.

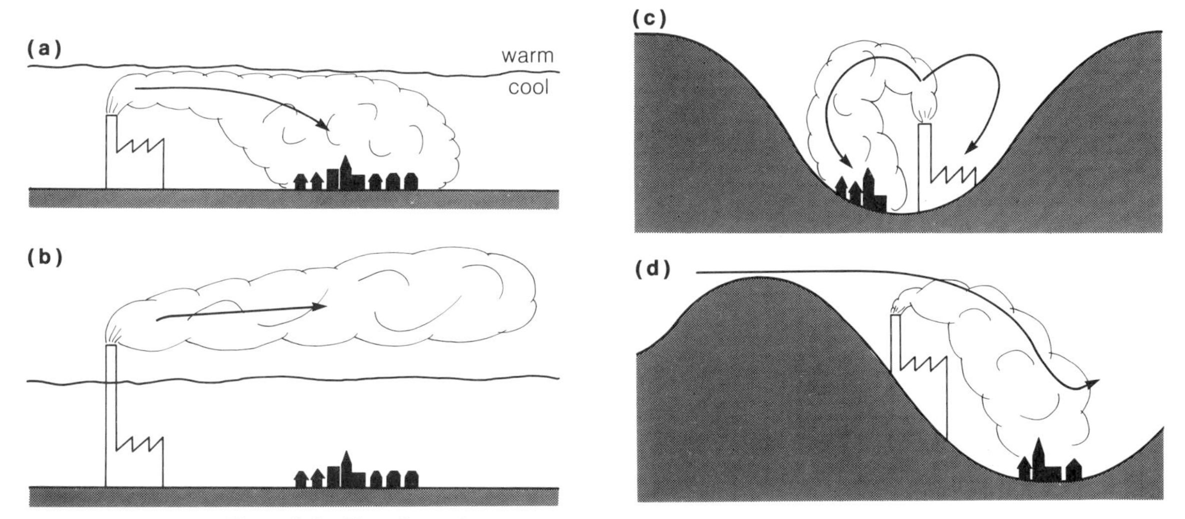

Figure 6.6 The effect of smoke stacks under various meteorological conditions.

Local climate and urban planning

In planning it is inadequate to use the simplistic rule of siting the residential section of a town upwind from industrial areas – defining 'upwind' by the prevailing (most frequent) winds – for it is not usually the prevailing winds that are important, but the winds associated with stagnant, highly polluting conditions (Figs 6.5 & 6).

In hilly areas, cold-air drainage will cause low temperatures, fog and the stagnation of pollution in hollows or along valley floors. These sites are therefore best suited to commercial (but not industrial) use as they are then occupied only by day when inversions are least likely and they do not need to be intensively heated at night. Slopes are best used for housing to keep it away from night-time frost hollows; houses are better sited on the sunny slopes (southerly aspect in the Northern Hemisphere) in high latitudes when insolation is low, and on the shady slopes in low latitudes where shelter from the heat is needed. Near lake shores, lake breezes reduce temperature regimes. In Chicago the most sought-after residential areas are within 2 km of L. Michigan, the depth of penetration of the lake breeze in summer.

Roof heights should be varied in the interests of increasing turbulence and providing adequate ventilation at street level.

6.2 Man's modification of regional climates

'Oh Lord, I'm about to round you up for a good plain talking. Now, Lord, I ain't like these fellows who come bothering you everyday. . . we want rain, Good Lord, and we want it bad; we ask you to send us some. But if you can't or don't want to send some, then for Christ's sake don't make it rain up around Hooker's or Leitch's ranges, but treat us all alike. Amen.' Thus prayed Big Dan Ming, Arizona, 1885, who was then sued by Hooker and Leitch for preventing rain falling on their land.

It seems hardly credible that man could affect the regional climates of the Earth significantly. The fact that he not only *can* but *has*, and that he will affect it even more dramatically in the future, is entirely due to the delicate balance that occurs between the components of the atmosphere.

Rain inducement

One of the few positive attempts to change the weather is 'rain-making'. Some clouds do not produce rain because they contain too few condensation nuclei, insufficient updraughts, or temperatures low enough for ice crystals to form. In theory it should be possible to induce rain in some cumulus clouds by increasing the number of condensation nuclei or introducing nuclei with a greater affinity for water. Attempts have been made to 'seed' clouds with very efficient nuclei – silver iodide or solid carbon dioxide in mid-latitudes to encourage the Bergeron process salt in low latitudes to encourage coalescence. These are delivered directly into the cloud, usually by spraying from aircraft. In areas of potentially high convective instability, such as the Canadian prairies in summer, seeding has been able to increase rainfall by over five per cent, but it is expensive. It is often carried out for a quite different reason: hailstones the size of golf balls frequently fall just as the crops are ripening; clouds are therefore seeded to encourage rainfall to begin sooner, before the hailstones have a chance to form.

Altering the global temperature and rainfall

People have only recently become aware of man's impact on the weather. However, as the operation of the atmosphere becomes better understood, it is clear that man can have a very powerful effect on climate in the long term. There are several possibilities, most primarily connected with burning fossil fuels.

CARBON DIOXIDE

If carbon dioxide (CO_2) emissions continue to increase at the present rate, the natural greenhouse effect may be strengthened and this may raise the global air temperature significantly. At present the rate of CO_2 emission will raise global temperature by 2–3 °C in 25 years.

In turn this may result in the complete melting of the polar ice caps, releasing sufficient water to raise sea level by several tens of metres and inundating the coastal plains where most cities are situated. At the same time the warmed oceans will release more CO_2 to the air, because CO_2 is less soluble in warm water, thus setting up a positive feedback system that constantly reinforces the greenhouse effect. This level of increase would be certain to change the global atmospheric circulation.

The change in circulation pattern that may occur in the decades ahead can be estimated by comparison with past warm periods, using the record contained in sediments; predictions can also be made by the use of computers. Most scientists agree that the changes to expect include increasing aridity for the mid-latitude interiors (Fig. 6.7). In turn this would cause a reduction in the yield of the prairies and steppes – areas that produce over half the world's wheat and maize and most of the surplus grain to feed the developing world. On the other hand tropical and maritime temperate areas would become wetter, possibly improving the output of rice and potatoes.

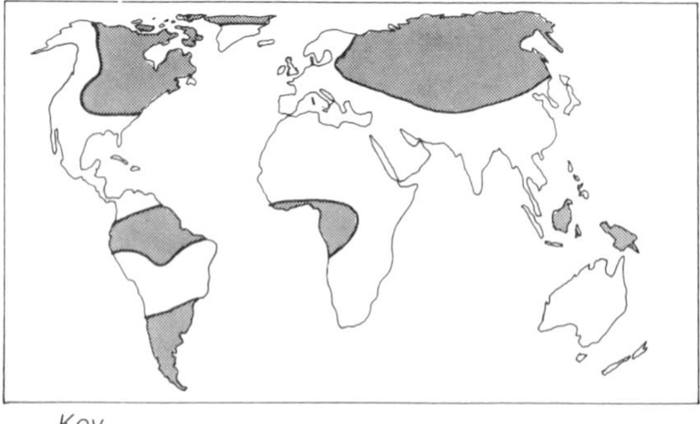

Figure 6.7 Possible changes in the global climate due to increasing atmospheric CO_2.

The implications for feeding the world and the global pattern of trade in agricultural goods are staggering. The US would stop being a grain exporter. The present US surplus is nothing less than the world's assurance against famine – the prospect that it might vanish is horrendous.

AEROSOLS

Increasing fossil fuel consumption has an effect wider than just producing CO_2; it also releases large numbers of particles – forming an aerosol – which scatter insolation and thereby *reduce* the global temperature. In addition, much more dust is being whisked into the atmosphere by winds blowing over areas recently exposed by desertification (see Ch. 29). How much this will reduce the impact of heating due to high levels of CO_2 is simply not known.

Acid rain

The atmosphere does not select between the natural particles and those added by man; both are carried aloft and moved as part of the general circulation, both act as condensation nuclei. However, the addition of large quantities of SO_2 and various oxides of nitrogen to the air has recently been discovered to have major environmental impacts, not in the immediate vicinity of the source of the pollution, but often thousands of kilometres away.

A large coal-fired power station can emit as much SO_2 each year as came from the Mt St Helens eruption (400 000 ft). At present the industrial countries of the Northern Hemisphere eject over 100 megatonnes annually of SO_2 into the air, along with oxides of nitrogen. Sulphur and nitrogen oxides react with water vapour to form dilute sulphuric and nitric acids, thereby creating acid rain. Even 'uncontaminated' rain is often slightly acid (pH 5.6) because it contains carbonic acid from the reaction of CO_2 and water vapour. Acid rain is anything with a pH less than that.

Mussels cannot survive in water with a pH of 6.0; lake trout die when the pH falls to 5.0. The rain falling in a zone of North America between Pittsburgh and Toronto often has a pH of less than 4.2, more than ten times more acid than normal rain. There are records of individual storms having yielded rain more acid than vinegar (pH 2.7) on parts of Pennsylvania.

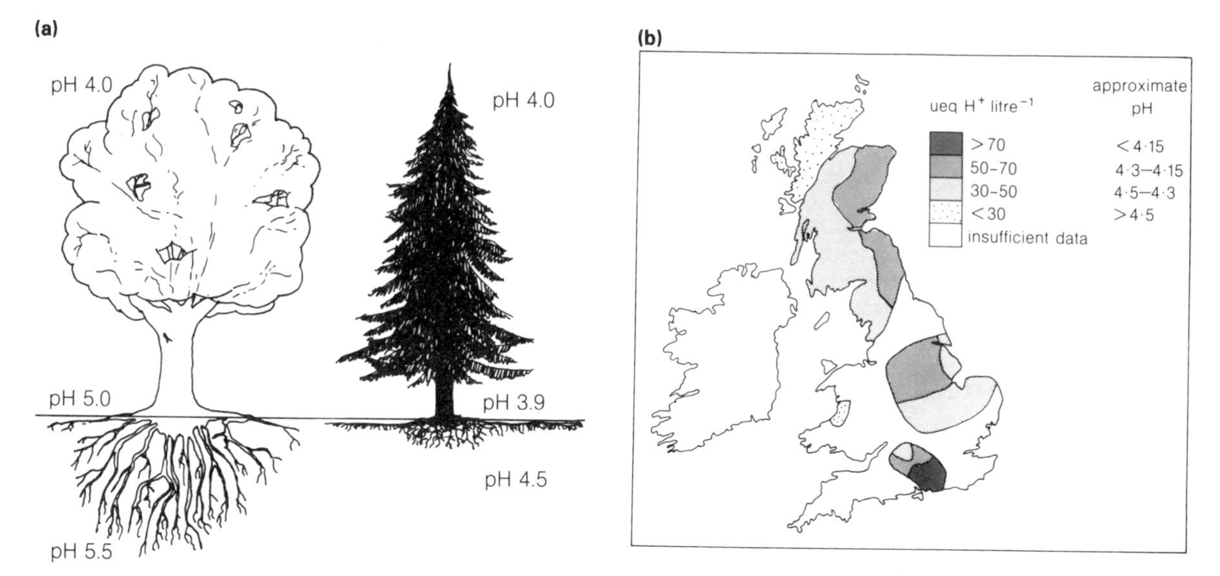

Figure 6.8 (a) The effect of acid rain on deciduous and coniferous vegetation (deciduous trees offer no buffering effect in winter). (b) The distribution of acid rain in Britain in 1983.

ACID RAIN AROUND THE WORLD

Canada: 50 per cent of all car corrosion is believed to be caused by acid rain and acid rain has now been confirmed in the high Arctic.
Poland: because of acid corrosion to tracks, trains at Katowice are limited to 40 km/h (25 mph).
Norway: the Norwegian stocks of Atlantic salmon are now largely extinct due to acidification of breeding grounds.
USA: soil acidification caused red-pine die-back in New York State; acidification was confirmed as an ecosystem stress in the major Hubbard Brook study; lake acidification has been recorded in Florida; while a White House science report concluded in 1983 that acid rain was probably causing crop losses of the same order as those due to ozone (5 per cent of cash value).
India: crop losses attributed to acid rain.
Sweden: 18 000 lakes are acidified, 4 000 severely acidified and 9 000 have fish stocks affected.
Italy: 50 lakes are acidified.
China: acid rain confirmed in Quinghai province.
Ireland, Soviet Union, Yugoslavia, Greece, Czechoslovakia and others: acid rain confirmed.

The environmental impact of acid rain varies considerably. If it falls on lime-rich soil or limestone rock, naturally occurring minerals can neutralise the rain; but on acid, leached, glacial materials there is no buffering capacity and here the effects are most severe. Much of Scandinavia, for example, has glacial soils, and water enters the lakes by throughflow without buffering, producing conditions too acid for the survival of wildlife. In Sweden 20 000 of the 100 000 lakes have lost their fish or are about to do so, and the remaining lakes can only be protected by pumping lime, an alkali, into them.

Acid rain also dissolves many toxic metals, including mercury and lead, iron and aluminium. Indeed, it is the aluminium leached from soils by acid rain that has killed the lake fish – they have died of metal poisoning.

In an apparent contradiction, acid rain can help to improve tree yields for a while, because the nitric acid provides a source of nitrogen for growth. Eventually, however, acid rain causes a decrease in the rate that organic matter decomposes in the soil litter and therefore reduces the rate of recycling, so in the long term trees suffer as well (Fig. 6.8a). Acid rain is, at present, killing most of Germany's northern forests.

Acid rain has actually been made worse by the very legislation designed to control the dispersal of pollution. Smoke stacks are now required to be very tall and are sometimes over 400 m; as a result, pollutants are carried away and become another country's problem. Scandinavia attributes its acid rain to power stations in Britain and Germany; Canada blames its acid rain on the industrial north-east of the US.

Today acid rain is also widespread in Britain. Between 1980 and 1983 Britain's rainfall was,on average, four to five times more acid than natural rainwater (Fig.6.8b). The concentration was greatest in south-east England, where the annual average was over seven times as acid as natural rainwater. Nevertheless, because the upland areas receive higher rainfall totals, they are the regions suffering the highest total deposition of acidity. They are also the areas where soils are naturally acid and where soil neutralisation of acid rain is least effective. Seventy percent of the acidity comes from sulphur dioxide emitted from power stations. Britain may export its acid rain to other countries but 80 per cent of it falls back on the nation that created it.

Part C

Water

7

The water cycle on land

7.1 The water cycle

The **hydrological cycle** (the water cycle) is the name given to the continuous cycling of water – as vapour, liquid or solid – between ocean, atmosphere and land (Fig. 7.1). The cycle is driven by the energy of the Sun. In just the same way as sunshine provides the energy which allows water to evaporate from wet clothes on a line, so the Sun provides heat energy to the oceans, lakes and other wet surfaces, causing evaporation. The rate of evaporation is, however, also dependent on the continuous replacement of moistened air with fresh dry air. Without such replacement, air will pick up only enough water to come into equilibrium with the wet surface. Furthermore, the capacity for air to hold water as vapour depends on the air temperature, warm air having a far higher capacity than cold air. Thus warm, windy weather causes far more evaporation than does cold, still air.

As air moves aloft it cools, and may become unable to hold all the contained vapour. At this stage condensation may therefore occur, leading to the formation of clouds and eventually to precipitation, as either rain or snow. However, the return of water to the oceans may not be immediate: some rain falls directly into rivers and returns rapidly to the oceans, but snow falling on the frozen wastes of Antarctica may be stored for tens of thousands of years before melting into the ocean. Furthermore, rivers only occupy one per cent of the land surface, and ice sheets only ten per cent; most precipitation falls onto rock, soil or plants and must follow tortuous paths even before reaching rivers.

You can get an impression of the processes involved in water transfer and storage whenever you are caught out in a storm. The first few drops of rain are absorbed directly by clothes, but within a few minutes continued rain saturates the clothes and water runs off as fast as it lands. At the end of the storm, clothes begin to steam as the stored water evaporates to the air. Nevertheless, because water drains downwards continuously under gravity, the lower parts of the clothes are the last to dry; they may even continue to drip for some time after the rain has stopped.

All these observations have their parallels within the natural landscape. The initial wetting of leaf surfaces is called **interception;** any surplus that flows over the ground and in channels is called **runoff**. The rain that wets the leaves never reaches the ground, being evaporated away at the end of the storm; it is only the surplus part of the rainfall that reaches the soil surface and proceeds towards streams. When there are many layers of leaves, such as with mature trees, interception storage (i.e. water withheld from the system) can be large – this is why it can be helpful to shelter under a tree during a storm. However,

(a)

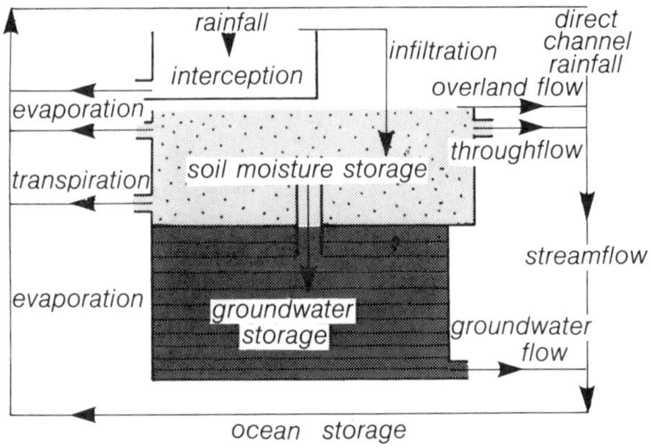

Figure 7.1 The hydrological cycle: (a) as a sequence of reservoirs, each of which must fill before spilling or leaking its surplus to others; (b) as a landscape model in humid temperate areas; (c) as a landscape model in semi-arid areas; (d) as a landscape model for a cold region.

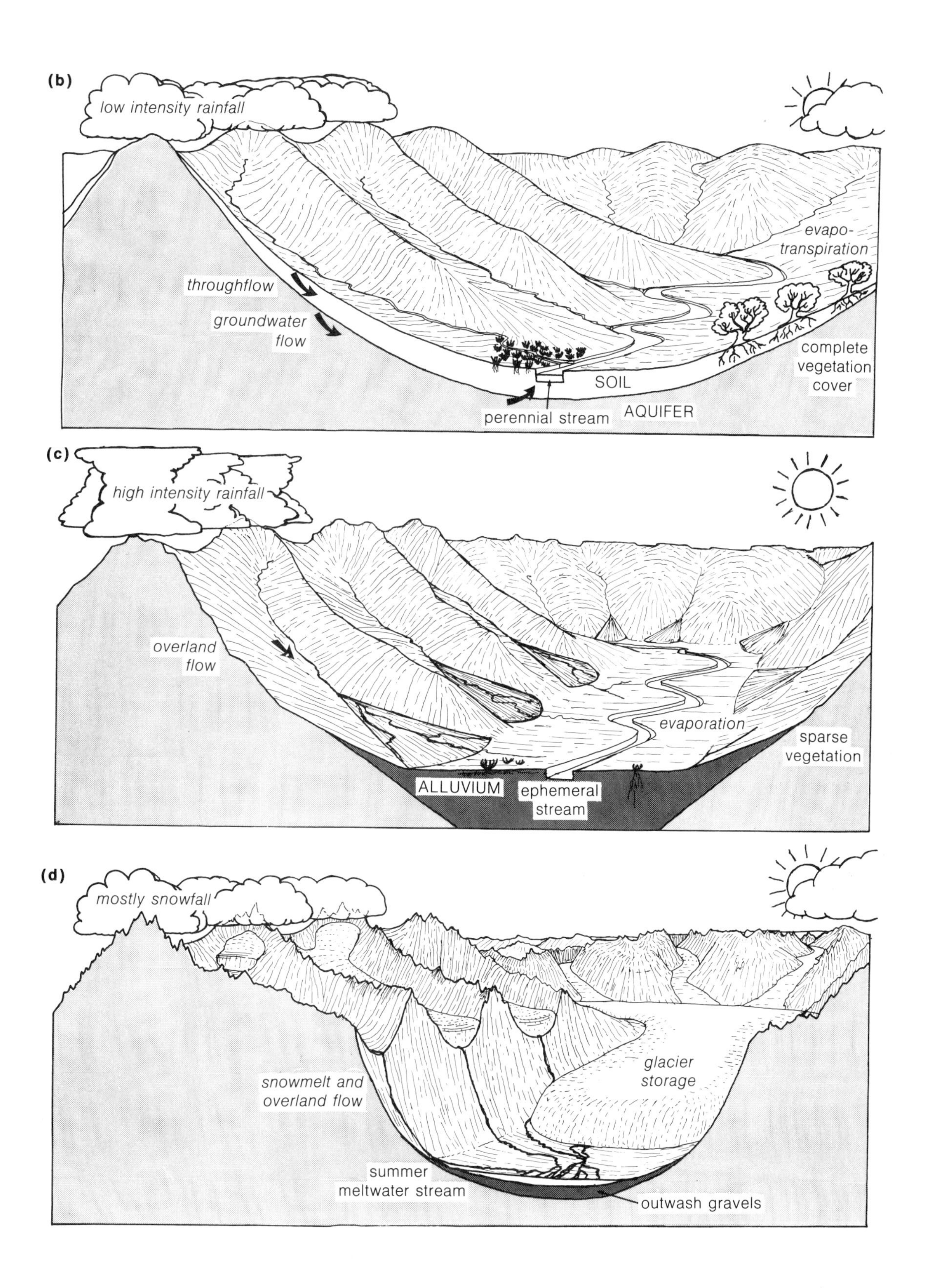

(b)
low intensity rainfall
evapo-
transpiration
throughflow
groundwater
flow
complete
vegetation
cover
SOIL
perennial stream
AQUIFER
(c)
high intensity rainfall
overland
flow
evaporation
sparse
vegetation
ALLUVIUM
ephemeral
stream
(d)
mostly snowfall
glacier
storage
snowmelt and
overland flow
summer
meltwater stream
outwash gravels

experience teaches us not to stand under trees after rain has ceased because drips will fall for some time. In effect some of the rainfall has been stored and returned to the air; the remainder has been slowed down, causing a delay or **lag** in its transfer to the ground.

7.2 Water in the soil

Usually water can infiltrate (soak into) the soil, making further progress by underground routes (Fig. 7.2). However, if the rainfall intensity is too great for the **infiltration capacity** of the soil – a rare event except in areas that experience severe convectional thunderstorms – water will also flow (as runoff) over

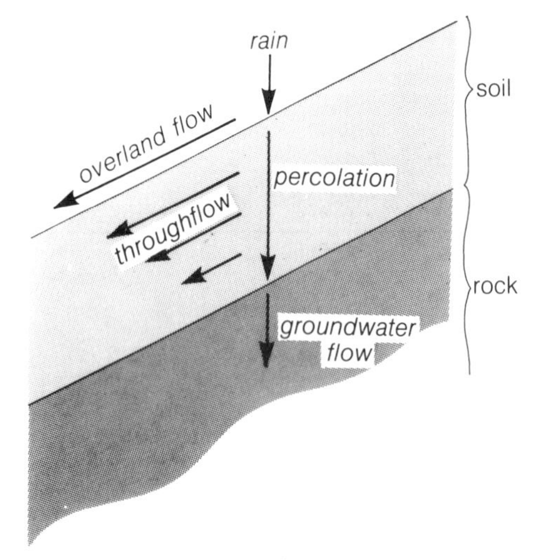

Figure 7.2 The components of soil-water movement.

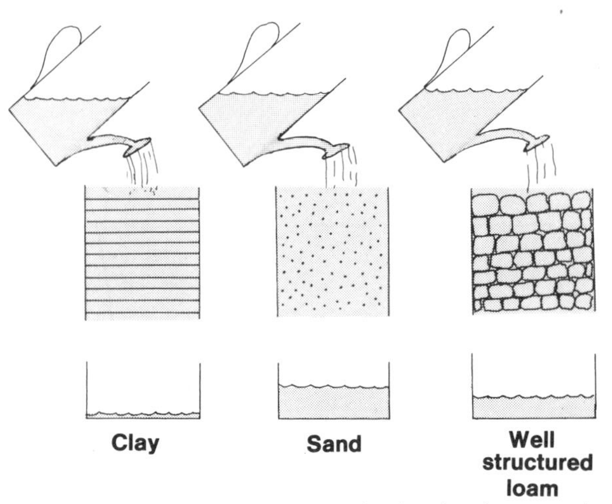

Figure 7.4 Percolation rates are closely related to natural soil permeability. Notice that well developed soil structures in the loam can give almost as effective permeability as sand.

the surface as **infiltration excess overland flow** (Fig. 7.3). Infiltration is closely connected with the further downward movement of water in a soil, called **percolation**, and with the lateral movement of water in the soil layers, known as **throughflow**. Indeed, the rate of infiltration decreases with time, slowing to a rate equal to water flow through saturated soil.

For hydrological purposes the soil can be regarded as similar to a giant sponge, holding water against gravity in minute (capillary) pores, while allowing a certain amount to percolate through according to the size and degree of pore connection. The total volume of spaces in the soil, the **porosity**, is a measure of how much water the soil will absorb, whereas the size and connection of the pores determines the rate of movement, a property called soil **permeability** (Fig. 7.4).

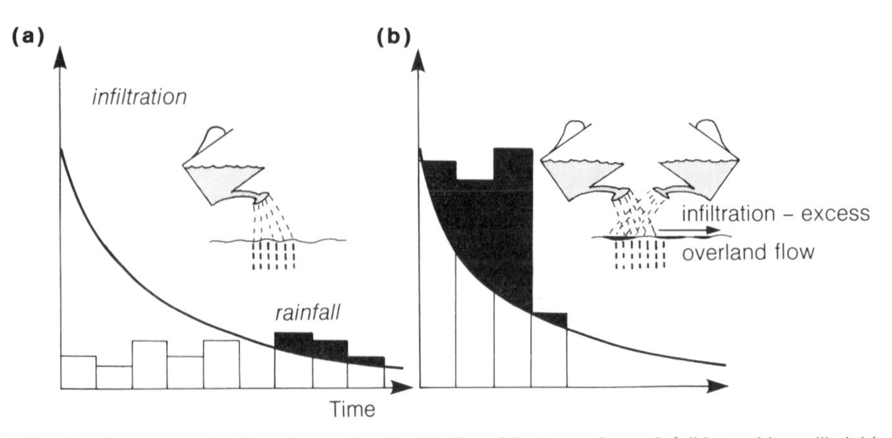

Figure 7.3 (a) Infiltration decreases with time, increasing the likelihood that even low rainfall intensities will yield overland flow. (b) Very high rainfall intensities associated with convective downpours will cause overland flow even near the beginning of a storm.

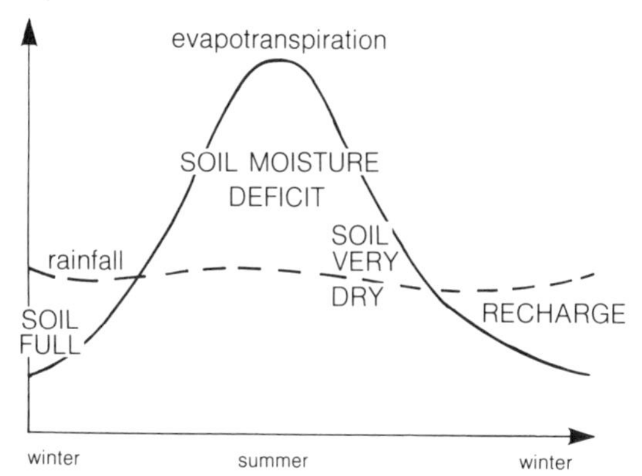

Figure 7.5 Changes in evapotranspiration through a year play a large part in controlling the amount of soil moisture storage. This diagram depicts the soil moisture budget over a year for a region of temperate humid climate.

Water held in capillary pores is used by plants for growth. For example, a tomato plant may use more than 2 litres of water a day – a tremendous drain on soil moisture storage. Roots draw up water from soil pores in response to loss by evaporation from leaves, a process called **transpiration**. In turn the rate of transpiration depends on the energy available for evaporation, varying both daily and seasonally (Fig. 7.5; Table 7.1). In summer, transpiration can remove

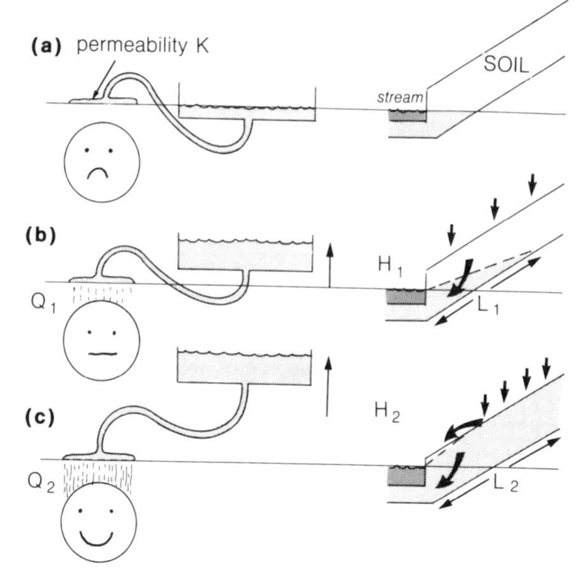

Figure 7.6 Throughflow contributions to streamflow depend on both soil permeability and hydraulic head: (a) levels the same – no water flows from soil; (b) levels in soil higher than stream – water begins to flow; (c) level in soil continues to build up and water flows correspondingly quickly from soil to stream.

Table 7.1 Interception and evapotranspiration loss for a spruce forest in Norfolk. The contrasts would be greater for a deciduous woodland because the leaves would be shed in winter and transpiration would be zero. (All measurements are in mm.)

Month	Rainfall	Rain falling through forest canopy	Total interception (evapotranspiration loss), *E*	Calculated transpiration, *T*	*E* + *T*	Water balance
January	39.1	22.6	16.5	4.5	21.5	+18.1
July	38.3	22.5	15.8	68.5	84.3	−46.0

ESTIMATING POTENTIAL EVAPOTRANSPIRATION

Evapotranspiration is extremely difficult to calculate accurately because it depends not only on the type of vegetation but also on the incoming solar radiation, air temperature, wind speed and air humidity. Nevertheless scientists have found by experiment that a reasonable estimate of **potential evapotranspiration** (P.E.) – the evapotranspiration that would occur if there were no soil-moisture restrictions – can be achieved using the formula:

$$\text{P.E.} = \frac{T - 32}{9.5}$$

where *T* is the mean monthly temperature in °F. This formula is particularly useful because it is based only on readily available temperature information. It is used throughout the world for estimating irrigation need.

moisture from a soil faster than it is replaced by rainfall. As a result the soil develops a **moisture deficit** and ceases to act as a source of moisture for streamflow. Indeed, changes of evaporation and transpiration (called **evapotranspiration**) control how much water flows in rivers in Britain, rather than any seasonal changes in rainfall.

Although evapotranspiration is a major source of water loss, in most areas some water will percolate beyond the reach of plants and begin to move through the soil towards stream banks. At this stage the pattern of movement depends on the natural variations in soil permeability with depth. Near the surface, burrowing animals and networks of plant roots disturb the soil and make it highly permeable, whereas at depth beyond this disturbance soils are more compact and less permeable. Futhermore, many soil processes deposit clays on the walls of pores in the lower levels of the soil thereby reducing the permeability even more. It is therefore quite common for soils to become saturated below the zone of earthworm activity (the subsoil) and, if rainfall is prolonged, water may even accumulate until it ponds to the surface, forcing further rainwater to run off over the surface as **saturation overland flow**.

On slopes, water can more readily move through the upper, more permeable soil horizons by throughflow, and saturation becomes less common. Nevertheless, in comparison with overland flow, throughflow is extremely slow – water may take several hours to flow a metre. Water flowing through a soil is subject to a time lag; this is why it arrives at channel banks to contribute to **streamflow** days and even weeks after a rainstorm, and in this way it contributes constancy and reliability to channel flow. However, on steep slopes throughflow is rapid enough to erode **pipes** within the soil. Such pipes can be several centimetres in diameter and, when they occur, they can considerably reduce throughflow lag time. When rain stops, the soil drains slowly. Large pores drain first, followed by progressively smaller ones (Fig. 7.7). The smallest pores (**capillary pores**) never drain, but hold water by surface tension until it is used by plants. When, about 48 hours after the end of a storm, the larger pores have drained, the soil is still very moist – a state called **field capacity**. Later,

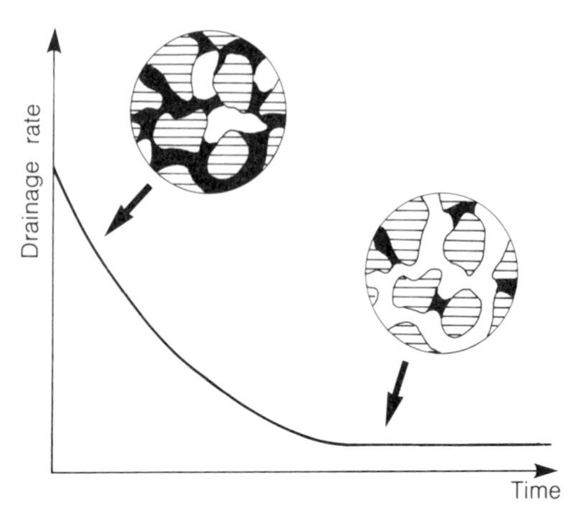

Figure 7.7 Drainage from a soil slows down as the larger pores empty and the small pores hold water by surface tension.

transpiration by plants will draw water from the top soil, but the subsoil will remain moist; the slow release of moisture from it may be enough to supply a stream during periods without rain.

7.3 Groundwater

The pattern of streamflow is only partly explained by the storage and permeability of soil. Where soils are underlain by permeable rocks (called **aquifers**) much water percolates directly into the pores or hairline cracks on the rock to flow as **groundwater** (Fig. 7.8). Groundwater reservoirs usually have a capacity many times greater than the soil and are important in sustaining a stream over long periods without rainfall. Many rocks can act as aquifers, not only chalk and limestone but also sandstone (which often has large pores) and in some circumstances even granite. The rate of flow in an aquifer depends on both the permeability and the pressure that develops as water accumulates and the **water table** rises. As with a pipe, if the head of water above the outflow point is increased, the rate of outflow will also rise; and as the head of water declines, so will the groundwater con-

THROUGHFLOW

The rate of throughflow (Q) depends on the quantity of water ponding up in the soil to provide the driving force (the hydraulic gradient, H/L), the soil permeability (K) and the area (A) of the stream bank through which the water can escape (Fig. 7.6). This relationship is called **Darcy's Law**:

$$Q = KAH/L$$

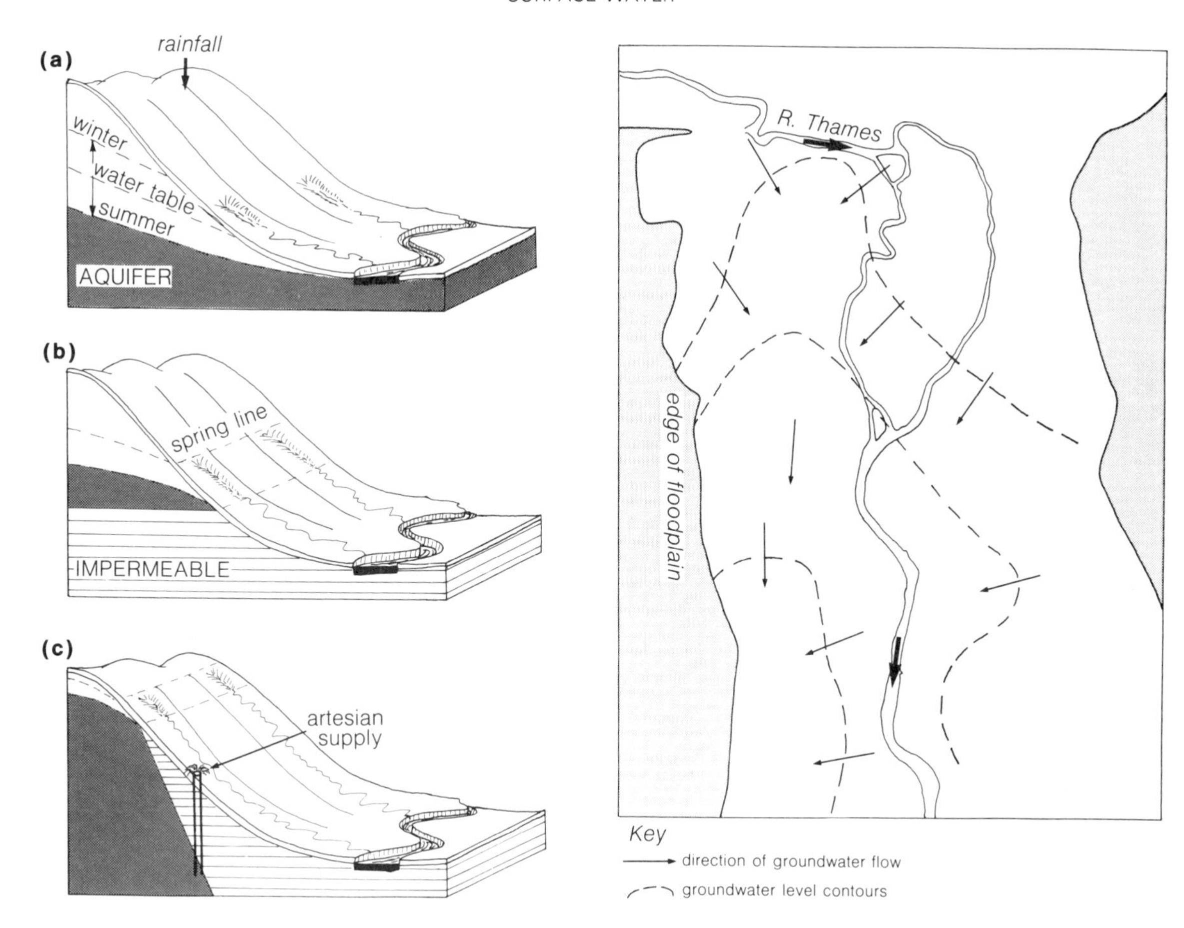

Figure 7.8 Groundwater. (a) Unconfined aquifers are very similar to thick soils, the spring line moving across the hillslope with changes in the height of the water table; the movement is primarily seasonal. (b) Unconfined aquifers underlain by impermeable strata have the lowest level of the spring line fixed. These springs rarely dry up in the summer, but increase in flow and migrate upslope in winter. (c) Confined aquifers contain a large amount of stored water which can be very useful to man. The flow of water from a well sunk to tap the artesian supply depends on the head of water in the aquifer. (d) Complex groundwater movement in the Thames floodplain at Oxford. Here the river and groundwater flows diverge.

tribution to streamflow. Like throughflow, groundwater flow can be described by Darcy's law.

7.4 Surface water

Despite the facts that most water runs through rather than over soils, and that throughflow and groundwater flow move water very slowly, river flows often increase rapidly soon after rainfall. This is because overland flow can occur in certain localities, even at low rainfall intensities.

Soils in some parts of a drainage basin always have a high moisture content. These occur most commonly within small folds or hollows (**concavities**) in hillsides or beside river banks, where they are commonly

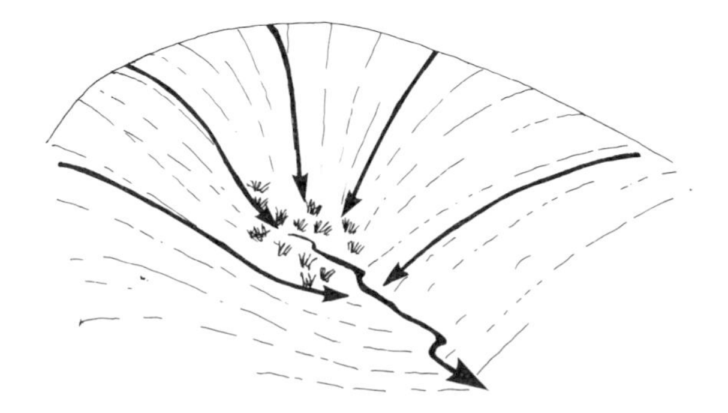

Figure 7.9 As water is focused into a concavity, saturation occurs and overland flow is initiated.

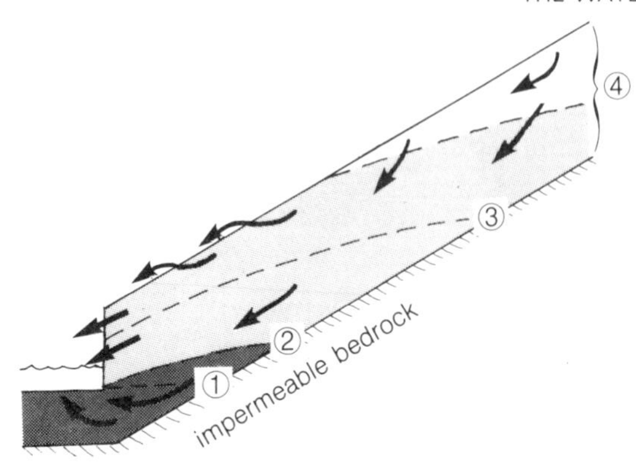

Figure 7.10 As a storm progresses, water flows through the soil and causes a rise in the water table and greater flow into the stream (1 to 3). If the water table rises to intersect the surface, overland flow will occur as water seeps out from the soil on the hillslope (4).

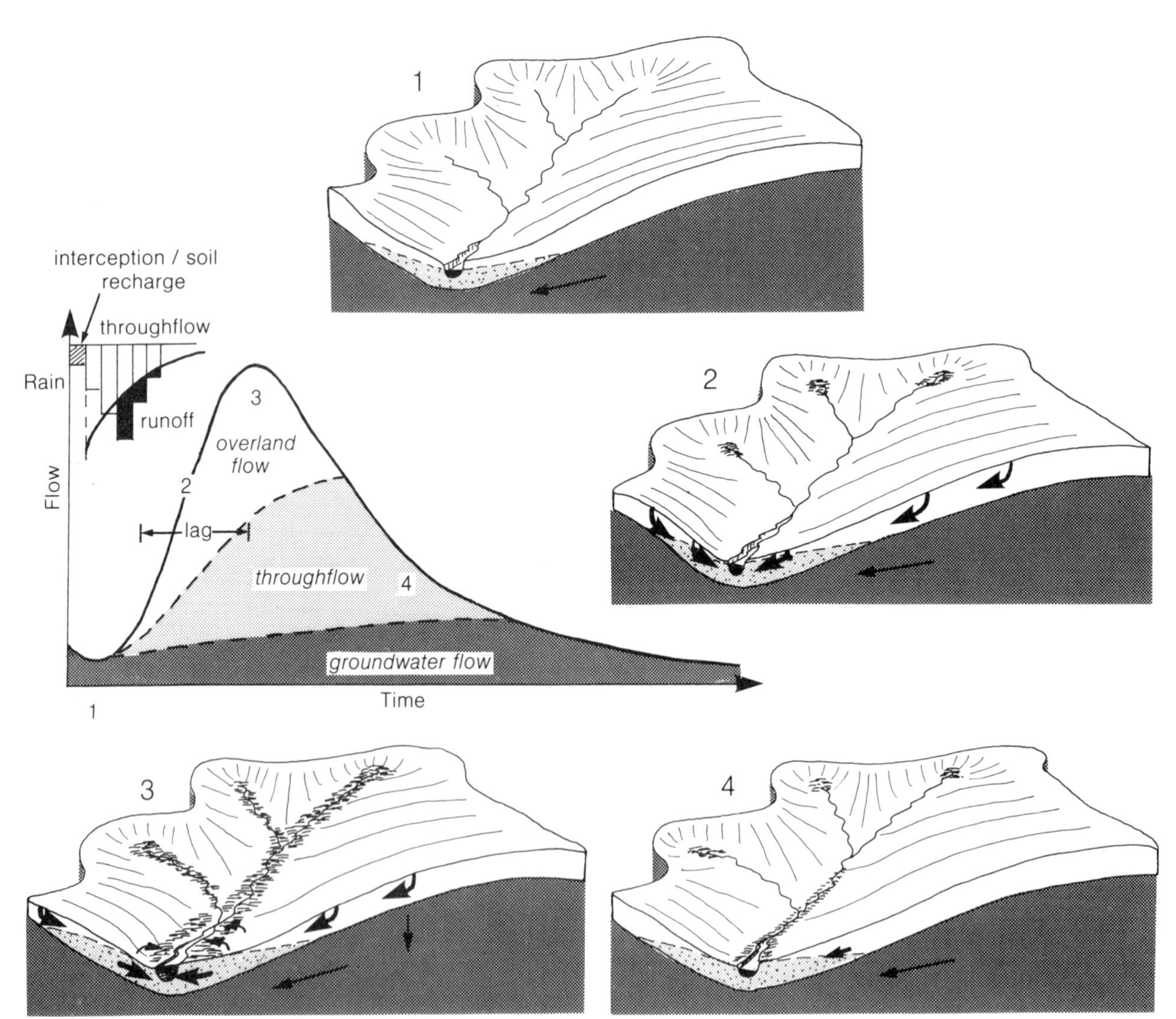

Figure 7.11 Changes in the contributing area at various stages of streamflow. The hydrograph is divided according to the contributing sources. Note that a hydrograph without overland flow will not have a sharp peak.

marked by growths of water-loving rushes or trees such as willow and alder. Hillslope hollows become saturated easily because water is funnelled towards their centres, directing more and more liquid into a smaller and smaller volume of soil. When the water cannot all be contained within the soil, some of it spills across the surface creating a spring or a region of seepage (**a bog**) from which issues a stream (Fig. 7.9).

The soil beside a river is always wet, and the lower part always saturated, because it is in contact with flowing water. However, water flows from bank to channel only if the water level in the soil is higher than in the channel (Fig. 7.10; see also Fig. 7.6). Soil beside a channel therefore always contains a saturated zone with a water table sloping up away from the bank. During a storm, water enters this zone as throughflow, raising the water table and increasing the head of water so that, in turn, more is forced through the bank. With a prolonged storm, the rise in water table within this zone may well force water out onto the soil surface as overland flow. Any water falling on the saturated zone will flow directly and quickly to the channel, thus enabling the stream to respond rapidly to rainfall. Those areas that become saturated in a storm (mostly banks and hollows) are together called the **contributing area** and are responsible for the size of the peak flow in the channel (Fig. 7.11).

The stream hydrograph

Streams and rivers gather their water from many sources. A tiny amount of rain falls directly into the channel; much more flows through the banks from the soil or enters the channel from aquifers either as hillslope springs or directly into the channel bed. In times of intense or prolonged rainfall there is a further contribution from overland flow.

The pattern of channel flow plotted against time is called a **hydrograph**. It represents all the contributions listed above from all parts of the drainage basin (Fig. 7.12). It usually looks quite different in shape from the pattern of any of its contributing sources. There are two reasons for this: first, a great many contributions are involved; and second, even after each contribution has reached the channel there is a considerable lag as it flows down stream.

The nature of each contribution and the effects of lag are easily demonstrated using some figures. For example, at any section of stream bank the contribution of flow through the banks may seem very small. But suppose each metre of channel bank contributed one drip of water each second and that 10 000 such drips are needed to provide a litre of water. This is a modest amount, taking nearly three hours to fill a

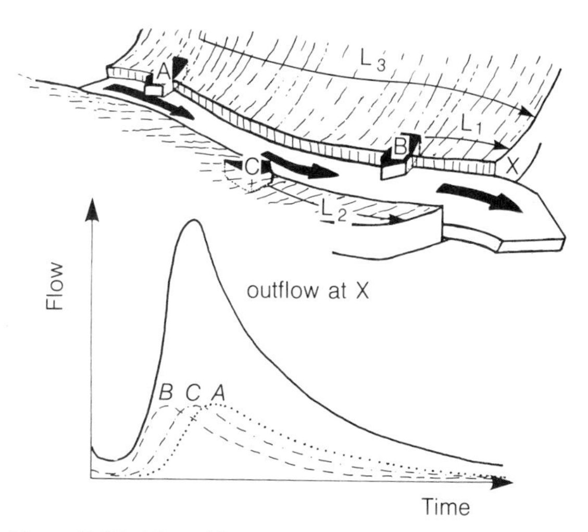

Figure 7.12 Three identical contributions, A, B and C each reach the measurement point X lagged by different times. B arrives first, having travelled distance L_1; C arrives soon after, having travelled distance L_2; while A arrives last after travelling L_3 and when the contribution from B is already declining. The hydrograph measured at X is the sum of many such contributions.

litre flask. However, there may be 4 km of channel in each square kilometre of landscape, which gives 8 km of bank. Multiplied up, the flow now becomes nearly 0.8 l/s/km^2 – a not-unreasonable amount to be expected for a small area during dry weather.

Although water flows rapidly once it has entered the channel, even travelling at 2 m/s a journey of 1 km will still take over eight minutes, one of 10 km nearly an hour and a half, while water entering the headwaters of a river as long as the Rio Grande (2000 km) could not reach the sea in less than twelve days. Therefore if rain fell evenly over the whole basin and each metre of bank contributed the same pattern of flow, lag effects produced within the stream would still ensure that the final hydrograph representing the mouth of the basin looked quite different from any of its contributions (Fig. 7.12).

Lakes

As rivers fill depressions in the landscape, they create lakes varying in size from ponds to inland seas. However, in principle their effect on the water cycle is the same. Just like water flowing from tap to bath, river water rushing seawards becomes lost in the vast stillness of the lake: a large body of water absorbs both velocity and volume. For example, a lake with a surface area of 1 km^2 needs 10^6 m^3 of water to raise its level by 1 m. It would take a river flowing at 1 m^3/s

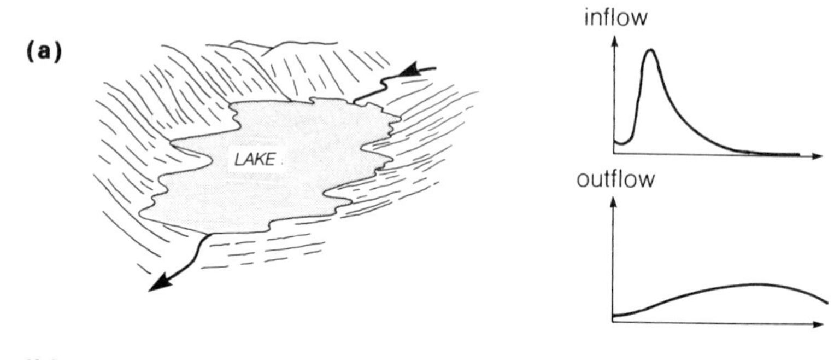

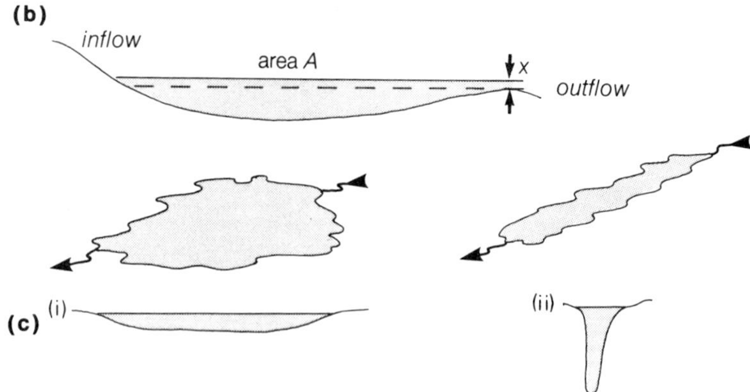

Figure 7.13 Lakes. (a) Lakes store water and can control a flood. (b) Outflow is proportional to depth (x), but x increases only slowly because it is related to the total lake volume (Ax). (c) Lakes can lose significant quantities of water by evaporation, especially if they cover a large area. Thus different shapes of lake serve different purposes: (i) much loss by evaporation, but good for flood control; (ii) less loss by evaporation but also less effective flood control.

more than ten days to achieve this. By contrast, the outflow of water from a lake is controlled simply by the height of the water level above the outlet. As a result, hydrographs of streams linking lakes have less peaked shapes than those of other streams; and they have especially long 'tails' as water is released ever more slowly from the lake as the water level above the outlet declines (Fig. 7.13).

8

Drainage networks

8.1 The origin of stream patterns

A river system seen on a map exhibits a remarkable degree of organisation of its tributaries, a pattern delicately etched in the landscape, which, it seems, might have been drawn by some guiding hand (Fig. 8.1). Indeed, the pattern of branching tributaries is so much like other patterns in nature such as the veins of a leaf that it is difficult to believe it has been formed by purely random processes. Playfair expressed this idea of the river as some kind of organic whole when, in 1802, he wrote:

> Every river appears to ... [run] ... in a valley proportioned to its size, and all of them together forming a system of valleys, communicating with each other and having such a nice adjustment... that none of them join the principal valley either on too high or too low a level.

Yet we need to be extremely careful in drawing conclusions about order or making comparisons with other natural structures, for they can be very misleading. Instead we need to begin by asking ourselves what the function of the river system really is. When we do this we realise immediately that rivers and river networks are simply by-products of water returning to the sea as part of the water cycle. The main point to remember is that there are many ways in which water can leave a drainage basin. One way is by flow over a bare rock surface. However, because water reacts with most rock, continual contact will eventually convert surface rock into soil; this will lead to less overland flow and more throughflow. With greater throughflow, fewer streams will be required to remove the water; water will be released from the soil over a longer period of time. Indeed, rivers would not be necessary provided that soils were deep and rocks sufficiently permeable to transmit any rain that fell.

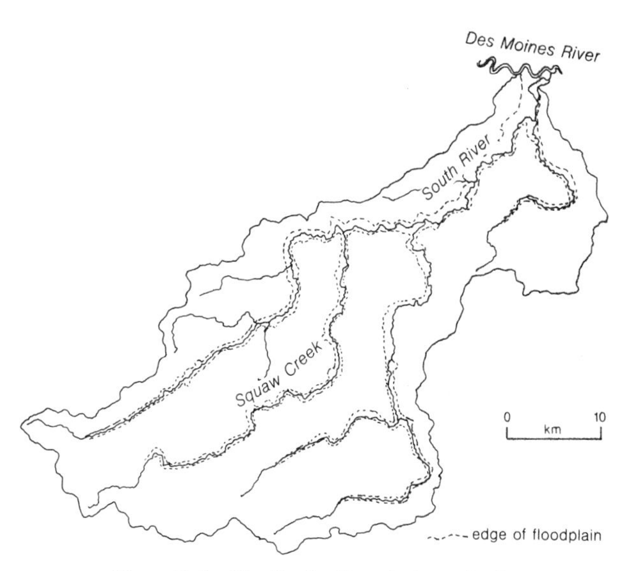

Figure 8.1 The South River drainage basin.

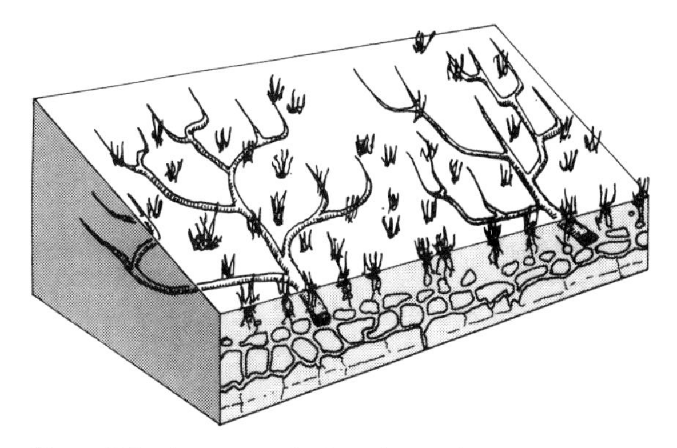

Figure 8.2 A network of sub-surface soil pipes in a well constructed soil may converge and lead to stream formation in regions with humid climates.

Most overland flow in humid regions results from soil saturation, perhaps because of the random convergence of water within a soil. However, where water seeping through soil pores moves fast enough to become erosive, converging flow leads to the development of subsurface soil pipes (Fig. 8.2). In regions dominated by convective downpours, on the other hand, infiltration rates are frequently exceeded and overland flow is common.

The impact of overland flow on the landscape depends on the amount of vegetation cover. In areas with a complete cover, vegetation may prevent surface erosion and channels may eventually form from the collapse of the roofs of soil pipes. By contrast, in areas without a continuous vegetation cover, converging overland flow will become progressively deeper and faster, and will be able to carry soil away and so initiate a channel.

At first channels on bare ground (**rills**) are nothing more than a few millimetres deep, converging and diverging to produce a multitude of intersecting waterways (Fig. 8.3). Nevertheless, the same ground irregularities that produced these rills will funnel more water into some than others, beginning a chain reaction: the more water reaching a rill, the more it will erode and the bigger the rill will grow. By this process rills capture water from one another until the most effective develop into deeper channels called **gullies**. Gullies do not cut deeply enough into a soil to be fed by throughflow during dry weather, so they usually contain water only during and soon after rain. But as the chain reaction continues, some gullies will cut deep enough to tap both throughflow and groundwater flow, thereby forming permanent **stream channels**.

The progression from rills or soil pipes to stream channels nicely illustrates the way one process takes over from another during landscape evolution. In the early stages the water movement is entirely random

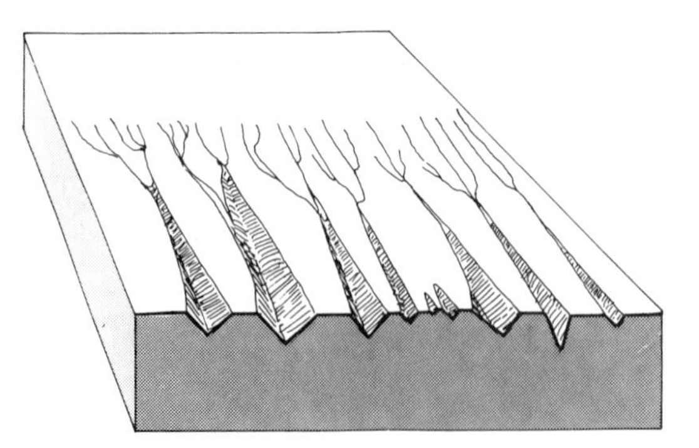

Figure 8.3 Initially, rills are so shallow they converge and diverge at random. However, the development of larger rills will tend to stabilise the flow, causing smaller rills to become feeders to the larger.

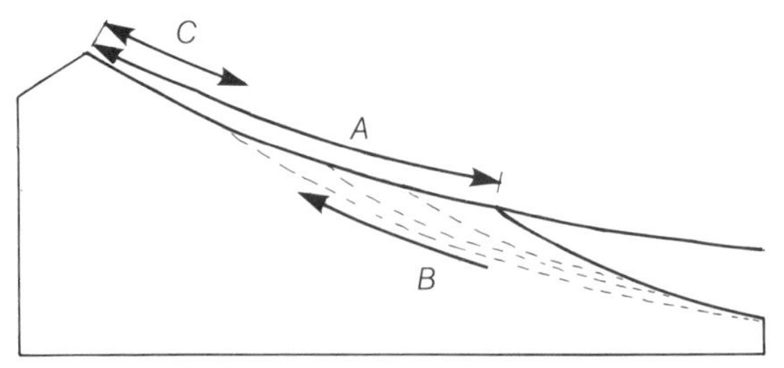

Figure 8.4 Even though there may be sufficient water to support a stream channel, because water flowing in sheets or through the soil is less effective at erosion, channels are only *initiated* at considerable distance from divides (*A*). Once formed, however, they promote water convergence, their erosion efficiency increases, and streams begin to cut back upslope (headward growth, *B*) until limited by water supply (*C*).

and vulnerable to deflection by chance irregularities in the land surface. However, in time the limited water supplies strictly determine the number of streams that can develop and so stabilise the evolving pattern in such a way that a reasonably even spacing of channels occurs.

The first stream channels to form will be a long way from the edges of the basin. However, as channels become more established they deepen, making the sides and the **head** (the upslope starting point) more liable to collapse. Thus, not only do the channels begin to develop valleys, but collapse at the headwater causes the channel to grow up slope into the region where sheetflow still dominates. So although the early stages of channel growth are a passive response to converging throughflow or sheetflow, the later stages reveal an active extension of the network – called **headward growth** or **spring sapping** – until there is simply not enough water to feed a stream (Fig. 8.4).

River-network analysis

The pattern of rivers that develops from a combination of random flow and competition for water was first analysed by Horton in the 1930s. He suggested that we should think of a drainage network rather as a collection of stream segments joined together in a systematic way. The most easily identified segments are the smallest 'finger-tip' tributaries with no tributaries of their own. They are most common at the edges of a drainage basin and are classified as **first order streams**. Those streams having only first-order tributaries are classified as **second-order**; those with only second-order tributaries are classified as **third-order**, and so on (Fig. 8.5). First-order tributaries are therefore the most common, with second-order tributaries fewer in number and more widely spaced, third-order tributaries even less common, and so on.

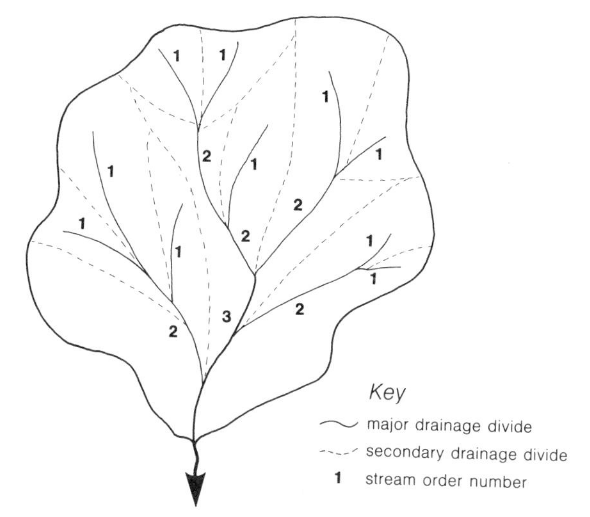

Figure 8.5 Description of a drainage basin using a numbering system.

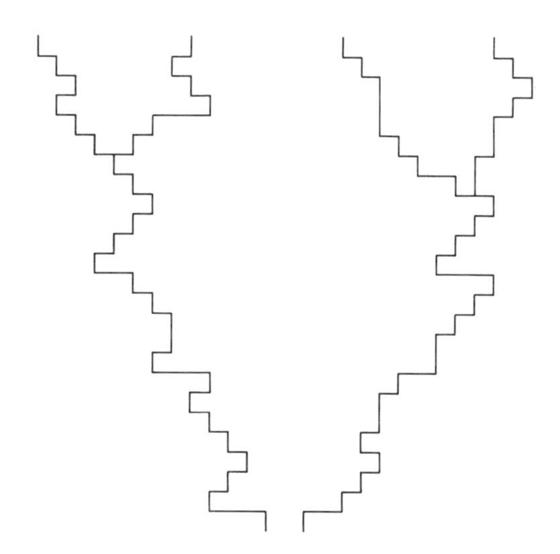

Figure 8.7 Computer-generated stream patterns have all the geometric properties of natural streams.

In this way stream segments within a basin can be shown to have a hierarchical pattern, much as Christaller has demonstrated a hierarchy of settlements in his 'Central Place Theory'.

Establishing a structure to the network shows that streams converge with a degree of regularity, even though they have formed randomly. Plots of the number of streams of each order against its order number (Fig. 8.6a), of stream order against drainage basin area (Fig. 8.6b), and of stream order against mean channel length for each order (Fig. 8.6c), yield simple logarithmic relationships in just the same way as 'stream' networks created by random walk computer programs (Fig. 8.7).

Although the basic stream network remains constant in *overall* pattern, it may vary in *detail* with changes in environment. For example, factors such as gentle slopes, climates with high-intensity rainfall, rocks with low permeability, and thin soils will all tend to encourage overland flow and a dense stream network; whereas low-intensity rainfall, thick soils or highly permeable rocks on moderate slopes will encourage throughflow and reduce the need for streams. This is illustrated by a comparison of granite basins in Dartmoor, England, and North Carolina, USA (Fig. 8.6). Here, as other factors are equal, it is the greater rainfall intensity of North Carolina that has created the more dense network. Similarly contrasts in stream networks within one climatic region

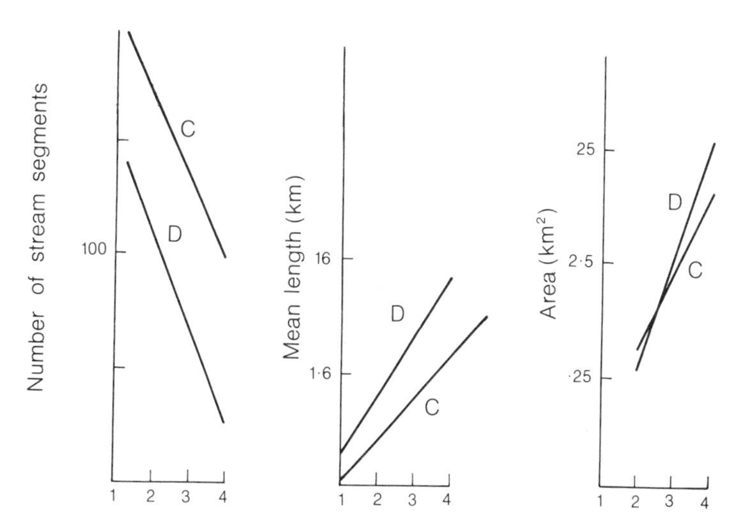

Figure 8.6 Stream segment numbering systems illuminate several interesting drainage properties. These graphs are for Carolina (C) and Dartmoor (D).

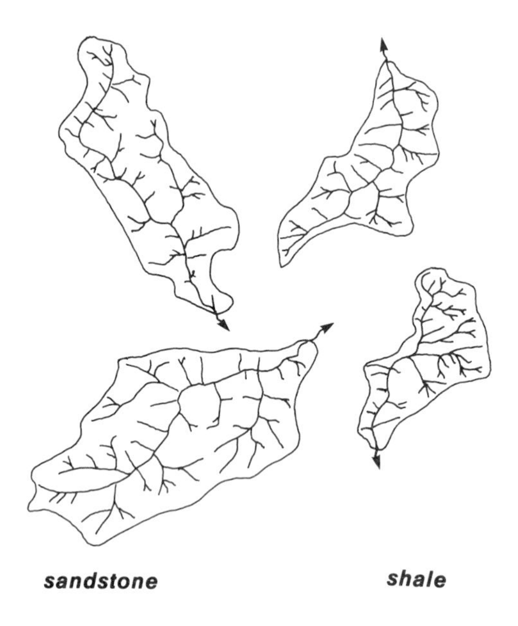

Figure 8.8 Contrasts between sandstone and shale basins in Devon.

result from variations in soil permeability. For example, in Devon the relatively impermeable shales produce a denser network than the more permeable sandstones (Fig. 8.8).

Stream orders are time-consuming to deduce. For many purposes the effect of environmental factors can be summarised by the **drainage density** of a basin, i.e. the total stream length in each square kilometre of basin. This is possible because the total length of streams within an area is a measure of the importance of overland flow.

8.2 Development of river patterns

Natural river networks may take millions of years to evolve and their development cannot be examined directly. Investigations into changes of river networks with time have therefore largely been made with scale models or computer programs. At Colorado State University one model consists of a sloping tray (9 m × 15 m) filled with a loam soil and provided with an overhead sprinkler to simulate rainfall. Such models are useful not only to show changes with time but also those that accompany changes in the environment, such as varying rainfall. Experiments with these models show that drainage density changes with both rainfall and slope and that there is a belt of no erosion near the divide. Similarly the growth of a single stream network confirms that fingertip tributaries stretch out towards yet unchannelled land where water still flows as a thin surface sheet.

But the model also shows a sequence of subtle changes that could not readily be observed in the slowly evolving natural landscape. While a dense network remains apparent in areas newly brought into the drainage net, the more established parts of the network begin a process of gradual simplification. Near the main channel the number of small streams decreases rapidly, and even near the margin of the model there are eventually significantly fewer tributaries. This proves that the high drainage density required at early stages in development or at the edges of the network is not necessary as the network matures, when water is able to move by subsurface rather than by surface routes. As rivers incise more deeply into a landscape and as valley slopes lengthen, the rivers are more able to transmit water without the need for large numbers of surface channels. Other changes in the channel pattern developed in the model include not only lateral migration of the main stem but also extensive capturing of the major left-hand tributary (Fig. 8.10). The main stem carries the most water, is most erosive and has the deepest 'valley'. Tributaries flowing into it develop steep gradients and rapidly erode headwards, in this case cutting across and diverting the flow of the major left-hand tributary. Thus simplification entails both channel elimination and major changes in the pattern of the network. Finally, the model helps illustrate two very important geographical rules: we cannot consider river channels in isolation from the rest of the landscape and the drainage network cannot become stable until slopes also have stabilised throughout the whole basin. In the meantime rivers change to keep in step with the continually varying pattern of water supply – an example of the dynamic equilibrium of a landscape discussed more fully in the following chapters.

Long profiles of rivers

Most rivers become less steep towards their mouths and long-profiles – vertical sections taken along rivers – or river and valley floors are, overall, concave upwards. There are several reasons for this shape. First, rivers moving coarse sediment must have steep gradients. Coarse debris is most commonly produced by mechanical weathering in uplands, but as the rounded nature of material downstream testifies, coarse material is rapidly worn smaller (**abraded**). Because of this downstream fragmentation, debris can be transported over progressively lower gradients. Secondly, rivers near their base levels have more constant gradients than those at higher elevations. The gradients of rivers are likely, on average,

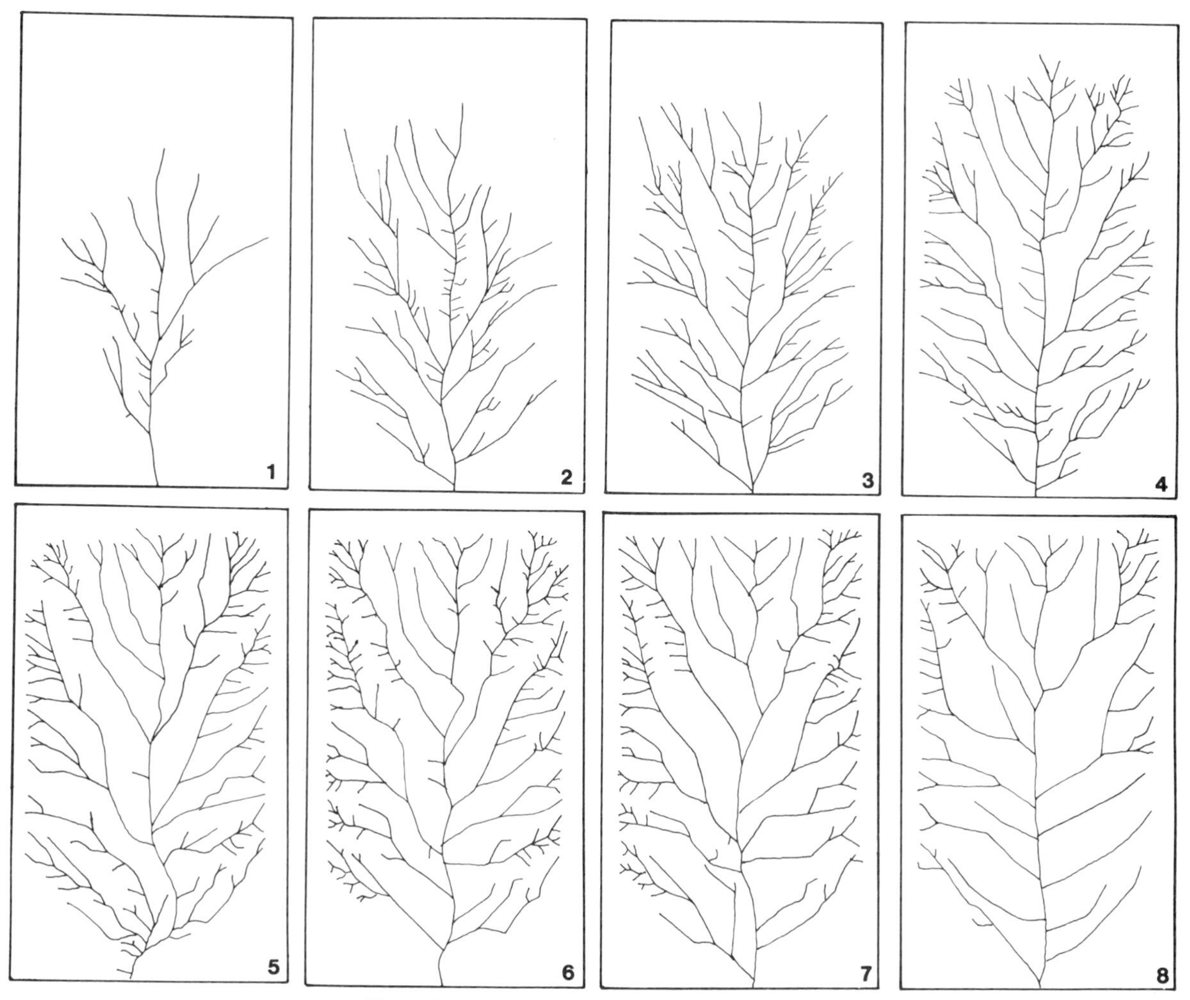

Figure 8.9 Drainage networks developed in REF.

to be lower near their mouths than near their sources. Further, as rivers gain more tributaries and their flows increase, more energy is available for sediment transport and erosion. This, too, encourages a long-profile concavity. Nevertheless river long-profiles are only *generally* concave upward. Looked at in more detail, local stretches of a river may be steeper or shallower depending on, for example, the sediment size supplied by the local rivers. But – which is more surprising – they may be steeper or shallower without any change of sediment size.

The tray model (Fig. 8.9) developed clear changes in its gradient such that in some reaches a convexity developed. The South River basin (100 km long) is formed in uniform material and thus has a pattern of drainage similar to the tray models (Fig. 8.1). Here again the long-profiles of the main stem channel and the alluvium are generally concave, but neither is concave throughout and convexities appear in several places. Moreover, the present channel and alluvial floodplain surface have their changes of surface out of step. Of course, changes in the long-profiles may be created by, for example, inflow of water from tributaries. However, Squaw Creek, for example, which effectively doubles the drainage basin of the South River, appears to have no effect at all on the long-profile.

If profile irregularities do not relate either to the different sorts of bedrock or to the entries of tributaries, they must be a natural part of river development. They may well be due to the variable way sediment accumulates in a floodplain. But it may take thousands of years for sediment to be flushed through a large river system and so changes in the

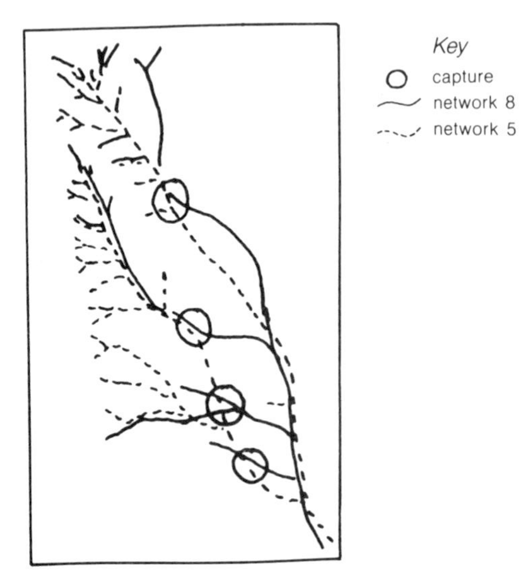

Figure 8.10 Part of the drainage patterns from stages 5 and 8 in Figure 8.9 have been superimposed and the points of capture circled.

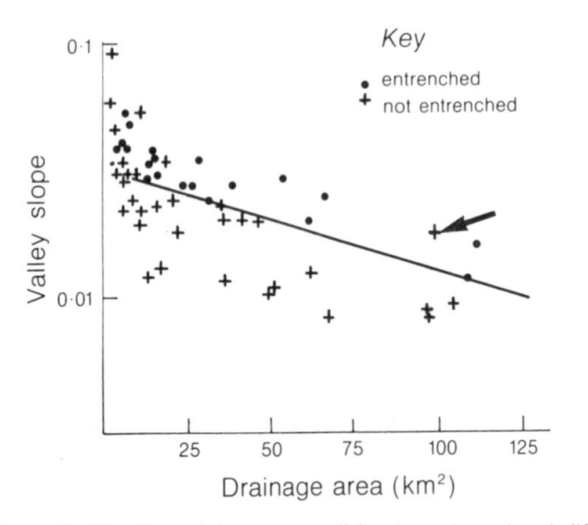

Figure 8.11 Except in very small basins where local differences in vegetation may outweigh other factors, there is a clear relationship between slope of valley floor and erosion. The example arrowed shows the only exception for large basins and suggests it is very likely to suffer trenching in the near future.

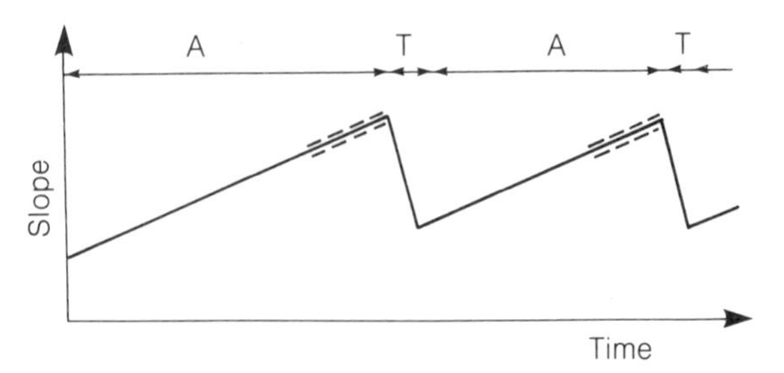

Figure 8.12 The changes in channel slope with time. If sediment from tributaries is too great for removal by the main channel it accumulates. (A) until the channel slope has been steepened to an unstable angle, when trenching (T) automatically occurs. However, a large flood can cause 'premature' flushing of sediment (indicated by the zone of dashed lines).

Figure 8.13 This valley in semi-arid New Mexico is being severely eroded by gullies. Cattle graze nearby. Is the gullying the result of overgrazing? Possibly, but in the adjacent valley where cattle also graze there is no major incision, so is the incision actually a natural part of valley development?

pattern of accumulation are difficult to measure. Only in rapidly eroding semi-arid uplands can the process be monitored. Here some of the basins are intensely gullied, while others, nearby and on the same bedrock, are not (Fig. 8.11). When these basins are plotted on a graph they fall into two groups, suggesting that sediment accumulates in the valley until some critical or threshold range of slopes is reached. Beyond this, instability develops, rivers entrench and gulleys form as sediment is flushed further down stream (Fig. 8.12). Therefore if the comparisons between humid and semi-humid regions are fair, at least in general terms, all drainage basins evolve in a series of stages. For much of the time only a little sediment is carried away by rivers and most accumulates as alluvium in floodplains. Slowly the sediment accumulates to such a depth that it becomes unstable. During this time large floods do not have much impact on basin development; but when the valley floor is unstable, floods may flush out large amounts of sediment.

At this time, and for a while thereafter, all river floods carry much more sediment than previously. When a new stable slope is achieved, accumulation begins once more. Therefore it is not surprising that, with such a **complex response**, river long-profiles, floodplain shapes and the sediment yields of rivers are difficult to interpret (Fig. 8.13).

8.3 The evolution of drainage systems

When we look at a small area of countryside, the arrangement of hills and ridges, rivers and valleys can often seem random and lacking in underlying unity. In an area of homogeneous rock this may be partly true, because a **dendritic** (tree-like) pattern evolves by random processes. Nevertheless, because of the contrasting physical and chemical properties of rocks, some are tough and resistant to mechanical weathering or inert and so removed only slowly by chemical reactions, wheras others both disintegrate and decompose readily. These contrasts are made apparent where earth movements have tilted a sequence of strata such that they lie edge-on to the land surface, yielding rapid changes of **lithology** (rock type) within a short distance.

The subtle nature of order and pattern that can develop in countryside in which rocks show marked contrasts in their resistance to erosion can be illustrated in South Devon where the plateau-land of the South Hams peninsula has been dissected by the rivers Erme and Yealme (Fig. 8.14). The first impression gained from the map may be one of a randomly produced dendritic pattern, yet a close inspection reveals that many of the streams are parallel, with the longer segments connected by short segments. Indeed, the drainage network more accurately fits a **trellis** (grid-iron) pattern. This has been produced as tributaries excavate the least resistant strata, leaving the more resistant to stand proud as divides. Furthermore, the tributaries often develop not along the centre of the weak strata but the junction of weak and tough, or lines of rock especially weakened and shattered by faulting. In South Hams, tributaries invariably flow across the dip of the rocks (they are **strike streams**); they flow on the tougher strata, gradually eroding the weaker rock at the junction. Nevertheless, the trunk streams of both the Erme and Yealme are little affected by changes in rock lithology, except that their valleys change width in proportion to the resistance of the strata through which they cut: in the weak rocks the rivers have cut wide, open valleys; in the tough rocks the valleys are narrow and steep-sided.

The patterns of trunk streams as observable now could not have developed under the present conditions: they must relate to an earlier phase when the rocks now exposed at the surface were buried beneath a more uniform covering rock (Fig. 8.15). In effect, therefore, the drainage pattern of the South Hams peninsula has been **superimposed** onto the landscape from some homogeneous cover rock which has since been eroded away. The trunk streams that developed on the cover rock flowed down the regional slope. Segments of them, called **consequent streams**, can still be identified, crossing tough and weak rock alike. By contrast the tributaries (called **subsequent streams**) have now altered their paths to etch out weak rock from tough and are now mainly **adjusted to the structure**. Indeed, the east-bank tributaries of the consequent streams have developed extensively across the weak rocks and look set soon to intersect the path of the consequents. When this occurs, water will divert from the headwaters of each consequent into the subsequent, making the subsequent stream the master and adding further shape to the adjustment to structure.

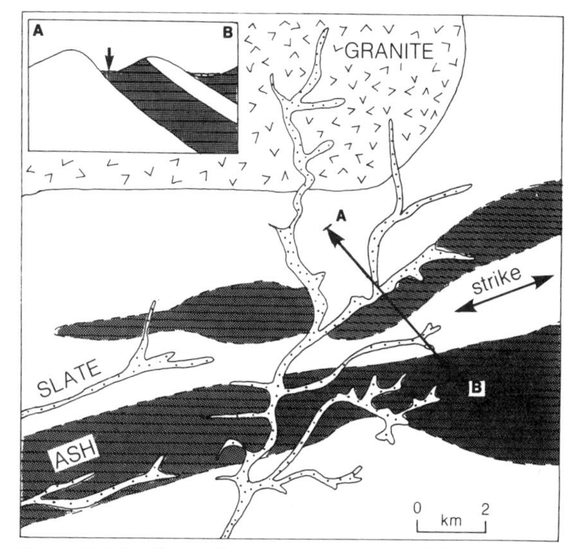

Figure 8.14 The drainage pattern of part of South Devon.

A few kilometres to the north of South Hams this process of **river capture** has gone one stage further. Here the River Lyd (a subsequent stream following the strike of the rocks) has captured the headwaters of the River Burn, a consequent stream held up by several bands of resistant igneous rock that lie astride its course (Fig. 8.16). Despite its longer path to the sea, the River Lyd has made a dramatic capture, creating a sharp **elbow of capture** and incising a **gorge** into the old valley floor of the Burn.

River capture is a normal part of adjustment to structure; it is not restricted to South Devon. Indeed, the more you look throughout Britain and the world, the more capture and other signs of adjustment to structure you can find. Such widespread adjustment has considerable implications for our interpretation of landscape development: much of Britain's drainage has been superimposed from some earlier covering rock. Some examples of capture are very impressive; for instance, the rivers Conway, Dee and Trent all contain elements of a former west–east flowing river – a proto-Trent – which has subsequently been captured by three strike streams.

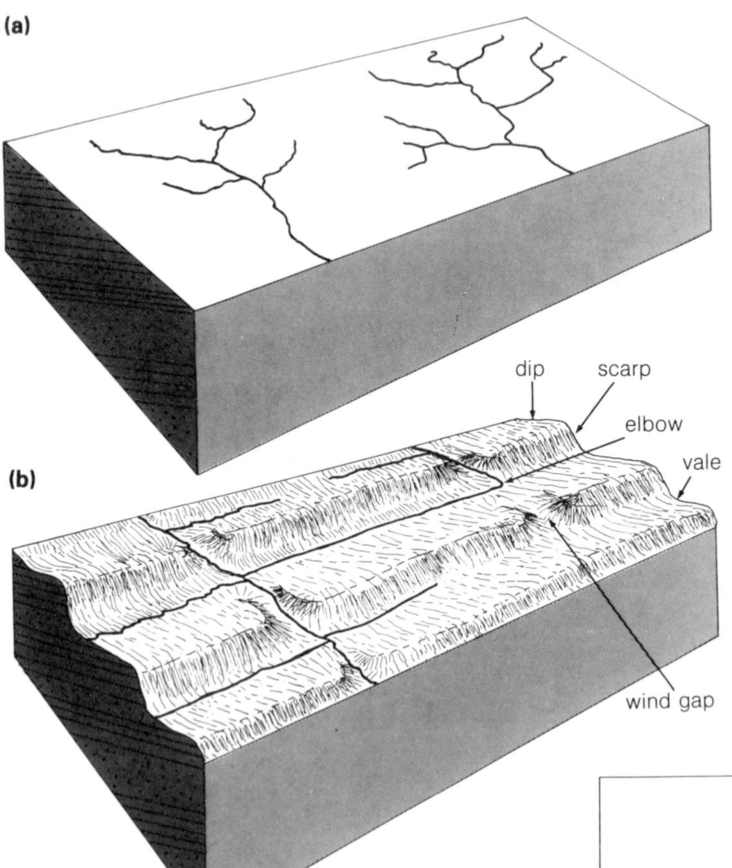

Figure 8.15 Progressive adjustment to underlying rock structure. River capture and valley formation in a region of dipping rocks invariably leads to the characteristic **scarp and vale** landscape.

Reconstructing drainage lines, as has been done with the proto-Trent, yields an extensive and coherent pattern of drainage for the whole of Britain – a pattern whose implications are even more wide-ranging. There is clear evidence that the former drainage of Britain was from west to east. This in turn suggests that the present Irish Sea was once the centre of a large fold on which the drainage was initiated. The centre of this fold may later have collapsed as a rift valley, producing the Irish Sea, perhaps as a result of the early stages of continental drift and the opening of the Atlantic Ocean. Whatever the cause, the centre of the fold has now gone; and only the upstanding western highlands remain, their fragmented drainage pattern still providing tantalising glimpses of past rivers that played a major part in forming today's landscape.

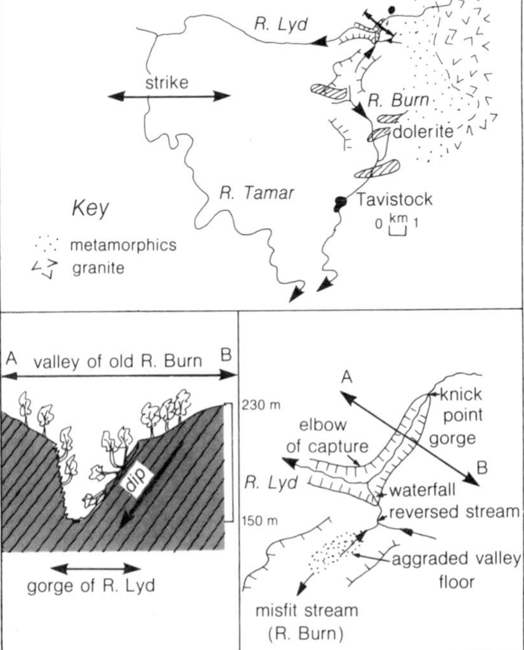

Figure 8.16 The setting of Lydford Gorge, Devon, illustrating the main features of river capture.

Channel shape and process

9.1 Channel shapes

The variety in river channels, as they return water from a drainage basin to the sea, is more apparent than real. There are only three basic channel forms (Fig. 9.1): a single **straight channel**; a single **meandering channel**; and a **braided channel** with many dividing islands. Straight channels are rare and mostly occur only on very gentle gradients: on moderate gradients rivers tend to meander. The greater the gradient, the more exaggerated the meander; beyond a certain threshold steepness, the channels become braided. The geometry of a meandering river remains constant irrespective of the river's width or length: this size is closely related to the drainage area (Fig. 9.2). These relationships show that rivers are closely adjusted to their environment.

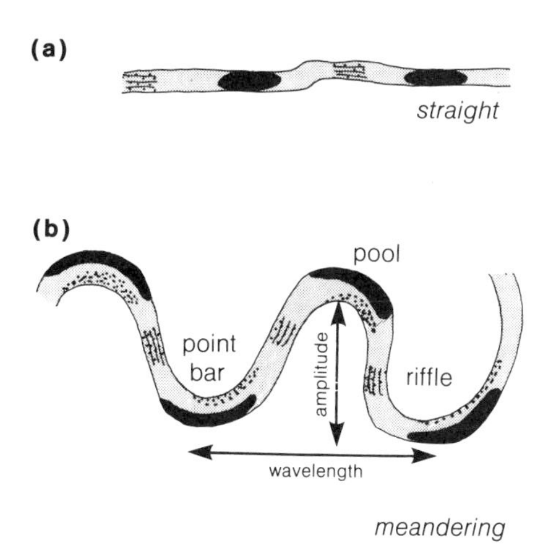

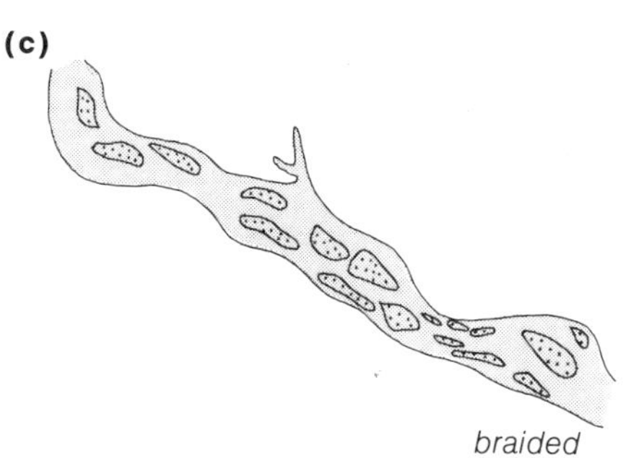

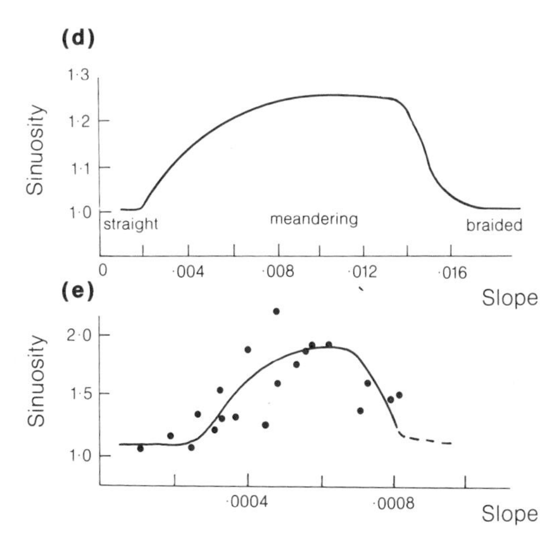

Figure 9.1 Basic river forms. (a) Straight forms are only found on very gentle gradients. On steeper slopes first meandering streams (b) develop, increasing their sinuosity with gradient until they change to braided forms (c) at every steep gradient. (d) In a laboratory model, changes from a straight to sinuous pattern and then to a braided pattern occur at two threshold slopes. (e) Field observations on the pre-channelised River Mississippi show a wide scatter, reflecting natural sinuosity variation due to meander growth and cut off. However, the line of best fit shows substantially the same pattern as for the laboratory model. The Mississippi does not have any braided sections.

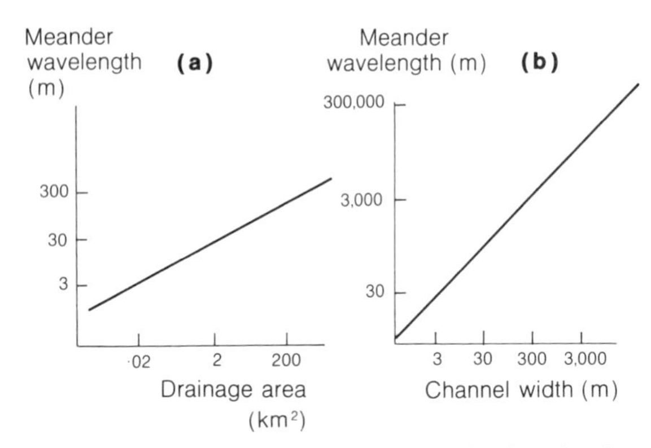

Figure 9.2 (a) Meander wavelength is closely related to drainage area (and hence river flow). (b) Meander wavelength is closely related to channel width, and thus the geometry of meanders remains the same for small creeks and large rivers.

9.2 Channel processes

As water flows through a channel it is retarded by frictional drag against the bed and banks (Fig. 9.3). The water does not all move at the same velocity: water near the channel boundaries (the bed and banks) moves more slowly than that near the centre of the river. Friction between water and air imposes a slight drag on the surface water also. The most swiftly moving water is therefore that occurring just below the surface and towards the centre of the river. The *average* velocity of the stream is similar to the velocity of water 0.6 of the distance from surface to bed. (This is why a stick floating near the surface travels faster than a fishing float that sits deeply.)

Stream energy depends on local gradient, velocity, and the mass of water. Most of the energy is dissipated in overcoming friction and is converted into heat. This is illustrated by meltwater streams which cut their channels into the surface of glaciers, melting the ice by means of frictional heat. The energy remaining after frictional loss is used in transporting sediment and in erosion.

The pattern of rivers, the shape of channels, the nature and amount of sediment carried, and the erosion performed are all related to the amount of energy that must be dissipated by the stream. In general the system works in such a way as to dissipate the energy as evenly as possible, a task that involves many subtle changes in river plan and section. The apparently endless variety of channels is thus the result of changes in frictional drag.

9.3 River velocity

River velocity is affected by variations in several parameters: channel plan, channel cross-section, channel slope, and the roughness of the channel boundaries (Fig. 9.3). However, a change in velocity between two adjacent reaches within a river system is not normally accompanied by variations in only one of the parameters listed above. In general all the parameters will alter a little in such a way as to accommodate the change in velocity with the minimum amount of work (the **law of minimum variance**). This means that changes in channel slope are rarely possible without complementary changes in channel plan, section and roughness. Nevertheless, for simplicity, they will be discussed separately below.

Table 9.1 Characteristics of the five stages in the development of alluvial stream channels (from E. Keller).

Stage 1	Stage 2	Stage 3	Stage 4	Stage 5
no pools or riffles	incipient pools and riffles spaced at about 3 to 5 channel widths	well-developed pools and riffles with mean spacing 5 to 7 channel widths and mode 3 to 7 channel widths	well-developed pools and riffles with mean spacing 5 to 7 channel widths and mode 5 to 7 channel widths	mixture of well-developed pools and riffles with incipient pools and riffles—mean spacing is generally 5 to 7 channel widths with a mode of 3 to 7 channel widths
the dominant bed forms are asymmetrical shoals	dominant bed forms are asymmetrical shoals	dominant bed forms are pools, riffles, and asymmetrical shoals (mostly point bars)	dominant bed forms are pools, riffles, and asymmetrical shoals (mostly point bars)	dominant bed forms are pools, riffles, and asymmetrical shoals
	pools and riffles are small	pools are about 1.5 times as long as riffles	pools are generally greater than 1.5 times as long as riffles	pools are generally much longer than riffles

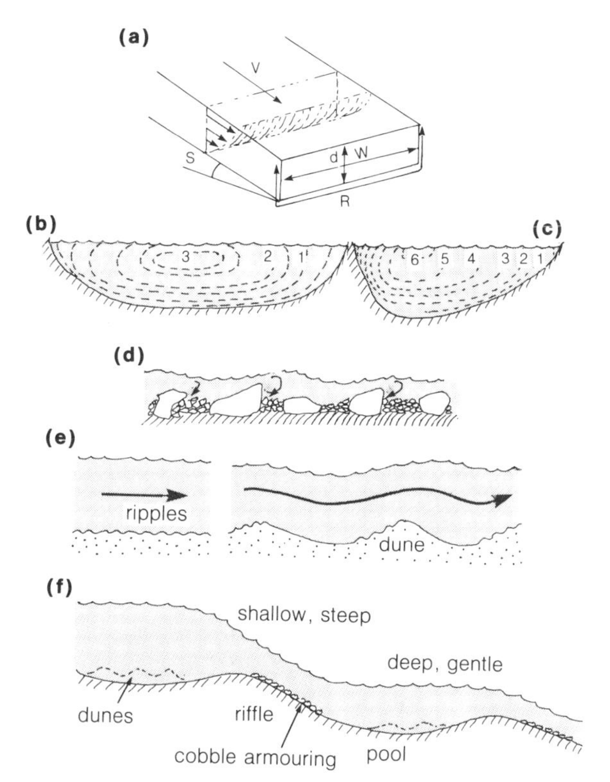

Figure 9.3 Characteristics of channels: (a) definitions; (b) velocity profile in a straight reach; (c) velocity profile in a curved reach; (d) the effect of bed armouring; (e) ripples and dunes; (f) pools and riffles.

Roughness

The size of channel material is largely determined by the type of bedrock over which the river passes. Nevertheless channels may become less rough, by deposition of fine material in between the coarse particles; or more rough by the winnowing out of fine material, the building of ripples and sand dunes, or, on a larger scale still, the development of a sequence of deeps and shoals (usually called **pools** and **riffles** respectively). Constructional features provide a particularly flexible means of changing channel roughness, building higher or being removed according to changes in velocity.

Channel geometry

An increase in the length of the channel cross section increases the frictional drag and therefore reduces stream velocity: a wide shallow channel and a narrow deep channel both impose more drag than a channel of more even proportions. The theoretically possible range of shapes, is in practice limited by the bank

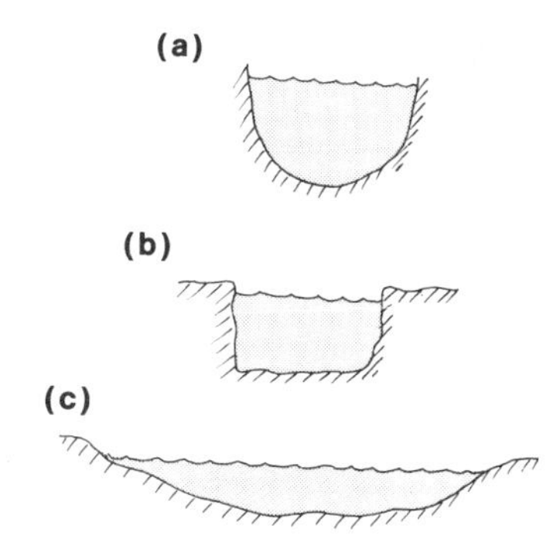

Figure 9.4 Channel cross sections. (a) The shortest cross-sectional profile is semi-circular. It is widely used by engineers but does not occur in natural channels. (b) In cohesive clay a rectangular channel forms. (c) This parabolic shape is typical for channels in non-cohesive material such as sand and gravel.

materials. Clays, which are cohesive, can stand as vertical banks; but sands and gravels are not stable at angles much over 30°, especially when constantly buffeted by moving water. Channels through sand and gravel are invariably wide and shallow (Fig. 9.4).

Water following a curved path travels a greater distance and is subject to more frictional drag than water moving in a straight line between the same two points. Meanders therefore affect velocity; the tighter the curve (the greater the **meander sinuosity**), the slower the water travels and the more energy is dissipated in getting round it. Indeed, a tight meander bend can dissipate as much energy as all the other variations in channel shape combined.

In a braided pattern, **dividing islands** cause energy to be dissipated. The more islands there are, the greater the channel cross-sectional perimeter and the greater the total drag.

Because of changes in external parameters the average velocity of the river changes little from source to mouth despite changes in gradient. In effect frictional drag is high in upland and low in areas of gentle gradient,which is one reason that coarse channel material and braided patterns are common near source regions, whereas fine channel material and low sinuosity meanders are typical of lowland regions.

9.4 Channel variation with time

River flow varies continually as the volume of water in the river grows after storms and later diminishes again. A channel that is in equilibrium with the flow at all stages has a very complex shape, for a channel that is suitable for high flows will not be suitable for low flows (Fig. 9.5). At times of high flow, more

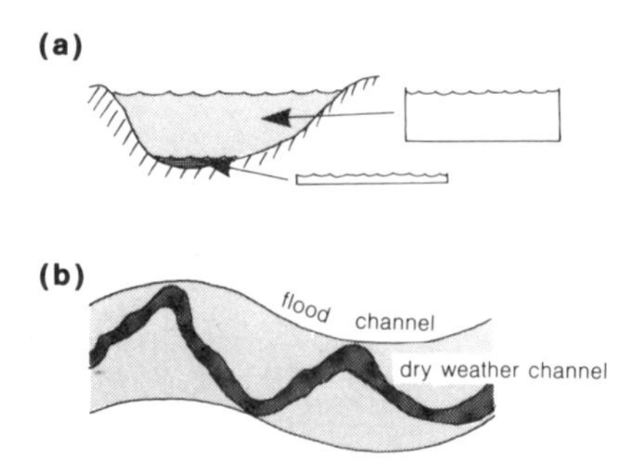

Figure 9.5 (a) The complicated river shape offers a low drag at high flow when water must move quickly, but at low flow water is confined to the flat bottom of the channel, a relatively inefficient shape. The result is to allow more energy for erosion and sediment transport at high flows and very little at low flows. (b) For the same reason, meander wavelengths are usually adjusted to relatively high flows. In regions where low flows are very small, a dry weather channel may be cut within the high-flow channel. This is very common in semi-arid regions or those with seasonal climates, but is not usual in humid climates such as in Britain.

Figure 9.6 Deep channels are not possible with sand and gravel banks. Instead the river widens and brings more distributaries into use to cope with storm flow.

energy is available for erosion so, in general, channels are adjusted to their larger, rather than their smaller, flows. In regions (as in Britain) where there is relatively little difference between high and low flows, the amount of water flowing in dry weather still fills the bottom of the high-flow channel. In regions where the difference is great, there are often dry-weather channels cut into the beds of the high-flow channels.

In non-cohesive materials, increases in discharge of water cause the channel to widen rather than deepen (Fig. 9.6). At low water only part of the channel network is used.

Braided channels develop islands of cobbles, gravel and sand that build and are destroyed with each storm. They do not have the single stable channel of a meandering river. At the onset of the flood wave, increasing energy allows the water to scour a larger

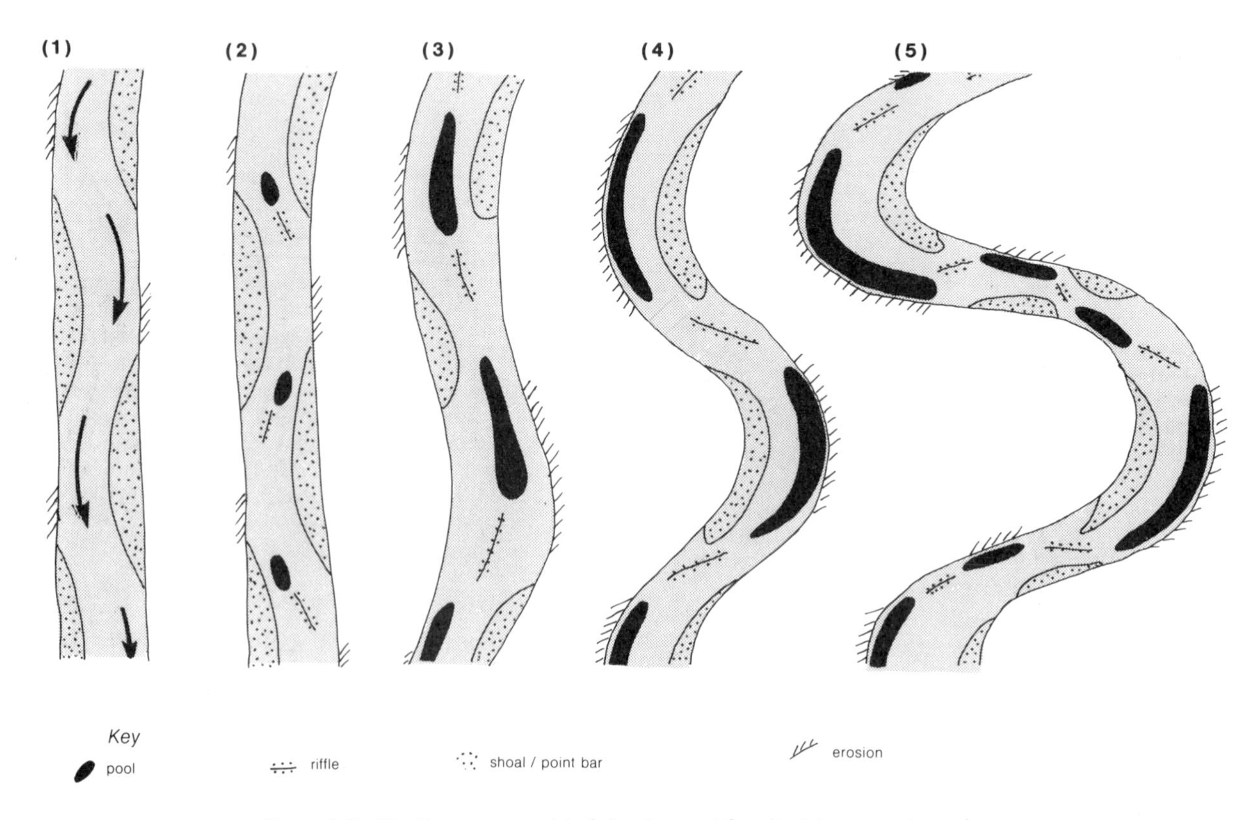

Figure 9.7 The five-stage model of development for alluvial stream channels.

channel, destroying the islands and setting their material in motion. Then, later, as the peak flow passes and the river has less energy for transport, the islands begin to reform. Each island begins as a local patch of deposition. In the protected lee of this patch further deposition occurs and the island grows. In this way islands change shape such that each flow is contained in a channel of appropriate size; a form of dynamic equilibrium made possible by the non cohesive nature of the material.

Channels can change shape over a long time but they rarely change quickly in response to storms. The 'fine tuning' necessary to keep dynamic equilibrium between channel and flow is achieved by means of changes in the pattern of sediment on the channel bed (Fig. 9.7). For example, sand dunes readily alter by heightening or flattening with changes in flow. In addition, most rivers have pools and riffles on their beds which control the flow at low water: pools normally occur on the outside bend of a meander and compensate for the high drag that occurs as water moves round a curve, whereas shallow riffles are found at points of inflexion between bends and increase the drag in regions where otherwise water would move quickly. The rivers that we see containing a balanced pattern of meanders, pools and riffles may be in dynamic equilibrium and as such they represent a stable 'mature' phase of river development. People have made many attempts to straighten rivers, usually with disastrous consequences, but the sequence of stages by which rivers re-establish their earlier meandering form is of considerable interest and may parallel the way in which natural rivers develop as the river valley widens and allows the formation of a floodplain.

9.5 Meanders and floodplains

River and channel are in **dynamic equilibrium**, a fundamental idea that explains for example why rivers do not keep eroding their banks and growing ever wider. Dynamic equilibrium does not, however, mean that conditions are unchanging and that no material is eroded. It is simply that where material is eroded from one bank it is balanced by deposition on the other.

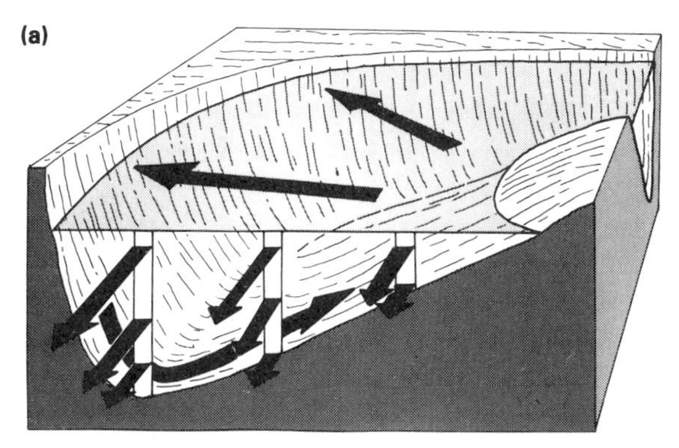

Figure 9.8 (a) Centrifugal force increases with the speed of flow and has therefore the most powerful effect on surface water, throwing it towards the concave bank and causing erosion. A balancing return flow is set up near the stream bed which transfers some of the eroded material to the convex band where it is deposited. (b) Bank collapse due to undercutting, River Severn, Shropshire.

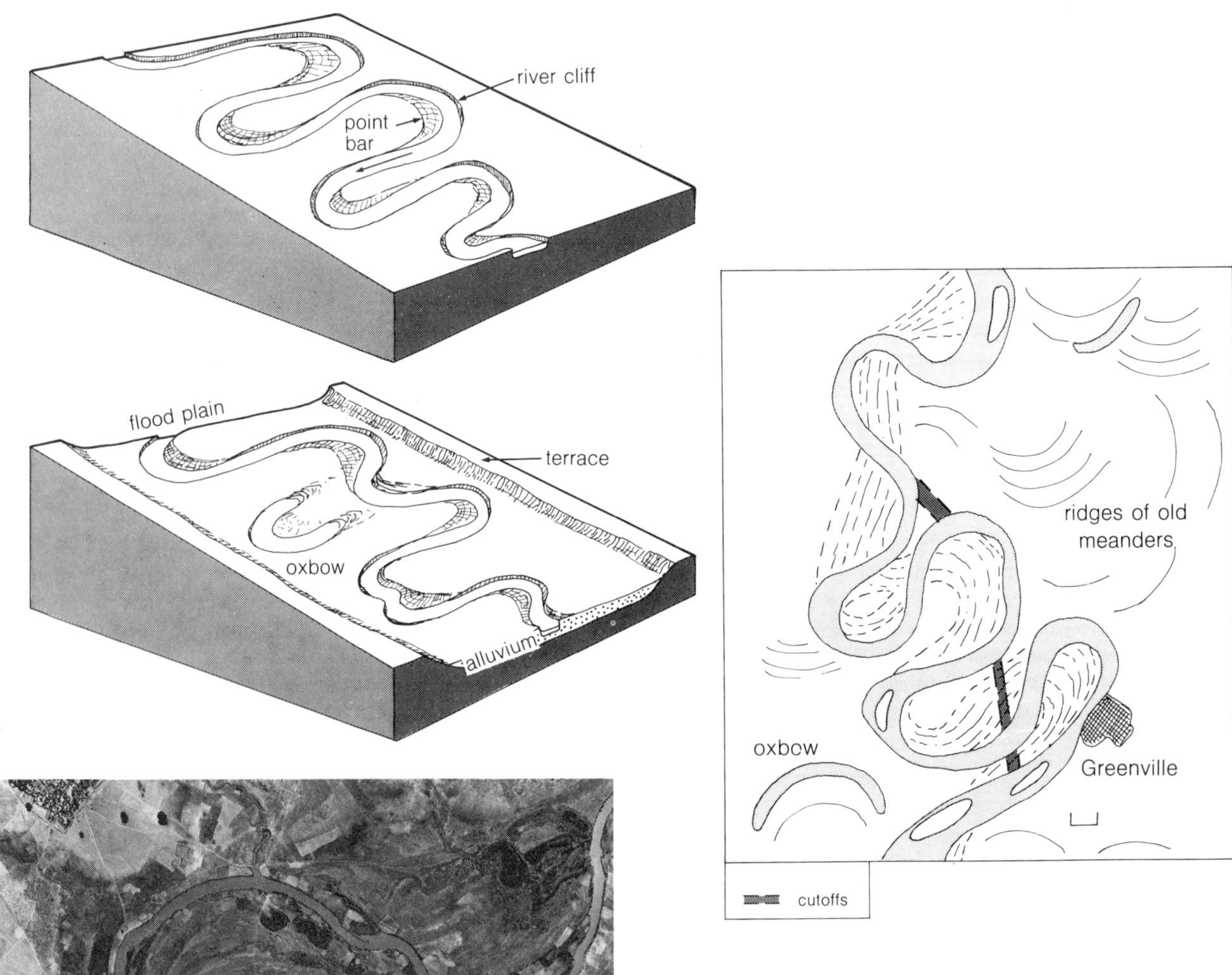

Figure 9.9 (a) Floodplain development occurs by a downslope shift of the meander belt. During this process random development of meanders can lead to meander cutoffs and oxbow lakes. (b) The shifting pattern of meanders shows very clearly on maps of the lower Mississippi. The floodplain occupies the whole of the map extract (notice the position of Greenville). The location of pools, rivers and other features are given in Figure 10.3. The 'cutoffs' refer to river modification as described on p. 94. (c) Meanders, point bar ridges (scrolls) and oxbows, Zaria, Nigeria.

Meanders are a clear example of dynamic equilibrium at work. As water moves in a curved path round each bend **centrifugal forces** throw it outward, forcing the line of maximum current, energy and erosion to the concave bank (Fig. 9.8). As the curve reverses at the next bend, centrifugal force again forces water to the outside bank. Coupled with a general downslope flow, centrifugal forces give the water a helical (corkscrew) twisting motion that carries sediment from the fast flowing outer bend where erosion occurs, to the slow flowing region at the next convex bank downstream where the sediment is deposited as a **point bar** (Fig. 9.8). Here erosion and concurrent deposition maintain the river channel shape while the river continues to shift across its floodplain (Fig. 9.9).

Similarly, dynamic equilibrium is demonstrated in river geometry (Fig. 9.10). For example,if a river increases its meander sinuosity too greatly the bed gradient will decrease, and this, in turn, will encourage deposition. Furthermore, the tighter curves will absorb more energy, causing water to build up during a storm, possibly overflowing its banks, spilling across a meander neck (an **avulsion**), cutting the meander out and eventually producing an oxbow lake. With a shortened channel the gradient steepens once again, causing scouring headward and deposition downstream until an equilibrium gradient is regained throughout the whole reach.

Changes within a river must always be viewed in these terms of *action and reaction*, deviations from an equilibrium position being restrained by in-built natural controls. However, both action and reaction may take longer than can be witnessed in a human lifetime. If any stretch of river appears out of step with the rest, it is probably acting to regain equilibrium but so slowly as to be imperceptible.

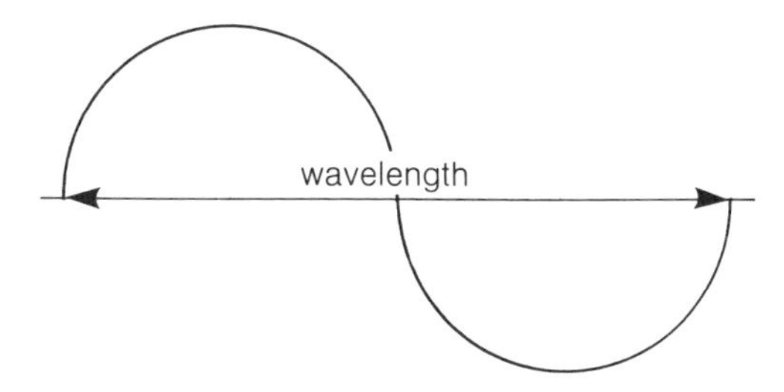

Figure 9.10 Meander patterns are stable in the long term despite cutoffs because the meander curve closely approaches the curve which has the smallest variation of change in direction, avoids any concentration of bend and minimises the total work needed for its formation. Thus, when erosion exaggerates a meander loop it is eventually cut off. The shape outlined above is most stable because it requires the minimum work; rivers rarely achieve it because their curves are at first formed by random flow.

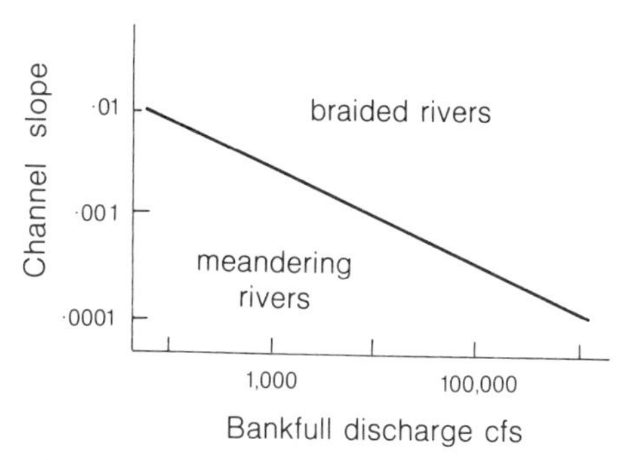

Figure 9.11 The relationship between channel slope and bankfull discharge.

Although rivers can respond to changes of environment in many ways, experience shows that it is usual for some combinations of plan and section to be most common. Thus meandering rivers on fine grained alluvium and braided streams predominantly found where gradients are steep and materials are coarser are typical combinations found all over the world. These **'quasi-stable' combinations** are common because they require the least work for their formation and maintainance. For this reason other combinations do not last for long and usually rivers change by 'flipping' directly from one stable form to another (Fig. 9.11).

9.6 Sediment-transport processes

The roughness of a channel may affect water speed considerably by increasing frictional drag. However, river channels are often formed from poorly consolidated material, so when water power increases, such as in localised eddies or during the passage of

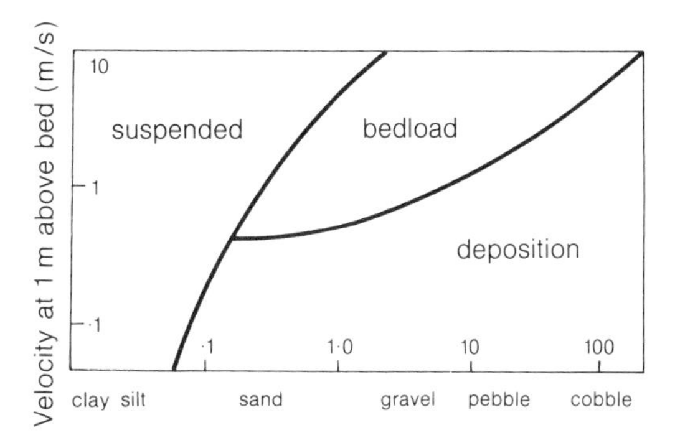

Figure 9.12 The size of material that can be moved depends on water velocity. For a particle to move, drag must exceed friction. When a moving particle comes to rest, the energy taken from the water is converted to heat.

a storm wave, the drag may become great enough to quarry material from the channel and set it in motion (Fig. 9.3d)

Each particle of sediment has a threshold below which water speed is insufficient to put it into motion (Fig. 9.12). Except for cohesive clays, less drag – and therefore lower speeds – are required to move small particles than large ones. Silts and fine sands are most readily moved because they are quite light, they are not cohesive and they project into the channel far enough to be brushed against by moving water. Medium and coarse sands are heavier and so present more of a problem. Gravel, although still heavier, increases bed roughness and turbulence and it is in fact moved more readily than large sand. Hardest to move, despite their large projecting areas, are cobbles and boulders, retaining their stability by weight alone. In a channel containing mixed sediment, sorting occurs: the finer material is carried away leaving coarse particles covering the bed and providing a protective shield to the sediment below – an effect called **bed armouring**.

Drag is not the only force on particles; stream water is always turbulent and continually produces eddies that spiral upward from the bed, lifting sediment into the regions of faster flow. Clays and silts are lifted from the bed and carried in **suspension;** the heavier sands tend to be bounced along the bed – a process called **saltation**. Only the largest particles, such as cobbles (which rarely receive enough lift from eddies) are rolled along by drag, or **traction** (Fig. 9.13).

There are limits to the total sediment that can be transported. For example, transport cannot exceed the available energy and therefore small flows cannot flush away large amounts of debris. There are also limits to the transport of each size of material. Thus the largest particles, which are rolled along, must form a queue when they move, each impeding other particles in the surface layer of the bed. Similarly, because sand particles move by saltation, the more sand moving the more the likelihood that moving grains will collide. Some will be knocked higher in the water, but many will also be knocked back against the bed, allowing only a 'skin' of sand to move, the sand below being knocked back against the bed by particles already in motion.

Rivers also carry a substantial **solution load** of dissolved material derived from the weathering of rock during soil formation and carried to the channel by throughflow and groundwater flow. Solution load is not dependent on the speed of water in a channel, and is often greatest after the high flow peak has passed, for this corresponds to the greatest throughflow contribution.

Because sediment transport processes mainly involve surface movement, even big floods rarely scour a channel to bedrock despite the enormous amount of energy available. Transport processes also determine the size of materials that make a significant contribution to the channel bed: beds rarely contain silts and clays because such low water speeds are needed to allow them to settle, and gravels are more readily moved than sands or cobbles. These differences may explain the predominance of either sand or cobble beds in natural channels.

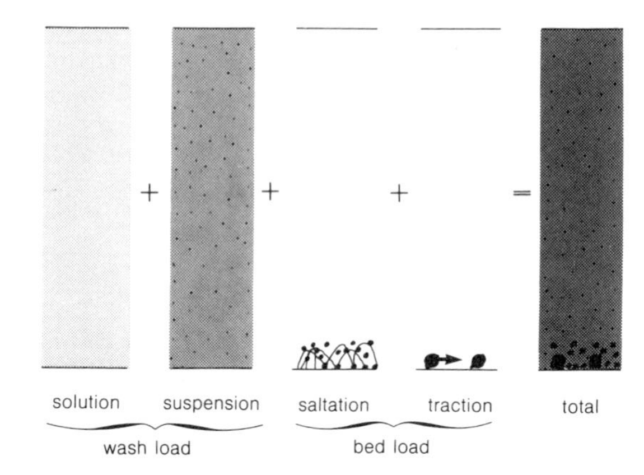

Figure 9.13 Sediment transport processes.

Bed materials of all rivers tend to fall into three categories: bedrock, sand, or cobbles, with downstream changes between these types very abrupt. Indeed, within a pool and riffle sequence the changes may occur within less than a metre.

Bedrock channels are easiest to explain. Usually only found in first order fingertip tributaries in uplands, they lack a coherent cover of alluvium because of their steep gradients. Bedrock channels have no consistent gradients, they must simply be steeper than that needed to transport any form of sediment. However, bedrock channels rarely occupy more than short segments of a total river system: natural channels are primarily formed in alluvium.

The material that dominates the bed of an alluvial channel is the size most difficult to move. Because the drag required to move sand is greater than that to move clay, silt or fine gravel, sand commonly dominates channel beds. However, because cobbles are the most difficult of all materials to move, when present in the channel they will always dominate, frequently armouring the bed. In channels with pronounced pool and riffle sequences, low-gradient pool reaches are sand bedded, and steep-gradient riffles are cobble armoured.

9.7 Transport patterns

Although river energy increases at times of high flow, there is rarely a close match between peak flows and peak sediment transport (Fig. 9.14). Not all major

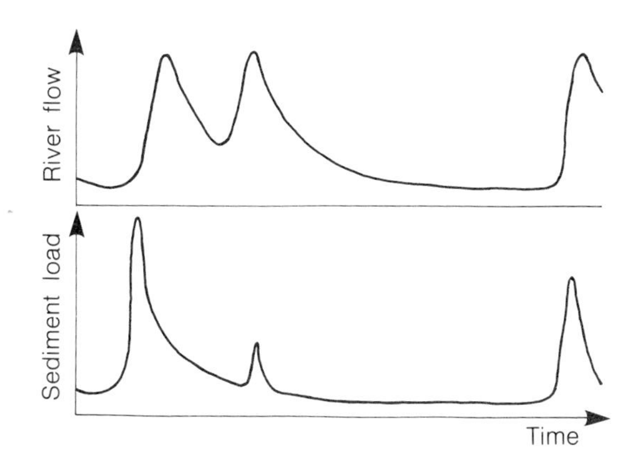

Figure 9.14 The relationship between the river flow and sediment yield is not constant.

floods move large amounts of sediment, and some quite modest flows are simply burdened with debris.

Sediment in a river can be compared to a queue of people waiting to travel by bus. Suppose enough people arrive at a bus stop to fill a bus, on average, every half hour. However, because of random traffic congestion, occasionally two buses arrive within 5 minutes of each other. At these times there will be few passengers for the second bus to pick up. Conversely, if a bus arrives only hourly, more passengers than normal will be left behind. In a drainage basin the rates of rock weathering, soil formation and slope transport produce a relatively continuous supply of sediment which is equivalent to the passengers arriving at the bus stop, But, because floods occur randomly, sometimes there will be much sediment to be moved, whereas at other times when storms succeed one another closely there will be little sediment for the second storm to carry. Sediment supply can also be strongly influenced by bed armouring; if the threshold velocity for the cobble material is not exceeded, very little material will move, whereas on those occasions when the threshold is exceeded there will be a very high sediment yield. Furthermore the whole pattern of river sediment transport depends on the pattern of hillslope sediment movement within the drainage basin.

10 River modification

10.1 Changing the shapes of rivers

Water within a drainage basin dissipates energy in many subtle ways. As a result not only is water effectively removed from the basin, but erosive energy is distributed throughout the channel network. Thus a change in any part of the river system causes compensating changes both upstream and downstream. As a result, there is in the long term a dynamic equilibrium. In the short term, irregularities in the river channel may occur because the self-regulating (negative-feedback) processes come into play only *after* the river has started to deviate from its established pattern. Deviations and compensating negative-feedback processes are a feature of a natural system. It is a widely known means of change called **order through fluctuation.** But when a man has built his house beside a river and the river then starts (or, as is more likely, continues) to change course, he wishes the engineer to correct the situation for him. Similarly, it is advantageous for industries to stand beside a navigable river. The owners of factories demand good footings for their factories and deep-water channels at both banks. And people don't like to be flooded – even if they have chosen to build their homes on floodplain land that inevitably *will* be innundated. They even demand a constant flow of water to their taps and fields.

All these conflicting demands present the engineer with a dilemma. Faced with pressure to make local and rapid adjustments, he must tinker with a system he doesn't really understand, and in ways that may later turn out to have been *opposed* to the development of the natural system. Nature may then respond in a way that requires even more modification by the engineer. The case studies outlined below describe in some detail the natural relationships between river and environment, and the results of a sequence of alterations by man to the natural flow. You should read these case studies as though they were a summary of river geomorphology, referring back to earlier chapters to see how the principles of river development are illustrated here. Work on the Mississippi represents the most far-reaching attempt to modify one of the world's greatest rivers; the implications of the results also are far-reaching. The alterations on the Rio Grande highlight how reservoir storage can lead to many difficulties.

Both case studies are from the United States rather than Britain. This is for several reasons:

(a) the US has undertaken far larger degrees of modification than any other country;
(b) highly variable river regimes are very difficult to modify;
(c) the US government has documented each study in great detail.

Some river modifications are made in Britain, but so far they have been on a relatively small scale; they mostly involve either channel straightening in urban areas (See Ch. 11) or reservoir construction. Major modifications in the future are most likely to be in developing countries, and there will be a great need to know the likely consequences. The US rivers in these case studies have regimes that match river environments in the developing world more closely than those of Britain.

10.2 Control of the Mississippi

It probably never crossed Huckleberry Finn's mind, as he and his friend Tom sat by the bank of the Mississippi hooking catfish, that the lower reaches of this mighty river were anything but a permanent and unchanging part of the landscape. Yet out across the river, which is two kilometres wide, nearly all the boatmen knew the ever-changing pattern of shoals and the dangers these presented. They knew that the most dangerous times were when they tried to cross from the deep water of one meander bend to the next. In the low waters of early spring before the arrival of the melting snows from the Prairies and the peaks of the distant Rockies, it was only the wide-bowed, shallow-draft, paddle-wheeled river boats that could ride these 'crossings' (shoals or riffles), and even then a man in the bows might need to plumb the route to ensure that there was sufficient draft. [and hence 'by the *mark... twain*' (2 fathoms)].

The lower river had always been extremely tortuous with its great goose-necked bends that nearly folded back and cut into one another (Fig. 10.1). And although this was inconvenient for the river traffic, it was part of nature and had to be endured. But then came competition from the railway, and the viability of the river traffic was brought into doubt. It became important to maintain sufficient depth for navigation at all times. Meander loops in the Mississippi have been being cut off at about seven per century in the 800 miles of the lower river. As a meander was cut off in one place, though, another would grow out, and the average length and gradient of the river was thus maintained. Each **cutoff** would take many years to become complete, during which time the back channel (**oxbow lake**) would fill with sediment. At the same time, the new channel would first take a braided form and then, as the slope decreased, gradually resume a meandering habit. Above the new cutoff, incision would occur as material was swept down the relatively steep new course, the excess sediment accumulating in the reach immediately downstream (Fig. 10.2). As the river became braided it would become temporarily shallower and wider, only to decrease again as a new sinuous form developed.

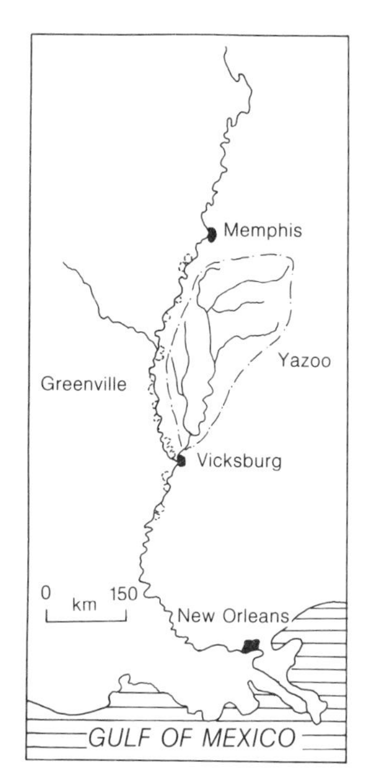

Figure 10.1 The lower Mississippi, illustrating the location of the places described in the case study.

When men first looked at the river Mississippi they saw it as both a blessing and a danger. They needed the water for navigation and irrigation, yet they feared the annual floods. During the 19th century, river flooding began to increase and navigation

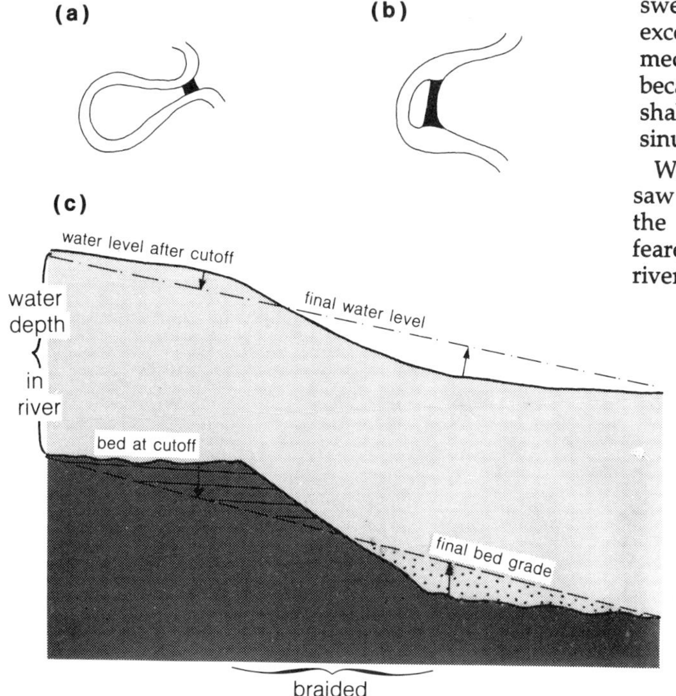

Figure 10.2 River cutoffs can occur naturally or be made by man either as (a) neck cutoffs or (b) chute cutoffs across a point bar. In both cases the cutoff causes a decrease in channel length and the channel changes as shown in (c).

became more difficult: elsewhere in the vast expanses of the basin (3.2 million km^2), other men were cutting down trees and ploughing soil, thereby increasing the proportion of overland flow and sediment yield. Eventually men were forced to build **levées** (artificial embankments, also called **dykes**) near the river to protect their farms and families. At first the levées were less than 1 m high, but slowly they were forced to make them higher – first to 1.5 m, then to 3 m, then to 6 m until today in some places they are 12 m high. At the same time the people were so afraid that a cutoff might cause a local disaster they did everything possible to prevent them. They were not aware that cutoffs were natural and essential. So gradually the river became corsetted in concrete banking walls called **revetments** while it was separated by the levées from its floodplain (Fig. 10.3). With a widely fluctuating regime, river geometry would normally have altered, but the concrete held the course firm. Revetments were undercut by the storm waters and some bypassed completely. This was met by determined resistance and the construction of newer and better walls. But although the river was held in place, the old pattern, now unable to change, did not help the movement of sediment: the flow of sediment became increasingly disorderly, leading to new problems for navigation.

It is said that soon after one period of flooding in the later 1920s the Chief of engineers in charge of the Mississippi navigation met with General Ferguson, a senior member of his staff, to ponder the problem of how to alleviate flooding. Both men became deep in thought and sat smoking their pipes until Ferguson, a practical man not given to extensive deliberation, said 'Well, do you want me to write a book, or fix the river?' After a period of silence and more pipe smoking the Chief quietly said 'Fergie, go fix the river' (US Corps of Engineers 1966).

Ferguson was not a man to be influenced unduly by the long-standing precedent of protecting the meander pattern at all cost. By the late 1930s 'Fergie' had indeed 'fixed' much of the river, reducing its length considerably. But, like a slow moving giant, the river was preparing to fix trouble for his successors in return.

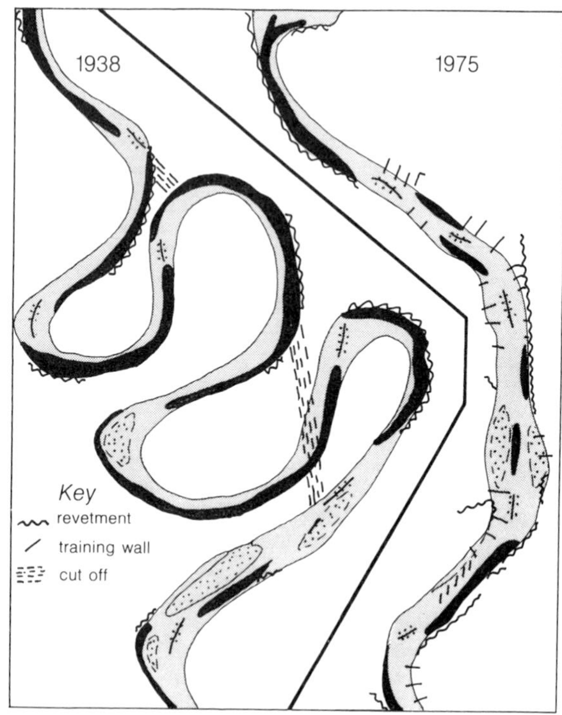

Figure 10.3 An example of river straightening on the Mississippi. This severe shortening reduced the channel length from 58 km to 6 km. Notice how many revetments and training walls have been needed to hold this unnatural alignment. Compare the 1938 and 1975 alignments with the model of meander development shown in Figure 9.7. The migration of the meanders is also shown in Figure 9.9b.

River response to alteration

Although the Mississippi is a huge, ponderous waterway, it is a dynamic and evolving system that sets its own rules and intends to play by them. Of course the rules are the same as those of any other river system. The Mississippi is just so large it takes people a long time to work out just what the rules are. Historical records show that before man revetted the meanders into place, the changes wrought by natural cutoff had taken between 30 and 80 years to complete. Change there will be, but with such a gentle gradient and such huge volumes of sediment to move, any change is bound to be slow. Not surprisingly therefore, in the two years of observation that were allowed by the US Corps of Engineers to examine the consequences, a trial cutoff showed little change in channel shape.

But perhaps instead of looking only at this deliberate cutoff, Ferguson ought to have cast an eye abroad to where the Germans had been struggling with the after-effects of a cutoff scheme they undertook in the first half of the 19th century. Between Basel and Mannheim the river Rhine had been reduced in length from 135 km to 85 km (37 per cent). So at the time Ferguson decided to fix the Mississippi, the Rhine engineers already had 100 years' experience. Their main conclusion was that any river took more than 100 years to stabilise. Severe **degradation** (erosion) still plagued the upper reaches, while **aggradation** (sediment accumulation) showed no signs of slowing down in the reach below the cutoffs. (This has remained true into the 1980s!) It was certainly clear that by 1927 the Rhine had responded very slowly to the cutoffs; if there *were* problems, they were not noticeable decades after the rectification (straightening). In the uppermost section (between Basel and Strasbourg) the river authorities finally had to give up the struggle and bypass the entire river with a canal (the Grand Canal D'Alsace), which, ironically, was begun at the same time as Ferguson commenced the Mississippi cutoffs.

The Mississippi is far bigger than the Rhine; its gradient is less and its time response correspondingly slower. But gradually it became clear that the whole of the cutoffs reach near Vicksburg acted as one giant cutoff, with the whole of the New Orleans reach aggrading and the whole reach near Memphis degrading. This parallels the Rhine experience. On the credit side it was certainly true that navigation was improved after the cutoffs and floods reduced for decades, but at some cost. In order to maintain the cutoffs and realignments (which included many chute cutoffs in the 1940s and 1950s that shortened the river by a further 88 km (Fig. 10.3)), a prodigious amount of revetting and training were needed. At the same time aggrading reaches and shallow braided sections have been kept to navigable depth only by dredging 1.7 billion (10^9) cubic metres of sand from the bed.

The situation today

You can view the revetting, training and dredging as an example of man's successful domination over nature. Or you can view it as one hell of a mess. It is a dubious success that condemns man to perpetual dredging on such a scale. A better way must exist, based on a thorough knowledge of the mechanics of water and sediment flow. Of course it is now impossible to re-establish the river in its earlier pattern, but even if this were possible, the river of the 1920s would not be suitable for the increased flows of the 1980s. With a greater flow the river needs a less sinuous course and a meander wavelength of about 16 km. If the river could be realigned to provide this sinuosity there might be a better natural control of sediment and thus river depths. To do this there must be some place in the river where the transported sediment can be temporarily stored. In effect, engineers must allow the river to recreate point bars. Unfortunately, if the river is to rebuild point bars the channel must become wider and this would mean demolishing and setting back some levées which at present are beside the channel. However, people live on and farm the land that would be needed, so the problem has become political. But whatever is done – or not done – will cost a lot of money.

River control for land drainage

The Yazoo is a tributary of the Mississippi whose drainage basin was cleared and cultivated for cotton and cereals during the 19th century. With clear-row cropping the land was soon suffering considerable overland flow; soil was quickly stripped away. Within a few years there were extensive areas of gullied land, and considerable valley-bottom silting wherever rivers were unable to carry the additional sediment away. As channels filled up, many piecemeal attempts at drainage were made and the stream bed was excavated. However, within a few years of each excavation, the downstream reaches were again completely filled with sediment: land drainage ditches ceased to be effective. Eventually the US Soil Conservation Service attempted an integrated plan for the whole basin including soil conservation measures such as revegetating badly eroded land and installing floodwater-retarding dams designed to reduce the sediment supply. At the same time the trunk streams were excavated and straightened.

From the time of the first plan in 1961, the Soil Conservation Service found themselves ever modifying the scheme, relocating dams and protecting more streambank with a **gabions** (rocks encased in wire mesh). Yet despite all their efforts stream erosion continues to be severe and the maintainance bill climbs ever higher.

Problems arise because channels have now been doubled in cross-sectional perimeter while straightening has doubled the gradient. In turn this has doubled the stream's speed and increased by a factor of 70 its capacity for transporting sediment. However, so successful have the soil conservation measures been that the carrying capacity of the rivers now greatly exceeds the sediment they receive from the fields. In times of flood this causes severe scouring of the bed and undercutting of the banks. Furthermore, the lower reaches of the trunk streams have been not only enlarged but deepened, creating 1 m-high steps. These mini-waterfalls are places of concentrated energy and erosion and have migrated rapidly upstream, scouring away even more channel sediment.

The Yazoo floodplain consists of unconsolidated sediment whose threshold angle for bank stability is about 35°; to match river trenching the banks are collapsing and the top width of the channels increasing to regain stability. At the current rate a channel cut with vertical sides and a top width of 9 m will widen to 46 m before a stable angle is reached. The process of bank collapse has been speeded even further by dumping excavation spoil on the banks to make levées; the weight of the spoil simply makes it more likely that the banks will slip. In addition,

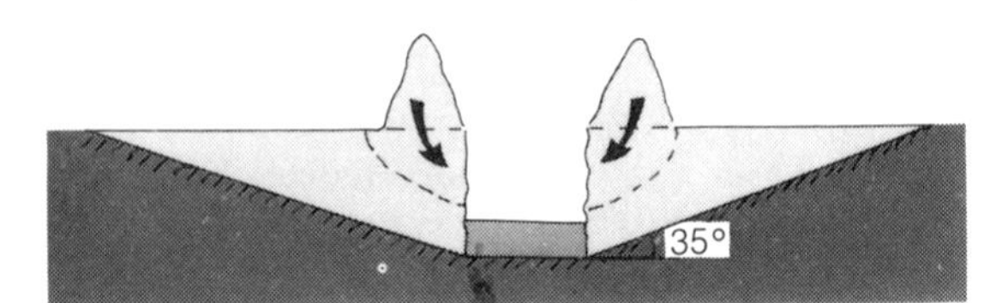

Figure 10.4 In the silty sands of the Yazoo basin, bank collapse will continue until a stable soil angle is produced. Channel widening will be fast at first because of the weight of levées on the unsupported banks.

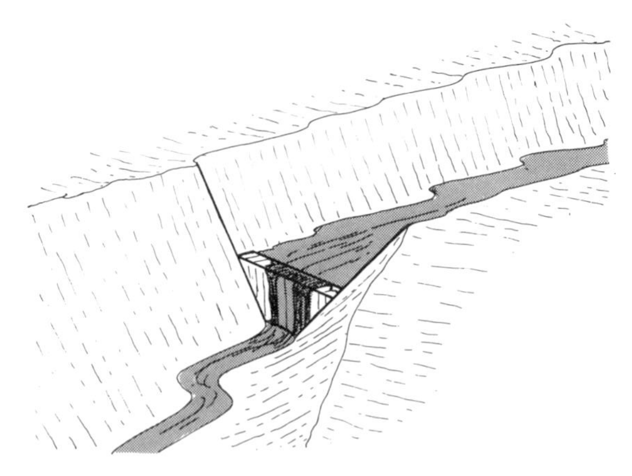

Figure 10.5 A concrete cill acts as a local base level, prevents trenching and means that the bank needs to widen less in order to decline to a stable angle.

Table 10.1 Rainfall, runoff and sediment supply for areas similar to North Mississippi (after Dendy *et al*, 1979). Notice that cultivation causes surface runoff to double and sediment supply to increase by 4000 per cent.

Land use	Runoff (% of rainfall)	Annual sediment yield (tons/acre/year)
cultivated	31	23.85
pine and hardwood	17	0.05

removal of river-bank trees during excavation has taken away any means of providing bank support (Fig. 10.4).

With hindsight we can see why the river responded to human alteration in such drastic ways. Flooding five to seven times a year may have caused major problems for agriculture, but it was a natural aspect of the drainage system. Channels in regions dominated by short-lived convectional storms are adjusted only to low flows, and whenever there was storm runoff, the entire floodplain became the high-water channel. When the channels were enlarged and flooding prevented, the river was divorced from its floodplain and channels here began to adjust to a size and shape more in keeping with that of streams of the humid areas. Until this is achieved the streams will not naturally have stable channels.

It is difficult to convince farmers who are losing valuable land by bank collapse that nothing can be done. In the past, conservationists have looked mostly at areas where erosion has been most severe (e.g. where channel top widths have increased most) and sought to stabilise these. But these channels are probably approaching their natural stability and could be left alone; it is those areas where widening has only just begun that will cause the most trouble in the future and where efforts at stabilisation should be concentrated. The cheapest remedial measure is to prevent further trenching by installing concrete **cills** across the channel beds (Fig. 10.5). However, 'they are not designed to produce a state of absolute stability, nor is it realistic to try to do so. Rather, any remedial strategy or tactic should be aimed at reducing the rate of channel change to an acceptable level or, in other words, to a condition of dynamic equilibrium' (Schumm *et al.* 1984).

10.3 Reservoirs on the Rio Grande

This is clearly demonstrated by the Rio Grande (called Rio Bravo in Mexico), one of North America's major rivers, gathering water from 11 per cent of the US and 44 per cent of Mexican territory of 860 000 km^2, to form a common bond in a region over three times the size of Britain (Fig. 10.6).

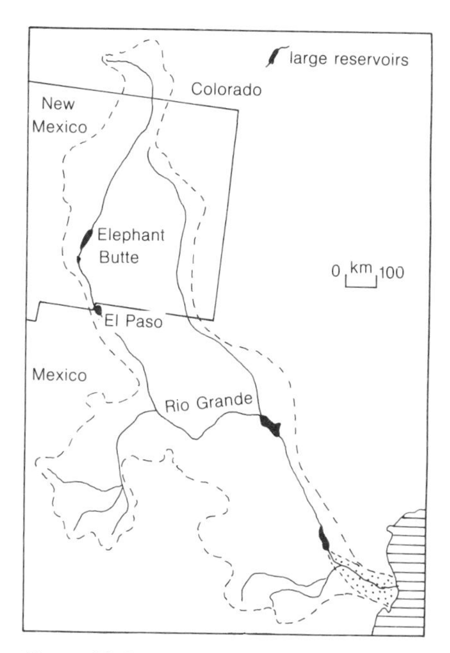

Figure 10.6 The Rio Grande river system.

Ever since the earliest settlements, the river water has been used mainly for irrigation. At first it was on a very small scale. However, as the 'Wild West' became part of history and life became more stable, more and more farmers arrived. Each farmer made his own arrangements to divert some of the river water to his land. As a result, these abstractions of water were local, uncontrolled and haphazard. At first this mattered little as there was plenty for all: but as more and more diversions were made without regard for the consequences downstream, the Mexican farmers, who had been drawing water from the river for over 300 years, gradually became unable to irrigate their land. The Texans fared as badly. It was clearly time for a planned approach to river use.

The planning and use of river water is an inextricable mixture of scientific knowledge, politics, economics, and social and environmental interests. But plans should be based on facts, and in this case facts were not available – river flows were unmeasured, precipitation values unknown and the possible causes of water loss unconsidered. The problems, though, were real: more water was needed, and reliably, for irrigation throughout the growing season; and new settlements needed protection from flood. The difficulty lay in devising a solution and deciding where to begin. In the border region between Mexico and the United States no progress could be made at all: Mexico was involved in its own internal struggles, and was a country with relatively little money to spend. Early attempts to control river flow and irrigation use were therefore limited to the upper reaches, which lie entirely within the US.

The Elephant Butte dam

The first plan was simple enough. A reservoir (called Elephant Butte) was to be built to store up spring flood-water that came down from the melting snows of the Rocky Mountains. The reservoir would help prevent downstream flooding. Then, throughout the summer, the stored water would be released for irrigation in a controlled manner. The problem was simple too. How big should the reservoir be and how much water could be released? The engineers suddenly needed to understand how water behaves. The first signs of trouble resulted from the building of the Santa Fe railroad. This crossed the Rio Grande upstream of El Paso on a bridge whose design reduced the river channel width from 1000 m to 250 m. Over the next fifty years the damming effect of this structure during floods caused so much deposition of sediment in the slack water ponded upstream that the railway track had to be raised 13 m to avoid being buried. Downstream from the bridge, the river, now relieved of much of its sediment, had plenty of energy to transport new material. Where better to collect it than from the downstream side of the bridge supports! So while sediment collected upstream of the bridge, severe erosion on the downstream side led to risk of collapse.

Elephant Butte reservoir acted like the Santa Fe railroad bridge on a much grander scale. Water flowed in and material settled out, reducing the storage capacity of the reservoir. Below the dam clear water flowed out. In a few years erosion had occurred over 220 km of river below the dam with the result that the channel became deeper and narrower, and the gradient much steeper. While the gradient steepened immediately below Elephant Butte, the river built up its bed near El Paso, locally halving the gradient. Sediment accumulated until the channel bed was higher than the city streets! Now El Paso was flooded *more* often than before the dam existed, even though the river contained less water. Nearby field drainage became impossible and irrigation water could no longer be flushed away. Toxic salts from the water accumulated in the soils and plants were killed. Something had to be done near El Paso beyond the short-term expedient of building levées (dykes).

Eventually 'brute force and ignorance' prevailed and the channel was 'rectified' (straightened and relocated), reducing its length from 250 km to 140 km and nearly doubling the gradient (Fig. 10.7). This expensive engineering solution was successful in flushing sediment out of the El Paso region and making irrigation possible once more, but also moved the problem downstream. As a result, once a year

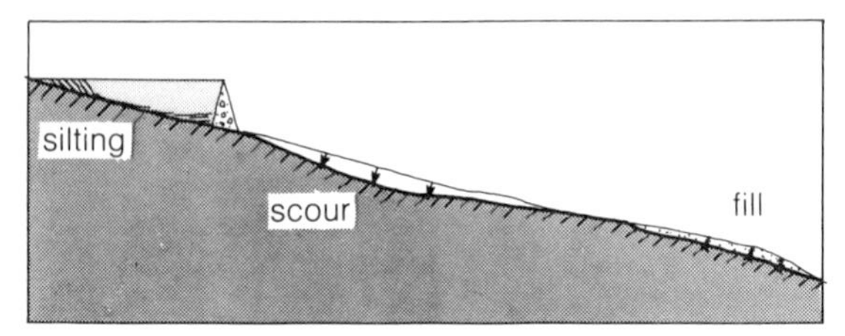

Figure 10.7 Problems that can develop near a reservoir.

large mechanical diggers have to move into the Rio Grande 140 km downstream of El Paso and clear away the accumulating bed load.

The Caballo dam

After the construction of the Elephant Butte dam it was decided to change the use of the reservoir water. In the 1930s it was decided to use Elephant Butte for hydroelectric power as well. This required release of more water than would be needed for the irrigation projects downstream. But how much water could be safely released? The only solution seemed to be another dam (the Caballo dam) constructed downstream of Elephant Butte for the express purpose of retaining the excess water released during power generation. Caballo could, for the first time, also prevent flooding downstream, which was a useful bonus because the power generation scheme has never proved economic. The new dam encouraged people to believe that they would be safe living on the floodplain. As this was not part of the plan, even more protection was necessary for the new, vulnerable houses.

Other control points

In the upper reaches of the Rio Grande there are now 13 major reservoirs designed to control the river. Most are in the Rockies, where they should have been built in the first place. Downstream, near the delta, development along the international boundary had been delayed by lengthy negotiation: construction should have benefited from the accumulated experience of earlier control schemes. However, the river near the delta is quite different from that in the upper reaches. So much water is abstracted from the upper Rio Grande that the river virtually disappears below El Paso and only flows again when fed by tributaries that gather water from the summer convectional storms that frequent the Gulf States.

In the delta region the water is in demand principally for irrigation, and, as in the upstream regions, reservoirs have been built. But flooding is more of a problem here than upstream because the delta surface slopes gently *away* from the river on both sides. Thus large areas are inundated whenever the river spills over its banks. The traditional solution has been to build levées. Levées are expensive and the land between them less useful than that in other parts of the delta. It therefore becomes important to know both the width that a **floodway** (an artificial relief channel) must be beween the levées and the height of levée walls.

The main dam near the delta is the Falcon dam (built in 1954) and again it has proved uneconomic to operate for hydroelectric power because it also has to guard against flooding and provide water for irrigation. But worst of all, lack of planning in the delta region has allowed land in need of irrigation greatly to exceed water supply; as a result both Mexico and Texas continue to suffer from water shortages. Compounding this problem has been a lack of proper drainage. When large volumes of river water are diverted to irrigate land any surplus must be drained away. This was not done, and now water levels in some places are within three metres of the ground surface; salts are not flushed from the soil, leaving large areas too saline for plant growth. Levées leave a lot of wasted land near to a river. Land in the floodway is ideal for farming. After a while, however, the farmers seemed to forget the real purpose of the floodways: they became angry when occasionally their crops were covered with water. Independent of spirit, they used their own machinery to build secondary levées in the floodways – thus interfering with the original flood-control scheme.

Perhaps somewhat ironically, El Paso and many other cities obtain most of their drinking water from wells rather than the river. This is partly because the townspeople failed to secure water rights when the river water was being allocated. Recently, with expanding populations, urban demand has exceeded supply from traditional wells. Wells have been pumped faster and new ones have been sunk. However there are often links between the water above and below ground. Near the Rio Grande, surplus irrigation water is now filtering down to the rocks in which the wells are sunk and it is being sucked up by the pumps. As this is the very water that farmers have recently flushed from their fields to prevent their crops dying, many of the Rio Grande towns today are finding that their water is rapidly becoming saline.

Floods

11.1 The causes of floods

Examination of the water cycle reveals the sources of water to a river, and shows why some types of flow are more rapid than others, but it does not directly reveal the reasons for floods. In most cases, however, it is floods that are the most important aspect of water flow, for it is during floods that rivers flow fastest, have greatest energy and modify the landscape most rapidly: floods affect man directly, bringing ruin to crops, damage to property and perhaps even loss of life.

In Britain, floods are uncommon. The largest flow, which recurs on average once a year (called the **annual flood**), does not even reach near to the top of the banks. Overbank spillage – true flooding – occurs on many rivers only about once every 2 or 3 years.

Floods result from special combinations of conditions – recipes for environmental disaster.

The magnitude and frequency of rainstorms

It may seem that, because rainstorms are produced by the three simple processes of frontal uplift, relief or convection, the pattern of rain that falls from any cloud should be similar.

But the situation is complicated because each air mass has a unique history of development and thus unique properties of humidity, surface temperature, environmental lapse rate and so on. As a result successive rainstorms can vary widely in the amount of water they release and the length of time over which it falls. However, a statistical analysis of rainfall records shows that the larger, longer or more intense the storm, the less frequently such a storm occurs (Fig. 11.1). As we have seen, the shape and size of a river channel is primarily adjusted to moderately sized flows produced from moderate rainstorms. Thus, while channels can easily accommodate lower flows, they are unable to cope with unusually large storms and overbank spillage occurs.

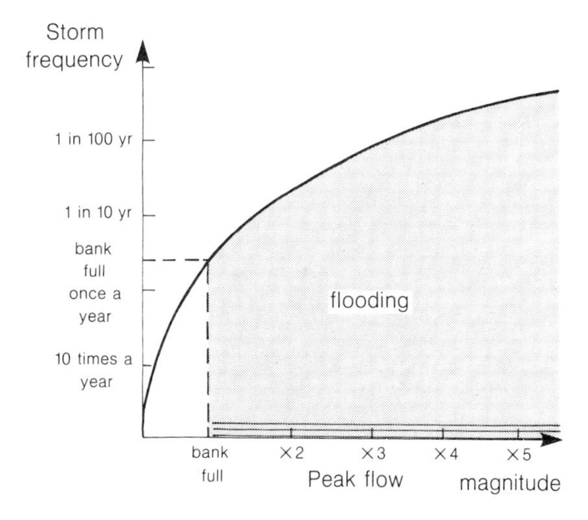

Figure 11.1 Flooding with large, infrequent storms.

Storm direction

Moving storms (Fig. 11.2) can have particularly serious effects on drainage basins that are elongated. There is little problem if the storm moves upstream: the main runoff from the lower part of the basin will have reached the outlet before the upper basin receives its rain and begins to contribute runoff (Fig. 11.2a). Conversely, if the storm moves down stream, all the runoff from different parts of the basin may reach the same part of the river simultaneously. Flooding then becomes very likely (Fig. 11.2b). Occasionally flood waves move at tremendous speed. In 1974 in Eldorado Canyon near Las Vegas, Nevada, a thunderstorm moved down the basin at 12 km/hr, a speed exactly matching the flood wave advancing down the channel. Rainfall and runoff were superimposed; water from the valley slopes reached the channel as the flood peak arrived. Picking up sediment from the channel bed, the flood arrived at the

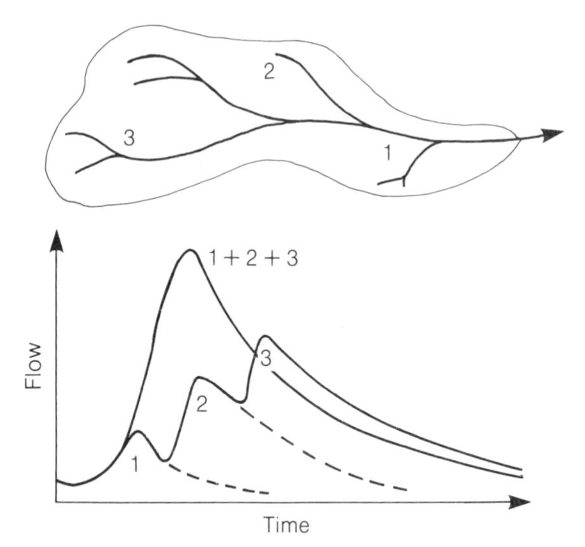

Figure 11.2 The effect of storm direction can be very important. A storm moving from mouth to source produces a many-peaked hydrograph (1, 2, 3), whereas a storm moving downstream combines these peaks into one of considerable size.

outlet as a wall of water 8 m high. This wall was said to have the consistency of ready-mixed concrete. At the mouth of the basin, a small tourist resort was instantly washed away: '... [the water] stacked with cars, trailers, etc., smashed into the coffee shop and post office and they exploded like there was dynamite inside...'.

Lack of vegetation

Floods are often associated with severe thunderstorms that produce infiltration excess overland flow. Such floods are common in semi-arid and arid regions where thin soils and sparse vegetation do little to promote infiltration or retard runoff. In these areas floods are more common than in areas of low rainfall intensity, such as Britain.

Stationary storms

As described above storms often cause flooding as they pass over a basin. Floods are even more likely if a cloud becomes stationary and is forced to release its entire contents over a basin (Fig. 11.3). Clouds are most easily trapped by mountain ranges. For example, in summer warm moist air drawn from the Gulf of Mexico towards the Front Range of the Rockies, USA, is further warmed and made unstable by intense daytime heating. Before this unstable air reaches the Rockies it has already developed towering thunderclouds; when its forward movement is blocked by the mountains it releases rain over the same small area for hours on end. On 30 May 1935 near Denver Colorado nearly 600 mm of rain fell in 4½ hours. On 31 July 1976 the cloud that became trapped over the Big Thompson basin in Colorado released 300 mm in 4 hours, causing water to cascade down every slope and producing a flood that killed nearly 150 people and washed away a road.

Storm duration

Heavy rainfall can be produced in the humid mid-latitudes by forced ascent of frontal air. However, in such cases it is the *duration* rather than the intensity of rainfall that is crucial. Floods result from saturation overland flow (see Fig. 7.2b).

One of the most famous examples of a long duration storm, causing ground saturation and inducing saturation overland flow, occurred on the northern slopes of Exmoor in Devon in August 1952. The River Lyn burst its banks and cascaded down into the seaside village of Lynmouth, tearing down the main road bridge and many houses with boulders weighing several tonnes. More recently, between 2 and 5 February 1977, a front moved slowly over the Appalachians, USA, releasing between 100 mm and 400 mm of rain onto land that had been saturated by a storm within the preceding few days. Contributing areas beside rivers and in hillslope concavities expanded until up to 30 per cent of the basin area was subject to overland flow, leading to widespread flooding in nearby riverside towns.

Flood waves

Floods may affect areas far beyond the confines of the basins in which rain falls, producing a **flood wave** that may travel downstream for hundreds of kilometres with water continually spilling over the channel banks. Rivers such as the Niger and the Nile experience annual flood waves following intense rainfall at their headwaters. The River Wye in Wales is also prone to flooding when water moves down the basin after storms over the mountain headwaters.

Melting ice and snow

Glaciers, even small ones, store tens of millions of cubic metres of water which may be released very rapidly by the warming effects of air or rain during the summer. Landscapes covered by winter snow are most liable to flood in early spring, when air temperatures rise above freezing and any precipitation is warm rain rather than snow. A warm spell or a period of heavy rain can cause widespread snowmelt and runoff over soil that is still frozen. Because there is no infiltration, all drainage is as rapid overland flow and severe floods are frequent.

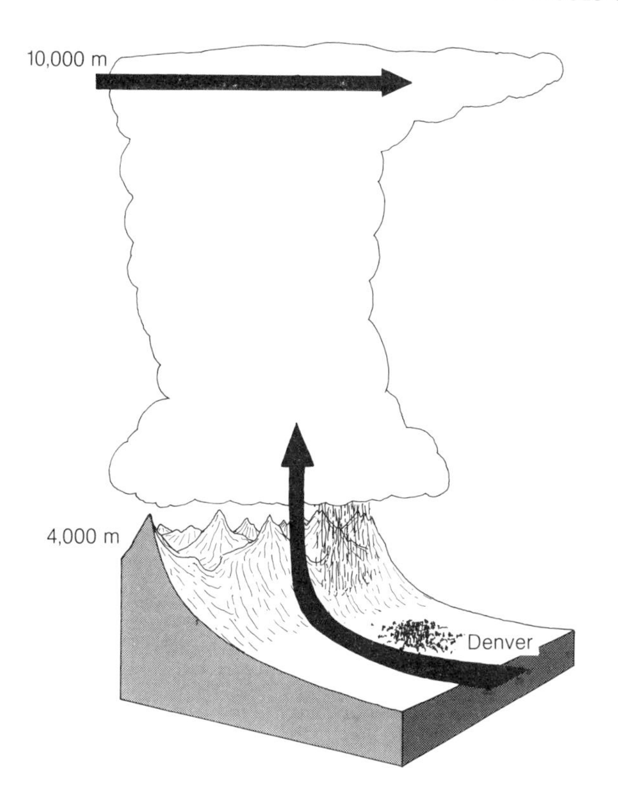

Figure 11.3 (a) In the Front Range area of the Rocky Mountains, summer days are usually hot and dry, but occasionally interspersed with intense and sometimes disastrous thunderstorms. When air becomes trapped against a mountain front it can only escape upwards, forcing it to cool and release even more water. Eventually the air becomes incorporated in the high-level westerly flow that brings it back again over the recently drenched mountain foot basins. (b) The highway through the Big Thompson canyon was completely washed away in 1976.

High tides

Sometimes a combination of onshore winds and high tides can drive water up river (Fig. 11.4). In funnel-shaped estuaries, such as the Severn and Amazon, high tide always builds water into a **tidal bore.** Whereas tidal bores when small are merely interesting, they become fearsome when enlarged by onshore winds.

On 2 February 1976 a depression drove gale-force winds against the coast of north-east USA, piling the sea against cliffs and causing water to surge into estuaries. Sea levels were already high because of the effect of the low pressure, but the level was heightened further by high tide and a wind-driven surge of water. The Penobscot river has a funnel-shaped estuary that faces directly into the teeth of these winds, and as water surged into the progressively narrower inlet, it was heightened further still until it was several metres higher than the river. As a result the water gradient in the Penobscot river was reversed and river water began to flow upstream, flooding the valley for over 80 km. In Bangor, Maine, the water rose 3 m in under 15 minutes; there was barely time for people to run to high buildings.

The severe coastal flooding of eastern England and the Low Countries in the winter of 1952–53 was caused by a similar effect. In this case a deep depression moved southwards driving the surface waters of the North Sea towards the narrow Straits of Dover. When the storm surge reached the Scheldt estuary in Belgium it was higher than normal by 13 m, more than the height of a three-storey house.

Whereas Maine has steep-sided valleys which restrict the extent of flooding, eastern England and the Low Countries have gentle-sided low valleys and almost flat coastal plains on which millions of people live. Coastal and river flooding left 50 000 people homeless and nearly 10 per cent of the Low Countries under water. This disaster led directly to major long-term plans for coastal protection, plans which include the Thames Barrage and the Delta Plan.

(a)

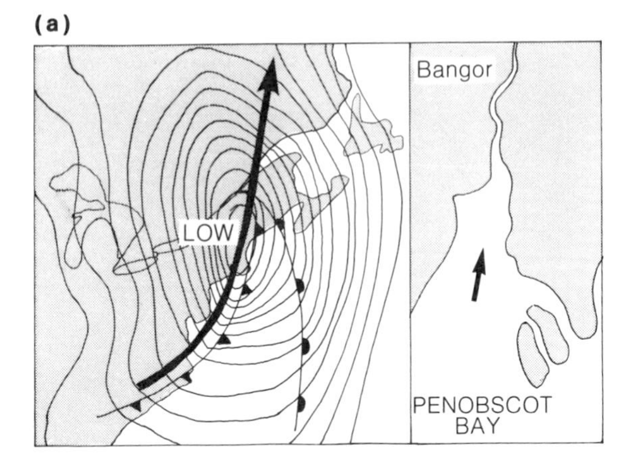

(b)

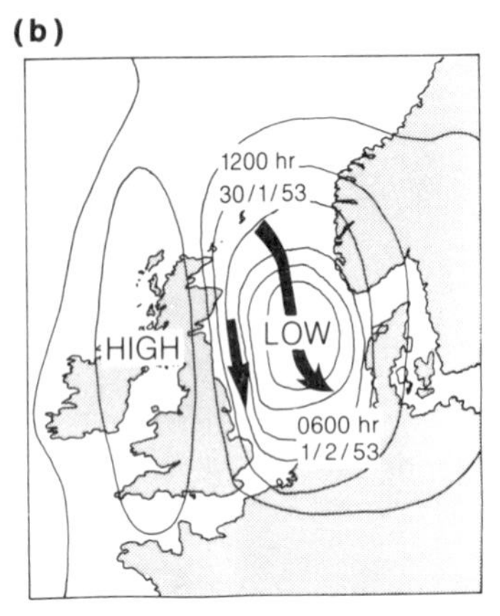

Figure 11.4 Storm surges: (a) Bangor, Maine, USA, 1976; (b) North Sea, 1953.

Urbanisation

People have increased the speed of runoff to rivers, largely by replacing permeable soil with impermeable brick, concrete and tarmac.

Urbanisation (Fig. 11.5) begins when rural lands are colonised by villages whose growth at first has little impact on the environment. As villages develop into towns and cities, their centres become more and more densely occupied until 100 per cent of the soil is covered with buildings or tarmac. With severe competition for central land, people begin to seek more space by moving to the outskirts and forming a surburban fringe where, although housing density is lower, there are still extensive areas of impermeable surface and a network of storm drains to carry the water quickly away. Finally rivers are straightened, enlarged and lined with concrete in order to receive water from the drains and speed its departure from the urban area even more quickly.

The installation of storm drains and the realignment of channels take place early in the urbanisation sequence rather than after houses have been built, so even a low level of urbanisation yields dramatic changes in storm flow, often increasing peak runoff, reducing the lag time to peak flow by a factor of four and making flooding much more likely.

Agricultural land drains

Agricultural land drains are buried beneath the surface to remove water that would otherwise pond up, causing stagnation and inhibiting soil aeration. Clay soils, with their fine pores and low permeability, can be improved considerably by installing drains, enabling land that would otherwise be relegated to poor pasture to be made suitable for arable cultivation. However, drains transfer water quickly to stream channels instead of leaving it stored in the soil. The effect is very dramatic because field drains, like urban storm drains, often increase the channel density within the drainage basin several hundredfold. In consequence the probability of the flooding increases alarmingly and may occur with even modest rainfall. In uplands that have been ploughed for afforestation drainage usually takes the form of a network of open ditches. Such ditches may be only a few metres apart and can have a dramatic effect on the speed of runoff from headwater regions. Improvement schemes in the Pennine headwaters of the Yorkshire Ouse, for example, have enabled water to flow rapidly towards York and Selby, with the result that these towns are now far more flood-prone and millions of pounds must now be spent by the Water Authority to protect them.

Deforestation

Tree roots help water to penetrate through the top soil and into the subsoil, so overland flow is unlikely where there are many trees. Furthermore, trees transpire more than any other form of vegetation. Cutting timber with heavy machinery that compresses the topsoil will therefore promote rapid overland flow and flooding. In the Fernow forest of the Appalachians clearcutting was found to increase storm-water flow six fold. Similarly, in two adjacent basins on Plynlimon, Central Wales, the river basin whose land is under forest always produces its maximum flow after the basin that has been deforested and used for pasture.

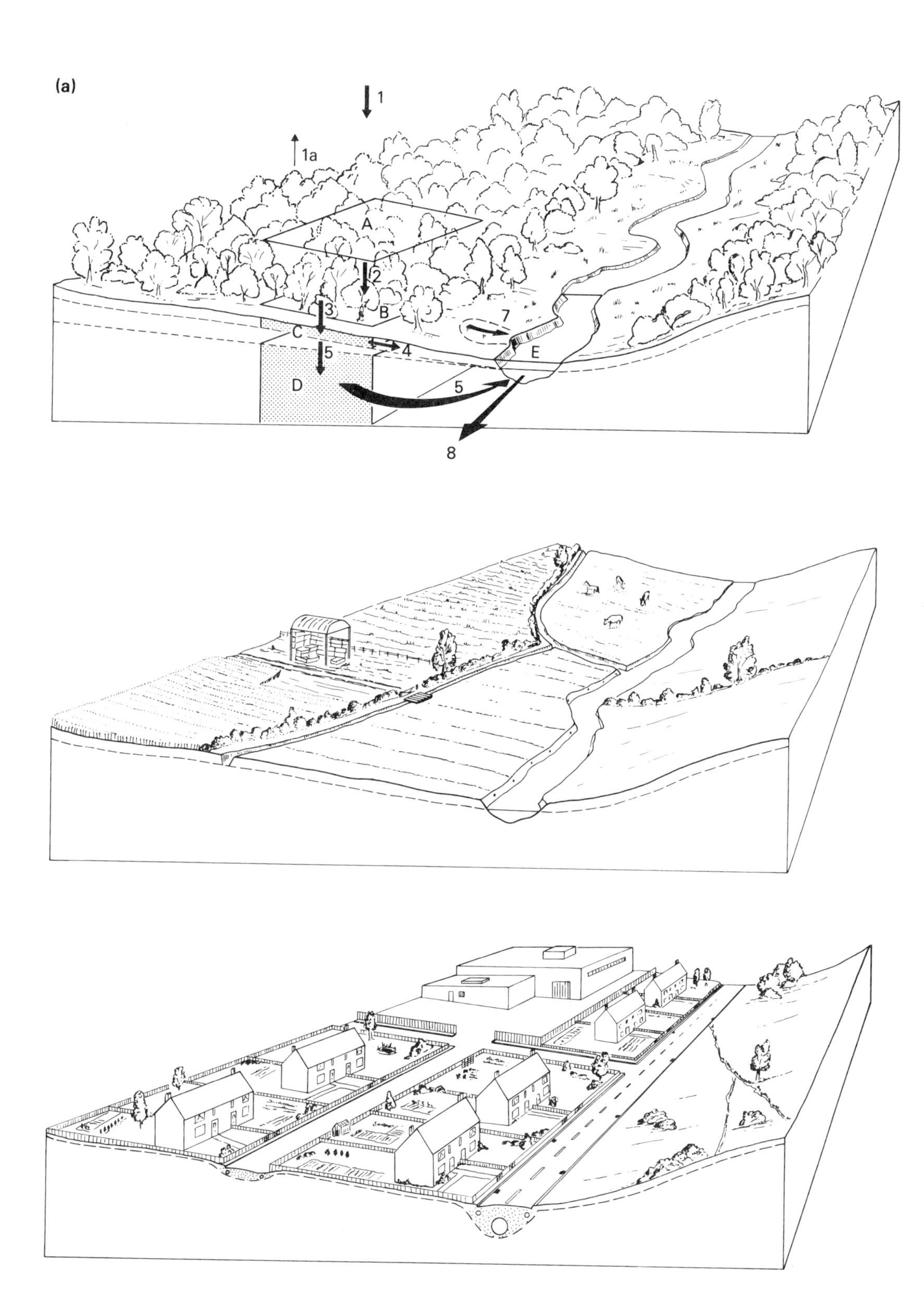

Figure 11.5 (a) Urbanisation of a drainage basin.

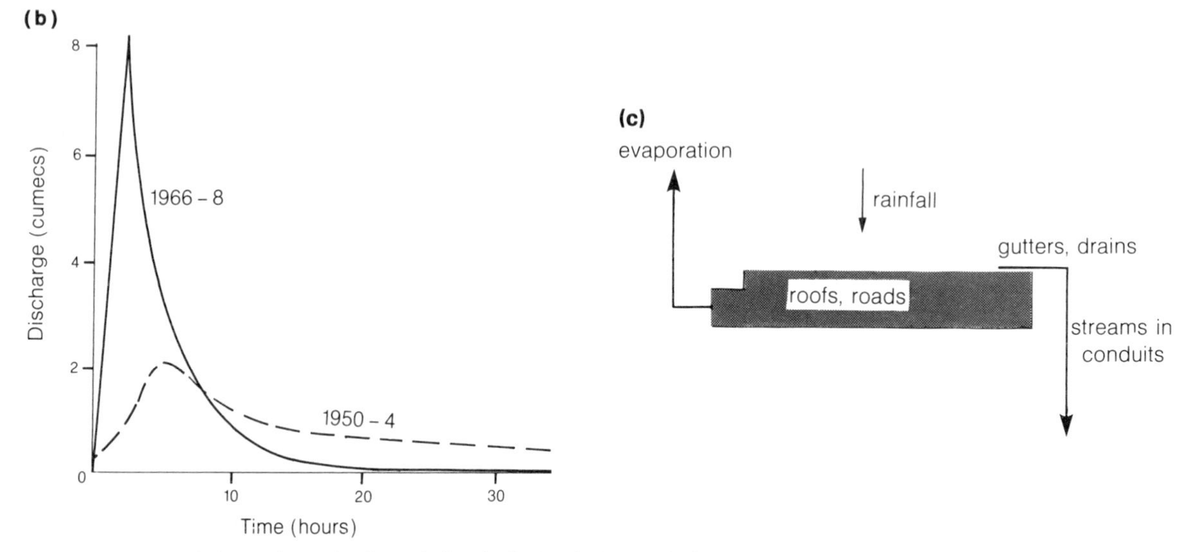

Figure 11.5 (b) Hydrographs at the Canon's Brook, Harlow before and after urbanisation. (c) Stages of man's alteration of the hydrological cycle.

11.2 Flood control

Floods are created by some unusual combinations of events. Nevertheless, flooding (**inundation**) occurs in all environments and can become a major hazard to those living on floodplains. The need for flood control depends on the frequency and severity of floods, and on the damage caused by them. It is possible to assess the flood hazard at any location by carefully recording streamflow. Areas liable to inundation can be drawn on a map and an estimate made of the cost of flooding for any chosen frequency of event (Fig. 11.6). Although a costing can be made on purely scientific and economic grounds, social and political factors may often be at least as important to the politicians deciding on flood-prevention measures.

Flood hazards are normally expressed in terms of the frequency of flood damage. Thus a town may on average be innundated only once every hundred years; any flood prevention measures must be designed to cope with this '100-year flood'. No special return period is crucial in deciding on action; measures taken follow a sliding scale. A return period of 1000 years makes the risk of damage seem too remote to worry about, but the return periods of 50 or 10 years tip the balance in favour of urgent action. A '50-year flood' does not mean that the next flood is 50 years away: it could occur tomorrow.

Small-scale controls

Suppose the area liable to flooding contains valuable property and that life as well as property is at risk. Under these conditions some form of action is almost certain. What options are available?

The first, and often the cheapest, option is to restrict the use of land liable to frequent flooding to functions that will not suffer too badly. Restrictions on house building, prohibition of basements, and building regulations that reduce flood loss are all easy to achieve. Such land would be suited only to factories and warehouses, and only then provided the warehouses were not used to store soluble harmful chemicals. It is now commonplace to designate floodplains as industrial development areas.

If the area is already settled and not due for redevelopment, this answer is impracticable: protection is needed. Traditionally the answer has been to build large embankments (**levées** and **dykes**) beside the river to contain the river. Examples abound: the Thames in London has a decorative containment wall (actually called The Embankment); in Germany almost continuous dyking along the lower Rhine prevents the flooding of tens of thousands of homes. However, dykes keep the river contained in the channel and isolate it from its floodplain. Consequently river levels become much higher than under natural conditions and, if a dyke breaks, potentially much more disastrous. Another result of containing storm water is to move the peak wave more quickly down stream – someone else must then cope with the consequences. For example, the 350 km of dykes on the lower Rhine in Germany leave the Dutch with quite a problem. Dykes may be unsightly if very high; and river crossings require bridges above street level.

In any case, dykes do not contain water completely.

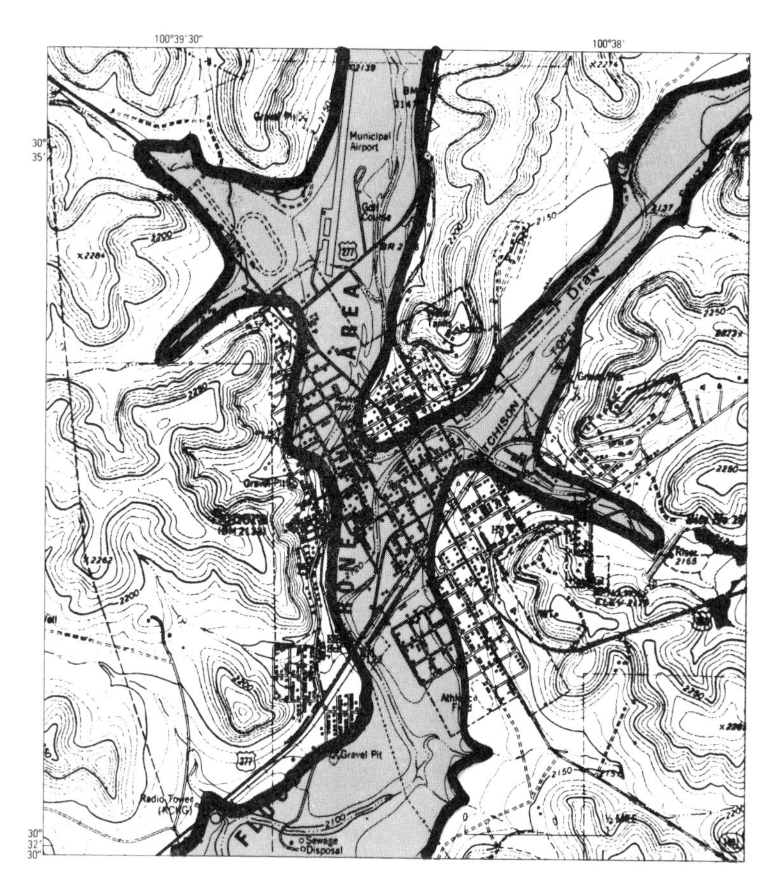

Figure 11.6 Portion of flood-hazard map of Sonora, Texas, showing approximate areas subject to inundation by a flood on Dry Devils River and Lowery Draw having an annual probability of 1 per cent.

Anyone living in New Orleans can testify that living below river level brings land drainage problems. For example, gravity cannot be used to get rid of surplus rainwater via drains. Water continually seeps through levées and needs pumping back up to the river. Embankments provide a solution that everyone can see, a strong political incentive to use these rather than some more subtle and effective – but less visible – means of flood control. One of the most popular alternatives is a **floodway.** This is a diversionary or overflow channel that bypasses urban areas; it carries water only when water in the main channel reaches a certain height. Some of the biggest floodways in use are on the Rio Grande Delta below Falcon Dam and on the Mississippi delta near New Orleans.

The construction of floodways allows some water-cycle theory to be put into practice. Figure 11.7 shows a floodway in Northern California, an area that experiences severe convective storms. It has been designed very carefully with wide sweeping bends of a wavelength appropriate to the expected flow. The channel is intended to conduct water efficiently while absorbing as much of the water's energy as possible. In this way erosion within and beyond the floodway

Figure 11.7 Floodway in California.

is kept to a minimum. Each step along the channel causes water to cascade and so dissipates its energy. Between steps the gradient is reduced to slow the water down. The danger of bed and bank erosion is thus confined to specific reaches – a desirable feature when it comes to repair. The bed and banks are made from pebbles and rocks which provide a rough, energy-absorbing surface and which are difficult to move.

Floodways and embankments are probably most suitable when it is impossible or uneconomic to control flooding in any other way. However, if floods are frequent and severe, the best remedy is to try to control the river flow.

Direct-control remedies vary from the simple and inexpensive to the sophisticated and enormously costly. It is often possible to reduce runoff at source. Urban areas are especially prone to rapid runoff, but where ground is totally paved, they can be designed to store water.

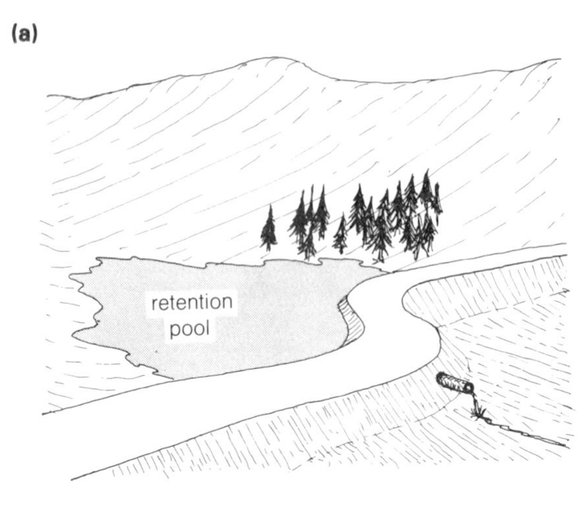

Car parks, for example, can easily store 6–8 cm of water without any damage to the cars parked. Flat-roofed buildings can safely store similar amounts of rainfall, and the water can be released slowly by drains made especially *small* – quite the reverse of established practice. Every little helps.

Bridges and **culverts** (pipes beneath roads) may restrict flow but this very 'disadvantage' can be exploited. If culverts are built *small*, heavy storm runoff will not be able to drain away and ponding will occur. But the area to be ponded can be designed specially and hollowed out (to form a **retention pond**, Fig. 11.8), providing a cheap way of withholding the flow of small tributaries until the main flood wave has passed.

Somewhat bigger **floodponds** and **soakaways** can form a dual purpose when aquifer recharge is also needed. In Long Island, New York State, many road drains are fed into special soakaway pits. When not in use for flood control these pits serve as baseball grounds – land is not wasted. In a totally different environment, the stream that drains part of Hemel Hempstead, England, can be diverted into a large pit: a flood gate opens automatically when the main river rises to a predetermined level. At other times the area is used as a municipal car park.

Larger schemes

In more comprehensive schemes, water is stored behind a dam across the main river, usually below the main headwater tributaries. With such schemes a permanent reservoir is created. However, not all

Figure 11.8 (a) Water retention pond created by an embankment and a small outflow pipe. Ponding encourages the greatest possible amount of infiltration. (b) Landscaping urban areas for flood control and amenity.

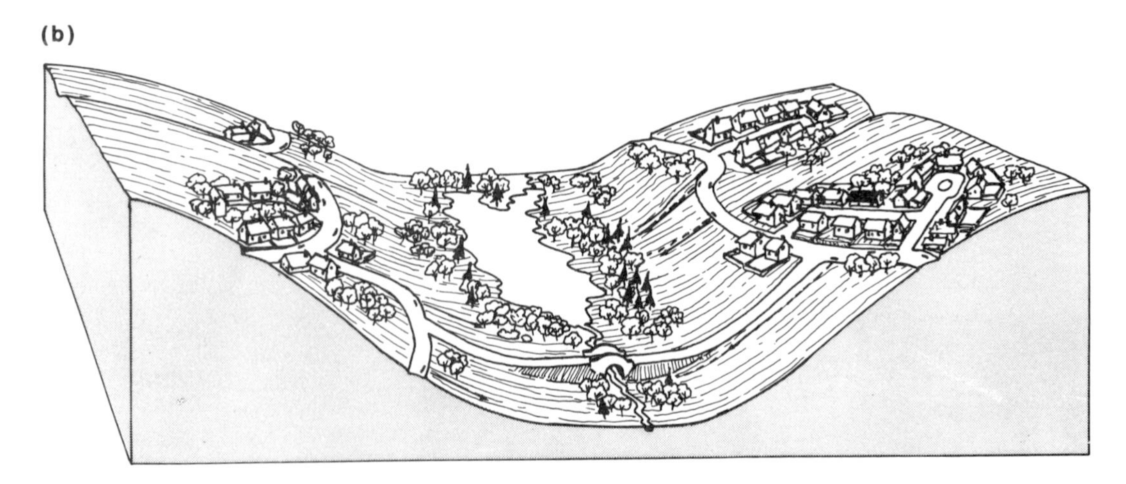

drainage basins have a suitable site for a dam of economic size and the people whose land would be flooded are likely to object. Dams and reservoirs are expensive and involve many social and political considerations. Flood control alone is rarely sufficient justification for a dam, but the use of the regulating reservoir in supplying water for urban areas may tip the scales in favour of its construction.

Dams can be very successful from the flood-control aspect. The Clywedog dam and reservoir on the main headwaters of the River Severn in Wales typifies the advantages (see Fig. 12.2). Before the dam was constructed, several towns beside the river (including historic Shrewsbury) were often flooded; since its construction, flooding has been eliminated.

The main dam is on a steep-sided, narrow valley which retains 49 million cubic metres of water in a lake 10 km long. Flooding is more common in winter when transpiration losses are small and more runoff reaches the river system, so a large capacity for storm runoff must be available by Autumn. This is achieved by drawing the level of the reservoir down during the summer: by 1 November each year there is about 8.5 million cubic metres' storage available. Because this reservoir is also used to supply water during a dry summer, it is never emptied; nevertheless, the level changes noticeably during the year. (Such reservoirs always seem to be half empty towards the end of a summer, leaving beaches around their edges.) The Clywedog water level normally changes about 4 m over a year, though a dry summer can result in a change many times more than this.

In general the larger the river, the larger the dam and reservoir required. One of the world's largest reservoirs is contained behind the Kainji Dam on the River Niger and is designed to control the annual flood that sweeps the length of this river. The most sophisticated schemes are in the USA, on rivers such as the Tennessee. This river was prone to severe flooding because of convective storms in summer and because farming methods left bare soil between crop rows, encouraging overland flow. Flooding, coupled with decreasing crop yields from eroded soil and the prospect of deriving hydroelectric power from stored water, encouraged the US Government to attempt an integrated scheme for the whole basin as early as the 1920s. In effect the Tennessee Valley Authority (TVA) turned the valley into a continuous string of lakes (Fig. 11.9). In addition, by grassing large areas and contour ploughing the remainder, everything pos-

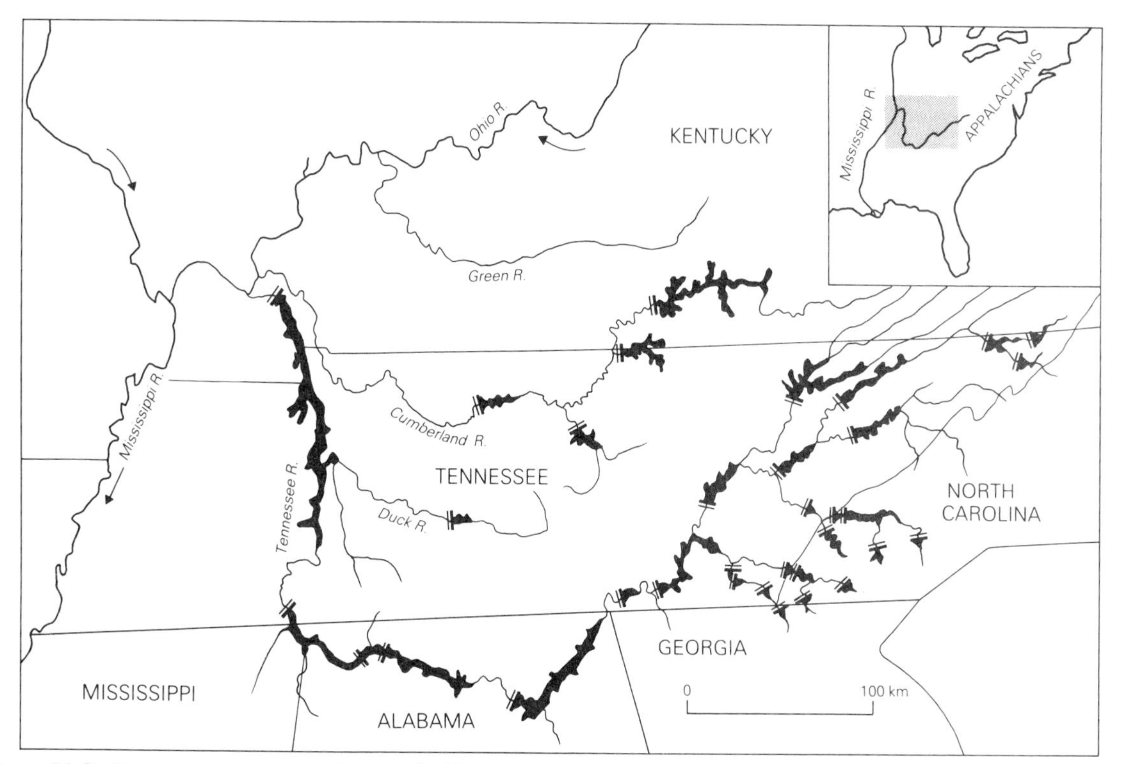

Figure 11.9 River management on a large scale. Multiple dams and reservoirs constructed by the Tennessee Valley Authority, USA.

sible was done to prevent surface runoff and promote throughflow.

As these examples illustrate, the theory and technology are available to prevent flooding. But the costs can be high and, unless the benefits are substantial, full-scale modification on the lines of the TVA is impractical. For this reason officials from overseas governments often go to look at the TVA scheme, but relatively few countries have attempted to copy it.

It is not enough to predict the flood hazard and outline protection measures that would achieve freedom from damage: who is going to pay? In the majority of cases, at least, a large part of protection costs would have to be met by local taxes. So any flood-protection scheme will rely on the willingness and ability of potential sufferers to pay. And as with many other things, the greater the number of people available to share the burden, the smaller the individual cost and the more likely the project will be adopted. The cost of flood protection is more easily absorbed into the budgets of large, prosperous towns than smaller communities, so in general the protection of small towns is piecemeal and haphazard. Local people may feel it is better to pay slightly higher insurance premiums and to suffer the periodic inconvenience of a flood.

12

Water supply: man's response to drought

12.1 Water demand

England. A country with some of the most reliable rainfall in the world. This green and pleasant land...

> 'This morning the water industry seems to be fairly good: the tap works, the sewage disappears, and if someone wants to go fishing one can get it somewhere. So why is everyone worried?' (Sir Norman Rowntree, as Director of the Water Resources Board, 1974.)

It seems strange that England should have a water problem, yet thousands of millions of pounds have been spent on schemes designed to alter the water cycle in ways that will allow water to be provided just where and when people need it. Although much water moves by throughflow or groundwater flow and is temporarily stored, even after a few days without rain many rivers are reduced to flows no more than 10 per cent of their average. Indeed, this **dry weather flow** can only supply a small fraction of the water needed today by a developed society (Fig. 12.1). From power stations to dishwashers, from lavatories to gardens the requirements for water are prodigious. In Britain, there is relatively little irrigation. Yet the national consumption, expressed as the amount per person, is over 300 litres each day; by AD 2000 consumption may be nearer 450 litres each day. Indeed, this century water consumption per head will probably quadruple, and since 1900 the

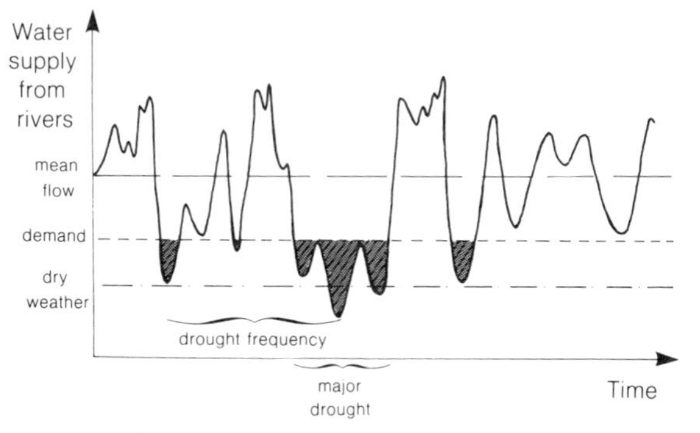

Figure 12.1 (a) An unusually long period without significant rainfall is called a **drought**. In Britain this is arbitrarily defined as a period of 15 consecutive days without significant rain; however, in regions experiencing a seasonal rainfall regime, drought has to be defined in relation to the expected rainfall for the whole wet season. A severe drought is an infrequent event but its effects may be catastrophic; because slight droughts are common provision is often made to compensate for them by storage in reservoirs. In any randomly fluctuating system such as river flow, it is unwise to rely on more than the small quantity called the **dry weather flow**. If agriculture, industry or people demand more than this, for part of the time arrangements will have to be made to compensate for the deficit. It is in those areas as the Sahel, where demand for water so often exceeds supply, yet there are no storage arrangements, where the greatest disasters arise. (b) The dry-weather flow of the River Wye, Plynlimon Central Wales. This size of flow could not maintain a developed society. Note that at times of high flow, water can fill and overtop the measurement flume.

total demand will have increased more than eight-fold. Demand varies with the season and the time of day. For example, domestic demands, which are small at night, increase rapidly in the morning as baths are taken, toilets flushed and breakfast crockery washed. These continually varying demands require a supply of great flexibility. Most of the water also needs to be purified.

12.2 Water storage

Every aspect of water supply requires careful planning because such important and far-reaching decisions are involved: structures built today may have a life-span stretching more than 200 years, the cost of each project is tremendous and the impact on the environment often substantial. Nevertheless, a reliable water supply can only be achieved by considerable storage, either by retaining it behind **dams**, pumping it from **aquifers** or abstracting it from the sea via **desalination.** Furthermore, the stored water must be distributed, often over many hundreds of kilometres, by aqueducts, pipelines or rivers.

A river is a natural aqueduct. If demand is from an urban area sited near to a river it is most convenient to store the water in headwater valleys and release it into the river as required, recovering it at the downstream demand points by pumping (Fig. 12.2). Natural lakes can be used to store water for public use if they are in the right place and of the right size. However, the changes in storage caused by varying releases during the year often result in the exposure of many metres of muddy foreshore. This opens an area of direct conflict between landscape protection and recreation on the one hand, and water supply on the other; a conflict which water supply usually loses, as the small number of regulated lakes in the Lake District illustrates.

With natural lakes either poorly sited or kept from water-supply use, water must be stored behind specially created dams and reservoirs. Before water-treatment plants came into general use, it was necessary to take water directly from uplands (where it was unpolluted), store it in reservoirs, and transfer it, still clean, directly by pipeline to urban areas. Reservoirs built then are easily identified by the 'Keep out – drinking water' notices and by the fences designed to keep people away. Treatment systems now allow reservoirs to be used for a variety of purposes – as well as storing water, they permit recreation and flood control.

Some reservoirs might seem big enough to cope with any demand, yet if water abstracted from rivers was never returned for further use down stream there would be an immediate and catastrophic shortage (Fig. 12.3). The supply system relies on each water-user returning as much water as possible after

Figure 12.2 The Clywedeog dam and lake across part of the headwaters of the river Severn, central Wales. This multipurpose dam is used to regulate the River Severn, providing a high degree of flood protection to the valley towns below; it is also used to release water on demand for the English Midlands and provides a large recreational lake.

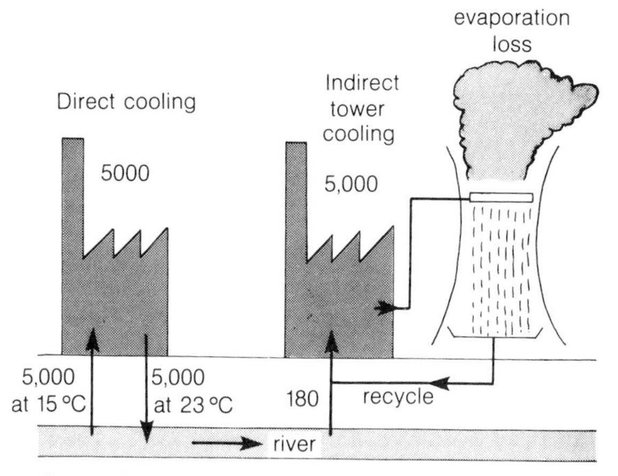

Figure 12.3 The gross water requirement by power stations on the River Trent in the English Midlands is 30 to 40 times the natural dry-weather flow. Cooling towers reduce the gross demand by operating their own miniature water cycles. Other industries must also return their water after use if at all possible.

use for treatment and further use down stream. Such a practice allows water managers to reduce the total release of water from reservoirs to the *net* consumption, for they could not cope with the gross consumption. This form of management allows a reservoir to supply several cities without the need for enormous storage. In practice only two per cent of the annual river runoff is stored, but its careful use allows periods of dry weather to be bridged most successfully. The cost of reservoirs simply to store two per cent of the flow is large, but it can readily be justified by the benefits gained, and reservoir construction on a small scale is not a very difficult problem. However, these reservoirs are not designed to guard against a 50-year drought; to be large enough for this would require colossal storage and could never be economically viable.

Many drainage basins lack suitable upland reservoir sites of large enough annual water surpluses. The River Thames, for example, drains a basin without upland storage but must meet the water demand from over 11 million people. Under these circumstances water has to be stored above river level in large embanked (**bunded**) basins, filled by pumping from the nearby river at times of high flow. Although they are vital, they are very expensive to construct, take up valuable space near urban areas and can only store limited amounts of water (Fig. 12.4). London's water engineers have to supplement the river and reservoir supplies with water pumped from the underlying chalk aquifer. In areas without aquifer supplies there would be no alternative but to build

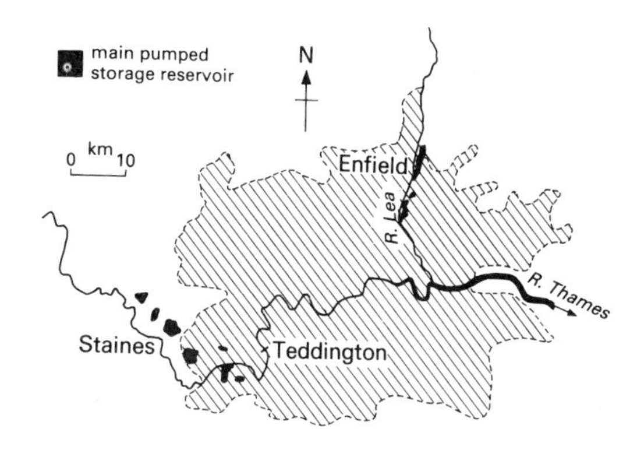

Figure 12.4 The major pumped storage reservoirs supplying London.

more surface reservoirs. A neat solution to storing water without using agricultural land is under trial in the Wash, where water is stored in an offshore bunded reservoir (Fig. 12.5). Schemes similar to this may all become realities in the early part of next century in such estuaries as the Severn (for Bristol) and the Dee (for Liverpool), and one day there may even be a bunded reservoir stretching across Lancashire's Morecambe Bay.

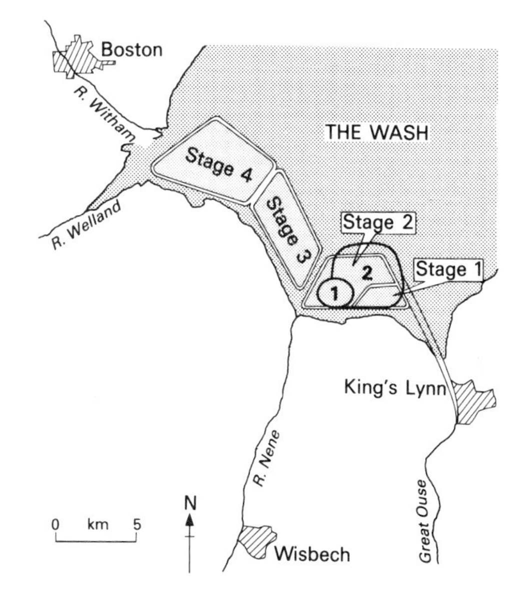

1 area : 6.5 km^2
storage: 65 × 10^6 m^3
yield: 180 000 000 l/day

2 area: 20 km^2 ?
storage: 200 × 10^{6}m^3 ?
yield: 150–300 million l/day

Figure 12.5 The Wash water-storage scheme. Early proposals for a four-storage bunded system have been revised based on a new demand forecasts to a two-stage scheme beginning after 2001.

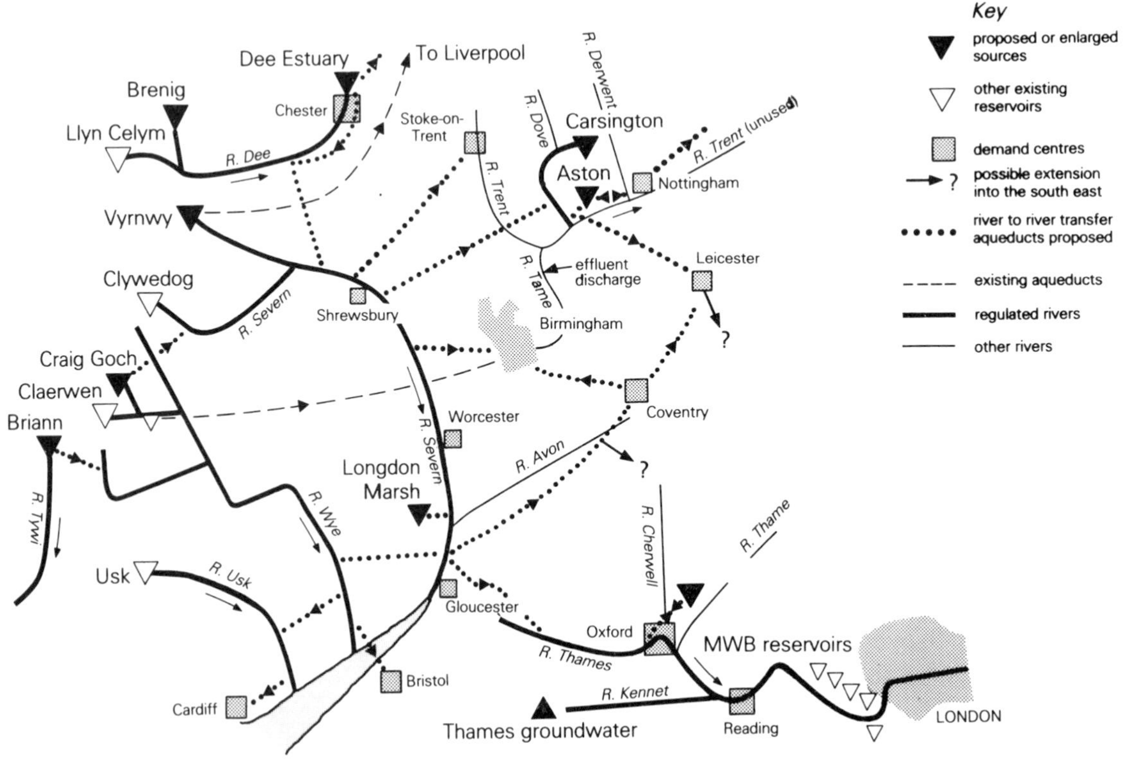

Figure 12.6 The face of water supply to come: a possible strategy for water supply in central England and Wales by the year 2001. Note especially the inter-basin transfers.

Storing fresh water in estuaries is expensive and before such reservoirs are constructed much more water will be obtained by rationalising the resources already available, especially by such measures as inter-basin transfers of water. In the 1970s the British water industry was reorganised into 'Authorities' (units) related to drainage basins, to facilitate just such rationalisation schemes. Inter-basin transfers are not new; the city of Birmingham, for example,has been supplied with water from central Wales for a hundred years. It is the *scale* of transfer that will show such a dramatic change in the future as more and more links between rivers are forged (Fig. 12.6). In Britain a national strategy has been adopted because supply and demand cut across so many local and regional boundaries (Fig. 12.7). The ability to transfer water and to regulate supplies over complete basins becomes particularly advantageous in times of severe shortage such as the 1975–76 drought as the graphic account quoted in the box illustrates.

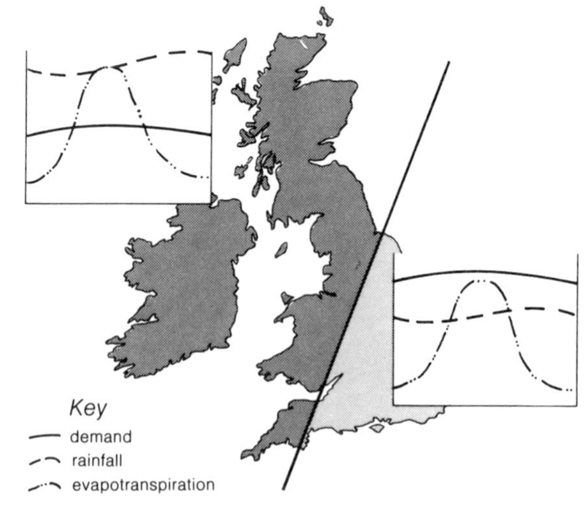

Figure 12.7 The reason for a national water strategy: imbalance of supply and demand.

THE 1976 DROUGHT

The North Central Division covers a wide stretch of the Dales . . . but the area most vulnerable to drought is that embracing Leeds and Harrogate. Supplies to these centres come from a chain of reservoirs in the Washburn Valley . . . As the drought bit deeper the engineers regulated the levels of these reservoirs so that the space to be filled was roughly in proportion to the catchment area of each: this avoided any chance of the lowest reservoir being full when it rained, and losing all its increment in overspill.

But it did not rain. The Leeds reservoirs were only half full in the middle of May. They held that level for a while as a result of additional supplies from the river Wharfe and consumer economies resulting from the publicity campaign, but by mid-July they were down to 41.7% of capacity, and got lower every week: 39.4%; 35.5%; 34.2% in early August. By the end of the month they were below 30%.

North Central Division was better situated than South Western Division, in that some augmentation was possible. The programmed measures to preserve supplies for Leeds and Harrogate for as long as possible ran as follows: Hosepipe restrictions; publicity campaign; reduction of pressures in the distribution system; drought order to permit reductions in compensation flow from reservoirs; drought order to permit temporary abstraction from the River Wharfe, subject to maintaining a prescribed minimum flow; on the assumption that the Wharfe would fall to that prescribed minimum flow, a drought order to permit temporary abstraction from the River Nidd, supported by releases from the Gouthwaite regulating reservoir higher up that river; further reduction of distribution pressures; drought order for provision of supply by standpipes; drought order for wider powers of rationing under the new Drought Act.

The hosepipe ban was placed on Leeds in January 1976 – it had been operating in Harrogate since the previous year. Reduction in reservoir compensations was authorised at the end of March, abstraction from the Wharfe at the end of April, and abstraction from the Nidd by late May. The publicity campaign and the first reduction in distribution pressures were started in April. In April also the Authority had budgeted for the expenditure of £600 000 on the purchase and erection of standpipes alone – not to mention all its other emergency operations throughout the region . . . . Yorkshire was lucky, or maybe just far-sighted, in that abstractions both from the Wharfe and from the Nidd could be fed into the Leeds supply system without any great expenditure on emergency mainlaying. There were existing trunk mains crossing both rivers. The Wharfe abstraction was pumped into the main that ran from Swinsty Reservoir to Leeds, and the Nidd water entered a trunk main from Leighton to Swinsty.

So, as Yorkshire inflexibly maintains, these operations involved no startling innovations, just thorough planning and hard work. It thought problems out and solved the practical difficulties as they arose. And the Yorkshiremen return again and again to the benefits accruing from reorganisation. Within all divisions they have brought about interconnections and transfers that would have been out of the question before (the Water Act of) 1974. Without that 'switchability', pockets of Yorkshire would have been dry by mid-summer, and local councillors would have been going round jingling their money in a quest for bulk supplies. Without an integrated command and transferable resources of labour and finance there could also never have been the acceleration of the capital programme that took place in the summer of 1976.

Andrews (1977)

12.3 Irrigation

There are many conflicting demands for water. For most urban dwellers public water supply has top priority, but in many parts of the world vital food production is possible only with the aid of irrigation (see also Ch. 33). On a world scale, irrigation systems are actually the largest single users of both river water and aquifer supplies, and most of the very large water storage schemes have been planned primarily for irrigation (Fig. 12.8).

Irrigation use is also unique because irrigation water is mostly lost water – at best only half returns by drainage through the soil; the bulk is used in evapotranspiration. In the USA, for example, the annual net consumption of water for irrigation is 144 trillion (10^{12}) cubic metres compared with only 2 trillion for power stations and 13 trillion for the whole of industry. Put another way, the water needed per

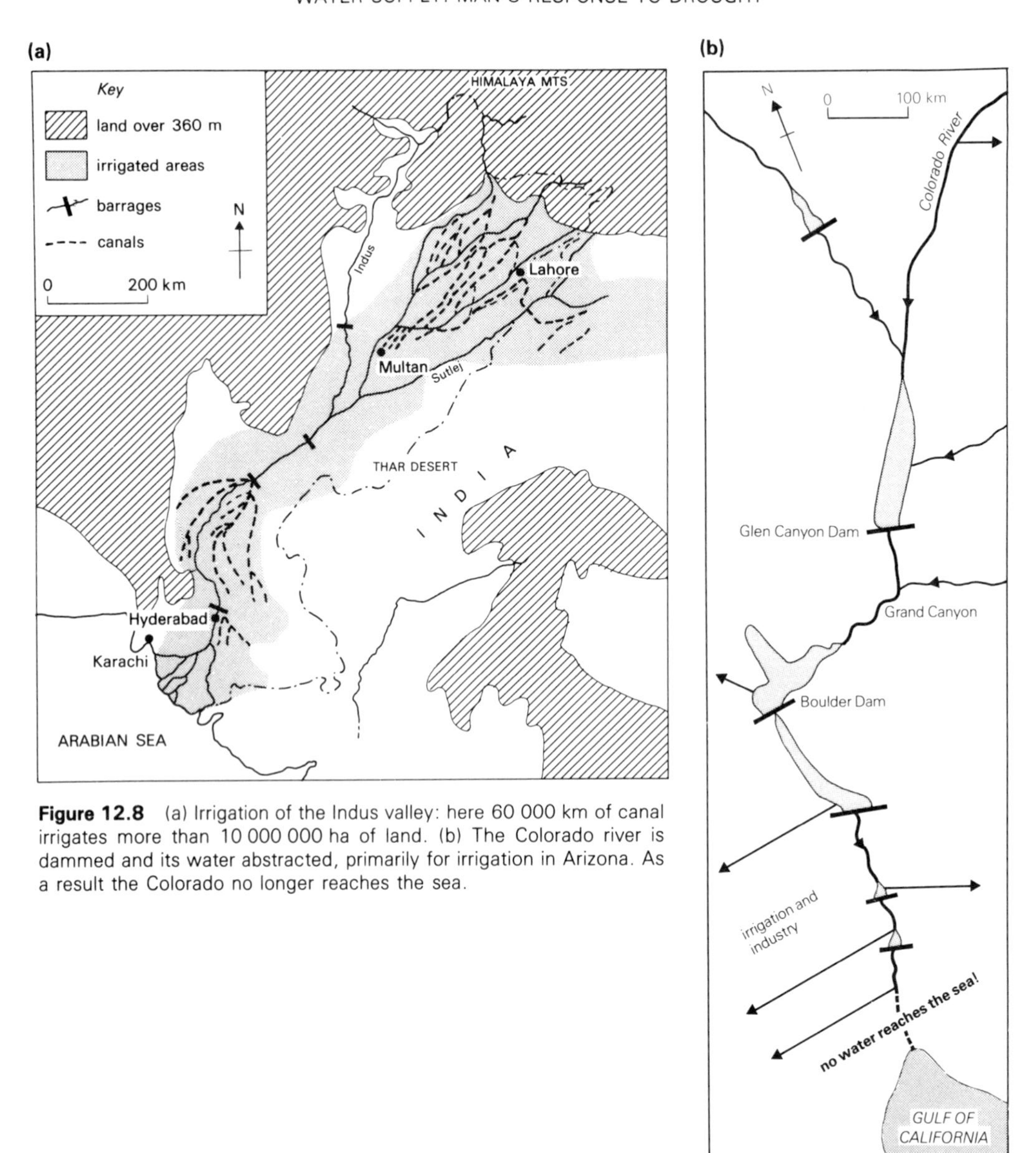

Figure 12.8 (a) Irrigation of the Indus valley: here 60 000 km of canal irrigates more than 10 000 000 ha of land. (b) The Colorado river is dammed and its water abstracted, primarily for irrigation in Arizona. As a result the Colorado no longer reaches the sea.

season to irrigate 0.5 ha of rice, 0.8 ha of cotton or 1.2 ha of wheat would supply water to 100 nomads and 450 head of stock for three years, and even 100 domestic consumers in a modern industrial city for two years.

Such large amounts of water cannot be taken from unregulated rivers, especially because those warm, dry areas with the greatest demand for water are almost invariably those with the more restricted natural river flows. Water cannot be delivered free: the more water used, the higher the cost. A balance must be struck between unrestricted supply of water, which would promote maximum growth, and restricted supply to reduce costs.

Experiments have shown that plants grow only at their maximum rates if soil moisture is kept within a very narrow range near field capacity. Even temporary dryness can cause severe check on growth. In fact most of the water is needed simply to keep the plants cool. Only one per cent of the water is used in growth and supplying nutrients from the soil. But no matter what the water is used for, it must be available quickly in the plant-root zone near the surface, so regular surface applications of water are essential. If the surface cannot be kept wet regularly, the range of crops that can be grown is reduced. For example, lettuce roots penetrate no deeper than 30 cm and potato roots 60 cm, yet cotton can reach water at 120 cm and most grains 180 cm. Lettuce will not tolerate any dry period; brussels sprouts may give 80 per cent of their potential yield (that with full irrigation) even if they receive no water for several weeks.

Soil texture is another important factor in determining water need. Coarse sands may only hold 0.5 mm of water equivalent for each 10 mm of soil; loams and clays will hold more than three times as much. Water moves very slowly in clay soils, so although the soil may be very wet plants may still wilt and die unless irrigated frequently.

On average the potential evapotranspiration rate is 1–3 mm/day in temperate areas; 5–8 mm/day in the humid tropics; and 9–12 mm/day in the arid zone. For effective plant growth, irrigation must match the evapotranspiration loss during the growing season. However, plants in hotter regions grow more quickly and have a shorter growing season than those in cooler areas. As a result a crop grown near the Equator often needs less irrigation water in *total* than the same crop grown in temperate regions.

Irrigation is a complex business, requiring an understanding of crop growth and yield as well as of water supply. Curiously, it is often possible to use fertiliser as a substitute for water (Table 12.1). These figures are based on experiments in southern England, where, incidentally, irrigation would improve yields in five years out of ten.

Because all water contains dissolved salts from chemical weathering, irrigation can cause severe problems to soil fertility if the amount supplied is only sufficient to balance evapotranspiration losses. With water in shallow circulation, salts that in low concentrations are nutrients, accumulate to reach levels that are toxic to plants. To avoid soil salinity, salts must be flushed (**leached**) down through the soil and back to the river, a process that demands yet more water and – equally importantly – a good drainage system.

Table 12.1 Irrigation and the use of fertiliser.

Fertilised with:	25 kg/ha N	50 kg/ha N
unirrigated	3.46	3.95
irrigated	3.83	4.08

In principle, water is needed to balance evapotranspiration use and salt leaching. Fortunately irrigation and leaching do not have to occur at the same time. Leaching can be achieved by a couple of large applications of water each year and this can best be done in the dormant season when plant demands are least. Alternatively, areas with a seasonal rainfall can use excess water in the wet season for leaching, leaving water supplied in the dry season strictly for the crops.

Water for irrigation can be diverted from natural streams, diverted from reservoirs or pumped from natural groundwater reservoirs (aquifers). It can be supplied to fields either by flooding furrows, by spraying or by drip from nozzles. Of these, flooding using natural streams is the cheapest but the hardest to control; drips to each plant using water pumped from aquifers give the greatest control but cost a great deal both in capital and fuel for the pumps. However, there are secondary advantages to piping water to the plants: evaporation losses are much lower than from water in canals; and pipes can be buried, allowing land to be worked more efficiently by machine. So the balance is cost against efficiency; simple against complex. Not surprisingly, the schemes used in developed and developing parts of the world are very different.

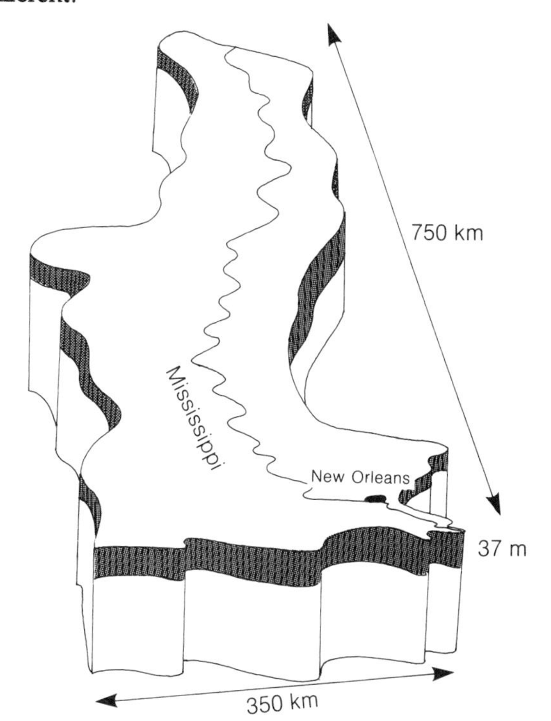

Figure 12.9 Aquifers contain an immense water reserve. Under the Lower Mississippi basin there is enough water stored to fill a reservoir the entire size of the basin to a depth of 37 m (i.e. 265 000 km^2 × 37 m).

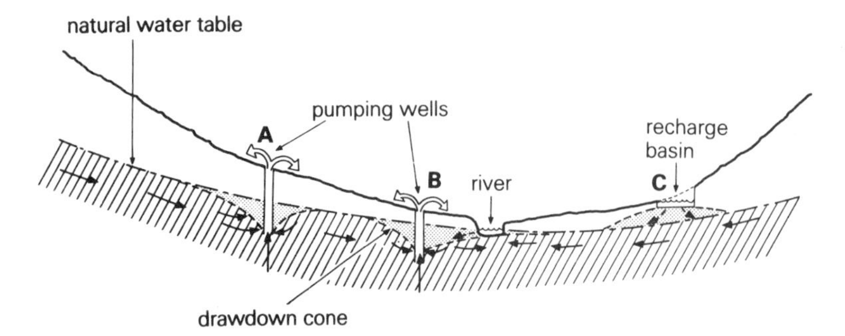

Figure 12.10 The nature of demand: the utilisation of groundwater resources. Pumping at **A** lowers the water table but has little other environmental effect. At **B** pumping drawdown depresses the water table below river level and causes abstraction from the river. The water table can be raised locally as at **C**. Such recharge basins are mostly sited near to pumping stations so that some of this water can be recovered in times of need.

The cheapest irrigation sites are diversions in the upper reaches of valleys where gravity will deliver water to fields downstream via canals (Fig. 12.8). Downstream a regulated river can be controlled with barrages that give lead-off points for canals. Alternatively water can be lifted from the river using anything from the primitive Shadouf (a swinging bucket with counterpoise operated by one man) to sophisticated diesel pumps.

If suitable rivers are not available to provide water, groundwater supplies become vital. Most regions contain large groundwater reserves (Fig. 12.9) and many farmers in developed countries prefer to use pumped ground water because they can control the supply more closely than if they flooded their fields from nearby rivers. However, demand may rise to a level at which it exceeds the rate of natural recharge, just as if one child drinks from a milk shake the level goes down only slowly, but when a dozen more place their straws in the same glass and begin sucking, the level falls dramatically and the glass is soon dry (Fig. 12.10). Aquifers are not inexhaustible. Many rivers gain part of their water from seepage: if pumping lower groundwater levels below river beds, the rivers will suffer reduced flows or may stop altogether. There is little point in increasing yields by pumping if the result is an equal decrease in yield to surface reservoirs.

Irrigation demand has resulted in important and serious side-effects: as water is abstracted and pores within rocks left empty, the aquifer begins to settle. Pumping from the aquifer that underlies California's Central Valley had to be stopped because, in places, the ground had subsided more than seven metres, and many buildings were beginning to collapse (Fig. 12.11).

12.4 Water quality

Water quality and supply are closely interconnected. One of the main uses for water is to dispose of waste, or **effluent**. It is very important, for example, to allow sufficient water to remain in a river at all times to dilute the effluent it may contain. In Britain an Act of Parliament even sets the minimum acceptable flows to deal with effluent dilution.

In the past little attention was paid to water quality and far too many of the world's rivers are contaminated by effluents to an unacceptable degree. The fact that in Britain the situation is not as bad as in many other countries is largely because most of our conurbations are near the sea: rivers flowing through the inland conurbations of the West Midlands and West Yorkshire suffer from severe pollution problems. For instance the River Tame, a tributary of the River Trent that flows through the West Midlands conurbation, carries more effluent than diluting water during times of low flow!

There are many effluents other than domestic sewage and industrial process wastes. The West Midlands is a metal-working area; there, even the runoff from streets following rainfall is polluted by the metals and hydrocarbons in the air and lying on the ground. No fish can survive in the Tame or in the Trent for 16 km downstream of their confluence. The Trent remains so polluted, even as far as Nottingham, that it cannot be used for drinking water even

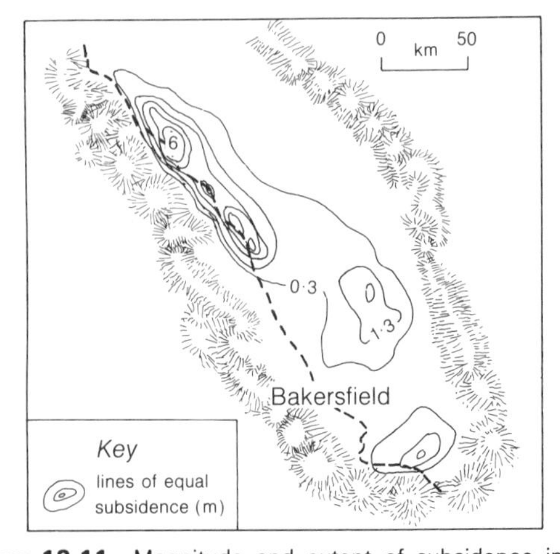

Figure 12.11 Magnitude and extent of subsidence in the southern Central Valley, California.

after normal purification. In an effort to alleviate the situation, the Tame is now diverted through a sequence of lakes (Fig. 12.12) where the worst of the pollutants have a chance to settle out.

Although industrial and domestic users are often thought to be the principal cause of pollution, environmental chemicals such as biocides and fertilisers are also a major problem. Fertilisers, for example, are leached through the soil to river waters where they may cause a rapid growth of algae (which turn the water green).

Many countries are less fortunate than Britain in that much of their industry is located inland. In West Germany the main concentration of resources and industry is along the banks of the River Rhine. Here 61 million people produce a total effluent load of 16 billion cubic metres each year, much of which finds its way into the Rhine. In general, more than half the surface water resources of West Germany are of poor quality or are grossly polluted (Fig. 12.13). Here the scale of the problem is much greater than that of the Tame – the whole Rhine cannot be passed through purification lakes. Maximum use must be made of **self-purification,** allowing natural bacteria the time to break down the effluent into non-toxic materials. Some bacteria are remarkably tolerant, but they do need sufficient oxygen and they are not able to live in

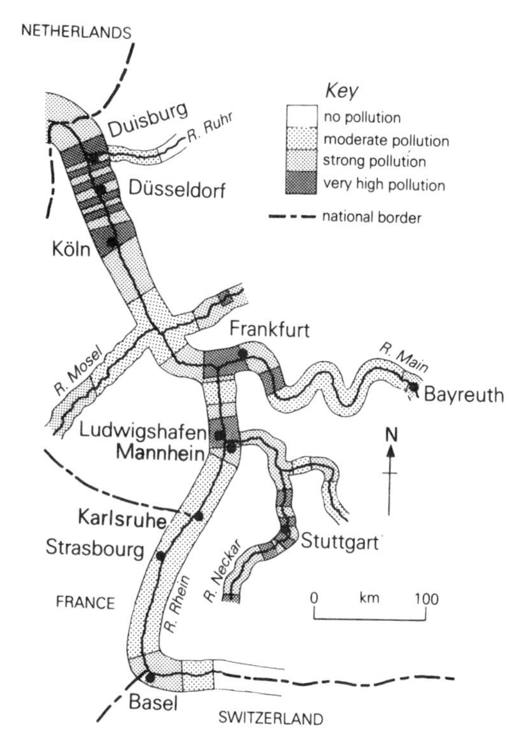

Figure 12.13 The degree of pollution in rivers in West Germany, 1967.

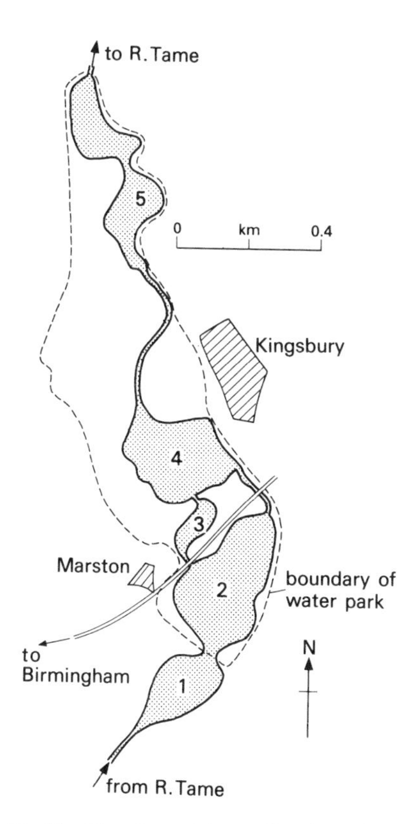

Figure 12.12 The River Tame lake purification scheme. The primary purification lake (1) has no recreational use, but the other lakes (2–5) can form the basis of a water recreation park.

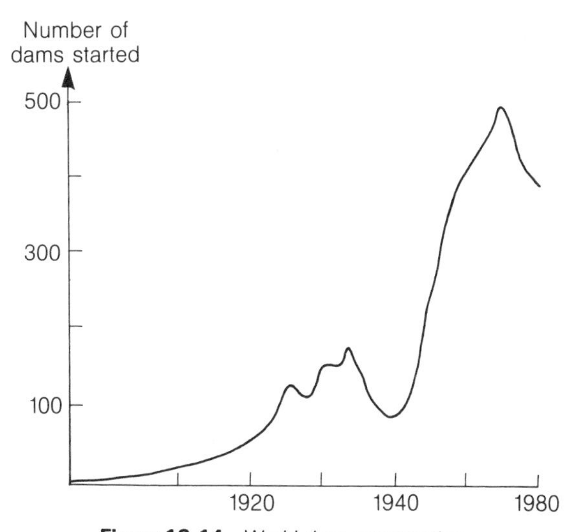

Figure 12.14 World dam construction.

environments polluted with heavy metals. Water can be oxygenated by making it flow down cascade ramps at weirs, or simply by allowing it to flow fast enough to become fully turbulent. But there is no easy way to remove toxic metal compounds, and legislation is needed to keep rivers free of them.

WORLD DAM CONSTRUCTION

In Africa and North America 20 per cent of the total runoff from all rivers is now regulated; in Asia and Europe it is only a little less at 15 per cent. Only South America and Australasia have made little progress in regulating their rivers.

The scale of dam construction is prodigious: between 1945 and 1971 over 8000 major dams were built (Fig. 12.14). The peak of dam building has now probably passed and there is considerable concern about the environmental impact and economic returns from these structures, especially as the profitability is often much lower than had been expected.

Part D

Slopes

Weathering

13.1 The significance of weathering

The term **weathering** is used to describe the changes that occur in the rocks of the Earth's surface, changes brought about by atmospheric agents such as water and varying air temperature.

The shape of a river valley reflects the result of direction erosion by the river, weathering of rock and the transport of weathered debris both down hillslopes and along the river channel. However, river erosion is confined usually to the river bed and exceptionally to the floodplain. By far the greatest contribution to valley formation is due to hillslope weathering and transport. Without significant weathering, river valleys would be no more than gorges in the landscape (Fig. 13.1).

Weathering involves two quite separate processes:

(a) **mechanical weathering**, whereby the rock is disintegrated without a change in its chemical composition;
(b) **chemical weathering**, whereby the rock decomposes, forming new materials (mainly clay minerals) and releasing other materials (mainly ions) in solution.

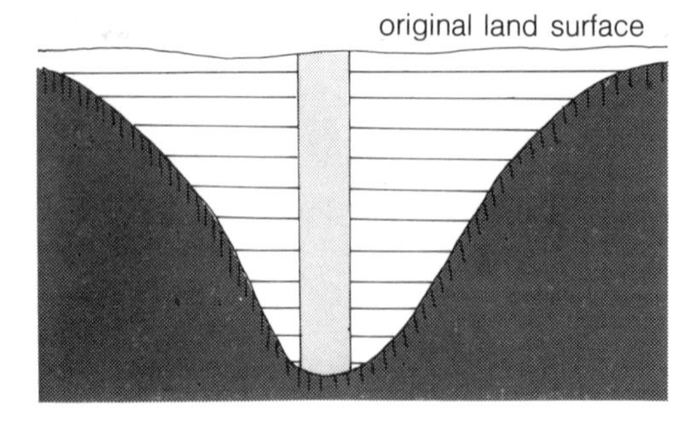

Figure 13.1 The contribution by direct river erosion to the shape of the valley is shown by the unshaded central region; the shaded regions indicate the far more substantial contributions from hillslope weathering and transport.

Although these processes are very different, they both act largely through the presence of water.

13.2 Mechanical processes

Geological processes

Rock can be broken into large fragments while it is still underground, by geological processes that cause disintegration without the use of water. Nevertheless, by providing deeply penetrating fractures in the solid rock they increase the pathways for water entry and subsequently more complete breakdown.

Most natural rock faces are criss-crossed with the hairline fractures called **joints** (which run across layers of rock) and **bedding planes** (which separate different layers of rock). This primary fracturing occurs because the great pressures that exist deep underground where most rocks form are not felt at the surface: as erosion gradually strips away the overlying rocks, the decrease of pressure allows the newly uncovered rocks to expand. However, because rock is brittle, expansion is nearly all achieved by fracturing (Fig. 13.2). The disintegration of rock in this way is called **unloading fracturing** or **dilation.** As rocks are unloaded they often fracture into distinctive patterns. For example, the pattern of granite fracture on Dartmoor determines the shape of the tors (Fig. 13.3a). In California, by contrast, the granite fractures parallel to the surface, producing a pattern of curved sheets (Fig. 13.3b).

Frost cracking

The initial fracture pattern caused by unloading may allow surface rock to fall away from the main mass. However, most rock remains in place after unloading

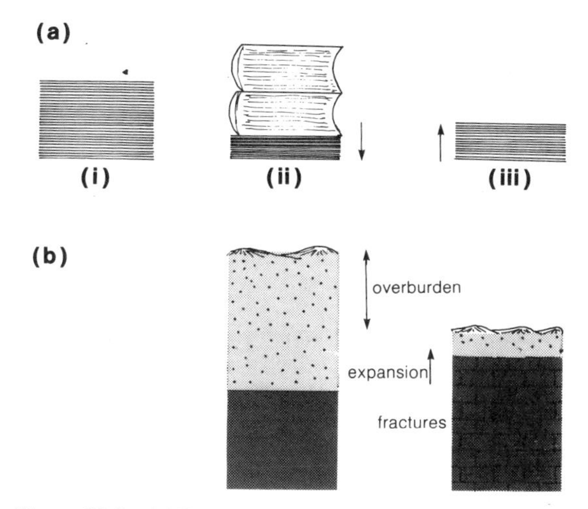

Figure 13.2 (a) The principle of unloading can be illustrated with a pile of paper: (i) stacked normally; (ii) loaded with an overburden pressure of many books; (iii) unloaded. (b) Unloading of rock by the erosion of overlying strata causes the rock to expand, primarily by fracturing.

Figure 13.3 Unloading fracture patterns at (a) Hound Tor, Dartmoor; (b) Yosemite, California.

and will not be transported away unless it is first reduced in size.

Freezing water expands by nine per cent in volume as it forms ice. Expansion alone does not always cause disruption. Milk freezing on a doorstep in winter will often prise off the bottle cap and extrude a cylinder of frozen milk; but if milk has already frozen in the bottle top, there is no outlet for expansion and the bottle then cracks. Something similar happens with rock that has become wet and is exposed to freezing conditions. In some circumstances ice is extruded; in others it is contained and forces the rock apart. The pressures involved in **frost cracking** are formidable: up to 2000 kg/cm^2 can build up as water freezes in a confined space, whereas few rocks can withstand pressure greater than a tenth of this (Fig. 13.4). Not surprisingly, therefore, frost cracking is one of the most potent of disintegrating forces. As over half the world's surface experiences frost, it is also one of the most widespread (Fig. 13.5).

Rock disruption by frost cracking occurs in many tiny steps and during each period of frost a rock fracture is wedged only imperceptibly further apart. The effectiveness of frost cracking is therefore closely related to three factors:

(a) the availability of moisture – frost cracking is more effective in maritime than in continental regions;
(b) rock porosity or primary fracture pattern – shale, chalk and sandstone are more vulnerable than crystalline rocks such as granite and gneiss;
(c) the number of freeze–thaw cycles – low-latitude mountains which experience a year-round diurnal temperature fluctuation about freezing (e.g. Mt Kenya in Africa) have far more freezing cycles

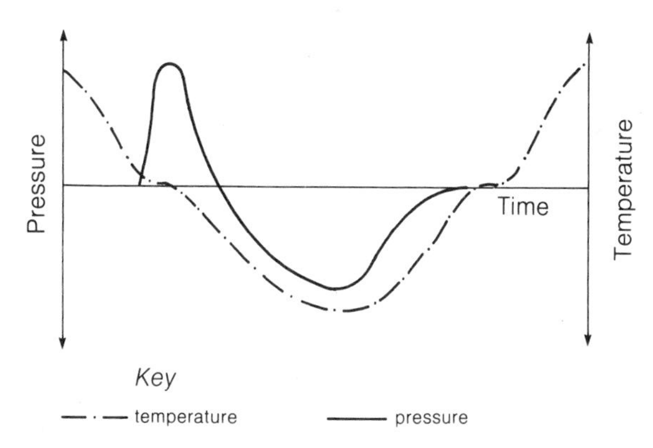

Figure 13.4 The changes in volume of water during a freeze–thaw cycle. At first there is a rapid rise in pressure on freezing, then contraction. Contraction is important in helping ice-wedge formation.

(a)

(b)

Number of rockfalls in 4 yr period
20
10
Jan
spring
autumn
Dec

(c)

Figure 13.5 (a) The results of frost cracking. (b, c) The seasonality of rockfall in Norway is clearly a reflection of the periods when air temperatures hover around freezing.

than those where temperatures only hover around freezing seasonally (e.g. Lake District); cold regions (e.g. Greenland) have the fewest freezing *cycles* of all because temperatures remain below freezing for much of the year.

Salt-crystal growth

Most groundwater contains abundant dissolved minerals. Rainwater, too, usually contains dissolved salts (derived from sea water). In warm arid and semi-arid regions rainwater penetrates only shallowly, even in porous rocks, before being evaporated. In time, however, repeated rainshowers leave more and more salt accumulating as crystals in the surface-rock pores. Salt crystals growing under these conditions can exert large pressures on the confining pore walls, causing sheets of surface rock to flake away (**spall**; Fig. 13.6). Laboratory experiments designed to

(a)

(b)

Figure 13.6 (a) This granite rock has been subject to flaking because of the growth of salt crystals below the surface. (b) Although the Sphynx lost much of its face from being used as target practice, its 'haircut' is probably the result of capillary rise of water in the limestone and sandstone strata and subsequent salt crystal weathering. The pyramids have not been affected in the same way because they are not carved from the solid rock.

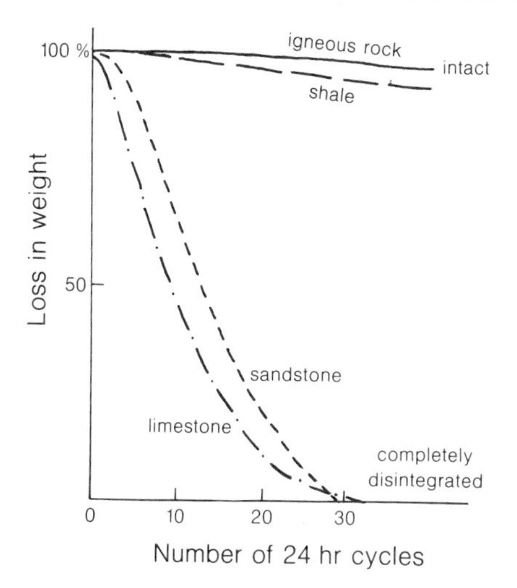

Figure 13.7 Blocks of rock subject to immersion in a saturated solution of sodium sulphate for 1 hour at 20°C; dried for 6 hrs at 60°C; and at 30°C for the remaining part of a 24 hr cycle, show that porous rocks such as sandstone and limestone are very susceptible to weathering by salt crystal growth. The temperature regimes were designed to simulate desert conditions.

simulate warm desert conditions show that very rapid disintegration is possible in highly porous rocks (Fig. 13.7), and suggest that salt crystal growth may be the most important mechanical weathering process in warm arid and semi-arid regions. The effect may be especially pronounced in rocks where the groundwater is close to the surface, for example near oases. Here, salt-laden water is continuously evaporating and salt crystals can grow rapidly in the surface pores, rupturing the surface and producing rock 'powder' that is readily blown away. Such a process may even be largely responsible for the material that allows deflation hollows to form in deserts.

Finally, and partly because crystallisation of salts can be brought about by cooling as well as by evaporation, a combination of a high salt concentration, groundwater and frost can have really spectacular results. Coastal wave-cut platforms, wetted by sea water and frozen overnight during low tide, may have their whole surfaces shattered within a few hours.

Plant roots

Many plants send tap roots deep below the surface in search of water. These roots begin as fine hairs which can therefore easily penetrate pores and hairline fractures in rock. As plants grow, their roots enlarge, exerting considerable pressure on the confining rock. The frequent examples of tree roots prising up tarmac and paving slabs on pavements is a clear indication of the mechanical force plants can exert. Their wedging effects are most important near exposed rock faces such as cliffs.

13.3 Mechanical and chemical weathering: the contrasts

The chemical processes that cause rocks to decompose are essential both for soil formation and for hillslope development. They complement the processes that cause mechanical disintegration, taking over in importance when climatic conditions do not favour frost cracking or salt-crystal growth, and on slopes where mechanically weathered debris cannot be transported quickly away.

Perhaps the easiest way to appreciate the comparative roles of mechanical weathering is to look at buildings. Buildings always decay for they are really only piles of rock, neatly arranged and cemented together. In general, the rocks in buildings fare better than most natural rocks because we take care to keep them dry. But although it happens slowly, weathering on buildings is similar to weathering everywhere else. For example, physical weathering occurs if rain makes a wall wet on a day when frost follows. Any

Figure 13.8 The disfigured form of this statue on the outside of a cathedral is clear evidence of the efficiency of chemical weathering.

water that has seeped into hairline cracks between mortar and bricks freezes; the mortar then shatters as does rock on a mountain. Every so often, the mortar in walls needs to be replaced (a process known in the building trade as 're-pointing'). Chemical weathering, by contrast, is more subtle in its effects. Although few new buildings show any effects of weathering, older buildings often need repair because their stonework is rotting. Fine blocks of stone, once tough and firm, now crumble to the touch, shedding dusty debris and revealing the corroded and pitted surface that is the hallmark of incessant chemical attack (Fig. 13.8). The stone of cathedrals is being rotted by one of the most common examples of chemistry the natural world has to offer. It appears more striking on cathedrals only because we see their sculptured shapes change, yet it happens unseen even more speedily beneath a layer of vegetation.

13.4 Chemical processes

Chemical weathering occurs because most of the Earth's rocks were formed at great depth under conditions very different from those found on the surface. Once exposed at the surface, rocks are chemically unstable and react readily, even with water.

All rocks are made from collections of **minerals**. Each mineral has a characteristic crystalline form and homogeneous chemical composition: in essence, it is simply a bundle of atoms of various kinds held together in a particular pattern by small electrical forces. Some crystals are held together very firmly: they are stable and hard to break apart. Quartz, made from silicon and oxygen atoms, is an example. Others, usually more complex and containing a greater variety of atoms, are held together less well and are more readily attacked. For example, feldspar,

THE REACTIONS OF CHEMICAL WEATHERING

Although in water there are free hydrogen ions (H^+) which can react with minerals, the presence of carbon dioxide gas (from air, soil, animals and decaying plant tissue) helps enormously. Together, water and carbon dioxide produce carbonic acid:

$$\text{water} + \text{carbon dioxide} \xrightarrow[\text{to produce}]{\text{combine}} \text{carbonic acid}$$

The rocks most easily weathered are often those held together by weak cements. Many sandstones have their grains cemented together by calcite, a mineral that readily reacts with water.

$$\text{calcium carbonate} + \{\text{water} + \text{carbon dioxide gas}\} \xrightarrow[\text{produce}]{\text{react to}} \underbrace{\text{soluble bicarbonate ions} + \text{soluble calcium ions}}_{\text{removed in solution by water cycle}}$$

Igneous rocks, such as granite and basalt, are made from minerals which have yet to undergo any weathering. Some of the minerals, such as quartz, are very stable; others, such as feldspar, weather rapidly under hydrolysis:

$$\text{feldspar} + \{\text{water} + \text{carbon dioxide gas (and plant acids)}\} \xrightarrow[\text{produce}]{\text{react to}} \underbrace{\text{clay mineral}}_{\text{precipitate}} + \underbrace{\text{ions}}_{\text{removed in solution by water cycle}}$$

Although hydrolysis and acid reactions are the most important agents of weathering, some minerals are also susceptible to other weathering agents. Oxides of iron, for example, may be reduced from the ferric to the ferrous form during weathering by stagnant deoxygenated water. Ferrous iron is soluble, and may therefore be removed in solution. Wherever iron oxide is a cementing agent, **reduction** may therefore be a significant weathering agent. By contrast, many iron compounds in rocks already contain oxides in their reduced form. If they take up oxygen (**oxidise**) from percolating water they expand and may break apart. Similarly, some compounds absorb water (**hydrate**) and swell, and in consequence split a rock apart.

a mineral common in igneous rocks, is formed of atoms of silicon, oxygen, aluminium, hydrogen, potassium, sodium and calcium. This variety of atoms only fit together in a loose open framework — a framework that is very vulnerable to chemical reaction.

Water contains oxygen and hydrogen atoms but many of these atoms become separated (**dissociated**). When detached, hydrogen is positively charged (written H^+) and is called an **ion.** Although ions in a mineral are held firmly together by electrical bonds, many ions will exchange with hydrogen ions. If a mineral is made wet chemical exchanges occur. One type of exchange is called **hydrolysis.** Hydrogen is particularly reactive and readily exchanges with atoms on mineral surfaces, causing the mineral structure to break down. This is the reason chemical reactions tend to result in surface 'pitting'. Carbon dioxide from the air accentuates the reaction when it dissolves in water, forming carbonic acid. Similarly, as water percolates through soils, plant roots and soil animal respiration not only add more carbon dioxide to the water but sometimes acids derived from plant decay are also released.

The result of acidified water reacting with minerals is to cause the reorganisation of the ions in the weathered mineral into arrangements that are stable in the presence of water. In effect the reaction causes the production of a precipitate, while leaving some components in solution. The precipitate comprises a group of minerals known as the **clay minerals** and these slowly increase in volume at the expense of the decaying rock. Clay minerals are a basic component of all soils.

Because chemical weathering acts on the surfaces of rocks, rocks that are highly fractured or porous are the most prone to decay. The rate of reaction depends on the chemical composition of the minerals, with quartz being the least reactive, and **ferromagnesian minerals** (iron and magnesium silicates, such as hornblend and augite) and calcium carbonate among the most reactive. Basalt, for example, weathers more rapidly than granite because it has a high proportion of ferromagnesian minerals. Some sandstones – those which have their grains cemented together with iron oxide – are very resistant to weathering. Those which are cemented with calcite (crystalline calcium carbonate) may be only poorly resistant.

Chemical weathering in a soil takes place slowly. Rainwater may be in continuous contact with soil or rock particles for many weeks before it ceases to be chemically active. Thus, as the bulk of rainwater percolates, it retains some capacity for reaction, a great contrast with mechanical weathering whose action is entirely confined to the surface (Fig. 13.9). Materials near the surface become completely weathered and only virtually inert stones of quartz remain. As a soil develops, therefore, the zone of most intensive weathering – the **weathering front** – moves downwards.

Biochemical weathering

The precise role of plants in assisting chemical weathering is very difficult to establish. Where dead plant tissue is rapidly decomposed by soil micro-organisms, the organic acids released are immediately stabilised to humus and therefore play no part in weathering (see soils). By contrast, when decay is slow – as under coniferous forest in temperate regions – organic acids are released that are not stabilised. Instead they percolate into the soil, where they are chemically active. It is believed, for example, that in regions with humid temperate climates organic acids may be responsible for taking iron and aluminium oxides into solution.

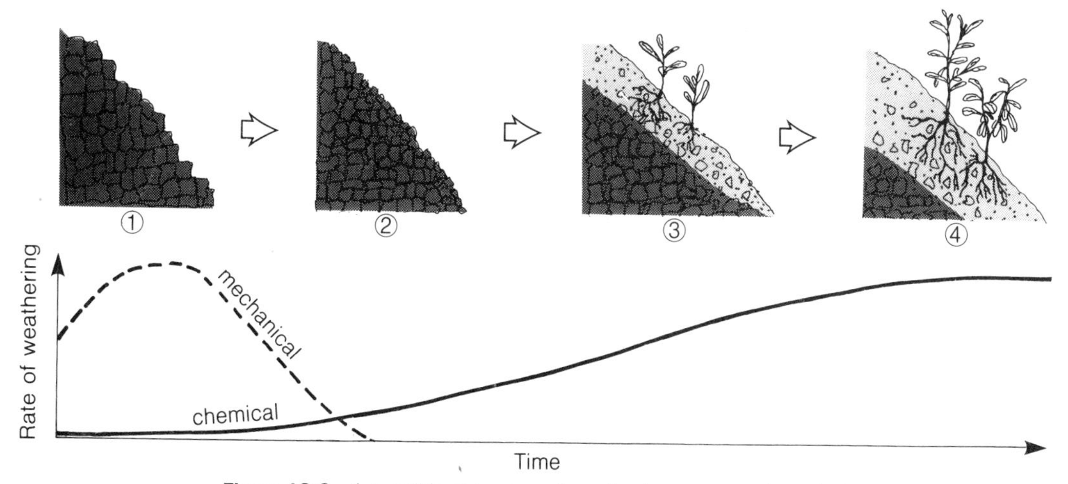

Figure 13.9 A possible sequence of weathering on a gentle rock slope.

Climate and chemical weathering

Many chemical reactions are faster in warm than in cool climates. Some reactions that occur in warm environments are simply not found in cooler regimes. For example, weathering reactions cause the breakdown and solution of silica in the humid tropics, whereas in temperate regions silica is virtually inert. The rate of weathering is further accelerated by continual percolation of rainwater that is still chemically active. So regions which have a high rainfall (and consequently high rate of percolation) *and* a high temperature are those where chemical weathering should proceed most effectively (Fig. 13.10). These conditions are most closely met in the humid tropics, where it is commonplace to find complete weathering

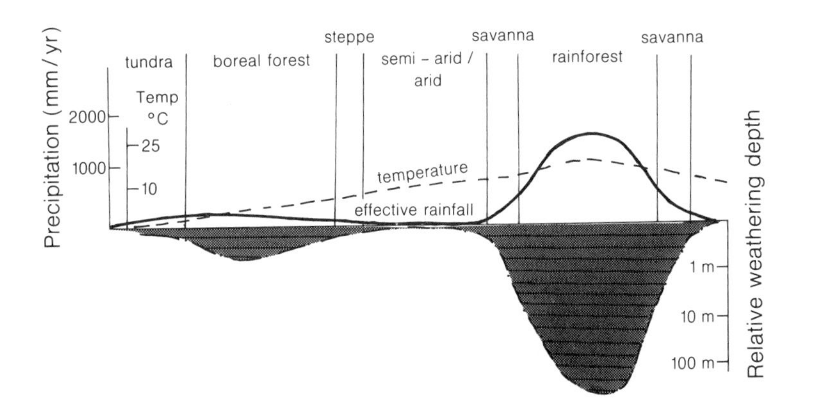

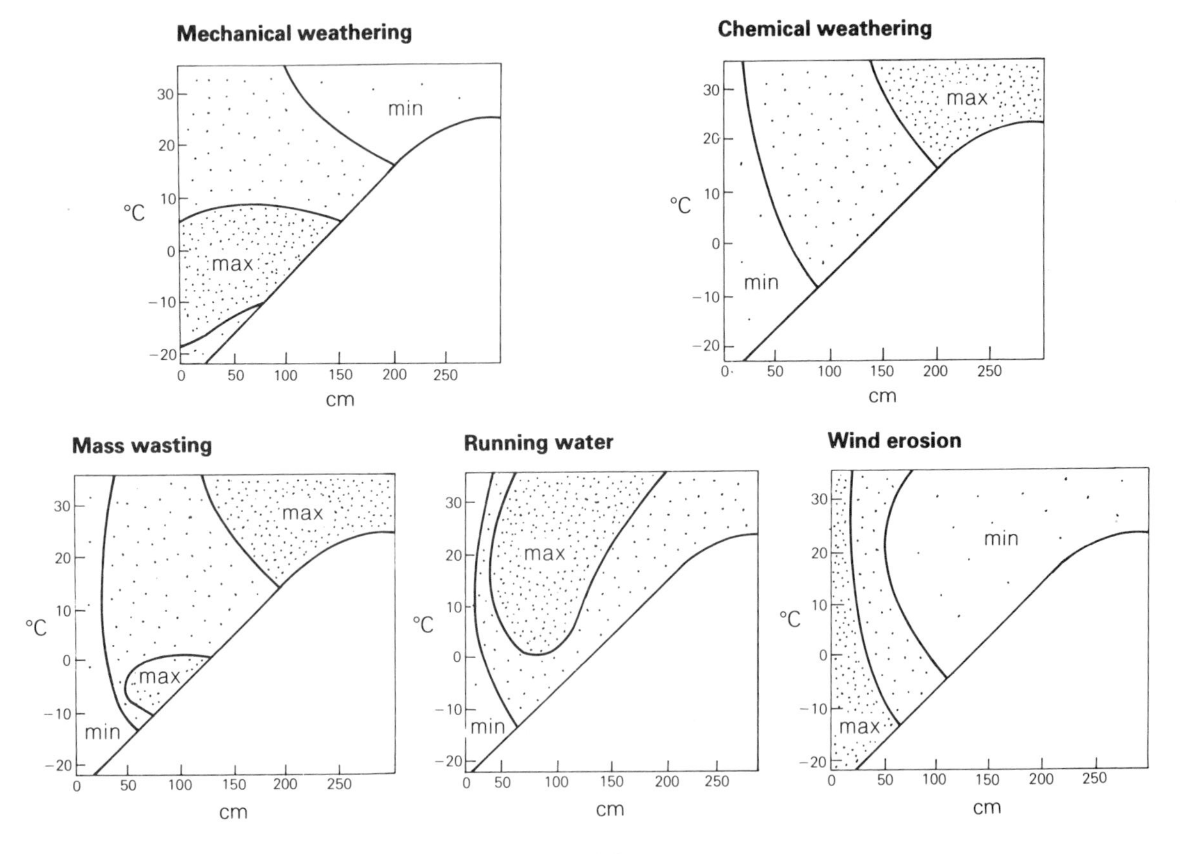

Figure 13.10 Some relationships between weathering, erosion and climatic factors.

(a)

(b)

Figure 13.11 The depth and intensity of rock weathering in the humid tropics is illustrated by the 'boulder' (a), taken from a rock face newly exposed by road construction. It was 7 m below the ground surface, yet it was so intensely weathered that it disintegrated when dropped (b).

of all rock to depths of many tens of metres (Fig. 13.11). Humid temperate regions also experience a substantial moisture surplus and consequently have high rates of percolation; but in these regions lower temperatures reduce the efficiency of the reactions, the depth of weathering rarely exceeds one metre, and iron, aluminium and silica remain largely immobile. Only carbonation proceeds effectively at low temperatures: carbon dioxide is more soluble in cold water than warm. Limestone weathering may thus sometimes be more effective in temperate than tropical environments.

Any conclusions about the rate and effectiveness of chemical weathering must take account of time as a factor. Many tropical regions have had a warm humid climate for millions of years. The depth and completeness of weathering may be related to time as much as to the character of reaction. Similarly, many mid- and high-latitude regions have only been released from ice cover for a few thousand years and weathering may still be increasing.

There have been few attempts to ascertain the rate of chemical weathering, but in one experiment some sets of small shale discs were buried at various depths in Welsh and Malaysian soils for two years. At the end of the experiment all discs had lost weight, but the Malaysian discs had lost over three times as much weight as the Welsh discs in comparable sites (Fig. 13.12). This difference suggests that weathering in the humid tropics is indeed much faster than in temperate regions.

It was assumed that the length of time rock is in contact with water affects its rate of chemical weathering. This was tested by leaving a shale disc on a piece of wood on the soil surface in Malaysia. At the end of the test it had weathered only half as much as the disc buried in the soil. Furthermore, all discs within the soil had weathered by the same amount; there was not a progressive decline with depth. This means that because the upper soil has already been weathered, water percolates without reacting. As a result the most active zone for natural weathering is

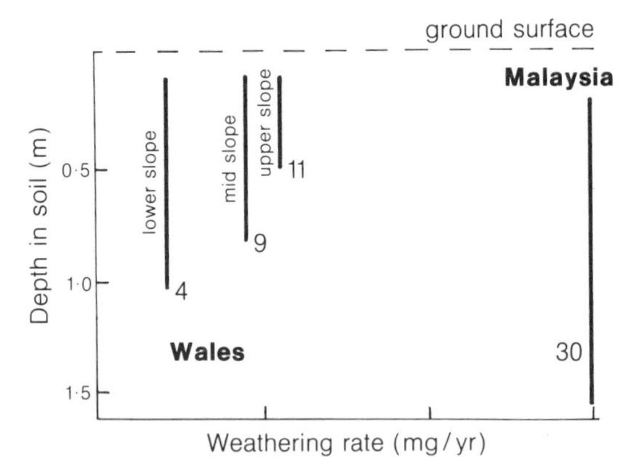

Figure 13.12 Contrasts between humid tropical and humid temperate weathering rates were very apparent for an experiment with shale discs. On average, Malaysian soils are 14°C warmer than Welsh soils.

near the base of the soil, simply because there is more unweathered rock there. But while the rate of weathering is fairly uniform within each soil profile, it appears that this rate decreases towards the base of the slope, showing that position within the landscape may also play a prominent role in determining the rate of weathering.

An experiment conducted on the drainage basin of the River Exe in Devon provides a further demonstration of the variability of weathering rates. In this case the dissolved load was measured for many small streams. By choosing streams that flowed entirely on one type of bedrock, it was possible to calculate the **denudation** (lowering of the landscape) as a function of rock type (Fig. 13.13). There is clearly considerable agreement between the geology map and the denudation map, with, for example, the Permian marls showing over eight times the rate of denudation of the Devonian slates. Results such as these support the contention that chemical weathering in a humid environment is closely related to rock type.

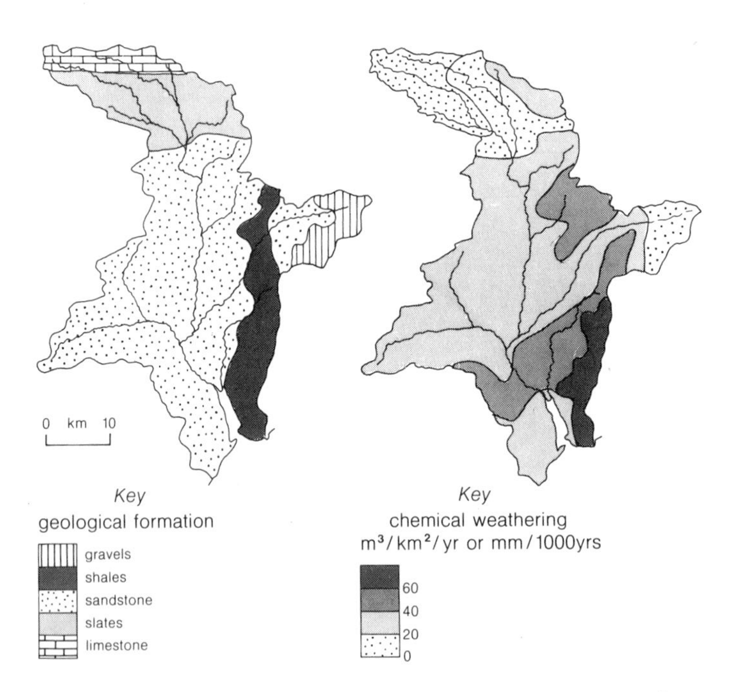

Figure 13.13 Rates of denudation (landscape lowering) compared with rock characteristics in south Devon.

14 Slope development

14.1 The Grand Canyon: an illustration

Hillslopes evolve through a combination of weathering and transport processes. In regions with humid climates the effects of weathering and transport are mainly camouflaged by a cover of vegetation; these are not easy places to study transport processes. In places such as the Grand Canyon, where the climate is too arid to support a plant cover, the movement of debris is readily seen.

In the spectacular Grand Canyon, Arizona, downcutting by the River Colorado has exposed a staircase of strata over two kilometres deep. Many tourists follow the switchback trail down the canyon wall, every turn revealing contrasts in weathering and transport of the varied rocks. Sometimes the trail plummets in man-made steps down the vertical faces; at other times it descends more gently on the ledges between the cliffs (Fig. 14.1).

Figure 14.1 The Grand Canyon.

Everywhere, debris litters the canyon ledges. Large blocks and small rest in loose, jumbled form waiting for something to push them from their shelves of temporary stability to crash down to the next ledge below. Some of the debris is large and angular, some small, but also with sharp edges as though it had been freshly plucked from the solid rock and laid beside the trail. It is easy to find stones for trail markers – the difficulty is in moving enough of them out of the way to make a reasonable trail to walk on. So a problem immediately comes to mind. The source of the material is obvious enough, but how does it move?

If you try to make a sand castle from dry sand you will soon discover that it simply can't hold slopes greater than about 30°. Rock rubble left by mechanical disintegration processes behaves in the same way. Characteristically **screes** (accumulations of coarse, angular debris) provide a fringing skirt to all steep slopes; the constant angle of many screes has a pleasing simplicity. A closer look, however, shows that materials of varied shapes yield slopes with different maximum angles of stability. If blocks falling onto a scree build it up beyond the maximum angle of stability, the whole surface collapses and gradually rebuilds back to this maximum angle. The Grand Canyon walls (Fig. 14.1) show each tread of the canyon staircase delicately adjusted in slope to allow the continual movement of the debris released from the riser above it. And as the risers and treads are made of different materials with varying joint patterns, so the angles needed to allow movement of debris on the individual treads vary also (Fig. 14.2).

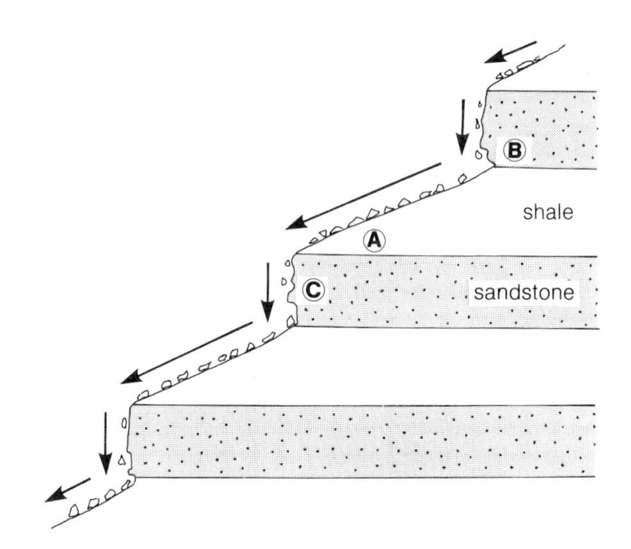

Figure 14.2 The small-calibre debris produced as shale weathers is only stable at a low angle (**A**). This gentle slope acts as one of the 'treads' in the valley-side staircase. As the overlying sandstone 'riser' is undermined at **B** it collapses, shedding debris over the tread. Eventually collapse of the sandstone block **C** will leave the toe of the shale unstable and the shale will retreat once more, transporting its surface debris to the tread below. The whole canyon side is an example of a **debris cascade**, a discontinuous conveyer belt with local control levels but with outputs from one segment providing much of the inputs to another. Such cascades operate on all hillslopes including those with soil and vegetation cover. The relationship can be written out as an equation:

debris in + weathering – debris out
= change in land surface

14.2 Transport processes

Detailed investigations have revealed that there are two kinds of transport processes of material down slopes: 'mass movements' and 'particulate movements'.

MASS MOVEMENTS

Where gravity is the most important moving force and where contained fluid is not the main influence on movement, **mass movement** occurs. Rockfalls, landslides, avalanches, debris flows, soil creep and gelifluction are examples of such movements. Each process involves one or more of several mechanisms: slide, creep, fall or flow.

PARTICULATE MOVEMENT

Where transport occurs within a fluid (such as ice, running water or wind), **particulate movement** occurs. Sheetwash, rainsplash, river flow, deflation, leaching and ice transport are all examples of such movements. Each process involves one or more of these mechanisms: suspension, saltation, traction or solution.

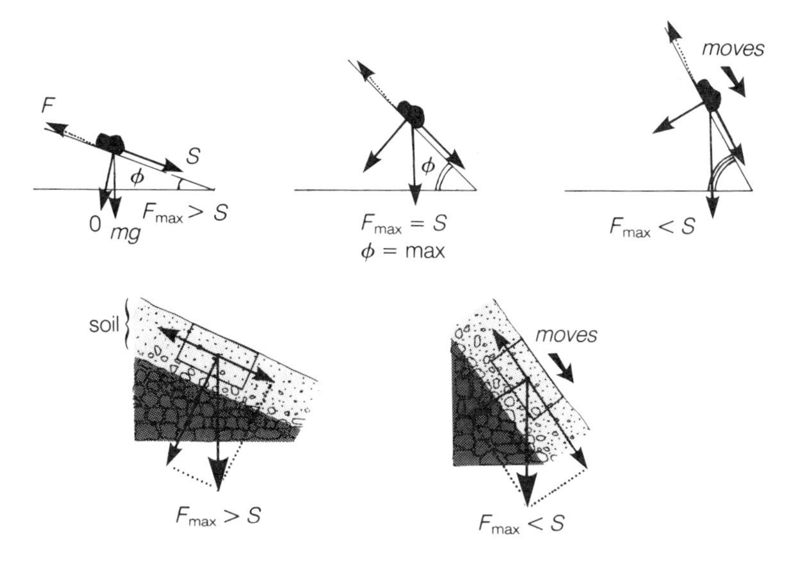

Figure 14.3 The stability of a piece of rock (above) or a block of soil (below) depends on the balance between effective stress (*S*) and resisting forces (*F*).

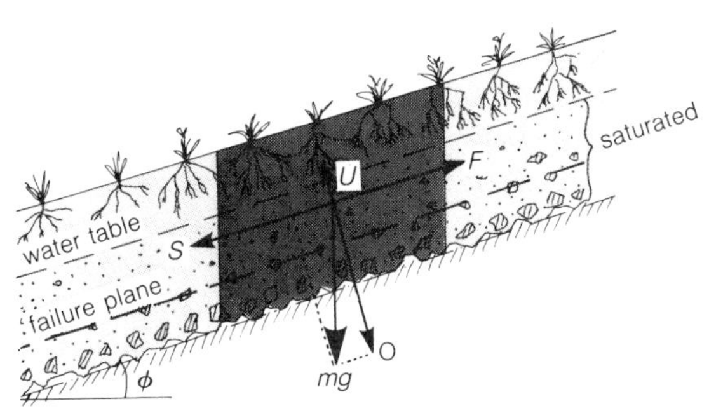

Figure 14.4 The forces acting on a partially saturated block of soil. Failure commonly occurs along a plane within the saturated zone.

COULOMB'S EQUATION

The forces acting within a soil can be expressed by the equation

$$S = C + (O - U)\tan\phi$$

where S is the sheer stress, C is cohesion, O is the overburden pressure, U is the upthrust due to water (buoyancy), and ϕ is the angle of internal friction for the material.

The equation can be illustrated by placing a hand on a table and trying to slide it. S is the force needed to cause sliding. Pressing harder down onto the table while sliding illustrates the overburden force O. The cohesive force, C, would be represented by making the hand sticky. Force U would be equivalent to the air under pressure that causes a hovermower to lift from the ground.

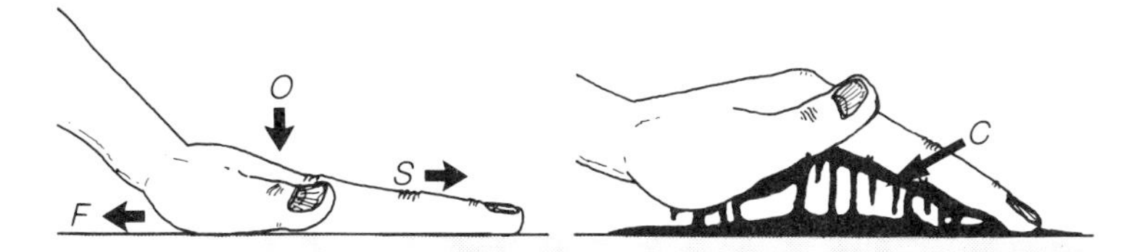

Figure 14.5

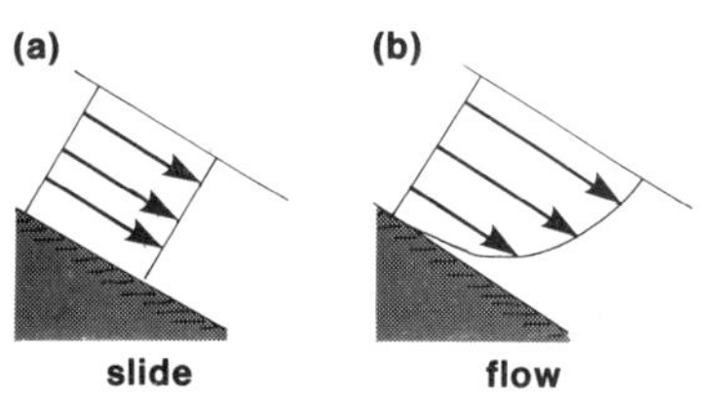

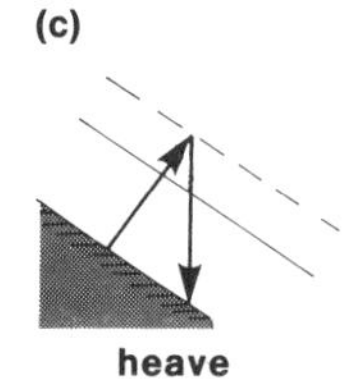

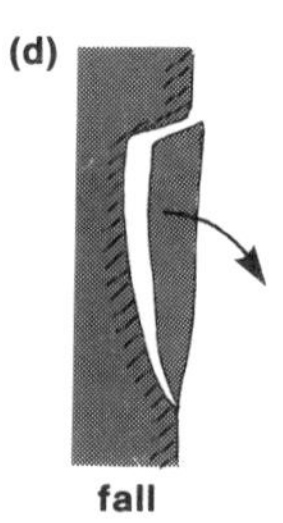

Figure 14.6 The basic forms of mass movement.

Mass-movement processes

There are four basic mechanisms of mass movement (Fig. 14.6):

(a) rock material may **fall** from steep slopes;
(b) material may **slide** if the internal cohesion of the material is strong in one part of the material but weak in another (usually a saturated) part: the material then fails as a block, sliding across a failure plane;
(c) material may **flow** if it has weak internal cohesion and if the threshold angle for stability is exceeded (friction with the bed retards movement in the lowest layers, the speed increasing upwards in the same way as in a river channel);
(d) material may **heave** if it comprises either clay sediments and soils (which absorb water on wetting) or porous materials: heaving occurs when the contained water freezes.

ROCKFALLS AND DEBRIS AVALANCHES

Rockfalls are produced by a variety of mechanisms but in all cases movement is caused by the factor of safety declining to below 1 (Fig. 14.3). Rock falls often occur after deep fractures have been produced by unloading.

Sometimes the rocks fracture into sheets tens of metres long, and then slowly flake from cliffs. Tall sandstone and granite cliffs often show such a failure pattern (Fig. 14.7). One of the most famous examples of flaking (known as **toppling failures**) was in Chaco National Monument, New Mexico. Threatening Rock, a slab 50 m high and 30 m wide, flaked clear of a cliff over a period of 2500 years. Finally it collapsed and broke into a pile of fragments. Rocks may develop a pattern of fractures that dip out from the cliff; in these cases slabs slide away. The coast near Watchet in Somerset has some spectacular examples of such **slab failure**. The steep sides of glacially eroded valleys are also commonly subject to slab failure because valley widening has caused rapid lateral unloading.

Not all rocks are strong enough to stand up in a massive way like the walls of a castle; most rocks become intensively cracked on unloading and tumble down as soon as a tall enough face is revealed. This process, called **rock avalanching**, is commonly found in cliffs dominated by shales. Some massive (joint-free) rocks resist all of these mechanisms and are only affected by mechanical or chemical disintegration. Such rocks vary from massive sandstones, from which the calcite cement weathers allowing individual grains to fall away, to metamorphic rocks, where hairline fractures are widened by frost cracking until blocks are prised away, creating **rock falls**.

Fragments released from a rock face may build at its foot to form a **debris scree**. Each piece of scree may be analysed for its factor of safety: the range of scree-slope angles commonly seen reflects differing threshold angles of stability consistent with the variations in size and shape of the scree material. The scree shape also depends partly on the height from which the debris is falling: debris falling from a substantial height may bounce and roll, dislodging and moving other scree material. Screes that build in

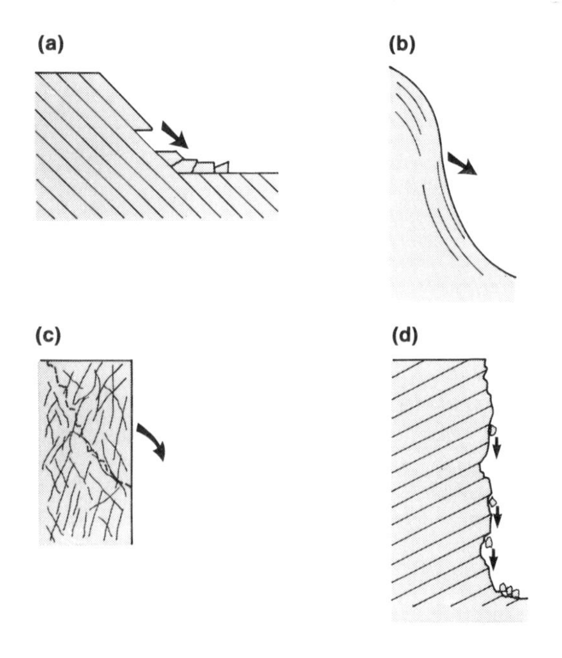

Figure 14.7 (a) Slab failure on dipping rock. (b) Toppling failure in glaciated valley. (c) Avalanching of fractured rock. (d) Rockfall from coherent cliff.

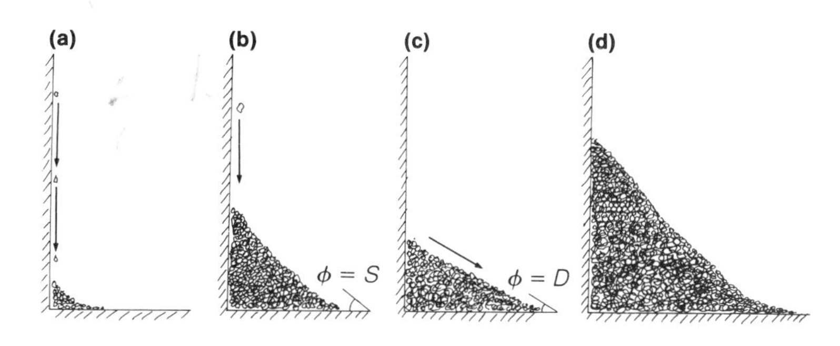

Figure 14.8 Scree formation: (a) large fall height, concave profile; (b) building to the threshold angle of static friction (S); (c) debris avalanching to the angle of dynamic friction (D); (d) final accumulation (after many episodes of (b) + (c)) to give a largely straight slope with a concave toe (see also Figure 20. 6).

this way at the feet of steep slopes tend to have concave profiles. As a scree builds, however, the average height of fall is reduced and the effect of bouncing becomes less pronounced; a straight slope develops, with material accumulating at the threshold angle of stability (Fig. 14.8).

Scree surfaces do not move every time a new block falls. Their angle of accumulation is slightly steeper than their angle of rest after failure. This is because the angle of internal friction for stationary material is greater than the angle of internal friction for moving material. It is common, therefore, to find adjacent screes standing at different slope angles even if all the particles appear to be the same shape and size. Many screes stand at angles of between 30° and 40°, the most common angles for dry, broken material.

LANDSLIDES AND DEBRIS FLOWS

Where chemical weathering dominates, hillslope debris contains clay particles, humic gums and plant roots as well as sand and stones. These fine materials give the soil some cohesion. Soils are further distinguished from screes by their ability to hold water in capillary pores.

Because cohesion is lowest and buoyancy greatest when a soil is saturated, most soil failures occur during or soon after a prolonged rainstorm. **Landslides** occur when internal cohesion is strong even in the presence of water. In these circumstances failure occurs either near the soil/bedrock boundary, in which case it is called a **shallow slide** (Fig. 14.9), or along a curved plane deep within water-saturated

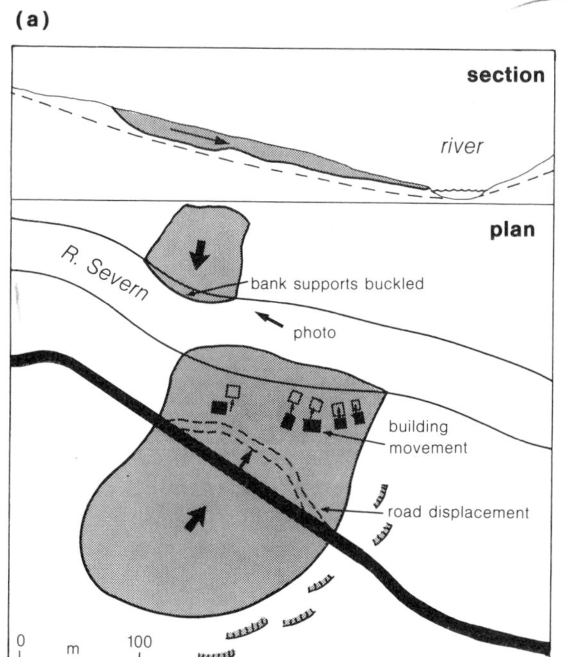

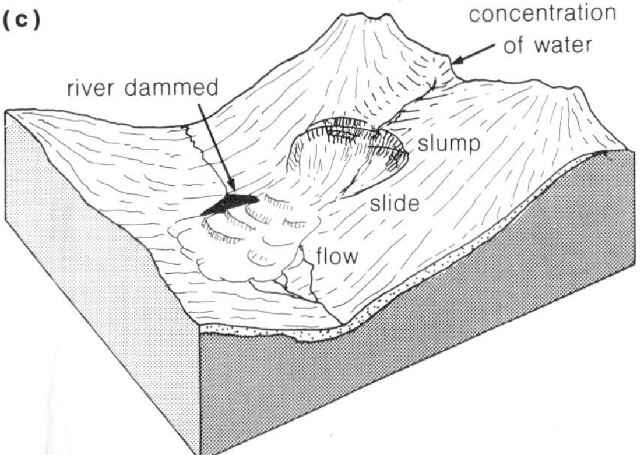

Figure 14.9 Examples of flows and slides: (a) The Jackfield slide, Shropshire; (b) the bulging bank of the River Severn opposite Jackfield; (c) the complex nature of most landslides.

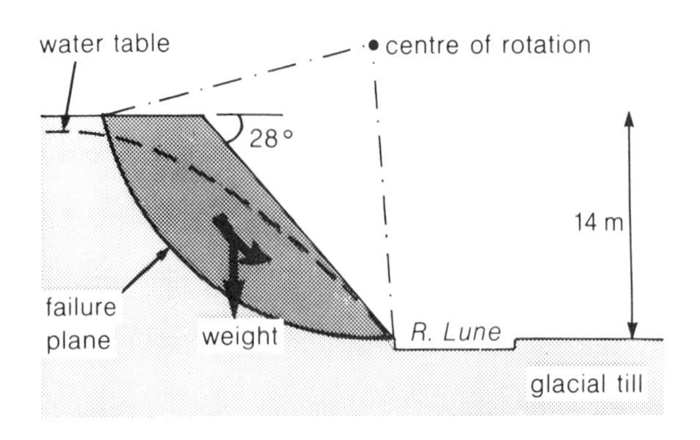

Figure 14.10 Undercutting by a river caused a 28° slope cut in glacial till to slip along a curved path at Selset near Middleton, Teesdale, Yorkshire. The slip moves in small steps very slowly and records show the site as a river cliff for over a hundred years.

sediment, when it is called a **rotational slip** (Fig. 14.10). Where the internal cohesion is poor, as is common in sandy soils when both clays and humic gums are scarce, instability results in a **debris flow**, whereby each particle slips over the one below and the debris behaves as a fluid.

Many landslides develop into debris flows because, as the soil slides over the rock below, intense vibrations are set up in the moving soil. When this happens the soil particles can move freely and begin to flow past one another.

Debris flows are most common in semi-arid regions: soils there contain little clay or humus and the plant cover is discontinuous and so ineffective at binding the soil. Landslides are most common in humid regions, as soils there are held together by clays, humus and a complete cover of vegetation.

The contrast between landslides and debris flow is well illustrated by two examples. The first relates to the Ironbridge Gorge, near the village of Jackfield, Shropshire. The 10° slope of the valley side here is too steep for thick soils to remain stable when saturated. In the winters of 1951–52 and 1952–53 the slope became locally saturated several times: the buoyancy effect that accompanied each storm enabled the soil to slide a few centimetres downhill. Over a two-year period sliding amounted to 20 m, eventually causing the destruction of a road and several buildings. The second example relates to the South Topanga Canyon in the hills outside Los Angeles, California. Prolonged heavy rain caused the sandy soil to become saturated: at 3 a.m. on 25 January 1969, the soil began to slide. With only partial vegetation cover and relatively little clay, the material was rapidly jolted into a suspension and the slide soon became a debris flow which demolished one of the houses and damaged several others.

Episodic movement of slides and flows Sometimes a soil that has previously seemed stable enough for people to build roads and houses on suddenly becomes unstable and fails. It is common in landscape formation for a catastrophic event to follow a long period of apparent inactivity. On a slope, failure is a natural result of weathering. Weathering produces fragments of soil that are more loosely packed than the parent rock. Even though this soil is bound together by clays, oxides and humus, these forces are small compared with the binding strength of solid rock. A time often comes when the thickness and weight of soil have increased sufficiently to cause failure without any external influence at all. The long period of progressive (but unnoticed) weathering changes instantaneously into the turmoil of a landslide.

Many slopes naturally develop by means of a series of catastrophic events separated by quiet periods. This is a main reason why slopes fail unpredictably, destroying houses and sometimes killing people and animals. In other cases, the final failure may well be triggered by saturation following a prolonged storm. Civil engineers today pay great attention to ensuring adequate drainage of slopes in the vicinity of important structures such as houses and roads.

Rotational slips Rotational slips are failures of saturated ground in which the failure surface is deep-seated and curved, and in which the movement has a pronounced rotational element. They may occur in thick unconsolidated sediment and are a feature of many glacial till cliffs (e.g. at Whitby, Yorkshire) and river cliffs cut into till (Fig. 14.10). However, slips can also occur in permeable rock underlain by weak impermeable strata. In these cases water percolates down to the impermeable material which then becomes saturated and fails (Fig. 14.11). Most coastal

Figure 14.11 Large rotational slips at Cain's Folly, Lyme Bay, Dorset.

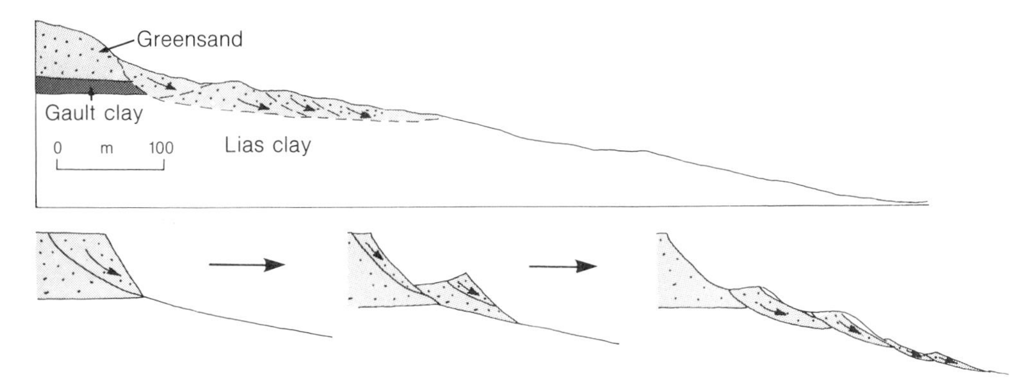

Figure 14.12 Steep valley slopes created by the River Char, Dorset, southern England, have exposed permeable Greensand rock over the impermeable shale. Water build-up in the Greensand has helped the upper cliff to slip along a curved path. Each new collapse has helped to 'shunt' older slipped blocks down slope where they break up into smaller pieces.

'undercliffs' have rotated into place, but deep-seated slips are also found on valley sides (Fig. 14.12).

SOIL CREEP

There is a critical minimum slope angle (usually between 8° and 15°) below which landslides and debris flow are not initiated. An 8° slope is quite steep – equivalent to a hill with a gradient of 14 per cent or 1 in 7. It would be unusual to find buildings on such steep slopes because of the amount of cutting and filling needed to prepare level foundations. Yet gentle slopes are not unchanging simply because they are free from *rapid* movements. Indeed, the absence of accumulated debris at the feet of many steep slopes indicates that transport and weathering keep in step.

One important process that operates on both gentle and steep slopes is the heave mechanism, which causes **soil creep**. Creep (Fig. 14.13) is produced by alternate swelling and shrinking of clay minerals as they become wet (during a storm, or through the winter) and then dry (between storms, or in the summer). Indeed, clay minerals are more like adjustable suitcases than rigid crystals. In tropical regions the amount of shrinking and swelling from montmorillonite clay is sufficient to make fields unworkable and to break up road surfaces. Even the less expansive kaolinite and illite clays present in temperate environments produce a substantial ratchet-like movement, particularly near the surface where the greatest wetting and drying takes place. Because soil creep depends on frequent changes in moisture content it is usually confined to the top 30 cm of soil.

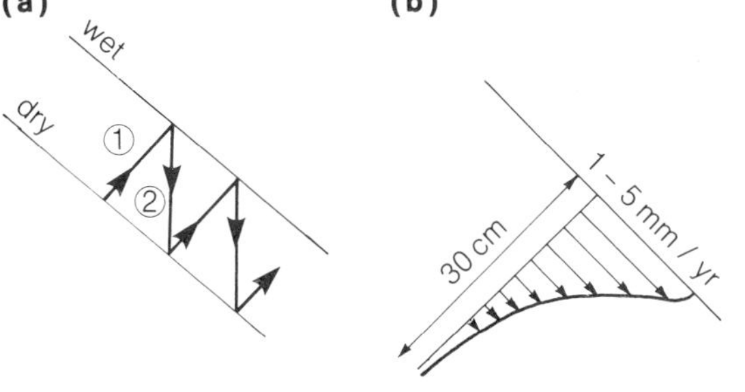

Figure 14.13 (a) As the soil becomes wet, the clay minerals expand and the soil surface heaves outwards (1). On drying, the clays contract downwards under the influence of gravity, giving a resultant ratchet-like downslope motion (2). (b) A velocity profile for soil creep. The actual surface movement in each year is variable and depends on the rainfall and drying patterns. (c) Soil creep on a steep slope near Sanquhar, Scotland. Notice that when the slope becomes too steep the turf tears away and creep is replaced by slide.

SOLIFLUCTION
Solifluction is the *slow* flow of saturated hillslope material. It can occur on slopes as gentle as 2°. It is most frequently experienced in cold climates when deeply frozen soil covered with snow begins to thaw. The shallow layer of soil that has thawed cannot contain all the snowmelt: at saturation, the upper surface begins to flow. This is called **gelifluction** and is more fully described on page 180.

Particle-movement processes

RAINSPLASH, SHEETWASH AND GULLIES
Because of the protective cover of vegetation, soils in humid regions are not normally exposed to the direct impact of running water. The most important hillslope processes are usually mass movements. However, in semi-arid and arid regions vegetation is discontinuous and overland flow therefore occurs almost every time it rains.

The effect of a raindrop falling onto bare soil is similar to that of a miniature meteorite crashing to the ground (Fig. 14.14). Its collision with the soil causes a deep impact crater and throws debris in all directions. Because warm semi-arid regions experience high-intensity rainstorms, the consequences of such **rainsplash** are more substantial than those of rain on bare soil in temperate regions: debris may be thrown over 20 cm into the air and several tens of centimetres laterally. But even in Britain the effects of rainsplash on bare soil can be significant, as demonstrated by flowerpots which may rapidly become encrusted with rainsplashed soil if left standing on a flower bed.

Rainsplash has several important effects. The impact of a falling raindrop breaks up cohesive clays and flings apart the clay particles. These are then carried into the soil by pecolating water, and eventually they clog up the pores. In time, the pores become sealed and an impermeable crust develops (Fig. 14.15). On slopes the majority of rainsplashed material is thrown down slope with the result that the surface of the soil is transported away.

As crusted soils become impermeable, infiltration rates decline and an increasing proportion of rainfall becomes overland flow. Initially water flows in sheets, much as it might flow over the surface of a road during a downpour. Shallow though it is, the water may entrain and transport soil particles, causing **sheetwash**.

As the sheetwash passes over the soil surface, chance irregularities in the ground rapidly cause the water to become concentrated. The water becomes deeper in these places and progressively less of the energy of the flowing water is used in overcoming friction; water then flows increasingly quickly. At some critical velocity the water can entrain sufficient soil to erode small channels, called **rills**. At the beginning these are mere traces on a slope that rearrange themselves with every storm; but they gradually undergo a form of piracy in which some rills become enlarged at the expense of others, focusing water into themselves and growing ever faster. These rills eventually become so big that they are not removed by subsequent storms; they become permanent features of the slope. At this stage the rills have become

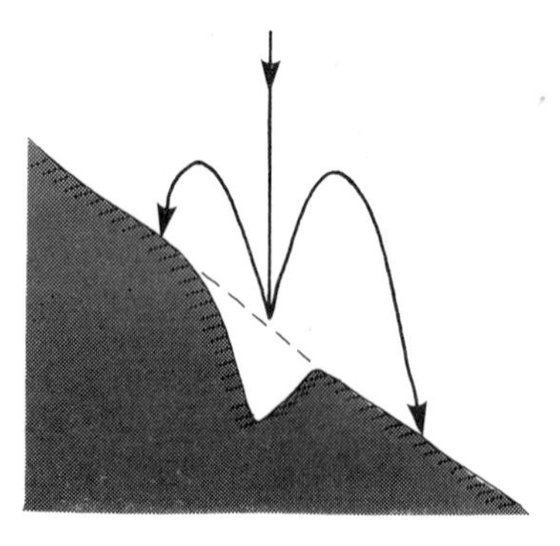

Figure 14.14 Rainsplash sends more material down slope than up slope. For uniform removal of material from a slope, a convex form needs to develop to provide a continuously increasing rate of soil removal.

Figure 14.15 A soil with severe crusting. Notice that the effect of rainsplash causes the upper slope to develop a convex shape.

Figure 14.16 Sheetwash on this newly formed road cutting has quickly been succeeded by an intricate network of rills and gullies. The material is glacial till.

gullies, effectively ephemeral river channels. Because they concentrate water, gullies widen their channels and erode headwards, all the time becoming more effective in transporting sediment (Fig. 14.16). The effect of this action is discussed in more detail on page 89.

LEACHING AND PIPE TRANSPORT

Overland flow is uncommon on hillslopes in humid regions. Most water infiltrates into the soil: it percolates first vertically, then downslope, parallel with the surface, as throughflow. Percolating water reacts with the soil and unweathered subsoil, producing a range of soluble materials which are carried away in solution. Transfer of soil materials in solution is called **leaching**. It is equally effective on gentle and steep slopes and may be the only transport process at work near valley divides. Indeed, because leaching occurs throughout the year its cumulative effect may be as important in removing soil as all mass movements combined. Only on the steepest slopes will mass movements predominate. Sometimes, and especially on steep slopes, rapid throughflow may enlarge soil pores and allow water speeds to increase. Many slopes have a network of enlarged soil pores that feed into one another, much as rills feed into gullies. The subsurface equivalents of gullies are called **soil pipes**. Both soluble and particulate material can be transported through soil pipes. Pipes come to the surface at the foot of a hillslope or in a hillslope concavity as a spring or as seepage. In many instances the roofs of enlarged soil pipes collapse, allowing pipes to become hillslope streams.

14.3 Slope form

Hillslopes are often quite remarkably complicated in shape, with variations in angle and curvature in both profile and plan. It is to be expected that a number of different transport processes will operate side by side or one above another on any hillslope. Yet despite this complexity it is possible to provide a simple classification of hillslope forms and to relate each part of the slope (each **slope element** or **slope unit**) to a characteristic set of weathering and transport processes. The classification in Figure 14.17 shows all possible slope characteristics, not all of which will occur on every slope.

Convex slopes

The upper part of a slope may be too gentle to permit rapid movement of materials by slide or flow. Here the dominant transport processes will be leaching and soil creep in humid regions and rainsplash with sheetflow in semi-arid environments.

Both soil creep and rainsplash produce convex slopes because progressively larger amounts of material must be removed downslope, and this is only possible with increasing curvature (Fig. 14.18). Leaching, however, is not dependent on slope angle and only causes uniform ground lowering; it does not influence slope shape. Similarly, sheetwash does not entrain particles until it reaches a critical speed, and this may not occur until the lower end of the convexity. However, as slopes steepen on the lower part of the convexity there will also be a tendency for soil to begin to slide. On slopes where the soil cover is thin but where there is a continuous vegetation cover, the soil slips away in strips, producing a broad staircase with small, irregular steps. Each step is called a **terracette**.

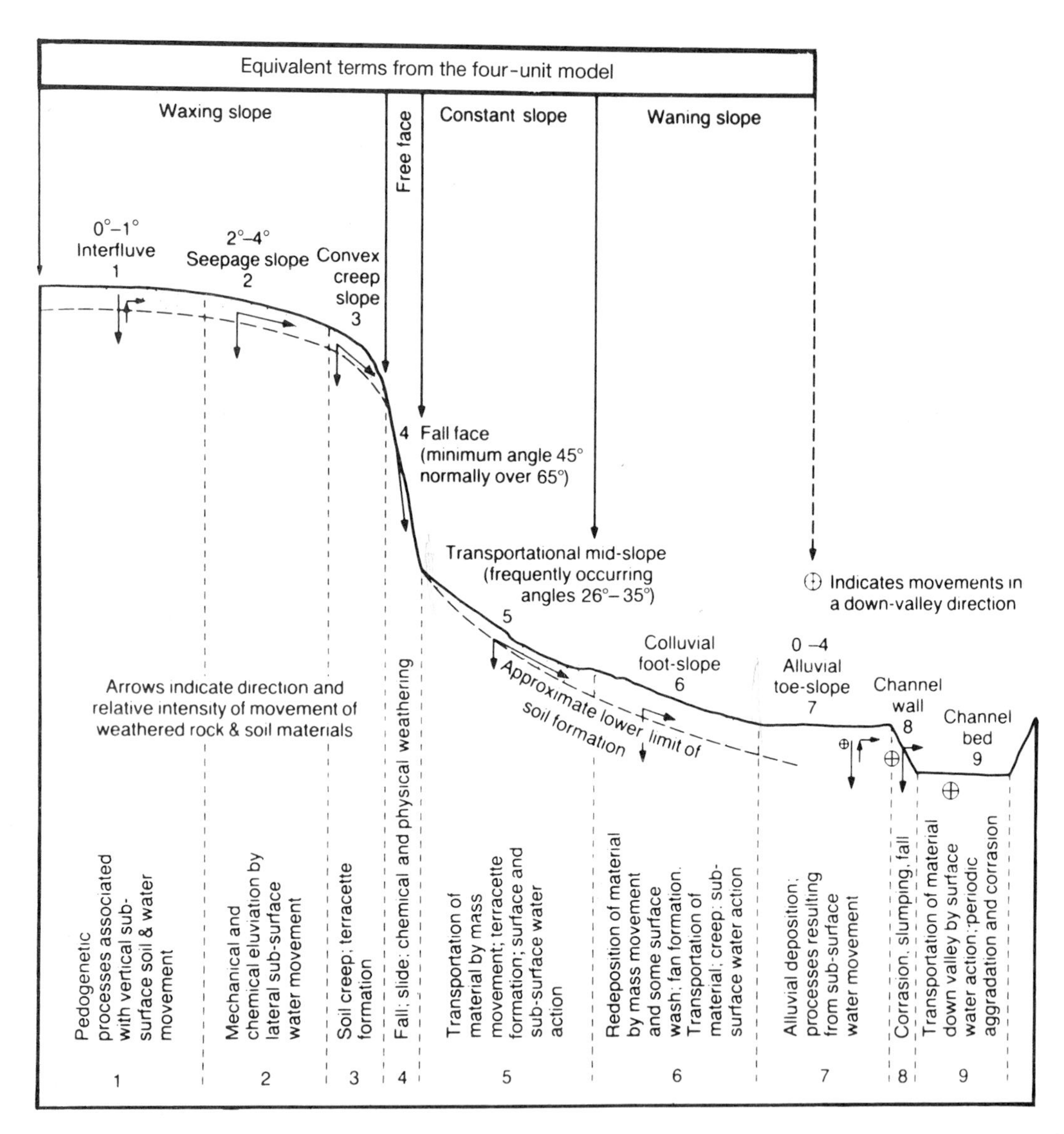

Figure 14.17 A nine-unit system for classification allows a wide range of slope segments to be described. A more simple four-unit system (shown at top) is also frequently used.

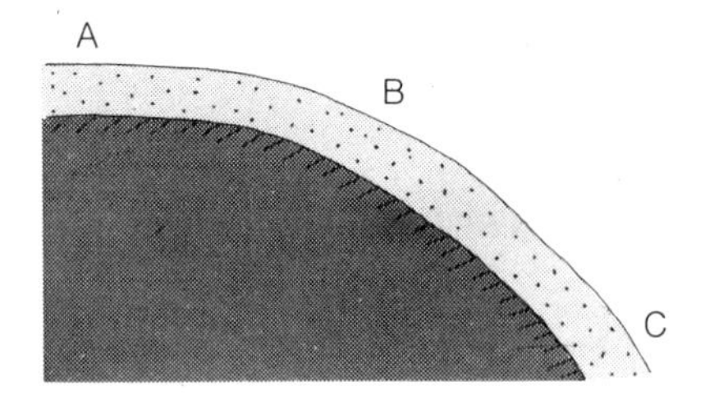

Figure 14.18 Processes on convex slopes. If soil weathered at B is to be removed effectively, the rate of movement must exceed not only the weathering rate at B, but also the rate of transport from A. To achieve this, the slope at B must be greater than that at A. The slope at C needs to be even greater if it is to remove its own weathered layer together with that from A + B. The only stable slope that will permit increased movement down slope is convex.

Straight slopes

Straight slopes vary widely in form from bare cliffs to gentle slopes with a deep soil cover. It is to be expected, therefore, that there will be an equally wide range of transport processes. In general there are two groups:

(a) **weathering-limited slopes:** slopes that are so steep that transport is at least as effective as weathering – under these circumstances slopes remain bare, and rockfalls and rock avalanches are often the dominant processes;

(b) **transport-limited slopes:** slopes that retain a soil and vegetation cover and where weathering is at least as effective as transport – these are slopes on which landslides and debris flows operate, giving slopes that may be very irregular and disturbed in detail even though they are, overall, straight.

Concave slopes

The upper parts of concave slopes may be sufficiently steep for landsliding, but in general, as slopes become less steep, so the range of processes changes to those involving overland flow. Rills and gullies are especially common in concave regions on slopes within the semi-arid zone, with pipeflow and leaching being important in the humid zones. When there is soil cover, soil creep can continue to work, although, for reasons stated above, it plays a minor role.

The lower concave slope (**footslope**) is distinguished from all the higher slopes because its form may be as much due to the accumulation of materials as to erosion. Many footslopes cannot transport the material delivered from upslope and it accumulates as **colluvium**. Slopes with colluvium do not, therefore, transport debris effectively and they are out of equilibrium.

14.4 Slope failure

When weathered debris is transported down a slope the forces causing collapse exceed those resisting movement. Scientists and engineers have long realised the need to predict slope failure, and have developed a basic measure of slope stability: the **factor of safety**. This is the ratio of the magnitude of the disturbing force (S) to the internal frictional and cohesive resistance of the weathered material (F). Failure occurs when $F/S < 1$.

The purpose of the factor of safety – to focus attention on the forces controlling transport – is most easily appreciated when one considers a piece of rock at rest on a plane slope (Fig. 14.3). Think of the piece of rock as having all its weight concentrated at its centre (at the centre of gravity). This weight gives rise to two forces. One, the **effective stress** (S), acts *down* the slope, tending to make the rock slide. The other, the **overburden pressure** (O), acts directly *into* the slope, holding the rock firmly in place. For as long as the rock remains at rest, S is balanced by an equal and opposite force – the **frictional force** (F) between rock and slope. Friction depends both on the surface roughness of the rock and slope and on the size of the overburden pressure. At rest, $F/S < 1$. However, as Figure 14.3 shows, as the slope angle increases, O becomes smaller while S becomes larger. At some angle, the **threshold angle for slope stability**, $F/S = 1$. At this angle, friction is only just able to balance S ($F = F_{max}$): any further increase in slope angle will cause the rock to move ($F/S < 1$).

The possible movement of a block of soil can be examined in the same way. The soil is held together by a number of **cohesive forces** (C), as well as by simple friction between soil grains. Cohesion can be thought of as the 'stickiness' of a soil. Cohesion in soils is provided by clays bonded together by ions, by humic gums released by soil micro-organisms and by surface tension. Surface tension – the force produced at the air–water surfaces in pores that are partly filled with water – differs from the other agents of cohesion in that it varies with soil moisture, disappearing altogether when the soil becomes saturated and there is no air.

Saturated soils are less stable than dry soils: not only is surface tension absent, but the water within the soil pores partly supports the soil grains, creating a **buoyancy force** (U). This force acts upwards in the soil and counteracts the overburden pressure (Fig. 14.4). Saturated soils therefore fail more readily than dry soils, and their threshold angle for stability is reduced to about half that of similar but unsaturated material. A soil on a convex section of slope which never becomes saturated may stand at a much greater angle than the same soil within a concavity where saturation is common. All soils are most prone to failure during a severe rainstorm, because it is then that the soil is mostly likely to become saturated.

14.5 Slope evolution

All slopes develop from the processes described above. If only one process operates one can predict the evolutionary course of the slope. For example, a steep, straight slope would evolve into one of the shapes shown in Figure 14.19, depending on the dominant process. On most slopes, however, several processes operate: each slope has a unique form that reflects the balance achieved between processes. For example, the common sequence of units – upper convex, straight, lower concave – shows that a different process dominates on each part of the hillslope even

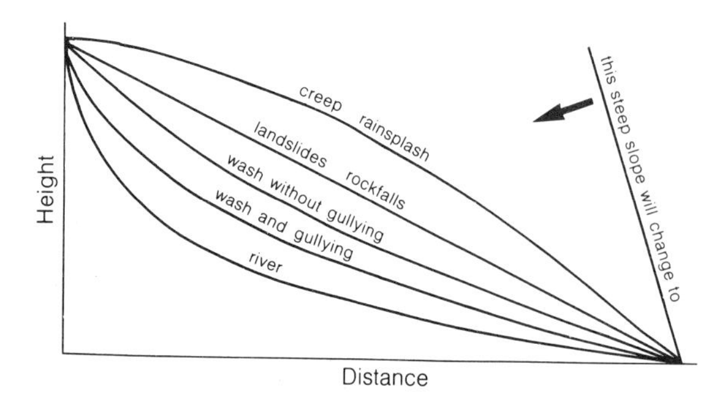

Figure 14.19 The characteristic profile that will develop under single erosion processes. Notice that this classification includes the slopes we call river channels. Those processes that become more efficient down slope result in concave profiles, whereas those that are most efficient at the top of the slope result in convex profiles. Straight slopes develop where processes are of uniform effectiveness.

though all processes are operating to some degree at the same time. Furthermore, hillslopes are only part of a constantly changing landscape and the slope shapes are also affected by processes operating within the river channel: the channel's ability to remove debris ultimately controls the hillslope debris cascade. A gradual lessening of the river's ability to transport sediment will lead to the accumulation of this material first on the river floodplain, as **alluvium**. Then, because material is no longer removed effectively at the slope base, there will be footslope accumulation, as **colluvium**. These changes, and the part they play in the evolution of the landscape, are discussed further in Chapter 16.

A case study

How quickly do slopes change shape, and how close to reality are the simple models described above? Most slopes do not readily yield answers to these questions, but the example given below describes a study of Hadleigh Cliff, Essex, in the past 12 000 years – using radio-carbon dating and historical records as evidence of change.

When Henry III granted a licence to build a castle on an isolated knoll of land beside the Thames estuary marshes, he must have expected its great stone walls to last forever. Yet even by 1240 the walls were cracking, and by the end of the century great sections of wall had collapsed as the land slid from beneath them (Fig. 14.20). Some four centuries later the castle was on the move again, as its southern wall began to slide towards the marsh – and it is still going. Even the northeasterly part of the castle has started to slip and the remains of one tower are kept in place only by a large slab of concrete.

Hadleigh Castle is built on top of a slope entirely made from London Clay. Steepened by river erosion at the base the slope was abandoned by the river about 10 000 years ago, leaving a slope of 26°. However, London Clay is only stable from landslips and slides below 8°, so the castle does not make a good insurance risk.

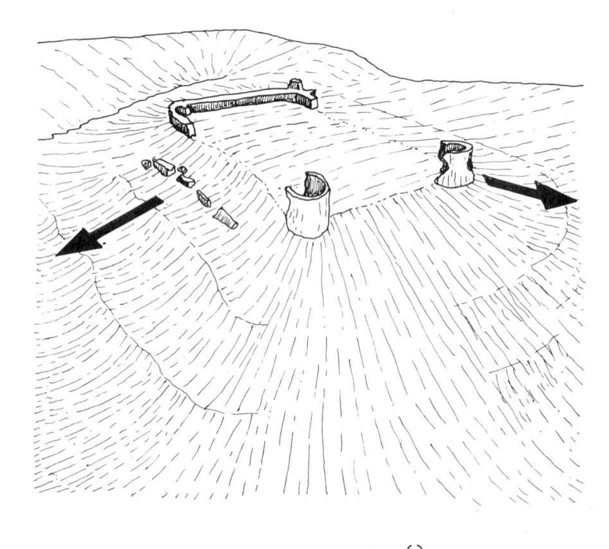

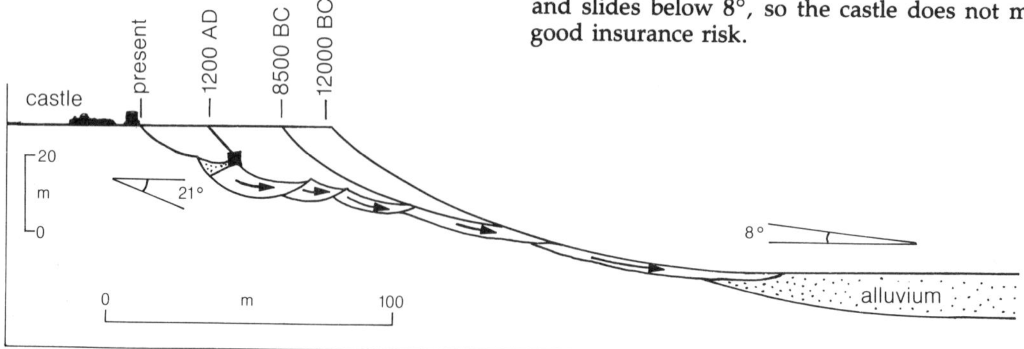

Figure 14.20 Hadleigh Cliff, Essex, showing the progressive landslipping since 12 000 BC.

It seems that two main periods of mudsliding occurred, about 9000 and 4000 years ago, each at a time of wet climate and high groundwater level. The more recent periods of rotational slipping may well be due rather to the extra weight of the castle. Whether wholly by natural process or speeded up by man, the effect on the development of the slope has been the same. Periodically material from the steep upper part of the slope moves downslope until it comes to rest at a shallower angle of about 8°. Once in place this material protects the lower slope, so further movement will occur near the slope crest. This causes the upper cliff to retreat parallel to its original profile while the lower cliff gradually assumes a smaller angle. In the long run the steep part of the cliff seems certain to be eroded away (as will the castle), but it has taken 10 000 years so far and there is still a large region of unstable cliff remaining. It is likely that, as the slope gradually changes shape, landslips and slides will be a feature of Hadleigh for thousands of years to come.

Slopes and man

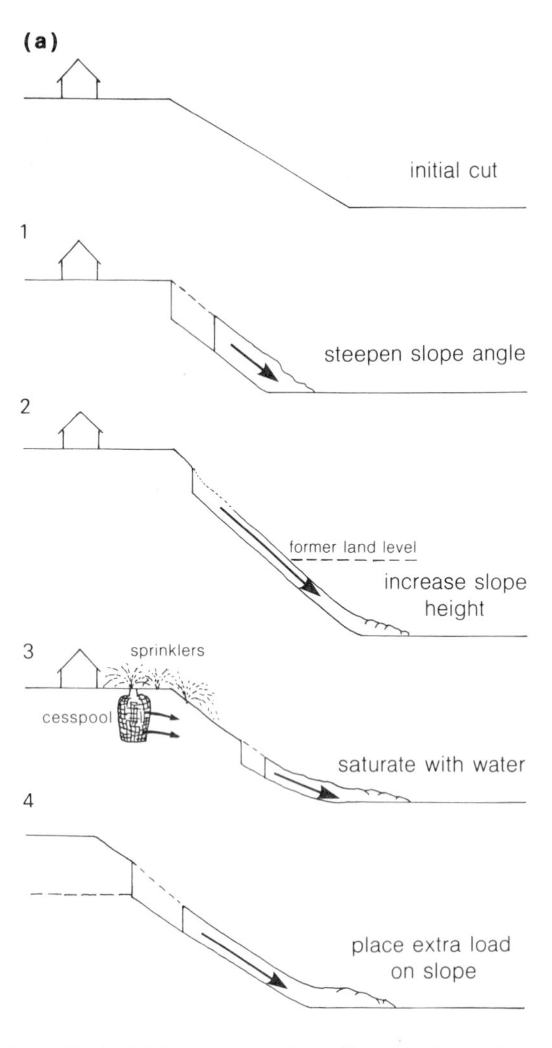

15.1 The importance of slopes

The vast majority of a landscape comprises slopes. Slopes – large or small, steep or gentle – have influenced the geometry of roads, towns and even nations. But slopes are so commonplace that they normally accrue very little interest. Yet they are part of an evolving landscape: slope materials are continually moving, whether it be by the subtle and unnoticed process of leaching or by the dramatic process of landsliding.

Slopes present the greatest hazard when people fail to appreciate certain basic concepts:

(a) the threshold angle of stability;
(b) the intermittent nature of movement;
(c) the buoyancy occasioned by water;
(d) the effect of increasing the overburden pressure.

Lack of appreciation of these basic geomorphological processes has led to many catastrophes. Most slope failures are a combination of several processes (Fig. 15.1), so an integrated approach must be adopted in slope studies.

(b)

Figure 15.1 (a) Four ways to destabilise a stable cut slope; (b) a landslide on a long steep slope, Redditch bypass, Warwicks.

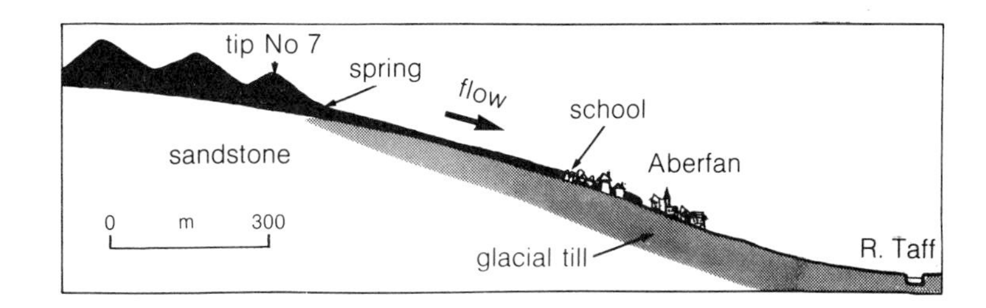

Figure 15.2 The disaster at Aberfan.

The case studies outlined below describe the natural relationships between slopes and their environments, and the results of alterations by man to the natural situation. You should read these case studies as summaries of slope mechanics, referring back to earlier chapters to see how the principles of slope formation and stability are illustrated here.

15.2 Some case histories

Aberfan, 1966

On the morning of 21 October 1966, in the village of Aberfan in the Taff valley, the Pantglas Infants' School was gathered together for morning assembly. High on the hill above, a spring continued to feed water into the base of one of the colliery spoil tips that loomed over the school and village. Heavy rain had saturated Tip No. 7; together with the spring water, this caused the toe of the tip to flow away, destabilising the rest of the spoil and allowing it to flow down the hill (Fig. 15.2). While assembly was in progress 107 000m^3 of colliery spoil engulfed the school and cost 147 lives.

The Coal Board had increased the soil overburden pressure by building a 67 m tip on the top of a steep slope, the lower part of which was of impermeable and unconsolidated glacial till. Further, it had introduced large quantities of water into the tip and promoted internal buoyancy by dumping over a natural spring. The 1966 flow was not the first sign of instability. In 1944 and 1963 small slippages had occurred nearby, but it was thought that all further movement would be slow enough for the villagers to be given warning. The reason the tip failed so dramatically on the last occasion was that the spoil had weathered, and its internal friction had changed.

Pwll Melyn, 1976

The A55 trunk road is the main east–west routeway in North Wales. In October 1970 an embankment was constructed across a small hillside valley to eliminate a bend and so improve traffic flow at Pwll Melyn. Soon after completion of the road a crack appeared along its centre. This was attributed to frost action and settlement and was simply filled in. In 1975 the crack reappeared and was again repaired. In October 1976, after a period of prolonged rain, the embankment failed completely, removing 30 m of the carriageway (Fig. 15.3).

The 1976 failure proved to be a shallow slide. Investigation showed that the subsurface culverts, installed during construction to drain the embankment fill, had been damaged and that water had saturated the embankment fill completely. The name 'Pwll Melyn' – 'yellow pool' in English – is perhaps a reference to a spring that had occasionally emerged at this point. The embankment was built on completely weathered shale: it was later found that, for a factor of safety of 1.0 on this material under saturated conditions, the slope angle of the embankment should not have exceeded 14°. This angle could only have been achieved with an embankment half as high as the one built. In other words, the embankment was built too high, adding too much overburden pressure to a slope made of weak rock which also contained a spring. Rebuilding the embankment could be a success only with much larger drains and a more permeable fill: the fill would otherwise become saturated again and probably fail.

Charmouth, 1960s

Near Charmouth in Dorset, sandy solifluction material up to three metres thick rests on silty clay bedrock (Fig. 15.4). The local slopes of 6°–10° are stable provided they do not become saturated. Slips had occurred in the 1920s, 1930s and 1950s when saturated conditions prevailed for short periods.

In the 1960s the cliff area was developed and a row of bungalows built. To provide a more level site, the small stream was contained in a culvert and its valley was filled in. The culvert was led to a soakaway pit in the middle of the site. This caused water to be fed directly into the solifluction material where, because of the underlying clay, it accumulated and saturated the surface material. Very soon this surface material

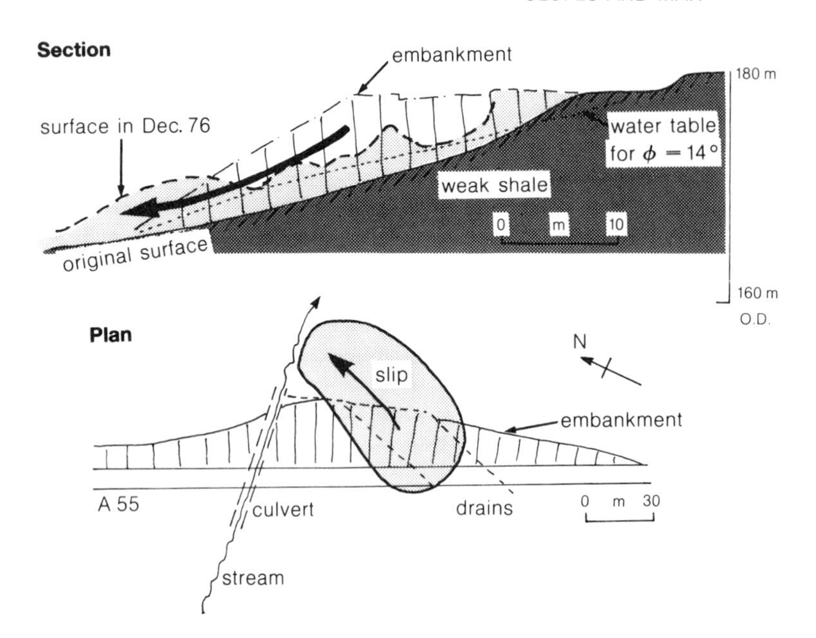

Figure 15.3 Features of the Pwll Melyn Slip, North Wales.

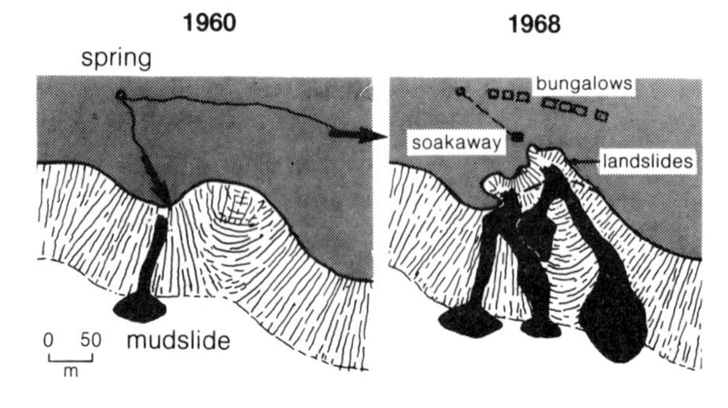

Figure 15.4 Debris flows caused by ill-informed drainage alteration, Charmouth, Dorset.

began to slide over the silty clay and down the cliff, causing the top of the cliff to retreat towards the new bungalows at 5 m/year. Only by diverting the stream completely away from the site was disaster averted.

Oakland, California, 1964

In order to build a major road, a steep slope with a thick mantle of weathered soil was 'benched' by making a cutting along the contour. This left the upper hillslope soil unsupported.

In 1964 there were minor earthquake tremors in the Oakland area and the hillslope materials were disturbed. With the slope cut away there was no restraining mass to prevent the soil moving. A massive shallow slide developed, completely covering the four-lane highway (Fig. 15.5).

Upland Britain, any time

Many roads are benched into steep hillslopes by cutting along the contour. When the slopes are increased beyond their angle of stability movement may occur without an earthquake. The frequent cautionary signs ('Beware falling rocks') along many cuttings bear eloquent witness to widespread instability. Debris may fall onto roads causing a hazard for drivers. As the slope begins to retain its angle of stability, extensive areas of soil and vegetation may be stripped away, making the slope less farmable (Fig. 15.6).

Blaenau Ffestiniog, 1981

Many mountainous areas once subject to glacial erosion are now experiencing the effects of lateral unloading and the possibility of rockfalls. As fractures open in valley-side rocks, frost action begins to prise them from the rock face. The problems (physical, social and political) of undertaking remedial measures to forestall a disaster are well illustrated by the article quoted in the box.

Figure 15.5 The Oakland landslide on Highway 24, California.

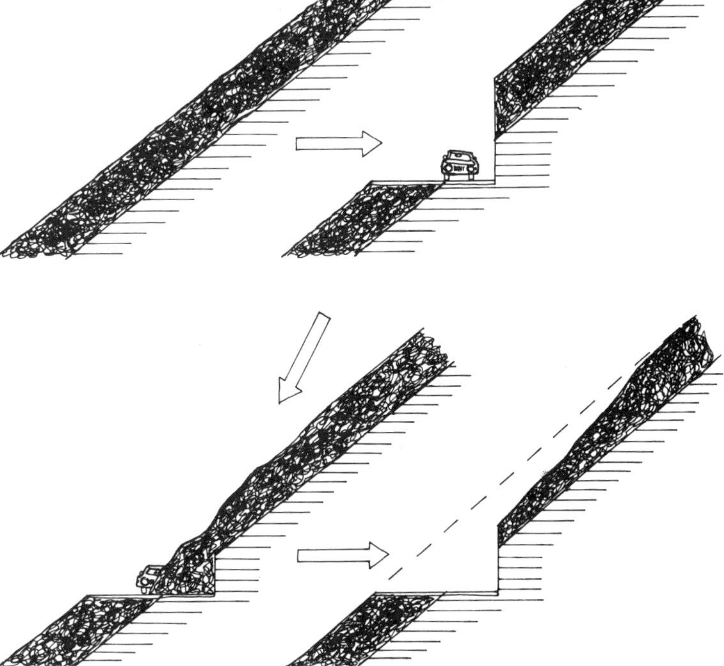

Figure 15.6 A road cutting leading to increased instability: see the text for explanation.

Problem of 100-ton boulders hangs over town

By CHARLES NEVIN

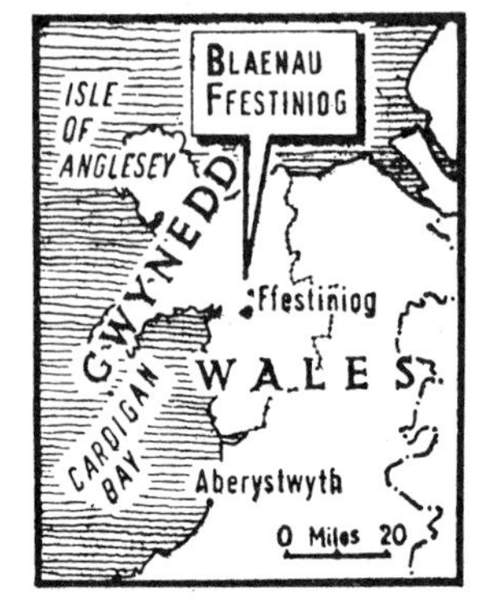

WHEN the good people of Blaenau, Ffestiniog lift their eyes heavenwards, it is not to check on the imminence of the next grey raincloud.

That is considered to be inevitable. In North Wales they say that "if you can see Blaenau it's about to rain. If you can't it is."

What they are scanning is the mountain they call Garreg Ddu (the Black Stone) and its cracked columnar outcrop of granite bwech gwynt (the windy gap) which looms 300ft above them. For it has just been confirmed that a distinctly non figurative threat is up there hanging over them — granite boulders weighing up to 100 tons which could crash down at any time.

The granite has been formed into long flaky columns or joints whose stability lies at the whim of ice, rain, drought, and the errant sheep who roam about it at giddy angles.

"Failure could occur at any time" is the Way James Williamson and Partners, consulting engineers, have put it to Gwynedd County Council and Meironnydd District County Council.

They say the instability of the boulders is increasing and add rather ominously that the failure is "not quantifiable."

Lying beneath in a relationship similar to an ant colony on a golf course is Church Street, Blaenau Ffestiniog's thoroughfare complete with church, pub, petrol station, shops and more than 100 houses.

Speedy action

James Williamson and Partners are the main consultants on the mammonth Dinorwic power station which is being hewed out of the mountains of North Wales nearby.

Their report commissioned by the two councils, and the local threatened National Westminster Bank for £20,000, urges speedy remedial action involving the use of steel nets and cables to anchor the boulders in place.

They recommend that a guard wall should be constructed between the houses and the mountain and suggest that part of the main road through the town should be diverted.

Mr Gerallt Hughes, chief executive of Meironnydd, estimates the total cost at between £500,000 and £1 million.

Neither Meironnydd nor Gwynedd have that kind of money in reserve. Approaches are being made to the Welsh Office and to the EEC, but Mr Hughes is not tremenodusly sanguine. The product of a penny rate in Meironnydd, he pointed out gloomily, is only £37,000.

But he says "the other alternatives, evacuation or no action, appear to be inconceivable in the circumstances."

Mr Hughes' sense of urgency however is not obviously echoed in Blaenau itself where most of the natives are reacting with the same phlegm displayed by residents in the environs of Etna or on the San Andreas fault line.

The sword of Dai-Mocles does not appear to be having the same concentrating effect as its near namesake.

Mr Hugh Jones, a former worker in the nearby slate quarry, turns from his pint in the packed threatened Commercial Arms and said: "I have been a rock man all my life in quarries. Those rocks have always been up there, and they always will be up there."

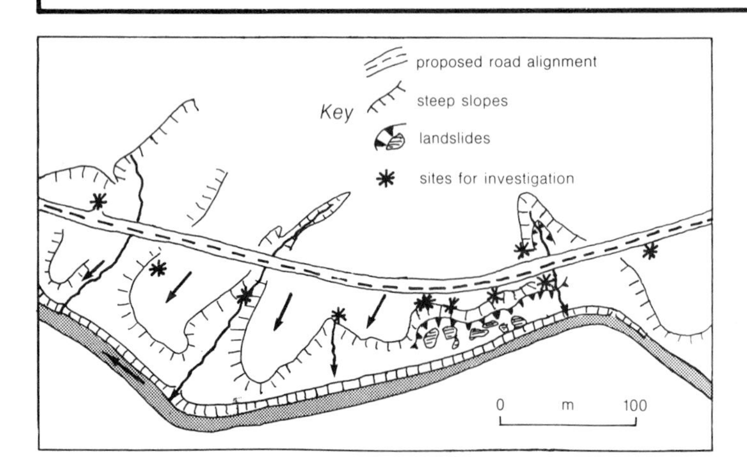

Figure 15.7 A geomorphological map of a riverside alignment on which the subsequent site investigation was based.

15.3 Identifying the hazard

The slope failures outlined above occurred on a variety of scales and over a variety of time periods. All could have been avoided if proper consideration had been given to the nature of slope processes. Urban planners can learn from the lessons of Figure 15.1 and zone their land for varying uses according to its stability.

Both urban planners and road builders can benefit from a geomorphological survey of the area on which development is proposed. Figure 15.7 shows an example of rapid geomorphological surveying for a road along an old railway track. This survey identified the slope processes that were occurring both from natural and man-made causes. Because each *process* was identified, its likely significance on slope stability could be better assessed and correct preventive measures undertaken.

Landscape development in mid-latitudes

16.1 The need for an overall model

In this chapter we will try to approach one of geomorphology's traditional goals – a description and explanation of the development of a landscape fashioned by running water. Perhaps because it is such a fundamental subject, it is also one of the most controversial. In the 19th century, scientists such as G. K. Gilbert began an intensive study of active processes; Gilbert is one of the fathers of the process geomorphology described in the previous few chapters. Process studies are essential if we are to understand the way the landscape develops, and in particular if we are to help people use the land effectively and safely.

If there is no model encompassing all the detailed investigations, geomorphology lacks a central integrating theme. It should be possible for a geomorphologist standing on a hilltop to look out over the surrounding countryside and answer these fundamental questions: Why is it like this? What was it like millions of years ago? What will it be like in the future? The lack of a model – a framework into which all the process studies would neatly fit – can also be an obstacle from a strictly practical viewpoint. As a famous conservationist, John Muir, said: 'When we try to pick out anything by itself we find it hitched to everything else in the Universe.'

16.2 The cycle of erosion

Towards the end of the 19th century, W. M. Davis attempted to provide the same kind of fundamental framework for landscape sculpture that Charles Darwin had presented for biology not many years previously. Just as Darwin had identified an evolutionary pattern from within the great diversity of natural forms that surrounded him, so Davis concluded that the immense variety of landscapes must be variations on a common theme. The basic processes of sculpture – water, wind and ice erosion – were always affected by the same simple laws of physics. It seemed to Davis that some parts of the landscape were at a more advanced stage of sculpture than others. If this was the case, different valley shapes could be interpreted as a *sequence* of shapes changing through time.

Using this argument as a base on which to build a theory, Davis proposed a characteristic sequence of events. He called this sequence the *cycle of erosion*, using the word 'cycle' in the limited sense that the starting and finishing points of the sequence were both landscapes of very little relief. The cycle is simple in outline: as with all models, the principles of development can be seen clearly only by reducing reality to its barest bones. Such simple events are rarely, if ever, found in the turbulent reality of nature, but the simplified model is a useful one. Many people have subsequently refined Davis's original ideas, especially so as to show that, as Gilbert suspected, some types of landscape can survive for a very long time without changing shape dramatically (Fig. 16.1).

The cycle in outline

Davis began by assuming that the 'cycle' could be initiated by a period of rapid land uplift *followed* by erosion. In many landscapes the two events happen simultaneously. On the newly uplifted land a

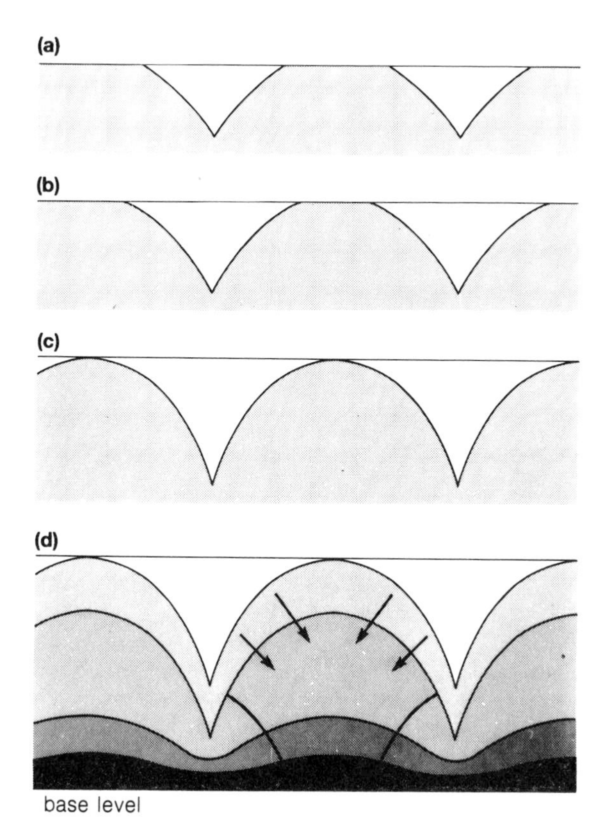

Figure 16.1 The cycle of erosion in outline: see the text for explanation.

drainage network would soon become established and erosion would subsequently cause the development of a pattern of valleys (Fig. 16.1a). The shape of these valleys would reflect the balance between river erosion, slope weathering and transport. At early stages slopes would be short and produce relatively little debris: rivers have surplus energy and so incise rapidly. Slopes would largely be steep and straight and the main transport processes would be landslides and flows. As the cycle develops, hillslopes would lengthen and release progressively more debris for delivery to the rivers (Fig. 16.1b). The need to transport more debris would steadily reduce the surplus energy available for downcutting; hillslope and river would move towards a position of balance. In time valleys would widen until they consumed the last of the original landsurface (Fig. 16.1c). Slopes could not then lengthen any more and the amount of debris reaching the rivers would stabilise. Rivers and slopes would thus enter a phase of approximate balance called **dynamic equilibrium**, whereby both slopes and valley floor are lowered at the same rate and the landscape would appear to undergo no change.

Eventually, however, river downcutting would be hindered as the valley floor is cut to near sea level. With river incision restricted, surplus energy would be used in lateral rather than vertical erosion, and the river would begin to develop a floodplain equal in size to the width of the meander belt. Without steady downcutting dynamic equilibrium would be lost and the landscape shape would change once more. Slopes would continue to be eroded and the material removed by rivers, but without downcutting the hillslopes would be progressively reduced in height (Fig. 16.1d). As slopes decline below the angle at which landslides could operate, the only processes for removal of material would be soil creep and running water and the rate of slope erosion would slow down. If the base level remains stable for a sufficiently long time, a landscape of low relief called a **peneplain** would develop and the 'cycle' would be completed.

Uneven development in the cycle

As Schumm (1979) has remarked: 'The numerous deviations from an orderly progression of the erosion cycle have led many to discount the erosion-cycle concept completely. Current practice is to view the evolutionary development of the landscape within the conceptual framework of the erosion cycle, but to consider much of the modern landscape to be in dynamic equilibrium. There are obvious shortcomings in both concepts. For example, although the cycle involves continuous slow change, evidence shows that periods of relatively rapid [...] adjustment result from external causes. This, of course, is equally true of geomorphic systems in dynamic equilibrium...Hence landscape changes and changes in rates of depositional or erosional processes are explained by the influence of man, by climatic change or fluctuation, by tectonics, or by isostatic adjustments. One cannot doubt that major landscape changes and shifting patterns of erosion and deposition have been due [to all these factors]. Nevertheless, it is the details of the landscape... that for both scientific and practical reasons of land management require explanation and prediction. These geomorphic details are of real significance, but often they cannot be explained by traditional approaches.'

Schumm is saying that because the cycle does not immediately allow the incorporation of short-term irregularities, and because there may be major interruptions due to climatic changes and so forth, the cycle concept has been abandoned by many people. Such complete abandonment, Schumm argues, is not justified. Instead, a closer look at the workings of the cycle is needed, and some of the previously unnoticed details given greater prominence.

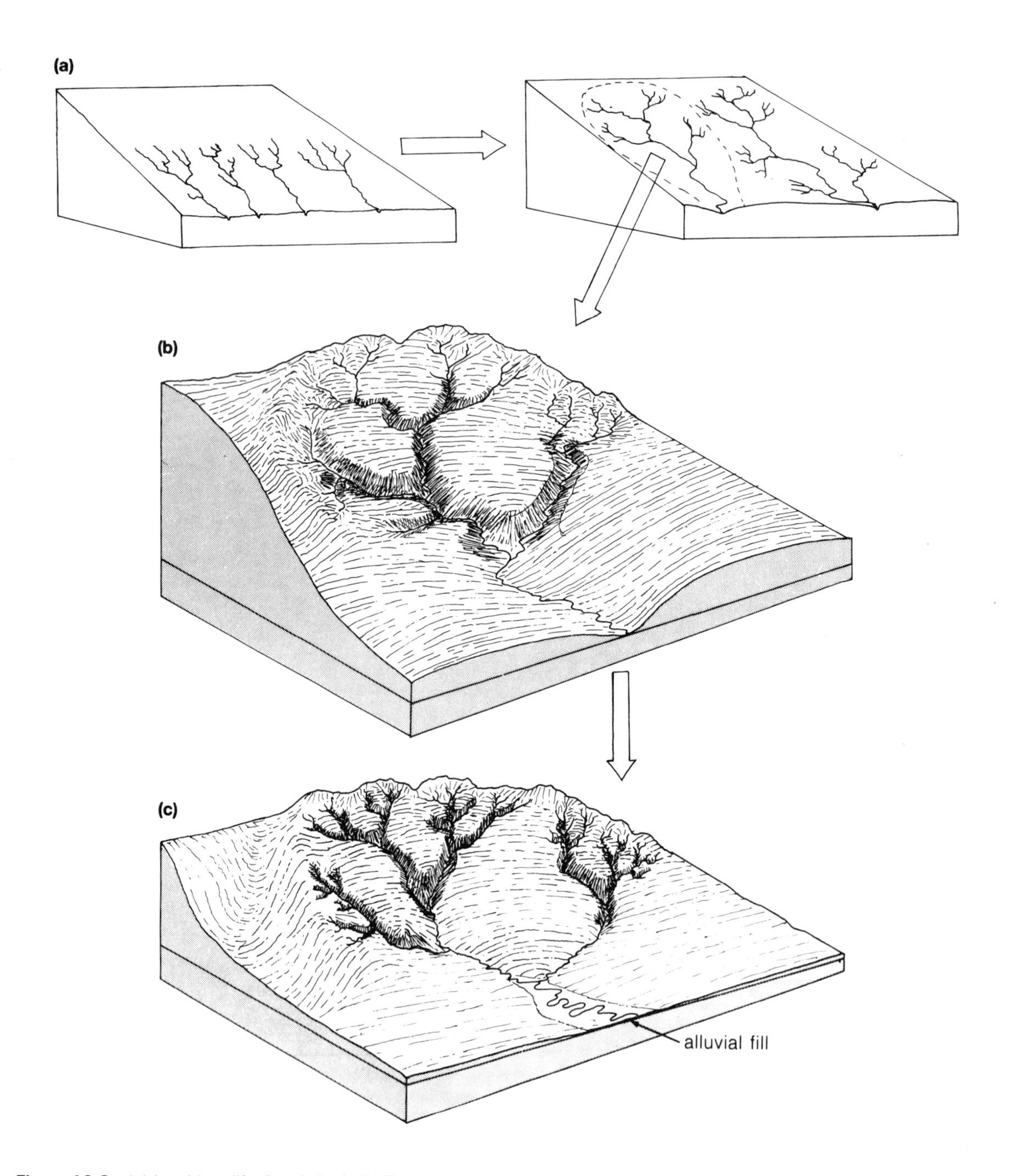

Figure 16.2 (a) Land is uplifted and tilted. (b) The drainage network extends headwards but most sediment is produced from the lower reaches, where hillslopes are the first to decline following a period of dynamic equilibrium. (c) The region of downcutting now proceeds independently and sediment reaching the main stem varies with phasing of contributions. Sediment stored in the lower valley floodplain builds up until it becomes too great: then it is flushed down stream, setting in motion a new pulse of downcutting that works its way through the system – a complex response.

Here is one example. Because the river is larger down stream, it is, in the early phases of the cycle, able to achieve more erosion here than in the headwater reaches. Thus the valley deepens and hillslopes lengthen near the mouth during a period when there is little sediment contribution from headwater regions (Fig. 16.2). Without the burden of extensive sediment from upstream, the river in its lower reaches can quickly cut down to near base level, develop a gentle gradient and begin to cut a floodplain by lateral erosion. As time proceeds, the upper valleys of a basin produce progressively greater quantities of sediment. This sediment is moved downstream to a region whose gradient is too gentle for continued transport. Deposition occurs and alluvium accumulates on the floodplain.

Thus, by considering the unequal rate of development of different parts of a drainage basin, it is possible to explain why deposition as well as erosion is a characteristic feature and why this might occur in an irregular way.

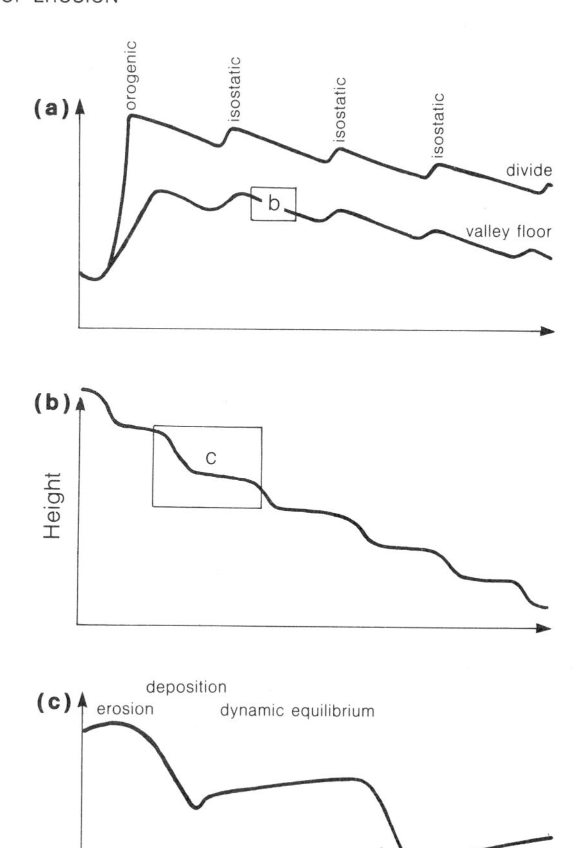

Figure 16.3 In the long term, a landscape is reduced to one of low relief (a) although there may be many interruptions. On an intermediate timescale (b) lowering may be episodic and river terraces may form. On a short timescale the valley floor alternates between short periods of erosion and deposition, and longer periods of dynamic equilibrium (c).

Progress within the cycle

It is possible to begin to see some of the complexities of the cycle by looking at the uneven development in *space*. It is equally important to look at uneven development through *time*.

To understand the complexity that an erosion cycle brings, it is helpful to examine it over different timescales. Over the whole cycle, involving a timespan of many millions of years, any uplifted landmass will be reduced to a landscape of low relief, unless it is completely disrupted by major mountain-building episodes. On a scale of many thousands (or tens of thousands) of years, isostatic forces will play a major role. Over the period of a complete cycle, unloading and episodic uplift will occur many times, each time causing a lowering of the base level and an interruption to the cycle (Fig. 16.3). Each period of uplift will, for example, rejuvenate the rivers, allowing them to cut down into their floodplains and so leave river terraces (Fig. 16.4). Similarly, global changes of sea level will cause either positive (downward) or negative (upward) movements of base level, resulting in either incision or deposition.

The intermediate scale of activity produces noticeable effects on the valley floor, but has far less influence on the development of hillslopes. Hillslope transport may be held up for a while and debris accumulate at the edge of the floodplain as colluvium. However, hillslope processes are not directly dependent on the work of rivers, and they can continue to operate unchanged for a long time. As a result slope processes only change when there are long-term fluctuations in river erosion (Fig. 16.5).

Although 'cut and fill' is the rule of landscape evolution on an intermediate timescale, there are also considerable fluctuations in the rate of erosion and deposition on the short timescale. These are produced by two effects:

(a) the irregular production of sediment from tributaries as each river erodes through rocks according to its available energy;
(b) the need for the trunk-stream gradient to become sufficiently steep to allow the movement of extra sediment (Fig. 16.3c).

Sediment moves irregularly within a river for much the same reason that movement occurs on hillslopes:

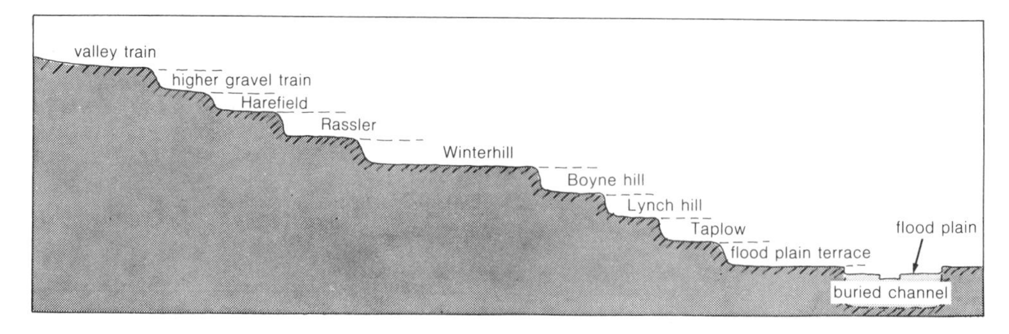

Figure 16.4 The middle-Thames terraces, showing the staircase of near-level surfaces produced by changes of base level.

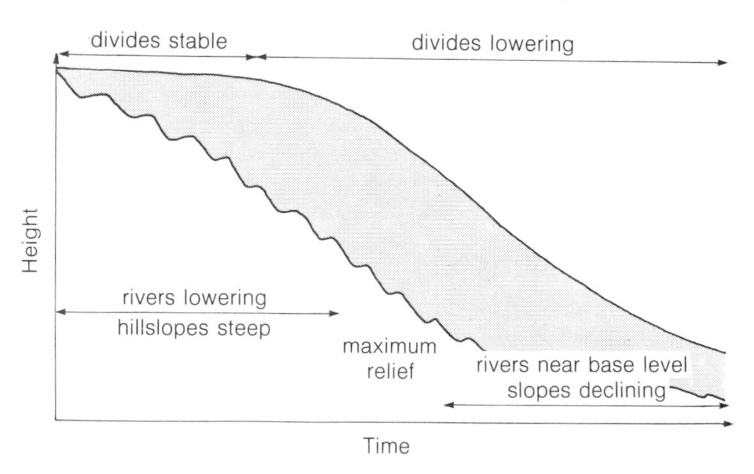

Figure 16.5 A model for valley evolution. The shaded region represents a stage of dynamic equilibrium and may be of long duration. Notice that valley floors are more sensitive to intermediate and short-term fluctuations than are hillslopes.

a steep gradient is needed to set material in motion but, once on the move, it will come to rest forming a surface that lies at a gentle angle. Material is progressively stored on the floodplain until some **threshhold gradient** is established, whereupon a phase of flushing will set in without any external influence (Fig. 16.3c).

The result of many processes operating on different timescales is to make landscape development appear very haphazard and without an underlying unity. But an examination of the cycle during different timescales helps to explain (for example) why apparently similar adjacent rivers produce different sediment yields. The yield simply depends on the point each basin has reached in its short-term cycle – one river may just have begun an accumulating phase; the other may be in the midst of an eroding phase. The concept of threshholds within a cycle further suggests that, when an accumulating phase is nearing completion, even a small storm can change a cycle from an accumulating to an eroding phase. This may explain why rivers sometimes move catastrophic amounts of debris during only moderate floods.

16.3 The influences of rock type and climate

The sequence of events outlined above has been described without reference to the influence of rock type or climate. In practice, this form of the cycle of erosion is most appropriate for a humid temperate climate, in which both physical and chemical weathering processes operate. In detail, the climate and type of rock can profoundly influence the angles of slope elements and the processes involved in their evolution. The main contrasts for slope development with three rock types and two climates are outlined in Figure 16.6.

Most landscapes are fashioned from a variety of rocks, which complicates the pattern of evolution further. For example, river capture will be a major factor during landscape development in regions of dipping rocks. It should always be remembered that we see the landscape today after a recent period of glaciation whose modifications will not fit into a model designed for river erosion. Throughout geological time, the continents have been changing shape and moving over the Earth; many landscapes now show the

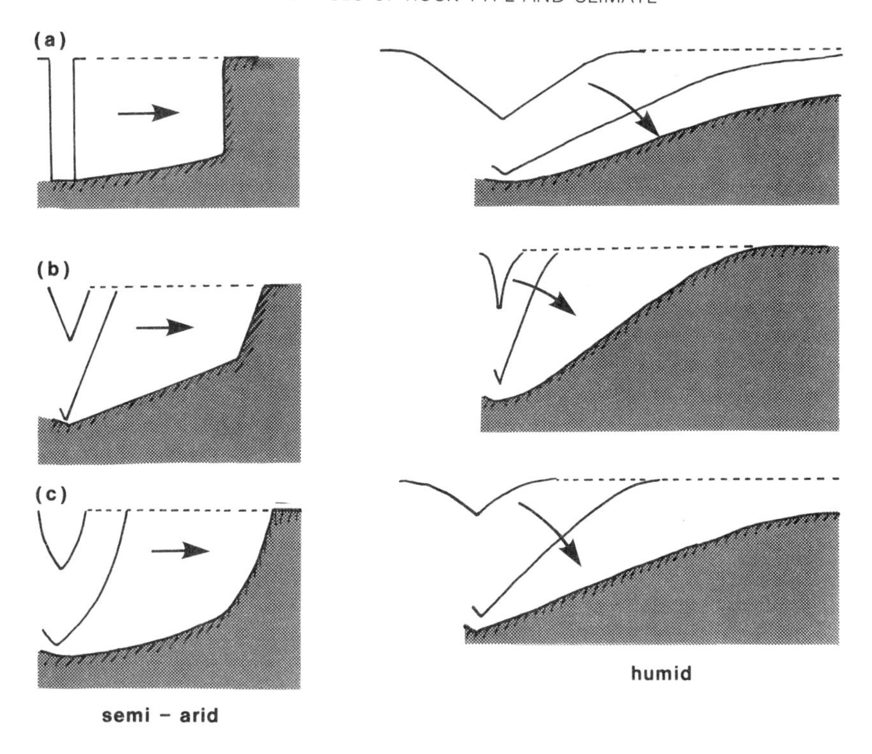

Figure 16.6 The evolution of slopes where rapid uplift is followed by a stable base level. *Under semi-arid conditions*, a vegetation cover does not develop and chemical weathering is restricted. (a) The sandstone will develop by parallel retreat as grains are weathered singly from the rock. These grains are easily washed away from the base of the slope by heavy rainstorms. The basal concavity is thus the result of surface wash. (b) The jointed rock is first kept steep by rock avalanches with a scree slope developing at its base. This scree slope gradually replaces the steep slope and is, in turn, replaced by a low angle 'pediment' (p. 162) resulting from wash of fines weathered from the scree. (c) The clay rock is much weaker and hence is subject to rill and gully erosion. This produces a concave profile which then retreats *parallel to itself* leaving a slowly extending wash slope at its base. Rainsplash will gradually produce an upper convexity. *Under humid conditions* a vegetation cover develops over a soil and chemical weathering dominates. (a, b, c) are very similar in form; all rocks will weather into soil and then the *slope will quickly decline* to its angle of stability by landslides/flows. Soil creep at the top of the slope and solution (and possibly some wash) at the bottom of the slope gradually shorten the straight part of the slope. As a result, the main contrast is in the angle of straight slope which varies with the lithology, being greatest for well jointed rock and lowest for clay.

imprints of more than one climatic regime. Davis's model, and the developments of it outlined above, serve to open our eyes to the great complexity the world has to offer; and to show how the short-term changes that are our immediate concern fit within a logical long-term framework.

Part E

Warm environments

Warm environments

17.1 Sources of variety

Although many of the processes that fashion the landscapes in warm environments are the same as those operating in cool environments, the distinctive landforms that develop in warm environments reflect changes in the relative importance of individual processes. Warm environments occur over a large part of the Earth's surface: the varied climates, rock structures and tectonic histories of the different regions have added to the variety of resulting landscapes. Thus the Amazon basin (which is a sedimentary basin of low relief) should not be lumped in with New Guinea (a mountainous region) just because they experience a humid tropical climate. Similarly, the intensely folded structures of the mountains of Iran yield a landscape quite different from the nearly level sediments of the central Sahara, even though both experience a desert climate.

In this brief discussion of the landscapes of warm environments, it is not possible to do more than dip into a plethora of landscape types and to try to pick out, identify and explain some of the more distinctive elements of each region.

17.2 Humid regions

The humid tropics are characterised by high temperatures and a moisture surplus in every month. This gives a predominance of chemical over mechanical weathering. Weathering is very intense and, except on quartz sandstones and quartzites, clay soils develop to depths of tens of metres.

Rainfall is characterised by intense convective storms which produce both throughflow and considerable overland flow. Most soils are saturated and landsliding is therefore common on steep mountain slopes. Rainsplash and overland flow occur in most situations. Massive erosion by sheetwash and gullying is being resisted only by a dense network of tree roots near the surface: each exposed root acts as a dam and stores sediment behind it.

Because of the dominance of chemical weathering, the slow rate of transport on all but the steepest slopes, and the reduction of rock to either quartz sand or clay, rivers have few tools with which to abrade their channels, and little more than a surface scouring occurs. Erosion of river channels in all except mountain areas is chiefly by chemical corrosion of the bed. Thus river long profiles are very different in the humid tropics from their counterparts in temperate areas. Rivers rarely come into any form of 'grade' (equilibrium), with each section being adjusted to match the sediment delivered from above. Instead of a smooth concave upward long profile, rivers have numerous rapids and waterfalls (Fig. 17.1) separated by quiet reaches of such small gradients that even meandering is suppressed.

Rivers do not incise into their beds, nor do they erode headwards. Waterfalls and their downstream gorges are *not* the result of waterfall recession. All the features result from differential weathering: the gorges are etched out along lines of fractures or other lithological differences in otherwise massive rocks. Rivers tend to flow over the *sides* of gorges rather than from their ends, in no way adding to gorge recession. For example, the waters of the Angel Falls in Venezuela drop some 800 m from a rock bar made of quartz sandstone which appears to be virtually immune to erosion.

In the humid tropics rapids are much more common than waterfalls, occurring wherever a resistant rock bar lies astride the river course. Lowering of the bar must await the relatively slow progress of chemical weathering, and thus the degree of stream erosion as a whole depends on chemical weathering of the most resistant material. Less resistant rock

(a)

(b)

Figure 17.1 (a) For all their apparent energy, rivers in the humid tropics do not erode waterfalls, the gorge into which the water plunges being a result of differential weathering. (b) Rapids are a much more common feature but again, there is little impact by fast-flowing water on river bed erosion.

weathers to great depth, but the river channels flowing over it incise only at a rate that matches the lowering of the rock bar. Vast areas of low-resistance rock thus weather and transport their debris to rivers without any matching incision. For this reason many landscapes evolve towards low plains, much like broad treads of a staircase, the short risers of which are made from the rock bars.

There is a large contrast between the landscapes of tropical humid mountains and plains. In the mountains, landslides carry unweathered rock quickly to the rivers and provide the coarse debris that is used in channel erosion. Braided rivers occur in the mountains of New Guinea, but elsewhere the alluvium is either fine-sand residue from granitic rocks or clay from weathering.

Many of the world's great tropical rivers (Amazon, Ganges, Mekong, Irrawaddy) flow across regions of tectonic subsidence, creating vast alluvial plains. During floods, the clay particles suspended in the water are carried great distances from the river channel, spreading out as a thin veneer to be trapped later by the roots of trees. Without a substantial coarse

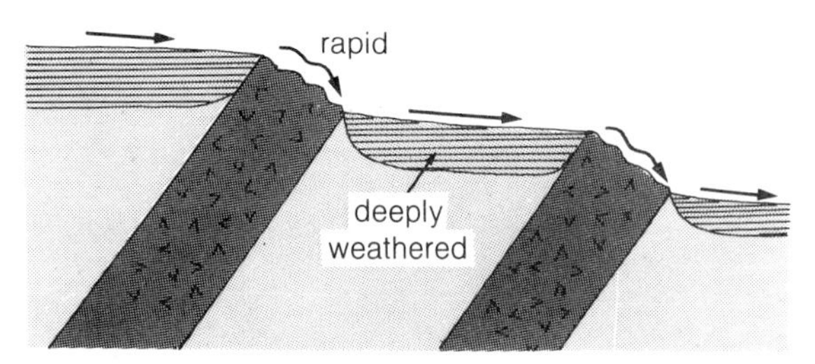

Figure 17.2 Long profiles of rivers in the humid tropics are characteristically irregular, rapids alternating with nearly flat stretches.

sand bedload, levées are not a common feature of river banks.

Because differential weathering is the key process in the development of landscapes in the humid tropics, it is important to consider both the rate of *weathering* of the *rock* (taking place many tens of metres below the land surface) and the rate of *erosion* of the *surface*. Differential weathering can occur even within a region of homogeneous rock, weathering proceeding more rapidly in those places where the jointing density is greatest. Variations in the depth of weathering are often not at all apparent on the surface, and rivers may or may not flow over the region of deepest weathering (Fig. 17.3). However, as the surface is gradually stripped away, those areas with the least frequent jointing (which are thus the least weathered) are gradually exhumed as **domes** (Fig. 17.4). Once the rock is exposed and water is no longer permanently in contact with the rock, the rate of erosion declines dramatically, and the continued lowering of the adjacent areas gradually allows the domes to gain greater prominence. The granite 'sugar loaves' of Rio de Janeiro display classic rock-dome shapes.

For domes to become exposed, surface erosion must accompany the weathering. At present the rate of surface lowering is very slow and it is believed that much of the surface stripping occurred in an early period of alternating wet and dry seasons (see below).

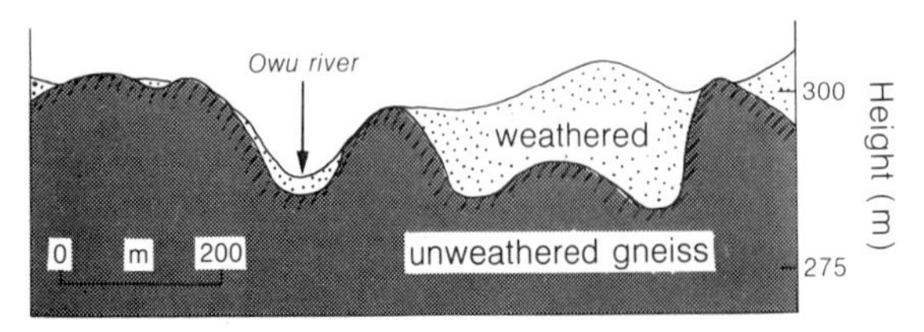

Figure 17.3 The Owu River valley in Nigeria has wide, gentle slopes except where it is contained between two resistant masses of gneiss. Notice how the land surface bears little relationship to the underlying bedrock.

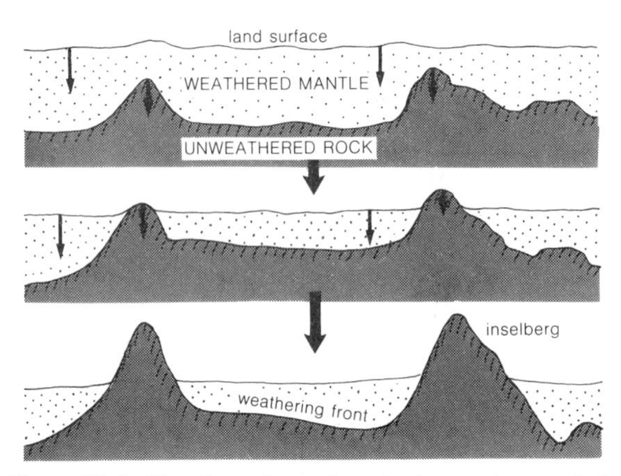

Figure 17.4 The formation of rock domes by gradual exhumation of the more resistant rock areas. Note that the arrows indicate relative rates of lowering.

17.3 Alternately wet and dry regions

Regions with a prolonged dry season are not subjected to the deep, intense chemical weathering of the humid tropics. Instead weathering is often incomplete; and although clays form, they are often washed out, leaving surface materials with a sandy texture. In the savanna, seasonal streams do not attain a graded profile because there is insufficient water to transport and sort the debris delivered from the uplands, and lowlands are often regions of considerable deposition. In many cases there is no distinct channel; when the rains come there is widespread flooding. Nevertheless, much fine-calibre material is removed, and many of the most important landscape features rely for their formation on both deep weathering and effective transport.

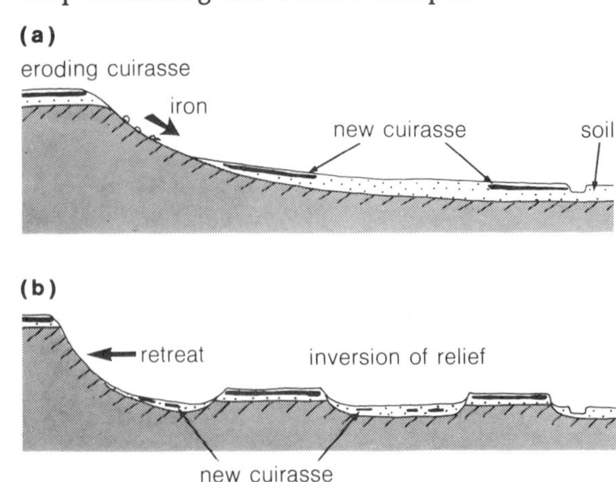

Figure 17.5 The formation of tableland landscapes in savanna.

Cuirasses – resistant sheets of rock-like material that mantle many plateaux – are the most characteristic feature of the stable plainlands. They develop from localised enrichment of certain soil layers in iron, aluminium or other 'cements' (Fig. 17.5). They are most commonly developed in sandy soils, where high permeability aids the transfer of iron and aluminium from slopes. When cuirasses have formed they are both acid and almost impermeable. As a result they become progressively less able to sustain vegetation and thus more easily exposed by erosion. As a cuirasse forms it tends to create the circumstances leading to its own erosion. When the cuirasse (in the soft state called **plinthite**) is exposed, however, it hardens into a rock-like material (then called **duricrust**). In the hardened form cuirasses play a considerable stabilising role. Although they are discontinuous crusts, they protect the underlying material from erosion and tend to cause inversions of relief. There are very extensive duricrusts in Africa, and they are especially prominent on the Jos Plateau of Nigeria.

Inselbergs (German for 'island mountains') are also common features of the seasonal tropics (Fig. 17.6). Their origin is similar to that of the humid tropical domes, but they form in a much wider variety of rocks because weathering transport in the seasonal tropics is sensitive to smaller changes in lithology. Under conditions of sparse vegetation, weathered debris is rapidly stripped from the more resistant elements of the landscape by high-intensity convective storms; it is spread as a veneer on the surrounding lowland. In turn weathering over the plains produces fine-calibre debris which can be transported over the gentle slopes. Thus it is probable that both inselbergs and the surrounding lowland plains are the result of differential weathering. They will continue to form as long as there is sufficient overland flow to remove material from the plains.

17.4 Dry regions

Truly arid lands are characterised by low rainfall of extreme irregularity, no rainy season and high daytime temperatures. Rainfall occurs as high intensity convective downpours of restricted extent; there is little opportunity for stream channels to become established except where there are very steep slopes.

Figure 17.6 An inselberg formed in gneiss rises from an etchplain in northern Kenya.

Chemical weathering is severely limited by the absence of water and surface materials are characteristically coarse. Little material weathers below fine-sand size or suffers abrasion during sandstorms to sizes smaller than silt. Fine material tends to be removed by wind and deposited (as **loess**) beyond the arid zone. The coarse surface materials have a very high infiltration capacity and thus inhibit prolonged overland flow (Fig. 17.7).

Overland flow begins when rainfall exceeds infiltration capacity in the area of the storm. With little vegetation to restrict movement, overland flow in mountainous areas usually has a very high sediment load. River flow in a desert often begins with a 'wall' of water which, because it contains very high levels of sediment, has the consistency of liquid cement. With constantly high percolation into the dry riverbed, rivers rapidly increase in viscosity, slow down and then stop flowing within a few kilometres of their source. As a result rivers rarely flow far from the confines of mountains, and mountain fronts are therefore regions of extensive deposition. Deposition usually takes the form of vast **alluvial fans** (Fig. 17.8). Sometimes fans may be sufficiently extensive to coalesce and form a continuous wedge of sediment (**a bajada**) along a mountain front.

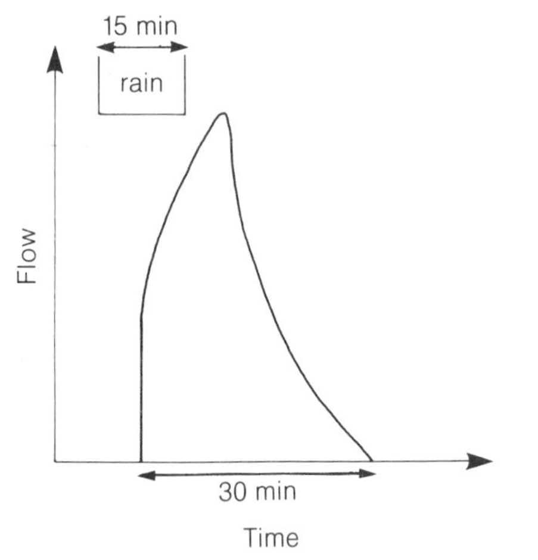

Figure 17.7 Hydrographs in the arid zone typically begin with a vertical rising limb (that represents a wall of water) and a speedily declining recession limb. Both effects are due to the high rate of bed infiltration.

There is no active mechanism for transporting coarse debris across plains, and the debris is therefore left littering the land surface. Some rocks weather to release sand grains directly; strong winds may bounce these along the ground (a process called **saltation**) causing sufficient abrasive collisions (**attrition**) for the grains to be reduced to silt size, after which they can move in **suspension**. Alluvial fans provide another source of fine-grained sediment.

Sand seas form about a quarter of all arid lands (Fig. 17.9). The sand, gathered from debris-strewn tablelands, river beds and alluvial fans, is fashioned by the wind and its eddy patterns into many delicate and symmetrical forms (Fig. 17.10a). Sand is produced both by rock weathering and by attrition of sediment load during flash floods. Its removal by the wind (a process called **deflation**) helps to explain the commonplace surface armouring on most hamadas.

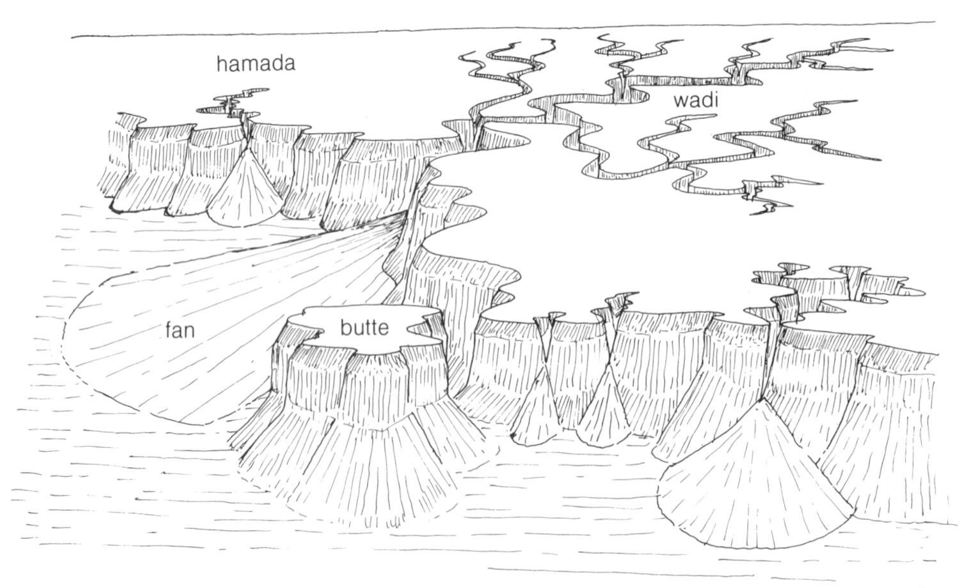

Figure 17.8 Sketch of a wadi system cut into an upland desert plateau south of Cairo, Egypt.

Figure 17.9 Desert landforms are often closely related. The sand sea in the distance derives its material in part from the weathering of the nearby hamada and wadi sides. The hamada, in this instance, is floored by a cuirasse of silica. It also forms the cap rock to the wadi side, collapsing and falling down slope as large boulders. An ephemeral stream channel between sand sea and hamada contains sufficient groundwater in its sediment to support palm trees.

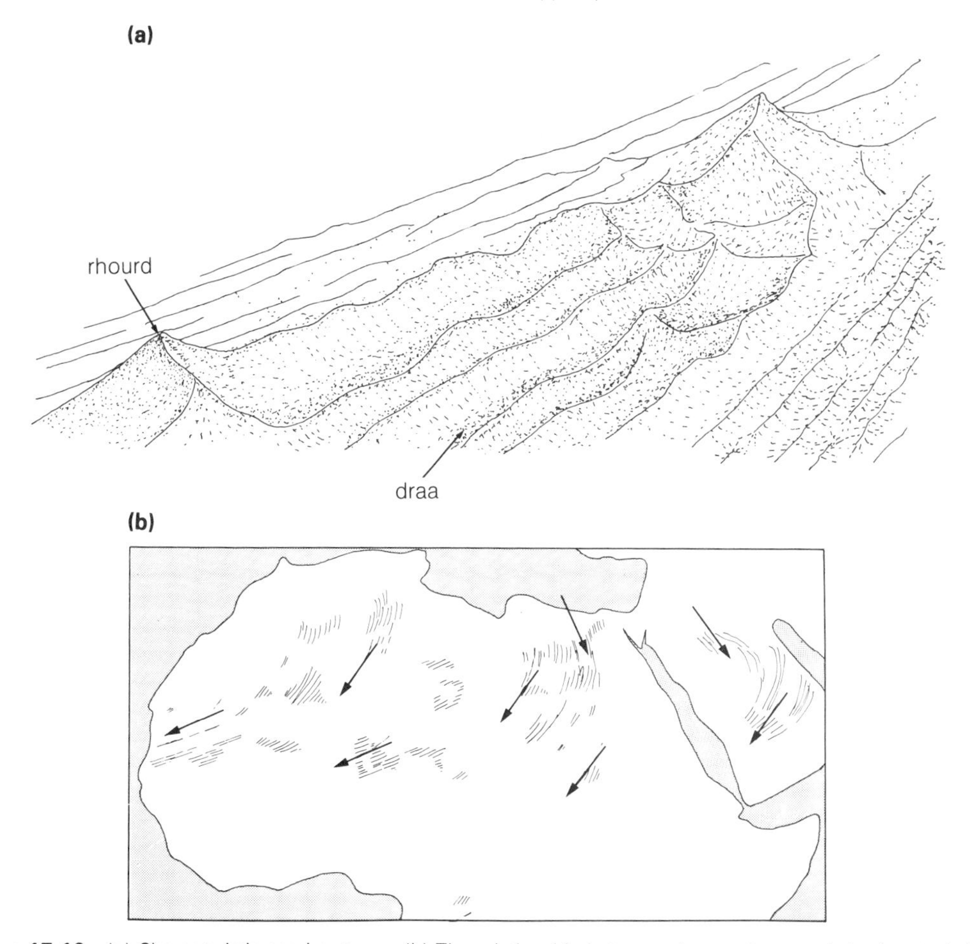

Figure 17.10 (a) Characteristic sand patterns. (b) The relationship between dune patterns and dominant winds.

(c)

Figure 17.10 (c) Yardangs etched into the rock surface of the Bahrain desert.

Sand seas depend on a continuing supply of debris from exposed rock surfaces or river alluvium.

It used to be thought that most of the landforms in the arid zone were the result of abrasion by sand grains carried by sandstorms. It is now realised that saltation can only modify features for a couple of metres above the ground surface. All the remaining curiously rounded forms are the result of an intricate pattern of differential weathering. Indeed, the main result of wind action is to reorganise weathered debris into the distinctive pattern of bedforms called, in increasing order of size, **ripples, sand dunes, draa** and **rhourds** (sand mountains). In restricted areas – in which level plains and constantly blowing (Trade) winds prevail – there are extensive areas containing wind-scoured elongated grooves in exposed rock surfaces, known as **yardangs** (Fig. 17.10c).

One can overemphasise the differences between the dry (desert) lands and the other parts of the tropics. Landforms in warm environments possess so many common characteristics that it may be more appropriate to think of them as parts of a sequence. Thus the domes of the humid tropics, the inselbergs of the savanna and the isolated hills of the desert (called **mesas** and **buttes**) may have many common formative elements.

Three quarters of the dry lands are bare rock with only a veneer of sediment. In tectonically stable regions, rock deserts contain prominent buttes, mesas and escarpments separated by extensive low-lying plains (Fig. 17.11). The most enigmatic of all these landforms is the **pediment** – a gently sloping rock surface of less than 5°, partly covered by a veneer of rock waste, that slopes from the base of a mountain front, mesa or butte and is developed on resistant rock in an arid region. Pediments commonly abut mountain fronts that rise abruptly at angles exceeding 20°, yet the mountains and pediments are frequently formed from the same resistant rock.

Traditionally the steep slopes have been viewed as forming by parallel retreat, leaving gently sloping surfaces in their wake. These gentle slopes were only sufficiently steep to allow the transport of weathered debris. Recent investigation has shown that the rock below the pediment debris cover is irregular, and that the apparent smoothness of the surface is entirely due to the debris veneer. Current rates of scarp retreat have been estimated at less than 8 mm/1000 years. Scarps are often many kilometres apart, so either pediments are phenomenally old or they did not form by retreat, but rather by differential weathering between regions of rock with contrasting

Figure 17.11 A rocky desert surface covered with a thin debris veneer in Arizona. Compare this photograph, which shows mesas in the distance, with Figure 17.12. From the ground, stream patterns are not apparent.

jointing patterns. Mesas and buttes may therefore share with inselbergs a common origin.

Many dry lands show patterns of channels formed by stream activity although, as with pediments, their origin is somewhat obscure. For example, high-density stream patterns are frequently etched into the thin debris veneer covering rock tablelands (**hamadas**) and pediments (Fig. 17.12). Indeed the debris veneer was probably transported by streams. Nevertheless, the encroachment of a sand sea (called an **erg**) over the stream network in Figure 17.12 suggests that overland flow is les effective than it used to be. Similarly, the deep gorges (**canyons** and **wadis**) that cut meandering dendritic patterns across some hamadas (Fig. 17.8) could be readily associated with a perennial river in a humid region. Flash floods that flow intermittently in these gorges carry huge debris loads and become so viscous that, if unconstrained by the landscape, they erode straight rather than meandering channels. This is further evidence that the arid land-forming processes seen today are not adequate to explain the formation of many prominent landscape features.

Figure 17.12 A high-density stream network suggests that the land surface has been subject to considerable overland flow. However, sand dunes have subsequently drifted across the rock plateau, confirming that the stream network is not currently effective at debris transport. The sand dunes show both linear and star-shaped patterns, characteristic of accumulation under varying wind patterns.

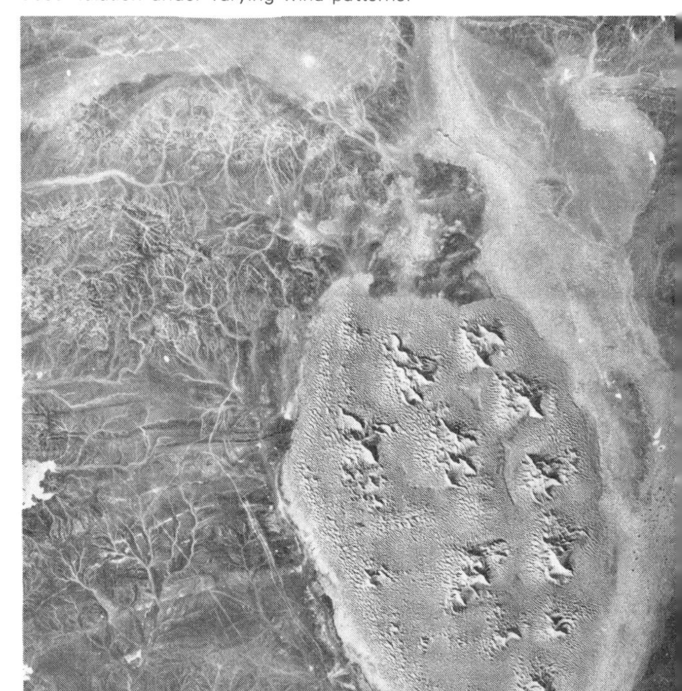

17.5 Warm coastal regions

Mountains and hills seldom reach the sea in the tropics, and most coastal margins are formed of river sediment and ocean-floor debris reworked into fringing coastal plains. In contrast with temperate regions, therefore, the majority of coastal sediment does not come from cliff erosion.

Deltas are more common than estuaries because rivers bring much sediment in suspension to the coast, where it rapidly accumulates, compensating for the postglacial rise of sea level. The gentle gradients of tropical deltas, however, do not allow any coarse particles to reach the sea: most of the beach sand derives either from erosion of coral or from offshore processes.

Coastal landforms in the tropics owe much to changes in sea level during the Ice Age. Sea levels fell at the onset of cold conditions and large expanses of seabed were exposed to intense tropical weathering. As sea level rose postglacially, the coarsest weathered material was driven ahead to accumulate as extensive offshore barriers.

Tropical seas have large swells generated by mid-latitude storms. Longshore drift is consistent in direction: it builds up major spits and in general tends to produce a smooth coastline. Vegetation and marine organisms further smooth out the coast. Mangroves, for example, grow on intertidal muddy ground, their long stilt-like roots reducing wave action and encouraging deposition (Fig. 17.13). Mangroves often colonise sufficiently to form a continuous coastal barrier and so help coastal plains and deltas to grow quickly. In many areas calcareous sands, often a mixture of quartz sand and coral sand, cement themselves together on the beach and become a rock-like mass which further protects the coast.

Limestones form the only significant sheer cliffs in much of the tropics, being undercut by an almost continuous notch (Fig. 17.14). Note that this is not a *wave-cut* notch, but primarily a *solution* feature.

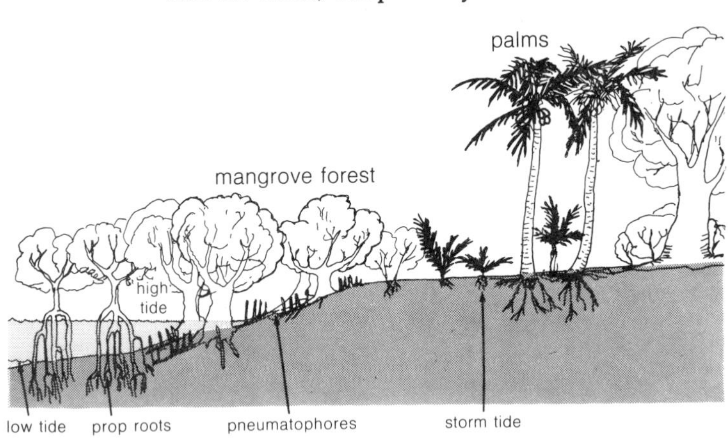

Figure 17.13 Mangroves colonise coastal margins, their roots helping to trap sediment (Malindi). Those nearest the sea have extensive downward-growing 'prop' roots; those in more protected situations have roots sticking up like the teeth of a comb.

Figure 17.14 A solutional notch eroded in limestone, Mombasa.

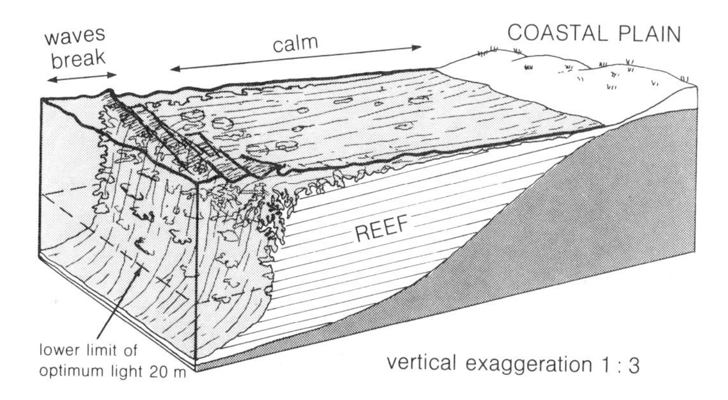

Figure 17.15 A typical fringing coral reef, showing outward growth and the important role of coast protection.

Most corals are restricted to tropical seas. They grow most readily in water temperatures between 25 °C and 30 °C and in water that receives adequate sunlight (less that 40 m deep). Corals cannot grow in muddy waters near deltas. By continuously growing outwards in search of the plankton floating in the ocean currents, successive generations of coral polyps build a **fringing reef** (Fig. 17.15). Fringing reefs are transformed into **barrier reefs** or **atolls** if there is a relative rise of sea level so that corals are forced to grow upwards to remain within the light. Barriers occur widely along the Indian Ocean coast of Africa and the Pacific coast of Australia (notably the Great Barrier Reef); and both barriers and atolls occur on slowly subsiding volcanic islands. Reefs protect the nearby coast and take the full force of the ocean waves, thereby encouraging deposition and the formation of coastal plains.

17.6 Difficulties in interpretation

At present many of the landforms of warm environments still provide more questions than answers. In all cases it appears that landscapes owe much to former climatic regimes and that they have to be interpreted as complex forms, in just the same way as many of Britain's landforms have to be interpreted partly as the result of processes active in the recent Ice Age.

Man and warm environments

18.1 The problems

Warm environments encompass so large a part of the world that a complete discussion of man–environment relationships would be lengthy. However, many of the problems of the humid tropics, such as extensive landsliding, are similar to those of temperate lands. In this chapter attention is focused simply on those problems that are unique to the warm environment. These problems, which are specific to deserts, are:

(a) the effects of advancing sand dunes;
(b) flash floods ;
(c) water supply.

18.2 Sand dunes

Few sand dunes move rapidly; and many sand seas occupy regions where seasonal changes in wind direction act to stabilise the area occupied by sand. **Barchans** (crescentic sand dunes) are possibly the most mobile sand feature (Fig. 18.1): these advance continually in the same direction as the sand, which is saltated up the windward ramp and then deposited at its threshold angle of stability on the leeward side. Although Barchan 'fields' are not common, they can be a great problem to airports and other structures. Indeed, any sand on a runway can be a major hazard. Furthermore, whereas the seasonal advance of sand may be of *relatively* small extent, this can still be sufficient to cover roads or fields and be a major problem.

In areas where rainfall is greater than about 150 mm/year it is possible to grow *Acacia* and *Eucalyptus* species on sand dunes. These species have deep rooting systems which help to stabilise the sand and to retard desert encroachment. For these species to become established, though, the surface must already be stable and active sand transport must be prevented. Short-term stabilisation can be achieved by spraying the dunes with a mixture of heavy oil and synthetic rubber, although when newly sprayed this is extremely unsightly.

Many areas have rainfall below 150 mm/year; these will not support stabilising vegetation. Sand control

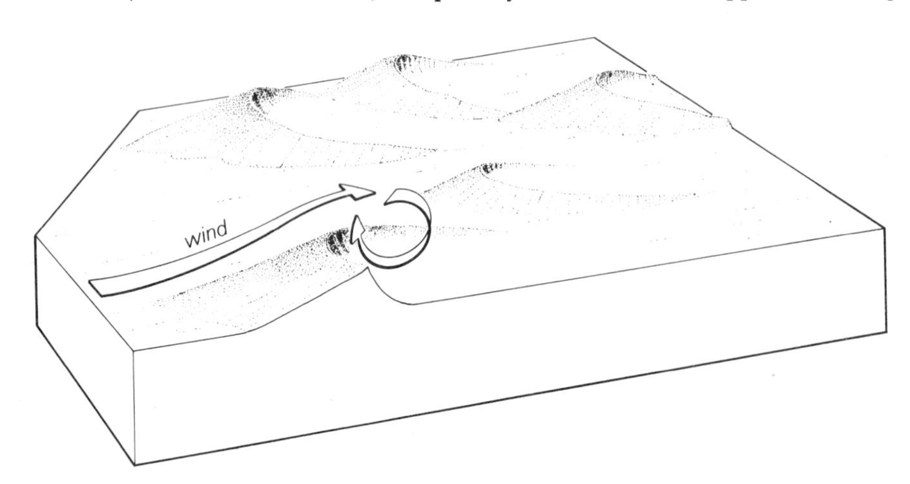

Figure 18.1 The method of barchan dune advance.

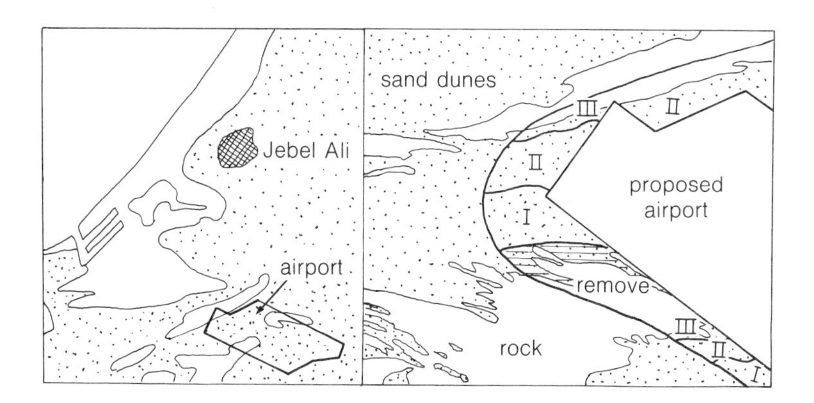

Figure 18.2 A geomorphological assessment of the sand hazard at Dubai airport revealed the areas that required stabilising and removing. Zones are marked I, II and III in order of increased hazard.

must then be achieved by artificial stabilisers, sand diversion or sand removal. Removing sand by bulldozer and lorry is not only an inelegant solution; it is expensive and is rarely successful long-term. Of longer term benefit is the stabilising of the sand with semi-permanent fencing, the area leeward of the barrier being paved to encourage wind movement fast enough to sweep the area clean. Sand can also be stabilised by repeated oiling. This is cheap, quick and effective, but has little else to recommend it.

Airports suffer particularly severe problems from blown sand. In the construction of airports considerable expense can be saved by locating them away from the most mobile dunes and by employing the stabilising techniques outlined above (Fig. 18.2).

18.3 Flash floods

Flash floods are a particular hazard if near mountain fronts or along paths used by intermittent streams. The effects of flash floods on man are far greater than those of sand-dune encroachment. For example, much of Los Angeles is built on shallow channels that are periodically flooded by storms in the Santa Barbara hills. At the time of the city expansion the problems of flood inundation were not fully realised; large amounts of money have been spent since on protection measures.

In the area near Palm Springs, California, many holiday homes have been built in the desert. Builders have sometimes constructed whole townships on alluvial fans, partly because these provide elevated positions and therefore improve the view from the houses, partly because they are away from the intense heat of the desert floors. These settlements suffer a considerable flood hazard, made more acute because, by definition, there can be no flash-flood warning.

Today considerable planning accompanies any large-scale desert development to minimise the cost of protecting against the flood hazard and to enable appropriate protective structures to be built. The proposed expansion of Suez City in Egypt, for example, would make many areas liable to flooding (Fig. 18.3). By carefully mapping the surface deposits and relief it was possible to identify the areas of major hazard and so provide the city planners with valuable information.

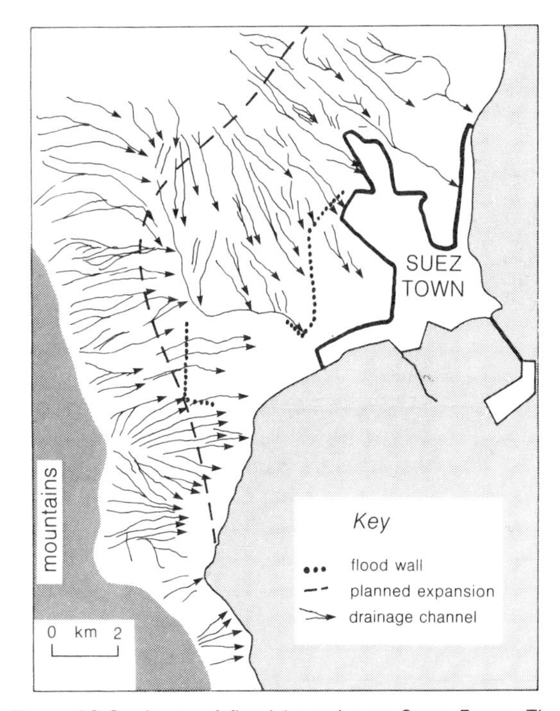

Figure 18.3 Areas of flood hazard near Suez, Egypt. The map shows the planned expansion of Suez and the hazards at each site. Notice particularly the flood protection walls.

Figure 18.4 The valley of the River Ziz, Morocco contains extensive date palm plantations, together with irrigated fields for cereals. Water is derived from the exotic river that receives water in the Atlas Mountains. Both settlement and agriculture are severely constrained by water supply in this environment and concentrate on the wadi floor.

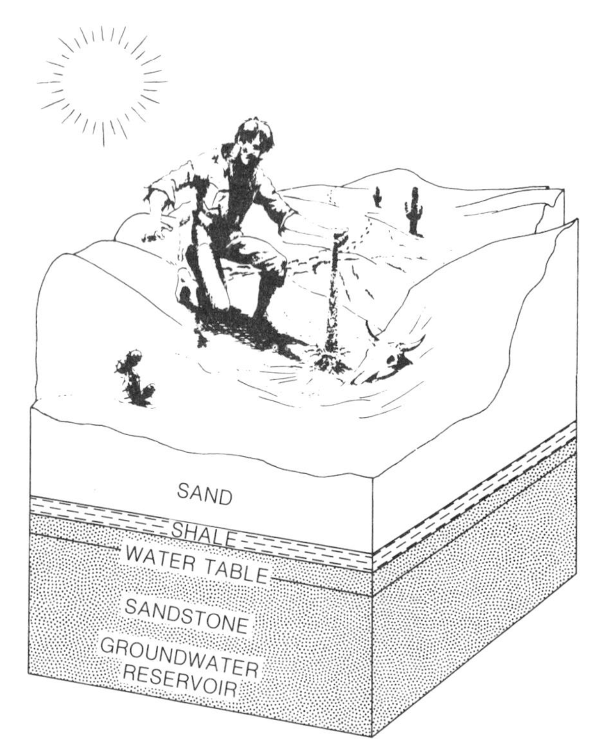

Figure 18.5 Groundwater reservoirs – the region's unseen waterhole.

18.4 Water supply

Achieving a constant and reliable supply of water is the most taxing problem for man's survival in the desert. Only rarely is it possible to bring water overland to the desert, and although the major world rivers that flow across desert regions (the Colorado, the Rio Grande, the Nile and so on) contain extensive diversion schemes and aqueducts, they can provide water only to relatively small areas (Fig. 18.4). In general, therefore, the only way to obtain water is to exploit the groundwater reserves (Fig. 18.5). Groundwater recovery is often a complex business involving energy-consuming drilling equipment and pumps, and often requiring desalinisation plants – many aquifers contain saline water.

Part F

Cold environments

Introduction to cold-climate environments

19.1 The effects of a cold climate

Probably no natural event stirs the imagination more than an **ice age**. For many people the words 'ice age' produce an image of ice first forming in distant mountains and then surging forward to spread as a vast blanket over the surrounding landscape – a true natural disaster of global proportions. Certainly there is clear evidence that, within the past two million years (called the Quaternary Period), ice managed to spread over 30 per cent of the world's land surface: and it clearly resulted in substantial changes to the landscape, such as the spectacular erosion of some valleys. The more carefully the ice age is studied, though, the more complex and subtle the events prove to have been. Even the term 'ice age' can be misleading: it conjours up a picture of a single long continuous cold period, whereas in fact there were really at least a dozen cold phases, of various degrees of severity, separated by warm phases, often more mild than the conditions of today. These continual fluctuations were vitally important; the major landscape changes occurred mainly as ice advanced or retreated. When deep ice actually covered the landscape it did not always erode the land beneath; in some cases it preserved the land from attack. Probably very little erosion is currently taking place in the central regions of Antarctica, even though these are covered by ice three kilometres thick.

In the relatively mild conditions of today it is difficult to appreciate the nature of an ice age climate. Sometimes a cold winter gives us the merest glimpse of what past conditions might have been like (Fig. 19.1) – conditions that could easily become a reality again.

Figure 19.1 In the winter of 1981/2, Northern Europe and North America experienced the type of cold conditions normally restricted to the Arctic. This photograph shows Chicago, USA at −32°C. In the winter of 1962/3 the sea froze at Blackpool, partly because outblowing winds from snow-covered Britain helped to push back the influence of the North Atlantic Drift!

19.2 The balance between temperate and cold conditions

In the winter of 1981/82 newspapers were full of stories about the severe weather suffered in northern Europe and North America. Overnight temperatures near Birmingham in central England fell to −27 °C – lower than those at the South Pole. In Chicago (Fig. 19.1) temperatures were −32 °C, and in New York they even fell to −36 °C. In many areas these low temperatures were accompanied by snow over a metre thick as depressions continued to bring precipitation that simply did not melt away.

But although severe weather occurs from time to time, such examples do not show towns in the grip of an ice age – the snow always melts away by the spring. Much longer periods of severe cold are required if the winter's snow is to remain throughout the summer. Yet although the warmth of mid-latitude summers may make us forget the severe winters, only small changes of average temperature are needed to tip the climatic balance and turn large areas into frozen wastes. Such changes can occur quite easily. For example, if the upper air waves moved on average just a few degrees farther south, their associated depressions would also move south and cause in Britain and the United States weather that at the moment occurs over Iceland or Greenland. Even now many areas are poised on a climate knife-edge between temperate and cold climates. A mere 1 °C drop in average temperature in the Cairngorm Mountains in Scotland, for example, would allow snow to remain throughout the year and ice to form. Perennially frozen ground (**permafrost**) already occurs in half of Canada and could easily expand further south (Fig. 19.2). Indeed, scientists have discovered that even in the depths of the ice age the average temperature was no more than 6°C or 7°C colder than it is today.

(a)

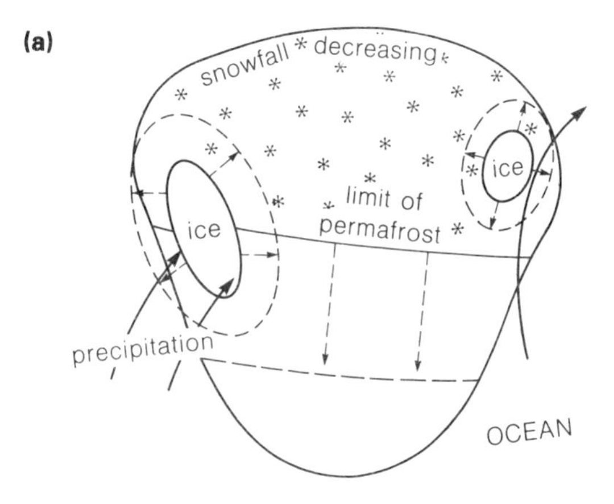

(b)

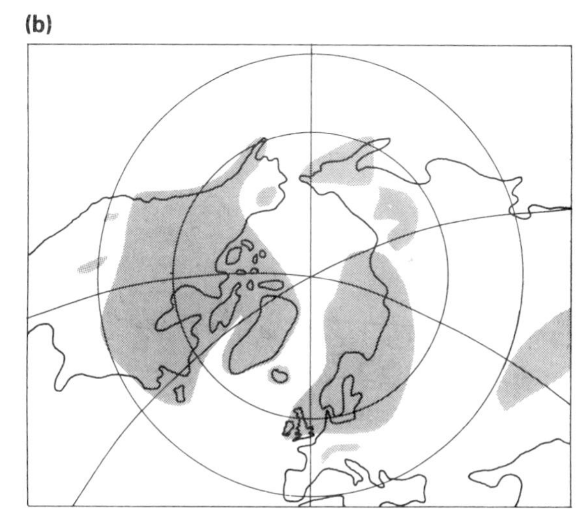

Figure 19.2 (a) The limit of permafrost and ice caps on an imaginary continent under present conditions and the expansion during a cold period (arrowed). Compare with (b) the maximum limits of the Quaternary.

19.3 The spread of ice

Whatever caused the change in climate, it is easy to visualise the first changes that occurred. As winters became longer and colder the ground was subject to prolonged and intense frost. In uplands the snow began to last throughout the year, packing ever harder until it turned to ice. Eventually even many lowlands also became ice-covered, although not necessarily by ice spreading from the uplands. Much ice seems to have formed directly on lowland; even at low latitudes, summers became so short that winter snow failed to melt. Once snow covered the ground it reflected the sun's rays and so encouraged further cooling.

The accumulation of snow to great depth would have been most favoured on the upland oceanic margins of continents. Some naturally 'arid' areas deep within continents must have remained so far from moisture-laden air that snowfall was only sparse and quite insufficient for ice formation. The common belief that ice sheets spread uniformly outwards from the poles is therefore far from the truth. Large areas of northern USSR, for example, must have remained then, as they are today, true cold deserts (Fig. 19.2).

Clearly the ice age must have produced two distinctive types of region: one (**glacial**) with a cover of ice, and the other (**periglacial**) experiencing deep freezing because it remained uncovered. Both types of region are still widespread, each covering about a

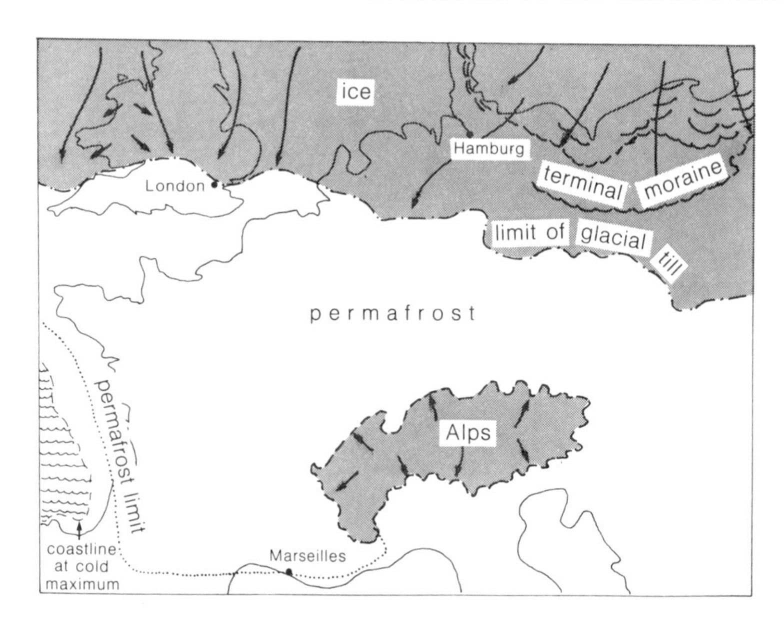

Figure 19.3 The widespread occurrence of glacial and periglacial phenomena in central and western Europe.

tenth of the world's landsurface. Furthermore, features remaining from the Pleistocene glacial period are a significant part of many mid-latitude landscapes (Fig. 19.3).

20

Periglacial environments

20.1 Types of periglacial climate

It is difficult to define the margins of regions belonging to cold climates (Fig. 20.1). The feature that distinguishes these **periglacial** regions from all others is the presence of ground ice and of landforming processes that depend on its partial melting. Periglacial processes are strongly influenced by the type of cold climate. There are many types of cold climate, each of which produces a different degree of frost, snow and wind action and which therefore yields landscapes which match the dominant process.

Professor Tricart has grouped periglacial climates in the following way.

(a) Cold dry climates with severe winters ('Siberian' type). These have very low winter temperatures, short summers, deep perennially-frozen ground, low precipitation and violent winds. Processes are influenced by intense freezing, little water erosion and enhanced wind action.

(b) Cold humid climates with severe winters. These are of two sorts. The 'Arctic' type is characterised by very low winter temperatures, but a more rapid rise of temperature in spring and longer summers than the Siberian type. Perennially frozen ground is widespread but there is appreciable snow and even some rain. Consequently freezing is less intense or prolonged. The effect of running water (meltwater) is significant and wind action is minimal. In the 'mountain' type, on the other hand, the climates give high precipitation which reduces the effects of freezing and of wind action. There is a higher mean annual temperature and summer melting is common, especially in valleys, giving considerable meltwater. Freeze–thaw cycles are more important in valleys than on summits where snow lies all the year. Frost is important but there is no perennially frozen ground.

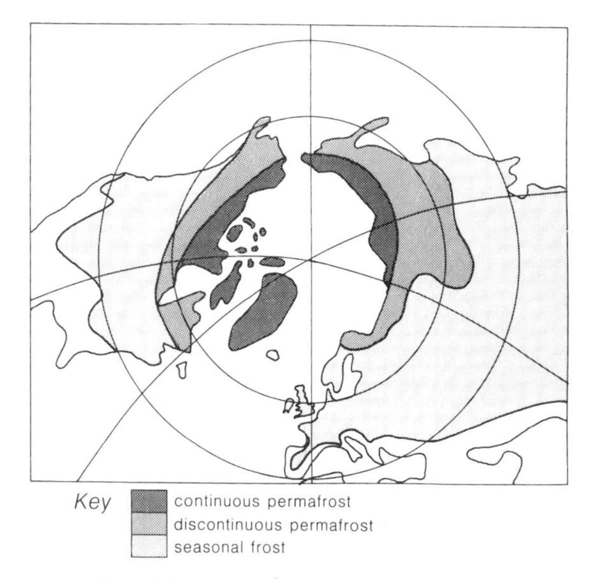

Figure 20.1 Although most people live in relatively mild regions where frost occurs only during a short winter season, the cumulative effects of frost cracking can be substantial. Further north, however, as winters become longer and harsher, perennially frozen ground becomes increasingly common.

(c) Cold climates with small annual temperature ranges are also of two sorts. The 'high-latitude oceanic islands' type have a mean annual temperature near 0 °C and numerous freeze–thaw cycles, but frost does not penetrate deeply. Considerable precipitation occurs and meltwater is significant. In the 'low-latitude mountain' type, the climates show no seasonal change but a large diurnal temperature range. They also experience considerable precipitation except at very high altitudes. There is a frost cycle virtually every day, but only slight frost penetration; there is no perennially frozen ground; and there is some wind action.

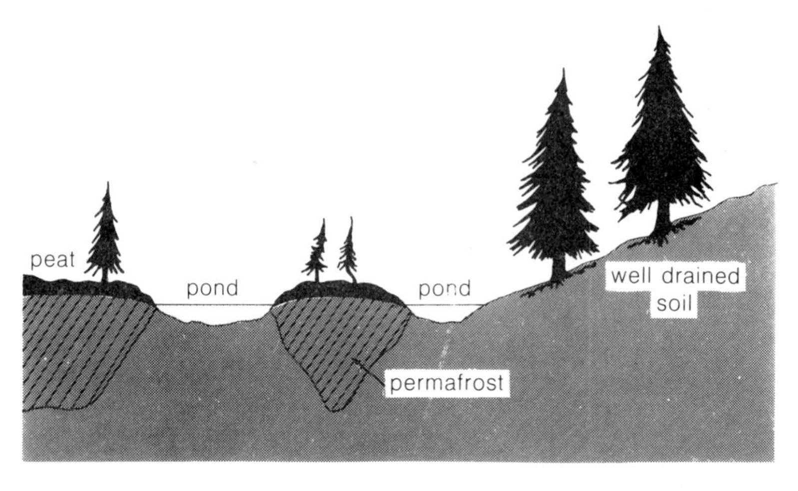

Figure 20.2 In the relatively flat areas of the Canadian Shield, small environmental changes control the distribution of permafrost. Because they restrict the depth of root penetration, areas underlain by permafrost are often marked by stunted trees.

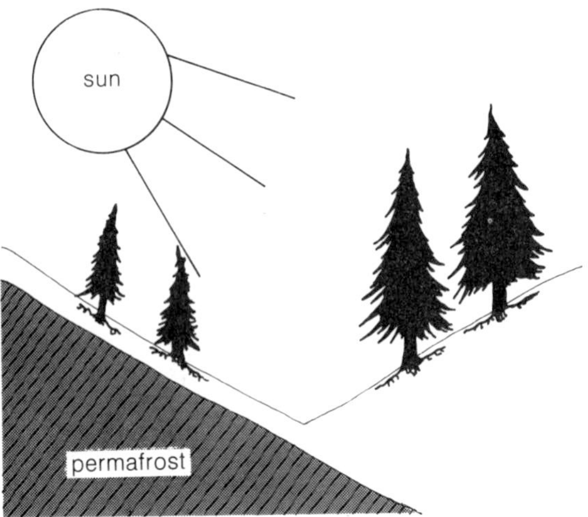

Figure 20.3 Along the margins of the permafrost zone, aspect can be very important. A seasonally thawed (saturated) layer on the north-facing slope can be expected to move more rapidly than on the unsaturated south-facing slopes. Eventually this may cause an asymmetrical valley form.

20.2 Permafrost

Frozen ground is central to periglacial processes. It can be of two types:

(a) *seasonally frozen ground*, where the ground remains frozen only during the winter – if the seasonally frozen ground forms above permafrost it is usually referred to as the **active layer**;
(b) *perennially frozen ground* or permafrost in which soil or rock is frozen continuously for a period of not less than two years.

The boundary of ground completely underlain by permafrost can be closely related to the areas where average air temperatures remain below −7 °C or −8 °C, conditions still experienced over large parts of northern Canada and Siberia. At its margins continuous permafrost breaks up into patches. In regions of **discontinuous permafrost** perennially frozen ground most commonly remains under peat bogs. Peat offers the most suitable environment for permafrost survival or development: because it gets wet in autumn and then freezes in winter, it allows frost to penetrate deeply. In summer the surface dries out and acts like an insulating blanket, keeping warm air at bay. In Canada patches of permafrost extend as far south as 53 °N (the latitude of Oxford, England; Fig. 20.1). However, in other areas at the same latitude permafrost is missing: spruce forests, for example, may shelter the ground from intense frost penetration, while ponds and rivers conduct sufficient heat during the summer to cause ground melting (Fig. 20.2). Aspect also is important, with south-facing slopes that catch the summer sun being commonly permafrost free (Fig. 20.3).

Regions with continuous permafrost are not necessarily barren. In many areas trees will grow by crowding their roots into the shallow surface layer of soil that thaws each summer. As a result it is rarely possible to detect the limits of permafrost from an aeroplane or simply by walking over the terrain (Fig. 20.4).

At present an average annual temperature of about −2 °C marks the southern limit of discontinuous permafrost. Nevertheless it is probable that much of the discontinuous permafrost zone is out of equilibrium with the present climate. Permafrost originated, and certainly reached its maximum extent, in the cold periods of the Pleistocene. It melts only slowly in response to atmospheric warming.

Figure 20.4 An ice wedge, Takutia, USSR. This deeply frozen ground has an active layer sufficiently deep to allow tree growth. The permafrost has only been revealed by river incision. Notice that the ice wedge is primarily a feature of the permafrost rather than the active layer.

20.3 Frost action

Frost action is by far the most widespread and important process in the periglacial zone. The term includes freezing and thawing; these processes result in frost-cracking of rock, frost-wedging of sediment and soil, frost-heaving of stones and frost-sorting of sediment in patterned ground.

Frost cracking

Frost action in rock (**frost cracking** or **frost shattering**) is the vital process that produces new broken material. Water expands by nine per cent when it freezes. Although rock may be very strong, the pressure exerted by ice forming in rock fractures or pores is extremely high, nearly ten times greater than the tensile strength of the strongest rock.

The size of broken material produced by frost cracking depends on the rock. An impervious rock like gneiss will shatter into large angular fragments, whereas granite can be reduced to its component crystals, sandstone to individual sand grains, and chalk to the size of silt. Frost action may be especially effective in porous materials – because ice crystals then grow in selected directions they enhance prising action (See also Ch. 13).

The primary influence on frost-cracking effectiveness is the *frequency* of frost cycles (rather than the intensity of the frost), so frost cracking is most important in climatic types (a) and (b) on page 00. Continuous shattering is possible only if the broken material is removed: material that stays in place acts as an insulating blanket, keeping the frost from the unbroken rock below. On the many flat mountain summits, such as those of the Cairngorms, the extensive sheets of shattered debris (called **blockfields**) have probably played a protective role, reducing the effectiveness of frost action. It is the highest, steepest or most exposed slopes, which are rarely covered with more than a thin brushing of snow, that are most prone to continual frost action.

On steep rock faces, summer rain or melting snow provides the necessary water. As rocks are prised clear of the face and fall away, then roll and bounce down the slope, their momentum may set other loosened blocks in motion, sometimes even causing a **rock avalanche**. Material from these avalanches often comes to rest on moderate slopes and is later removed by snow avalanches (Fig. 20.5). Elsewhere it accumulates as a **scree** (Fig. 20.6), or falls onto the surface of glaciers and is thus carried away as lateral moraine.

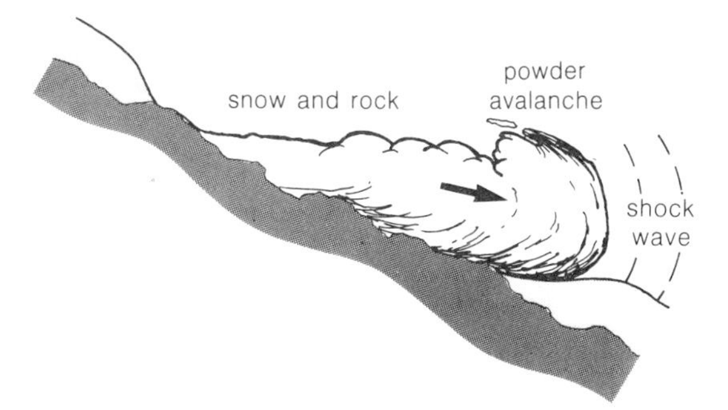

Figure 20.5 As snow gathers speed on a mountainside the surface, powdery fraction lifts away and then moves down hill as a high-density gas, often reaching speeds of 200 km/h, pushing air ahead of it and creating a shock wave. It is the shock wave that blows a house to bits before the main mass of snow even arrives. However, it is the momentum of the main snow mass that helps transport rock debris down hill. Avalanches are most common on slopes between 20° and 40°.

Figure 20.6 The Wastwater screes, Lake District, a fringe of angular debris shattered from the glacially steepered valley wall.

The amount of frost cracking depends not only on the number of freeze–thaw cycles but on the continued presence of water. Permeable or porous rock, in which water is already present, is liable to greater frost cracking than is fractured impermeable rock. But as water supply depends also on position within a landscape, hillslope concavities (which focus both surface and groundwater flows) and valley bottoms are likewise liable to frost cracking. Indeed, increased frost cracking in hillslope concavities is the main cause of the growth of hillslope hollows; while cracking of stream beds that dry up in autumn may promote the rapid channel deepening that is often characteristic of a permafrost region.

NIVATION HOLLOWS

Nivation hollows are enlarged hillslope concavities that show the combined effects of frost action, mass movement and meltwater at the edges of, and beneath, long-lying snowdrifts (Fig. 20.7). The snowdrift prevents frost action at the centre of the hollow, but prolongs it at the edges. Melting snow provides a continued supply of water. Each day, solar radiation melts some of the surface snow, releasing water which tends to percolate into the surrounding rock. At nightfall, water in the rock near the (cold) snowdrift is likely to freeze and initiate cracking. Cracked rock then falls onto the snow and slides away (Fig. 20.8). This process of 'countersinking' snowdrifts into the hillside cannot but enlarge the hollow, year by year, allowing it to retain more snow.

The sizes and shapes of nivation hollows depend on both the structure of rock in which they form (the lithology) and the frequency of frost cycles. If the rock readily cracks into small fragments, as do shale or chalk, then a wide, shallow hollow forms. Other rocks, including granite, allow a steep-sided deep hollow to form, and this shape may enable cirques to develop. In northern Canada there may be fewer than forty frost cycles each year, a number which makes the development of nivation hollows extremely slow.

Figure 20.8 Frost cracking and snow transport, Switzerland. The rock is metamorphic gneiss whose fractures have been produced by unloading.

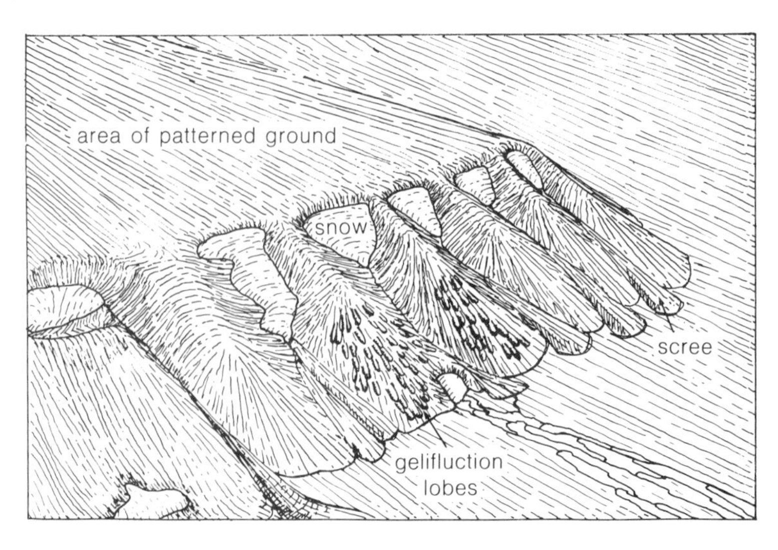

Figure 20.7 Nivation hollows bite deep into the edge of uplands on Axel Heiberg Island, Canada. Below them lie the scree slopes made from frost material transported across the snow patch surfaces.

Ice growth in sediments

When ground contains a large amount of water, the effects of expansion by freezing can be dramatic. **Ground ice** may take many forms; the principal ones are pingo ice, ice lenses, ice veins and ice wedges.

A **pingo** – the Eskimo word for 'hill' – is a large, domed mass of sediment with a core of ice (Fig. 20.9). Pingos are very distinctive features of some regions in the continuous permafrost zone. It is thought they may form in two ways:

(a) by the inward-freezing of water-saturated sediment – the open-system type (Fig. 20.9a);
(b) by the upward flow of groundwater from beneath a thin permafrost cover, rupturing it and forming an ice lens in the uppermost part – the closed-system type (Fig. 20.9b).

The ice sediment mounds so formed may be up to 100 m high – true hills of ice (Fig. 20.10). The final melting and collapse of a pingo leaves only a small circular ridge of sediment as evidence of its existence.

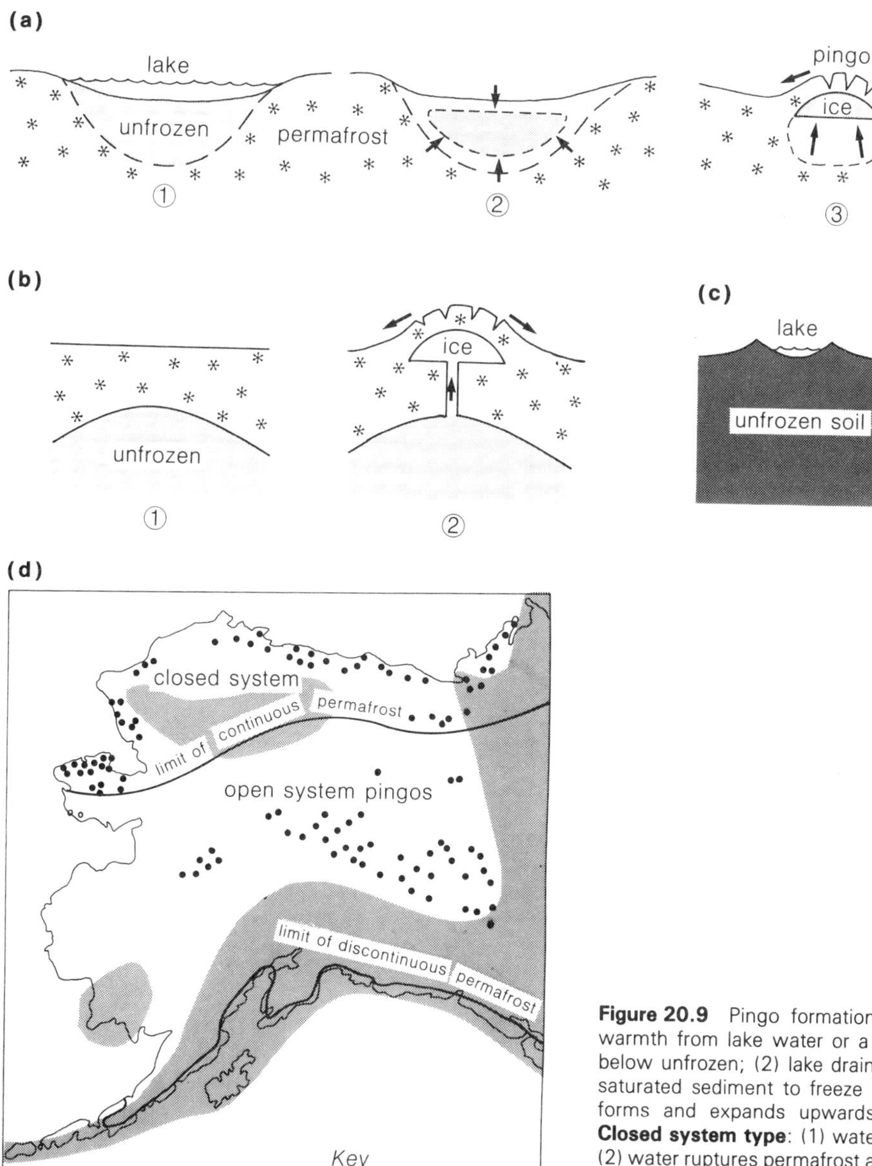

Figure 20.9 Pingo formation. (a) **Open system type**: (1) warmth from lake water or a river keeps saturated material below unfrozen; (2) lake drained or river diverted allows the saturated sediment to freeze from all sides; (3) an ice lens forms and expands upwards, causing pingo growth. (b) **Closed system type**: (1) water trapped beneath permafrost; (2) water ruptures permafrost and forms a surface ice lens. (c) The form of a pingo after melting. (d) The distribution of pingos in Alaska.

Figure 20.10 A pingo, Victoria Island, Northwest Territories, Canada.

Nevertheless, pingos are important indicators of the extent of former periglacial environments; their evidence was used in constructing the map in Figure 19.3.

Ice lenses are irregular masses of ice within soil or sediment; they form mainly as the freezing front advances downwards at the onset of winter. They are extremely variable in size, the smallest being less than one millimetre across, the largest several tens of metres. Under freezing conditions (liquid) water is attracted towards regions of ice: once centres of ice growth have formed they grow quickly into lenses, displacing soil or sediment. Ice growth causes the soil surface to heave up.

Ice-lens growth has particularly important effects on slopes, where it causes **frost creep** (Fig. 20.11). As water in soil pores begins to freeze and develop into lenses, the soil heaves away from the slope. Later, with the onset of a summer thaw, the ice lenses melt and the soil slumps to a new position slightly down slope from that occupied in the previous summer. In this way frost creep can move a sheet of surface materials each year. Furthermore, without the confining pressure of overlying soil layers, ice near the surface grows upwards, sending columns of ice sometimes several millimetres out of soil pores (Fig. 20.12). The growth of **needle ice** can lift surface stones, even boulders.

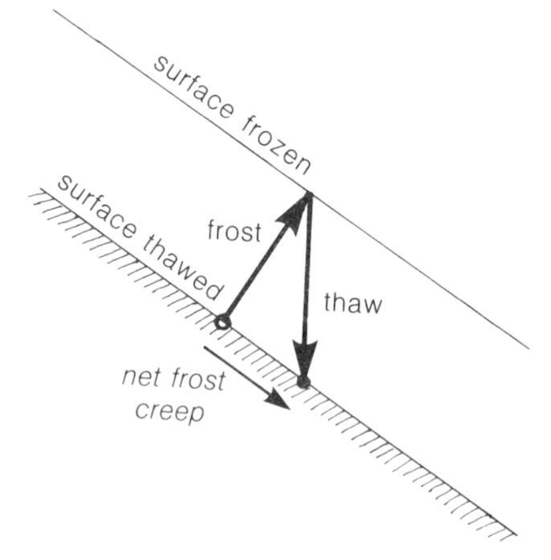

Figure 20.11 The mechanism of frost creep, resulting in a saw-tooth downslope movement of surface material.

Ice can grow not only horizontally, but also downward into the soil, producing **ice veins** which often develop large V-shaped **ice wedges** (Figs 20.4 & 13). Two mechanisms are involved:

(a) a period of *rapid* intense freezing, causing the ice in the ground to contract quickly (as shown in Fig. 13.4), thereby opening up a network of

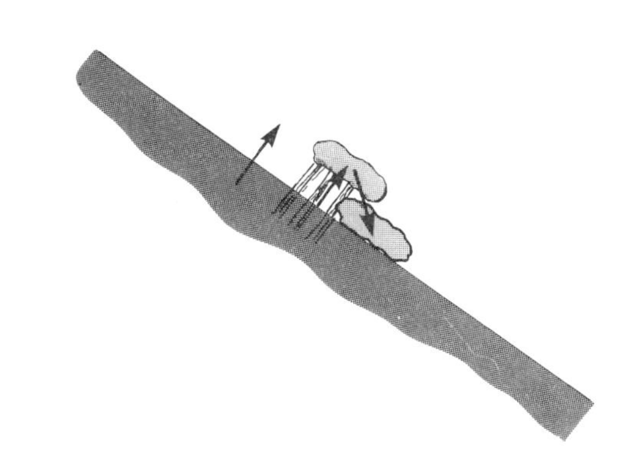

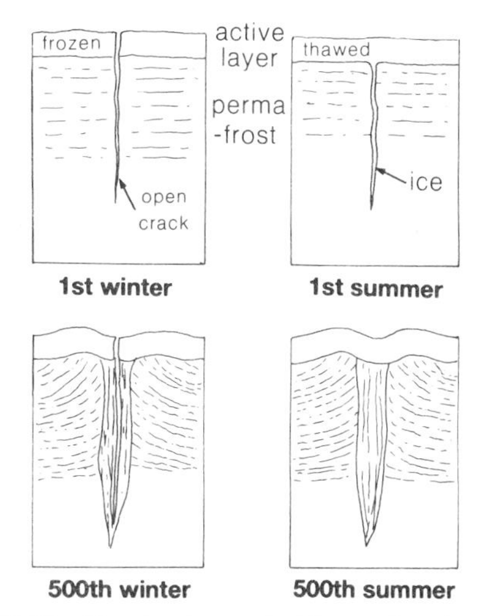

Figure 20.13 The development of ice wedges by repeated contraction freezing.

Figure 20.12 The growth of needle ice.

polygonal cracks which often extend into, and become imprinted on, the permafrost;

(b) the growth of ice in the cracks using meltwater, groundwater or even water vapour.

The development from veins to ice wedges occurs as soils crack each year along the same lines. Wedges can grow sufficiently large to squeeze and dome up the cracked soil. The semi-permanent pattern of polygonal cracks and ice wedges produces the **patterned ground** so common in type (a) climates on gentle slopes (Fig. 20.14).

Figure 20.14 Patterned ground. A polygonal pattern of cracks is common in flat periglacial regions. Individual polygons are up to 40 m across and they rise in a gentle swell to the middle. In this photograph snow still picks out the troughs around the polygon margins: they are 2–4 m wide. There is a large variety of patterned ground apparently forming in an equally wide variety of climates, so the features probably result from several different processes.

Frost sorting

Patterned ground may sometimes be associated with particle sorting. It is common, for example, to find the depressed margins of polygons filled with coarse sediment, while the centres contain only fine material (Fig. 20.15).

Larger stones conduct heat more quickly, so as a freezing front penetrates the ground it advances to the bases of large stones before reaching those of adjacent small stones. Any water at the bases of large stones will be frozen first, causing expansion and heaving individual stones upwards (Fig. 20.16). Subsequent freezing of water below fine material pushes the small stones into the cavity left by the heaved stone. During thawing fine stones tend to collapse into the spaces below the larger stones. Both of these processes prevent the return of the larger stones. Once they have reached the surface their progress to the edges of the polygons may be assisted by needle-ice growth.

(b) during freezing

ground surface

stone

ice vein

frost table

① ② ③ ④

⑤ ⑥ ⑦ ⑧

during thawing

Figure 20.16 A suggested origin for the upward movement of large stones during freezing.

20.4 Meltwater

Water melting from snow patches and ice lenses saturates slope soils and sediments as they begin to thaw in spring. Slow flow of soil from higher to lower ground in these circumstances is called **solifluction** or, more precisely, **gelifluction** (Fig. 20.17).

Gelifluction may occur as sheets, lobes or terraces, depending on the distribution of drifting snow and on the terrain. Where extensive snowbanks occur,

Figure 20.15 (a) The pronounced particle sorting associated with some patterned ground. Polygons tend to merge into stripes as slope increases. (b) Small sorted polygons on Mt Kenya at 5000 m.

(a)

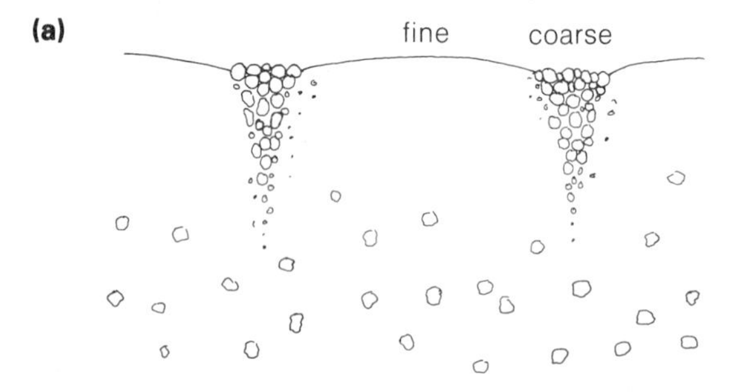

(b)

Figure 20.17 (a) A section through a gelifluction lobe. (b) A gelifluction lobe, Isle of Arran, Scotland.

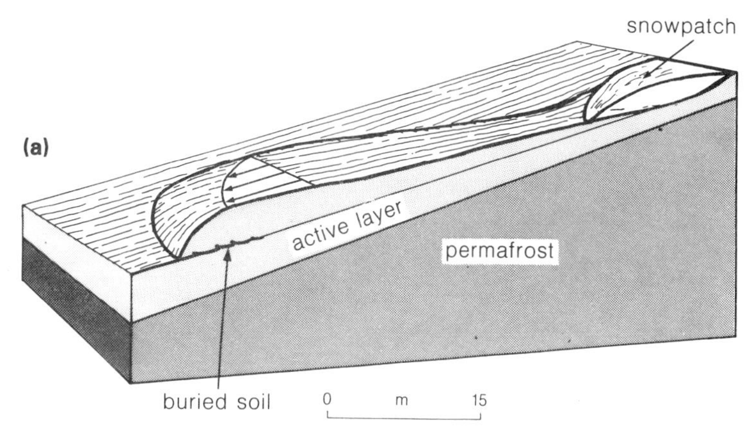

(b)

sheets and terraces are most common; where snow is confined to patches gelifluction mainly occurs in lobes.

Gelifluction is a true flow, continuing as long as snowmelt feeds water into the soil at such a rate that it cannot easily drain away. For a few months each spring, when only a thin surface layer of soil has been unfrozen, saturation produces buoyancy within the soil and allows 'waves' of soil to edge a little farther downslope. Gelifluction can occur on slopes as gentle as 2°, so prolonged gelifluction tends to smother breaks of slope and merge hillslope into plain (Fig. 20.18). It is probably enhanced by the presence of permafrost because the large depth of cold ground reduces the rate of soil thawing. However, gelifluction may also occur on ground without permafrost during the period when the surface ground is thawing.

20.5 Periglacial activity and the landscape of Britain

Periglacial activity has played a significant, if mostly unrecognised, part in forming the landscape of Britain. During each cold period, periglacial conditions reduced the steepness of sea cliffs, escarpments and exposed slopes of glacial valleys towards their

Figure 20.18 Characteristics of periglacial landscapes. The steep wall in the distance is fringed by a scree slope, in contrast to the gelifluction lobes that mantle the more moderate slope in the middle ground. The gently sloping foreground contains a braided stream and patterned ground (Axel Heiberg Island, Canada).

threshold angles of stability, forming extensive screes or producing finer material that could be moved by gelifluction to infill valleys and spread as a mantle over the edges of plains. On gently sloping land the soil was intensely frozen and formed into patterned ground. Because northern Britain was mostly ice-covered in the last ice age, the time over which periglacial activity lasted was shorter than in the south. Nevertheless the extensive till sheets that still mantled many slopes after the melting of ice sheets provided just the right type of water-retaining, unconsolidated material that allows gelifluction to work most effectively.

Today gelifluction material cloaks most moderate slopes in areas as far apart as Cornwall and Scotland. Nevertheless the irregular sheets, lobes and terraces that must have been common while gelifluction was active have now been smoothed out by soil creep and other processes; gelifluction material is now very difficult to see. Perhaps the best places to gauge the impact of gelifluction are along the coast. Here the sea has trimmed back many valleys, exposing them in cross section and revealing gelifluction infill (Fig. 20.19).

The widespread scree slopes of the Lake District (including the famous Wastwater screes (Fig. 20.6) and the screes of the Yorkshire Dales (best seen in Gordale valley) all testify to a long period of periglacial activity when frost shattering was much more intense than it is today. Many of the screes are now becoming covered with grass – a sure sign that the form is a relic from the past.

Material removed from nivation hollows now forms extensive lobes and sheets on the adjacent low ground. Gelifluction on the chalk scarps in southern Britain was particularly effective in moving chalk rubble (called **head**). Although the head forms a virtual 'skirt' along the scarp foot, its extent is not easily seen. A considerable thickness of gelifluction material was only recently discovered near Godalming in Surrey when a new bypass unknowingly cut across a lobe removing the material at the 'toe' end and initiating a landslide.

The recent widespread use of aerial photos has been of great assistance in identifying areas where patterned ground once formed. Because the material in the edges of the polygons has a texture different from that in the centres, crops grow at contrasting rates and give a pattern that is readily seen from the air. Such photos lend support to the belief that, perhaps 12 000 years ago, the view over Cambridgeshire would not have been unlike that of Axel Heiberg today.

Figure 20.19 Periglacial features at the coast of Start Head, Devon. Gelifluction material mantles the lower slopes and is being eroded to reveal a buried raised beach.

21

Glacial environments

21.1 Ice on the move

There is often no simple regional division between periglacial and glacial environments – for example,on the mountainsides high above the valley glaciers in the Alps, processes of frost cracking release from a periglacial environment debris which falls and bounces onto the surfaces of glaciers. However, there is a fundamental distinction in processes: in periglacial regions the ice, although it is important in causing localised movement of soil or shattering of rock, is not itself part of a coherent moving mass.

Glacial processes are entirely dependent on the effects of moving masses of coherent ice. Streams or sheets of ice may be no more responsible for shaping the whole of a landscape than is a river for shaping the whole of a valley; other processes may be equally important(see glacial valleys). For example, streams of ice usually follow and modify pre-existing river valleys; much less commonly do they create entirely *new* valleys (Fig 21.1). But when ice overwhelms a landscape the results can be extensive and spectacular.

Figure 21.1 This before and after reconstruction of the Yosemite Valley in the Sierra Nevada mountains of California illustrates the classic view of glacial erosion *modifying* a V-shaped valley formed by fluvial processes and transforming it into a 'U'-shaped valley. Notice that both *widening* and *deepening* are involved.

Figure 21.1 contd.

21.2 From snow to ice

Some of the most important processes that occur in glacial regimes result from the change from individual crystals of ice ('snow') to a solid mass ('ice'). It is therefore important to understand both the changes and their implications.

Catch a snowflake in winter and you can see its delicate ice-crystal form against the dark background of a glove. Look at snow piled high on the branches of trees and you can see the star-shaped crystals locking together in a light, rigid framework which would be the pride of any civil engineer. Yet these delicate crystals of snow are doomed to be crushed and broken as soon as they leave the cloud; they were not designed to take the weight even of one other snowflake. Indeed, where flakes touch they melt, continuously transferring the point of pressure to another part of the structure and then refreezing (Fig. 21.2a). In this way crystals shrink back on themselves and delicate branches are replaced with thicker, part-melted stumps of ice. The process of 'spot welding' not only compacts the snow but gradually fuses the crystals into a more rigid mass.

Further metamorphosis of snow may happen in one of two ways. In very harsh climates, such as in Antarctica, temperatures remain below freezing throughout the year and snow compaction is simply the result of adding more snow and waiting a long time. Snow may change to the more dense crystalline form called **firn** or **névé** (with a density of 0.4 g/cm^3) at a depth of several metres and finally to **ice** (density 0.8-0.9 g/cm^3) at several tens of metres (Fig. 21.2b). But with only a few centimetres of snow added each year, this process takes centuries to accomplish.

In regions where summer temperatures rise above freezing, warm air or rain can achieve in a few days consolidation of snow that would take decades in Antarctica. Mild weather and rain both cause the surface

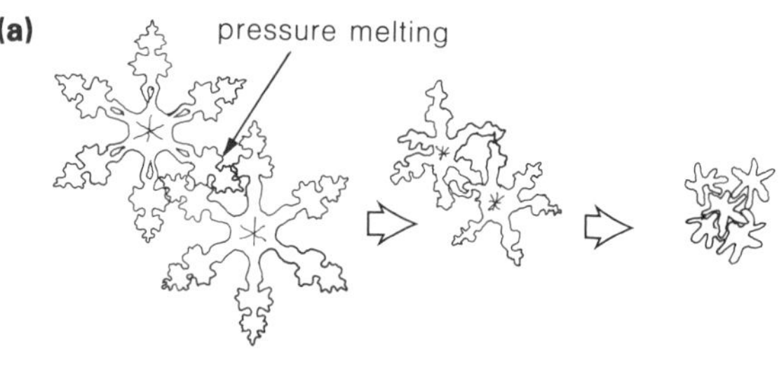

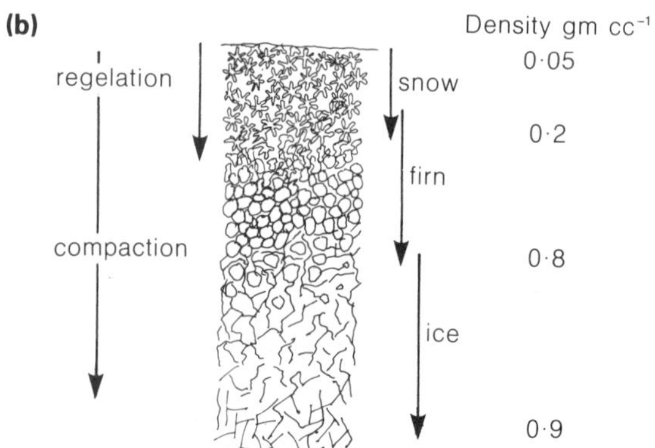

Figure 21.2 (a) Changes from snow to firn by pressure melting. (b) Changes from snow to ice (not to scale).

snow to melt and percolate downward; it later refreezes. Day by day additional melting consolidates the snowpack, gradually filling in the air spaces and producing névé that looks and feels not unlike poorly made cut glass.

21.3 The properties of ice

Looking at an ice cube you might think that ice really shouldn't move at all. An ice cube is clearly a very brittle solid; no amount of finger pressure can distort it but it shatters into tiny pieces when struck with a hammer. Even large pieces of ice seem to behave in much the same way: ice breakers need to ride right up onto an ice flow and bring their entire weight to bear before the ice will finally crack and break into large slabs. Icebergs hundreds of metres high ride like stately galleons in the water, maintaining a sharp, angular outline much as when they first broke away from the parent ice sheet. One might suppose that valley and cirque glaciers or ice sheets behave in much the same way – as large, rigid slabs of ice. The surfaces of glaciers and ice sheets give some support to this idea, for they are often criss-crossed with great open cracks, or crevasses (Fig. 21.3).

If we treat ice as a rigid slab we can estimate its chances of moving, in just the same way as we did for clay soils on a hillside (Fig. 21.4). A lowland ice sheet should be perfectly stable because it is lying in a horizontal plane with its centre of gravity as low as possible. A sloping ice slab in a glacier will be much less stable and should slide down slope if friction between ice and rock can be overcome.

However, ice is not totally rigid and is able to advance over lowlands and move down curving valleys. In Figure 21.5, for example, the curved track of the ice has been further highlighted by streamlined trails of fallen frost-shattered debris (**lateral** and **medial**

Figure 21.3 A crevasse on a steeply sloping section of a glacier. Notice how the crevasse width decreases with depth.

(a)

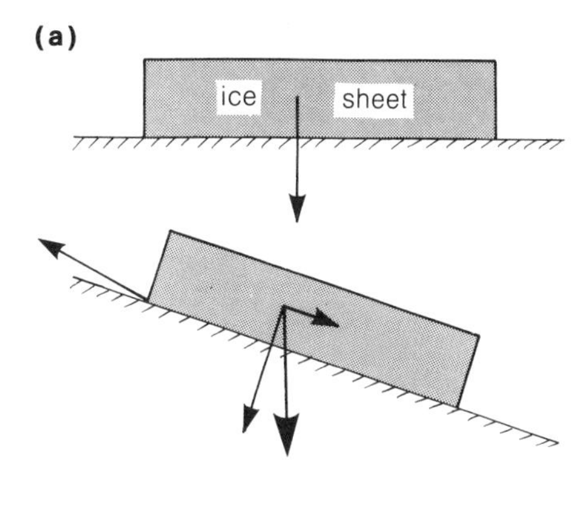

(b)

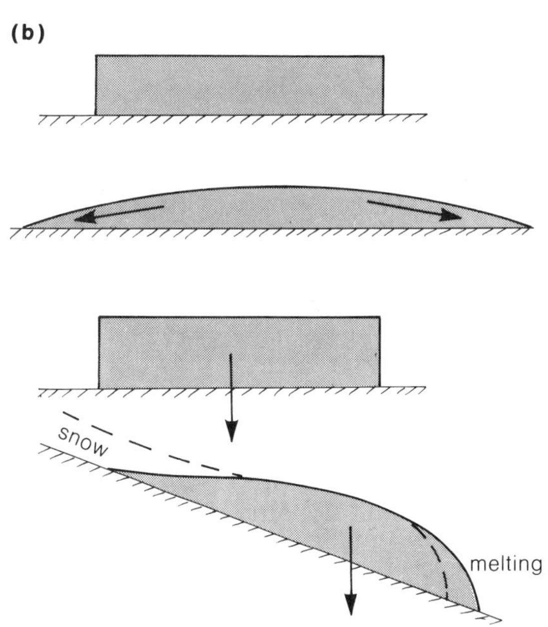

Figure 21.4 How to think about ice on the move. (a) As a **brittle solid:** a rigid horizontal ice sheet is stable, but a glacier on a slope will tend to slide (symbols as for **Fig 14.3**) (b) As a material that can **deform under pressure** from its own weight: a horizontal slab spreads out and lowers its centre of gravity, while a glacier flows down slope due to both accumulation and melting.

Figure 21.5 A valley glacier in Greenland. Notice the surface moraines and the crevasse pattern.

moraine) that sweep in snaking curves on the glacier surface.

Deformation in ice below its melting points is achieved by slippage within and between ice crystals. This allows ice to creep forward at speeds which may reach hundreds of metres a year (Fig. 21.6). Clearly ice behaves in two quite different ways. It does behave as a rigid solid which moves with great difficulty across an irregular landscape or through small valleys; yet over time it also behaves as a slow-moving fluid, continually moulding into new forms (Fig. 21.7). It is these dual roles, of solid and liquid, that provide the key to the distinctive landscapes sculptured by ice. Near the surface there is relatively little pressure on the ice; it simply cracks into crazed patterns of crevasses. Deep in the body of the ice stresses force it to deform.

At the base two situations are possible:

(a) ice may melt in heat either from friction as the ice moves across the rock floor or from geothermal sources within the bedrock. Meltwater thus formed provides a kind of buoyancy, reducing friction with the floor and helping the ice slide over its bed. When this happens ice sheets and glaciers are said to be **warm-based**; a large amount of sliding occurs and glaciers become highly erosive.

(b) in regions where melting does not occur, the resulting **cold-based** ice remains stuck to the floor, all the movement is by internal creep and erosion is minimal.

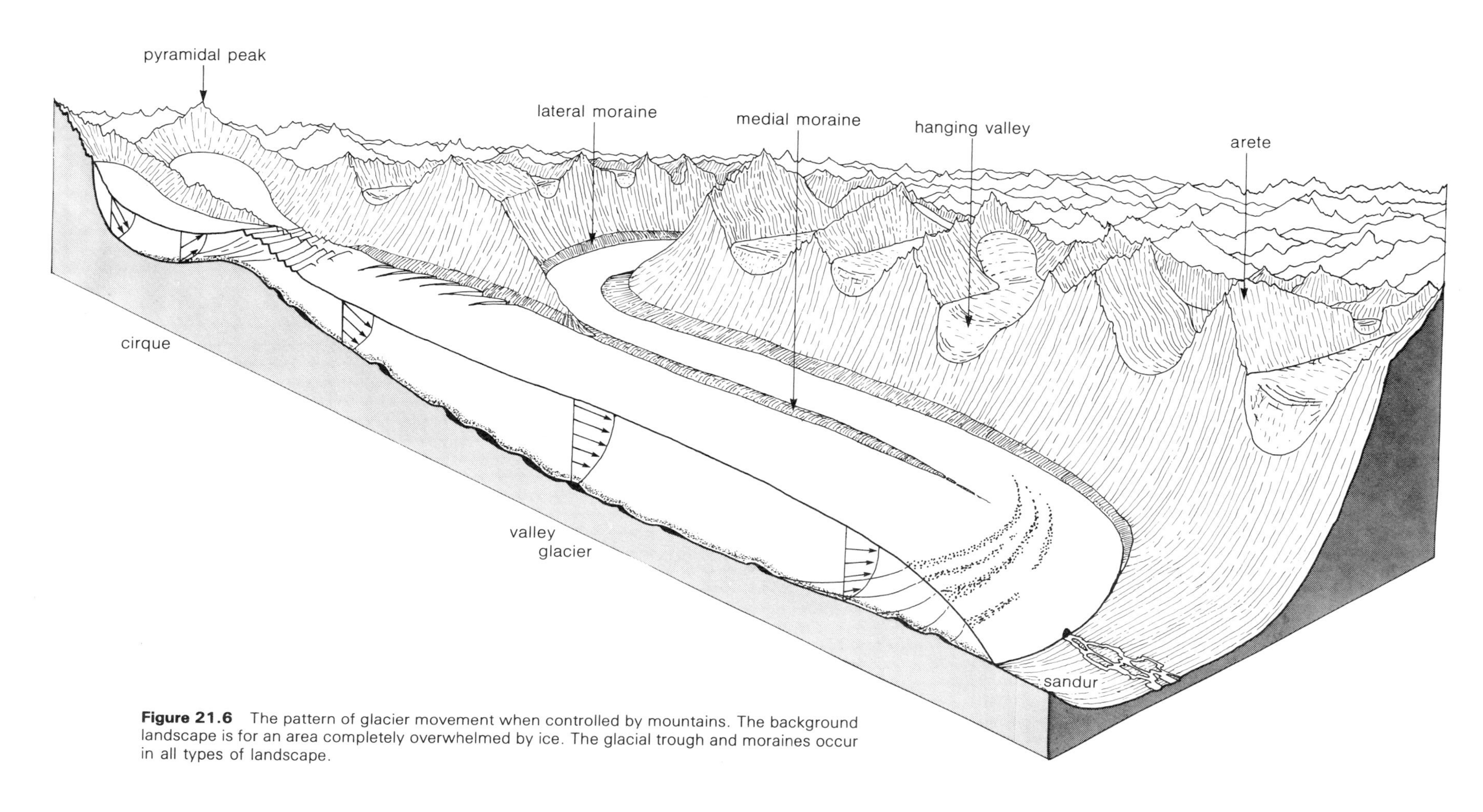

Figure 21.6 The pattern of glacier movement when controlled by mountains. The background landscape is for an area completely overwhelmed by ice. The glacial trough and moraines occur in all types of landscape.

sand when making ice cubes. You will easily be able to scratch a piece of glass if you rub the ice cube back and forth like a file. The pressure of sand grains on glass can be substantial, especially if the embedded sand happens to be corner-on (Fig. 21.8). Domestic examples help us to speculate on the role of ice. To observe equivalent processes under an ice sheet tens of metres thick we must enter an ice cave formed where the ice sheet bed arches across a bedrock hollow (Fig. 21.9). Here, in the blue twilight world beneath a creaking mass of ice, both quarrying and abrasion are clearly realities. The most prominent

21.4 Glacial erosion

Glaciers and ice sheets can only erode where the ice touches rock. Although ice will mould itself to the general outline of the rock, it is too strong and rigid to flow into every small irregularity. As a result, all the erosive activity is concentrated onto regions of contact between moving ice and rock.

The ability of ice to melt under pressure is the key to its movement, but this also limits its ability to erode. Ice is incapable of eroding rocks by direct pressure; in the same way that people need strong tools to break rock, so ice can erode effectively only when supplied with debris.

Supply and transport of debris are part of the vital interplay between the glacier sole and bedrock. The effects are very complex but fall into the two categories of **quarrying** and **abrasion**. You can gain an impression of quarrying whenever you scrape ice from a car windscreen or pull a package stuck to the side of the freezing compartment in a fridge. No matter how smooth the surface appears to be, freezing water seems to act like contact adhesive. By contrast, abrasion can be illustrated by adding a few grains of

GLACIATED VALLEYS ARE REALLY CHANNELS

It can be very difficult to relate the small streams that flow apparently aimlessly across the floor of glacially-enlarged troughs (glaciated 'valleys') to the large streams of ice that once filled them. But a stream flowing in a channel 1 m wide and 50 cm deep at 2 m/s can remove the same amount of water as a glacier 1 km wide and 300 m deep because the latter moves at only 100 m/year. We need to think of glacial troughs as *channels* rather than valleys (Fig. 21.7).

Erosion of the bed (valley floor) and banks (valley hillslopes) of a glacial trough initially proceed until a shape is formed that allows an efficient transport of ice, just as the bed and banks of a river are adjusted to water flow.

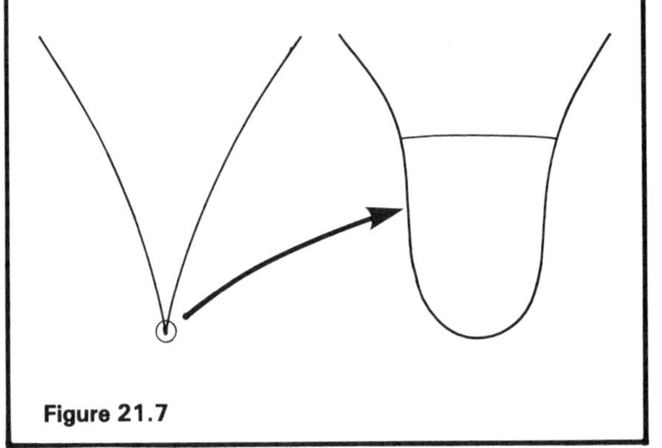

Figure 21.7

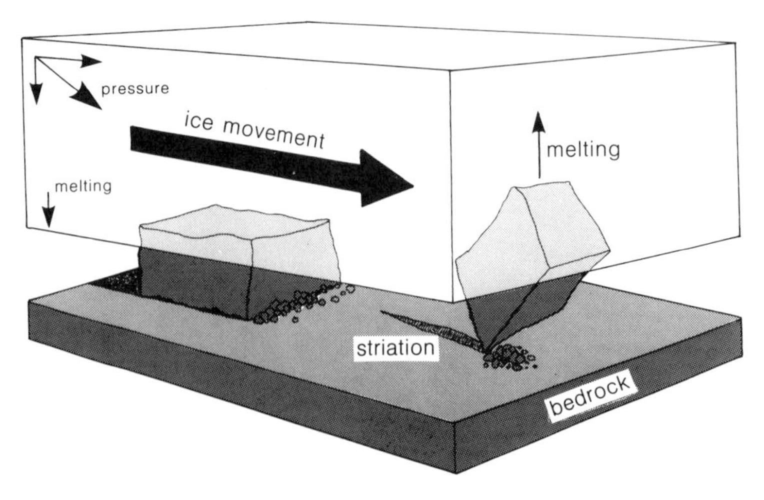

Figure 21.8 Ice can press a piece of debris hard against bedrock until it is absorbed by pressure melting. A piece of debris corner-on is particularly effective because the pressure is very concentrated. Notice how the debris will tend to have its point of contact flattened as it abrades the bedrock.

Figure 21.9 (a) Detail of quarrying. The shattered rock is frozen to the ice and is being carried across the roof of the cave. (b) A chunk of debris partly protruding from the base of the ice: compare with Figure 21.8. (c) Striations produced by abrasion on a rock now beyond the ice margin. (d) The continued abrasive and crushing power of ice sheets is clearly demonstrated here. Scientists investigated the change in size of material eroded from a dolomite rock band as it was carried forward in the till. To begin with, material is very large (quarried fragments), but these are gradually crushed to rock flour (fine sand and silt). After a journey of 300 km, over 70 per cent of the dolomite has been reduced to this small size (data from Ontario, Canada). (e) Glacial till, Whitby, Yorkshire.

feature of the upslope side of the cave is the frozen 'waterfall' of ice that has formed: for years meltwater has seeped away from the zones of high pressure and built much like the flowstone of a limestone cavern. Here, then, is proof that ice melts under pressure and refreezes (a process called **regelation**) at the base of a glacier. There is also evidence of erosion: where ice and rock diverge, some meltwater has refrozen around fractured rock, sticking it fast to the glacier sole. In the roof, dark streaks mark the trails of such debris in transport from the site of quarrying to the place where the ice closes back down on the bedrock on the downslope edge of the cave. You can picture the abrasion that must occur as soon as ice, debris and bedrock crush together.

In a single ice cave the whole theory of debris quarrying and abrasion becomes reality. To confirm the interpretation in an experiment, two blocks of rock, one basalt and the other marble, were bolted to bedrock in a place where ice is normally in contact with the rock. After only three months both had typical scratch marks or **striations** (Fig. 21.9). The basalt had lost one millimetre from its upper face; the marble had lost three millimetres and had had a chip torn (quarried) off one corner.

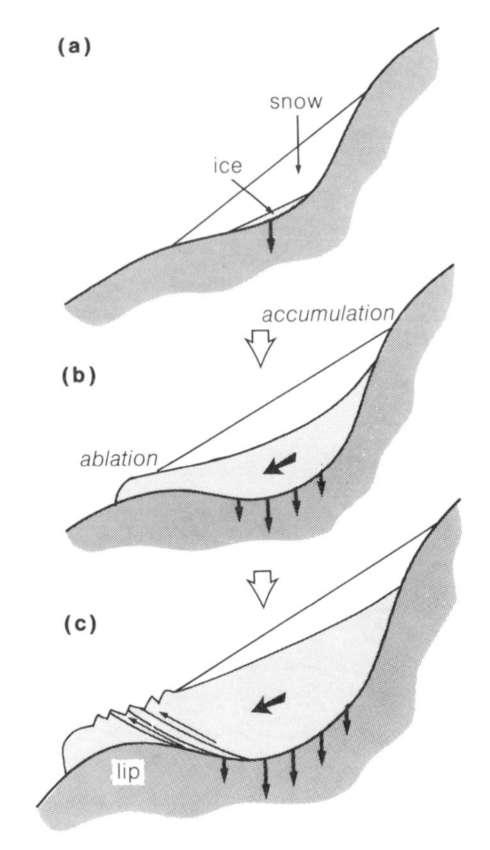

Figure 21.10 Cirque glaciers are never deep enough for the weight of ice to prevent erosion. In general, therefore, the thicker the ice, the greater the abrasion rate. When ice forms in a nivation hollow (a) abrasion will gradually cause overdeepening in proportion to the ice thickness (b). The main reason for any glacier flow is to restore an equilibrium profile determined by the capacity of the ice to support a slope. Each year most snow is added to the top and most ablation occurs at the toe. The ice then deforms to regain its equilibrium shape by creep and slide. However, on a cirque glacier (c), this movement superficially resembles a rotational movement. Notice that the ice gets thinner near the toe and thus deforms less readily. This leads to the development of shear planes and movement by overriding. In turn, the lack of abrasion in the shear zone allows the cirque lip to form.

Variations in ice erosion

Many measurements are needed to establish how much erosion occurs. Usually the degree of quarrying depends on the presence off well fractured rock and an irregular rock surface over which many caves can form. But rock fractures are not all inherited – some will have been produced by previous periglacial action; others must be the result of rock unloading as erosion replaces high-density rock by much lighter ice. The rate of abrasion depends not only on a continuing supply of debris (from quarrying) but also on the speed of the moving ice: the faster the ice moves, the more debris is scraped over the bedrock.

Abrasion also depends on ice pressure. Try sliding one hand forward over the surface of a table while pressing down on this hand with the other one: to begin with sliding is easy, but, as the pressure increases, motion becomes increasingly difficult. Ice pushing rock fragments over its bed behaves in the same way: very thin ice will press down so lightly onto the debris that it will slide easily but cut into the bedrock hardly at all; very thick ice will press down so heavily that the debris will be prevented from moving and no erosion will occur. At some intermediate value forward movement occurs with the pressure of ice causing maximum abrasion. With increasing ice thickness, abrasion first tends to increase to a maximum value; it then declines again until under very thick ice it does not occur at all. Nevertheless, in warm-based glaciers the sheet of water produced at the base can take some of the weight of the glacier. This is the condition that allows some erosion to take place beneath even thick ice.

These variations in abrasion rate have important consequences for cirque, valley-glacier and ice-sheet erosion. For example, ice in cirques is never very thick, so the greatest abrasion always occurs under the thickest ice – a process that eventually leads to the formation of a bowl shape (Fig. 21.10). Valley glaciers

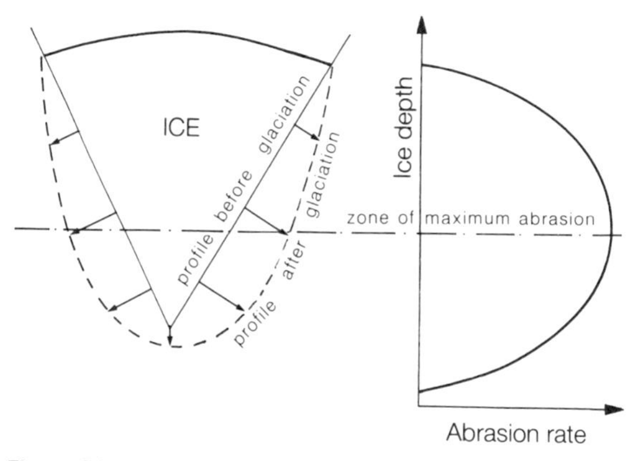

Figure 21.11 The rate of abrasion of a valley partly depends on the ice thickness. In some cases, the zone of maximum abrasion rate will be above the valley floor, leading to the development of a widened valley profile.

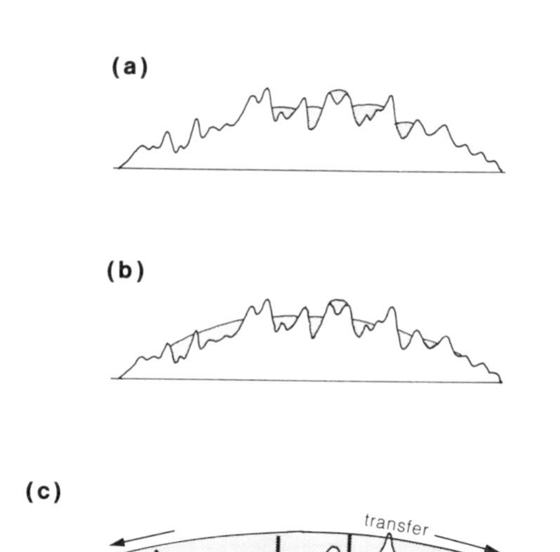

Figure 21.12 (a) In the early stages of a cold period, ice is confined to small ice caps, cirque and valley glaciers. This is the period during which valley shapes are transformed (Alps; Rockies at present). (b) The maximum phase of ice-sheet development in high mountains. Lower peaks are overwhelmed and abraded. Upper peaks remain ice-free and frost shatter continues (Alps; Rockies at maximum ice; Scandanavia/Josterdalsbreen at present). At this stage there are well developed arêtes, cirques, and pyramidal peaks. (c) The maximum ice-sheet development in northern Europe, when ice overwhelmed virtually all the landscape. The weight of ice depresses the land. Much ice in the valleys remains stationary and large areas are cold based. At this stage there are subdued peaks, arêtes are rare, and cirques remain from early and late phases of ice cover.

are often so deep that the zone of maximum erosion may be some distance above the trough floor, causing widening and the formation of a U-shape (Fig. 21.11).Ice sheets may be so thick that valleys are overwhelmed and ice within them may stop moving altogether (Fig. 21.12). Cirques are therefore likely to develop at early or late stages of a glacial phase; in periods of maximum ice advance, all but marginal areas will be overwhelmed and put into cold storage by the growth of an ice sheet.

Ice sheets in mountains

The change from valley glacier to ice sheet is significant in many ways. Essentially it shows that valleys have failed to remove the snowfall effectively from an area, much as rivers sometimes fail to cope with rain from a storm. Consequently the ice begins to flood out from its rock channels (valleys) and adopts a different means of movement. Progressively the valleys fill: ice coalesces into an **ice cap** and then an **ice sheet**. The physical restriction of the valleys, their width, is gradually lost; the surface ice can flow out at a rate that eventually balances the rate of new snowfall. However, in the first stages of this sequence, as ice spills over the valley divides, it cuts deep troughs much as flooding rivers cut trenches in their bank levées. Such trenches, called **glacial diffluence troughs**, are an important feature of a glacially eroded landscape: they provide vital new routes for ice through otherwise difficult terrain (Fig. 21.13). Without them, many of the valleys of the Alps, Rockies and Western Scotland would never have received enough ice to become overdeepened and later be transformed into fjords.

Ice sheets on lowlands

The earliest phase of a cold period may well be characterised by ice forming in mountain regions and the development of the classic pattern of cirque and valley glaciers. Advancing valley glaciers will subsequently coalesce into an ice sheet as they spread out onto the surrounding lowland.

Very soon, however, the increasing cold that prevents snow melting in the mountains will have widespread effects in lowland areas as well, for if snow does not melt it will accumulate into ice sheets on lowland and upland alike. As a glacial period becomes established ice sheets will form on low-lying areas and spread out quite independently of ice flowing from mountains. Indeed, as Figure 19.2 shows, the area covered by ice in the last glacial periods was so great that not all the ice could possibly have come from mountains. We therefore need to think of lowland ice sheets as partly or even mainly self-nourishing; although some ice may travel long

(a)

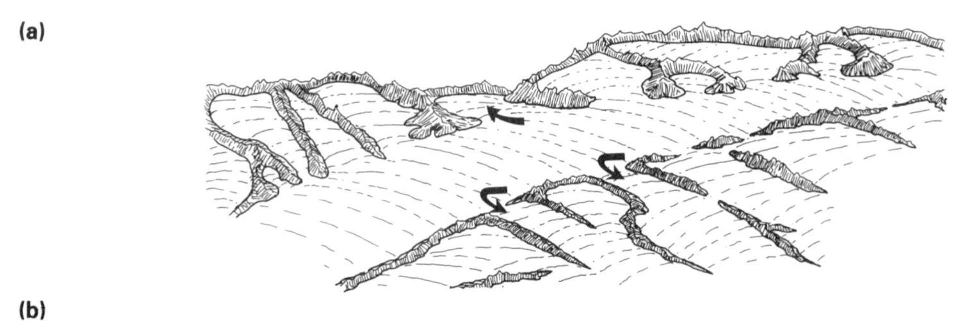

(b)

Figure 21.13 (a) At the maximum of the Alpine ice sheet, only the tops of the mountains would have been visible. Ice spilled across divides, breaching many river headwater regions and creating glacial diffluence troughs, which have provided the passes so vital for through transport in mountain regions. The sketch represents the Engadine region of Switzerland. Similar conditions created many famous passes such as the Pass of Glencoe, Scotland; Llanberis Pass in Snowdonia, Wales; Dunmail Raise pass between Ambleside and Keswick, Lake District, England; and the Pass of Tuolumne, Yosemite, California. (b) The rugged scenery of the granite peaks of Arran, Scotland shows that the peaks survived as isolated mountains (**nunataks**) above the general level of the ice. Notice that the headward retreat of cirque backwalls has led to the development of pyramidal peaks.

distances, much of it must continually form near to the spreading ice margin in the same way that rivers continually gather water from tributaries spaced out from source to mouth. This marginal growth is also made more likely by the changes in climate that occur as ice sheets grow (Fig. 21.14). The intense cold of a large ice sheet will cool the overlying atmosphere sufficiently for a permanent anticyclone to become established. This will block most moisture-laden winds, greatly reducing the rates of snowfall and ice formation towards the centre of the ice sheet.

Whether ice spreads from a mountain or lowland source, its effect on the lowland landscape will be immediate and substantial, for ice is capable of a great deal of erosion even on gently sloping terrain. The energy for movement, transport of debris and erosion depends on the product of mass and velocity. As a river near its mouth can often carry far more sediment than its component headwater mountain streams, so the vast bulk of an ice sheet can more than offset the slow flow. Because most lowlands of once glaciated regions are today buried under many metres of glacial debris, one may imagine that lowland ice sheets only deposit. There are no tangible results of erosion, such as deeply excavated glacial troughs. But the fact that an ice sheet leaves debris behind when it finally melts away does not mean it was incapable of erosion during its advance (Fig. 21.15). Indeed, the floors of glacially eroded troughs usually owe their flat form to deposition as the ice melted. Even an ice sheet that has scoured tens of metres of rock from thousands of square kilometres of ground leaves a landscape that looks far from spectacular when compared with glacial valleys (Fig. 21.15). Only occasionally can the true erosive power of lowland ice be estimated – in eastern England, for instance, where an ice sheet near the limit of its advance stripped back the line of a chalk ridge farther in a few thousand years than fluvial processes could have managed in millions. Where the ridge was overridden, the junction of upstanding chalk with the underlying clay rock is now twelve kilometres in front of the scarp edge compared with the relative position of the scarp where ice failed to override (Fig. 21.16).

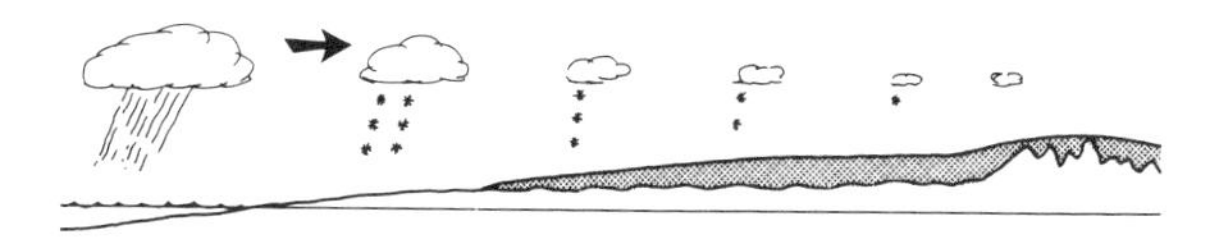

Figure 21.14 As ice sheets spread, the cold high surface induces a high pressure over the central ice sheet and ice accumulation slows down. Nevertheless, ice sheets continue to grow, nourished largely by snowfall on marginal areas produced by relief effects.

Figure 21.15 Many lowland landscapes near to the centre of ice dispersal show clear marks of glacial scouring. Many develop a very distinctive knobbly terrain. This example is from Sutherland in Scotland.

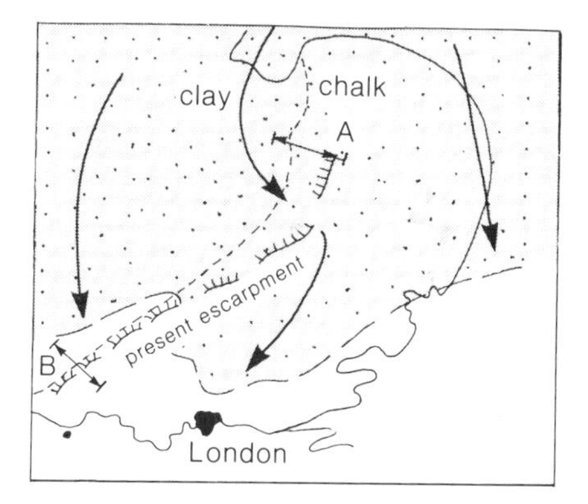

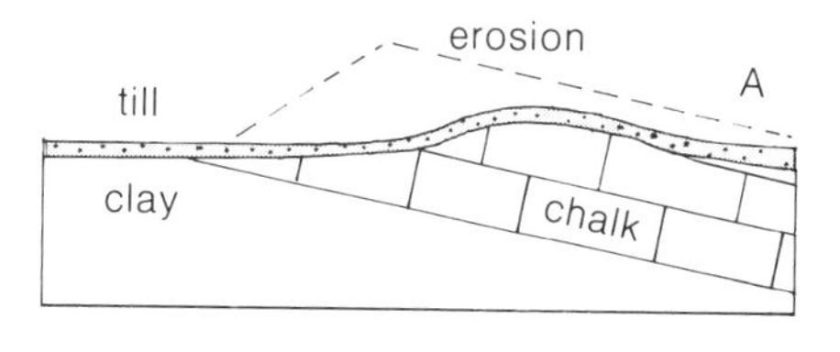

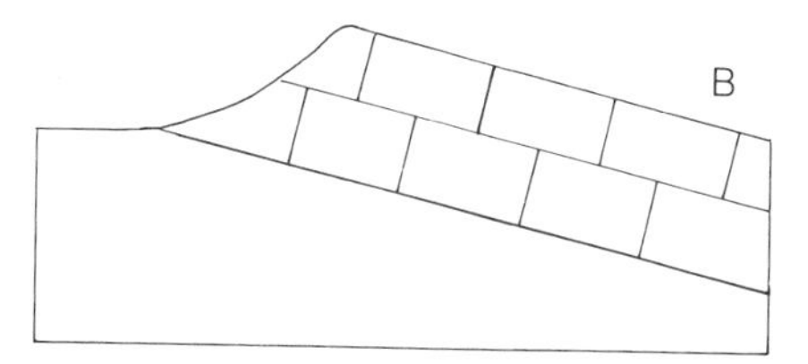

Figure 21.16 In Eastern England, a large part of the chalk outcrop has been eroded so greatly by an ice sheet that the chalk hills are in some places more than 12 km from their preglacial positions.

21.5 Glacial deposition

In the long term, despite widespread erosion, it is lowland that is most vulnerable to deposition, partly because the ice sheets move so slowly and partly because enormous debris loads acquired from upland and lowland sources make ever greater demands on available energy. The delicacy of the energy balance that controls transport, erosion and deposition shows very clearly in areas where 'drumlins' occur (Fig. 21.17). **Drumlins** are streamlined mounds of glacial debris deposited from, and reworked by, the base of a moving ice sheet. Their presence at first suggests that insufficient energy was available for debris transport, yet their reworking shows that at least enough energy remained to change their shapes. Drumlins often occur in groups of many hundreds in just those places where ice sheets may be thought to have considerable energy reserves. Perhaps, therefore, drumlin patterns may have the same relationship to an ice sheet as sand dunes to the bed of a river. Just as alluvial rivers continue to transport debris as they build up alluvial floors, so drumlins often rest on a sheet of basal debris (**till**) that is deposited simultaneously (Fig. 21.18).

The frequent change beneath an ice sheet from erosion to deposition often results in extremely complex landscapes, with scoured bedrock, hummocky till and upstanding ridges of moraine all occurring within short distances (Fig. 21.19). Deposition is rarely simple, even at the ice margin. Only if ice advances at the same rate at which its margin melts will the margin remain stationary long enough for debris melting out of the ice to build into a substantial ridge or **terminal moraine**. Many margins are not marked by any moraine at all (Fig. 21.20), while even those with moraines show a number of ridges that indicate repeated changes of marginal position.

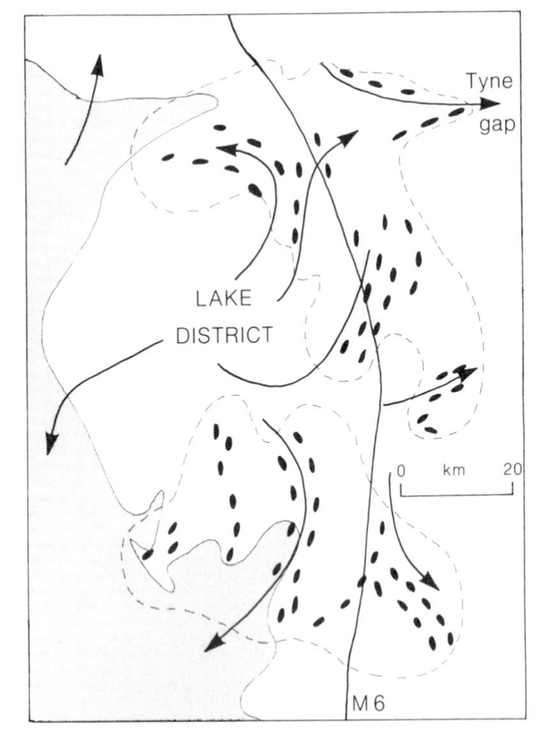

Figure 21.17 The orientation of drumlins indicates the direction of ice movement. In areas such as north-west England, drumlin fields are very extensive. Notice how they are closely associated with areas marginal to upland ice-sheet sources. These drumlins also show a good correlation with the areas of permeable bedrock. This is because any water formed by basal ice melt drains into the rock, considerably increasing the energy needed for debris transport.

(a)

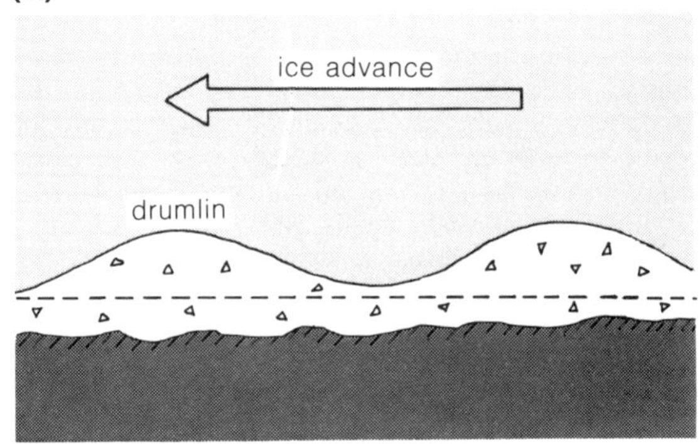

(b)

Figure 21.18 (a) Drumlins are often only surface irregularities on a widespread till sheet. They are features of active ice movement and rarely found near areas of maximum ice advance. (b) Drumlins half submerged in Newport bay, Co. Mayo, Ireland.

Figure 21.19 Deposits near the margin of an ice sheet.

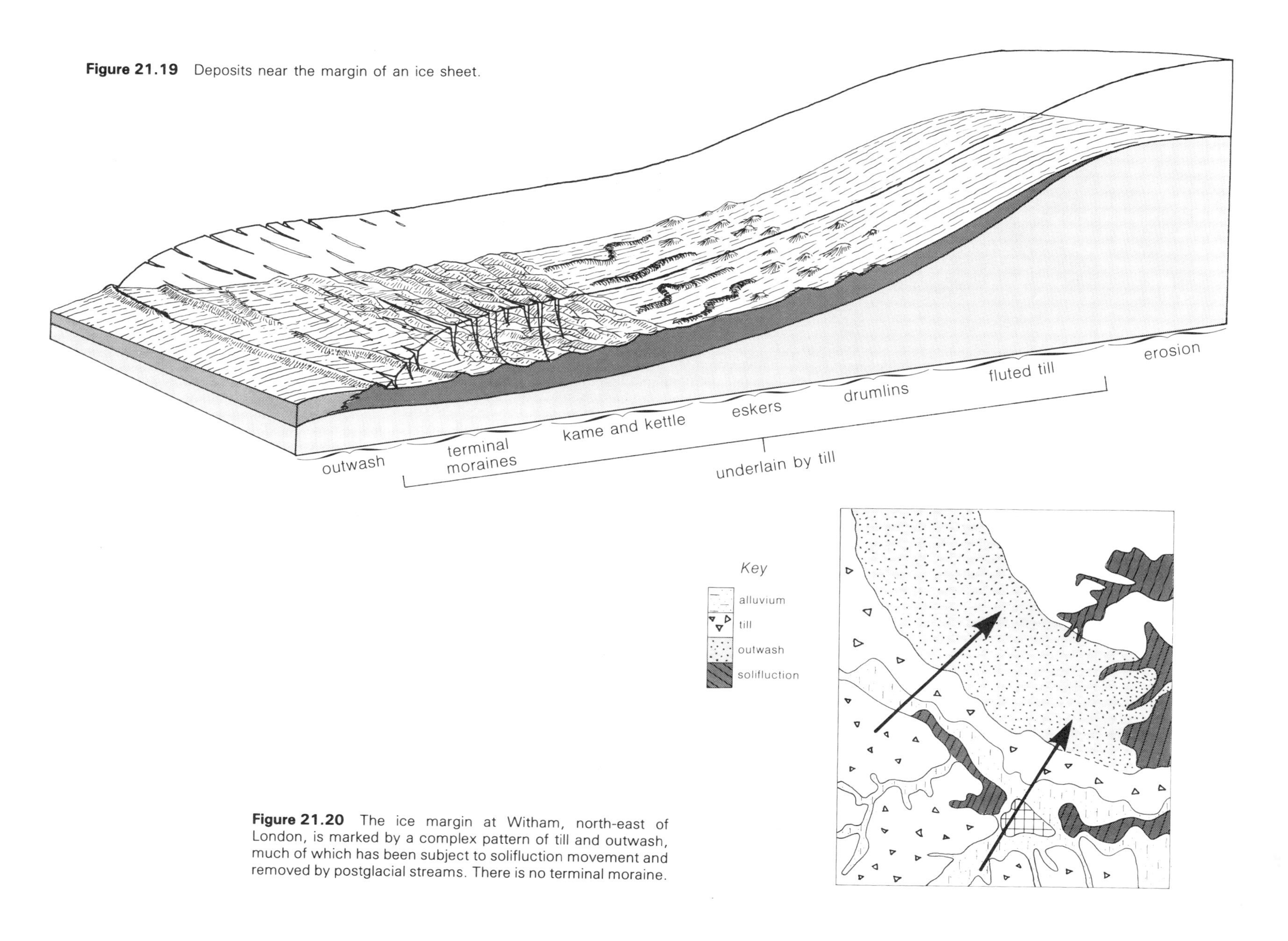

Figure 21.20 The ice margin at Witham, north-east of London, is marked by a complex pattern of till and outwash, much of which has been subject to solifluction movement and removed by postglacial streams. There is no terminal moraine.

21.6 Glacial ablation

Although some till is deposited beneath advancing ice sheets and terminal moraines build up during equilibrium conditions, the majority of deposition occurs at the end of the cold period, when the climate warms and melting (**ablation**) can set in over large areas simultaneously. At this time there are major contrasts between areas near the ice margin and those hundreds of kilometres further away. Only near the margin will ice melt back (**retreat**) faster than it melts down (**downwastes**). But even here everything depends on the rate of retreat. Slow retreat will give time for meltwater reaching the ice margin to criss-cross the area newly uncovered and reduce any ice-deposited features to a uniform sheet of sorted rubble (an **outwash plain** or **sandur**) (Fig. 21.21). Rapid retreat will give meltwater streams little chance to rework glacial material, and deposits may well be left intact (Fig. 21.22).

Warmer conditions spell an end to net snow accumulation except in the highest and most poleward regions. Deprived of snow, the reduced ice sheets simply cannot maintain the steep equilibrium profile needed to push ice near the margin forward. Ice sheets wane and the ice comes to rest. In this confused situation much surface water is created. Some water cascades into crevasses and forms icy ponds which gradually fill with debris transported by meltwater streams across the ice surface. Other water reaches the bedrock below the ice and flows as

(a)

outwash plain

0 km 5

Würmsee

(b)

Figure 21.21 (a) In southern Germany, ice advanced from the Alps, finally stopping along a wavy line now marked by a complex of moraines. When the ice finally melted, the terminal moraines acted as barriers behind which lakes have formed. Among the lakes is the Würmsee, after which the most recent cold period has been named. Beyond the moraines is an outwash plain. (b) A sandur plain beyond the Morteratsch glacier, Switzerland.

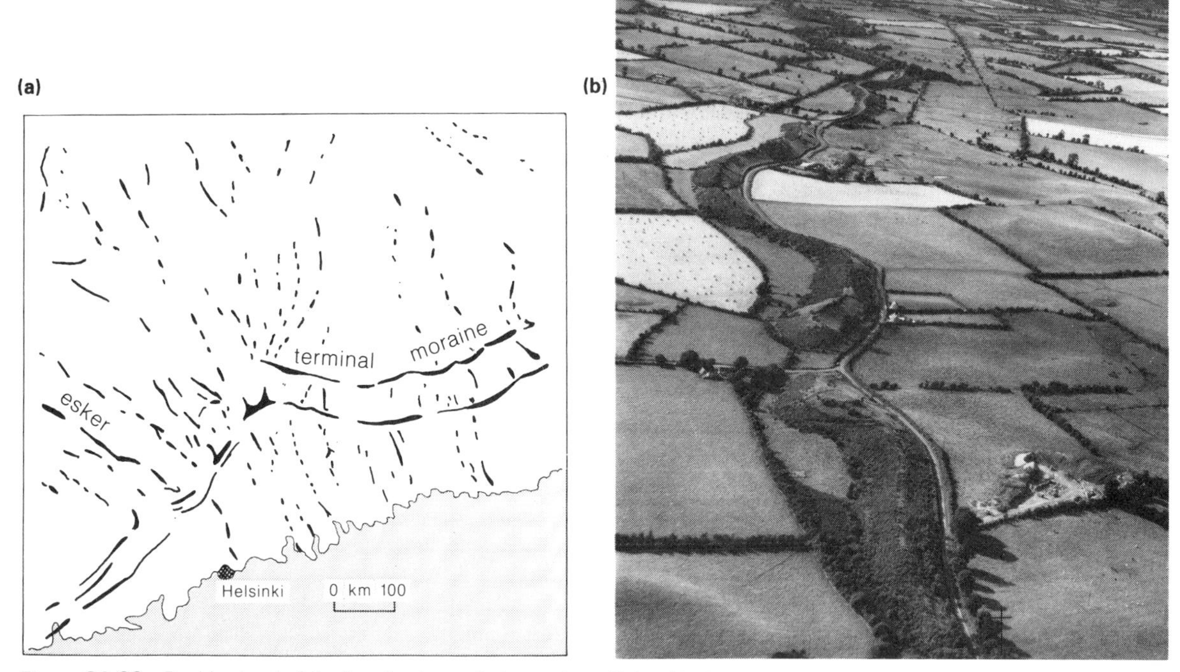

Figure 21.22 Rapid retreat of the ice sheet margin in southern Finland laid bare and preserved not only the terminal moraines but also the beds of subglacial meltwater rivers (eskers). (b) An esker south-east of Trim, Ireland.

subglacial rivers, sometimes cutting deep trenches into the bedrock (called **subglacial gorges**), at other times depositing a bed of coarse gravel and boulders.

Although some water flows in an organised manner, much may be held up in the chaos of hummocks and troughs formed by the debris melting out of the ice-sheet base. Debris has become thoroughly mixed up into the lower ice layers: how much there is at any place and how much melts out is therefore a matter of chance. Where ice contains much till, deposition may produce a hill. The till deposited during ice advance will contain many blocks of embedded ice that

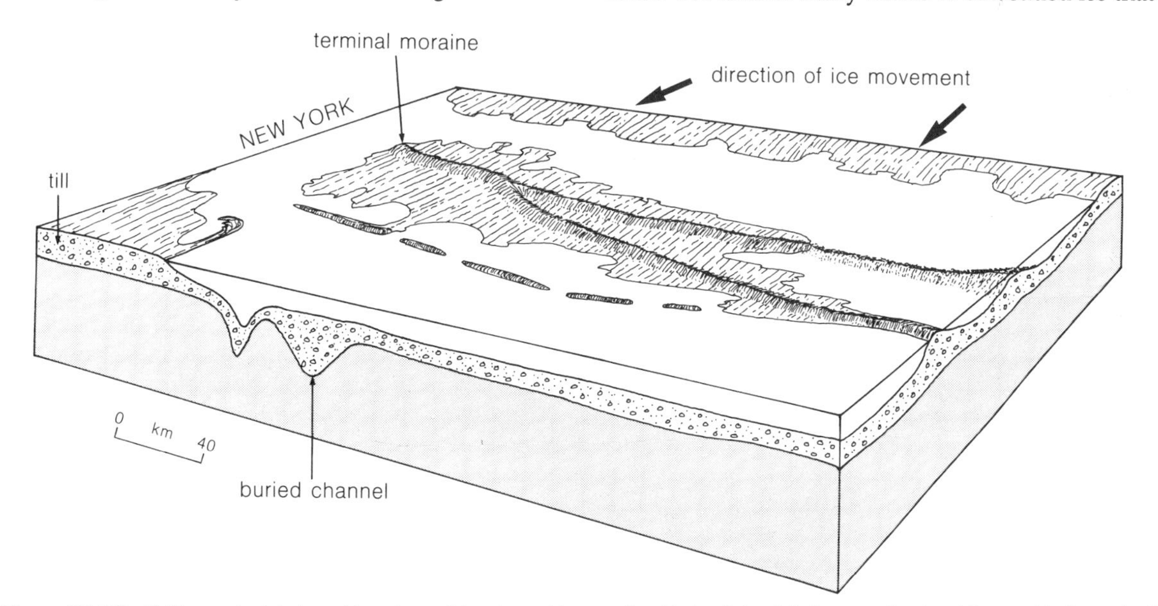

Figure 21.23 Without glacial deposition, Long Island would not exist. Bedrock is all below sea level, while an old channel of the River Hudson cuts a deep trench across what is now the western end of the island. The barrier islands have been built from outwash sands pushed up by the sea during the postglacial sea level rise.

became detached from the main ice sheet. When these ice blocks finally melt depressions (**kettle holes**) will be left which add diversity to the till surface and become a place of storage for meltwater.

The confused and hummocky nature of a till sheet surface tends to aid its preservation as it hinders the development of an organised drainage system. It may also help the survival of piles of sandy sediment (**kames**) from crevasse ponds or winding ridges of sand and gravel debris (**eskers**) that are the last remains of subglacial river beds (Fig. 21.22).

Today many coastal areas of northern continents, and even some entire countries, exist only because of widespread glacial deposition. Denmark and the Netherlands are perhaps the best known examples, but the East Anglian and Lincolnshire coasts of England are almost entirely made from till, while New York's famous Long Island would otherwise be far below sea level (Fig. 21.23).

21.7 The cycle of ice ages

Today the widespread advance of ice sheets during the **ice age** is a familiar story even though the idea is only a little over a century old. Before then most scientists did not even suspect that ice could advance at all, or play any major role in landscape erosion. As evidence for ice advance and erosion was found it became possible to show that ice sheets had inundated tremendous areas not just once but at least four times. With an eye to painstaking detail, despite each successive ice advance having eroded and reworked much of the debris of its precedessors, scientists were able to find enough remnant deposit to help them spell out a fascinating legacy of change.

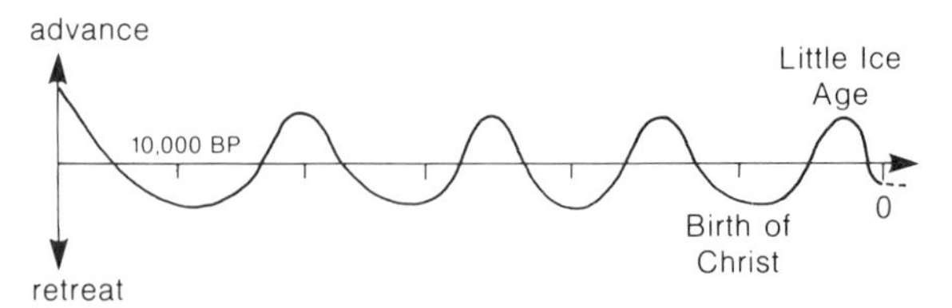

Figure 21.24 Since the end of the last major cold period, some 12 000 years ago, there have been a regular succession of cool phases when glaciers advanced worldwide. Cold periods had climatic fluctuations just as great as those at present, although they were all much colder than today. Glaciers advanced and retreated in sympathy, sometimes leaving a complex pattern of terminal moraines as silent witnesses (e.g. Fig. 21.20).

Four Alpine advances were established, called Günz, Mindel, Riss and Würm after a selection of local rivers and lakes near which each displayed its best evidence. Later, workers in northern Europe and America were able to find much the same sequence, although it appears that not all advances have been as important, nor have they all occured simultaneously in every region.

Today, evidence from deep-sea sediments shows that during the two million years of the Quaternary Period (the Ice Age), the global climate has been first falling into the icy grip of a cold glacial period and then curling back into a warm 'interglacial' period repeatedly. The continuing fluctuation of climate (Fig. 21.24) has led to widespread speculation of a possible return to a prolonged cold period with the prospect of ice sheets advancing once again to reach New York and London. The ice age may not be over yet.

Man in cold-climate environments

22.1 The permafrost zone

Those of us living in the temperate zone are used to thinking of cold conditions as features of the past. Where patterned ground once formed there are now vast fields of wheat; glacial troughs that once held silent rivers of ice now echo to the noise of highway traffic. Yet an increasing number of people live in the ten per cent of the world in which cold conditions survive and in which permafrost reigns supreme.

When you live and work in a region underlain by permafrost main everyday activities become a challenge. The undisturbed ground may feel as hard as a rock, but if you place a heated house upon it melting ground ice will soon lead to ground settlement and structural collapse (Fig. 22.1). An unheated warehouse on similar ground will protect it from the summer thaw; the consequent rise in permafrost may be sufficient to heave up the floor. In permafrost regions the remedies are to build either on very thick gravel pads or on piles. Structures kept well clear of the ground permit a natural air flow and so leave the permafrost undisturbed. Even tarmac roads conduct heat in a manner so different from that of natural cover that permafrost can melt, allowing the road foundations to flow away. Gelifluction is no respecter of man: any building or road on a slope is liable to move each spring. The safe rule is never build on a sloping ground unless piles can be driven into solid rock or permanently frozen ground.

Permafrost brings other problems. For example, normal sewage pipes and cess pools cannot be built, nor can conduits for electricity and telephone lines be placed underground. All must go in heated tunnels (called **utilidors**) above the ground (Fig. 22.1d). Water supplies are another problem, for shallow lakes are often frozen to their beds and service pipes are liable to continual rupture by freezing.

The same freezing force that brings large boulders to the surface and creates patterned ground will even succeed in lifting building piles if these are not frozen in the permafrost to at least twice the thickness of the active layer. Without these precautions even rail and road bridges can be heaved into the air in the most spectacular manner (Fig. 22.2).

Over the past few decades probably the greatest demand for exploitation of the permafrost zone has come from those concerned with oil and gas extraction. The vast reserves of Alaskan and Siberian oil and gas, and the need to transport them over thousands of kilometres of permafrost, have brought many new problems besides the old ones of building construction. Oil from the wells may rise at temperatures over 50°C and must be kept from contact with the permafrost. Oil pipelines cannot be buried underground nor be stored in conventional surface tanks because of the risk of rupture (Fig. 22.3). There is nothing easy about survival in the permafrost zone.

Figure 22.1 Problems of building on permafrost. (a) An unheated warehouse causes the permafrost to thicken and the warehouse is heaved up. (b) A heated house melts the permafrost and causes settlement. (c) In College, Alaska the church was built on frozen sand and gravel of low ice content, but the adjacent church school stretched over ice-rich silt with ice wedge polygons. The school collapsed as heat from the building melted the underlying ice. (d) A utilidor buried in ice-rich permafrost. Heat from the utilidor has thawed the ground and allowed the irregular subsidence of the utilidor that resulted in broken pipes. This is the reason many utilidors are now built on stilts above ground. ▷

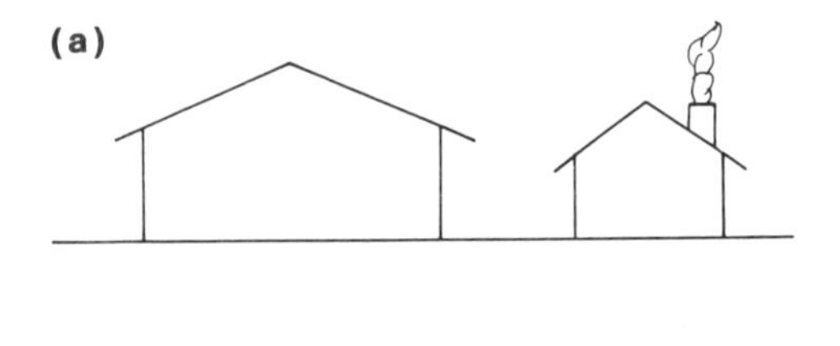
(a)

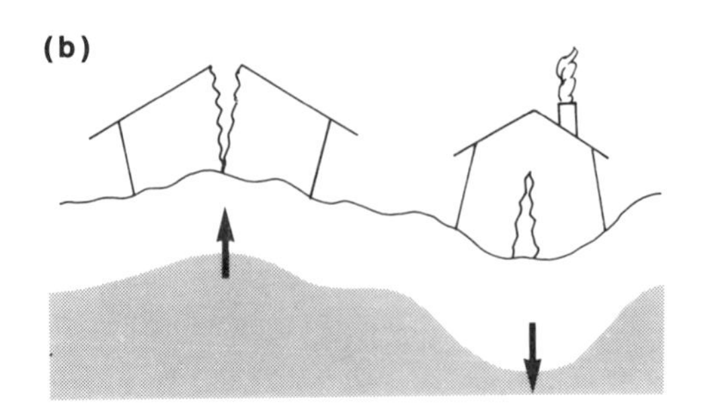
(b)

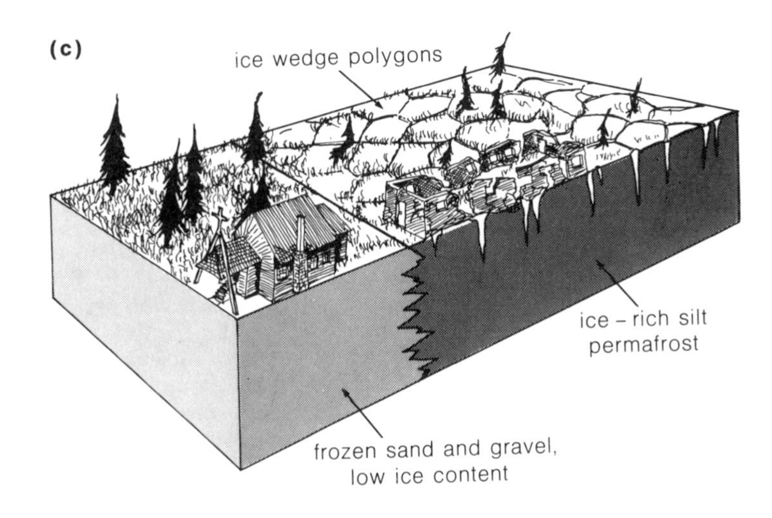
(c)
ice wedge polygons
ice – rich silt
permafrost
frozen sand and gravel,
low ice content

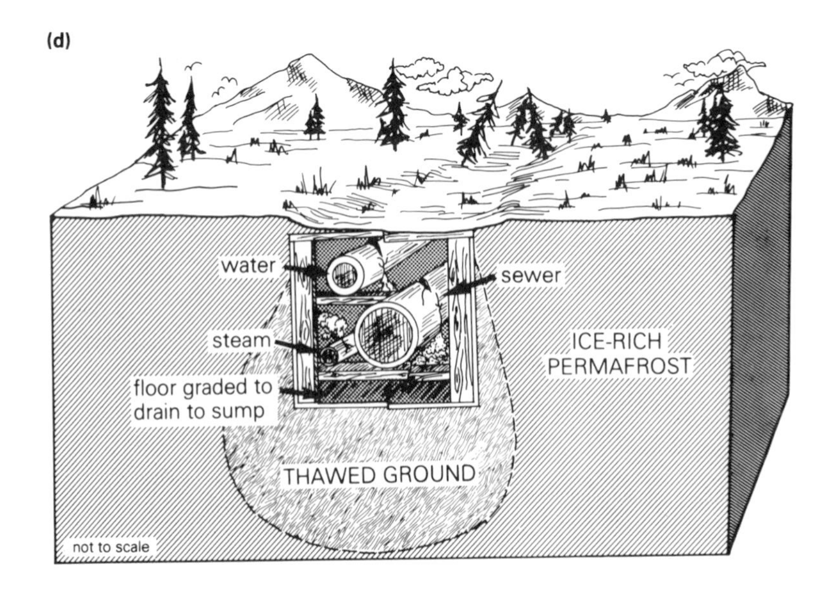
(d)
water
sewer
steam
ICE-RICH
PERMAFROST
floor graded to
drain to sump
THAWED GROUND
not to scale

Figure 22.1

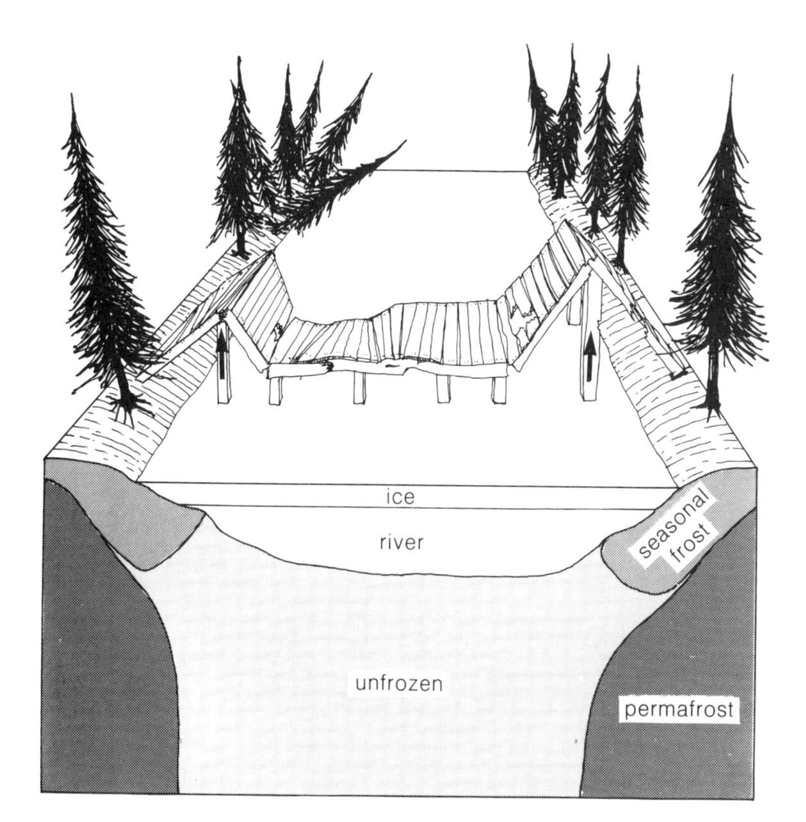

Figure 22.2 Frost-heaved piling of bridge spanning outlet of Clearwater Lake 8 miles south-east of Big Delta. Piles are of wood.

Figure 22.3 The danger of rupture with buried oil pipes has resulted in their construction above ground.

22.2 The influence of glaciation

If you live in a lowland area you might feel that glacial periods are irrelevant to your way of life – there are no deep glacial gorges or fjords, no terminal moraines, no prominent esker ridges to influence man's activities, but unseen below your feet may lie tens of metres of clay, sand and gravel: the hidden evidence of the recent all-pervading influence of ice.

A sample area south of Shrewsbury in the English Midlands, for example, shows hardly a spot unaffected by glacial deposition – by till, outwash sands and gravels of a former ice-margin lake clays (Fig. 22.4). Yet although these gently undulating landscapes may appear to be underlain by solid rock, many are formed entirely of glacial deposits.

Farmers are probably the most widespread users of land and they are acutely aware of the subtle variations of texture and fertility that accompany glacial debris. For example, till often yields clay-textured soils; in flat areas this tends to give imperfect or poor drainage. These types of soil must either be improved with expensive sub-surface drains or left as low-profit permanent pasture. Where till has been deposited in gentle hills the same soils are much more useful, for slopes allow surplus water to drain away and clay textures store moisture which can be used by plants during a summer drought. With these advantages cereal or fodder crops are most common and land can be coaxed into producing high yields. Where till gives way to outwash sand and gravel, soils drain too

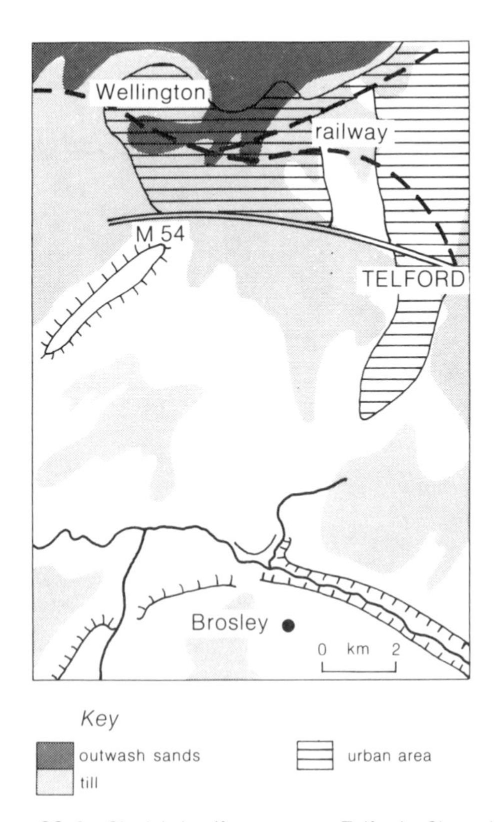

Figure 22.4 Glacial landforms near Telford, Shropshire, England.

quickly, leading to wilting problems in the summer. They are also less productive and so need more expensive fertilizers. Too dry for grass and too poor for cereals, these soils can be a severe test to farming.

But the farmers' problems are a resource for the building-materials industry: outwash sands and gravels are needed for concrete and for road bases in an area where river gravels are very scarce. The spreads of outwash north of Telford are dotted with gravel pits old and new – you can almost map the outcrop of outwash sands by mapping the pits that exploit it! Though till may help the building-materials supplier, it has always been a problem for builders; slopes of more than a few degrees are very prone to landslides. Many of the older villages seem to have avoided areas of till cover altogether, preferring either the drier outwash gravels (as with Wellington) or solid bedrock (as with Broseley). However, when work began on Telford New Town in the 1960s, there were no sites completely free of till. Telford's planners therefore designed shops, houses, industrial estates and bypass roads in patterns that made maximum use of flat land, leaving unstable steep slopes as grassed open spaces. The nearby motorway was forced onto till by the need to bypass the Wrekin hills. Because of the instability of till slopes, the several long cuttings were fashioned at a very gentle angle in an attempt to forestall landslides. The railway runs through several steep-sided cuttings, excavated in the days before the long-term instability of till was known. The railway engineers chose to cut through till because it was an easy material to excavate, but, as one would expect, landslides have since become a problem.

The nearby Ironbridge Gorge is the birthplace of the modern iron and steel industries, located at one of the few places where solid rock is found at the surface. Here the advantage gained from exposed coal seams was further enhanced by river transport. But even Ironbridge Gorge is a product of the ice age: it was formed at a time when the original northern outlet to the sea was blocked and what is now the River Severn diverted, spilling southwards (Fig. 22.5). Wherever you look in Shropshire, ice has left its mark.

In New York State, USA, the ridge-and-valley landscape of the northern Appalachians may also seem unaffected by past ice sheets. Yet the whole form of

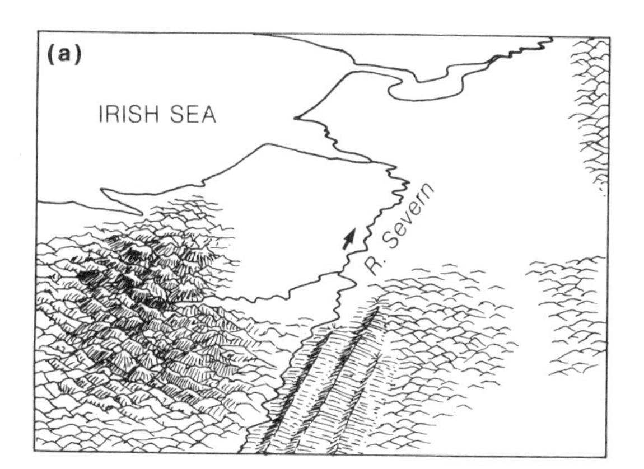

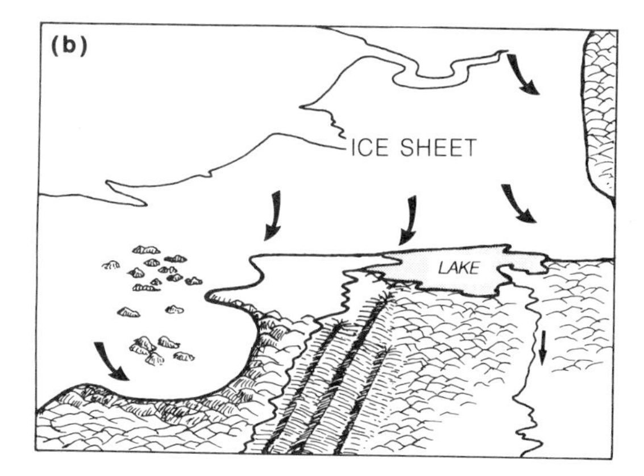

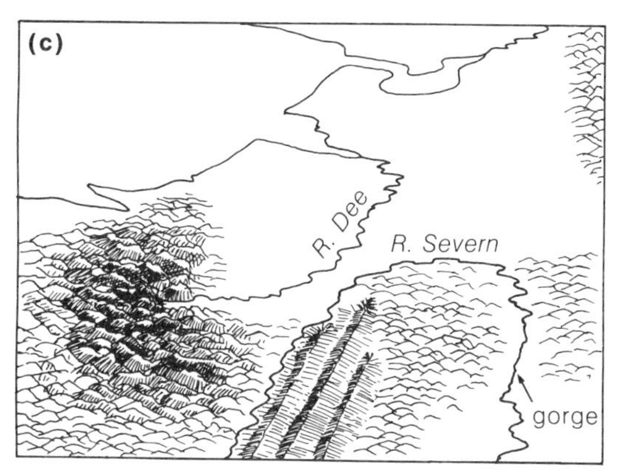

Figure 22.5 The glacial deposits of the Telford area come from an ice sheet whose source lay far to the north. At one stage ice covered all, but in a subsequent advance the ice limit lay near Telford, damming the River Severn into a lake and diverting its waters south, creating the Ironbridge gorge.

the original valleys is masked by till which is, on average, 20 m thick. In this area ice flowed *across* rather than along the valleys, leaving large quantities of till on the lee sides of the ridges, while keeping the sides facing the ice advance relatively till-free (Fig. 22.6). When the ice finally melted away, preglacial valleys of a symmetrical form had been replaced by valleys with pronounced asymmetry. Without a close investigation of the terrain, it is difficult to build up a clear picture of deposition in this 'till shadow' region. Some farmers discovered water within a few metres of the surface, others had to drill at least 30 m before they reached through the till to the underlying aquifer (at double the cost).

In many areas the complex pattern of deposition that occurred near to the ice margin has partially trapped outwash deposits between layers of till (Fig. 22.7). Because the outwash is water-bearing it can be used to supply water and to supplement surface streams in times of drought.

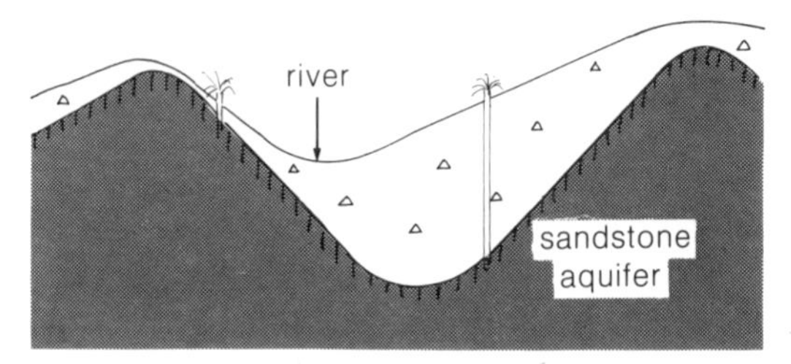

Figure 22.6 Cross section of 'till shadow' hills, Appalachians.

In upland environments glacial erosion has provided a pattern of through valleys or passes which have facilitated transport in mountainous terrain. Even **spillways**, channels created as meltwater cut into hill ranges, can provide vital transport routes. For example, the traditional routes from the east to the west coasts of America go through spillways, cut by the old rivers flowing from ice-marginal lakes.

Erosion has also influenced the sites of power plants and reservoirs. A glacially scoured trough, such as Thirlmere in the Lake District, in northern England, may require only a small dam to create a sizeable reservoir. Small dams are significantly cheaper than the large dams usually needed for reservoirs in unglaciated valleys. The Rocky Mountain storage lakes on the upper Colorado, USA, exploit the same advantages. In Snowdonia, North Wales, the natural cirque lake at Dinorwic was easily enlarged by building a relatively small dam. The cirque is now used as a convenient high-level pumped-storage reservoir in a power system, releasing water to generate hydroelectric power at times of peak national demand.

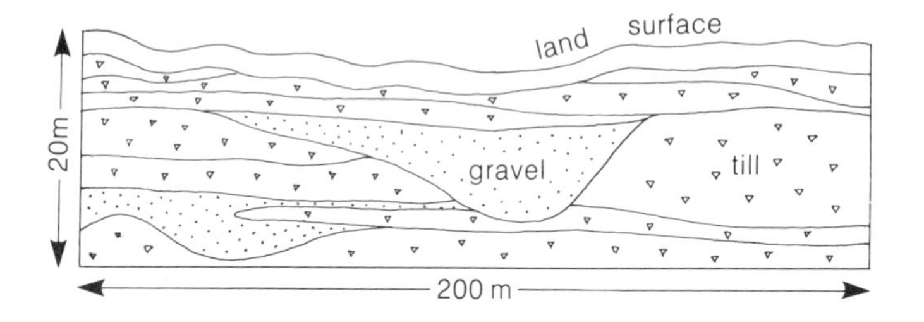

Figure 22.7 A section from the wall of the Harmattan open cast mine, Danville, Illinois. The tills of 3 glacial periods are shown. The large gravel feature in the centre is a buried channel, and is not only water bearing but provides valuable aggregate materials.

Part G

Coasts

23

Wave processes

23.1 The products of wave action

Coastlines are truly remarkable. Seen from the air the sweeping curves of beaches appear merely as thin, insubstantial lines, yet these beaches absorb the considerable energy of the waves. Beach materials may be nothing more than sand, a material mobile enough to be whisked about by wind, yet able to cling tenaciously to a beach; rarely are waves able to sweep it away.

The coast and its materials are resilient because both their shape and their reaction to waves are ideally suited to the nature of attack. Yet to explain how the coast is adapted to resist waves or how it changes with time is surprisingly difficult. Brief visits for holidays and news reports (Fig. 23.1) give little more than 'snapshot' impressions; they do not provide a balanced view of coastal processes. News reports tend to focus attention not on the long-term resilience of the coast but on its apparent short-term frailty – on the destruction of sea walls or the 'loss' of beach. But most of these examples in fact reflect man's interference with the coast rather than its own natural response.

The responses to waves of natural and man-modified coasts are strikingly different. Man, for example, has often favoured the building of sea walls made from vast slabs of reinforced concrete. Nature, on the other hand, uses loose beach sand or shingle.

The contrast between a stable beach of loose sand or shingle and a sea wall reduced to rubble is very striking. Rigid barriers are rare in nature, being found only as massive cliffs. In effect, natural coasts survive because there are reserves of sand for use in times of wave attack; and even when reserves have been exhausted and beaches severely depleted, wave processes rebuild beaches during the much longer calm periods. As we shall see, man's determination to yield not one centimetre of land to wave erosion has no part in natural coastal processes. Coasts, like other parts of the natural environment, are in long-term dynamic – not static – equilibrium. Without the long-term view, we will only be struck by the dramatic, isolated events where losses occur during storms and we will fail to notice the equally important gradual rebuilding and long-term equilibrium. The only way to gain the necessary perspective, both to understand the development of the natural coasts and to enable man to modify the coast successfully, is to examine coastal processes systematically, beginning with the nature of tides and the formation and movement of waves out at sea.

23.2 Tides

Tides are crucial in the study of coastal processes; although it is the waves that have the energy to act, the tides vary the range of heights over which waves operate and in part they determine the distribution of material on a beach. Tides are also responsible for some important currents in the ocean.

Tides are a result of the gravitational pulls on the Earth of both the Moon and the Sun. Because gravitational fields are directed toward the centre of a planet, the pull of the Moon is different at every point on the Earth's surface. (Fig. 23.2). At most places the pull is oblique, so there is an upward component (an **attractive force**) and a horizontal component across the Earth's surface (a **tractive force**). The Earth's gravity counteracts the upward pull, but there is no little natural terrestrial force to counteract the sideways pull. Water is constantly dragged towards the position at which the Moon is overhead, eventually accumulating as a high tide. Because the Moon orbits the Earth as the Earth spins on its own axis, the high tide appears twice a day at every place on the Earth's surface.

Figure 23.1 The Penlee lifeboat disaster of December 1981, when a lifeboat was smashed to pieces on the rocks off Cornwall, England while attempting to rescue the crew of the 1400 t *Union Star*, here seen capsized and in the process of breaking apart.

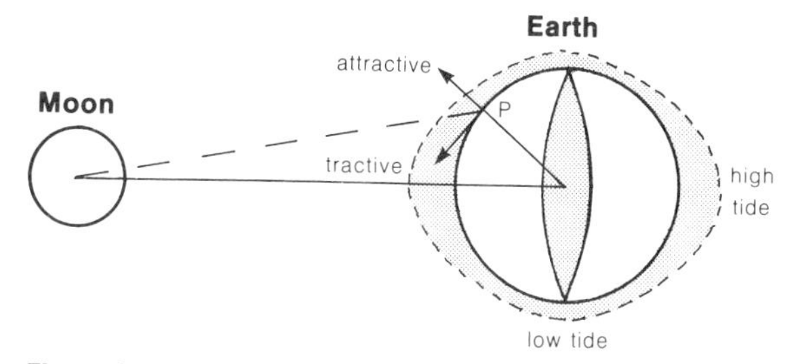

Figure 23.2 The gravitational forces of the Moon primarily responsible for tides.

The Sun has a gravitational effect similar to that of the Moon, but because it is so far away its effective pull on ocean waters is only half as great. This is still quite sufficient to complicate the pattern of lunar tides. For example, when Sun, Moon and Earth are in line, the solar influence reinforces the lunar high tide giving an extra high tide or **spring tide**. When the order is Sun, Earth, Moon, the solar and lunar pulls are in opposition and a minimum or **neap tide** is produced. The varying orbits of Sun, Earth and Moon produce constantly varying sizes of tide, with neap and spring tides each occurring roughly twice a month.

Tides in the oceans

If there were no land masses, water would be dragged continually eastwards around the Earth's surface, following the Moon's orbit (Fig. 23.3). However, because oceans are divided by continents into separate compartments, the pattern of tides is dramatically altered. It is further modified by the tendency of moving water always to veer to one side. In the Northern Hemisphere, moving water veers to the right; in the Southern Hemisphere, it veers to the left. Winds veer also. To illustrate the resulting change imagine an ocean bounded on every side by continent. As the Moon appears at the western horizon (in the Northern Hemisphere) water moves towards it. As the flow begins the water is deflected to its right, with the net result of a high tide in the north-west (rather than in the west). As the Moon rises and moves across the ocean, the water continually moves also, again deflected always to the right. Over a full day the combined effect of the tidal motions is thus to create a wave which slowly rotates around the ocean basin in two complete sweeps. This largely explains why high tide occurs at a different time at each part of a coast.

You can make a mini-tide for yourself. Next time you have a drink, give the container a small swirling movement and watch the resulting wave (tide) run round the liquid inside the container while the centre of the liquid stays still.

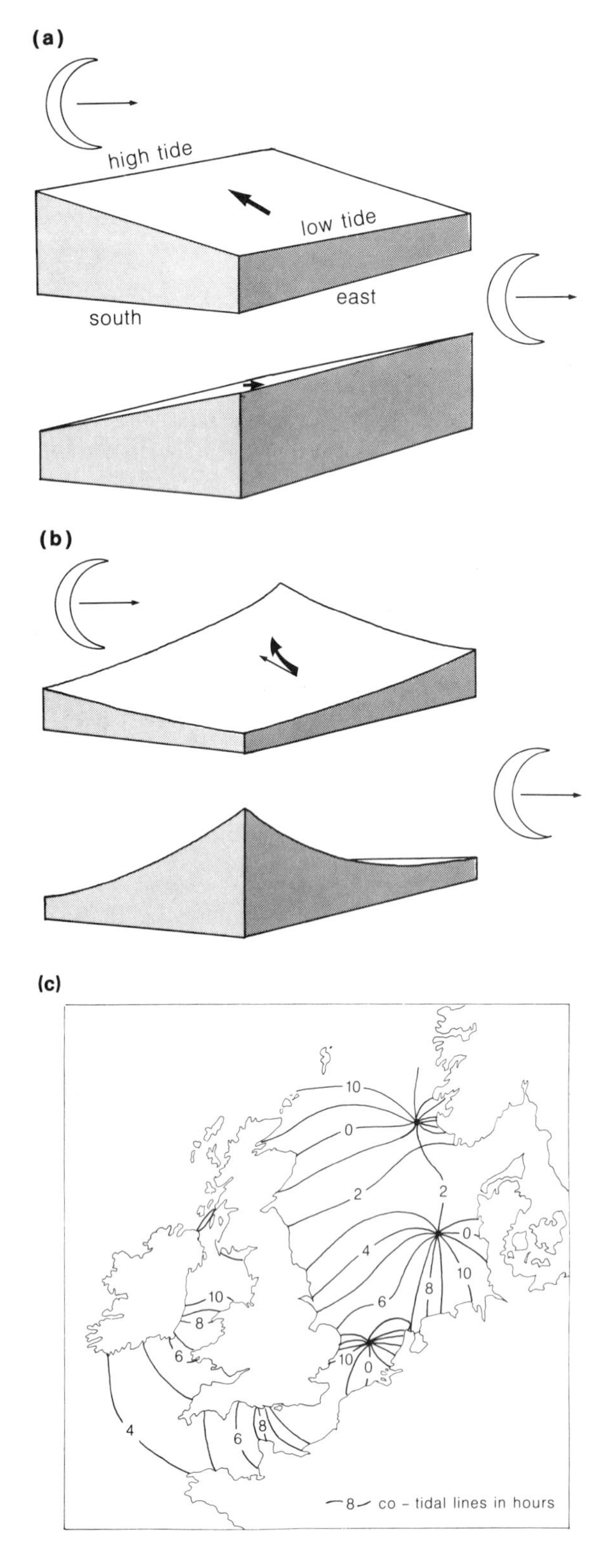

Figure 23.3 (a) As water follows the movement of the Moon, tides move across an ocean basin from west to east. (b) The Coriolis effect that deflects moving water to the right prevents a simple east–west wave motion. Instead, it forces the tide to run round the outside of the basin. In this case the centre of the ocean experiences no tide and is called the **amphidromic** point. (c) Tides around the British Isles.

23.3 Waves

The growth of waves

There are two discernible types of wave created in seas and oceans. Of these the pressure waves called **tsunamis**, which are generated by earthquakes, are the most spectacular. Although almost imperceptible in deep oceans, they grow many tens of metres high as they reach shallow beaches, their momentum causing them to 'climb' out of the sea and move on shore to limits well above those of the most severe wind waves, wreaking havoc on the coastal lands. Fortunately tsunamis are also very rare. Dropping a pebble into a pond of still water produces a similar, if far smaller, instantaneous transfer of energy, creating a simple sequence of ripples that spread outwards from their source. However the second and more common type of water waves are those created by wind blowing across a water surface (Fig. 23.4). In

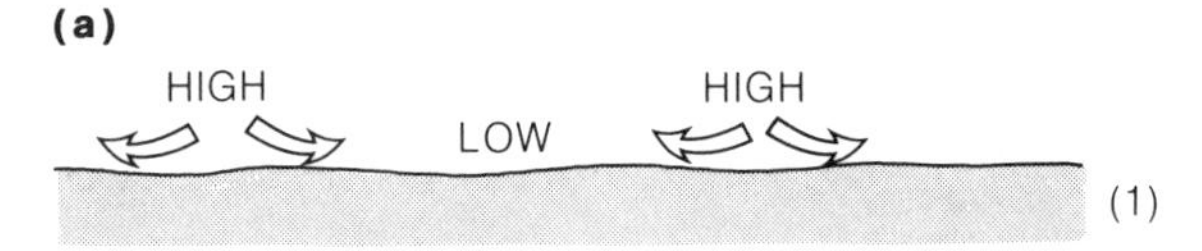

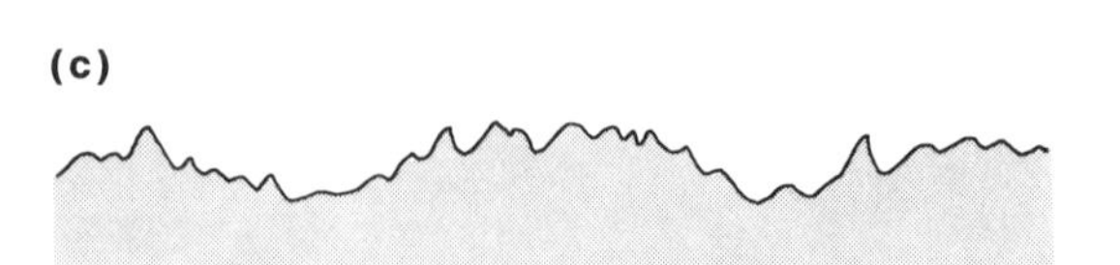

Figure 23.4 (a) The wind simultaneously generates waves of many heights, lengths and periods as it blows over the sea. Wind waves (a), caused by changes in air pressure, catch the exposed windward face of disturbed water (b), and a complex wave train develops (c).

this case energy is transferred from wind to water continually rather than instantaneously. As the wind blows across a motionless water surface, frictional drag causes the air to develop eddies, some of which spiral air downwards onto the sea surface. Where they press the water is pushed down; where compensating eddies spiral upwards water can rise. This causes the water surface to undulate. The upstanding places 'catch' the main thrust of the wind, as do the sails of a ship and are pushed forward as waves.

A wind actually 'pushes' many series of wavelets (**wave trains**) as it blows, each train moving away at an oblique angle to the direction of the wind. Most wavelets decay, but those moving downwind are reinforced and grow. Most of these waves are closely spaced and steep-sided; they require little water for their formation. Widely spaced waves take far longer to form because much more water must be set in motion. If winds prevail for long enough they can set up waves whose bases allow them to grow to heights of many metres before they break (Fig. 23.5). These waves can travel faster than the closely spaced waves, devouring them and absorbing their energy. For this reason widely spaced waves eventually become dominant during a prolonged storm.

Winds produce such a variety of waves that a method of description is needed more precise than simply 'closely spaced' or 'tall'. Experience shows that the most useful way to describe a wave is by its **wavelength** (distance between succeeding crests, *L*), **wave height** (vertical distance between crest and trough, *H*) and **wave period** (the time taken for succeeding crests to pass a given reference mark, *T*) (see Fig. 23.6). 'Closely spaced' waves, for example, might more accurately be described as having (say) a wavelength of 40 m, a height of 1 m and a period of 5 s.

Ocean waves

You can appreciate the way waves develop by blowing across the surface of still water in a dish. Even

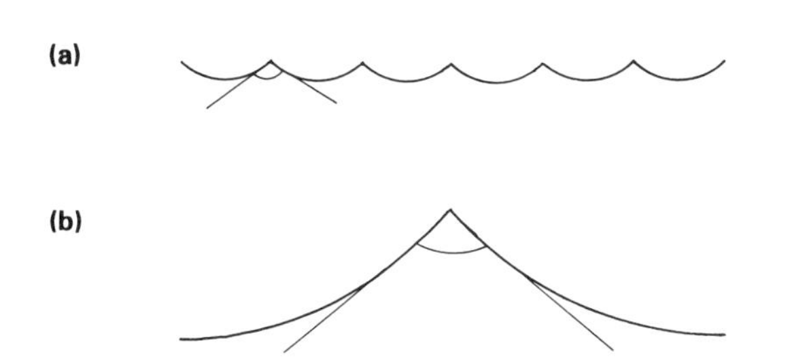

Figure 23.5 (a) Closely spaced waves form quickly, but are restricted in height, as the crest angle soon reduces to 120° and breaks. (b) Widely spaced waves form slowly, but can become towering giants before they break.

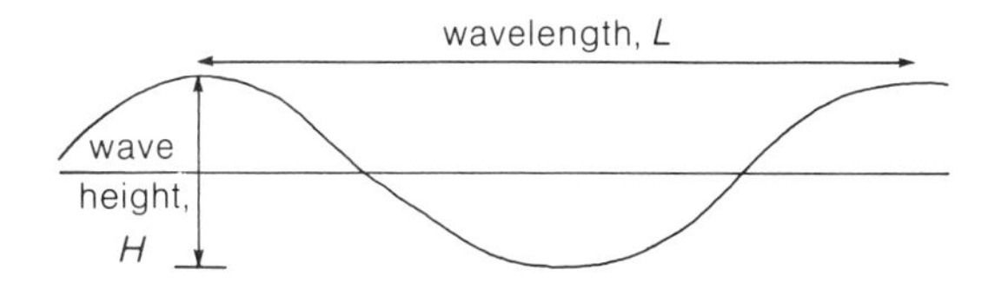

Figure 23.6 The dimensions of a wave.

blowing gently you can produce a pattern of air pressure waves that will set the water rippling. If you blow harder, feeding more energy into the water, the ripples will grow into waves that change from a rounded shape to a sharp-crested one; eventually they break. The longer you blow, the greater the area of disturbed water and the greater the complexity of the wave pattern as newly created wave trains are ruffled by the continuing flow of air.

The pattern produced in a dish models the formation of waves over open oceans, with wavelength, height and period all determined by the nature of the wind and the amount of energy that can be transferred to the water. For example, the longer the wind can blow constantly over a wave (the longer the **wind duration**) the more energy can be fed into it and the larger it will grow. Similar effects result from a greater **wind speed** and from a greater distance (**wind fetch**) over which the wind can blow (Fig. 23.7).

If a strong wind blows constantly for many hours there is time enough for all the many types of wave

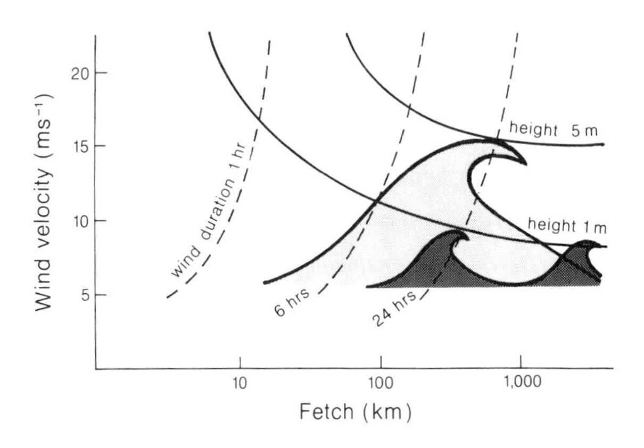

Figure 23.7 Wave formation. A 1 m high wave can be produced from many combinations of wind speed, fetch and wind duration. It could be formed by an 18 ms^{-1} wind in 1 h or a 7 ms^{-2} wind lasting 24 h. But for a wind to blow over the same water for 24 h requires a fetch of 400 km. By contrast, a 5 m high wave cannot be formed in a short time without hurricane force winds: even at 20 ms^{-1} such a wave would need 5 h and a fetch of over 60 km to develop. In general, therefore, it is rare for large waves to develop anywhere except in the open ocean.

WAVE ENERGY

Every wave is a parcel of concentrated energy moving across the ocean.

The amount of energy in each wave (E) increases with the square of the wave height (H) and its period (T). Put as an equation, this is

$$E \propto H^2 \times T^2$$

In other words, tall, broad waves carry the most energy. Storm waves often have a considerable height but a short period; many swell waves have a small height but a long wave period.

It is not easy to tell exactly which waves have the most energy and it becomes vital to measure all the dimensions of a wave. In Figure 23.8, for example, the energy in the storm wave train would be $H^2 \times T^2 = 5^2 \times 9^2 = 2025E$, while that in the swell wave $0.5^2 \times 12^2 = 36E$. Thus the storm wave train has 56 times as much energy as the swell wave train. However, if the storm wave had a height of 1 m and a period of 5 s then the energy ($1^2 \times 5^2 = 25E$) would have been only two thirds that contained in the swell wave half as high.

to reach full height. At this stage the energy passing into the waves is balanced by energy expended in friction. These conditions produce a **fully arisen sea** in which waves vary from towering giants the height of a house to the smallest ripples.

Local storms may create waves so near to coasts that the complex wave train of a fully arisen sea reaches the shore in almost its original form (Fig. 23.8). Most of these **storm waves** are steep, with wavelengths ten to twenty times the wave height. Some waves formed in the open ocean may be driven for long distances under constant winds such as the Trades. But even waves formed over a long fetch and over a long time eventually move away from their source area and receive new energy. Once out of the generating area, the waves begin to decay as energy is used in keeping the wave moving. Because long waves travel faster than short waves and contain more energy, only the long waves travel far from their source areas before fading away (Fig. 23.9). These are the broad **swell waves**, for example, that reach the shores of south-west Britain having travelled from the south Atlantic. They characteristically have wavelengths thirty to over five hundred times their height.

Waves near the coast

Waves are propagated by the frictional drag of wind, in the way we might spin a top by striking it a glancing blow (Fig. 23.10).

Individual water particles move only in circular orbits; they do not travel *with* the wave. This type of wave, called an **oscillating wave**, is similar to that produced when, for example, a rope is flicked or a whip cracked: the wrist is moved up and down, and the progressive transfer of energy along the rope or thong occurs as a further sequence of up-and-down motions staggered in such a way that a wave form is perceived to travel forwards. It is this difference between the particle movement (up and down) and the wave form itself (forwards) that explains why a ship can ride up and down on waves but is not carried along with them, and why a fishing float merely

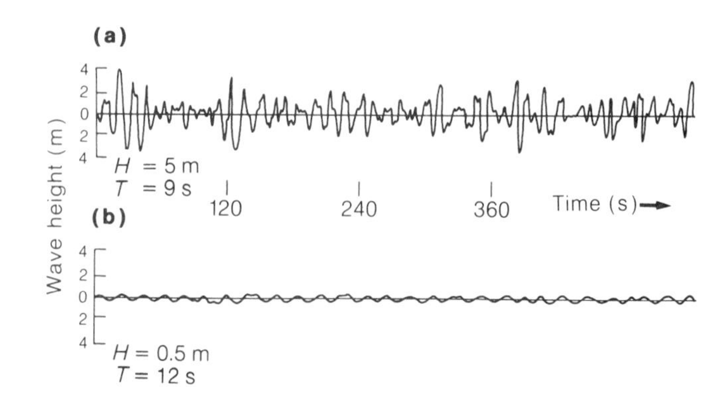

Figure 23.8 Characteristic wave trains. (a) A complex of storm waves. (b) The much more regular shape of swell waves. It is useful to be able to represent these trains by simple values. For this purpose we use characteristic wave height, H = two thirds the maximum wave height, and characteristic wave period, T, found by dividing the number of axis crossings into the total wave train time.

Figure 23.9 Swell waves on the coast of England, subdued after travelling long distances.

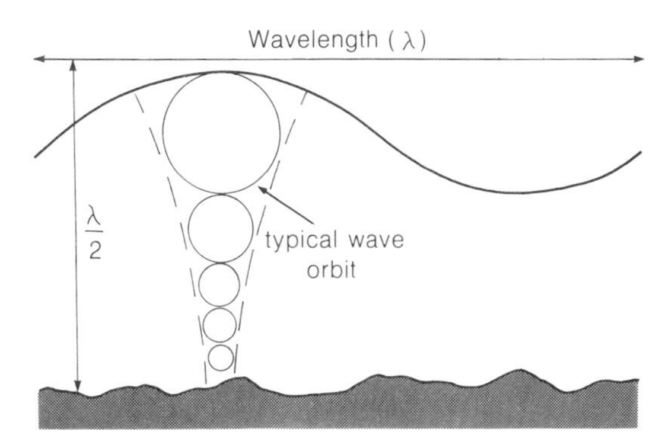

Figure 23.10 Surface disturbances only affect water to a depth of half a wavelength.

bobs up and down in the waves created by a passing boat.

Although we see the wave move along a rope, not all the energy applied by the wrist reaches the far end. Some is used in overcoming friction in the rope itself, and more is used to displace the surrounding air. Similarly water waves cannot travel over the surface without disturbing the water below, so using up some of their energy: this is why waves decay. However, the shock-absorbing properties of water are so great that no motion from the surface reaches depths below about half the surface wavelength (Fig. 23.10). Out in the open ocean, therefore, surface waves never disturb the bottom and they decay only slowly by dissipating energy in movement. In shallow water the situation is dramatically different. Some ocean waves, for example, may have wavelengths of over 100 m, and these therefore need 50 m depth of water for full damping to occur. Whatever the critical depth may be, when a wave 'feels bottom' it begins to change shape, as the circular orbits of its water particles become distorted into ellipses. Increasing frictional resistance with the sea bed causes the waves to slow down; in effect they begin to concertina and develop shorter wavelengths.

Wave refraction

Only rarely will a wave crest arrive over a uniformly shallowing sea bed in a direction exactly parallel to the coast. When this happens, all parts of the wave 'feel bottom' at the same time and begin to slow down exactly in step. The wave crest finally breaks, in a completely even way, in a single straight line parallel to the shore. But few sea-beds are smooth and evenly sloping; most are irregular in form, containing frequent ridges and troughs. And wave crests rarely arrive parallel to the coast, but rather at some oblique angle to it. Wave speed is very sensitive to water depth. Because each section of wave crest is in part independent of the rest, an uneven sea-bed may slow down one part of the wave crest before another. The wave does retain its overall unity, however, despite variations of speed in its individual sections. It accomplishes this in much the same way as a line of soldiers who execute a turn (Fig. 23.11a). These changes of speed along the wave crest cause the whole crest to swing round – in effect it bends towards alignment with the sea-bed contours.

The consequences on coastal processes of wave-crest distortion (called **wave refraction**) are far-reaching. The convergence and divergence of waves affects the distribution of energy and therefore the power each section of wave can bring to bear on the coast. These effects are easily illustrated using wave-refraction diagrams such as those in Figure 23.11b. In diagram (i), for example, waves approach obliquely an evenly shoaling coast; wave crests bend in such a way that they become much more nearly parallel with the shore. The direction of movement of the crests is more clearly shown by lines called **orthogonals.** These are normally drawn from points equally spaced along a deep-water wave crest and they represent the sequential positions of these points on the original wave crest. In diagram (ii) of Figure 23.11 a complex bay and headland coast cause the waves to converge onto the headland and diverge within the bay – a pattern clearly shown by the orthogonals. However, although in this case (as well as the first) the sea-bed contours follow much the same pattern as the land above sea level, it is the sea bed that controls refraction, not the coast that we see. Diagrams (iii) and (iv) illustrate how hidden features of the sea bed will cause waves to diverge or converge along what, from the land, appears to be a perfectly straight, even beach.

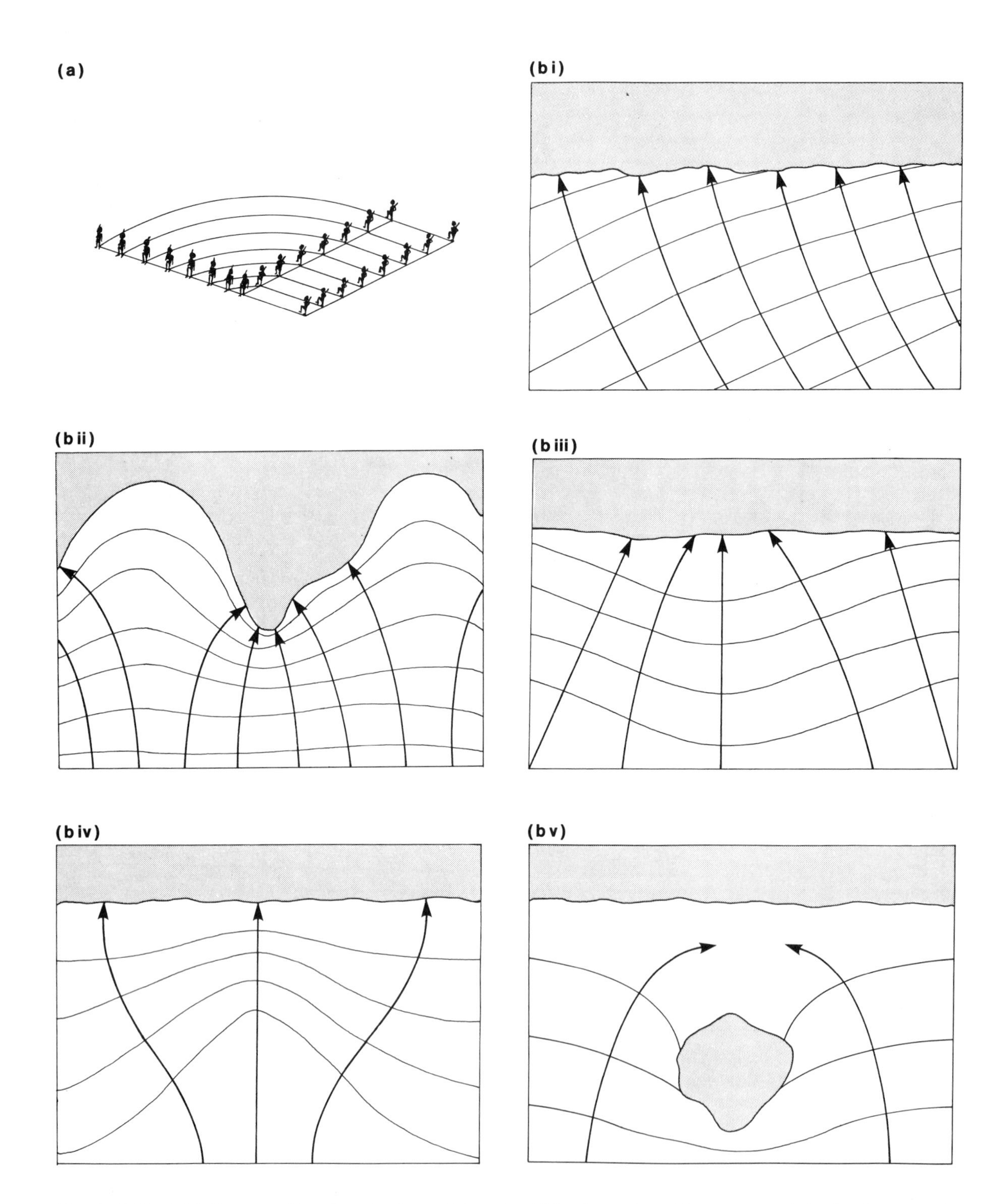

Figure 23.11 (a) Soldiers marching abreast only maintain a straight line by moving at the same speed. As they turn a corner, the inside man marks time, while the rest slow down in a progressive manner. In this way, the integrity of the line is maintained although it swings in a curve. (b) Wave refraction patterns: see text for explanation.

One further example of wave distortion occurs in the lee of islands. As wave crests approach an island they swing round in the same manner as when approaching a headland. However, as diagram (iv) shows, toward the rear of the island the parted wave crest is drawn together again in such a way that the two sections of wave come face to face. The ensuing conflict consumes much energy, and such regions often become sediment traps.

REFRACTION AND COASTAL ATTACK

Wave orthogonals are much more useful than wave crests because they divide waves into segments containing equal amounts of energy (Fig. 23.12). If wave refraction causes orthogonals to converge, the energy contained between them must be concentrated; if the orthogonals diverge, energy must be more widely spread. In real terms energy-concentration changes are seen in terms of varying heights along a crest: regions of concentrated energy have crests that grow in height while regions of energy dispersal seem to grow lower. These changes of energy along a wave underly the common observation that large waves occur out at headlands while only small waves are found in bays.

CASE STUDY: THE COAST OF SOUTHERN ENGLAND

Sufficient is now known for wave motion and refraction to be modelled mathematically in computers. To study waves approaching any stretch of coast the only new information required each time is detail of the sea-bed contours, and the height, wavelength and period of the expected waves.

Off the coast of Bournemouth surveying reveals an irregular sea bed. Although these irregularities occur far below sea level and remain permanently out of sight, they will have a major influence on wave refraction: when sea-bed data are fed into the computer, the predicted modifications to a storm wave approaching from the south-west are very severe (Fig. 23.13). Notice, for example, how orthogonals are concentrated at the headland between Poole Bay and Christchurch Bay. But while this result is to be expected (see Fig. 23.12) simply from the coastal plan, the concentration of orthogonals in the centre of Christchurch Bay is quite unexpected. A study of local records, however, clearly shows that this area (Barton-on-Sea) is currently experiencing severe erosion problems. The orthogonal pattern also shows

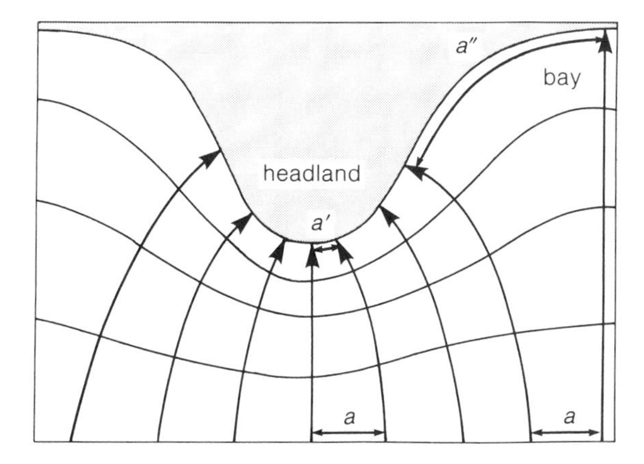

Figure 23.12 Wave orthogonals reflect wave energy. Wave energy, represented by wave length *a*, is concentrated into length *a′* by refraction. On entering a bay, the same energy is spread over a wide length *a″*.

(a)

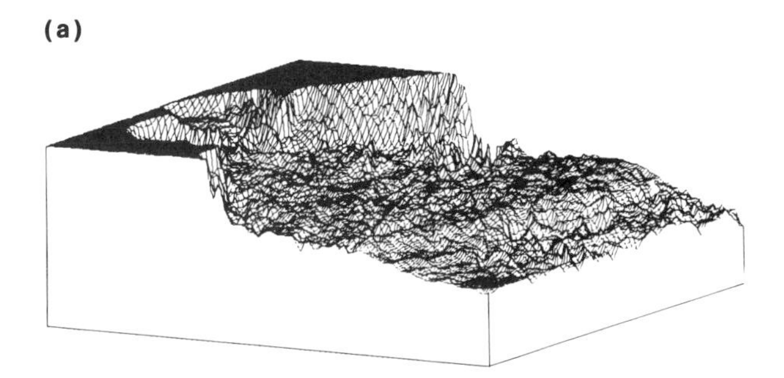

(b)

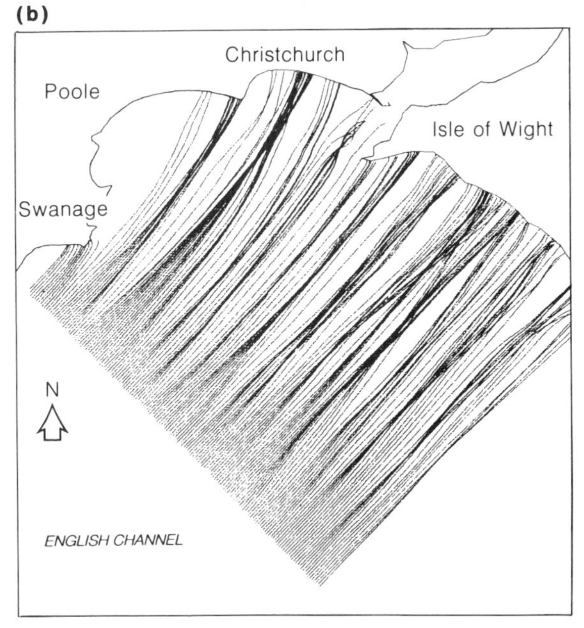

Figure 23.13 Wave refraction in practice on the south coast of England. A natural sea bed is an extremely irregular place: seen as a three-dimensional reconstruction, the coast near Bournemouth (a) is full of shoals and deeps. It is hardly surprising, therefore, to find that waves are greatly refracted (b) even far away from shore.

that there is no easy way to prevent the concentration of wave energy and its associated coastal erosion, for the source of the problem is shoaling of the sea bed more than 30 km offshore! This is a salutary reminder that coastal processes involve more than the narrow strip of land we call the shore.

The pattern of orthogonals produced in Figure 23.13b is, of course, for only one type of wave coming from one direction. Poole Bay and Swanage Bay may be sheltered from such waves, but both bays would be far more exposed to southerly and southeasterly waves, and these too would need to be studied if we wished to make a more complete study of coastal processes in this region.

24

Coastal landforms

24.1 The impact of waves

Think of the coast as the buffers at the end of a railway track. When waves reach the coast all of the contained energy must be dissipated quickly.

Figure 24.1 The characteristics of the coastal zone. Here apparently calm water out to sea rears up and breaks as a plunging wave on the beach. Ahead of the breaking wave is the backwash remnant of its predecessor. In the foreground, the beach sediment shows quite a range of particle sizes, while in the distance there is little beach, and waves break directly against the cliffs.

As they approach the coast, waves lose energy in two ways. At an early stage frictional drag with the sea bed slows the wave down and begins the long process of wave refraction. It also uses up some energy. The transfer of energy along the ocean surface is rather like a wave travelling down the thong of a whip: although some energy might be used up along the way, the main kick is at the end. The water ultimately reaches the shore: whatever energy is left in the wave is dissipated against the cliff or beach. However, it is not just the *amount* of energy that is important: just as vital is the way it is applied to the coast. Every time a wave breaks against the shore it delivers a short sharp shock – a rapid transfer of energy with an impact as dramatic as the crack of a whip. Consequently the coastal (or **littoral**) zone contains many of the most rapidly changing landforms on Earth (Fig. 24.1).

Types of littoral zone

Those of us who live inland may view the coast in somewhat black-and-white terms, often classifying it rather casually by the presence or absence of a beach. But, of course, there are many types of beach-free coastline and many types of beach (Fig. 24.2). For example there are also many stretches of coast, especially near river mouths, dominated not by rock or by beach sand but by fine 'mud' deposits. The more closely we look, the more variety we find. Some rocky coasts, for example, contain pockets of sand in small coves or bays between headlands; conversely some sandy beaches extend for many kilometres, broken only by the occasional bank of rock.

Yet although there is tremendous variety in coastlines, the underlying processes are less varied; they depend directly on the way energy in one form can be changed into other forms. Energy dissipation can be as **heat** produced by wave turbulence; by pushing sediment into **motion**; and by **erosion** of solid rock.

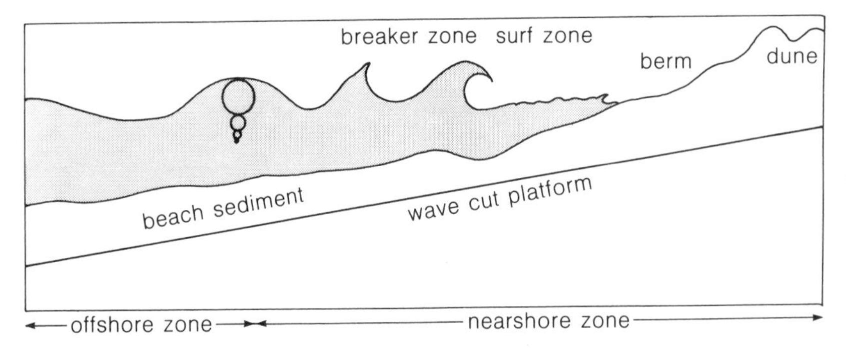

Figure 24.2 Terminology used to describe beaches and coastal waves.

As waves move towards the beach the water becomes shallower. When the depth is less than 1.3 times the wave height the wave begins to break. The area between the point of first breaking and the point at which the wave finally loses its orbital form is the **surf zone** of broken water.

In calm conditions many small waves break on the shore without disturbing any sediment. They spill forward and dissipate their energy by gliding gently across the beach as a thin sheet of broken water; the rough sand surface acts as a brake. Larger waves, with a greater height or period, have more energy to dissipate; simple friction with the beach does not exhaust the wave energy. As a result, the churning water in the surf zone, the forward movement of the broken water up the beach (called the **swash**) and/or the return flow (called the **backwash**) all set sediment moving. Storm waves may contain so much energy that, in addition to moving sediment up and down the beach, they even carry some of it away, exposing the bare rock at the foot of a cliff to direct attack.

Coasts without beaches

Wave refraction often causes energy to be concentrated at certain places along a coast. These places can usually be identified because they lack beaches; any sediment produced by erosion is immediately swept away. Most headlands are high-energy environments where waves transfer their energy directly to the cliff rocks.

At least at high tide, most headlands experience reasonably deep water where short-period (small) waves may arrive without breaking. These waves are often reflected back to sea by steep cliffs, much as light waves are reflected in a mirror (Fig. 24.3). In some cases incoming and reflected waves match so that their wave forms interact – a sort of stroboscopic effect such that the waves stand still. You can produce a **standing wave** (or **clapotis**) in a bath, by moving the water rhythmically back and forth. With this type of wave the water simply appears to rise and fall without any to-and-fro movement. However, because there are two waves involved, each reinforcing the other, the final wave height is much greater than that of the incoming wave alone. For example, a wave moving towards the coast with a height of 1.5 m can be transformed into a standing wave 3 m high.

Because of the continual change of their height against the cliff, standing waves apply constantly changing pressure. Water can apply a lot of pressure even when calm – imagine a bucket of water on your foot. The constant change of pressure associated with rising and falling water is often enough to weaken and break a well-jointed rock.

For a wave to reach the cliff without breaking the depth of water must be more than 1.3 times the wave height. This depth may occur only at high tide: it is quite common to find that waves that produce clapotis at high tide start to feel bottom and break as low tide approaches and the water becomes shallower. Long waves will probably break at all stages of the tide and will never produce clapotis.

The pressures that build up against a cliff when a wave *breaks* are quite different from those produced by a standing wave. Instead of changing smoothly, **breaking wave** pressures are brief 'shock' waves acting only in the region in which the wave crest hits the cliff. In the case of fractured rock a breaking wave can act rather like a pneumatic drill, forcing water rapidly into the fractures (Fig. 24.4). However, the shape of

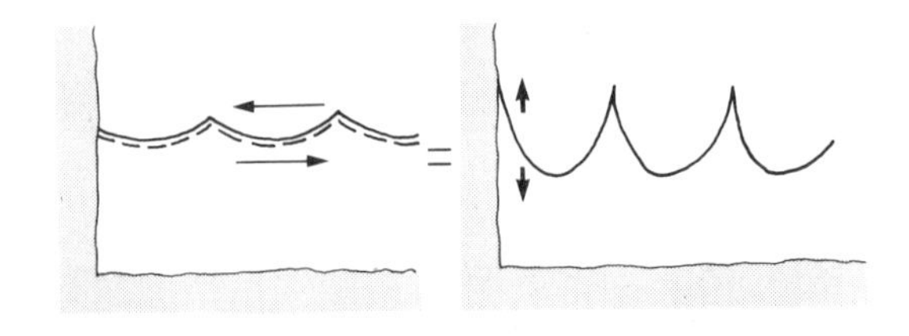

Figure 24.3 Standing waves at a cliff impose continually varying pressures.

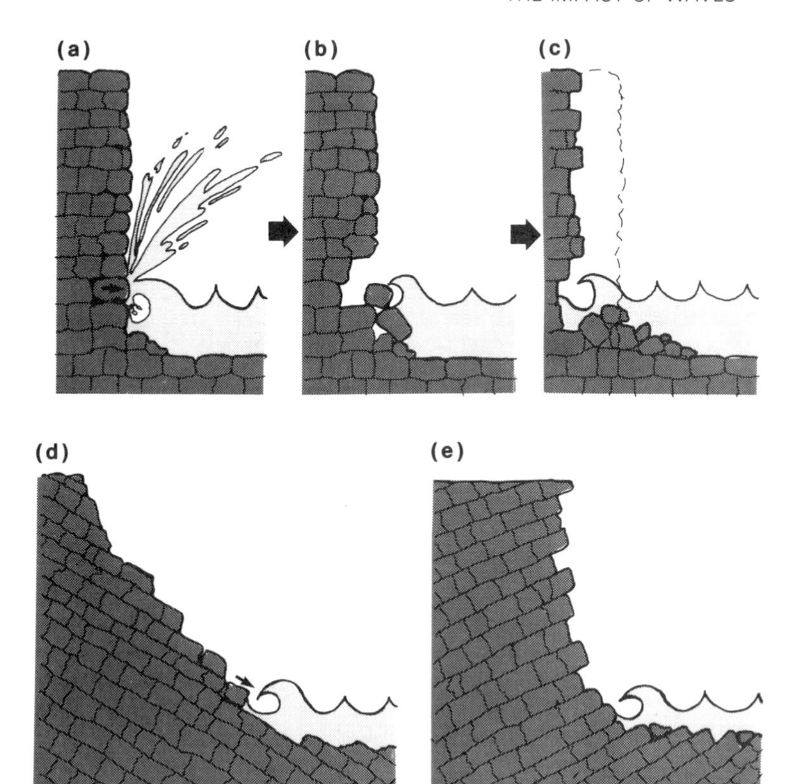

Figure 24.4 The hydraulic action of breaking waves (a–c) is most effective with plunging waves and fractured rock. The resultant cliff profile (d, e) is very sensitive to rock structure.

the breaker is critical: 'plunging' crested waves are by far the most effective for they trap a cushion of air against the cliff, allowing the body of the wave to drive the air deep into fractures and behind surface blocks. Breaking wave pressures may be over 15 times those for a non-breaking wave, even on a cliff without fractures; and air compressed into fractures will create local pressures that are even higher.

This is the **'hydraulic' action** of waves. It is the main agent responsible for cliff undercutting between tidal limits and this in turn leads to the collapse and retreat of many cliffs of fractured materials.

Headland scenery

Of all natural processes, marine action is probably the most sensitive at etching out variations in rock strength. Almost any headland will show a diversity of **coves** or **caves**, **arches** and **stacks** where hydraulic action has picked out faults in the weakened rocks or places where joints are more frequent (Fig. 24.5). This difference from erosion on land is due to the ease with which weathered debris can be removed. Every particle of debris on land has to join the slow cascade that will carry it down hillslopes and along river

Figure 24.5 Handfast Point, Dorset with the stacks of 'Old Harry' (at the end) 'and his wives'. Notice the contrast between marine and sub-aerial erosion, the former producing a crenulate pattern of bays in two directions, suggesting that a pattern of rectangular joints is being exploited. The surface landscape shows none of this variation.

channels to the sea; waves in a high-energy environment are able to carry sediment away instantly, leaving the rock exposed for further immediate attack. Under such continual onslaught by the sea, many rocks yield ground very quickly, leaving only the most resistant as headlands (Fig. 24.6). Even resistant headland rocks eventually retreat, leaving behind only the occasional remnant arch or stack to rise from a gradually widening rock bench cut down to wave level.

In this way cliffs are slowly replaced by **wave-cut platforms**. Wide platforms cause waves to run into shallow water a long way from the cliffs, forcing them to break and so dissipate most of their energy in a mass of foaming surf before reaching the shore. As a result platforms grow ever more slowly until a stage is reached when the material eroded from the cliff cannot immediately be taken away. From then on such material remains permanently covering the landward side of the platform, and a **beach** is formed.

Coasts with beaches

Even a casual observation of waves moving on shore shows that all waves dramatically change shape as they reach shallow water. At sea, wave crests may be rounded; but as they move into water less than half a wavelength deep, increasing friction with the bottom causes a distortion of the waves and they become progressively more sharply crested (Fig. 24.7). A wave remains unbroken only as long as its crest does not sharpen to less than 120°. At a depth of about 1.3 times the wave height, the crest finally breaks. As

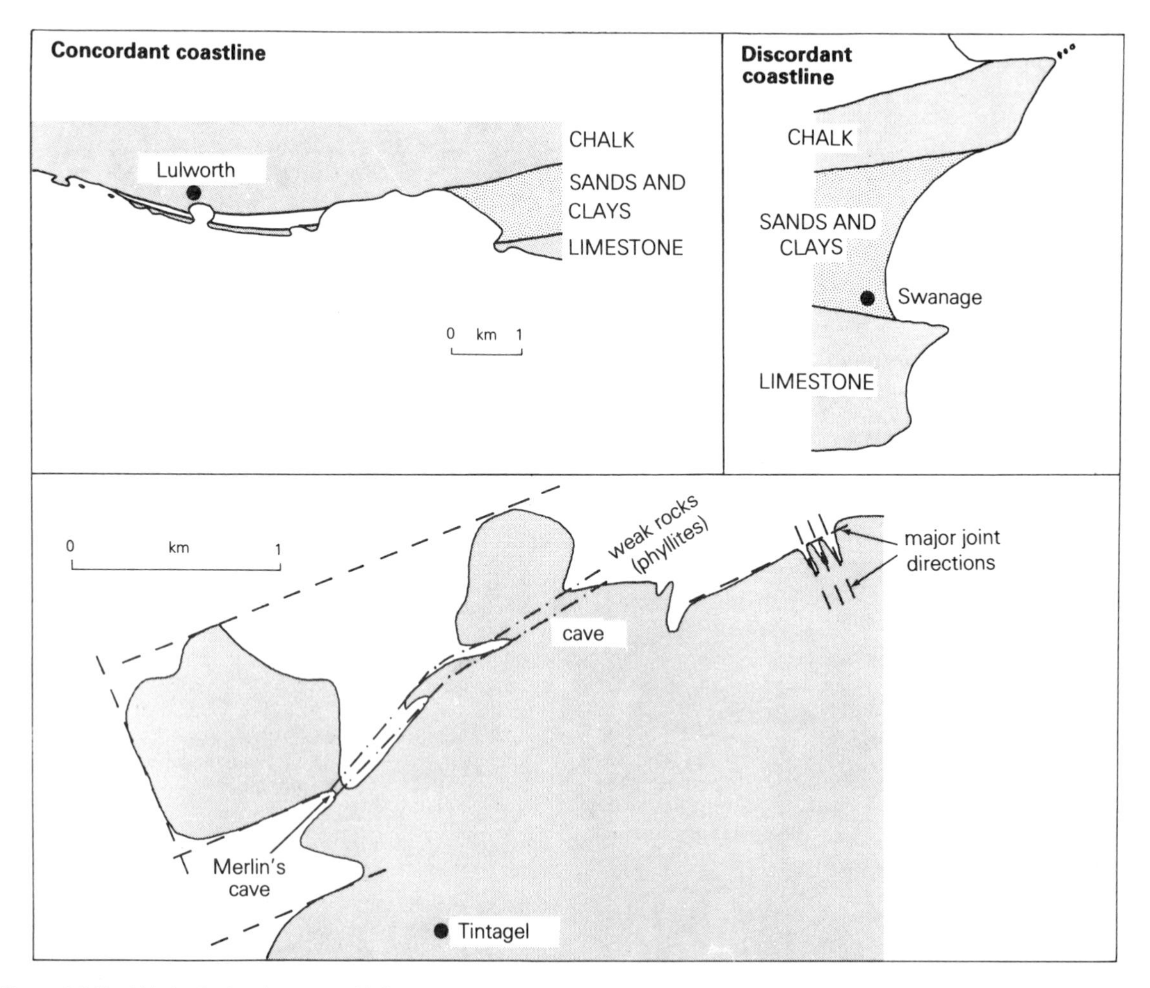

Figure 24.6 Lithological and structural influences on coastlines: (top) broad patterns of headlands and bays, Dorset; (bottom) an intricate pattern of coves, Tintagel, Cornwall.

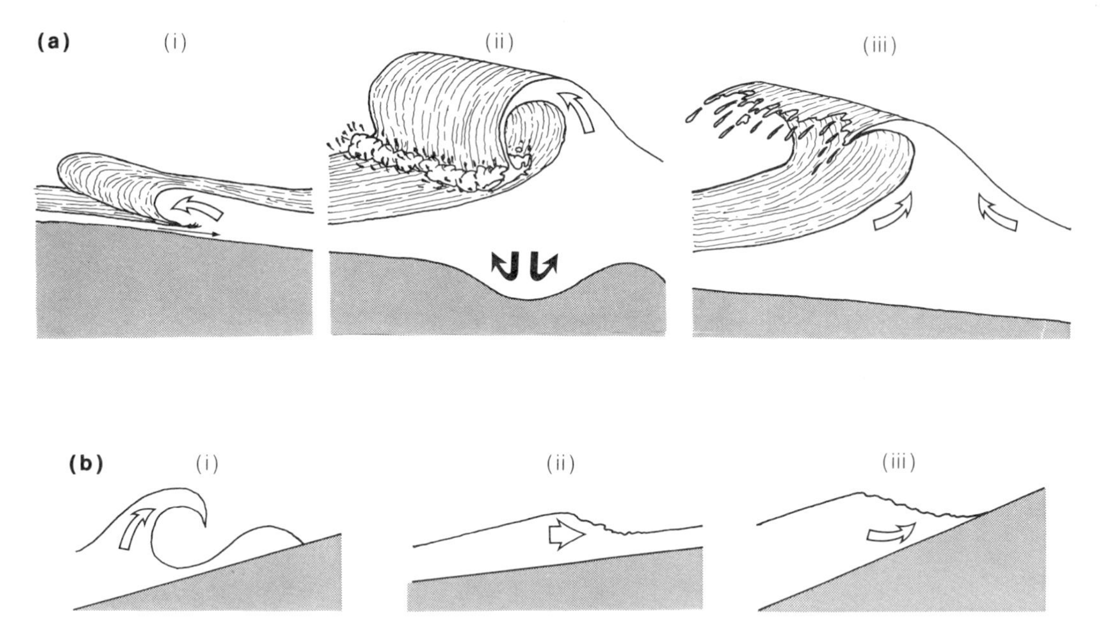

Figure 24.7 (a) Three stages during the formation of a plunging wave. (b) The three main wave types: (i) plunging; (ii) spilling; (iii) surging.

there is no longer enough depth for the water to sustain even a distorted elliptical orbit, water begins to fall away from the crest line and a **breaker** is born (Fig. 24.7).

The type of breaker that forms along a shore depends both on the wave shape and the beach slope. For example, short-wavelength waves approaching a steeply sloping beach will break only close to the shore. With little width of beach on which to dissipate their energy, these waves will rear up (i.e. grow higher) before finally breaking in such a way that water **plunges** vertically from the crest line and down onto the beach, disturbing much sediment. The same wave on a gently sloping beach will feel bottom slowly and gradually, the wave motion will distort progressively, and energy will be sapped from the wave over a large distance. When the wave finally breaks, the wave orbits are elliptical and the crest **spills** forward, riding over the water ahead of it. In contrast with the plunging breaker, this form of movement causes little disturbance at the breaker line. Instead it tends to thrust water ahead as a powerful **swash**, and it is in the long swash that most sediment is moved on shore.

Waves of large wavelength will feel bottom far from the shore and will invariably produce spilling waves, whereas on really steep beaches a wave may reach the shore before it breaks, the base of the wave **surging** forward to balance the **backwash** (return flow) of the previous wave.

24.2 The movement of sediment

Beaches

Beaches are merely veneers of sediment (usually sand or shingle), resting on wave-cut platforms. They are continually reworked by waves into quite complicated shapes, controlled not only by the type of wave that breaks but also by the size and shape of the material from which it is made. For example, waves can pile up cobble-sized (shingle) beach material into a much steeper slope than is possible for sand.

Some parts of a beach are more exposed than others. The effect of variations in exposure shows particularly clearly in bays in which some beaches face directly into the dominant wave direction while others remain relatively sheltered. Along the Californian coast, for example, the dominant waves come from the north-west. In Half Moon Bay (a pleasure beach for San Francisco) the greatest shelter occurs in the lee of Pillar Point (Fig. 24.8), with exposure gradually increasing southwards. In the most sheltered region (shown by sample line A) sediment is fine (on average 0.17 mm diameter) and the beach is gently sloping. However, sample lines B and C show progressive changes, with coarsening sediments and steepening beach. At the most exposed location (D), not only is the sediment more than three times coarser than at A (on average 0.65 mm diameter) but the beach is also twice as steep.

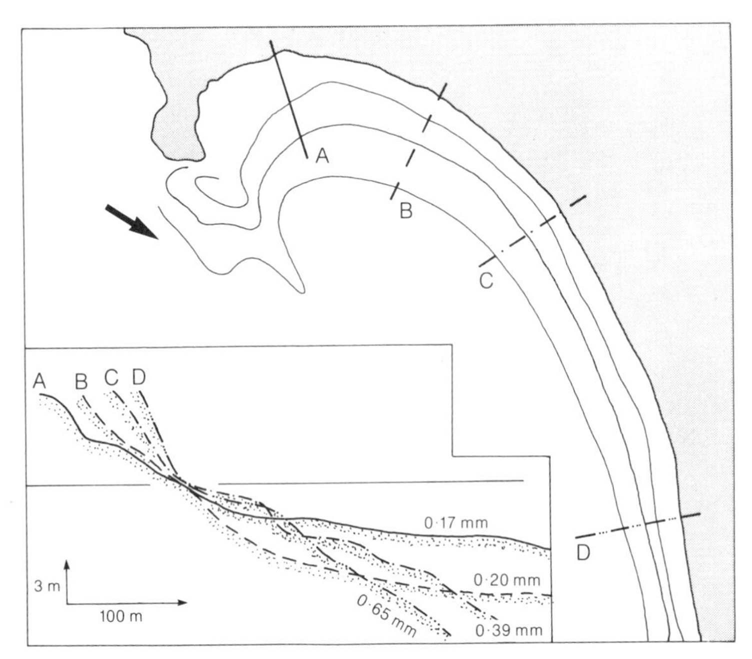

Figure 24.8 Half Moon bay.

The type of beach change found in Half Moon Bay is typical of the pattern for beaches within bays. For example, near Bournemouth on the English Channel coast (Fig. 23.13) three successive bays show the same plan as Half Moon Bay (a curving shape known to mathematicians as a 'log spiral'), with sediments becoming coarser and slopes steeper towards places of greatest exposure and highest wave energy.

Beaches in profile

The study of Half Moon Bay puts the scientific edge on a property of beaches we all know and take for granted – beaches are composed primarily of sand. More rarely beaches are dominated by shingle. In all cases the material is coarse and non-cohesive. Because of this it is very easily moved, either by small children making sandcastles, or by the rhythmic action of waves. Near the breaker zone, where the beach is saturated with sea water, sand is partly buoyed up and its movements made easier still. Thus there is a constant interplay between wave and beach: as the slope of a beach may influence the way a wave breaks, so breaking waves can readily change the beach form.

As they break, plunging waves on a steep beach drop a curtain of water. Because energy is dissipated in a confined zone the swash is relatively small and it quickly comes to rest, only to rush back down the beach under the influence of gravity. In consequence plunging waves tend to dig out a trench where they break, to comb back material from the surf zone and to throw sand into a **bar** offshore of the break line (Fig. 24.9). But this also produces an offshore transfer of sediment which eventually *reduces* the beach slope, and tends to convert plunging waves into spilling waves. Spilling waves mostly form on gently sloping beaches and are much less dramatic in their action. Characterised by forward-thrusting breakers, although they disturb far less sediment and therefore

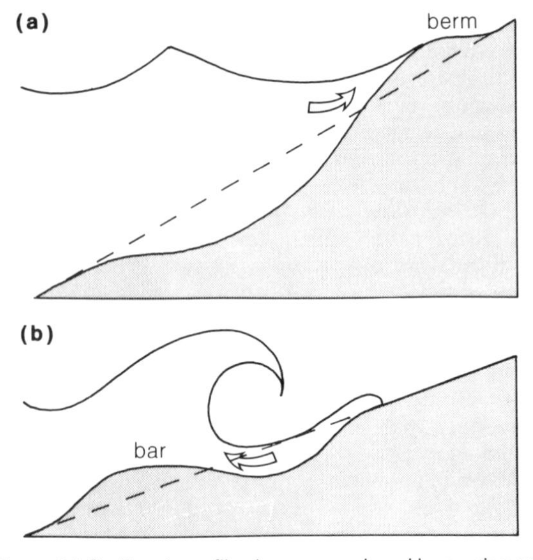

Figure 24.9 Beach profile changes produced by surging and plunging waves (note that these features change position during a tide).

alter the beach less speedily, the powerful swash nevertheless constantly takes material on shore. However, gentle beaches give little help to the backwash, which remains weak and is made weaker still by the large beach area in which percolation can occur. This means that spilling waves tend to push material in shore, and so *increase* the beach slope.

The effect of surging waves is very much more difficult to decide upon without a close study. They form only on very steep beaches and then the shore swash and the powerful backwash seem finely balanced. Yet measurements of beach profiles clearly show how these waves thrust much more material forward than is combed back down again by the backwash. Indeed these are the waves that give beaches, and especially shingle beaches, their distinctive **steps** or **berms** (Fig. 24.9) So, clearly, surging waves steepen beaches.

TIDES AND BEACH PROFILES

If waves constantly worked over the same piece of beach, the shore would be a much less interesting place. On the Mediterranean coasts, for example, virtually tideless holiday beaches have to be cleaned of litter and smoothed over by machine. Contrast this with the beaches that front the open Atlantic or Pacific oceans, where tidal ranges of several metres allow waves to sweep clean and refashion broad swathes of beach each day. Yet tides do more than help waves clean beaches; they control *where* the waves will work and therefore have as much influence on beaches as the waves themselves.

The effect of a combination of varying waves and tides is best illustrated by an example. In any two-week period between spring and neap tides, we can expect at least one storm to comb down material from the upper beach, to create an offshore bar and to produce a simple concave upward profile (Fig. 24.10) comprising a steep upper beach (possibly with eroded sand dunes) and a much flatter and wider lower beach. After the storm swell waves will replace plunging waves: their spilling action on the flat lower beach will gradually drive material back onshore. In particular, the offshore ridge or bar will slowly be pushed landward and welded back on the upper beach. At the same time each high tide will see surging waves form against the steep upper beach, washing material forward to create a step or berm. As a result of this rebuilding activity coupled with a gradually falling tide, the high-tide berm created each day will be lower than the one preceding it. In theory, therefore, the period between spring and neap tides

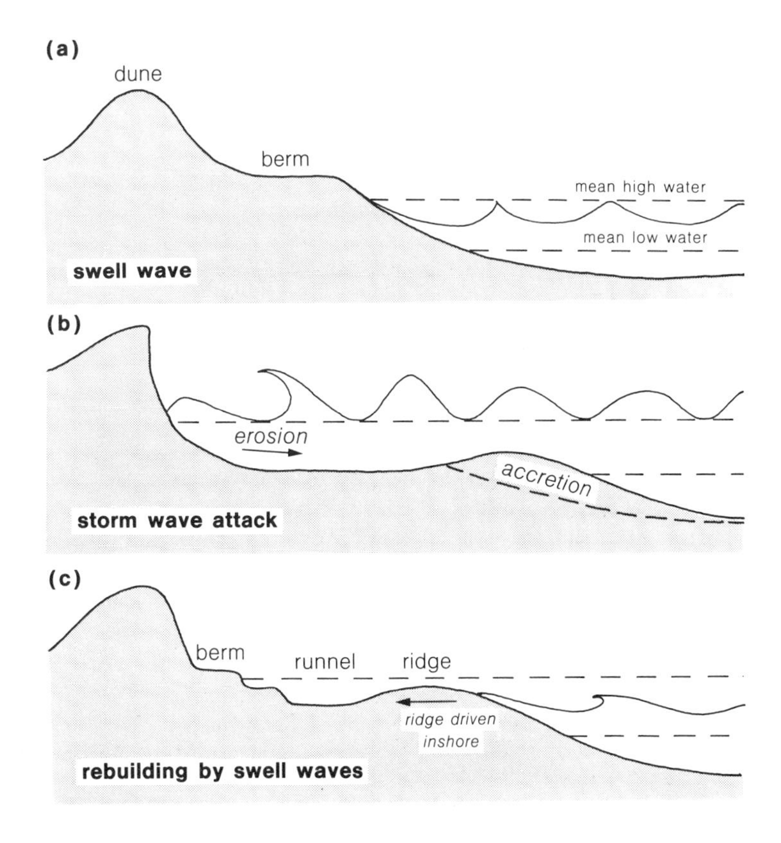

Figure 24.10 Beach profile variation. In this sequence, a storm causes severe erosion of the dunes and upper beach. The beach is subsequently rebuilt by driving material back from offshore, while the dune front reforms by catching blown sand.

will yield a sequence of berms like a broad flight of stairs.

Changes are much less dramatic on the lower beach. By the time neap tide arrives and reveals the offshore bar formed by the earlier storm, it will still be separated from the upper beach by a wide trough or **runnel**. Many weeks will be needed for the bar to come back on shore, although even after a few days it will probably have been altered to a broader form and have been pushed slightly nearer land.

The time between spring and neap tide will always reveal the greatest variety of beach forms. By contrast each new tide between neap and spring progressively covers the beach, pushing all the berms into one and forcing this combined feature higher each day. However, this idealised sequence of rebuilding may in practice be disrupted at any moment by a change in the weather: renewed storm activity might comb back down the beach in one day the constructional efforts of months. But in general terms mid-latitude regions see much downcutting and storm action in winter and rebuilding in summer. Beaches are very often at their lowest in spring and at their highest in later autumn. Like the weather, though, beaches come in endless variety. With so much wave energy at work on material that is so easily moved, a beach is rarely the same from one day to the next.

SAND DUNES

It is really quite rare for waves to reach to the back of a beach. Except in times of storm or spring tide, large areas remain exposed to onshore winds. Most of this sand is dry, loose and easily moved by the wind. Even sand exposed between tides can dry out enough to bounce and skip across the beach surface (a process called **saltation**) under the driving effect of an onshore wind. When not protected from winds by high cliffs, beach material readily blows off the beach (**deflates**) and is carried inland where it may be stabilised by special types of vegetation (Fig. 24.11).

A dune will continue to grow in height until sand is blowing off the dune at the same rate at which it is arriving. The presence of a line of dunes backing the beach (called **foredunes**) sets up a wave motion in the wind: the wind carries any surplus sand inland, producing a spaced sequence of smaller dunes. After a time quite a complex network of dunes may form, but it is the high foredunes that are the most important from the viewpoint of beach stability. They act as a *barrier*, preventing waves and high water from moving inland, and as a *reservoir* of sand, feeding the beach during times of erosion by storm waves. When the storm is past, foredunes are replenished by slow deflation from the beach. In this way beach and dunes often remain in a long-term dynamic equilibrium.

Figure 24.11 Beach dune sands being trapped by marram grass roots.

Beaches in plan

From a cliff vantage point, wave crests and beaches can be seen in a new perspective. This reveals that beach materials are not simply moved on and off shore; they are also moved laterally. Sometimes these **longshore movements** are emphasised on a small scale by the formation of beach **cusps** – crescent-shaped hollows in berms that focus swash and backwash in a spectacular manner (Fig. 24.12). Cusps are normally associated with surging waves and berm building, but involve a pronounced rhythm of longshore movements as well.

Waves approaching the coast are, inherently variable. A wave train may be dominated by one set of wave forms, but there are always many waves approaching the shore, moving at different speeds

Figure 24.12 Beach cusps on a steep beach.

and having different heights. Sometimes they reinforce, making one part of the dominant wave crest higher than others; at other times they cancel each other, reducing the crest height. So as the wave moves on shore, water also moves *across* the crest, creating a cross current from places of high water to low.

Together, onshore and offshore currents produced by swash-backwash and cross currents divide up the water into **cells**. The dominant feature of these cells is the offshore return or **rip current** (Fig. 24.13). When waves break parallel to the coast, both swash and backwash move directly up and down the beach without any consistent transfer of water along the shore. Under these conditions local variation in breaker height can take on some importance by setting up the local transfers of water that result in both **longshore** and rip currents. However, most waves approach, most breakers plunge, and most swash moves forward *at an angle* to the beach. Some waves arrive on the beach at an angle even of 45°, despite the effect of wave refraction. In these cases there is a consistent flow of water in one direction along the shore; this flow is also associated with a longshore current. Where a breaker plunges to the beach and temporarily throws a lot of sediment in suspension, longshore currents (usually moving at speeds about 0.7 m/s) are able to transport at least the finer particles.

Longshore currents operate within the surf zone, usually being strongest about halfway between the break point and the swash limit. But because they occur beneath breaking waves they remain permanently out of sight to an observer on shore. By contrast, the final thrust of swash up the beach produces results that are very easy to see.

If you walk along the shore within the surf, the powerful movement of sediment is very apparent. First, the swash rolls and bounces particles over your feet and sends them shoreward at the angle of the wave advance. Then, as soon as the wave energy is dissipated, the backwash begins, this time directly down the line of steepest beach gradient, pulling particles from under your feet, rolling and bouncing them along until they are halted by the advancing swash of the next wave. One result may be a net movement of material up or down the beach, but in

(a)

(i) swash limit
breaker zone
rip current

(ii) longshore current

(iii) beach drift

(b)

Figure 24.13 (a) There are three wave-induced movements in the nearshore zone: (i) a cell circulation of rip currents and associated longshore currents; (ii) longshore currents produced by the oblique wave approach to the shoreline; (iii) the saw-tooth motion produced by the alteration of swash and backwash. Normally all three types of movement are present. (b) Pebbles accumulating on the updrift side of a beach groyne.

every case the saw-tooth movement of particles resulting from different directions of swash and backwash will have caused a **beach drifting** of material. Together longshore currents and beach drifting account for the movement of sediment along the shore, a phenomenon called **longshore drift**.

Because longshore sediment by beach drift and longshore currents is controlled by the direction of wave approach, drift can vary both in importance and direction daily. In the long term the net direction of drift will depend on the direction of the dominant waves. This can be determined only by careful measurements: it can be misleading to draw conclusions about longshore sediment transport from observations over just a few days. A better estimate of the long-term net direction of the 'coastal conveyor belt' can be made by looking to see where sediment has built up – against one side of some rocks, say, or against a jetty – although even here there may be changes from one season to another, with seasonal changes in the dominant wind direction. The movement of beach sediment is prodigious. It is common for a large storm to comb down 30 000 m^3 of sediment (about the load of 1200 large trucks) along every kilometre of beach. Longshore sediment transport of over 500 000 m^3 per year (equivalent to the load of 20 000 trucks) past a given point is not unknown.

The pattern of deposition

Measurements show that longshore sediment transport is rarely continuous even on long uniform stretches of coast. Instead coastal processes appear to confine sediment to cells consisting of a **source region** of eroding cliff or river mouth, a **conveyor reach** of shore, and an offshore area of deposition (a **sediment sink**). Sometimes the transfer of sediment offshore is even sufficient to form **sandbanks**, although more usually it is reworked by waves and possibly even used to supply another cell. Within a sediment cell, however, smooth transfer of sediment along the shore is ensured by the natural momentum (the impetus to keep going in the same direction) of the waves. Waves have such enormous momentum that sudden changes in wave speed and direction are impossible. The direction of advancing waves therefore changes only slowly, with wave crests sweeping in graceful curves rather than making awkward turns in response to variations in water depth or change in coastline shape.

With such resistance to rapid change it is clear that we should not expect an immediate modification to the pattern of wave crests simply because there is an abrupt alteration in the direction of the coast. And because the transport of sediment is so closely related to wave action, the slow refraction of wave crests is matched by slow changes in the direction of sediment transport. Thus at abrupt changes in the orientation of the coast, sediment direction and shore become temporarily out of step; sediment drifts not along the shore but in a gently sweeping curve out into deeper water. In the deeper water the waves do not transport the sediment so it begins to accumulate, forming a sort of natural beach extension or jetty called a **spit** (Fig. 24.14). The accumulation of sediment gradually reduces the depth of the water; wave action transports material in the direction of approaching wave crests; and the spit is thereby extended further out from the shore.

Sand spits and wave fronts interact in a manner that always encourages the growth of the spit. Wave fronts bring in the sediment that provides shallow water to enable the wave fronts to resist a change of direction, which in turn encourages further deposition on the spit.However, complete closure of a bay is hindered by the flushing effect of tides rushing through an increasingly narrow sea connection (Fig. 24.15). Spits may therefore be expected to grow

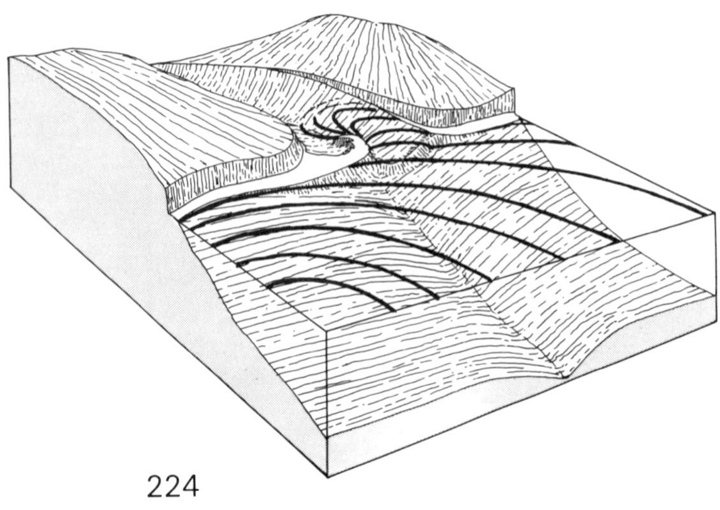

Figure 24.14 As a spit builds out into deeper water more sediment is required and the rate of advance slows. Tidal ebbs and flows make the final bay closure more difficult by carrying sediment offshore where it forms sandbanks.

Figure 24.15 (a) The recurved spit of Spurn Head, on the North Sea Coast. Notice that there are extensive mudflats in the shelter of the spit and submarine sandbanks on the outer side which have caused waves to break. The arcs of wave crests produced by wave refraction show very clearly. Notice also several lighthouses that indicate constant changes in spit position. Today the very existence of Spurn Head is threatened because man-made groynes along the shore to the north have cut off the supply of sediment needed for nourishment. In 1982 the narrow neck was only kept intact by using bulldozers after erosion by a severe storm. (b) Holy Island, Northumberland. Two drift-aligned spits are receiving sediment from the south. Notice that the spits seen above high tide are merely the 'tips of the iceberg' – a better impression of the quantity of deposition is given by the intertidal sand flats.

(a)

(b)

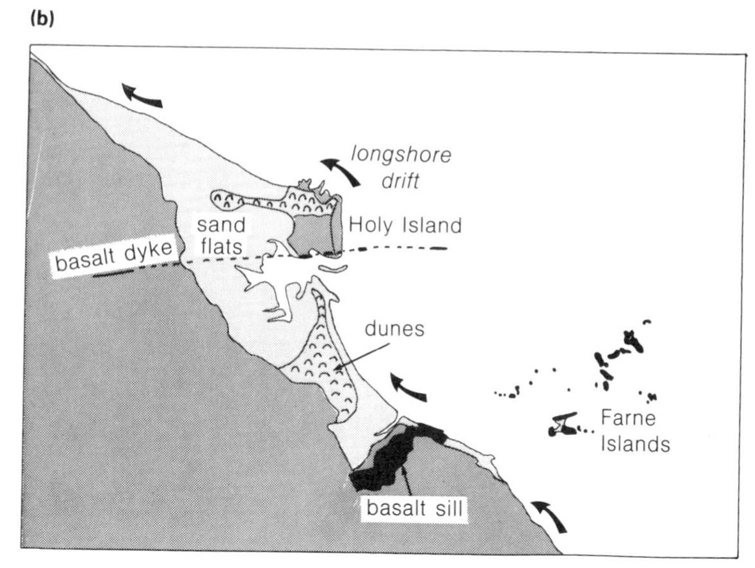

across only *small* bays, closing them with complete sand ridges called **barrier beaches** (formerly called **bay bars**).

Further into the shelter of an estuary, where wave energy is low, littoral drift does not occur; deposition here is mainly of river sediment (Fig. 24.16). Rivers bring in a wide range of sediments in contrast to the uniformity of beach materials. Some of the mud may be carried out to sea by the remnant current of the river, but much will usually settle out and flocculate in sheltered places close to the estuary shore, where the mud is rapidly stabilised by salt-marsh vegetation. Elsewhere the formation of a network of **intertidal islands** reveals a remnant kind of stream braiding being extended as far seawards as the river current can prevail.

Considerable shelter is provided in the lee of offshore islands where interacting wave fronts use up

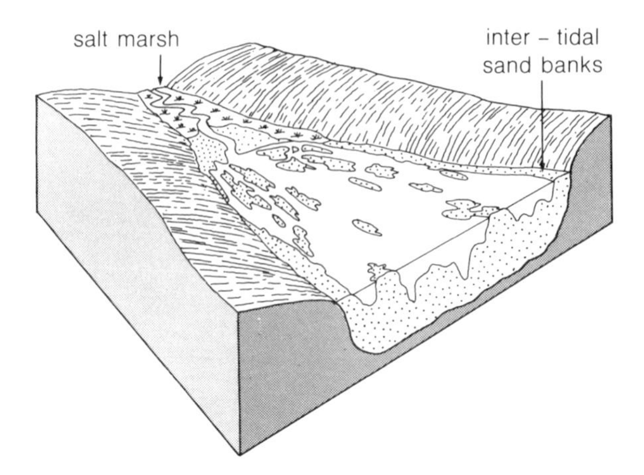

Figure 24.16 A typical sediment arrangement in an estuary or ria at low tide, when intertidal sandbanks are exposed.

Figure 24.17 (a) A typical sediment sequence along a barrier island coast. (b) The Barrier Island coast near Corpus Christi in Texas is over 160 km long.

(a)

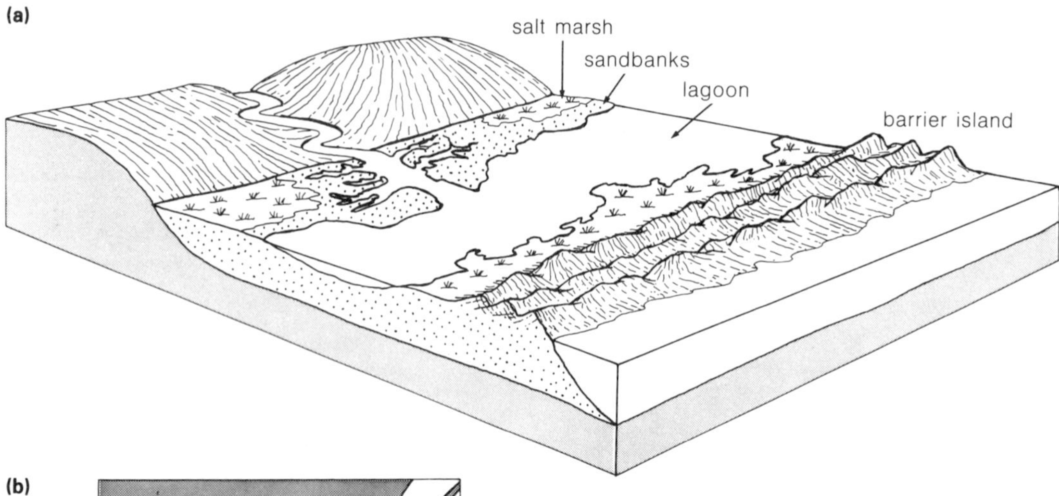

(b)

Corpus Christi

GULF OF MEXICO

0 km 25

their energy in battle and deposit a sedimentary causeway to form a **tombolo** (Fig. 23.11). Similarly, widespread shelter is also provided behind the **barrier islands** so typical of the Atlantic seaboard of the USA and elsewhere (Figs. 24.17 & 18).

Deltas

Deltas are coastal landforms of deposition produced when more sediment is brought by a river than is taken away by the sea. Deltas are therefore found on high-energy coasts only where large rivers can provide considerable volumes of sediment. Deltas will be more common on low-energy coasts, along which longshore drift is weak.

Rivers transport very varied sediment to the sea, although usually by far the greatest proportion is silt and clay; sand is often only a minor constituent of a river's sediment supply to the sea.

Figure 24.18 The barrier beach at Slapton Sands, Devon is thought to have been driven onshore by the rising sea levels at the end of the Ice Age. It has not been formed by longshore drift.

The shape of a delta continually varies as in-flowing rivers break down their confining levées and change direction in times of flood (Fig. 24.19a). However, the general tendency for material to be pushed outwards from the coast gives a dominant region of accumulation directly beyond the river mouth: this forms the apex of the delta. The final shape is a balance of deposition and erosion by wave action, as the Mississippi delta clearly illustrates (Fig. 24.19b).

Sediment cells and sediment budgets

As we have seen, the coastal zone is more than just a collection of places where erosion, transport or deposition occur. Like any other parts of the natural environment, coastal processes are interlinked: a change in one place causes changes elsewhere. A study of the littoral zone is therefore only half complete if we simply identify (and keep isolated) features of erosion or deposition.

This concluding section therefore looks at coasts in the same way as an accountant studies a profit-and-loss account. In effect we shall try to examine the 'budget' of complete **sediment cells**. In a coast of bays and headlands, sediment cells are reasonably easy to identify: each cell is formed between two adjacent (source) headlands, with wave refraction constantly causing littoral transport to carry sediment toward the intervening (sink) bay, where a broad beach forms. In other cases, such as an open beach, the cell it not so clearly defined; long-term, detailed measurements of cliff retreat, of longshore-transport rate, and of spit, bar and sandbank growth are the only way to clarify the situation (Fig. 24.20).

CASE STUDY: THE HAMPSHIRE COAST

The English Channel coast in the shelter of the Isle of Wight (Fig. 24.21) has evolved in a complex manner. Atlantic Ocean wave fronts are refracted in two directions around the island, causing longshore drift to converge near Portsmouth. This region is underlain by poorly resistant sand and clay beds so that, despite the sheltered position, some stretches of coast are retreating by five metres each year.

Drift here does not proceed smoothly along the coast: sediment accumulates as spits and sandbanks. Some of the sediment sinks can be identified from maps: the **cuspate foreland** of Gunner Point, for example, is clearly the sink caused by the meeting of two drift directions. But measurements show that these are not the *major* sinks – these major sinks lie hidden as sandbanks off shore. The pattern of sediment movement is made even more complicated by three natural harbours whose tidal ebb and flow keep the entrances flushed clear to bedrock. They alternately draw sediment into the harbour on the flood

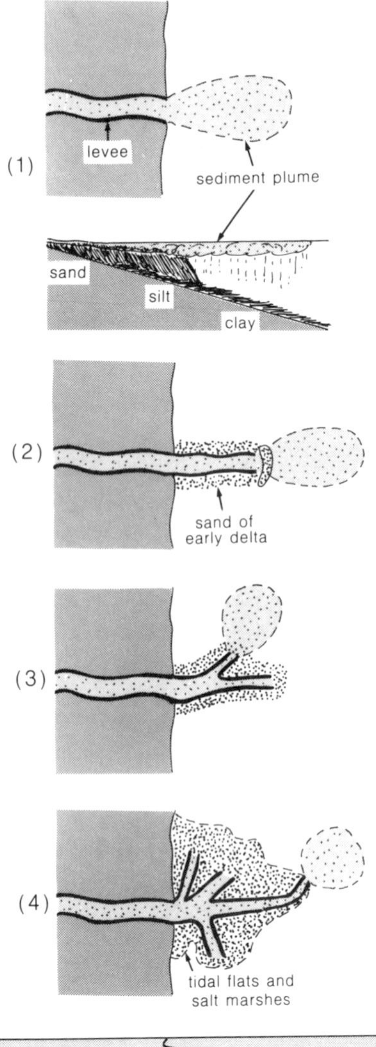

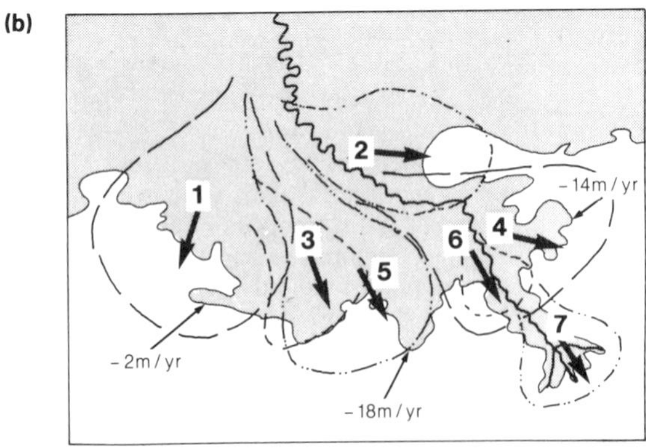

Figure 24.19 (a) The sequence of delta formation. (b) The Mississippi delta: the numbers (m/yr) refer to erosion rates by wave action. The bold numbers indicate the stages of delta growth.

tide (where it has formed recurved spits) and flush it out on the ebb tide, to form offshore bars. It is impossible to discover the way this section of the coast is evolving without drawing up a detailed budget.

CASE STUDY: THE YORKSHIRE COAST

The coast around Whitby provides an example from a contrasting environment. Here the coast takes the full force of the north-easterly winter gales (Fig. 24.22a). Wave trains approaching from this direction are refracted, which concentrates their energy on the headlands at either end of the bay. But even though energy in the bay is dispersed, the unconsolidated cliffs of glacial till are still being actively eroded. The headlands, which consist of horizontally bedded shales, sandstones and limestones, are clearly subjected to wave energy in excess of that required to remove sediment: there is no beach; a large wave-cut platform skirts the cliffs; and despite active cliff collapse, relatively few boulders remain at the cliff base (Fig. 24.22b). The sediment produced from these source areas and transported into the bay by longshore drift accumulates as a wide beach, but is insufficient to dissipate wave energy in the bay. This is why the till cliffs are actively being eroded and, indeed, supply the majority of the beach debris. The beach itself is subject to considerable changes each season – downcombing by plunging waves in winter, followed by rebuilding by spilling waves during the summer (Fig. 24.22c).

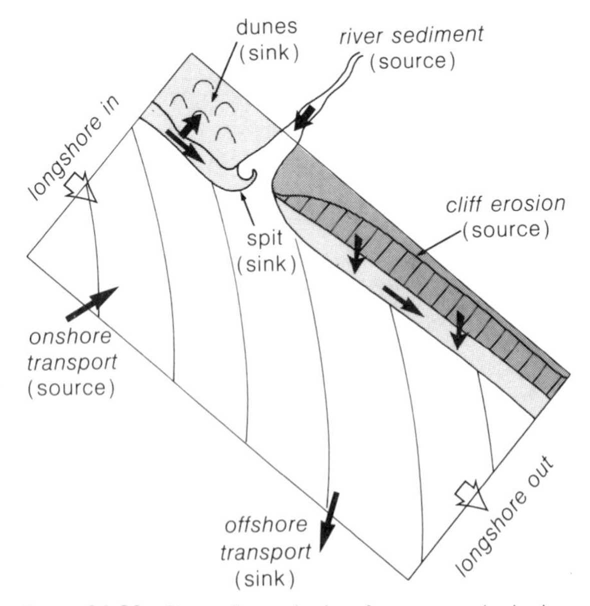

Figure 24.20 The sediment budget for a coast: the budget is calculated for the area inside the box (it need not be a complete natural sediment cell).

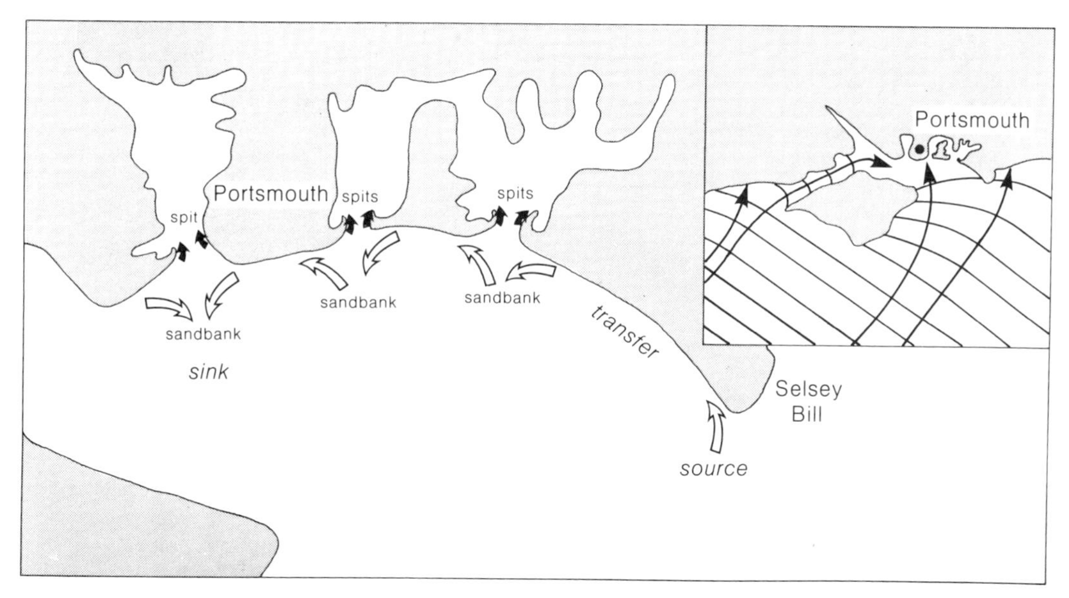

Figure 24.21 Drawing the sediment cells near Portsmouth shows the complex pattern of opposing spits and offshore bars to be closely related to the erosion of Selsey Bill and the growth of the Horse and Dean Sands (for detailed budget data see Ch. 25).

(a)

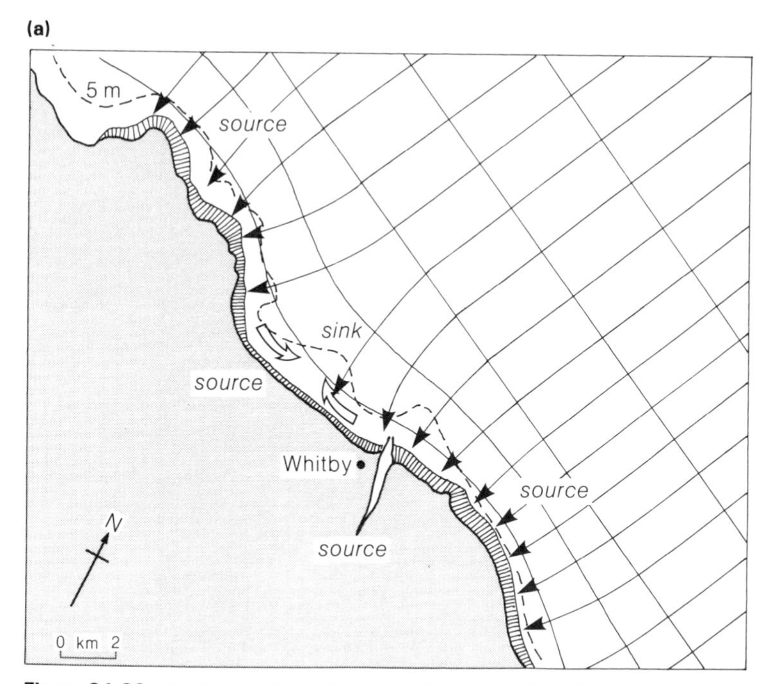

(b)

Figure 24.22 (a) The pattern of wave refraction and sediment movement at Whitby. (b) A wide wave-cut platform below the cliffs on the southern (source) headland at Whitby. (c) Whitby beach in June 1982. (d) Rotational slumping on the till cliffs.

(c)

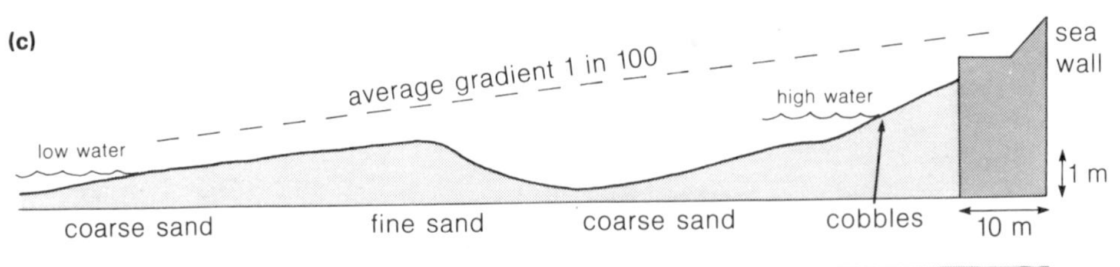

(d)

Figure 24.22 contd.

24.3 Sea-level changes

As we have seen, wave energy is distributed in a way that tends to produce a smooth, evenly curving coastline. However, many coasts are far from achieving it.

Part of the reason is that a poorly resistant rock will always retreat faster than one of great strength. And whether a spit can grow across a bay and smooth out the coast depends on the size of the task, the amount of available sediment and time. As a result, in those areas of hard rock in which sediment supply is low, spits are very rare. But even where spits are common, they rarely grow across large bays. Nevertheless many bays are completely cut off by barriers. This kind of disparity between currently developing landforms, spits and finished landforms such as bars, suggests that many features must have formed under conditions that no longer prevail.

Possibly the most difficult problem concerns the evolution of barrier islands and long barrier beaches. In Australia, for example, there is a linear coastal offshore beach nearly 150 km long. It is hard to believe this grew as a simple spit from a headland at one end. Similarly the barrier islands off the southern US Atlantic and Gulf States stretch for many hundreds of kilometres without touching land. It is even difficult to explain small features such as Slapton Sands in Devon (Fig. 24.18) in terms of simple beach extension.

The recent large fluctuations of sea level caused by global cold periods (ice ages) and the widespread advance of ice sheets were of especial importance to coastal landforms. The most recent cold period, for example, probably extracted over 80 m depth of water from oceans in the process of building the ice sheets. With so much less water, seas retreated until they were very close to the margins of the continental shelves; with the resulting lowered base level, rivers cut deep trenches into their valleys and extended out across land that was once below the sea. Adjustment to the changed conditions soon took place: trenches widened out to form valleys; the long-abandoned cliffs, no longer kept steep by wave action, gradually became gentler (Fig. 20.19). The most profound changes probably occurred in areas where ice spread out over former sea beds, especially near to mountains, where glaciers were able to sweep down, eroding troughs way below the former sea level.

The end of the cold period was marked by equally dramatic changes. As ice melted, water began to return to the oceans and flood over the long abandoned sea beds. These were very different sea beds from those the water left behind: by now many northern areas were covered with thick glacial debris, and even in those areas not directly affected by ice, weathering had formed soils on the old sea beds. As sea level rose, therefore, there was much loose debris for waves to form into beaches. Gradually, over thousands of years, this material has been driven against the present shores as long barrier beaches. The abandoned cliffs, which had been reduced to gentle hill slopes, have since been retrimmed; deepened valleys have been flooded (to form estuaries, rias or fjords) and the process of valley infilling has begun (Fig. 24.23).

Figure 24.23 Loch Ranza, Isle of Arran is a glacially enlarged valley now partly flooded by the sea. Extensive intertidal sands show where sediment from the small river, together with littoral drift is gradually filling in the loch. The houses skirting the loch shore are built on a platform (raised beach) eroded at a stage when the sea was at a slightly higher relative level than today. The castle ruins may stand on a small bay head spit.

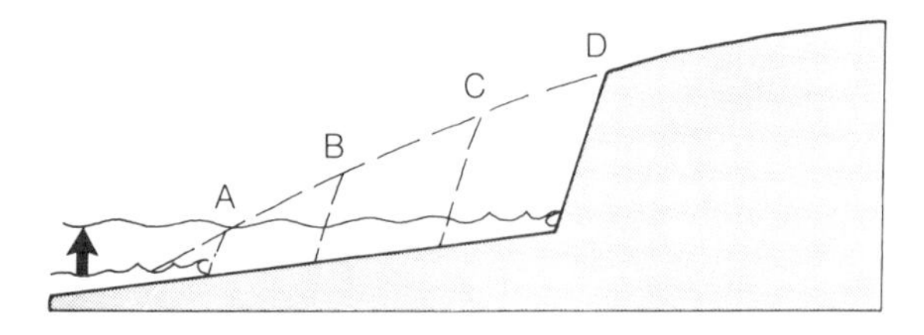

Figure 24.24 Wave-cut platforms can only become extensive if sea level is gradually rising.

Ice sheets did more than cause sea levels to fall. The tremendous weight of ice slowly depressed the land until this land lay tens or even hundreds of metres below its former level. Land recovers its equilibrium only slowly: when the ice melted and sea level began to rise again, the sea in many places rose over land still far below its equilibrium level. Into this land the sea cut cliffs; and it developed wave-cut platforms and covered these with a veneer of beach deposits. In time the land began to recover its equilibrium height – not all at once, but in a series of jerks, each rapid rise of a few metres being followed by a long period when uplift was much slower. Each time the land was relatively stable the sea cut new cliffs and developed new wave-cut platforms (Fig. 24.24). In the area where most uplift has occurred these old shorelines now rise like broad flights of stairs from the present coast (Fig. 24.25), although even today we have certainly not seen the last of the isostatic recovery.

If such past reconstructions are correct, many coastal landforms are effectively 'fossil' features that may or may not be in equilibrium with their present environment. If they are not, sand and shingle bars, for example, must certainly be among the most fragile landforms of all. Before we tamper with beaches (as sources of building materials) or alter them in any substantial way, we need to understand which of our coastal features are capable of rebuilding themselves and which are not. Indeed the equilibrium of coastal landforms can be disturbed remarkably easily, as we shall see in the next chapter.

Figure 24.25 If sea level rises quickly, the coast will be drowned and sedimentation will occur. On the other hand, a rapid fall of sea level will leave the coastal wave-cut platform and its beach high and dry. Both types of movement can be seen around the coasts of the UK, illustrating the complexity of sea-level changes. These photographs show a staircase of raised beaches, Kildonnan, Arran (top) and a close-up of raised beach material, Holy Island, Northumberland (bottom).

25

Coasts and man

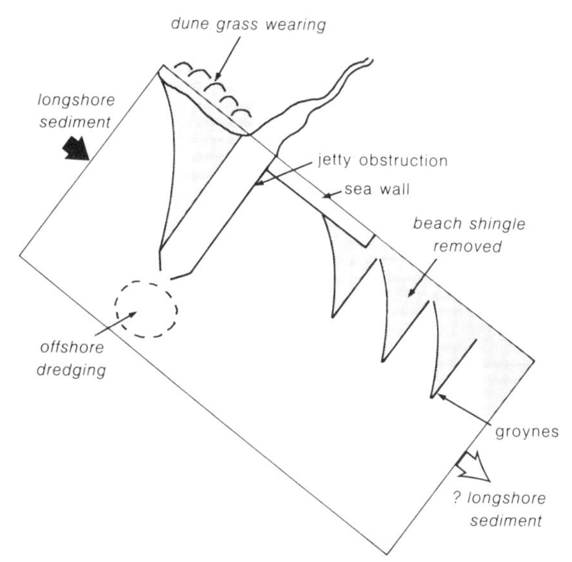

Figure 25.1 The major forms of man's interference with the coast. Because there are so many changes, the impact of any one structure on longshore drift is often difficult to assess.

25.1 Man's intervention

People have always lived by the sea, defending their shores against invaders, building harbours for fishing and trade, or hotels for tourism. But, as the previous chapters show, coastal environments are liable to rapid short-term changes.The collapse of a piece of cliff, the shift of a sandbank during a storm, or flooding over sand dunes are perfectly normal parts of coastal evolution – but to people with houses on the cliff top, a harbour near the sandbank or farms behind the dunes, such a natural event could be a disaster. Man may want the coast to remain unchanged, but this is completely contrary to the natural processes of dynamic equilibrium. When short-term changes occur (such as the loss of a holiday beach during a winter storm), man feels the need to step in with some dramatic action instead of waiting for natural processes to regain equilibrium. The result usually disturbs completely the dynamic equilibrium of a whole stretch of coast, causing repercussions far and wide (Fig. 25.1).

25.2 Problems created by longshore drift

Whenever waves break on a beach, they move beach material. But although you can see sediment being washed in and out with every breaking wave, you cannot easily see longshore drift: how are you to tell that a sand grain moves 70 m a year along the shore when all grains look alike? So it is not surprising that, time after time, widespread coastal disruption has followed the building of even a small jetty or harbour wall.

Looked at on a map, a jetty is a trivial feature compared even with a beach. Yet because it is built out *across* the beach, it interrupts the natural longshore movement drastically, causing accumulation on the updrift side, and starving the downdrift side of a beach supply. (Fig. 25.2a).

Deep water created in the lee of the jetty is useful for ships, but the downdrift beach is also starved of a sand supply to replace that drifting away. As a result downdrift beaches become smaller and their 'shock absorber' effect is lost. This leaves the coast open to direct wave attack and erosion until a new sediment supply is produced which can replenish the beach and absorb the wave energy once more. Building a jetty may cause the loss of a potential holiday beach and may open sea-front homes increasingly to attack by the sea.

Remember that the natural shape of bays in the lee of headlands is a sweeping curve of special shape (Fig. 24.8). This is the shape that wave action will tend to create in the lee of a jetty, harbour wall or pier

(Fig. 25.2). If widespread erosion is to be prevented, engineers must protect the coast – or knock down the jetty! Traditionally engineers have tried to compensate for the downdrift erosion caused by a jetty by installing wooden or steel shutters (called **groynes**) across the beach to act as further sediment traps. But groynes can only stabilise material – they cannot create it. Often their installation is as effective as closing the stable door after the horse has bolted. In many cases not enough sand remains to protect the coast, and engineers have had to build **sea walls** to protect homes. Unfortunately sea walls *increase* erosion problems if their design does not enable them to absorb wave energy. Vertical walls, or those with slight curves, act as cliffs: they reflect wave energy, causing more turbulence in the surf zone, greater disturbance of sand and an even greater loss of beach.

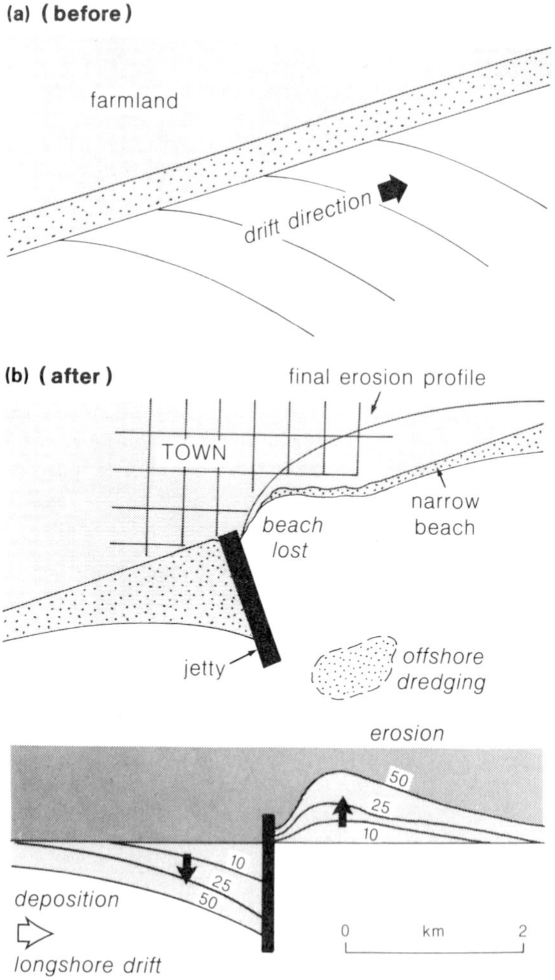

(a) (before)

farmland
drift direction

(b) (after)

final erosion profile
TOWN
narrow beach
beach lost
jetty
offshore dredging
erosion
50
25
10
10
25
50
deposition
longshore drift
0 km 2

Figure 25.2 (a) How a simple structure like a stone jetty can cause drastic changes in the coast both up drift and down drift. (b) A laboratory model of a proposed breakwater at Cotonou, Dahomey showed clearly the changes in coastline that might occur within 50 years.

There are alternatives to smooth-fronted sea walls. What is needed is some form of structure that will dissipate wave energy – a rough surface on which waves can break but which will not itself be carried away. As it happens, the cheapest and best method is the one employed by nature and seen at the foot of most cliffs – a **pile of large rocks**. But piles of large rocks are unsightly on holiday beaches and they have not been widely used.

In the long term the only way to be sure of keeping a sandy beach for holidaymakers and of protecting coastal homes is to maintain natural longshore drift. Today, with the huge cost of coastal-defence structures and their limited success, more and more engineers are looking seriously at either pumping sand into a beach from an offshore sediment sink (**beach nourishment**) or dredging sediment from the updrift side of the obstacle and pumping it ashore on the downdrift side (**sediment bypassing**). No one pretends these are cheap or elegant solutions, but they do keep sand on a beach without unsightly groynes, they protect the coast using natural materials, and they do not cause trouble elsewhere.

Today coastal 'protection' (a word that still conjures up the old idea of static equilibrium) is big business, with problems and solutions unique to each area. Below are a number of case studies that illustrate the way engineers have worked with or against natural processes and the outcome of their endeavours.

25.3 Case studies

Sandy Hook, New Jersey

New York City needs as many beaches as possible to provide recreation areas for its inhabitants. One widely used beach, on a sand spit called Sandy Hook in neighbouring New Jersey (Fig. 25.3), is in danger because it is starved of sediment by an extensive groyne 'field' and many sea walls to the south. So complete is this ultimate horror of continuous concrete that coastal engineers now use the term 'Newjerseyisation' to describe extensive man-made coastal modifications. The whole coast is now out of balance, but the groyne fields and sea walls cannot simply be ripped up. More groynes would limit the use of the beach even if they were restricted to the threatened area; they would also move the erosion problem further along the spit.

If the beaches of Sandy Hook are to be stabilised without further unsightly structures and the erosion

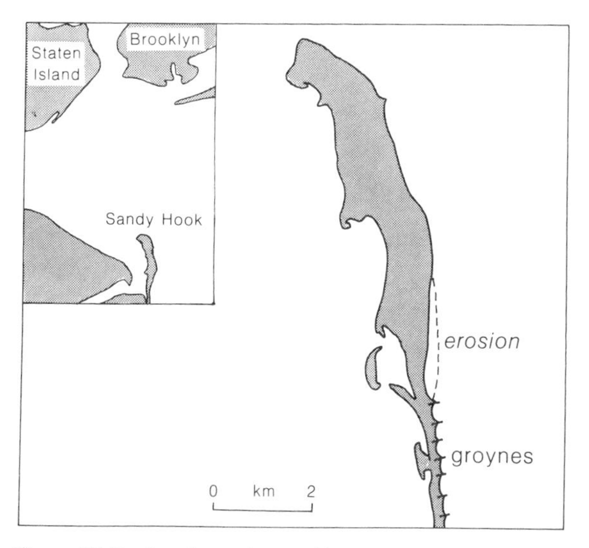

Figure 25.3 Beach erosion problem areas, structures, and erosion control operations at Sandy Hook, New Jersey.

spiral is to be broken, an approach based on wave processes has to be found. In this case investigation revealed that the 50 km of groyne field at the southern end of the spit is trapping 60 per cent of the sand that would otherwise be moving into the beach area. In this exposed Atlantic seaboard location with high energy waves, 60 per cent is a huge amount of sand – something like 270 000 m^3 a year. No wonder that about 23 m is being lopped off Sandy Hook's beach each year by erosion, or that a large part of the beach car park and two bath houses have already been lost. To restore the smooth spit profile 1.5 million cubic metres of sand needs to be put onto the beach, a vast and costly amount of material. In this region of high beach demand the only way to get the beach back is to nourish it with new sand. But nourishing beaches is quite different from building structures: adding to the beach, rather than stabilising an already deteriorating situation, means that more sand will be needed every year – a continuing process that will cost a lot of money. At Sandy Hook, they are fortunate in having a nearby supply of material dredged from the (downdrift) navigation channels of New York Port: if this is placed in shallow water, swell wave action will do the rest by pushing the sand on shore and reworking it into a smoothly sloping beach.

This long-term strategy can be aided by dumping material to make an offshore ridge that will absorb some of the wave energy before it reaches the coast. Of course, in areas where there is no nearby supply of sediment, beach nourishment may be too expensive; but near seaside resorts in which a pleasant, structure-free beach is the main resource, natural replacement measures should make economic sense.

East Anglia

Sometimes dredging for port navigation channels may be inevitable, and the material can be usefully replaced on nearby eroding beaches as is the case for Sandy Hook. However, both rivers and dredging can interrupt the natural flow of material along the shore. The river and the coast may reach a state of long-term balance, but when the dredged sand is taken from the coast and used for building material on land, it is lost from the coastal system for ever. Great care has to be taken to ensure this does not lead to disaster.

East Anglia owes most of its existence to deposition in the Ice Age: its rocks are poorly consolidated and readily eroded. Moreover, much of the northern region juts out into the high-energy wave environment of the northeasterly gales that seep down the North Sea. It is a combination that must inevitably give rise to much sediment movement and cliff retreat. Like Sandy Hook, there are many large sand spits to prove that movement by longshore drift is important. For example Orford Ness only became 18 km long because of a massive contribution of material from the cliffs. At present the main feeder cliffs (sediment-source regions) are in Norfolk, where 400 000 m^3 of cliff tumble into the waves each year (Fig. 25.4). These cliffs are natural beach nourishers and the beaches in turn are natural energy absorbers, keeping down the rate of cliff erosion elsewhere and providing the East Coast resorts with their famous sandy beaches.

As with all coastal regions, the sediment movement around East Anglia is divided into separate cells of activity: material eroded from one place may not move all the way along the coast. Thus not all of the vast sediment supply from Norfolk reaches Harwich. With losses to offshore regions, what was 300 000 m^3 of southward drift at Norfolk is reduced to only 100 000 m^3 at Harwich. But here, to keep the port open, the dredgers remove 200 000 m^3 each year. Of course some of the dredged material is supplied by the river whose valley provides the site for the port, but dredging allows so little sediment to drift southwards along the coast to Clacton and nearby seaside resorts, that the beaches there are in danger of being stripped away completely. Already the beaches near Clacton are steeper and narrower than they used to be, and the coast is deprived of its buffering sand. The eventual loss of the beach and consequently of the holiday resort trade would be an economic disaster to Clacton. The sooner the dredged sand is diverted back to the coast the better, or short-term expediency will be overtaken by long-term catastrophe.

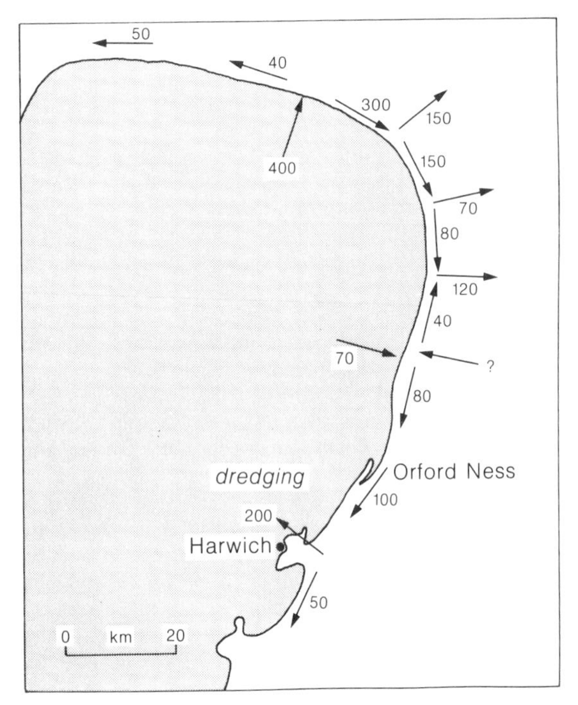

Figure 25.4 Sediment movement along an apparently uniform shoreline such as East Anglia, is a complex affair, consisting of a system of sediment 'cells' with much interlinking between nearshore and offshore regions. Unless this pattern is fully understood, an action such as dredging or sea wall construction in one place may cause havoc elsewhere.

Sediment comes all the way from the north of Norfolk and it will continue to arrive. If sea walls are built to protect farm land near the eroding cliffs, they will cut off the supply all along the coast. It is better to accept that some cliffs must be lost as part of the natural balance of nature than to see all of the coastal beaches lost. It is simply the price that has to be paid for maintaining long-term equilibrium; compensating farmers for lost land is much cheaper than finding new jobs for the workers in the seaside towns.

Hallsands, Devon

The classic case of ill-advised dredging occurred a century ago in South Devon. The Royal Navy needed material to extend the Devonport dockyard. A supply of suitable cobbles was found off shore in nearby Torbay. Half a million tonnes of these cobbles were taken from the bay just off shore of the fishing village of Hallsands. Even while dredging was under way, the beach below the village started to become steeper and narrower; it disappeared at an accelerating rate. The onshore–offshore sediment cell had been interrupted and the waves were seeking a new equilibrium with the available sediment. Soon, high-energy waves were able to pound unrestrained against the narrow rock ledge (a raised beach) on which the houses sought to retain a foothold. In just a few years the ledge was retreating rapidly and the houses were in ruins. Today only a few broken shells of the houses still remain, ready at any moment to collapse into the sea. While they remain they are a silent memorial, a reminder to those who would alter one part of the coastal region without first ensuring that they understand the consequences.

The Portsmouth coast

Selsey Bill and Gunner Point are the natural limits of a balanced sediment cell (Fig. 24.21): here offshore bars make the sediment source regions. But offshore bars are a fine source of building sand, as are the sediments at Selsey Bill. Quarrying the bars therefore causes a loss of sediment input to the cell, which, in turn, has caused severe erosion. Unfortunately, engineers have reacted to this loss simply by building a groyne field on the Bill. As a result longshore sediment transport has been cut from 70 000 m^3/year down to only 7000 m^3/year (Fig. 25.5). No wonder down drift Bracklesham Bay is losing 5 m off its coast each year (Fig. 25.6). More surprising, perhaps, is the fact that, despite the reduction in sediment supply, there has as yet been little erosion elsewhere. But this happy situation is soon to change; with no new sediment supplies and with constant dredging, the offshore reservoirs that have been 'buffering' the other beaches (by continuing to supply sediment) are fast running out of material. Even if dredging is stopped the problem will not be solved; in the long term engineers will either have to turn to the Horse and Dean Sands – the cell's natural sink – and start a programme of beach nourishment, or let Selsey Bill disappear.

Cliff modification at Llantwit Major

The balance between cliffs, beach and waves is often ignored, even today. This is particularly the case when small-scale measures are to be undertaken and no expert advice is sought. Many problems still arise at the local level, as, for example, at Llantwit Major between Cardiff and Swansea in South Wales.

The South Wales coast is exposed to a variety of storm waves. Only the most resistant cliffs survive without rapid retreat. At Llantwit Major the cliffs are far from resistant, being formed of alternating bands of shales, sandstones and limestones. Like their counterparts in Dorset (Lyme Regis) and North Yorkshire (Whitby) they have a wide skirt of wave-cut platform free from beach sediment (Fig. 25.7).

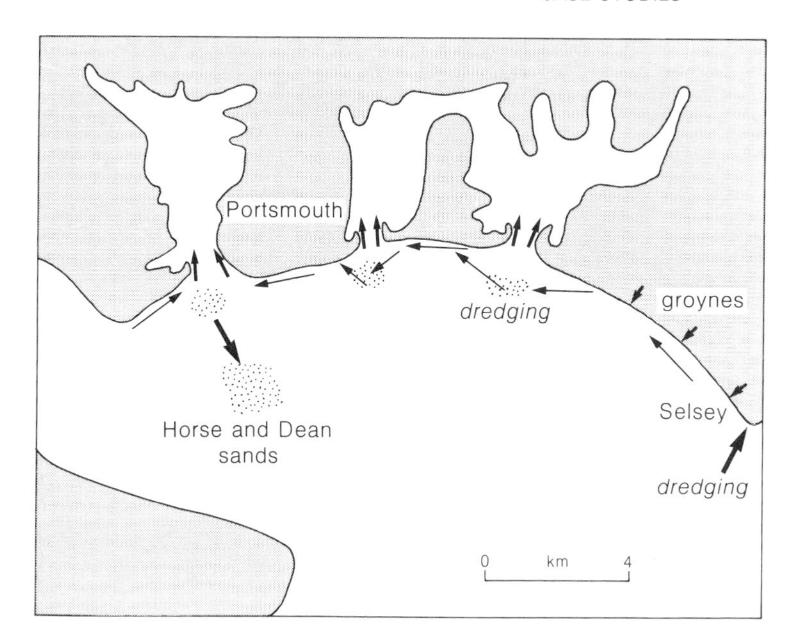

Figure 25.5 Sediment transfers near Portsmouth (see also Fig. 24.21).

Figure 25.6 Severe changes in the natural pattern of beach drifting are illustrated in Bracklesham Bay, Hampshire. Some attempt to restrict the erosion of the sand-dune coast behind the beach has been made, using wooden slatting. However, it is likely that such action will result in further undermining and beach removal. Notice that the groyne has had to be both heightened and strengthened in recent years and that broken flint cobbles have been dumped at the foot of the 'sea wall' in an attempt to absorb some of the wave energy.

Figure 25.7 Blasting was designed to produce a protective ridge of rock debris at the foot of the cliffs at Llantwit Major, South Wales.

This is clear evidence of a high-energy wave environment and poorly resistant rocks where progressive undercutting causes cliffs to collapse periodically, avalanching rock onto the wave-cut platform. Once there, the debris is quickly broken up and driven to the east by longshore drift. By this means the sea at Llantwit Major keeps its doorstep very clean, despite cliff retreat by up to 0.5 m a year.

Llantwit Major is a pocket beach on an otherwise cliffed coast. Being close to urban areas it provides a good holiday beach, except that near the ends there is a constant danger from cliff collapse. In an attempt to forestall the possibility of danger to families as they sit with their backs against the collapsing cliffs, the local authority blasted the top of the cliff with the intention of providing a more stable angle and at the same time protecting the cliff base with the pile of blasted rock. It was hoped that the protective skirt of rock would be effective for up to 20 years and at the same time make the beach a safer place.

Unfortunately, the blasting had two undesirable effects: first, it weakened the cliff by shaking the rock; and second, it threw down onto the beach mostly small material which was easily removed by the waves. In only ten years more than half the protective rock had been swept away and rockfalls from the cliffs had begun again. In retrospect it seems that, far from retarding erosion, blasting has actually speeded up cliff retreat by presenting the waves with material more quickly than could have occurred under natural conditions. It illustrates just how much thought and care must precede even the most minor coastal improvement project.

Part H

Ecosystems and farming systems

26 Man's impact on environmental systems

26.1 The resilience of natural systems

Natural systems involving the living world at first appear very diverse, but they all fit into simple basic structures. For example, all living creatures occupy food 'chains' of no more than three or four links, or **trophic levels** (e.g. plant, herbivore, carnivore). The variability within each system is therefore primarily related to the adaptations that each species makes in response to stress (energy, water, nutrients).

Adaptations follow two basic strategies. There are:

(a) organisms that have a high reproductive rate, produce many young, take little care of their young, and multiply rapidly under favourable conditions: these so-called 'boom-and-bust' organisms (such as the locust) rapidly exploit favourable ecosystems but suffer a considerable death rate when the food supply nears exhaustion;
(b) organisms that have low reproductive rates, ensure the care and survival of their young, have long lifespans and many means of surviving fluctuating environmental conditions, and maintain populations at around the level appropriate for average conditions.

Under normal circumstances man falls into the second group; when stressed, he may tend towards the first.

No organism is able to respond to environmental changes immediately. If the time between stress periods is long, it may permit recovery; if it is insufficient instabilities can develop. Thus a **shifting-cultivation farming** system (whereby people use a piece of land to exhaustion, then move to fresh land) is stable provided sufficient time is allowed for natural recovery (see Fig. 33.5), but it becomes unstable and degrades if increasing use forces natural recovery times to be shortened.The tragic cycle of poverty in the Sahel is a classic example of this form of instability.

Many insect pests, on the other hand, have a short recovery period and adopt the first strategy, so they are very difficult to exterminate; often they quickly adapt to pesticides. Desert species too tend to adopt the first strategy because the environment conditions are unpredictable and impose considerable stress, whereas tropical-rainforest species mostly adopt the second strategy because they have a stable environment. When man interferes with such systems, therefore, he tends to have a far greater destabilising impact on those with long cycles, such as rainforest species, than on those with short cycles, such as desert species and pests.

26.2 Man's search for food

Man is partly a vegetarian (**herbivore**), collecting seeds, tubers, and fruits; and partly a meat-eater (**carnivore**), taking easily captured game. As a result people have progressed using the twin strategies:

(a) increasing the effectiveness of predation (i.e. by engaging in animal husbandry);
(b) improving useful plant productivity (by **cultivating**).

In turn these strategies have had three consequences:

(a) to *exploit* the ecosystem, placing stress on most species to allow a few to prosper;
(b) to *disrupt* the habitat for use by cultivation, building and so on;
(c) to *pollute* the environment with concentrated wastes.

In each of these strategies people have found that if they put too much stress on the natural recovery cycles of the species they are eating, then instability will result and the **law of diminishing returns** will apply. For example, in order to maintain the catch of whales (which fall into the long-recovery-cycle group) people have resorted to ever-increasing levels of sophisticated technology (e.g. harpoon gun, radar). This has applied so much stress to the whale population that catches are declining despite the increased technological investment.

Some would argue that it can sometimes be appropriate to engage in a short-term crop maximisation strategy, overcropping the population, reducing it to low limits in order to recover investment rapidly, then abandoning the resource for decades (or even centuries) in order to allow it to recover naturally. Thus whaling and many forms of fishing are equivalent to shifting cultivation of the land. This strategy assumes that, given time, nature can restore the original situation.However, as a low population is unstable, such recovery cannot be guaranteed. This is why land-based strategies have evolved to give long-term increases in the chosen species numbers, by providing more food and nutrients, reducing competition, and reducing consumption of it by species other than man.

Many scientists now aim to help bring about better food production by using the principles of short- and long-term adaptive cycles. For example, pest control is cheapest and most effective when the life cycle of the pest is understood. Similarly, it is better to adopt a selective logging policy in a large forest, and thereby disrupt as little as possible the long-term nutrient cycles and ecosystem, rather than adopt a clear-cutting policy which may lead to instability (soil erosion and leaching) and make recovery impossible.

26.3 Urban and industrial development

Because people are becoming increasingly urbanised, they are disrupting ecosystems in a further fundamental way. A natural ecosystem operates a closed nutrient cycle: nutrients are retained within the system. Crop harvesting, in contrast, breaks this cycle: it removes the nutrients to cities and sends the wastes in over-concentrated form to pollute the sea or other disposal areas.

Within the last few decades the expansion of industrial and urban development has resulted in social stress to curb the worst effects of the environment. Conservation measures are increasingly based on a proper understanding of the environment cycles. They include:

(a) retention of natural ecosystems of sufficient size to support a natural genetic stock for all trophic levels;
(b) protection of soil fertility;
(c) retention of visual landscape patterns;
(d) the retention of some historic land use practices (e.g. mixed farming);
(e) the restoration of degraded environments.

These are the topics that will be discussed in the following chapters. However, as has been demonstrated, they can only be discussed in the light of a proper understanding of each natural system.

Soil processes

27.1 Properties of soil

Some of the most valuable parts of our environment are taken for granted. Perhaps the most abused of all is soil.

The non-specialist may see a soil only as a featureless, uniformly brown material in his garden, but the apparent uniformity of a soil is largely the product of plough or spade: it is rather like a layered trifle that has been stirred with a large spoon to give an even brown mess. Yet dynamic processes are continually at work to restore the natural layers, which is one reason gardeners find it necessary to turn the soil over so frequently.

Today it is difficult to find naturally developing soil in much of the world because agriculture has encroached over nearly all accessible areas. However, it is vitally important to appreciate the natural processes of soil formation: even under man's husbandry, such processes are at work attempting to fulfil nature's design. It is surely better to work with nature than against her.

Table 27.1 Soil-horizon nomenclature

O	O organic horizon above the mineral soil, divided into: L litter layer F partly decomposed or fermented layer H well-decomposed or humified layer	accumulating organic material
A or E	Ap ploughed surface horizon (in cultivated soils) Ah mineral + organic surface horizon Eb brown-coloured (eluvial) horizon from which clay has been removed Ea ashen-coloured (eluvial) horizon from which clay and sesquioxides have been removed	'topsoil': may be a region of loss
B	Bk horizon containing illuvial carbonates (often as krotovina) Bf horizon containing illuvial iron as a pan Bh horizon containing illuvial humus BL indurated clay–sesquioxide horizon in tropical soils (plinthite) Bs horizon containing diffuse accumulations of illuvial iron and also clay (i.e Bs includes Bt) Bt horizon containing illuvial clay Bw horizon of alteration, differentiated from Ah and C by colour and/or structure and to which there has been no significant clay eluviation (w = weathered) little-altered horizon (the parent material)	'subsoil': may be a region of gain
C	The suffix g indicates a gleyed horizon; (g) indicates mottling; G indicates a horizon so subject to gleying that no other soil-forming process can be identified. For transitional horizons the convention Ah/Bw, Bt/C, etc. is used.	solid rock, alluvium, colluvium, loess, glacial till, wind-blown sand etc

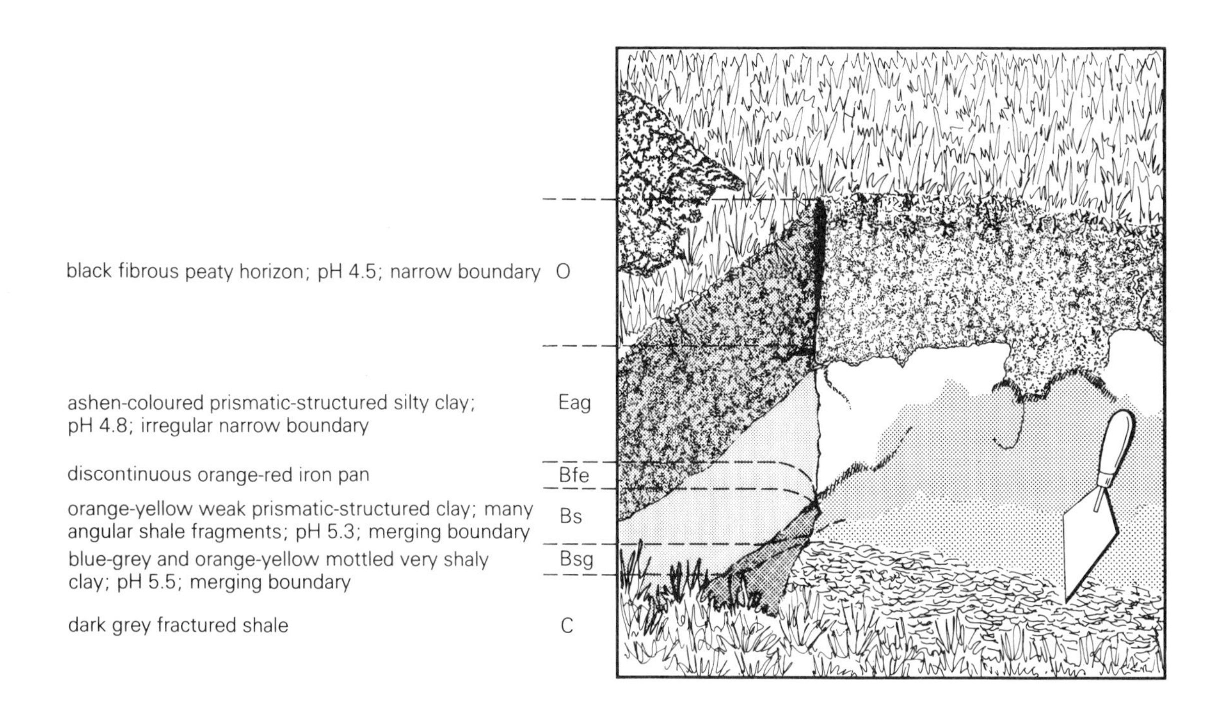

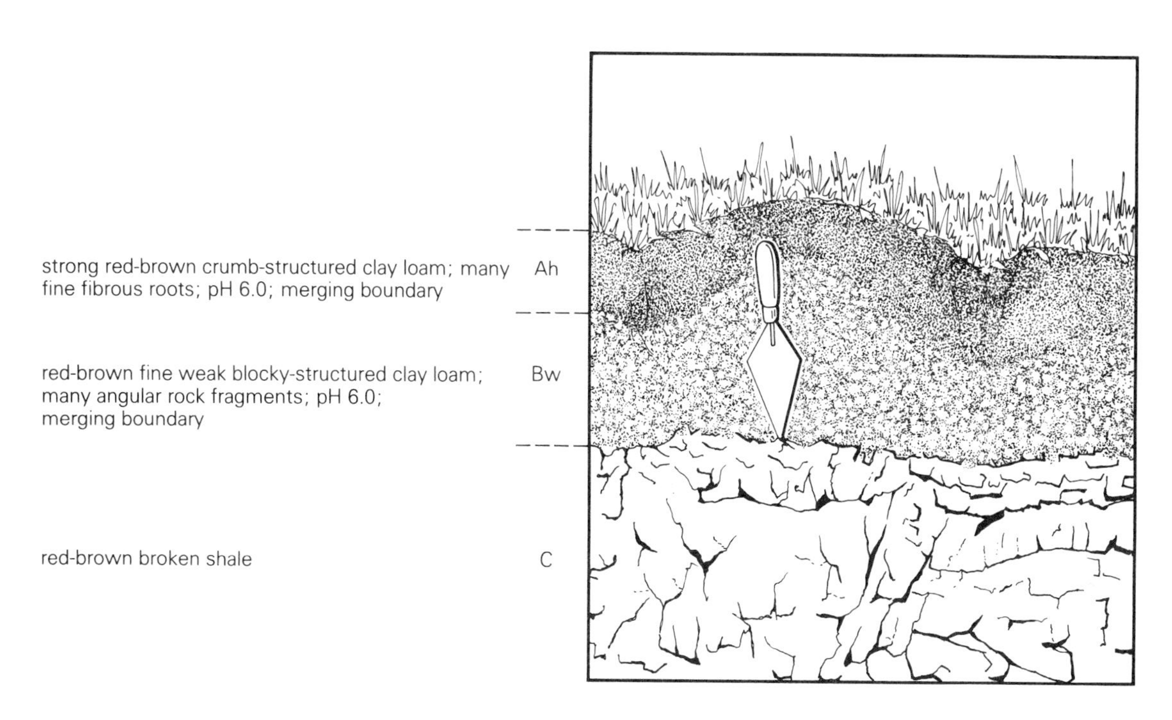

Figure 27.1 Two soils common in Britain. The upper soil, a *podzol*, is frequently found in uplands; the lower soil, a *brown earth*, is more characteristic of lowlands. Descriptions of each horizon and the labels used to represent the dominant processes involved are also shown.

Colour

The most obvious property of many soils is their patterns of colour. The soil in Figure 27.1a, for example, is clearly divided by its colour differences into several distinctive layers (**horizons**). These horizons are described in Table 27.1.

Soil colour is produced by:

(a) iron oxides weathered from the mineral content of the soil, which produce a range of red and orange stains in freely draining soils, and grey, blue and green stains in waterlogged soils;
(b) decayed organic matter (often humus), which imparts a black stain;
(c) the natural, usually pale, colours of the bulk of the soil-mineral matter.

The colour pattern enables us to identify regions of the soil dominated by particular constituents and thus the regions where particular processes are most effective (Fig. 27.2). For example, in the podzol of Figure 27. 1a, the black colour of the uppermost profile indicates a dominance of humus. Below this the pale colour horizon (labelled Eag) is simply that of the unstained mineral matter. The orange colour of the subsoil indicates a concentration of iron oxides. It seems, therefore, that at least three different soil-producing processes are at work within this soil; each process is replaced by the next within a short distance. By contrast, the red-brown colour of the brown earth topsoil (Fig. 27.1b) reflects a combination of iron oxides, humus and natural mineral colour. In this profile there are no striking colour changes, a clear indication that the processes at work in this profile are different from those operating in the podzol.

Texture

Texture is another basic soil property. It relates to the proportions of various sizes of mineral matter in the soil. Small particles, between 2 mm and 0.06 mm are called **sand**; those in the size range 0.06 mm to 0.002 mm are called **silt**; and those below 0.002 mm in diameter are called **clay**. The gravel and larger fractions are grouped under the heading of **stones**.

Variations in texture are most easily noticed by the 'feel' of moist samples when rubbed between the fingers; texture is not readily apparent simply by visual observation. Sandy soils have a rough, gritty feel and do not hold together well (are not cohesive). Silty soil feels slippery and soapy, and is more cohesive than sand. Clay soil is sticky, plastic and very cohesive; it can easily be rolled into a ball.

Usually soils contain a mixture of these fine sizes and are therefore referred to by combined terms such as 'sandy clay' or 'silty clay'. A fairly equal mix of sand, silt and clay, which is called **loam**, is recognised by growers as a most desirable texture: it yields an easily worked, fertile and well drained soil.

Often horizons that show colour differences also show variations in texture. Thus, whereas the Eag horizon of the podzol in Figure 27.1 is silty-clay textured, the lower, Bs,horizon has a 'heavier' clay texture. Textural changes can occur without changes of colour, and many soils show an increase in the proportion of clay in their lower horizons – a signal of movement processes with the soil.

Structure

Most soil particles are not packed loosely together as individual grains, but occur as organised groupings. A piece of soil rubbed gently between the fingers may break apart, but recognisable clumps of many particles will usually remain. Clods of soil left lying on

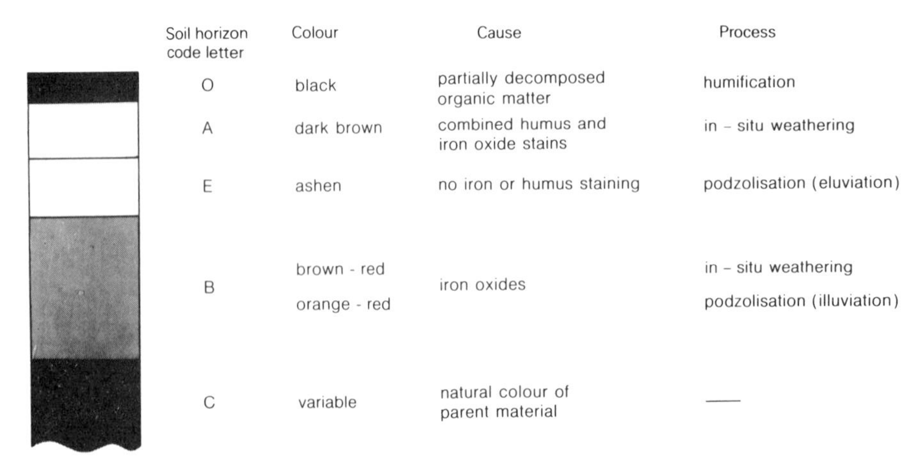

Figure 27.2 The relationship of soil colour to processes of formation and soil constituents.

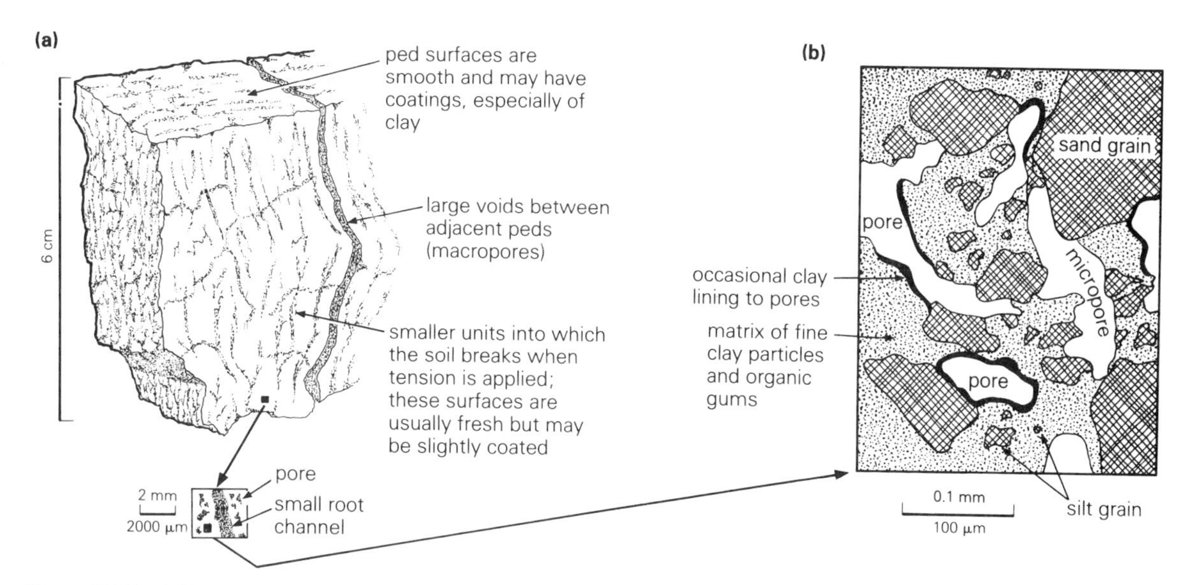

Figure 27.3 Soils can be examined at several levels of detail. The easiest unit to identify in the field is a ped (a), whereas the intricate composition of a soil is more readily seen under an optical microscope (b) or an electron microscope (Fig. 27.5). Notice that some clays are deposited on pore walls much like scum left on a bath after it has been drained. They are a clear demonstration of movement processes in operation.

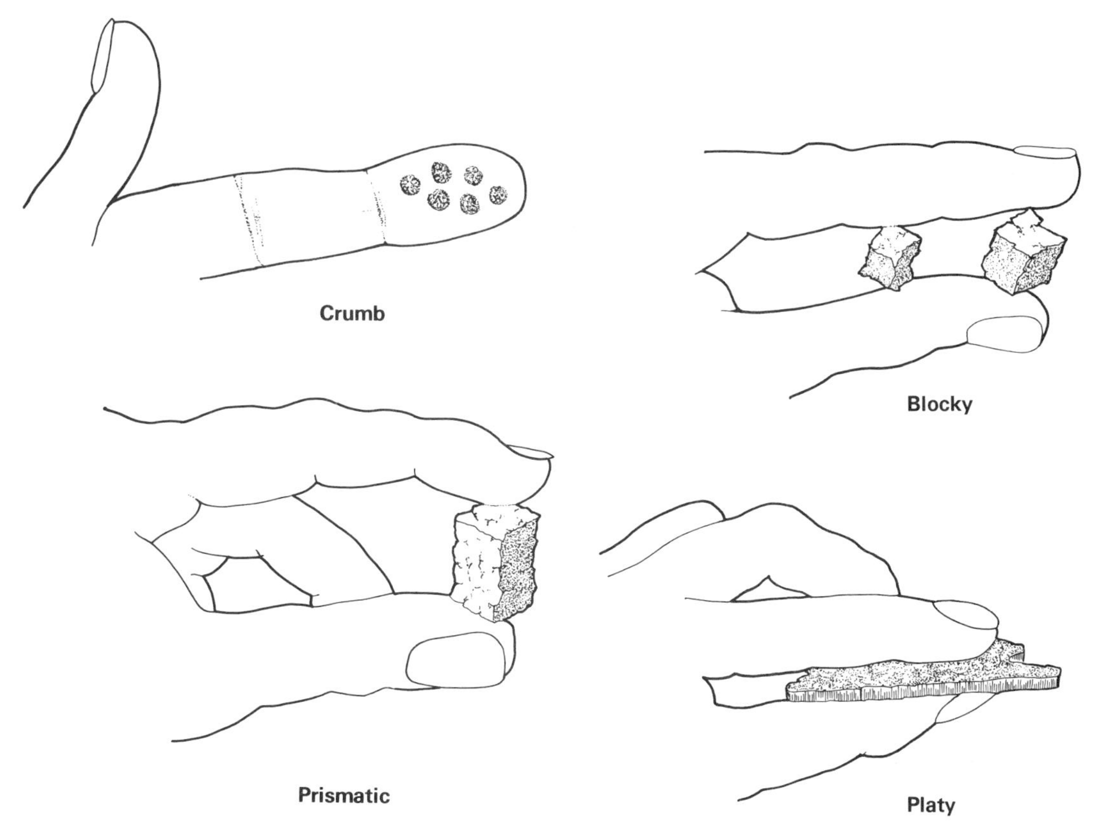

Figure 27.4 The sizes and shapes of peds.

a field after it has been harrowed are another clear demonstration of the tendency of soil material to hold together. Natural soil groupings are colloquially called 'clumps' or 'clods' but technically called **peds**. Sometimes they are strongly formed and very stable (have a **strong structure**); others fall apart when handled (have a **weak structure**). Peds occur in a variety of common shapes (Fig. 27.3). The smallest shape, **crumbs**, are roughly spherical, whereas the larger peds are more angular, either being **blocky**, elongated (**prismatic**), or flat (**platy**); (Fig. 27.4).

The differences in size and shape of the peds is an indication of which soil-forming processes are dominant. Soil structures are largely formed by soil wetting and drying and by root and animal activity. Such structures are thus most prominent in the topsoil.

Porosity and permeability

Natural soil structures are separated by well defined 'fractures' that provide relatively large pathways for air and water movement. Much smaller spaces occur *within* each ped. These are caused primarily by the poor packing of the irregularly shaped grains (Fig. 27.5). All the shapes and sizes of the soil are collectively known as **pores**: the smallest pores that can hold water against gravity drainage are known as **capillary pores**.

The amount and size of pores determine the **porosity** of a soil, which is affected also by the texture and structure of the soil. In clays the size and degree of linkage of the pores within each ped may be very low and such that they can transmit very little water. Thus in well structured but fine textured soils, the pores *between* peds can be of greater value to soil drainage and aeration than the pores *within* peds. Soil porosity is not a reliable measure of the ability of the soil to transmit water, nor of its ability to retain water. The greater the proportion of capillary pores, the greater the amount of water that can be held against gravity (and thus tide plants over a dry spell); but the term 'porosity' does not distinguish between large and small pores and it should be treated with care. Nevertheless, one can assess the relative effectiveness of the large and small pores by simple observation. Furthermore, one can measure the water-transmitting capacity (the **permeability**) of the soil.

Organic matter

Organic matter is the essential constituent which, together with mineral matter, defines a soil. Soil

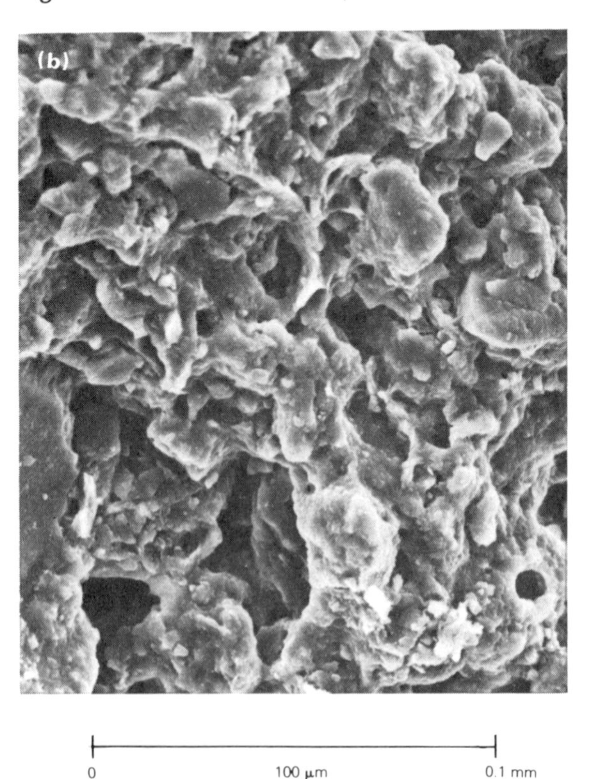

Figure 27.5 All soil particles exhibit poor packing, leaving spaces (pores) for air and water access.

organic matter consists of the dead leaves and stems lying on the ground and the live roots below it. Roots investigate the soil in search of water and nutrients. They spread out in a complex net which both anchors and stabilises the plants.

As roots tend to follow paths of least resistance, they mostly penetrate a soil via its pores. Initially only fine root hairs will penetrate pores: as the root grows, the pores are mechanically widened. This causes general topsoil heaving and promotes widespread fracturing of the soil, increasing soil permeability. When the root dies and is decomposed by soil organisms, the abandoned root channel provides yet further paths for percolation.

At depth soils become more compact and so more difficult to penetrate. Nutrients are less common and aeration is less readily available. These factors limit the depth of root exploration and tend to confine root activity to the topsoil. Furthermore, the vast army of soil organisms that decompose and incorporate dead leaves into the soil are similarly limited by their need for aeration and food. Organic matter therefore tends to be concentrated near the soil surface.

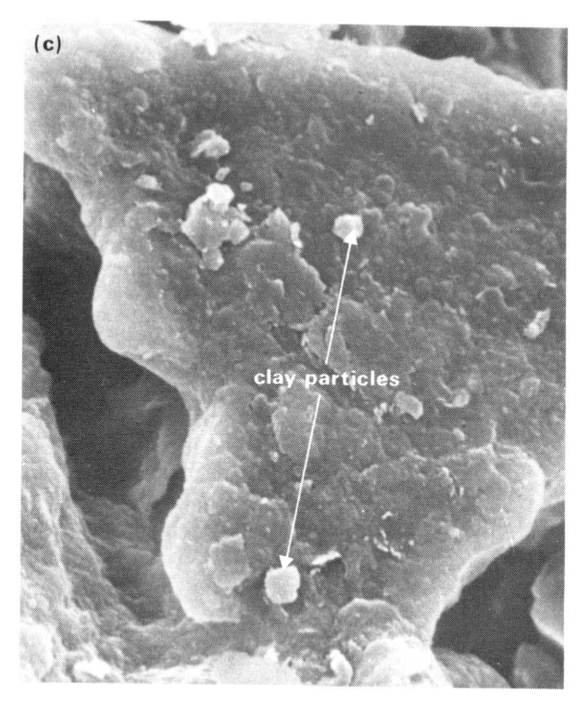

Table 27.2 Soil reaction.

pH value	Reaction description
8	alkaline
7	neutral
6	slightly acid
5	moderately acid
4	strongly acid

Soil acidity

Soils are described as acid, neutral or alkaline according to the proportion of hydrogen ions among the various positively charged ions in the soil water. If the proportion of hydrogen is greater than that of the other ions, the soil is said to be **acid**; if the proportions are equal it is said to be **neutral**; and if the proportion of hydrogen is smaller, the soil is **alkaline**. There are, however, only about 10^{-4} grammes of hydrogen ions per litre of solution even in very acid soils, and fewer than 10^{-8} in some alkaline soils.

It is convenient to denote soil acidity by giving only the power term (e.g. '4' instead of '10^{-4}'). As a reminder of the real significance of the number – the proportion of hydrogen ions in the solution – this number is prefixed by the letters **pH**. Neutral soils have a pH of 7. The more commonly encountered range is shown in Table 27.2.

Although soil acidity is not an observable soil property, it can be measured easily. Like the other properties, it helps enormously in identifying the major soil processes in operation.

27.2 Processes in soil

A **soil** is *a mixture of mineral and organic fragments organised into horizons by physical, chemical and biological processes*.

There are three basic and quite separate sets of processes involved in producing a soil (Fig. 27.6), which together control a sequence of events that produces all soil profiles:

(a) the breakdown (weathering) of soil parent material;
(b) the decay and incorporation of plant and animal matter;
(c) the organisation of soil matter into soil horizons.

Weathering of mineral matter

As discussed in Chapter 13, weathering can be divided into:

(a) *mechanical* weathering, which produces a layer of shattered and angular rock;

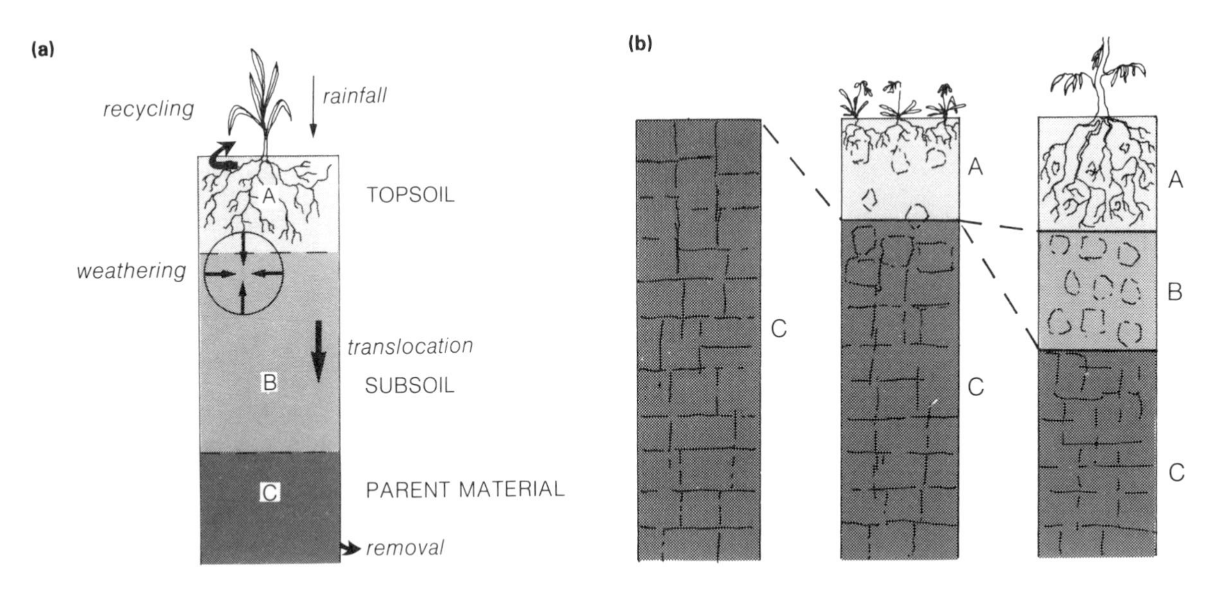

Figure 27.6 (a) The basic soil-forming processes. (b) The sequence of horizon formation during soil development.

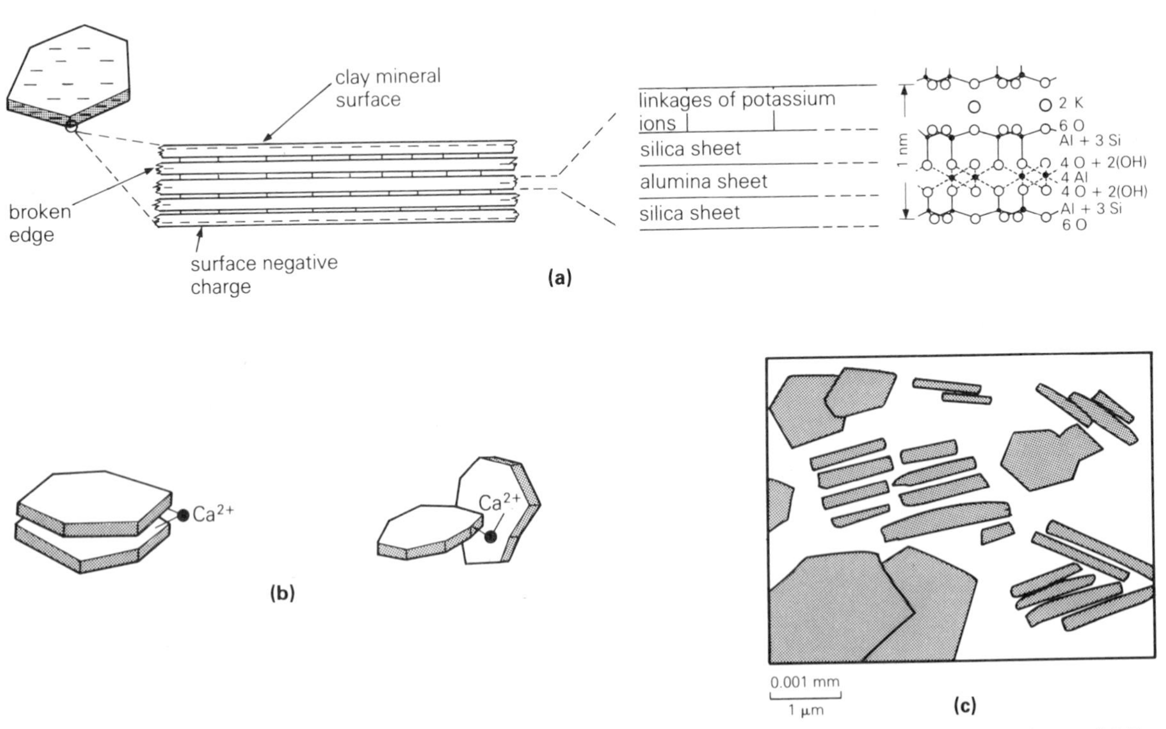

Figure 27.7 The nature of a clay mineral. (a) Its sheet-like form, showing the position of charges and internal linkages. (b) The way in which clay minerals can aggregate using cations in face-to-face contact and edge-to-face contact. (c) A natural grouping of clay mineral particles seen in thin section under an optical microscope and showing both types of aggregation. Pores are created because of irregular packing of grouped particles although they are much smaller than those shown in Figure 27.5.

(b) *chemical* weathering, which (using water, carbon dioxide and organic acids) decomposes rock to produce clay minerals, iron and aluminium oxides and to release nutrients (ions) for plant growth.

The main products of chemical weathering, the **clay minerals**, are very small (colloidal) particles less than 0.002 mm across (Fig. 27.7). They are very thin platey crystals which have an enormous surface area compared with their volume; most importantly, they carry at their surfaces a significant negative electrical charge. When some minerals are weathered they also release **iron** and **aluminium oxides** (collectively called **sesquioxides**). (Iron oxide alone is the familiar rust.) In many soils these oxides are evenly dispersed through a soil horizon as coatings on the mineral grains.

In addition to these insoluble products of weathering (the precipitate of a chemical reaction), many substances remain in solution: these may be carried away as the water percolates through the soil or they may be used by plant roots as nutrients. Most of these substances are ions that do not readily combine with each other in the soil solution. Nevertheless, because the ions are electrically charged, the positive ones (**cations**) are attracted to the negatively-charged surfaces of the clays, where they are loosely held.

(a)

Figure 27.8 Soil micro-organisms. (a) Cross section of a twig amongst leaves (also in cross section) (A) in the F layer of a moder type of organic matter. Fungi (dark filaments at F) are actively breaking through the outer regions. Mites (C) have been consuming the less resistant inner parts of the twig, leaving faecal pellets (D) behind. (b) An electron-microscope photograph of fungi attacking a leaf surface; (C) An electron-microscope photograph of a protozoan (the smallest of soil animals) amongst fungal filaments.

The decay of organic matter

Organic matter is a generic umbrella term for all the organic tissue, live and dead, within a soil. Much of it is in the form of simple chain sugars called **cellulose**; there are also harder, woody substances (in which the cellulose structure has become impregnated with other substances) called **lignin.**

As plants build up their tissues, they use not only carbon and hydrogen from the air and from water, but

(b)

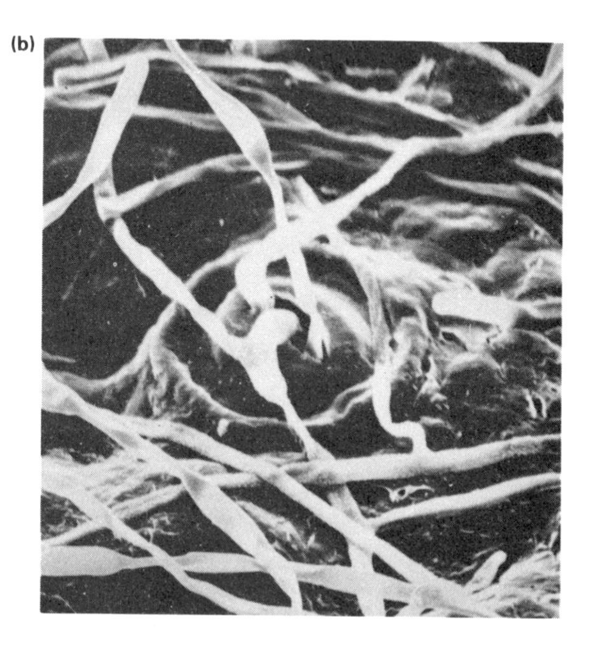

(c)

Visual form of organic matter	Stage of development
	leaf on tree begins to die, dehydration occurs
	leaf in litter is broken down by micro-organisms and larger fauna
	leaf in fermenting layer loses identity and only the most resistant tissue remains
	humus is encorporated into the soil body

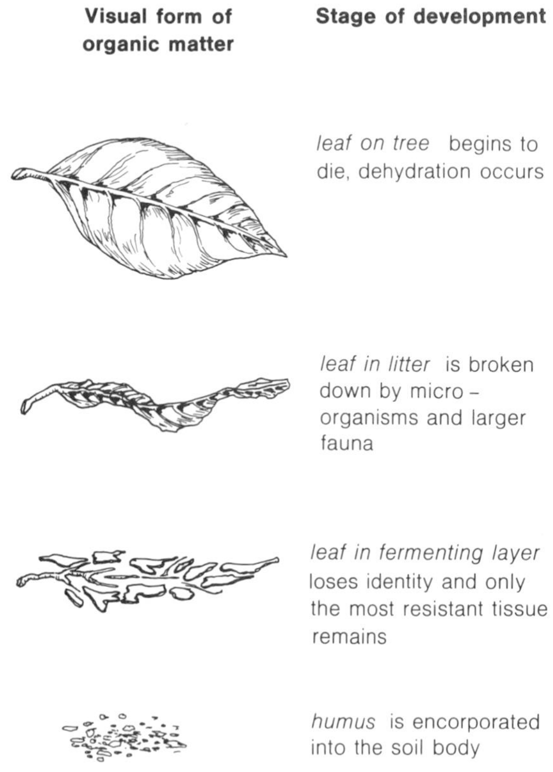

Figure 27.9 Stages in the decay of organic matter.

a whole range of ions released by weathering. The absorption of ions by plants is only a temporary loss to the soil (see nutrient cycles, p. 280) provided that the ions are quickly returned by rapid decay of tissue when the plant dies. Rapid decay occurs most readily in plants such as grass which contain a large proportion of simple cellulose. It is retarded in plants such as trees that contain lignin. This difference has major implications for soil acidity and fertility.

Organic tissues are mainly decomposed by **soil micro-organisms** (Fig. 27.8) rather than by direct contact with rainwater. The most populous of these micro-organisms in neutral and alkaline soils are **bacteria**. There may be as many as 1000 million bacteria in each gram of soil, although the larger **fungi** are possibly the most important decomposers because they can tolerate acid environments. In acid soils many fungi form direct links between decaying tissue and plant roots, giving very efficient recycling.

In addition to the micro-organisms there is a vast army of larger animals including **mites**, **earthworms**, **insects** and **mammals**. These consume, break down and incorporate dead organic tissue; they also help to mix soil materials together. In temperate regions earthworms are the most important at soil mixing; in the tropics this role is played by termites, millipedes and ants. Without their help the only direct addition of organic matter would be from the *in situ* decay of roots; the bulk of the plant residues would be left lying on the mineral matter surface as **litter** (Fig. 27.9).

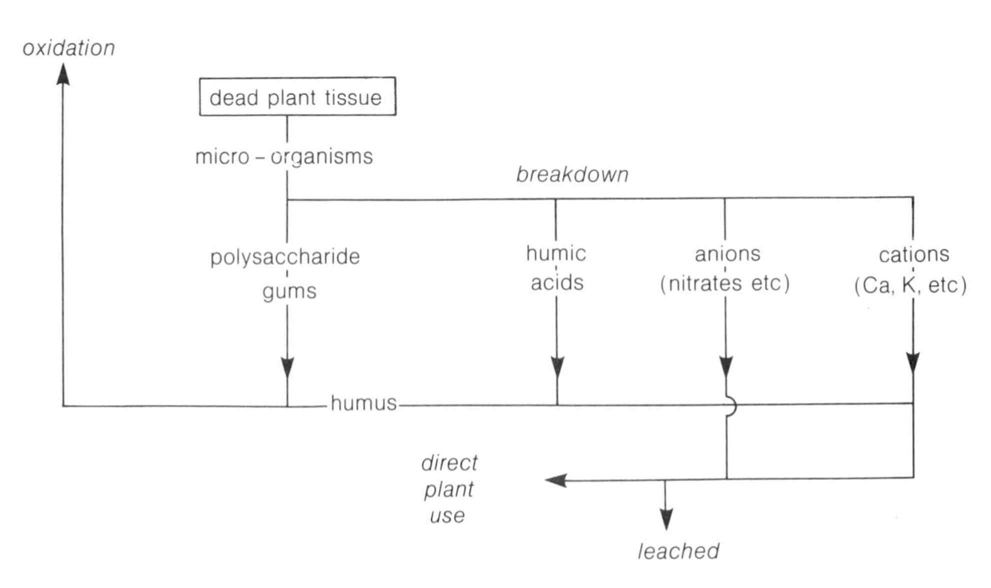

Figure 27.10 The process of plant decay.

Organic matter is decomposed in two stages, again largely by soil micro-organisms or by micro-organisms within the guts of the larger soil animals (Fig. 27.10). In both cases break down occurs in two stages:

(a) **mineralisation**, in which organic matter undergoes chemical reactions that change it into simple organic substances (mainly acids) and release a range of compounds (such as nitrates) and ions (such as calcium, Ca^{2+}) into solution;

(b) **humification**, in which some of the acids immediately recombine (polymerise) to give an insoluble precipitate called **humus**.

Mineralisation yields simple substances which form the building blocks for the next stage and at the same time provide nutrients (ions) for further plant use. Humus, the result of humification, is the organic equivalent of clay minerals; like the clays, it has a negative charge and a large capacity for storing cations on its surfaces. In addition, micro-organisms

THE IMPORTANCE OF CLAY, HUMUS AND IONS IN A SOIL

The great importance of clay minerals and humus in a soil is their capacity for absorbing ions released by chemical weathering of rock or dead tissue. They therefore act as an invaluable *reservoir* of nutrients for further plant growth and as a *buffer* against immediate loss (**leaching**) in the soil water. Nevertheless, the cations do not become permanently fixed to the clay or humus surfaces, but are continually being displaced by others in such a way as to maintain a balance between the proportion on the surfaces and within the soil water.

Hydrogen (H^+) ions are steadily produced within the upper soil as part of plant respiration and also arrive at the surface with fresh rainwater. Non-hydrogen ions are replenished by rock weathering, the continued decay of plant tissue, and in rainwater.

An exchange of ions will occur between soil water and clay–humus surfaces every time rain falls. Percolating water will wash away (leach) the old soil water together with its cations and replace it with new rainwater. If the non-hydrogen ions are not replaced sufficiently quickly there will be a gradual increase in the proportion of hydrogen ions on the clay–humus surfaces and therefore an increase in soil acidity. There will also be major changes in the stability of soil structures.

The amount of charge that can be held on the clay–humus surfaces varies with the different types of clay and humus. It is referred to as the **cation-exchange capacity (CEC)**. From an agricultural point of view the clay–humus role is paramount, for without its storage capacity, fertilisers applied to the soil surface would be rapidly washed away and wasted. Thus sandy soils with a low clay–humus content have a low CEC and a correspondingly low fertility. However, even a clay soil can vary in its potential fertility depending on its humus content, for humus has a CEC several times that of the equivalent amount of clay: clays with a high humus content are thus much more fertile that those without.

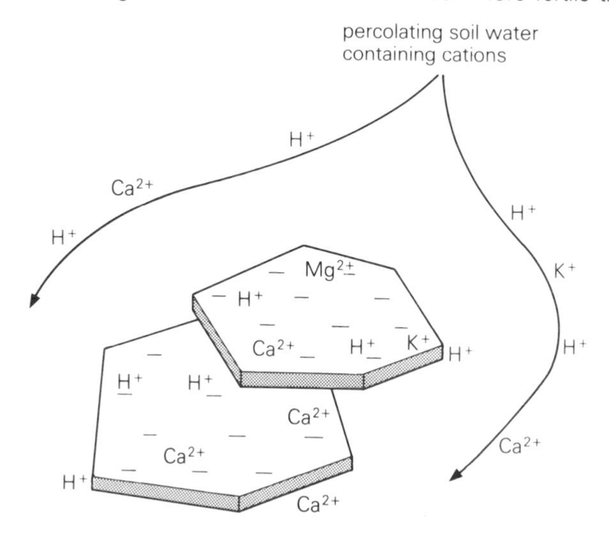

Figure 27.11

convert part of the organic matter into new forms of long-chain sugars called **polysaccharides**. These do not have a significant surface charge but are physically sticky. Mostly produced in the gut of animals, the polysaccharides act as long sticky threads, binding sand, silt and clay together.

Sometimes, under acid conditions, these conversions are not accomplished. In particular, organic acids are not converted to humus and polysaccharides do not form. In these circumstances the acids remain reactive (aggressive) and can weather new mineral matter or rapidly remove ions from the surfaces of clays and humus.

EFFECT ON SOIL HORIZONS

The type of organic matter decay is reflected clearly in the appearance of the upper soil horizons. Under grassland, and with neutral or alkaline soil conditions, complete break down of organic matter occurs. A multitude of soil animals mix the material evenly into the the upper part of the soil. This type of organic-matter distribution is called **mull**. If soils are more acid (as might be expected under deciduous woodland) decay is slower because there are fewer micro-organisms at work. As a result a thin layer of **litter** remains clearly visible on the surface with a layer of partly decayed material beneath it. This arrangement is called **moder**. Under extremely acid conditions (as under coniferous forest or heath) there are so few soil animals that a thick layer of black, greasy, partly-decayed organic matter lies below the surface litter. This arrangement is called **mor**. Finally, under conditions of extreme waterlogging, most micro-organisms cannot get sufficient air to survive: plant tissue is little decomposed, instead accumulating on the surface to form **peat**.

The aggregation of soil particles

There are several independent processes that tend to bind soil particles together: electrical forces; stickiness; cements; and surface tension.

ELECTRICAL FORCES

Some of the cations held on clay-humus surfaces have sufficient electrical charge to be able to link with the negative surfaces of more than one clay or humus particle and thereby form a bridge (Fig. 27.7). These are cations such as Ca^{2+} (called **divalent cations**). Other cations, such as Na^{+} and H^{+}, have only sufficient charge to bond to one surface at a time. They are therefore nutrients but not stabilising agents. The process of **bridging** to build clay–humus groupings is called **flocculation**: it can build packages of particles up to the size of sand. At this larger size they become much more stable and less likely to be washed or to blow away. Farmers therefore frequently add lime (calcium hydroxide – a source of divalent cations) to their soils both to provide nutrients and to encourage flocculation, which in turn gives better soil aeration and drainage.

HUMIC GUMS

Cation bridges are vital, but they only build small soil packages: larger packages are formed by sticking particles together with polysaccharide gums. No electrical charges are involved in this case and therefore all kinds of particle, including sand, can be stuck together. The difficult problem of wrapping up the particles with the polysaccharides is usually accomplished in the gut of earthworms or other soil fauna which eat their way through the soil.

Humic gums are an unstable product of tissue decay: in time, they become oxidised and lost. Nevertheless, because they are often the most important stabilising agent in a soil there must be constant replacement by the decay of fresh organic matter.

A soil needs only about 0.02 per cent by volume of polysaccharides for a high level of stability – an amount that can be provided from a raw plant content of only 3 per cent. A low level of polysaccharides is one of the main reasons for soil instability and accelerated soil erosion. Low polysaccharide levels are common in cultivated land where the organic matter is removed as the harvest. There is no *artificial* substitute for polysaccharides, which is why some form of **manuring** (fresh organic matter) is so vital to soils throughout the world, no matter how sophisticated their farming practices.

CEMENTING AGENTS

Iron and aluminium oxides are usually unstable in a soil and are quickly precipitated after their release by weathering. They form coatings on and between soil particles and can help soil stability considerably, especially in acid soils where cations are leached away and humic gums are in short supply.

SURFACE TENSION

Water held in capillary pores under surface tension can considerably increase the stability of a soil, which is why dry, rather than wet, soils rarely blow away. Because surface tension depends on water being held in small pores, its effects are most important in fine-textured soils.

The movement of soil materials

Complete stability of soil materials is impossible. To achieve even a high level of stability, divalent cations such as Ca^{2+} have to be returned to the soil at a rapid rate, while polysaccharide gums need continual renewal. These favourable conditions only occur in humid climates with plentiful rainfall when soils are developed on cation-rich rock such as chalk and under rapidly recycling vegetation such as grassland. Even then, the upper humic horizon will be subject to leaching. In semi-arid regions stability is easier to achieve because rainfall rarely percolates completely through the profile. In all other situations continued percolation leaches cations, reduces the effectiveness of flocculation and allows clays in particular to be washed downwards.

The general term for the movement of soil materials within a profile is **translocation** (Fig. 27.12). There are two clearly separate transfer components of translocation: the movement of material in solution, called **leaching**; and downward washing of particles in suspension called **eluviation**. The general term used to denote the precipitation of leached materials and/or the deposition of eluviated particles is **illuviation**. This is a basic division which allows all soil horizons to be categorised as:

(a) those subject to *no translocation*, only *in situ* weathering;
(b) horizons of *loss* (eluviation and/or significant leaching);
(c) horizons of *gain* (illuviation).

IN SITU WEATHERING

Soil profiles that have experienced no translocation rarely have sharp boundaries between horizons and show only a gradual decrease in the intensity of weathering with depth (Figs 27.6 & 16). These profiles normally have a pH greater than 6.0, which encourages not only a high level of microbial activity but also a large earthworm population, whose reworking activities mix the organic and mineral particles to give a dark coloured surface horizon with a structure of small, crumb-shaped peds.It is given the identifying label Ah.

The only visible changes to the profile occur as the organic content decreases with depth and its influence on soil colour lessens. At the same time, an increase in stones may herald the parent material. This (subsoil) horizon containing weathered mineral matter but little organic material is labelled Bw. Below it lies the little-altered parent material, labelled C. In Britain and other humid temperate climates, a soil with a horizon sequence Ah–Bw–C is called a **brown earth**. In the humid tropics weathering produces not a brown colour but a brilliant red, but the horizon sequence remains the same.

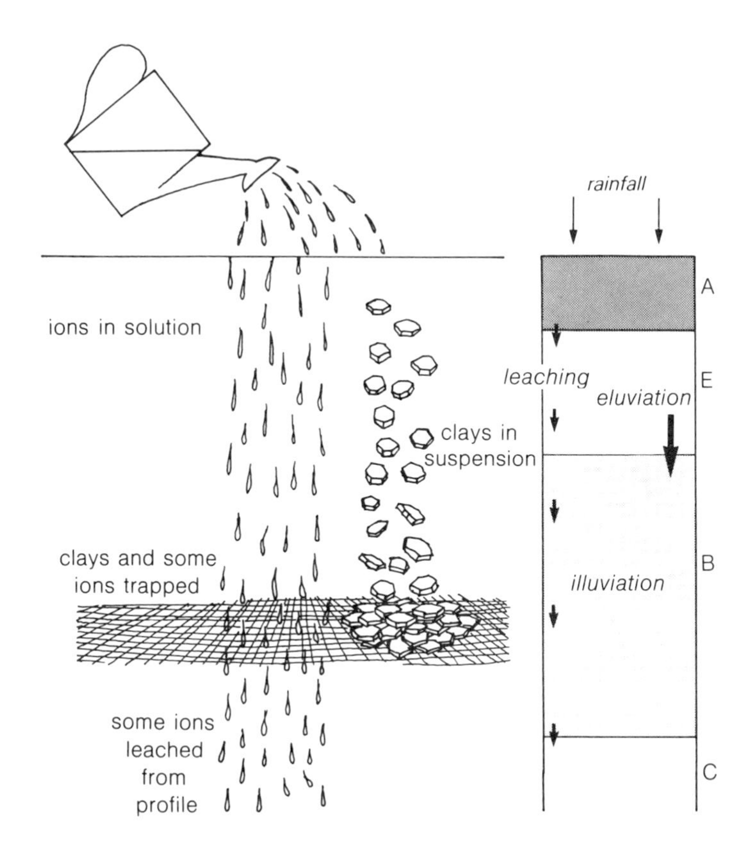

Figure 27.12 Translocation processes. Translocation (movement from one part of the soil to another) involves leaching (in solution) and eluviation (transport in suspension). Illuviation is the process of redeposition of both leached and eluviated material.

PROCESSES INVOLVING WEAK TRANSLOCATION

The rapid recycling of cations that occurs under grass and helps maintain soil stability is less efficient under deciduous forests, where many cations are stored in the wood of trees for perhaps several hundred years. Under these conditions the soil is often slightly acid (pH 6 to pH 5) and the microbial population consequently less abundant. In turn, this slows down the rate of plant decay, leading to a moder of organic matter, fewer divalent cations and a lower rate of polysaccharide formation. In these circumstances the soil is less stable: some clays become dislodged and are eluviated (washed) through the soil pores of the upper horizon to become trapped in the finer pores of the subsoil. This leaves a clay-depleted near-surface horizon (labelled Eb), either below or instead of the Ah. Below the Eb lies the illuvial accumulation horizon, its pores lined by thick clay coatings (see Fig. 27.3). This illuvial horizon is designated Bt. Many brown earth soils show evidence of clay eluviation, especially under woodland or when developed on cation-deficient parent materials such as sandstone.

PROCESSES INVOLVING STRONG TRANSLOCATION

Eluviation is only one of the processes that cause movement and it must, of course, be preceded and accompanied by a moderate degree of leaching. However, some materials – notably the oxides of iron and aluminium – remain stable, with the consequence that there is no noticeable change in soil colour. Nevertheless, in soils where the parent material is exceptionally poor, where coniferous forest grows, or where the throughflow of water is large, soils become severely depleted in non-hydrogen ions and the pH falls below 5.

Many micro-organisms cannot survive in very acid conditions: as a result, organic matter is decomposed slowly and incompletely. Many humic acids do not recombine to form humus, but instead remain soluble and able to react with the sesquioxides, bringing them into solution. Once in solution the sesquioxides can be leached through the profile – a movement immediately recognisable by the striking colour contrasts that develop in the profile. This severe translocation, which is produced by cation leaching, eluviation of clays and finally sesquioxide leaching, is called **podzolisation**. It leaves a near-surface horizon (called Ea) depleted of all constituents except quartz grains, and the horizon pales to an ashen colour (that of the grains).

Translocation involves both regions of loss and regions of gain. The sesquioxides lost from the topsoil must therefore either be carried completely away in the percolating water or redeposited lower in the soil. In practice both occur, but sufficient illuviation occurs in the subsoil to produce a number of distinctive colour bands. Uppermost is any translocated humus, shown as a black horizon, labelled Bh; below lies an horizon of iron accumulation, often so intense that it forms a sheet-like **iron pan** (labelled Bf); and below this lies a more diffuse accumulation horizon containing not only redeposited clays, but also further iron and aluminium oxides that stain it a dull orange colour. This last horizon is by far the thickest and is labelled Bs.

PROCESSES INVOLVING TRANSLOCATION BECAUSE OF WATERLOGGING

Waterlogged soils are quite common. In some cases this is because of a high water table close to a stream; in others it is the result of a clay-textured soil or an iron pan having such a low permeability that water is forced to pond on the surface.

Most soil micro-organisms need oxygen to survive, and under saturated conditions of slow water movement they may remove all of the oxygen from the water. In their search for further supplies, micro-organisms extract oxygen from the (ferric) iron oxides, reducing them to ferrous oxides. This change is accompanied by a change of soil colour from brown or orange to grey, blue or green. The ferrous compounds are also soluble and can be leached away. The process whereby oxygen is depleted is called **gleying.** As soil that is waterlogged only occasionally (usually for less than two months a year) does not become fully gleyed only irregular patches of blue-grey reduced soil form, giving the soil a mottled appearance. Fully gleyed horizons are labelled G or g according to the severity of the gleying; mottled horizons are labelled (g).

OTHER MOVEMENT PROCESSES

In tropical areas there is often a strong seasonal contrast in rainfall, giving a water surplus in the wet season followed by severe deficit in the dry season. Under the conditions of high temperatures and plentiful water that accompany the wet season, considerable leaching occurs and many soils become very acid. If the site is well drained a podzol may even develop. However, where drainage is less adequate, as at the foot of an escarpment, the iron and aluminium oxides released by weathering in the wet season are not removed from the profile but remain within it, drying out and re-precipitating during the succeeding dry period. Such areas also usually receive large additional contributions of oxides leached from the better draining soils. Together, these sources concentrate massive amounts of iron and aluminium oxides in the B horizon (designated BL) –

a unique process called **ferrallitisation**. Sometimes these sesquioxide accumulations can become doimant within a soil, and such soils have traditionally been called **laterites**.

In semi-arid and arid areas there is rarely sufficient rainfall to leach the soluble products released by weathering completely from the profile. Semi-arid areas may have sufficient rainfall to flush away the highly soluble sodium (Na^+) and potassium (K^+) ions (and thus prevent the soil becoming saline), but not enough to remove calcium totally. Soils with carbonate deposition within the profile (as a Bk horizon) are called **chernozems** and the process of accumulation of calcium within the profile is described as **calcification**.

In extreme, rare circumstances there can be a net *upward* movement of water through a soil. This occurs, for example, beside coasts or lakes in arid areas,where the soil water table is near the surface and capillary action, encouraged by evaporation from the surface, can draw water upwards. Surface evaporation removes the water, leaving the sodium and potassium salts to crystallise out as crusts – a process called **salinisation**. However, this is *not* a typical desert process – it can only occur locally under special water table conditions. The majority of soils in the hot deserts are little more than collections of poorly weathered mineral fragments devoid of much real development: they are called **skeletal soils** or **lithosols**.

28

World soils

28.1 Soil development

Combinations of the soil-forming processes described in Chapter 27 are responsible for the world pattern of soils. These *internal* processes are controlled by five *external*, or 'environmental', **soil-forming factors**: climate, parent material, relief, vegetation and soil organisms, and time.

Climate

Climate controls the range of temperatures, the rainfall and the moisture regime of a soil. Climate therefore exercises a fundamental control on both weathering and translocation. For example, cool wet climates discourage plant growth, transpiration and soil organisms. As a result dead tissue is only poorly decayed, humification does not occur properly, and percolating water becomes very acid. Widespread leaching and eluviation occur, and eventually the removal of iron and aluminium oxides may lead to podzolisation. By contrast, in hot, dry climates water rarely percolates through the soil profile, leaching and eluviation are reduced, and soluble materials largely remain in the soil. Hot, seasonal climates result in complete weathering and most profiles become very deep and dominated by clays.

The role of climate is best illustrated by examining a large area with relatively uniform relief and parent materials. This is the case for Saskatchewan in Canada. Most of the soils are developed on glacial till and the area has only low relief. The climate is partially controlled by latitude and distance from the ocean, and partly by the rainshadow effect of the Rocky Mountains, whose effect becomes more pronounced towards the south.

The controlled effects of latitude, distance from the sea and rainshadow give a climate that becomes both hotter and drier towards the south-west. Thus there is a large water surplus in the north-east and a severe deficit in the south-west.

There is a wide variety of soils in Sakatchewan (Fig. 28.1). Their pattern shows a clear relationship to the soil-water budget. Where there is a large water surplus, soils vary from being totally waterlogged (tundra gleys) to showing signs of severe translocation (iron podzols). As percolation decreases, the degree of translocation becomes less pronounced, and leached brown earths form. When percolation is insufficient to reach through the profile, calcium is deposited as krotovina, *in situ* weathering is the dominant process and chernozems form. As soil moisture declines even further, vegetation becomes more sparse, the organic matter available for incorporation is reduced, and only a thin form of chernozem, called a **chestnut soil**, forms.

Parent material

Soils develop on a wide range of parent materials, from solid rock to wind-blown sand. Most soils in Britain form on unconsolidated materials such as glacial till, river-terrace gravels, silty alluvium, loess, hillfoot colluvium, and solifluction material. It is for this reason that the term 'parent *material*' rather than 'parent *rock*' is used.

Parent materials usually influence soil development within the overall limits imposed by the climate. For example, parent materials exercise considerable influence on the availability of cations and on the soil texture. If the parent material is quartz sandstone, there will be few calcium ions to neutralise any hydrogen ions in the percolating rainwater. In humid climates, sandstone parent materials are associated with acid soils. Furthermore, sandstone does not weather into clay; the soil will have a dominantly sandy texture and be freely draining. A combination of poor nutrient status and rapid drainage will encourage severe translocation. Shale-derived soils,

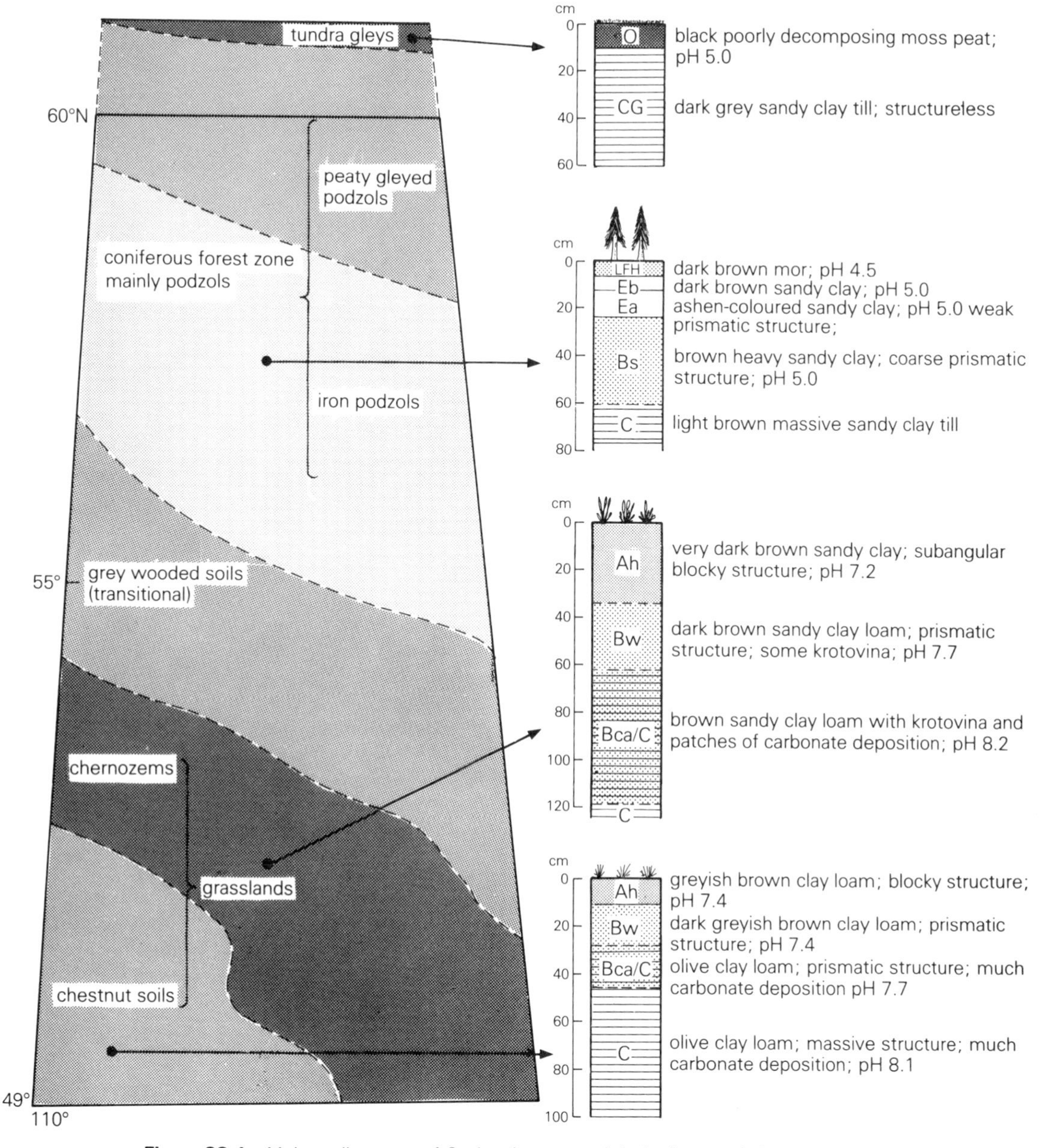

Figure 28.1 Major soil groups of Saskatchewan and their characteristic soil profiles.

and/or those formed on glacial tills, will weather to clay and the soil will develop a heavy clay texture and be poorly draining. Slow percolation rates will inhibit severe leaching but will promote gleying. At the same time clay is available to retain cations as a buffer against leaching and frequently the soil remains nearly neutral.

The influence of parent materials is clearly illustrated in the uplands of the Pennines, where a cool humid climate produces a large soil-moisture surplus and encourages translocation. On the limestone uplands, the recent retreat of ice sheets has left patches of glacial till, wind-blown loess and outwash sands and gravels, although bare rock is occasionally exposed as limestone pavements and very steep slopes (Fig. 28.2). Where the rock is at the surface, soils form in the weathered joints (**grykes**) between the blocks (**clints**) of limestone pavements. With little

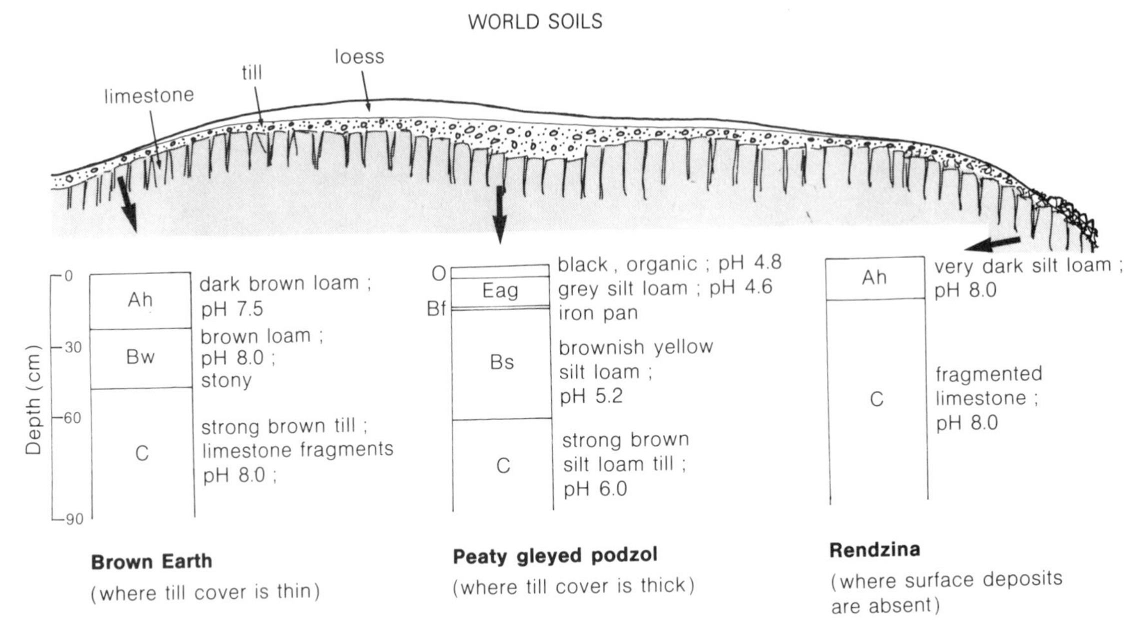

Figure 28.2 Characteristic soils of the Pennine limestones.

mineral content, soils are rich in organic matter and continual solution of the limestone keeps the material very alkaline. The soils have an Ah–C profile and are called **rendzinas**.

When the limestone is overlain by a shallow thickness of till, there is sufficient mineral matter to form a soil that is almost independent of the solid rock. Nevertheless, because the till is derived in part from the limestone, the fragments of rock within the till continue to release large quantities of calcium and maintain a high pH. This Ah–Bw–C soil, with a high pH throughout the profile, is called a **brown calcareous soil**. Where the till is thick, on the other hand, there may be sufficient depth for translocation to be a major factor. An acid moorland flora grows on these soils, dominated by mat grass (*Nardus stricta*). The till has a high iron content and severe leaching causes the formation of an iron pan. But because it has a fine texture and soil particles are not easily washed through the pores eluviation is suppressed and horizons do not show much textural variation.

Relief

Relief influences soil development through differences in drainage conditions. Soils on slopes are generally well drained, but on flat divides, in valley bottoms or on gently-sloping plainlands, water cannot drain away quickly and waterlogging occurs. Without good drainage, translocation does not operate and gleying becomes a major soil-forming process. Because most soil organisms do not thrive in stagnant conditions, organic matter tends to decay slowly, accumulating on the surface and frequently forming **peat**.

Rapidly changing water conditions are commonly encountered in till plains, where the uniform parent material is formed into swells and troughs (Fig. 28.3). The major soil process on freely draining sites is *in situ* weathering. This produces a brown earth. However, towards the troughs the fine-textured soil becomes increasingly unable to shed the water delivered by throughflow from the ridges, and gleying becomes apparent. In each profile the Ah horizon is the most permeable, primarily because of the effects of earthworm activity, and gleying is therefore most apparent in the Bwg and Cg horizons. In the troughs draining is so poor that water builds to the surface for some months each winter and the whole profile then exhibits gleying. Soils in the troughs are therefore described as **surface-water gleys** or **stagnogleys**.

Vegetation and soil organisms

Although vegetation is primarily controlled by climate, many types of plant have the same climatic range. Some (such as grasses) are readily recycled and help to build humus, polysaccharides and a supply of nutrients for more plants. In doing so they inhibit translocation. Others (such as heather) are slow to decompose, discourage soil organisms and take out more from the soil then they replace. Heathers can even make acid a soil formed on limestone.

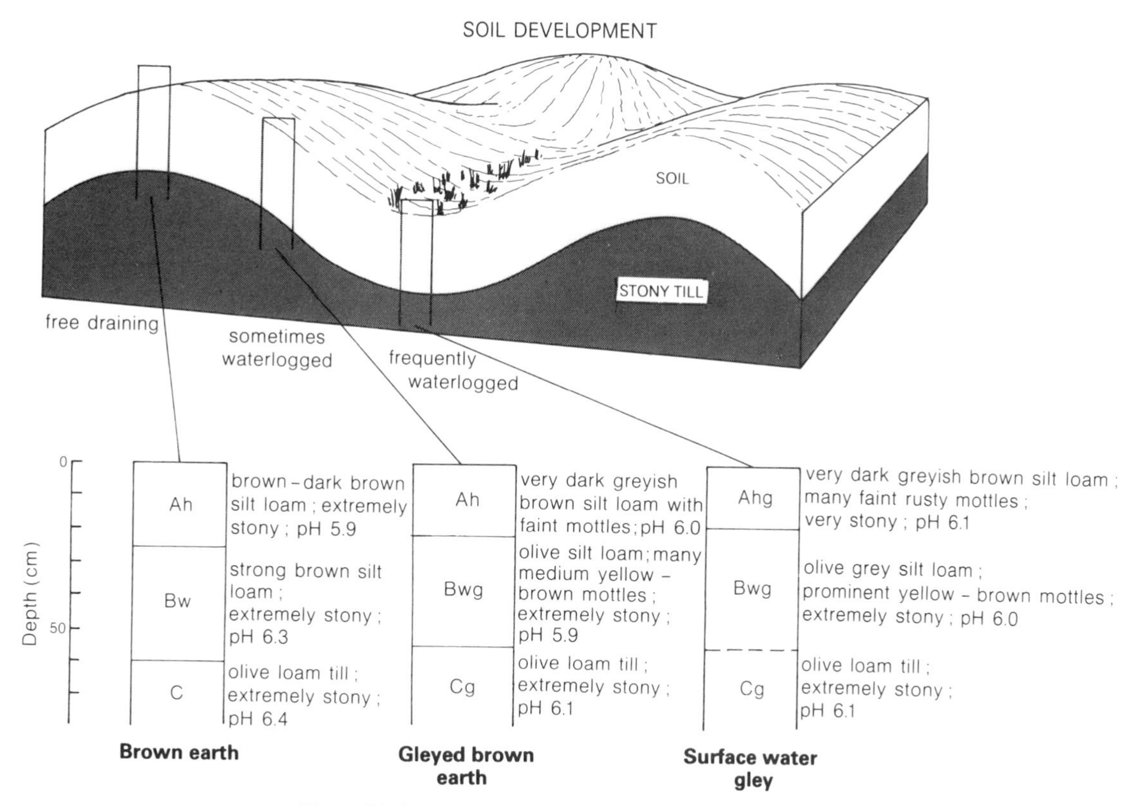

Figure 28.3 The effect of relief on soil development.

Vegetation and soils have such an interdependent relationship that it is usually very difficult to determine the exact role of vegetation. Plantations of conifers adjacent to stands of deciduous trees offer the best opportunity to examine the way different recycling rates affect soil acidity and translocation (Fig. 28.4). Simon's Copse in Surrey has extensive conifer plantations set within an oak woodland. The soils are all level and developed on sandstone. Although the conifers have not yet reached an equilibrium with the soil, already they have had a dramatic effect on the pH and an embryo ashen (Ea) horizon has been observed.

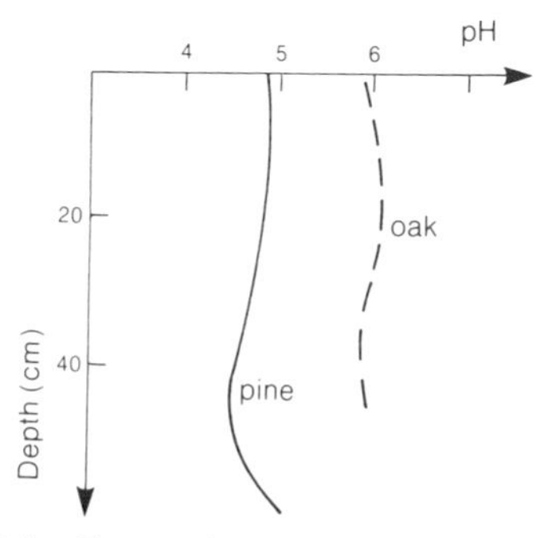

Figure 28.4 Changes in soil acidity produced by conifers.

Time

Soils develop slowly. The processes of leaching, eluviation and *in situ* weathering form soil horizons imperceptibly over hundreds or thousands of years. Indeed, most world soils have probably not fully come into equilibrium with their environment since the last cold period of the ice age ended 12 000 years ago. Even the soils that were beyond the reach of ice sheets have suffered much colder conditions and major changes in soil-moisture balance; many soils in middle and high latitudes have developed on glacial or periglacial materials entirely within the postglacial period. Only soils in tropical regions have had a long period of continuous and relatively undisturbed development. Most of these are very deep and highly leached.

Most valley-bottom soils are developed on alluvium and receive new material with every flood. These are new soils, and show only weak signs of soil development. Raised beaches and river terraces provide other places where some impression of the time needed for soil development can be assessed, but the initial stages of soil formation can most clearly be seen on developing sand spits. The South Haven Peninsula in Dorset, for example, consists of three parallel sand-dune ridges. On the foredunes, which still receive

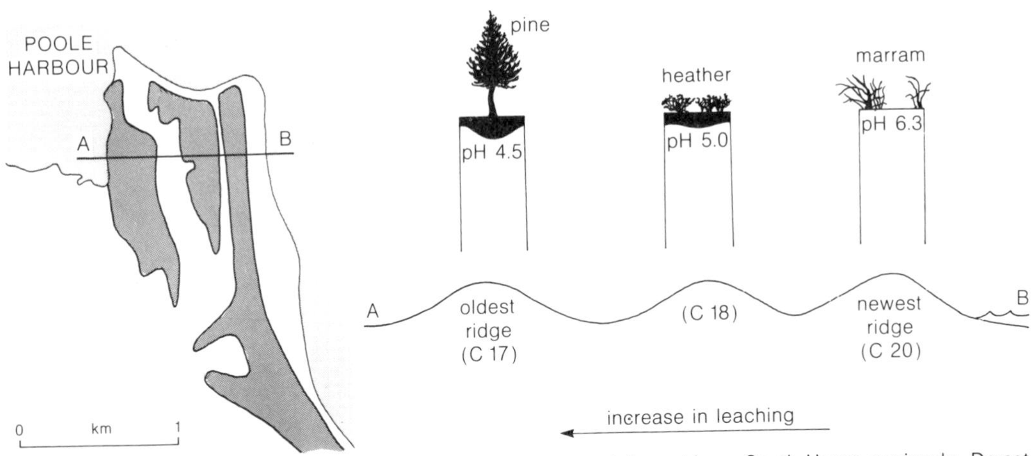

Figure 28.5 The changes in soil development with time on sand-dune ridges, South Haven peninsula, Dorset.

blown sand, marram grass has colonised and the first stages of leaching can be seen (Fig. 28.5). On the second dune ridge, formed in the 18th century, leaching has proceeded further and a wide range of vegetation is growing. Pine colonisation has occurred on the oldest ridge which was established by the 17th century, an O horizon is forming and continued leaching has reduced the pH to 4.5.

28.2 Soil classification

In the previous sections soils have been given names such as 'podzol' and 'chernozem'. These names are a form of shorthand nomenclature which quickly identify groups of soils that have distinctive soil forming processes and patterns of soil horizons. However, soils are part of a continuum and often there are no well marked boundaries separating soil groups. Nevertheless, in an effort to map soil distributions, many detailed soil classifications have been developed.

In Britain most soils fall into the two **major soil groups** called **podzol** and **brown earth**. However, it is apparent to most people examining soils in the field that each major group contains a wide range of soils that can readily be distinguished in terms of colour, texture and parent material. The basic soil classification usually adopted is called a **soil series** and is defined as *a group of soils developed in parent materials yielding soils with the same or a similar texture and mineral composition, and the same or similar succession of horizons in the profile*. Some idea of how soil series are differentiated can be gained from Figure 28.3. In this case till is the parent material for all profiles, but the colour of the upper horizons and the nature and depth of gleying are different. All three soils are therefore classified as parts of separate series. Because soil series are closely related to local parent-material influences, they are often given a reference name derived from the local area – the Denbigh Series, the Southampton Series and the Thames Series are examples.

Soils scientists have built up a hierarchical system by grouping the soil series into the larger, more variable groups called major soil groups. Major soil groups in turn have often been grouped into:

(a) those associated with climatic and vegetation zone, called **zonal soils** (e.g. brown earth, podzol, chernozem);
(b) those, strongly influenced by parent material, that develop in more than one climatic zone, called **intrazonal soils** (e.g. rendzina);
(c) those that show little horizon devlopment, called **azonal soils** (e.g. alluvial soils).

Table 28.1 Soils: simplified terminology used in this book.

Major groups	Major environment of occurrence
brown earths	cool and warm temperate deciduous woodlands
chernozems	temperate grasslands
chestnut soils	semi-arid margins of temperate grasslands
desert soils	hot deserts
ferrallitic soils	tropical humid climates
gleys	poorly draining sites
lithosols	poorly developed soils
peats	sites with excessive water supply
podzols	cold and cool temperate woodland and tropical rainforest
rendzinas	limestones
tropical ferruginous soils	Tropics with distinct wet and dry seasons
tundra soils	arctic, subarctic, mountains

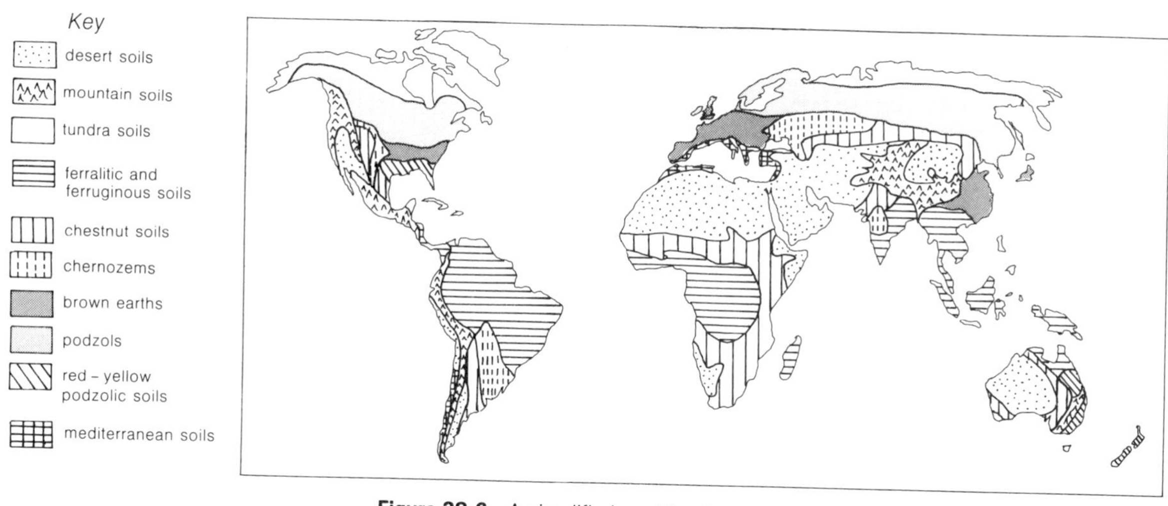

Figure 28.6 A simplified world soil map.

Some people believe this scheme overemphasises the influence of climate and vegetation as soil-forming factors, and there have been attempts to avoid the use of the terms 'zonal', 'intrazonal' and 'azonal'. British soils are not grouped beyond major soil groups. The simplified terminology used in this book is given in Table 28.1. The simplified world distribution of soils is given in Figure 28.6 and detailed descriptions of some of the most important climate–vegetation–soil patterns are given below.

The cool and cold temperate climates

These are regions where there is a moisture surplus throughout the year, resulting in a tendency for severe translocation on all freely draining sites (Fig. 28.7). However, many of these soils have only been forming for the short period since the retreat of ice sheets and, on parent materials with a high calcium content, they have only suffered a loss of ions (brown earths). On less nutrient-rich parent materials eluvia-

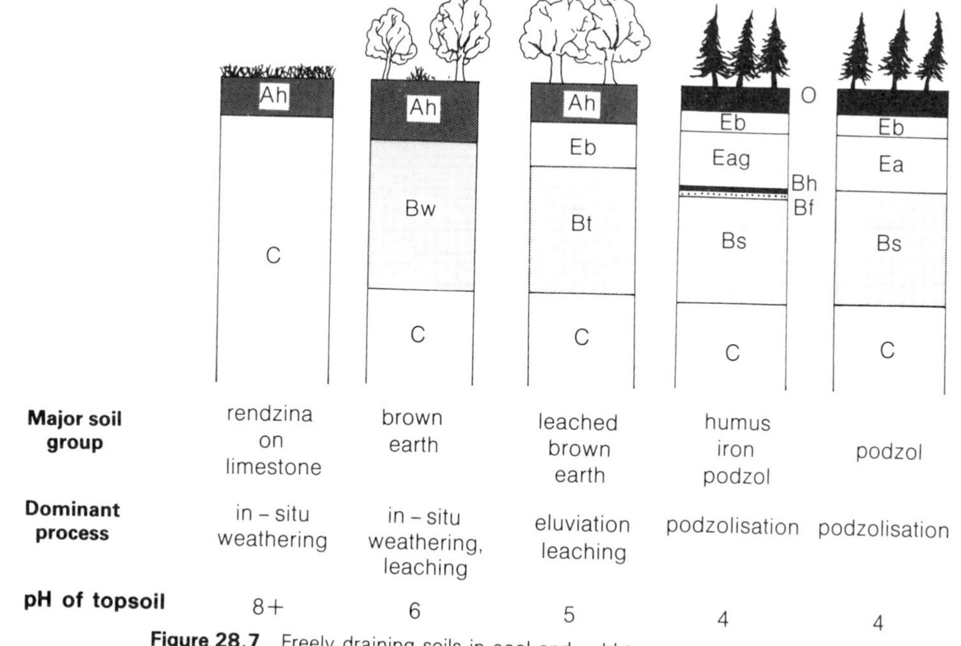

Figure 28.7 Freely draining soils in cool and cold temperate regions.

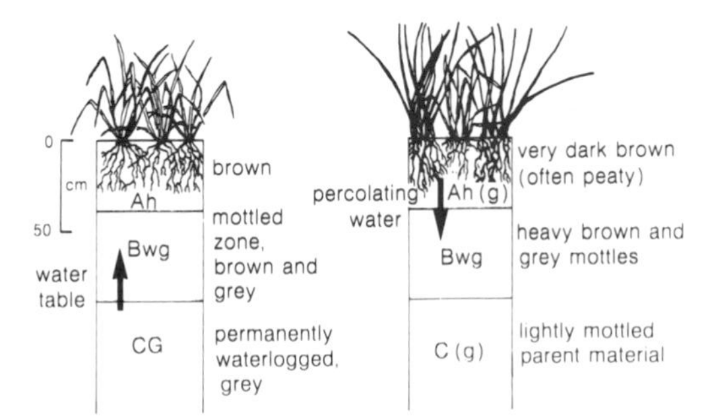

Figure 28.8 Poorly draining soils in cool and cold temperate regions. Groundwater gleys form on floodplains, and are due to the water table being near the surface. Surface-water gleys are found on gentle slopes, mainly on parent materials with a high clay content (clay vales or lowland tills).

tion has begun to dominate the profile and soils show clay translocation (leached brown earths). On nutrient-poor parent materials (such as outwash sands and gravels), where conifer or heath vegetation grows (as in much of Scandinavia and Canada), or where the climate is cold and wet and discourages soil organisms, the brown earths give way to podzols. In the less freely draining sites gley soils develop, irrespective of the type of parent material (Fig. 28.8), and on areas where limestone is the parent material, rendzinas form.

The temperate sub-humid plains

These are regions of low precipitation, dry hot summers and cold winters. All soils display a classic deep Ah horizon, formed by decomposition of the deep root mats of the prairie grasses (Fig. 28.9). In the warmest, driest regions, such as in the rainshadow of the Rocky Mountains, there is little opportunity for weathering or translocation of soil materials, and even the most mobile constituents rarely move far down the profile. In addition, the widely spaced, slowly growing vegetation forms (bunch grasses and sagebrush) yield little organic matter, although their deep rooting systems do ensure a relatively deep Ah horizon.

The most widespread soils of these zones are the **chernozems**. These soils occur in the better watered parts of the grasslands where growth is more prolific. Soils have a deep black Ah horizon mainly produced by *in situ* decay of the deep and extensive grass rooting systems, and the work of the soil organisms. However, there is little leaching, and only the most mobile constituents are lost. Calcium remains entirely within the profile, sometimes as concretions and infillings of old root channels. Yet despite the presence of calcium, because many of the soils are developed on predominantly silty parent material (loess) they do not develop a stable structure, the surface horizons being retained against wind erosion largely by the intricate grass rooting system. These are very fragile soils when used for cultivation.

The **prairie soils** also have a deep Ah horizon, but increasing rainfall causes a greater degree of leaching and a transition to the brown earths of the woodland biome.

The sub-humid Tropics

These soils are very different from those of the mid-latitudes because great lengths of time have been available for *lateral* as well as vertical transfers of soil constituents. Soils in these regions regularly occur in

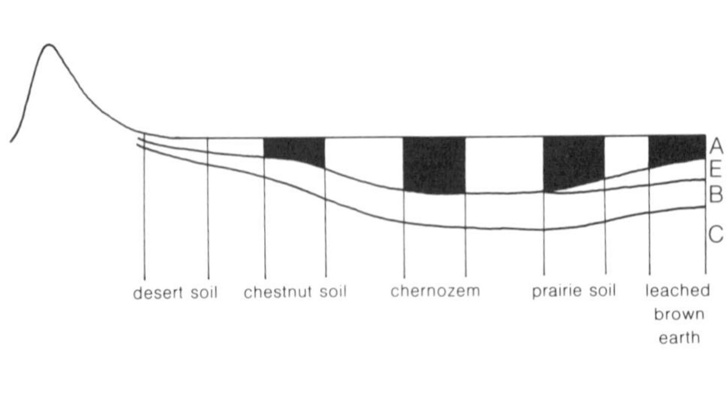

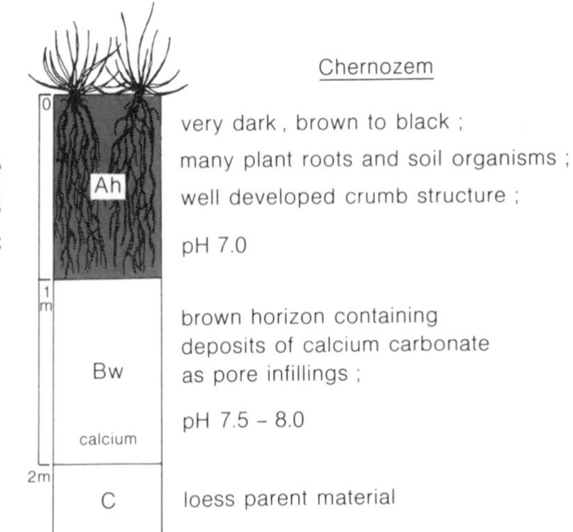

Figure 28.9 Soils of the temperate grasslands in North America. The chestnut soils and prairie soils are shallow versions of chernozems on the arid and humid margins, respectively.

the same topographic relation to one another, and an association of such soils is called a **catena.** Furthermore, large areas are developed from acid igneous and metamorphic rocks. Typically the landscape consists of wide plateaux separated by short, steep escarpments (Fig. 28.10).

In the more humid savannas there is sufficient water available in the wet season to cause deep percolation, intense weathering and clay formation. Soils also develop a deep red colour as iron oxides are released during weathering. The upper soil is continuously turned over by termite activity, the depth of their activity being marked by a layer of stones (Fig. 28.11).

Soils near the base of escarpments experience a seasonal rise and fall of the water table, together with much throughflow from upslope. In the wet season, iron released by weathering is mobilised and moves in the groundwater, but it is precipitated in the soil as nodules or sheets (called **plinthite**) as the soils dry out in the dry season. This is a very slow process, taking hundreds of thousands of years. In areas where these soils become dissected by rivers and exposed to air, they harden irreversibly to a material called **duricrust** (formerly **laterite**). The extensive duricrust caprocks that fringe most escarpments in the humid savanna are believed to originate as plinthite horizons.

Figure 28.11 This upper ferruginous (tropical red) soil has been worked by termites. The line of small stones may mark the depth of their activity.

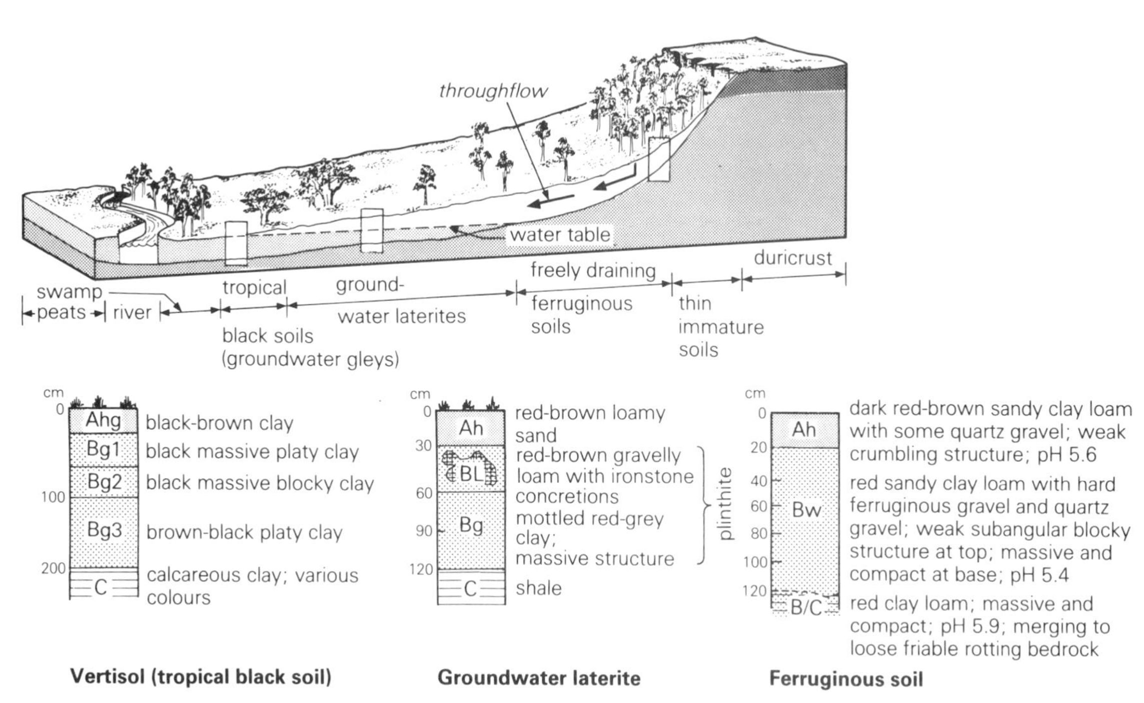

Figure 28.10 A soil catena developed on acid igneous and metamorphic rocks in the 'moist' savanna zone.

In the dry savanna soils experience a shorter rainy season and water rarely percolates through the soil. With less intense chemical weathering soils have a sandy texture and most of the finer particles come from dust storms that blow off the adjacent deserts. Many of these soils are alkaline and can be cultivated successfully if irrigation water is available.

The humid tropics

These soils form under conditions of constant high temperatures and moisture surplus. Intense chemical weathering and translocation are therefore the dominant processes, grouped under the term **ferrallitisation**. This involves the total breakdown of rock and the leaching of all ions and much silica. In these conditions iron and aluminium oxides are the least mobile and, together with clay minerals, form the bulk of the soil. All soils are therefore deep red in colour. The relative proportions of clay and oxides depend on the parent material: on acid parent materials, clays dominate; on basic materials, clays are only poorly developed and oxides are the major soil constituent.

On drainage divides and steep upper slopes (especially on quartzite) soils are sandy and still dominated by quartz gravel, developing similar profiles to the podzols of the cold temperate regions.

It is important to realise that all the soils are very old, weathering and leaching have progressed to such a stage that the weathering front is now *tens* of metres below the surface on all but steeply sloping soils, and that *all* soils are very acid. Thus, even under luxuriant forest vegetation, pH values of topsoils are commonly 4.0–4.5. These are poor soils for agricultural use. Only the **valley-bottom gley soils** contain a high nutrient content and are suitable for cultivation.

29

Soil use and abuse by man

29.1 The protection of soil

Today, most of the world's soils are not forming under the natural conditions described in Chapter 28, nor are they in equilibrium with their environment. This is because they have been subjected to hundreds and perhaps thousands of years of cultivation (Fig. 29.1). In some cases the soils have been resilient enough to withstand this treatment and appear little altered, but elsewhere they have been irrevocably destroyed. Soil erosion – in the sense of soil removal by leaching, eluviation and mass movement – when it is on a modest scale, is a natural process and part of landscape formation. However, man has often *accelerated* the process and caused rapid soil loss, loss of fertility and a change in the water-flow properties, producing saline or waterlogged soils.

With such intensive use it becomes vitally important to understand how to find the right level of soil fertility for sustained crop growth; how to identify those soils most vulnerable to erosion, and for which special treatment may be needed; and how to conserve already eroded soil and upgrade existing soils to make them more productive. Such studies are especially important if forecasts of an increase in world population (by 2 per cent per year) are correct: it may be that soil will be the first resource to be totally used – possibly within the next 50 years.

Figure 29.1 Modern cultivation methods include stubble burning in autumn and heavy application of biocides to suppress 'weeds' and 'pests'. The resulting sterile landscape becomes very fragile and liable to accelerated erosion.

Maintaining soil fertility

Soil erosion is rarely massive and spectacular. The majority of eroded land has simply become less fertile. Nevertheless, under cultivation, a reduction in fertility (and the consequent loss of structure-forming cations), a decrease in soil micro-organisms, and a reduced production of polysaccharide gums go together. As structure-forming processes become weaker, the pores between the structures become less distinct, water percolates through or infiltrates the soil less well, and overland flow and erosion begin. Individual particles are no longer bound tightly into larger, heavier groups, and are much more liable to blow away. Loss of fertility is thus of concern not only because crop yields become smaller, but because soils become more prone to erosion.

On cultivated land soil fertility can be maintained by applications of cations (applied as fertilisers such as lime) and from applications of decaying organic matter (manure or organically derived fertiliser).

Land-use capability

Soils which have a high nutrient status may still be unsuited to intensive cultivation for purely physical reasons. For example, they may be too shallow for large plants to keep a good anchorage; or they may be 'droughty' and dry out quickly in summer, increasing the risk of wilting. They may be deep but of a clay texture, causing water to pond on the surface after storms. They may have so high a proportion of stones that they become difficult to cultivate with machinery. It is often easier to improve the nutrient status of soils than to change their physical properties to make them more easily workable. For example, poorly drained clay soils can only be improved by extensive use of expensive artificial subsurface drains.

The purpose of predicting land-use capability is to assess the *potential* of soils under reasonably careful management. This provides an important guide not only to farmers but also to the planners who make land-use decisions. The strength of a capability survey lies in discovering the true potential of farmland. Land's use for maximum profit may be incompatible with the need to conserve the soil resource in the long term. Farmers can be shown that the best, most versatile flexible soils (called class 1 and class 2) should not only be used for undemanding crops such as cereals just because these yield well. Root crops are much more demanding and so should be grown on the best land, leaving cereals to class-3 and class-4 soils, for if these soils can be drained, they can still produce high yields. Similarly, root crops are much more difficult to harvest from class-3 soils: in a wet period these soils become heavy and stick to the crop, adding considerably to cleaning costs. Because of the need for a long growing season, areas to be used for high-density grazing are best allocated to high-class soils, and only low-density grazing allowed on low-grade land which may be more liable to trampling damage.

In the past cereals have often been grown on the best land and animals left to graze on the land that cannot be used for anything else. From a consideration of soil use alone this is often an unsound practice and may not maintain the soil in its best condition or yield the highest returns. But other factors also have to be taken into account when choosing land use:

WORLD SOIL POTENTIAL

There are about 13 billion (10^9) ha of land free from ice in the world, but most of this is unsuited to cultivation: 2.6 billion ha is mountain or desert, where soils are too thin or poorly weathered; 1.7 billion ha of chestnut soils are in the dry semi-arid zone; 1.6 billion ha consists of acid podzols; 0.7 billion ha consists of cold tundra gleys; 1.4 billion ha consists of 'laterites'.

The largest group of potentially cultivable soils (1.0 billion ha) are the tropical red soils (ferruginous soils), but these are highly leached and need massive fertiliser applications. Thus the highest concentrations of arable land are at present found in the semi-arid to sub-humid grassland zones, where 0.5 billion ha of chernozems, chestnut and prairie soils are cultivated, often with the help of irrigation. The great river valleys contain 0.3 billion ha of fertile alluvial soils; and a further 0.2 billion ha of fertile black soils occur in some tropical areas. Some desert soils and podzols are cultivable, together yielding 0.8 billion ha. The other soils (amounting to 0.4 billion ha) are mostly brown earths. In total there are 3.2 billion arable ha – 24 per cent of the land area of the earth and 2.3 times the area currently under cultivation. Of this only 0.3 billion requires irrigation to yield any crops at all, but most needs enormous inputs of fertiliser to get reasonable returns.

An additional 3.6 billion ha can be used for grazing on land that could not – or should not – be ploughed.

capability takes no account of relative market prices and other economic factors (see von Thunen's model, p. 334). In practice farmers tend to make a pragmatic choice that lies in the direction of economic pressures (see, for example, the South Devon case study, p. 316). The intervention-price system operated by the EEC has been particularly important in encouraging much wider growth of cereals and sugar beet than under the previous 'free market' system.

Land improvement and reclamation

Land-capability classification points to the most useful ways of using and improving land. But of course farmers have tried to improve their lands for many generations, using the means available to them at the time. Many thousands of cartloads of sand have been taken to clay soils to try to lighten clay textures, and many thousands of loads of clay have been moved to sandy soils to improve the sand's water retention and fertility.

Land improvement is a subtle process, often taking so long to accomplish that it goes unnoticed. Only when dramatic land improvement schemes are undertaken are they brought to our attention. These large schemes are usually called **reclamation projects**. Some of the most famous have occurred in the Netherlands. In Britain the fenlands were the largest

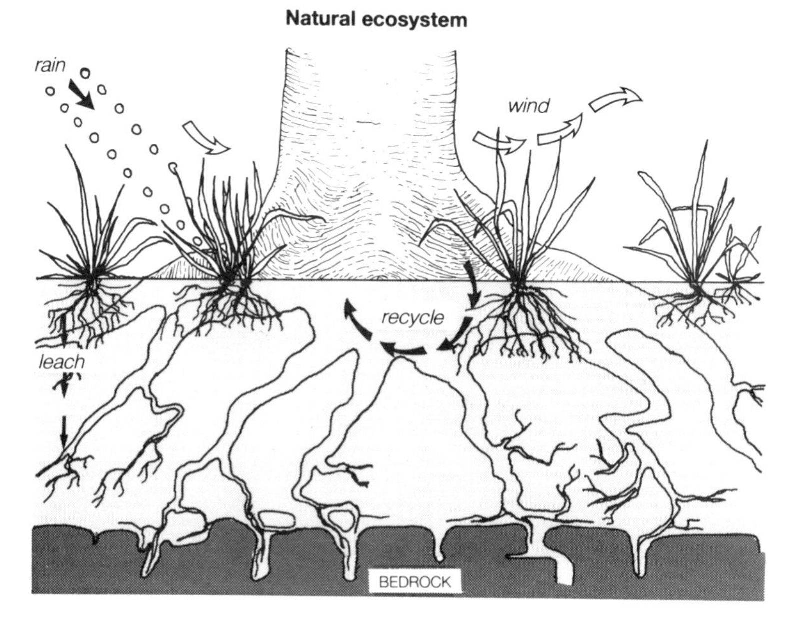

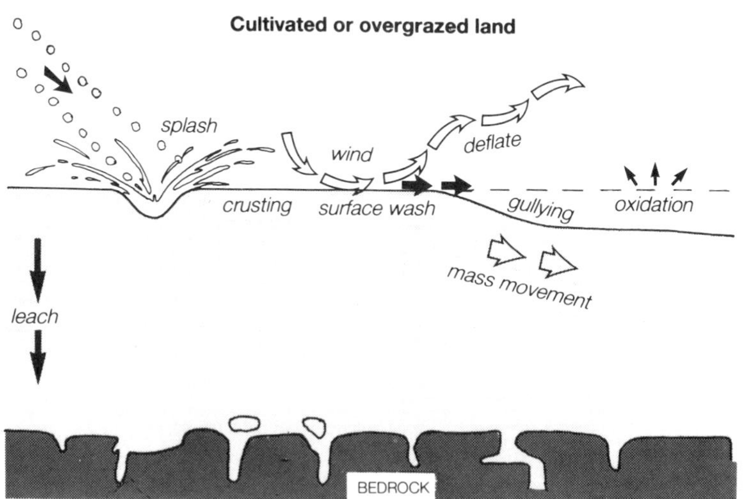

Figure 29.2 The possible accelerating erosion mechanisms that can occur when a natural ecosystem is replaced by cultivated land.

area of concerted reclamation. But whatever the size of the project, land improvement should always be preceded by capability analysis to ensure that the scheme will provide an economic return.

29.2 The destruction of soil

Accelerated soil erosion

All soils suffer erosion as part of landscape development through leaching, soil creep or other mass-movement processes. However, if the *rate* of erosion is increased beyond that at which weathering can provide new soil material, the soil will begin to degrade. Erosion mostly affects the topsoil, where the majority of nutrients are held. Accelerated erosion therefore leads to a loss of fertility and reduces productivity of the vegetation that grows on it (Fig. 29.2). Accelerated erosion is always most serious on cultivated land, because:

(a) the soil is exposed to raindrop impact and high winds;
(b) there is no permanent, deep root system to help retard mass movements;
(c) there is often insufficient application of fertiliser or manure to replace cations and polysaccharides, leading to a weakening of the soil structure.

It is important to remember that the natural vegetation is in equilibrium with its soils in a particular way. For example the speed of organic matter decay in a tropical rainforest may leave little protective litter: rainsplash erosion may therefore occur, but tree roots still help to stabilise the soil, and the rate of weathering is very high. By contrast the rate of organic matter and weathering is much lower in a temperate deciduous forest, but a deep leaf litter protects the soil from raindrop impact. In the long term the most important requirement is that a balance is achieved between the vegetation and the soil.

The developed world

WATER EROSION AND CONSERVATION

In most places, water erosion is a far more serious cause of erosion than wind. It has several effects:

(a) it removes the more productive topsoil exposing the subsoil and resulting in lower crop yields;
(b) water running over the land surfaces (overland flow) does not percolate into the soil and so cannot be stored by the soil against times of drought;
(c) water reaches streams more rapidly via overland flow than via throughflow, leading to increased risk of flooding;

Figure 29.3 Water erosion across a steeply sloping field on the Isle of Wight. Extensive sheetwash and gulley action can be seen. The eroded material is being deposited within the nearby wood (shown as light tones).

(d) river waters receive more sediment and fertiliser, causing increased silting of reservoirs, burying of soils and crops on floodplains.

Water erosion first breaks peds into individual particles by the direct impact of raindrops (rainsplash). This creates a **surface crust**, allowing the transport of fines. Later, if sheetwash is very extensive, channel flow may develop. Initially this takes the form of shallow drainage channels (**rills**), only a few centimetres deep, which change their pattern between storms. Eventually larger permanent channels (**gullies**), tens of centimetres to metres deep, can form (Fig. 29.3). From a farmer's viewpoint, gullies are a much more serious handicap than rills, for they cannot be ploughed out. Water erosion is greatly influenced by the size of a storm. Because the energy for erosion comes from falling raindrops it is important to note that, for example, a 4.5 mm diameter drop has 500 times as much energy as a 1 mm drop, and that a 7.5 cm/hour storm has 100 times as much energy as a 5 cm/hour storm. Erosion capability thus increases dramatically with the intensity of a storm, and it becomes essential to know the critical 'threshold' size of storm that will produce significant erosion, and how often such a storm occurs.

Water erosion is more frequent on soils with a weak structure and/or bare surface. However, the organic matter content does not help to protect against splash because polysaccharide gums have a low strength against impact when wet. It is growing vegetation, cations and sesquioxide cements that maintain peds against impact and keep the soils permeable. Plant roots also stabilise the soil against mudflows or landslides.

The basic principles of conservation against water erosion involve keeping the ground covered as well and for as long as possible. Protecting the soil from raindrop impact can be accomplished with a surprisingly low quantity of organic **mulch**; for example, 1 tonne/ha can reduce rainsplash effects by half. Additionally, a **rough tilth** (surface cultivation), **terracing** or **grass strips** planted along the contour all help to retard water movement.

In most developed countries steps towards tackling the problem of severe water erosion were made some decades ago in special areas (such as south-east USA) when catastrophe was at hand. In general soil erosion is probably more prevalent in Europe and the USA today than it was one or two decades ago. The trouble is that farmers often do not notice topsoil loss or take measures to protect against it until there is a marked reduction in their yields. At this stage it is often too late. Despite large resources in the developed world, and the technical advice available to farmers, there is still too great a tendency to rely on the natural buffering capacity of the soil to withstand abuse, and insufficient attention directed towards the long-term implications of soil erosion by water.

WIND EROSION AND CONSERVATION

Wind erosion occurs on undeveloped or poorly vegetated surfaces exposed to strong winds. Clay and silt-sized materials picked up by the wind are carried in suspension, whereas sand-sized material is moved by saltation. Usually saltation is the dominant process, moving **friable** (easily broken) material such as fine sands, loamy sands and peats. During a 'blow', saltation carries material from fields and drifts it as dunes against hedges and buildings or fills in ditches. The finest materials are carried away completely. As a result the topsoil texture gradually coarsens. Eventually the stones exposed by the removal of fines will armour the surface, but only after the soil has lost most of its agriculturally useful materials.

In Britain blowing can occur on any arable field, but it is most common on the flat lowlands of eastern England where sandy-loam or peaty-topsoil textures prevail. The most common time for blowing is from March to the end of May, when the fields are bare or have only small seedlings and when the weather is characterised by strong winds and dry spells which dessicate exposed topsoil. Indeed farming practices make soils liable to blowing. For example, to grow root crops on a light-texture soil, a fine seedbed is needed in spring. Thus soil clods are broken down to a fine tilth. The topsoil rapidly dries to yield particles that can readily be moved by the wind (Fig. 29.4).

The effects of a blow can be spectacular. In the United States major periods of uncontrolled blowing laid waste a vast area of the dry western interior which today is still called the **Dust Bowl**. In Britain, although less disastrous, the effects of a blow can still be important. Drainage channels in the fens, often two metres deep, are filled with soil, seed and fertiliser; seedlings are twisted and bruised; leaves are shredded by sand abrasion to such an extent that total or partial redrilling is needed.

Figure 29.4 Wind erosion on bare cultivated land in Nottinghamshire.

Wind erosion can be partly reduced by maintaining a good soil structure so that particles do not become dispersed on cultivation. Liming the soil helps to provide the calcium ions that encourage flocculation; adding manure or ploughing in a grass or clover crop provides the raw material that enables soil organisms to form polysaccharides. As an emergency measure, blowing can be prevented by ploughing the soil just before the blow is expected, thereby bringing moist soil to the surface.

For longer-term control, **windbreaks** can be planted. These often consist of tall poplars intergrown with hedge shrubs to provide a semi-permeable break at all levels. These are effective at reducing wind speeds for about five times the tree height to windward and up to twenty-five times to leeward. But despite their many ecological advantages, trees are expensive to plant and slow to grow; they use up valuable land, they shade growing crops and they compete for soil moisture in the summer. Today such shelter belts cannot protect more than a fraction of the large fields needed for mechanised farming and are rarely used. Alternatives have been adopted in some places. For example, barley is sometimes drilled in alternate rows with crops such as sugar beet, celery and carrots (**intercropping**): the barley protects the soil and the other crops while they grow. The barley is then removed by rotavator in late May when the root crop can protect itself. In a few places **hessian screens** of one-metre high hessian tied to wire-mesh netting have been employed, but this method is labour-intensive and can only be an economic proposition to protect tender high-priced crops such as lettuces. Unless very valuable crops are being grown such techniques are often regarded as simply not worth while.

The traditional way to obtain a seedbed, kill weeds, break up poached (trodden) land and allow better root penetration is by ploughing in autumn and harrowing in spring, but this creates an environment that is very prone to blowing. The use of herbicides has increased the opportunities for replacing the mouldboard plough with the less disruptive chisel plough or with machines that drill seeds directly into the ground without first making a traditional seedbed. These are known as **minimum-cultivation** farming methods. When the land does have to be left bare, straw can be spread over the field and/or incorporated with a special cultivator, or the ground can be loosened below surviving stubble using special reciprocating harrows (**stubble mulching**).

In climates more arid than that of Britain (such as that of the American prairies) conservation measures are taken more seriously because the economic consequences of erosion are much greater. Water conservation is vital, and the drier areas are only cropped once every other year to allow the soil to become moist to the full root depth. Fallow land is arranged in alternating strips with the cultivated land, so that it can also perform the useful role of a windbreak. Such practices give the prairies a distinctive striped appearance.

The developing world

SOIL EROSION AND CONSERVATION

Soil erosion is a major problem in many parts of the developed world despite, or perhaps because of, high levels of technology. It is interesting to note that farmers in the developed world are often *less* aware of erosional problems than farmers in the developing world. The difference is that whereas in the developed world there are the resources to combat erosion if there is the will to implement them, in the developing world the resources are limited and there are immense obstacles – traditional land-use practices – to be overcome. Awareness of a problem and action to solve it do not always go hand in hand.

There is an immense variety of environments in the developing world, but those areas where soil erosion is a major problem can broadly be divided into:

(a) mountain and hill areas with high rainfall and steep slopes;
(b) plainland areas with low, seasonal rainfall.

Each type of area has its own distinctive mixture of physical and social problems. They will be illustrated by the following two case studies.

Case study: The Uluguru Mountains, Tanzania The Uluguru mountains of Tanzania rise to over 2600 m and have a rainfall of up to 2850 mm/year (Fig. 29.5). There is no dry season, at least 50 mm of rain falling in every month. Although valley sides are steep, intense weathering allows deep soils to form even on slopes as steep as 60°, the soil being retained largely by tree roots. Soils are nevertheless liable to erosion from rainsplash, sheetwash, landslides and mudflows even under rainforest vegetation. Rainsplash occurs because the forest canopy is so high that raindrops falling from the leaves regain terminal velocity by the time they reach the ground. There is little leaf litter on the soil surface to prevent raindrop impact and roots are frequently exposed by rainsplash and sheetwash. Nevertheless, the roots act as 'dams', retaining soil and reducing erosional loss.

Forest clearance began at about the beginning of the 19th century as population pressure increased on the surrounding plains. However, the plains people knew little of steep-land cultivation and they continued to use the methods they had learned on the plains. Understanding the degree of potential soil

Figure 29.5 Part of the Uluguru Mountains. The steep cultivated slopes are visible in the foreground with rainforest on the upper slopes.

erosion as a result of cultivation on steep slopes, the (German) colonial power declared the upper slopes a forest reserve. Caught between over-used plains and the forest reserve, the people had no alternative but to use the remaining slopes even more intensively, bringing into cultivation even the steeper land they had previously left alone. At the same time they reduced the period of fallow between cultivations. As soil became exposed more frequently under these farming techniques, severe rainsplash and sheet erosion accelerated to such an extent there was a danger that the very soil needed for cultivation would be removed entirely.

In the 1930s the (by then British) administration sought to introduce terracing techniques. They tried a simple technique: vegetation and crop residues were spread in lines across the hillside and covered with soil hoed from an area immediately upslope (Fig. 29.6). In this way flights of small terraces could be created that would become more definite year by year. At the same time organic matter dug into the soil would improve the soil's fertility. However, not only were such techniques more difficult; they went against the traditional shifting-cultivation system of

Figure 29.6 Creating new terraces on a steep mountain slope in equatorial Africa. Notice that vegetation is spread in lines, then covered with soil hoed from up slope – the task is very arduous.

the people. In the end the technique was abandoned. Today rainsplash and sheetwash are up to 100 times greater on cultivated soils than under forest (equivalent to 0.3 mm loss of soil each year) and gullies up to 7 m deep have formed near the mountain footslopes. Landslides are also more frequent: one storm alone caused landslide losses equivalent to a 14 mm soil loss over the whole area. In the Ulugurus conservation measures are now needed more urgently than ever, but they will have to be introduced slowly and with careful plans to educate the local people to understand their benefits. It would probably be more effective to adopt first those measures which appear to interfere least with traditional life. For example, trees can be replanted along the upslope margins of roads (where artificially steepened slopes exist), or in lines along the contour 30 m or so below ridge crests (a common place for landslides), or along stream lines to restrict bank erosion. This kind of tree planting has the added advantages of providing a local supply of timber without destroying more forest; it uses very little land; and can be recognised by the local people as effective, cheap and feasible.

Case study: The Dodoma Plains, Tanzania The Dodoma region of central Tanzania is characterised by plains surrounding isolated hills (inselbergs). The rainfall of 500–600 mm/year is concentrated into one rainy season whose convective storms often produce rainfall intensities far above the infiltration capacity of the soil. Under these conditions soil erosion is a potential menace and can only be resisted with a continuous vegetation cover. In this region the savanna ecosystem has a carrying capacity of 2.5 ha/animal, but at present there is only 1.9 ha/animal so overgrazing is widespread (Fig. 29.7).

To help offset the lack of water in the dry season many small dams have been built, but the reservoirs they impound have been silting up badly due to accelerated soil erosion brought on by overgrazing (Fig. 29.8). Overgrazing is severe near the reservoirs simply because this is where animals must come to drink. The many worn trails they make are particularly prone to develop into gullies. Conservation is difficult because it involves a reduction in stock density, a concept totally alien to, and resisted by, the local tribes. As a result the Dodoma region is on the verge of falling into the trap of desertification.

Recently the face of such areas is being transformed by new, communally constructed terraces. Often initiated by aid agencies in time of famine, the food for work campaigns have allowed farmers to see real benefits from conservation for the first time.

Figure 29.7 The results of overgrazing near Dodoma. (a) The savanna trees have helped to reduce rainsplash and sheetwash; gullying has been initiated between the trees. (b) Severe sheetwash and gully erosion have made this land unusable.

Figure 29.8 Erosion patterns near a reservoir, Zaria, Nigeria.

TROPICAL DESERTIFICATION

Desertification denotes the extension of a typical desert landscape to areas in which it did not occur in the recent past. These areas are mostly those that border the true hot deserts, receiving an annual rainfall of 100–300 mm. The lower limit of 100 mm corresponds to a situation of permanent water deficit in which the potential evapotranspiration is greater, on average, than the precipitation every month of the year.

Under undisturbed conditions the zone bordering the deserts has a thorn scrub vegetation consisting of between 20 per cent and 40 per cent ground cover of perennial shrubs and grasses. It is just enough to protect sandy soil surfaces from net loss by wind erosion. (Removal of soil from in front of plants is compensated for by deposition behind them.) The critical separation of plants to allow net loss is about five times their height. With a wider spacing wind erosion (deflation) rapidly increases, removing fine soil and leaving the surface armoured with pebbles. Nearby, the deflated sand accumulates as dunes, covering, and rendering unusable, yet other soil areas. At this stage, with the surface continually moving, most permanent plant life becomes impossible.

In areas dominated by clay soils, land depleted of sufficient vegetation cover is eroded by rainsplash and the bare ground surface sealed. This causes extensive overland flow, widespread gullying, a reduced supply of moisture to plant roots and further plant death. In some areas even over-irrigation or lack of drainage from irrigation causes desertification. In areas with rainfall below 200 mm, desertification may be irreversible. Such conditions exist in large areas of North Africa and the sub-Saharan Sahel zone. The many processes leading to desertification are displayed in Table 29. 1.

Table 29.1 Physical and biological processes causing desertification.

Processes	Effect
Water	
scarce precipitation; uneven and erratic rainfall distribution; mismanagement of irrigation water; over-exploration of aquifers and surface reserves; water losses	scarcity of water
inadequate drainage systems; uncontrolled surface runoff	mismanagement of rain water
ignorance of usage; deficient levelling of land; inadequate distribution; inadequate irrigation methods	mismanagement of irrigation water
Soil	
reduction in vegetal cover; uncontrolled runoff; sedimentation and silt; degradation of soil structure; inadequate farming practice; wind erosion; soil profile depth decrease; loss of fertility in surface soil; leaching; reduction of capacity to retain moisture	erosion
excessive accumulation of salts; flooding; excessive irrigation; water quality; deficient leaching practices; inadequate drainage	salinity and deficient drainage
Vegetation	
shifting agriculture; deforestation; overgrazing; weed encroachment; uncontrolled fuel wood collection; excessive tree felling; fire; drought	reduction in vegetal cover

The desertification spiral There is considerable difficulty in establishing the cause of desertification. Some researchers claim that it is largely the result of climatic fluctuations that sometimes leave these fragile lands with many years of continuous drought. If this is so the desertified lands should recover with time.

Other people point to the huge expansion of the population of countries in the arid zone. The average growth rate of most of them is about 3.0 per cent, which means that they are doubling their populations about every 25 years; the consequences will include over-grazing, over-cultivation and the over-use of any woody plants for cooking fuel. Indeed, because people in the arid zone are primarily pastoralists, the growth in livestock numbers is roughly in proportion to the growth of the human population: for subsistence, peasant farmers need to rear between two and four standard stock units (1 SSU ≡ 10 adult sheep or goats ≡ 2 cows ≡ 1 camel). Naturally this immense upward pressure has led to stock densities far beyond the carrying capacity of the land. Under these conditions the plant cover is lost and the ground is trampled down and compacted, leading to increased runoff and acceleration of the desertification process.

In an attempt to compensate for overgrazing, many people have turned to cultivation: for this they need about 250 kg of cereals per year for subsistence. And because people are not able to increase their yields, an increasing population requires that more land is cultivated. As cultivation is driven into areas of lower and lower rainfall, the average crop yield decreases to between 100 and 200 kg/ha/yr. Even this is not a reliable yield because arid-zone rainfall is notoriously irregular: in reality farmers get a good crop one year in three or four, and three or four complete failures in five years (Fig. 29.9).

The result is a vicious downward spiral. As crops fail malnutrition occurs, making the farmers less able to work for the next growing season. In turn farmers cultivate the land less well and receive lower yields. Cultivation by primitive methods traditionally requires land left as fallow, but as pressure increases fallow is reduced, soil fertility is no longer restored and yields fall even further. For example in part of the Sudan the area cropped for groundnuts and sesame was 0.3 million hectares in 1960, yielding 110 000 tonnes; by 1973 the area cultivated had increased to 1.6 million hectares but the total yield had fallen to 88 000 tonnes! But worse still, the soil left barren after crop failure suffers wind erosion – rates of up to 10 mm of topsoil loss in a month have been measured in Tunisia. Of course loss of topsoil and exposure of low-nutrient subsoil makes cultivation all the more difficult.

Each person needs at least 1 kg of wood each day just to cook food. Given that the natural undisturbed vegetation in the arid zone yields about 500–1000 kg of woody tissue per hectare, every person would destroy at least half a hectare a year if there were no regeneration. But even with regeneration it is

Figure 29.9 Desertification in Kenya. Here, wind and water erosion following overgrazing has removed the topsoil and created virtually irreparable damage near to the waterhole.

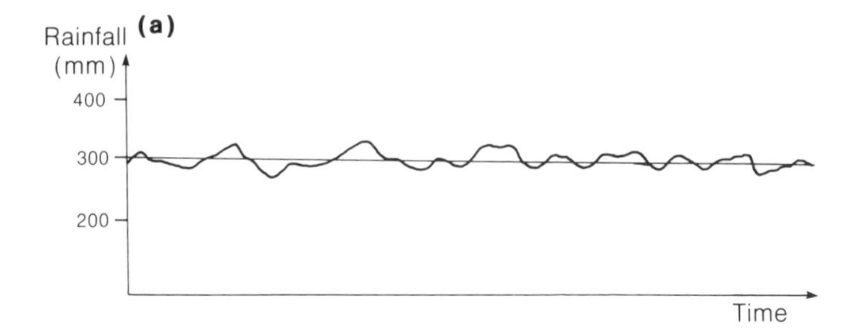

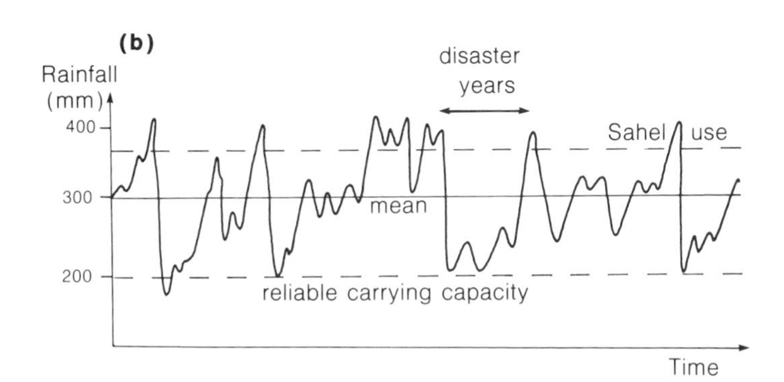

Figure 29.10 If rainfall in the arid zone were distributed about the mean in the way shown in (a) it would be much easier for subsistence farmers to survive. However, rainfall is most irregular (b). For a reliable yield, even in low rainfall years, the population density (carrying capacity) must be low (lower dashed line). Recent population pressure has brought more people to rely on the same land: now they only get enough to eat when rainfall is high (upper dashed line); when rainfall is low they are vulnerable to starvation, malnutrition, and disease.

estimated that a family of five will destroy over a hectare of woody vegetation each year. With over 100 million people depending on this type of fuel in Africa and the Middle East alone, they must destroy something like 25 million hectares (an area half the size of Kenya or the whole of the UK) each year. It is clearly an enormous additional pressure producing desertification.

There is – almost unfortunately, some might say – a means of curing the water-supply problem for people and livestock in the arid zone; boreholes can be dug to groundwater supplies. The result has been a great increase in livestock numbers irrespective of the carrying capacity of the land. In consequence there has been *total* destruction of vegetation for many kilometres around each borehole. In the great Sahelian drought of 1968–73 most animals actually died of hunger, not thirst.

In all, the problem is one of great complexity. 'The image of desertification progressing along a lengthy line at a more or less constant speed of x km per year is childish.... In fact the rate of advance of desertification is not known: there are only very coarse estimates' (Le Houérou, 1975).

The difficulty of desertification stems from population pressure and the great vagaries of rainfall. In the arid zone average figures are meaningless for agricultural use. To combat desertification there is a need to reduce the livestock numbers to an ecologically realistic carrying capacity for, as Landsberg (1976) exclaimed: 'I wish people would quit blaming climate or climate change for failure of agriculture in marginal lands. It's a scapegoat for faulty population and agricultural policies' (Fig. 29.10). This means carrying capacities need to be held down to the levels needed to allow survival in the *poorest* years: at present they are geared to the best rainfall years, which is why the Sahelian drought was so catastrophic. Finally, it is worth noting not only that climate in marginal lands can have a severe impact on the people, but that the people can also affect the climate. Removing the vegetation by overgrazing, for example, will increase the albedo (reflectivity) of the soil, and less solar radiation will be absorbed by the ground. In turn this may lead to less ground heating, less convective cloud formation and less rain. One computer-run model has even suggested that rainfall might be decreased by up to 40 per cent in this way, reducing still further the rainfall where it is needed most.

THE SAHEL DROUGHT

Between 1965 and 1974 there was a period of intensified drought, a natural part of the long-term climatic cycle. Droughts had been known before and they had caused hardship but, although climatically the 1965–74 drought was no worse than many in the past, its effects were far more devastating. In the six Sahelian countries of Mauritania, Senegal, Mali, Upper Volta, Niger and Chad better medical facilities have led to a rapid rise in the population: in 1961 there were 19 million people in the Sahel; by the end of 1973 there were 25 million. Thus, as the drought began to take effect, although the population continued to rise steadily, the subsistence economies declined in yield and the gap between demand and supply rose dramatically. In 1972–73 the grain shortage in the Sahel was over 1 million tonnes, while the number of cattle lost by starvation was over 3.5 million.

But whereas in the past droughts caused large-scale southward migrations from these fragile lands, the neighbouring countries, which were similarly overstocked and already have high populations, have become increasingly reluctant to allow any movement across their borders, thus making the plight in the Sahel even more severe.

Faced with the inability to move, the Sahelian peoples overgrazed the dry savanna to the point of exhaustion. Subsequent recovery of the range has been very slow: it can only occur if the carrying capacity of the land is held well below maximum. Unfortunately no Sahelian government has been able to withstand the pressure to restock and the savanna thus remains severely degraded.

30

The basis of biogeography

30.1 The importance of biogeography

Biogeography is concerned to describe the distribution of plants and animals over the Earth's surface, and in particular it tries to show the interactions between the multitude of species and their environments.

Too often the vegetation or the wildlife of an area is dismissed with hardly a thought, yet plants and animals play a crucial role in the world. They are important even from a purely humanocentric point of view because they provide the diverse genetic stock from which all our domesticated crops and animals have developed. Today, for this and other more generous reasons, people are becoming increasingly aware of the great threat that exists to the survival of the natural world. If this natural world is to be preserved, we must understand how it works. Furthermore, agricultural scientists are turning more and more to studies of the natural world in an effort to increase the productivity and stability of farming systems. Any study of the world of man is intimately connected with the study of both plants and animals.

30.2 Plants, animals and life

If we want to understand how living things are distributed, we must understand how both plants and animals function. All living things have certain functions in common: they feed, grow, respire, move, reproduce, respond to their surroundings and excrete waste substances. However, plants and animals also exhibit fundamental differences.

Plants build their tissue and grow using inorganic molecules obtained directly from the air, water and soil and the energy from the sun in a process called **photosynthesis**. This process produces starches and sugars from carbon dioxide and water. These materials are then combined with other minerals to form proteins which are the building blocks for tissue membranes: celluloses, waxes, fats and resins. Because of their direct dependence on soil, water and solar energy, the distribution of plants is closely related to the global variations in the availability of these resources.

Animals, on the other hand, rely on plants for their food because they cannot synthesise organic material directly from inorganic substances. Most animals are fairly mobile: because many are able to use several plants as sources of food, they are not as closely tied to one part of their physical environment as are the plants.

30.3 The distribution of living things

The Earth's rocks and climates, slopes and oceans present an extremely varied range of environments for living things. Over the great expanse of geologic time – at least 3300 million years since the first organism in the fossil record – life has evolved which exploits every conceivable range of environments, from the barren harshness of Antarctica to the apparent luxuriance of the humid tropics; from the hot desert to the equally barren centres of cities.

The main process that has led to the present stage of evolution, **natural selection**, involves continual adaptation to new environmental factors. Nevertheless, major steps in evolution seem to occur spasmodically rather than progressively and continously. For long periods of time, all species of plants and animals are bounded by certain **environmental barriers** beyond which they cannot normally flourish. The factors that limit the distribution of a species are usually complicated and subtle in their effects. As a result there is rarely a fixed line beyond which a species will not flourish; rather there is a broad zone across which growth or reproduction become progressively more difficult until eventually

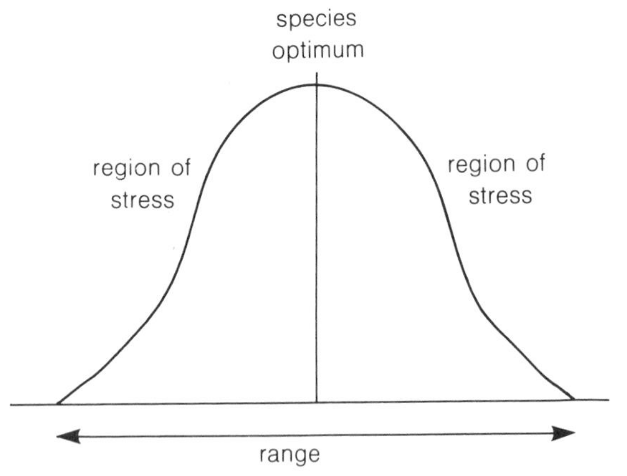

Figure 30.1 The concept of range for a species. There are rarely fixed boundaries, but more often a gradient of stress away from a region of optimum growth.

the species cannot compete effectively with others for the available space or food (Fig. 30.1).

The most commonly important restrictions on the distribution of a species are temperature, moisture and light. In a typical garden there are many species that illustrate these limitations well. For example, geraniums were originally brought from tropical regions where frosts are unknown: geraniums here would perish each winter unless taken indoors and kept warm. The daisy, on the other hand, which gardeners call a weed when in their lawns, is native to Britain: it would not grow in warmer regions because it requires night temperatures to fall below 10 °C. Marigolds need a lot of light and will not flourish in the shade whereas a clematis grows poorly in the sun. Tomatoes wilt if the soil is not kept moist; petunias usually die from stem rot unless they are kept fairy dry.

Within the environmental limits that control the geographical distribution of a species, there may well be competition between members of the same species for available resources such as light and water. Long dry periods will often see the weaker seedlings die back and thereby leave more room and moisture for the stronger to grow on vigorously. In this way species tend to control their own density, a feature particularly apparent in semi-arid regions where water supply is always a limiting factor and plants adopt what at first sight might be thought unnecessarily large spacing. Nevertheless, below the surface an extensive root system will usually occupy all the intervening space. Some species, such as sagebrush, have deep rooting systems that do not spread far. Others flower or grow at different times of the year. In these circumstances species can grow much closer because they do not compete for the same soil moisture or nutrients (Fig. 30.2).

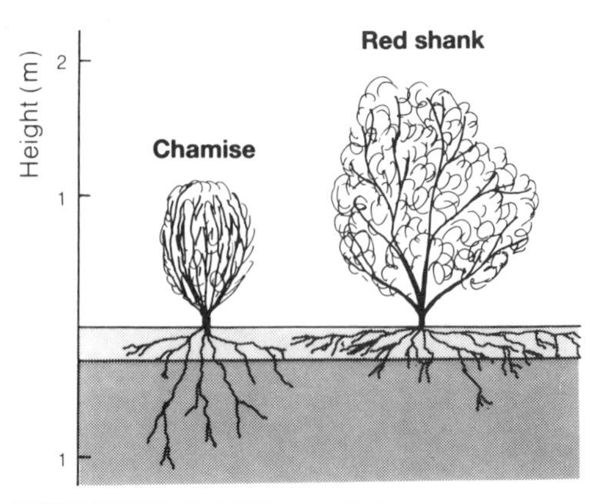

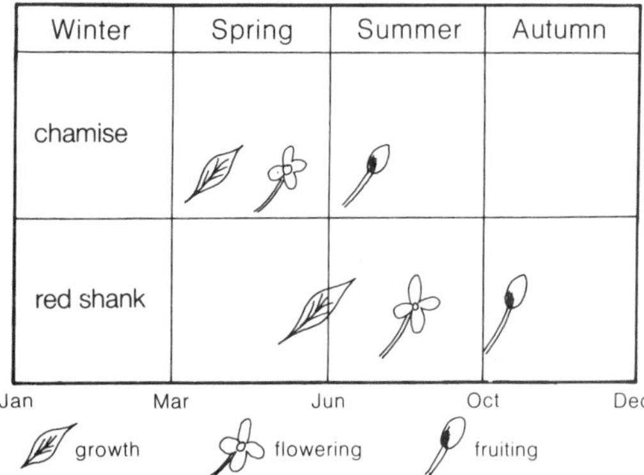

Figure 30.2 The chamise and red shank minimise competition by occupying different root zones in the soil and by growing, flowering and fruiting at different times of the year.

Competition between individuals of the same species may result in greater spacing, but competition between different species may instead lead to the destruction of one of the species. This is readily seen in a temperate woodland clearing where a wide variety of quick growing herbs such as *Ranunculus* (buttercup) and grasses (Gramineae) soon develop in bright sunshine, but then die back as the more slowly growing tree species, such as birch (*Betula*) or beech (*Fagus*), spread out their leaves and shade the ground.

Overall, however, evolution must equip all species to survive. To do this the stress of competition must be low and the available resources must be shared out among many species. It is therefore usual to find a varied community of plants and animals each having a food requirement and adaptation to its environment so arranged so that no one species is dominant enough to displace the others. This yields a stable home or **habitat** for many species living in a closely interactive system (an **ecosystem**) where the limited resources of light, nutrients, moisture and shelter are as fully used as possible. When the ecosystem is mature, such full use tends to keep out potential invaders simply because there is little surplus food. The tropical evergreen rainforest is the prime example of a fully integrated ecosystem which only man has been able to disrupt.

The pattern of species

Although plants have very complex distributions, when they are grouped into major units, or **biomes**, the close correlation between world maps of vegetation and climate clearly suggests that climate is the main controlling factor. By contrast, studies of plant distributions within biomes show that soil, drainage and slope play major roles.

Many plant species are very sensitive to the amount of available moisture and aeration to their roots. Plants with a high soil-moisture tolerance (or need), such as Sitka spruce, are therefore usually found in valley bottoms or on poorly drained plateau soils; those with less tolerance, such as Scots pine, occur only on steep, will drained slopes or sandy soils. Similarly, soil acidity (pH) often exercises a major control. Heather, for example, can survive on acid soils because iron cations are freely available, but it is never found on soils with a pH greater than 5 because iron is stable in near-neutral soils and the plant does not have a ready means of extracting it. On the other hand some plants, such as certain saxifrages, do not have a mechanism for controlling the input of the aluminium cations mobilised under acid conditions and they therefore die of aluminium toxicity.

30.4 Ecosystems

Just as a drainage basin is a natural unit for studying the pattern of individual rivers in the landscape, so the ecosystem is a convenient unit for studying individual plants and animals in their natural environment. An ecosystem is a visually distinct unit within the environment and may be very variable in size: it may comprise a single tree, a complete wood, or the climatic zone of broadleaved forests. But whatever the scale, emphasis is always placed on the way species interact with their physical environment and with each other – by succession, competition, predation and symbiosis – and on how energy is distributed and nutrients cycled. Only by studying these is it possible to get close to understanding the reasons for the behaviour and distributions of the species within the ecosystem.

Energy

All plants need light energy to enable **chlorophyll** (the green pigment) to begin the complicated process of creating new tissue (Fig. 30.3). However, plants

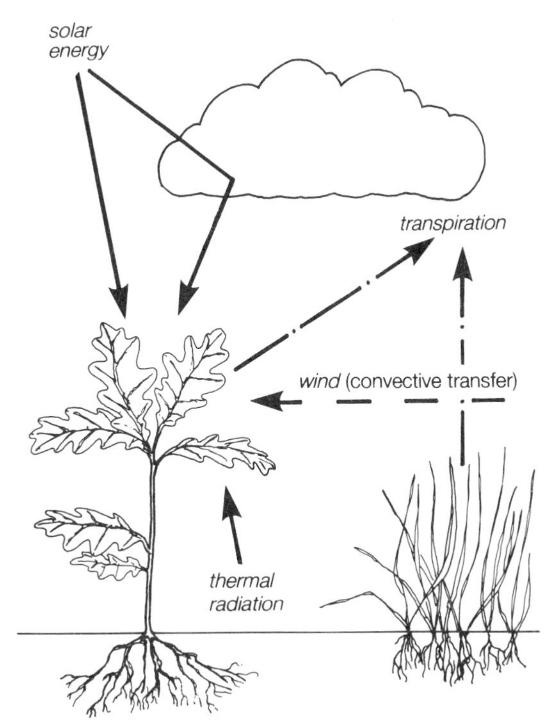

Figure 30.3 The energy environment of a plant, showing the various physical factors which influence the amount of energy it receives and the temperature which it attains.

are able to make good use of most light levels and their growth is only restricted by dense shade; their rate of growth is therefore usually controlled by many other factors.

Plants are also tolerant of a wide temperature range. For example, most plants will grow in temperatures between 6 °C and 35 °C even though their optimum for growth is 25 °C. Only when temperatures rise above or fall below the level of tolerance does the stress on the plant become intolerable. Plants have several mechanisms for coping with high leaf-surface temperatures, the most important being **transpiration**, whereby large quantities of water are exuded through **stomata** on the leaf surfaces, and evaporation cools the leaf. Indeed, over 99 per cent of all plant moisture requirements are to allow adequate transpiration – less than 1 per cent is used for growth. Low temperatures result in gradually slower rates of biochemical reactions (**metabolism**) until some essential reactions cease entirely and growth stops. For many plants 6 °C is the lower limit for growth, while frost can cause cell walls to rupture, so killing the plant.

The energy gained by a living organism is used to grow, to move and to continue its normal life processes. By far the greater proportion, however, is used in 'living' as opposed to growing: the example of transpiration given above is a good illustration. Nevertheless, the increase in living tissue weight that occurs as an organism grows – the increase in **biomass** – is an important indicator of the productivity of an ecosystem. In practical terms this is of especial significance to farmers.

Animals cannot directly use sunlight as a source of energy. Instead, they rely on its conversion by plants to stored chemical energy. Only a few animals (herbivores and decomposers) can make direct use of plant cellulose; others have either to eat plant protein (nuts, fruit, and so on) or become carnivores.

All energy gained by plants and animals is eventually returned to the environment. For example, when people move rapidly they begin to convert chemical energy to heat energy and they become hot. Even when at rest, humans lose heat at a rate equivalent to an 80 W light bulb. The same pattern of conversion from chemical to heat energy is true for all living things, although it is often less obvious in plants. It is the conversion of chemical energy into unusable heat that makes all animals such inefficient users of the food they eat.

Organisation

A stable ecosystem is in balance: it contains systems which provide for the continual renewal of life. These complicated systems of interdependent organisms can be viewed as an intricate network of **food chains** (Fig. 30.4), usually called a **food web**. Each chain has a pyramidal structure with the mass of each organism higher up the pyramid being less than the total mass of the one consumed by it. For example, when herbivores (such as cattle) eat grass, they cannot utilise all of the energy in the grass. Cattle must therefore eat a large bulk of grass leaves to survive. When man eats beef he cannot utilise all of the energy in the meat: if his diet consisted only of beef, there would have to be a much larger number of cattle than humans. The numbers, density and distribution of higher members of the food chains are dependent on the distribution and biomass of lower members. The need for the greatest living bulk to be of vegetable matter also explains why most ecosystems are referred to by the plant association rather than the animals that depend on it.

Changes within an ecosystem

In 1980 the Mt St Helens eruption devastated an area of many hundreds of square kilometres in a matter of minutes. Where once there had been large stands of mature coniferous trees dominating a stable ecosystem, suddenly there was matchwood and ash. All the trees had been blown down and their leaves and branches burned away. Somewhere, in almost every country every week, a piece of hillside finally

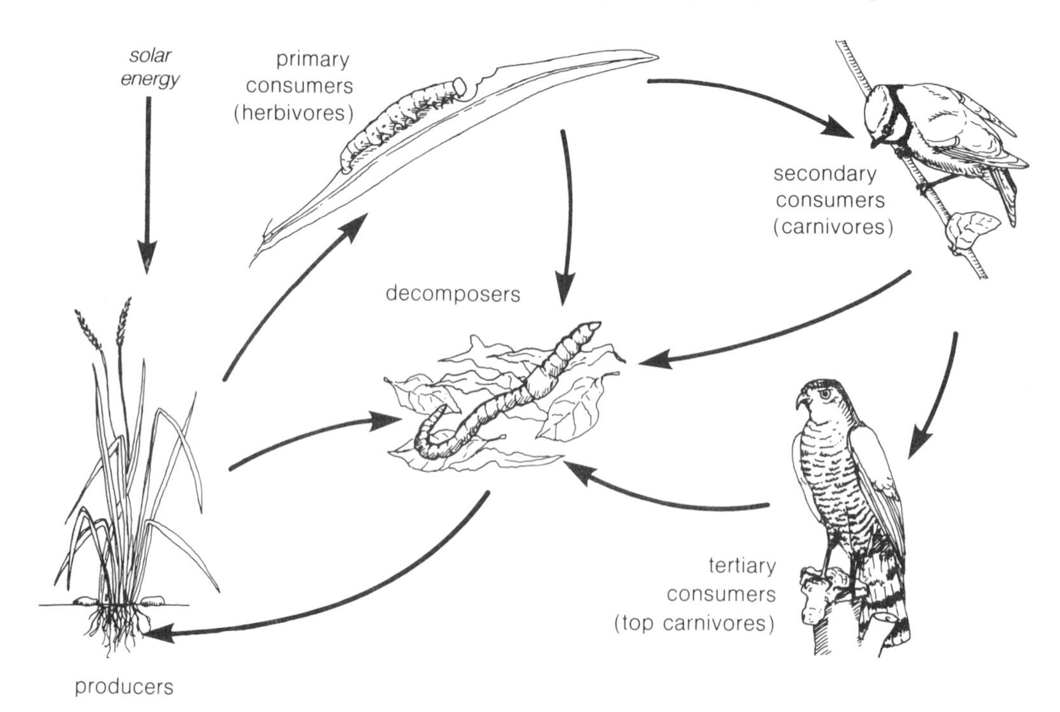

Figure 30.4 An example of a food chain fueled by the continuing energy of the Sun.

gives way and slides, leaving a large, bare scar in the place it had been. Occasionally there are devastations on a scale even greater than from volcanoes. The last was when ice spread over 30 per cent of the land area of the globe during the ice age. But whatever the scale, such examples illustrate how commonly ecosystems can be disrupted and how often they must rebuild anew under purely natural conditions. Recently man has considerably increased the occurrence of ecosystem devastation, so that rebuilding is now a widespread and common phenomenon. Rebuilding does not always lead to a return of stable conditions (Fig. 30.55). The ecosystem is a *dynamic* rather than a static entity, and has a limited self-renewing capacity.

The **colonisation** of a new site and the development toward a mature, stable, and only slowly changing ecosystem called a **climax** is a more or less continuous process, but it can be viewed as a series of overlapping steps, or **successions**. Each succession is dominated by a particular group of plant species which controls all other life forms (Fig. 30.5). For example, in a site in a temperate humid area that has been newly cleared to bedrock, there are relatively few nutrients available; only the **pioneer** plants (such as lichens) that can obtain their mineral requirements directly from rocks or rainwater, can colonise. However, as individual lichens and mosses complete their life cycles, die and decay, nutrients are released in a form available to higher plants; while chemical weathering, soil formation and rainfall gradually add to the nutrient supply allowing the biomass per unit area to increase with time.

As a direct consequence of the increasing availability of nutrients, both the mass and the variety of species also increase, allowing the next phase of succession to begin – **competition** among colonisers for space and nutrients. In this phase open ground is gradually replaced by herbs whose seed is blown in from surrounding areas. Most of these plants have similar nutrient and habitat tolerances. They also provide the moist, temperate microclimate that allows other seedlings to flourish. Thus in time larger and sturdier perennials begin to take over and crowd the annuals out. They in their turn are replaced by shrubs and quick-growing short-lived trees such as birch. The succession is completed when the slow-growing but larger trees, such as oak, dominate the ecosystem. Another common succession takes place on land that is being built by rivers into lakes or the sea.

Whatever the circumstance, the initial rapid rate of increase in biomass gradually declines until, with the final climax, it ceases to change significantly at all, each year's new growth on average being closely matched by the annual decay.

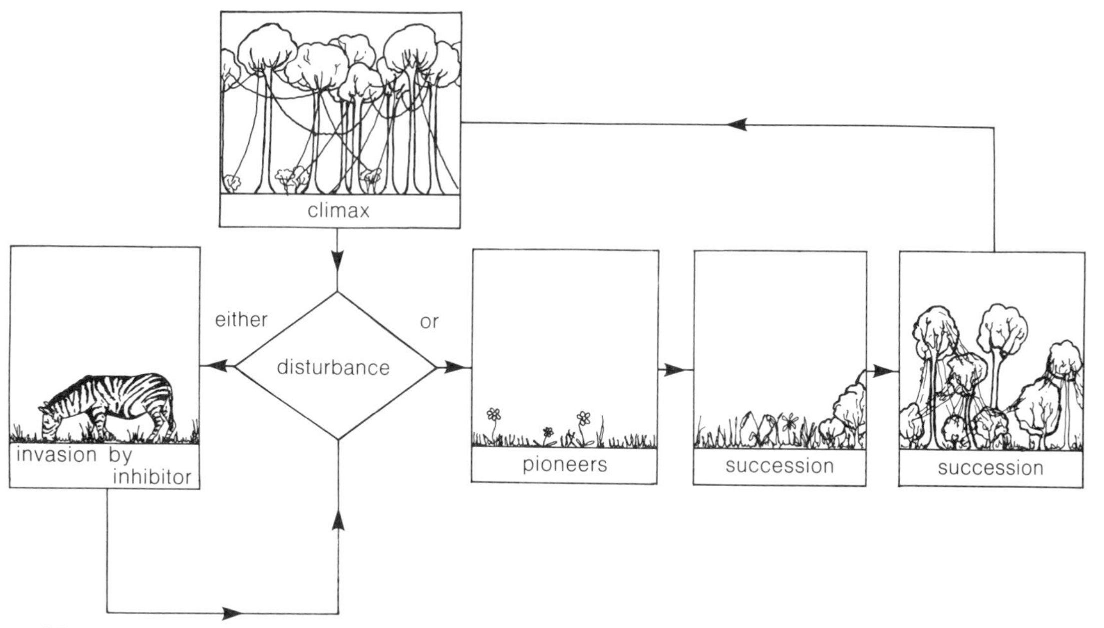

Figure 30.5 When a stable community is severely disrupted by clearance, a series of successions may occur that eventually lead back to a climax community. Thus a tropical rainforest may be self preserving. However, under certain circumstances, such as when fire repeatedly destroys a stable community and herbivores then enter, the succession is inhibited and the colonising pioneer species never create the environment suitable for succession to proceed. This has probably been the case for the world's grasslands (steppe/savanna).

30.5 Evolution and equilibrium in ecosystems

Once we understand the basic concepts of succession to a climax we can understand the continued survival of the climax in the face of the evolution of the individual species of which it is composed. Although ecological systems cannot evolve by natural selection in the same way as an individual species, the constant interplay between the species that make up an ecosystem means that, as each species evolves in sympathy with those on which it depends (e.g. hummingbirds on plant nectar; flowers and their pollinator insects), so great groups or 'constellations' of species change together and constantly produce new patterns within the landscape.

The question of evolution can therefore be widened to involve the way that even a biome (p. 296) evolves within its physical environment. But although we may seek to understand the evolution of large ecosystems, their complexity presents formidable problems. Scientists have only so far made tentative steps toward some of the answers. The difficult task they face is easy to appreciate when faced with problems such as these.

(a) The number of species in ecosystems varies widely – for example, evolution has restricted the species of butterflies resident in Britain to about 60 whereas nearer 1000 have evolved in New Guinea.
(b) Some species are relatively more abundant in some regions than others.
(c) There are many more small species than large ones.
(d) Terrestrial food chains typically have only three or four links (plant, herbivore, first carnivore, second carnivore) though there are many variations in the amount of energy transferred and the physical details of the organisms involved.

The gradual drift of the continents within the last 200 million years has had a profound effect on the distribution and variety of all living things. (The pattern of continental movement is described in detail in Chapter 1.) As the continents parted, so land connections that had allowed plants to propagate and animals to wander freely over Pangea were severed. The consequences were particularly important to evolution; despite the natural diversity of living things, many share a common origin. The many orders of mammals, for example, evolved from a common origin sometime between 300 and 200 million years ago, just before continental drift began. From their point of origin the mammals radiated outwards, adapting to new ways of life and evolving new forms as they moved into new environments. Such **adaptive radiation** is paralleled by the phenomenon of **convergence**, in which animals and plants that have very different ancestries take on much the same adaptation in similar environments. After continental separation each continent became a separate centre of adaptive radiation, leading not only to the rapid world-wide diversification of species but also to convergent evolution, as forms developed to fit the available environments (Fig. 30.6). Thus although the tropical rainforest in each continent may look superficially the same, each continent is composed largely of different species of plant and inhabited by different species of animal.

It appears from the fossil record that there always have been consistent patterns in the number of species associated with a given region, which suggests that each region has been 'ecologically full' for a long time. For example, when North and South America finally joined by continental drift about 2

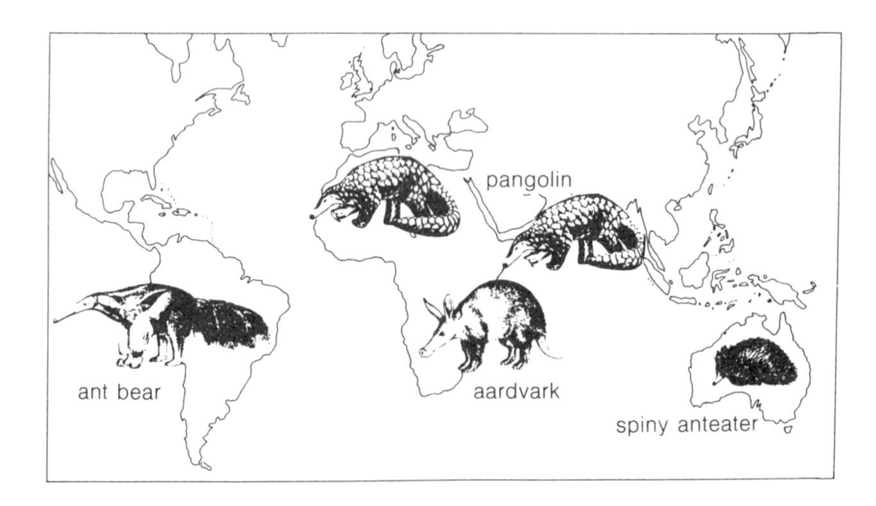

Figure 30.6 Convergent evolution is exemplified by the four ant-eating mammals, each of which belongs to a separate Order and has been evolved on a separate continent.

million years ago, each continent had about 25 or so families of mammals. At first there was a lot of interchange across the new land bridge and the number of families on each continent rose markedly, but gradually competition caused the extinction of some, such that now the number of species is the same as before.

Observation suggests that the total number of species present in any region increases with the area and decreases with its isolation: small, remote islands therefore have few species, while continents have many. Thus the small island of Peninsula in the Aleutian chain near Alaska has 6 species on its 156 ha; larger Attu has 247 species on 9057 ha; and the largest island, Kodiak, has 423 species on 897 400 ha. Islands that have been decimated by natural catastrophes such as volcanic eruptions have recolonised with the same number of species as before the eruption, although many of the new species may be quite different.

Europe has many more species than Britain, which in turn has more species than Ireland. During the periods in the ice age when Britain was temporarily attached to mainland Europe the number of species rose to equal that of the continent, but since the rise of sea level and the consequent isolation the number of species has been gradually dropping again to a level proportional to its area. The same decrease in the range of species has been observed in the mountain 'islands' of the Sierra Nevada in the US since they became isolated from the main area of Boreal Forest after the decline of the ice age.

All these observations strongly suggest there is an underlying community structure to an area although exactly *which* species are present is entirely unpredictable. Thus within an ecosystem there are a certain number of ecological niches that must be filled. There are many species similar enough in form and habit to fill these niches (Fig. 30.6); which ones actually arrive first is a matter of chance. However, once any one species within a competing group becomes established, competition will tend to ensure the extinction of the other potential colonisers. This in turn helps to explain why there are different numbers of species in different regions: the larger the region (and the less restricting its environmental controls, such as frost or drought) the greater the number of similar species that can survive and find a niche.

These relationships hold only for a 'mature' ecosystem. As we have seen, as colonisation proceeds into new areas, the number of species gradually becomes larger and the proportion of any one species smaller. The first to arrive take up most of the limited niche space, but gradually become partly displaced by the later, more slowly growing arrivals whose development provides a greater variety of habitats. Grasses colonising a bare field, for example, leave relatively few niches, whereas the trees that gradually colonise the same area and rise above the grasses create many more niches in their crown space, in the shade they cast and the protection they offer.

Within an ecosystem small animals can subdivide any habitat to a greater extend than large ones. Thus cattle may eat several species of grass indiscriminately, yet the herbs may act as host for a wide variety of small arthropods. The length of any food chain is limited by the poor efficiency of conversion of energy (about 10 per cent) that takes place at each level. It still seems strange that the number of levels should be three or four so consistently throughout the wide variety of the Earth's environments unless there is some other more powerful control. At present no such control has been found.

Limits to growth

All organic materials are made from four basic elements of the atmosphere: carbon, hydrogen, oxygen and nitrogen; and thirteen elements from the Earth's crust: potassium, calcium, magnesium, phosphorus, nitrogen, sulphur, iron, copper, manganese, zinc, molybdenum, boron and chlorine. All of these elements exist on the Earth's surface but they can only be mobilised either by photosynthesis or by chemical weathering of rock. The release of elements by chemical weathering is always slower than the potential requirements of the biomass growing in it. A continual extra supply of nutrients from rainwater is vital, and it is essential that the vast majority of these hard-won elements are recycled (Fig. 30.7).

Although nitrogen is the most common element of the atmosphere, it is useless to plants in its 'free' form. It must first be 'fixed', as ammonium ions (NH_4^+) or nitrate (NO_3^-). Nitrogen fixing is mainly accomplished by soil bacteria; the nitrogen is released as nitrates when the bacteria die. Despite the vast numbers of these microorganisms in a soil, nitrogen is required in such large quantities that it is the element that most commonly limits plant growth. Nitrates are negatively charged, so they are repelled by clay and humus surfaces and are never retained in the soil mass. Without such natural stores they are rapidly leached by percolating waters unless they can be recaptured by plant roots. The high demand for nitrogen and the ease with which it is lost from the soil explain why nitrate fertilisers are among the most important of the farmer's aids.

Sufficient supplies of all the elements listed above are needed for maximum growth. In a natural ecosystem one or more will probably be in short supply, especially in an ecosystem undergoing succession and therefore an increase in its biomass. Plants are not able to speed up the release of elements

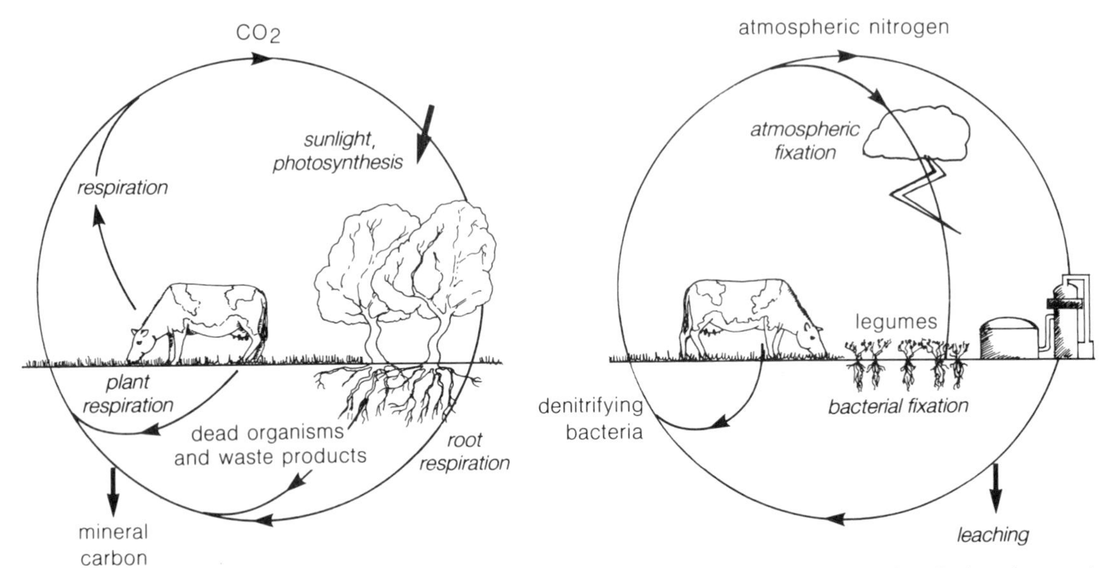

Figure 30.7 (a) The *carbon cycle* involves two competing processes: photosynthesis and respiration. During photosynthesis, plants convert carbon dioxide and water into carbohydrates and free oxygen. The latter combination of substance represents a rich store of energy, which is utilised in respiration as the carbohydrates and oxygen are recombined to yield carbon dioxide and water again. Respiration is common to all organisms that can live in the presence of oxygen, and thus all contribute to the return of carbon dioxide in the atmosphere. Some carbon is stored in minerals such as coal and petroleum, but that too is returned to the cycle when it is burned. (b) The *nitrogen cycle* traces the circulation of the nutrient that is often the factor limiting the growth of plants. Atmospheric nitrogen is useless to plants; the element must be supplied in 'fixed', or combined, form, as in ammonium ions (NH_4^+) or nitrate ions (NO_3^-). A little nitrogen is fixed by lightning and other processes in the atmosphere, and a more important contribution is made by bacteria, notably those that live in the root nodules of legumes. Nevertheless, the pool of available nitrogen in most soils remains small. The element is removed by leaching and by bacteria that return it to the atmosphere, and it is lost through the harvesting of crops. To compensate for these losses, nitrogen fixed industrially is applied to the soil as fertiliser; industrial fixation has become a major component of the nitrogen cycle.

from a rock: whatever supplies are available must be shared between the growing plants. In most ecosystems, therefore, the plants are growing at less than their full potential, which is why man is often so successful in improving plant yields by applying fertilisers, by decreasing competition through adequate plant spacing and, if necessary, by irrigating.

In a mature ecosystem, the vegetation remains stable by conserving its nutrients as completely as possible. This is achieved by the patterns of roots that are spread out as a net within the surface soil to catch ions from rainwater or released by the decay of dead organic tissue. In temperate regions, neutral soils contain clay and humus that help store ions until they are needed by plants. In tropical environments where higher temperatures and rapid oxidation prevent the survival of humus and where high rainfall leaches ions from the clays, plant roots and mycorrhiza (fungal threads connecting decaying matter to roots) must be able to retain all the ions directly.

Because of the restricted availability of ions and soil moisture, a stable ecosystem limits the number of each species in such a way as to keep within the effective food supplies. This applies both to the plants and to the animals that depend on them. In effect, therefore, the land imposes a certain **carrying capacity** on the numbers of each component of the ecosystem. Because some of the resources on which the ecosystem depends vary from year to year, the carrying capacity of the land will also vary. For example, extra biomass may be produced in a year of plentiful rain; this in turn will allow an increase in the number of animals. Similarly, drought causes a

Table 30.1 Net primary productivity (source: Whittaker and Likens, 1975, and others).

Biome	Productivity ($g/m^2/year$)	% of Earth's land surface
tropical rainforest	2200	12
deciduous temperate forest	1200	8
northern coniferous forest	800	11
savanna	900	} 19
temperate grassland (prairie)	600	
semi-arid scrub (including desert)	90	29
cultivated land	650	24

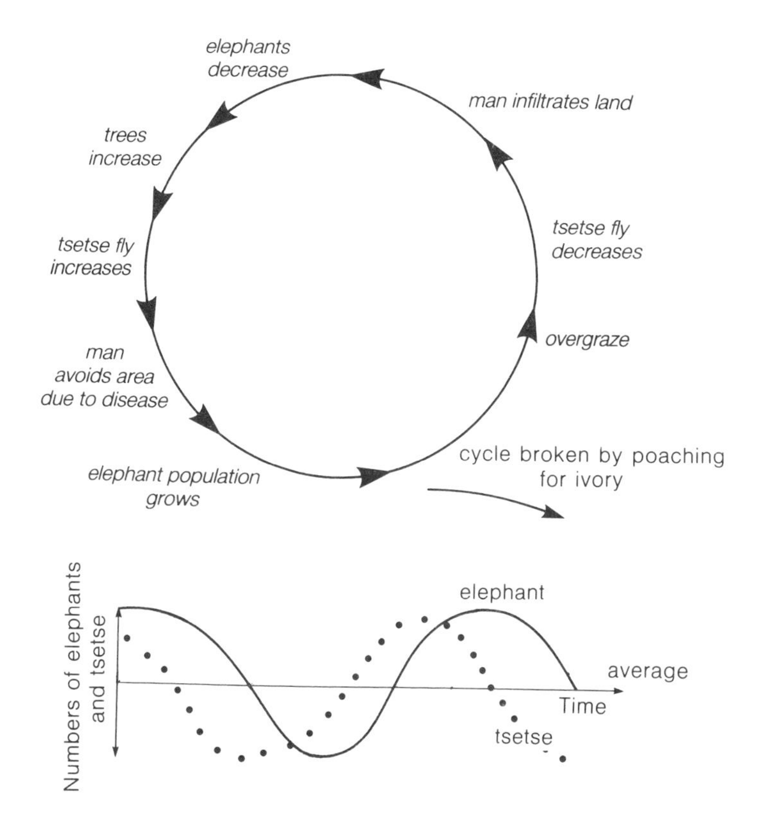

Figure 30.8 Dynamic relationships in elephant numbers.

reduction in biomass and a consequent fall in animal numbers, due to starvation.

We can compare the net primary productivity of different types of ecosystem (Table 30.1). This shows that the biomass productivity is much greater for forests than for grasslands or crops. When man replaces forest with crops, a complex and stable ecosystem is replaced with a simple, unstable system having a lower primary productivity. But the purpose of agriculture is to increase the *edible* food content of the area. Because the edible component of many natural ecosystems is so small, agriculture is a very effective means of providing man with more food.

Periodic fluctuations in ecosystems

People in East Africa worry about the periodic large increases in elephant herds and there is strong pressure to have many animals killed. Yet there is nothing unstable about the elephant population: the size of the herds merely responds to the availability of forage. The difficulty arises because the long lives of elephants enable them to get out of phase with their food supply: elephants born in a time of plentiful supply may live for 50 years.

The availability of the elephant's food is partly dependent on the tsetse fly. These flies inhabit acacia thorn savanna at altitudes between three and four thousand feet. Elephants do not suffer from the fly-borne disease as cattle and humans do. Humans therefore normally keep away. As the undisturbed vegetation grows, elephant food supplies increase and the elephant population increases. Eventually a stage is reached when the elephants are eating the forage faster than it can be produced. Now both elephants and tsetse fly are deprived of food and they both die back – a natural cycle of stability. Of course the reduced tsetse-fly population at these times has encouraged men to come in to take advantage of the low phase of the cycle and destroy the remaining scrub, denying the tsetse a home. Such action upsets the natural balance, thereby requiring man to cull the elephants as well.

Urban man and ecosystems

31.1 Man and the natural environment

There are a number of ways in which man can affect the natural environment. He may do so

(a) *directly*, by releasing pollutants into the air, water or soil (leading to everything from metal poisoning of birds to acid rain destroying trees),
(b) *indirectly*, by upsetting the natural balance of an ecosystem, exterminating higher trophic levels such as birds of prey and mammals and thereby allowing lower trophic levels such as insects to destroy plants.

People are more aware of direct and acute effects than of indirect or chronic effects, simply because the damage in the former case is immediate and striking. It is difficult to assess the impact of indirect effects because changes are slow and insidious.

Most basic of all the services supplied to mankind by the natural environment are *food production*, a storehouse of *genetic information*, and waste *recycling*. Man cannot replace any of these vital functions. The whole intricate network of biological cycles that provide our food allows the survival of the system if one species fails. The natural order works on the basis that diversity ensures stability.

For the past ten thousand years man has been clearing woodland and grassland alike, replacing diverse ecosystems with the most simplified and unstable ecosystem of all – the cultivated field. Similarly, rangelands have been overgrazed and many species killed off in the search for more edible grazing. A natural pasture, with its variety of summer flowering plants, may be a delight to the eye, but it is not the most economic way to feed animals: most of the flowering plants are inedible. The difference between man and nature is that, whereas man attempts to maximise the productivity of the small number of species that he or his animals can eat, natural ecosystems operate in such a way as to maximise stability. These two goals are incompatible because stable ecosystems have zero net productivity; all is used and recycled. Clearly man is obliged to create a degree of instability in order to provide food for the world's population. However, the more disruption that occurs the smaller are the remaining source areas of plant diversity which, because of the genetic information they contain, are in turn the potential sources of new variability. Similarly, the use of biocides (chemicals that kill specific organisms) leads to a drastic reduction in animal and insect life, often disrupting the natural food webs.

The most fundamental relation between human population size and the environment is that population size acts as a *multiplier* of the impact of each member of the population. Thus:

environmental disruption = population x consumption per person x damage per unit of consumption.

For example, in the USA the population increased by 40 per cent in the 20 years from 1946, while the amount of lead emission per vehicle-kilometre, and the number of vehicle-kilometres per person nearly doubled. Thus 40 per cent (1.4 fold) x 2 x 2 = 5.6, or 460 per cent. Because the factors are multiplied up, *no factor is unimportant*.

Not only is there a multiplier effect but also a law of diminishing returns operating on farming. Each increase in yield from an area can only be achieved by a larger increase in fertiliser, biocide and energy. These increasing changes cannot proceed unchecked for ever; sooner or later negative feedback will occur. It has been estimated that mankind's demands on the environment increase at a compound rate of 5 per cent per year, resulting in a doubling of demand every 14 years. This means, for example, that there will be a fourfold increase in demands on the environment between 1986 and 2014.

One of the major problems in slowing down or halting this rate of impact stems from the momentum involved: it is difficult to get people to act until they perceive their environment deteriorating, and by then it is often too late. Some substances, such as DDT (or the extreme case of plutonium), remain active for very long periods after they have been introduced into the environment, and may be progressively concentrated as they pass through the food web. Without more effective long-term planning, the chances of widespread disaster continue to increase.

31.2 Natural vegetation in urban areas

Natural vegetation tends to be used in very limited ways within urban areas because it is often regarded as difficult to work with and costly to maintain. However, natural vegetation can provide both cost-effective and aesthetically pleasing alternatives to unrelieved areas of brick, tarmac and concrete. For example, a fully grown deciduous tree with a crown diameter of 14 m has a leaf area of about 1600 m^2. The leaves absorb carbon dioxide and release oxygen, they help to break down sulphur dioxide pollutants and they trap up to one tonne of dust each year. Furthermore, trees and hedges provide sound-absorbing layers of leaves. These are all major benefits to the urban environment. If major roads were built within 'tunnels' of trees or hedges, the vegetation would reduce both noise and air pollution; if planted with some sensitivity, they would also provide interesting vistas for drivers (Fig. 31.1).

Natural vegetation can be successfully planted in areas where there is little competition for land within the urban environment. For example, floodplain land is unsuitable for residential building but can successfully be used as a **linear park**. Other spoilt ground, such as that near flooded gravel pits, quarries, and spoil tips, provides areas where natural vegetation can grow, and much needed recreational opportunities. The Lea Valley country park in East London and the Irwell Valley (Moses Gate) country park in northern Manchester are recently developed examples (Fig. 31.2).

The controlled growth of vegetation on the banks of water courses is one of the best and least expensive ways to stabilise soil, reduce bank erosion, provide a diverse biological environment to help preserve wildlife, and yield a pleasant environment for people to enjoy (Fig. 31.3). All of these objectives are achieved using planning based on detailed knowledge of both the river-flow pattern and the environmental conditions needed for the growth of each plant species. For example, a meandering river provides the most effective means for distributing river energy,

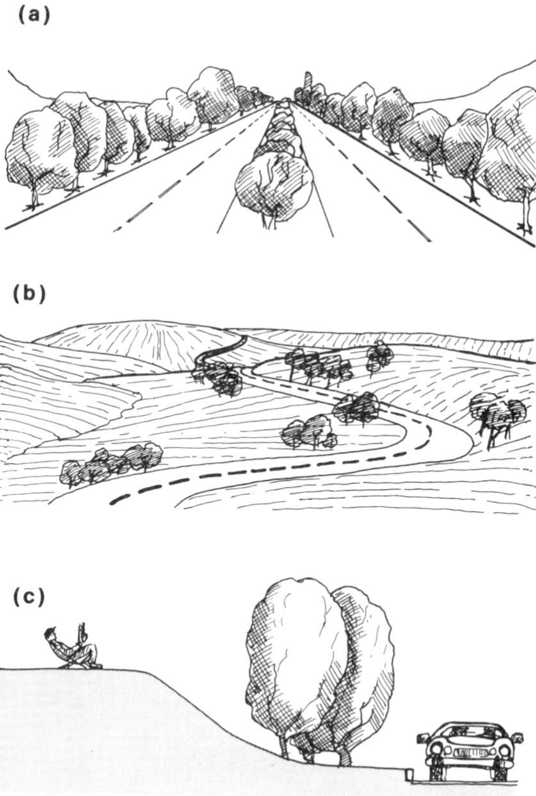

Figure 31.1 (a) Tunnels of trees help preserve the environment away from the road but they can be claustrophobic for the driver. (b) Curved routes are better, as they are in an open landscape, where natural vegetation can still be used to great effect in producing variety. (c) Shelter belts including shrub–tree screens add to privacy, and reduce traffic noise and pollution.

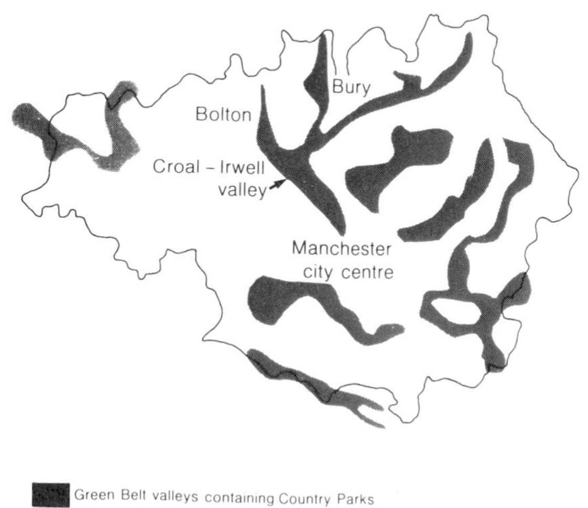

Figure 31.2 The linear park development in the Irwell valley, Salford, Manchester.

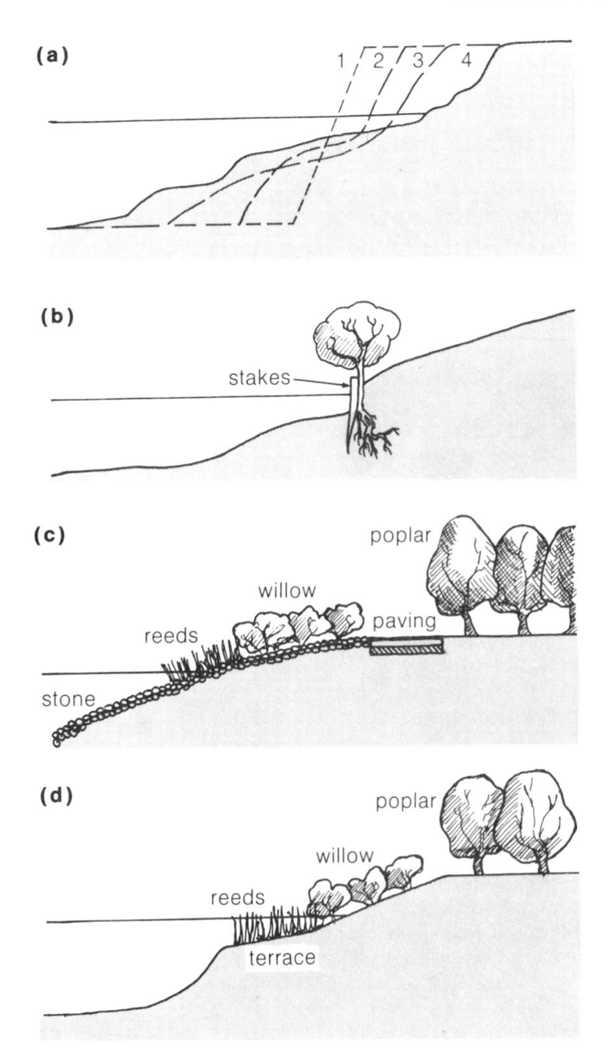

Figure 31.3 (a) Stages of natural erosion from a steep bank to a stable profile. (b) Erosion barrier of willow, alder or other tree which will take root and form a natural barrier. (c) A bank of gravel and stone pitching with reed and willow cuttings. (d) A reed bed planted on an underwater berm as a form of shoreline protection.

and therefore reduces erosion-maintainance costs. It also provides the widest variety of ecological conditions for both land and water plants and animals.

The pattern of natural vegetation that lines a river bank can readily be reproduced, from the rushes that survive in areas of fast-flowing water below river cliffs, to wet banks above the water line where alders and willow can grow (Fig. 31.4). This stable riverside vegetation protects the banks from excessive erosion and provides a pattern of shelter for insects and shade for fish. It allows the river to become **self-cleansing** (Fig. 31.5).

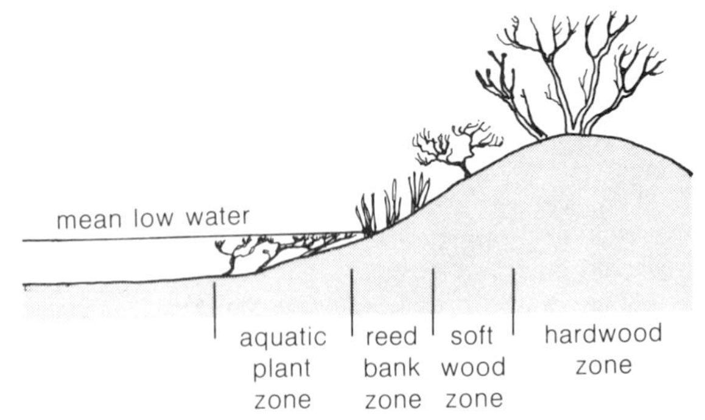

Figure 31.4 Vegetation zones on freshwater river and pool banks.

31.3 The changing British countryside

The British countryside is changing fast simply from the individual imaginations of many thousands of people, each of whom alter the countryside just a little but whose combined effects alter it a lot. New buildings, quarries, reservoirs, roads, hedge-row removal, woodland reclamation, marsh drainage, river straightening, pollution, recreation and ploughing of permanent pasture all result from attempts by people to get more from a limited land resource.

When, in 1977, the Countryside Commission made a survey of leisure activities within England and Wales, they discovered that the countryside is one of the most attractive forms of outdoor recreation, with over half of the population visiting it at least once a month, and over three quarters (37 million) spending at least one day in the country during a year. Many of these people (33 per cent) visit the countryside for drives, having outings and picnics within short distances of the road; 20 per cent go on long walks, hikes or rambles, and 18 per cent go to visit the coast. The managed attractions – stately homes, historic buildings, ancient monuments, safari parks and the like – attract another third to the countryside. Overall, therefore, most people prefer to visit unmanaged, easily accessible areas on a fairly casual basis.

How can all of these activities be made compatible with the preservation of wildlife, and the variety in a landscape that is so treasured by those who enjoy the countryside? One solution has been sought in the formation of **National Parks**, large tracts of land recognised for their outstanding natural beauty (Fig. 31.6). National Parks are not intended to fossilise the inefficient and quaint farming systems of a bygone age. They are intended to be areas where planned progress can go ahead with especial care to the protection of the natural countryside. For example, if a

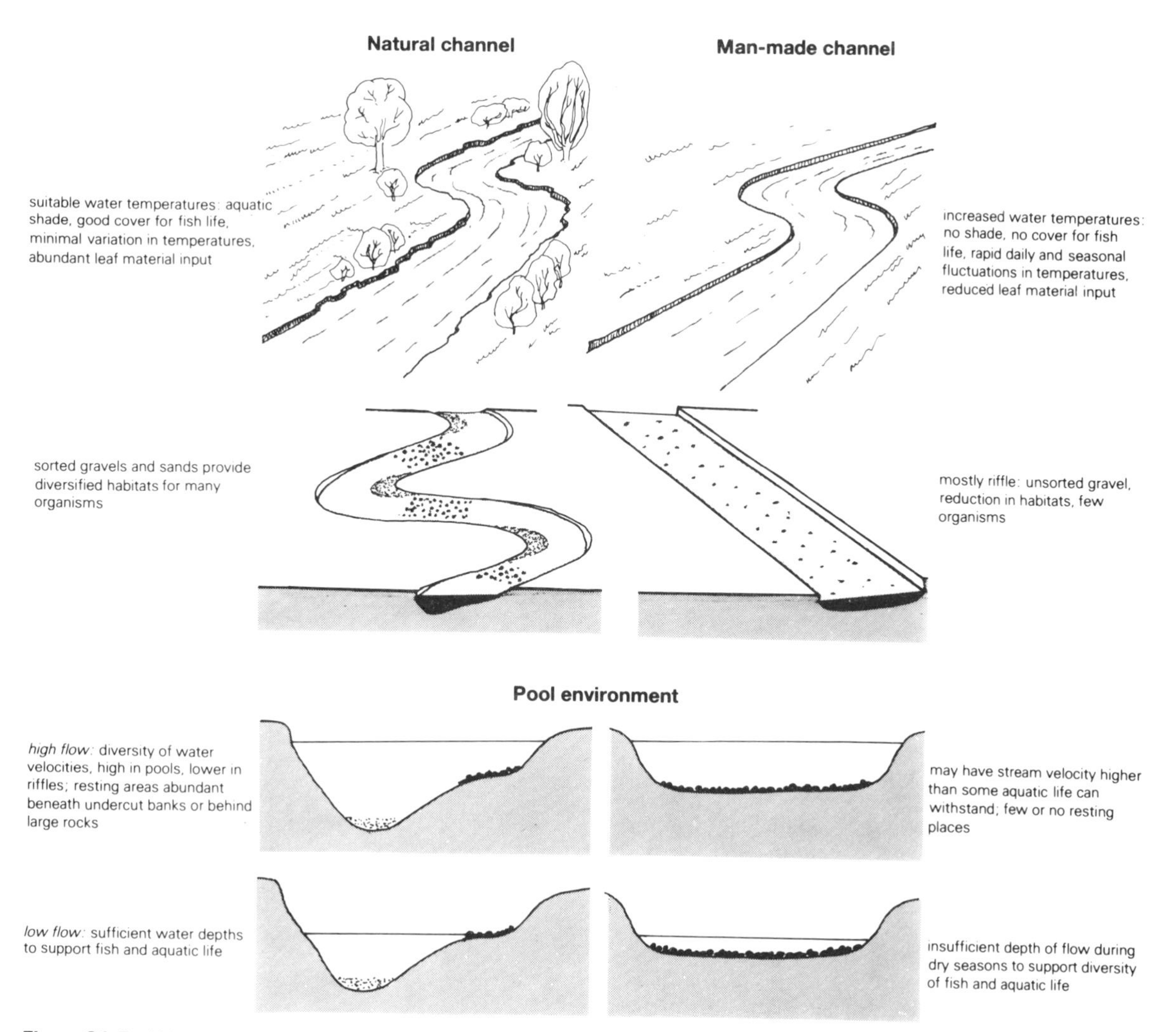

Figure 31.5 When river channels are altered, the environment for living creatures changes. Fish, for example, use the varied habitats of pool and riffle sequences in a river for food and protection. When man straightens river channels he destroys much of the variety, reduces the habitats, and the number of species and individuals that can survive.

landowner in a National Park wants to plough moorland or heath, he has to give notice to the local planning authority so that they may have a chance to buy the land or compensate the owner for the loss incurred by restricting its improvement.

Beyond these areas protection of the countryside rests largely with the sympathies of farmers. But the farmer cannot be expected to be too sympathetic to wildlife; he wants his high-yielding plants to grow without the effects of competition from other species. The days of the floristically rich meadow or the wild plants adding a dash of colour to cornfields are nearly over.

Perhaps the landscape should be recognised as serving two separate functions, one *protective*, the other *productive*; with the protective area confined to the least productive parts of the country – in effect to the highland north and west. Indeed, the National Parks are in this highland zone. But this would leave many people a long way from any natural amenity. So would it be possible to protect areas for amenity use in the midst of a sea of intensive farmland? The highest levels of the food chain, predators such as marsh harriers, need large territories for survival. A pair of marsh harriers needs something like 120 ha of reedbed over which to hunt for food. Reserves of

(a)

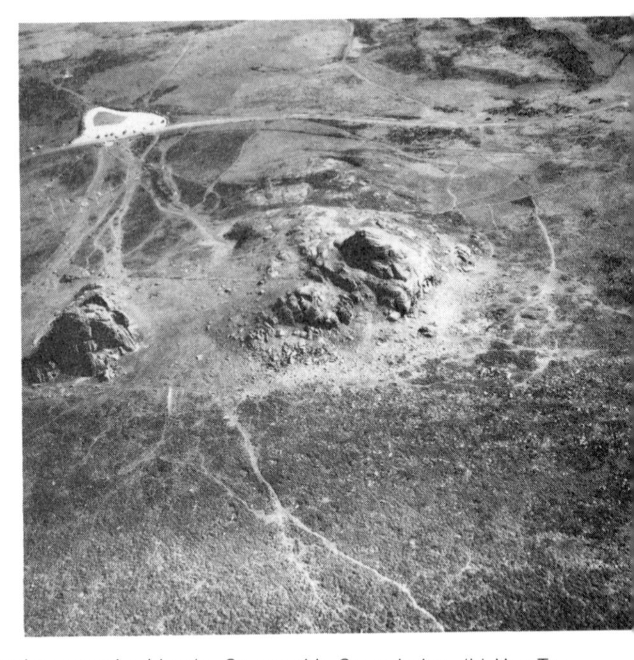

Figure 31.6 (a) National parks, country parks and recreation paths recognised by the Countryside Commission. (b) Hay Tor, Dartmoor, showing a popular region of a National Park subject to overuse.

lowland need to be substantial before they can yield stable ecosystems. Even then their likely isolation from other similarly protected areas will lead to problems of breeding and maintenace of the species.

Where will further large viable units of land be found? Mainly they will include such areas as golf courses, country estates and managed forests. They will have to be used wherever they may be rather than built around areas of special scientific interest which may lie in land otherwise quite unsuited to amenity development.

If sufficiently large areas can be zoned from agricultural use, they can be further zoned in a concentric way, with the outer, most visited, areas acting as buffers to the remoter central areas, which can be used as sanctuaries. This is a pattern already employed in large National Parks in other countries.

Some features of the zoning concept have been adopted in Sussex, where the East Sussex County Council has bought 280 ha of chalk cliffs adjacent to the Cuckmere estuary and declared it 'The Seven Sisters Country Park' (Fig. 31.7). Next to it are 800 ha of Forestry Commission beechwood, National Trust and local council lands. Within this area, called the 'Sussex Heritage Coast', an access zoning policy is helping to preserve the most vulnerable wildlife. It is a conservation blueprint for the future. Such areas provide a protected place for the urban dweller to walk after a drive in the productive countryside.

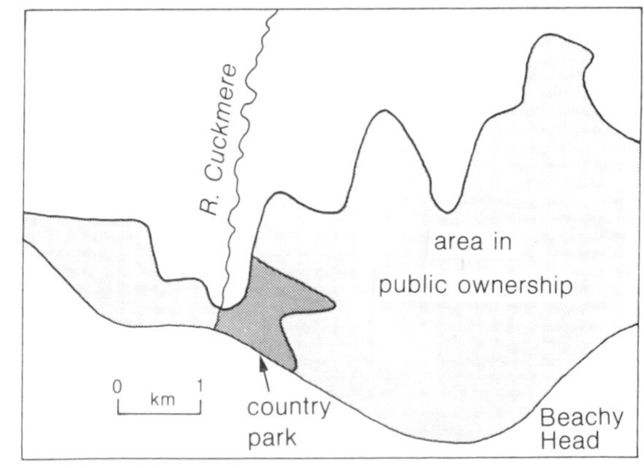

Figure 31.7 The Sussex heritage coast.

On the other hand, the pressure for short-distance country visits, highlighted by the Countryside Commission's survey, cannot be satisfied on a purely ecological basis. Many country parks are much smaller than necessary to provide a viable ecosystem containing the higher trophic levels, but they can offer elements of the countryside close to urban areas. In effect they are a 'safety valve' for the urban dweller, providing 'real' countryside virtually on the doorstep. They are designed as 'natural' enclaves within agricultural land for people who wish to spend a few hours in the countryside. They offer the additional advantages of car parks, toilets and refreshments.

For similar reasons, it is necessary to provide recreational 'corridors' that thread their way through the countryside. There is a network of public waymarked recreation and long-distance paths in Britain, designed to cater for those 20 per cent of tourists who want a more adventurous walk.

Clearly, making way for the urban dweller in the countryside requires many compromises. Providing access on a variety of scales, with plans based on surveys of recreational demand and ecological considerations, offers the best prospect of success.

31.4 The impact of tourism

When areas such as National Parks become major protected centres of amenity, their use may increase to a level where environmental quality is threatened. It is therefore important to know the level of possible

Trampling feet put a blight on honeypots

by Roger Ratcliffe

BELOW the shining limestone cliff at Malham Cove in the Yorkshire Dales, a surprise awaits walkers who return this spring to this dramatic beauty spot – a stone staircase cut into the 240ft sheer precipice.

They were completed just before Easter, convenient for those who wish to reach the superb viewpoint above, but as officials of the Yorkshire Dales national park readily admit, are "rather drastic in an unspoilt area".

Such artificial features are becoming more common at the most popular beauty spots, known as "honeypots" because they attract the greatest swarm of visitors. The question worrying national park and other countryside planners is: how can the honeypots' natural beauty be preserved while providing facilities for the holiday hordes?

In 1976 the national park put in 100 timber steps up the cliffside at Malham; but, says Baker, it was quickly realised that this was not enough. Hardened paths were laid at the foot and new steps were made out of limestone to merge with the background. "In an area like this it was the only way it could be done", he says. "It would have been destroyed otherwise."

Malham's problems have been tackled with just £12,000. But another of the country's most visited honeypots, Snowdon, requires an estimated £1,500,000 to put right the ravages of popularity. The mountain is the most popular tourist destination in Wales, with at least half a million visitors a year. On a busy summer day it is reckoned that 2,500 people make it all the way to the 3,559ft summit, the highest in England and Wales. Officials of Snowdonia national park decided to take action when they found that the six main paths up Snowdon were being stripped bare of vegetation and soil and they got down to the job of hardening paths and persuading people to keep to them.

Every British honeypot has similar problems:

- At Tarn Hows, a picturesque lake that is the most visited beauty-spot in the Lake District (500,000 visitors a year), hardened paths had to be laid to put a stop to erosion and some parts of the area were fenced off to control car parking.
- At Hadrian's Wall, the dramatic Roman structure stretching across northern Britain, the grinding of an estimated 250,000 pairs of feet a year has caused damage to the wall and its complex of ditches and earthworks.
- At Loch Lomond, Scotland, hundreds of thousands of trippers have made it necessary for local councils to get together and form what will be virtually a national park authority for the area, so that the scenic beauty can be preserved.
- At Land's End, Cornwall, the 60ft cliffs where England tapers into the Atlantic have a new private owner who is now engaged in a legal dispute over his plan to charge visitors (at present a million of them each year) £1.50 a head. He plans to use some of this income to tackle honeypot erosion.

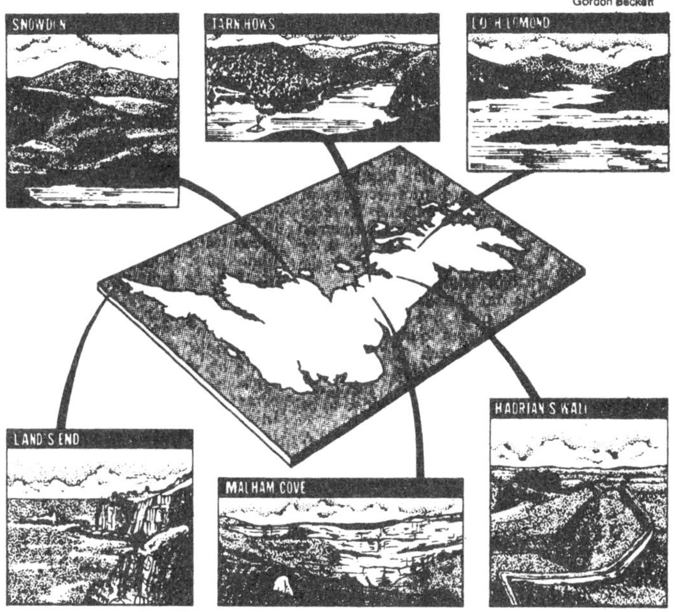

Beauty in peril: some of the famous sites suffering from popularity

Low-cost Rural Housing (best served by an integrated scheme

by Thuthuka Yeni

Low-cost housing for the urban low-income group in developing countries has posed an array problems behind which can be found an interlocking network of cultural, organisational and social economic variables. As focus shifts from the urban to the rural setting, the pressing need for shelter remains constant whilst the impact of the individual variables flactuates in degree.

What may constitute a major constraint in the low-cost housing effort in the urban areas, say the cultural component for an example, gets relegated to the back ground in the rural context where cultural homogeneity is the rule rather than an exception. The list of cross-sectoral items that are candidates for comparison in influence and effect is naturally quite long. For our immediate purposes, the environmental and resource considerations suffice to illuminate the discussion.

The land that interlinks the environment and the natural resources is such a strong and persuasive one that the ecologists have argued persistently for its preservation with the belief that the balance and stability of the ecological systems depends upon it. Lessening the diversity of both simple and complex natural ecosystems have a likelihood that they may lead to their destabilization, whose consequences might be undesirable from the human views-point.

The conservationist element embedded in the diversity-stability quality has its aim at limiting the destruction of the habitat that suplants the flora and fauna some of which are species in systems important to people. Uncontrolled forest clearing thus deprives plants and animals of a supportive environment essential for the balance in the ecosystem.

The farmers prime asset and the backbone of the rural substance is the soil, whose richness depends on the supplies of nutrients from decaying organic material. For a continued nutrient cycle, the fundamentals of soil ecology can therefore not be ignored. Bearing this in mind then, the crucial question becomes how can the ecological and the economic considerations of rural housing be reconciled if the short-term urgency for shelter poses an environmental peril in the long-term.

The two main natural resources that form the base of rural housing construction are the soil and wood. By generating a discreet awareness on the part of the rural potential builder of the mutual dependence between the forest trees and the soil, consequently his own survival and prosperity, a point of equilibrum can be reached between ecological interests and subjective economic means.

As the farmer undertakes to procure resources to construct a dwelling, he should be guided by an awareness that trees have stored energy and nutrients good for the soil and are protection from wind and torrential rain.

If cleared, there is a danger of run-off, denudation, formation of gullies, alteration of climate and inducement of desert conditions, all together not conducive to a healthy crop yield. Thus, the watch word is moderation.

Construction materials for rural house building requirements can be composed of locally available materials which in addition to the soil and wood for mud blocks and framework include crushed stone which can be fired or unfired, rocks used for foundations, raffia and dried grass for roofing materials need to be readily available in the local itself but can be obtained from the nearest major trading and commercial centres access to which is facilitated through a transportation infrastructure even at the most rudimentary level.

Building one's own home is a natural human urge and rural communities still cling to this age old custom where family groups pool their labour resources together to set up settlement agglomerations, unit by unit, literally in the manner of the admonishing the "you scratch my back, I scratch my back".

However, no matter how vigorous and forward looking communal team work and participation to can seek to be, some complex building requirements may and often do require task specialization and division of labour where handling of certain materials is entrusted with specific workers. This sort of expertise can be solicited through a contractor. Assuming that the scale of the building projects and economic resources of rural dwellers are of a limited nature, they have recourse only to the services of a small contractor.

The modern rural builder is no longer satisfied with stereotype rural home wood frame, mud walls and thatched roof, hastily fabricated' with little incorporation of skilled planning. Nowadays he is willing to undertake home building under conditions that will permit that for the economic resources available him, social and aesthetic need together with durability are simultaneously met in the construction.

If such an aim has to be furthered, the rural builder needs both the effort of himself and unskilled community helpers on one hand and small contractors with the skilled labour force of foremen. Carpenters, rooffitters, painters, masons and plumbers on the other.

The small contractor can assist the home owner in minimizing the costs of materials and job site expenses in a manner that is consistent with target accomplishment. In the final analysis, the foregoing would be served by an integrated scheme within the rural physical planning programme which can promote a well directed and co-ordinated developmental effort in that sector.

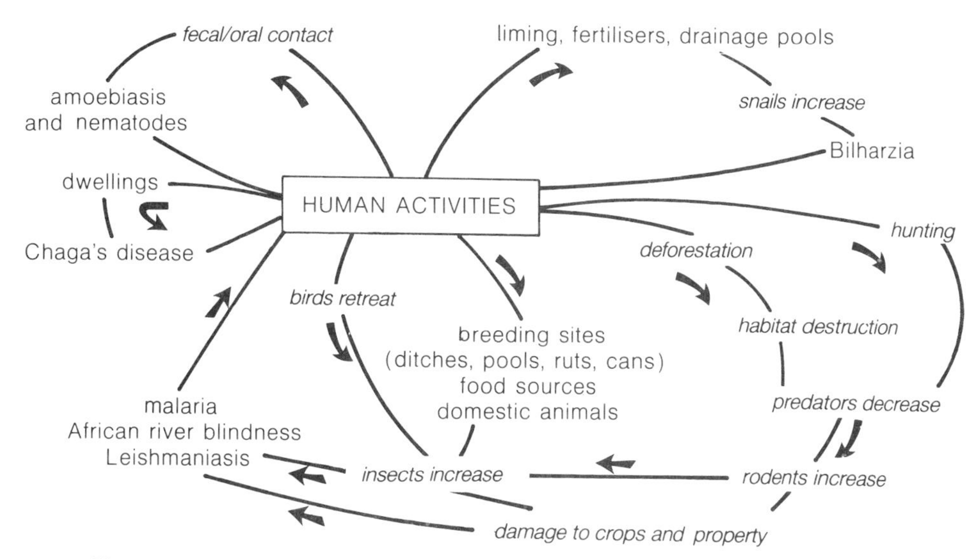

Figure 31.8 Disease and health risks increased by man's activities in the Tropics.

use that can be allowed without severe damage – in effect, the 'tourist-carrying capacity'. This would be the maximum number of visitors that a park could carry for a specific level of development and environmental quality. In Britain, no charge is made for entry to National Parks, nor control exercised on numbers of visitors. Other countries, and many of Britain's private landowners, do charge an entry fee and use part of the resources so obtained to increase the carrying capacity of the land, a valuable socio-economic trade-off.

31.5 Disturbed ecosystems and the risk to health

Many forms of human colonisation are sedentary, tending to concentrate disruption. Man-managed habitats destroy those of animals, causing natural predators to retreat and many rodents to increase unhampered (Fig. 31.8). Rodents not only damage crops and buildings but they also spread disease. The retreat of birds encourages an increase in insect numbers. Puddles and vehicle ruts provide breeding grounds for insects, while household refuse attracts both rodents and insects. This explains a common observation of travellers to the tropics – that undisturbed forest is relatively insect-free, whereas human habitations are commonly infested.

The unforeseen result of the long-term use of biocides has been to concentrate poisons in birds, mammals and fish. This, in turn, allows insects to multiply. At the same time both insects and rodents have become resistant to many biocides, thus strengthening further their dominance within an ecosystem that is already unbalanced.

World vegetation

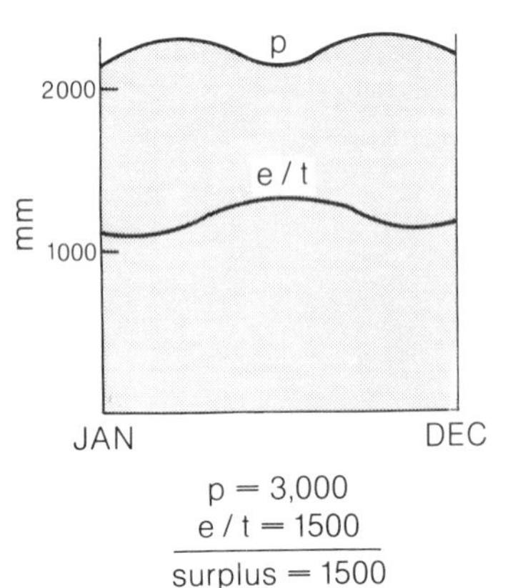

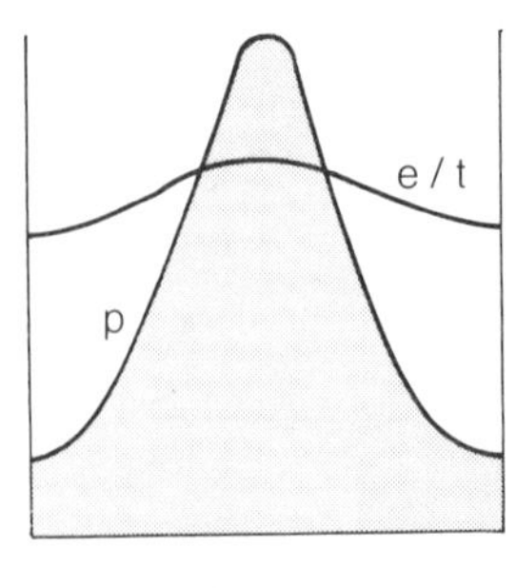

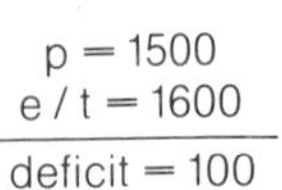

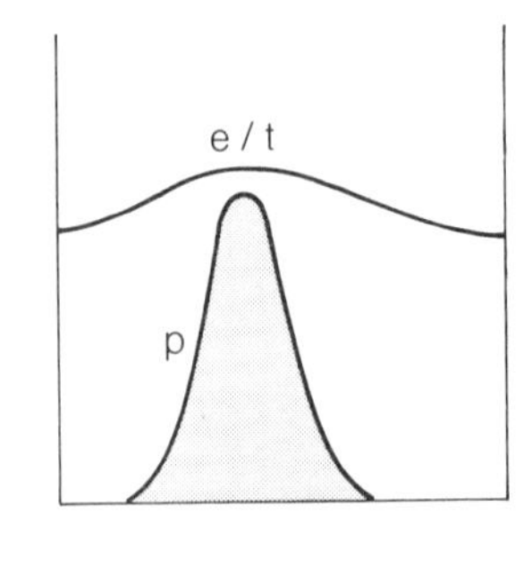

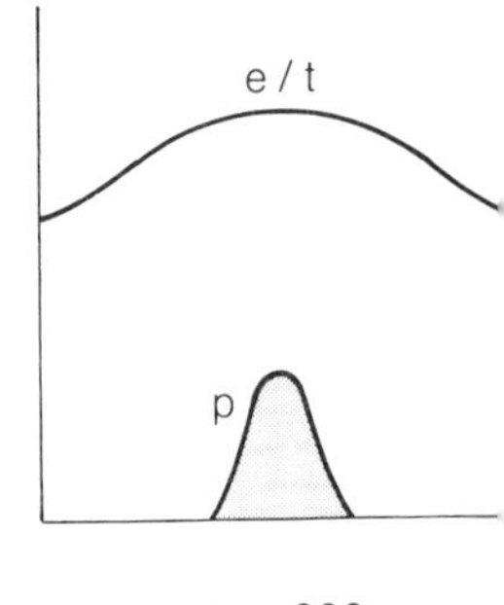

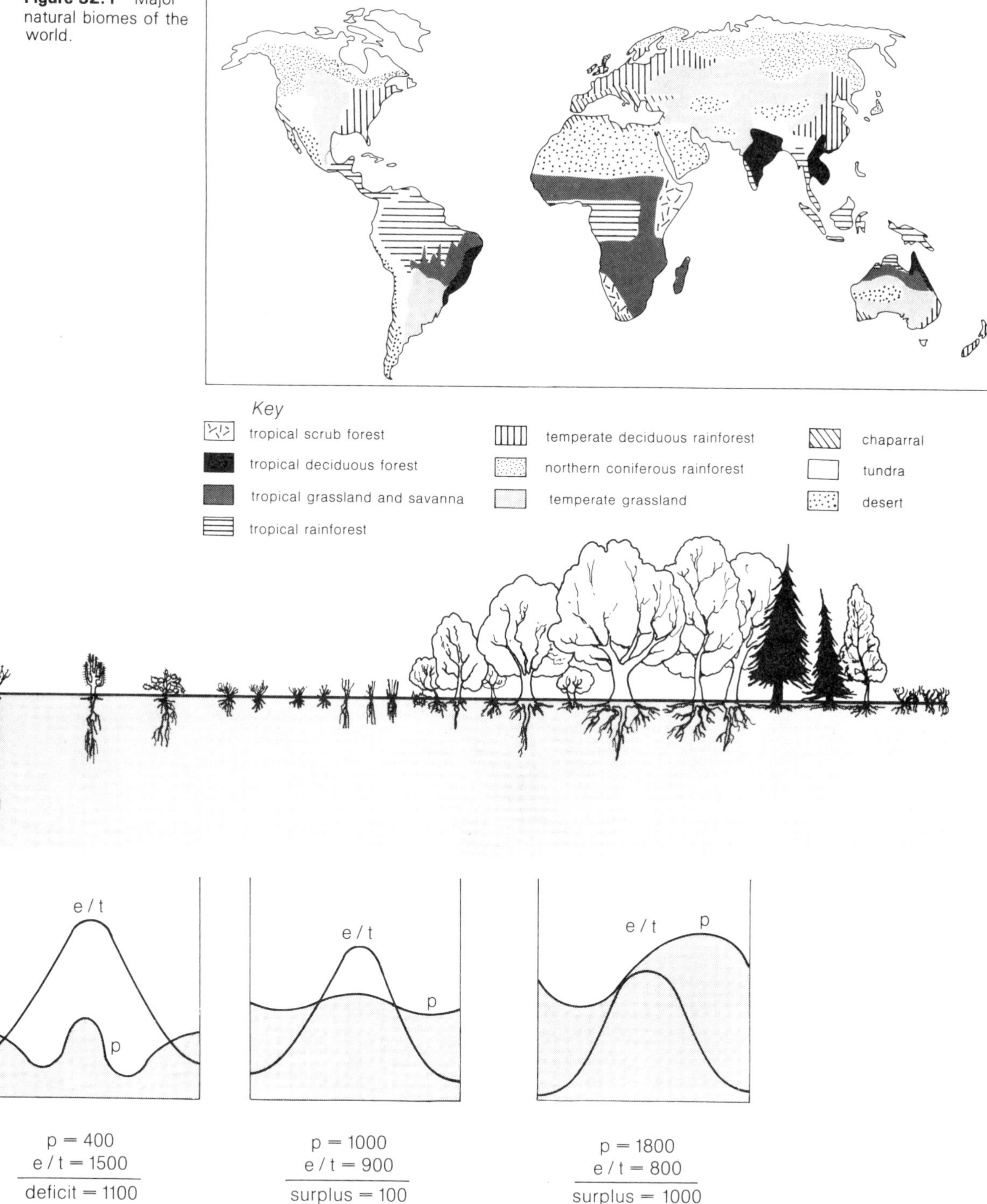

Figure 32.1 Major natural biomes of the world.

Figure 32.2 The changing pattern of biomes for the Northern Hemisphere. Note the extreme rooting depth or other adaptation to extreme moisture stress in deserts. Towards the poles moisture surplus returns, this time due to a reduction in evapotranspiration rates.

32.1 Biomes

Much of biogeography is concerned with the study of the patterns of flora and fauna that cover the world's surface. The most general unit of this study is the **biome**, the largest ecosystem that retains an overall homogeneity. Each biome is primarily recognised by the characteristics of the mature vegetation, because vegetation provides the habitat for animals. Thus each biome describes a specific assemblage of natural vegetation, such as a coniferous forest or a savanna grassland. Climate exercises the fundamental control over biomes and a combination of climate and natural vegetation strongly influences soil formation. Because of this relationship, broad patterns of climate, natural vegetation and soils are generally coincident (Fig. 32. 1).

The two principal climatic factors influencing the character of biomes are moisture supply and temperature (Fig. 32.2), although the widespread occurrence of each biome is a reflection of the ranges of tolerance exhibited by plant communities. Nevertheless, each biome has its own particular limiting characteristics (Fig. 32.3): the evergreen rainforest biome, for example, is present in regions with a wide range of moisture conditions, but does not occur where low temperatures are found; the grasslands biome tolerates a wide variety of temperatures but not wide variations in moisture supply. Taking the world biomes together, temperature is a markedly more significant controlling factor than moisture, and this refects the relationship of temperature to the energy available to the ecosystem. Thus most biome patterns broadly parallel changes in latitude.

What is now natural vegetation has not been 'natural' over most of the world for thousands of years. Man has cleared vast tracts by fire and plough; small vestiges of many natural biomes survive. It is important to identify biome patterns and to explain the natural ecosystems involved, for these represent natural dynamic equilibria and provide a set of reference points against which we can judge man's agricultural efforts. A knowledge of biome equilibrium will also help the farmer to improve his yields in a way that will cause as little long-term harm to the environment as possible.

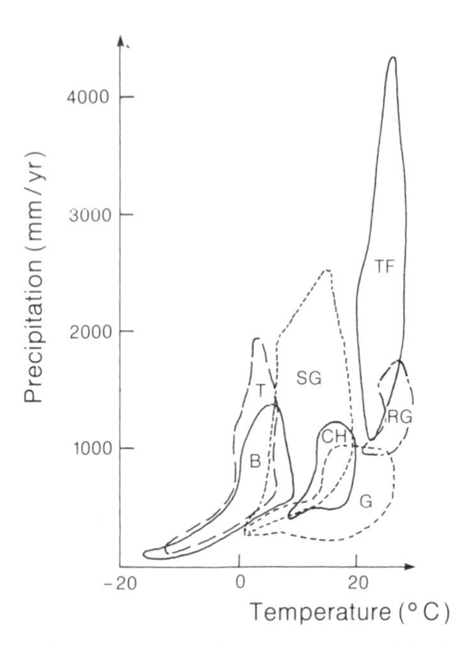

Figure 32.3 Relationships of temperature, precipitation and biome type. Continuous lines enclose evergreen types: B, boreal forest; CH, chaparral; TF, tropical rain forest. Pecked lines enclose seasonally green vegetation types: T, tundra; SG, deciduous forests; RG, temperate rainforest; G, grassland; D, desert and semi-desert.

32.2 Biomes of the Tropics and sub-Tropics

The regions of the tropical and subtropical world display a wide variety of climates and vegetation patterns, but they are all linked by limits imposed by seasonal variations in rainfall rather than by temperature. Divisions between the biomes reflect the lengths of the dry season in each: as evapotranspiration exceeds rainfall, various adaptations are necessary or plants will wilt and die. In the **tropical rainforest** biome there is a moisture surplus in every month, and a bewildering variety of species. The presence of even a short dry season quickly reduces the species variety: **tropical seasonal forests** contain only trees adapted to the more stringent conditions. Modifications to withstand drought become increasingly apparent as one looks at areas with increasingly long dry seasons. There is first a transition from woodland to various forms of **savanna** (mixed woodland and grass), then to **thorn scrub** and finally to the extremely specialised vegetation forms of the **deserts**. Each of these biomes is discussed below.

Rainforest

The main features of a rainforest include a monotonously green of the foliage; a scarcity of large, brightly coloured flowers; the lianas and other natural 'ropes' that hang in graceful curves between the tall, slender trees with their buttressed bases; an apparent scarcity of animal life except for occasional birds and omnipresent ants and other insects. Is this what makes a 'jungle' ?

(a)

(b)

Figure 32.4 (a) The tropical rainforest; (b) A typical buttress root.

Mature rainforests are relatively clear and free from shrubs because the canopy of tall trees shades the ground and prohibits most undergrowth (Fig. 32.4). The ground is also remarkably free from rotting vegetation, there being no equivalent of the deep tree leaf litter that is so characteristic of temperate woodlands: with high temperatures and moisture, and the help of the myriad insects and fungi, decomposition is very rapid, conversion from leaf to humus often taking no more than a few weeks. By contrast, a **jungle** is rarely a natural form of vegetation. Rather, the dense, interwoven thicket of impenetrable shrubs and saplings belong only to a phase of growth after clearance by man or to river margins where light penetrates the ground.

The tropical rainforest is, as with all biomes, a self-supporting ecosystem: it contains an association of producing, consuming and decomposing organisms all of which ultimately derive their energy from sunlight. The producers are the woody plants, from the giant trees which provide the leaf canopy over 45 m above the ground to the tiny trees that are no more than shrubs in height, and the lianas that hang between them. The consumers are the animals – the birds, small mammals, monkeys, elephants and the like that inhabit every level of the forest. The decomposers are the fungi, safari ants, millipedes and mites that break down dead organic matter into the minerals, nitrogen and carbon it contains, thus enabling plants to use these in building new cells (Fig. 32.5).

The characteristic of the tropical rainforest most apparent to anyone who walks through it is the amazing variety of species present. For example, a 2 ha sample of lowland rainforest on Malaysia has yielded nearly 200 species of trees 30 cm or more in diameter;

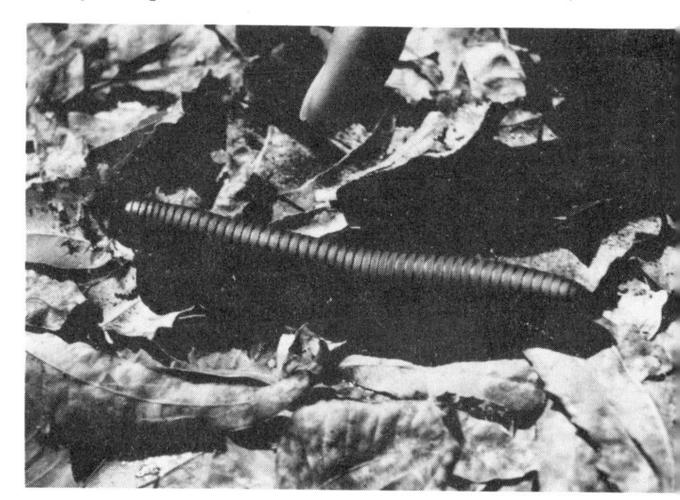

Figure 32.5 This millipede, over 15 cm long, is a major decomposer in tropical forests: it is black with red legs.

in a similarly-sized piece of temperate woodland there would rarely be more than half a dozen. The undergrowth, too, is rich in species of ferns, lianas, orchids and bromeliads – the red-centred, broad, spiked-leaved plants that grace many a home window-ledge – that grow **epiphytically** (non-parasitically) on the trunks and branches of trees.

The tree canopy contains trees of many differing heights and there is no uniform tree-top level (Fig. 32.6). Nevertheless, there are major contrasts between the top of the **canopy**, which is exposed to direct sunlight and air movement; the **understorey** which is in partial light, has a high humidity and still air; and the **forest floor** – again with still air and high humidity, but with no direct light and a lower and uniquely constant temperature. It is this variety of environments that provides the different foods and living conditions that encourage the diversity of animals that inhabit each layer. Above the canopy multitudes of insects provide food for a wide variety of birds; in the canopy tree fruits provide a staple diet for monkeys, sloths and other creatures whose limbs are especially adapted to climbing, swinging and jumping from tree to tree; and on the forest floor the hogs, elephants and rhinoceros depend on the rain of fruit and other plant materials that continually drop from the canopy above.

Over 500 resident-species of birds have been identified in just 800 km^2 of lowland rainforest. The same species are not found evenly scattered over the world's forests; rather there is considerable variety from area to area. Rainforests may look alike but often they have very few species in common.

Although there is a great variety of species in a rainforest, each species (except termites and ants) has a very low population *density*: this is why so few are seen. Furthermore, many are nocturnal. The large number of species may be due in part to the great age of the rainforest – it has probably remained much as it is today for tens of millions of years. Here, in a constant environment without drought or cold, plant and animal reproduction can occur at any time of the year and the evolution of species proceeds unchecked. On the other hand, the low population density may be due to the incessant pressure from the plant-eating animals, many of which feed exclusively on a single plant species. Seed scattered from such plants near to the parents may produce seedlings that are readily consumed by the herbivores, whereas those (fewer) seeds that are scattered over larger distances, perhaps by birds in their droppings, are more liable to survive predation.

Many plants have developed specific defence mechanisms to ward off their predators. The success that they have achieved, mainly with various exuded poisons, may well account for the small mass of animal life in the forest compared with the mass of

Figure 32.6 A rainforest has several strata, each creating a distinctive environment. Notice that many trees take on an 'umbrella' form with no branches near the ground where leaves would not be able to photosynthesise effectively.

vegetation. Despite the profusion of vegetable matter, there is a shortage of edible plants in a rainforest. And just as there is little food for animals, so primitive hunter-gatherer peoples of the forest find that the restriction of their food supply means that of necessity they, too, must maintain a low population density if they are to survive.

THE RAINFOREST NUTRIENT CYCLE

In the agricultural advisory offices in coastal Brazil, prospective farmers are still often told of the riches in soil that lies astride the newly developed Trans-Amazonian Highway which cuts through the heart of the rainforest. They are told of the rich humus that has accumulated in the soil, undisturbed over the centuries and just waiting for farmers to use. Yet the reality of the rainforest is exactly the opposite. Essentially, the soil in which the luxuriant vegetation anchors its roots is extremely acid and infertile: the rainforest exists only by maintaining an almost closed nutrient cycle.

The overwhelming proportion of the nutrients needed to maintain the rainforest ecosystem are held in the vegetation itself: the rest mostly comes as

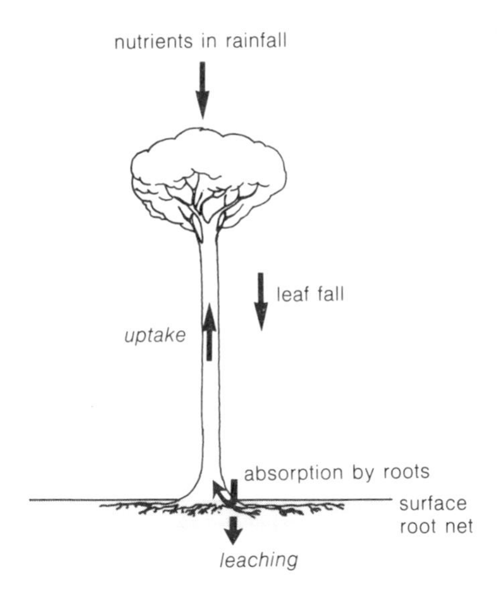

Figure 32.7 The rainforest nutrient cycle. Note that when the rainforest is replaced by cultivated plants this system breaks down, partly because a large part of the biomass is removed as food and partly because most food plants do not have sufficiently efficient means of rapidly absorbing nutrients.

'liquid fertiliser' with the daily rainfall (Fig. 32.7). The rain has also, for millions of years, chemically weathered and leached all minerals from the bedrock to leave only either iron-impregnated clay (plinthite or laterite, p. 255) or white quartz behind. Over most of the rainforest, the active weathering front, and therefore the place where nutrients are released from minerals, is tens of metres below the surface – tree roots rarely go deeper than a metre or so.

All rainforest trees have roots that spread laterally just below the soil surface: a network of large roots and a multitude of fine rootlets that literally form a net to catch the nutrients released by the decomposers before these are leached away by the percolating rainfall. In the acid conditions of humid tropical soils, bacteria are not able to survive in large numbers and the main decomposers are the fungi, whose hyphae branch everywhere, carrying the nutrients directly from dead tissue to the plant roots so effectively that almost nothing escapes.

The nutrients leached from the ecosystem are replaced by minerals contained in rainfall: in the central regions of the Amazon a year's rainfall may contain 0.3 kg of phosphorus, 2 kg of iron, 10 kg of nitrogen (as ammonia and nitrates) and 3.6 kg of calcium per hectare. Leaf fall yields just over 18 kg of calcium per hectare. Nutrient recycling is so effective that the concentration of calcium in the nearby streams remains too small to be detected.

It must be stressed that only the *natural* vegetation is adapted for effective recycling. Cereal crops, for example, have a root system which is incapable of trapping nutrients. This difference underlies the apparent contradiction between luxuriant forest and farming failure.

RAINFOREST SUCCESSION

The natural rainforest is in a dynamic equilibrium in which every ecological niche is used; no outside invader is able to survive. There are no invasions by 'weeds', and no epidemics of diseases. Yet sometimes the natural stability of the forest is locally disrupted. Suppose, for example, that a large tree topples over and brings down others with it. Sunlight immediately floods over the forest floor and soon seeds, already brought in by birds but left dormant under the previously unfavourable shade of the forest, begin to sprout and flourish. Within a matter of weeks the forest floor is a mass of grasses and herbs all fighting for light and space and rising in continuous cover to about 2–3 m. However, they do not survive long: soon tree saplings rise above them, providing within two or three years a higher-level cover sufficiently continuous to shade the ground and cause the initial colonisers to die back. These new trees form a secondary forest – a real 'jungle' of fast-growing softwoods. But they, too, have a limited life and die within about 30 years, leaving the taller but slower-growing hardwoods to rise up and begin to close a new canopy. The repeated succession of trees, each type yielding ground to another until a final and stable *climax* vegetation assemblage has reformed, may take centuries. It is not easy to replace.

Seasonal forests

Not all parts of the humid tropics receive an equable distribution of rainfall through the year; many areas, such as India, southeastern Asia and western Africa, receive much of their rainfall in short monsoon seasons between which monthly totals may fall below 50 mm. Plants adapt to the seasonal availability of rainfall, either by shedding their leaves in the dry season (developing a deciduous habit) or by developing hard, glossy evergreen leaves that lose water slowly. With a lower moisture budget, trees in tropical seasonal forests grow less vigorously and may reach only half the height of those in the equatorial rainforests (Fig. 32.8). Stress also reduces the species variety; the many-layered rainforest canopy is replaced by one containing only a single understorey, and large areas become dominated by single species (such as teak in south-east Asia).

The tropical seasonal forest is a form of transition between rainforest and temperate forest. In the seasonal forests, soils are less heavily leached than in

Figure 32.8 A tropical seasonal forest has many fewer species than a rainforest: many trees are deciduous.

the rainforest and more humus remains in the upper horizons. This allows trees to be more dependent on the soil for nutrient supplies, just as they are in the temperate forests. In the dry season leaves accumulate on the forest floor and become liable to catch fire. When fire plays a major role then the forests often become more open and grade into the savanna biome.

Savanna

The **savanna** has an extremely wide global distribution but is perhaps most extensive in Africa. Here it extends in a wide belt both north and south of the equatorial rainforests and also occupies the high plateau regions of East Africa. Low seasonal rainfall and high temperatures contrive to make water supply the main limitation on plant growth. Tussock grasses form the major element of the natural vegetation, often growing 3.5 m and even, in the case of the elephant grass (*Hyparrhenia*) to 5 m. Yet despite their vigour in attaining such heights during the wet season and despite their deep rooting systems (Fig. 32.9), grasses die back in the dry season. Nearly all savanna grasses grow, ripen and set seed in the short period from the onset of the rains to the early part of the dry season.

Although moisture is a limiting feature, unbroken grasslands are not a characteristic of the savanna: trees dot the landscape, their frequency being greatest on the moist margins that grade into tropical seasonal forest. On the margins of the deserts, savanna trees decline and become no more than thorn scrub (Fig. 32.10). Trees tap water from deeper within the soil than do the grasses and can therefore grow

Figure 32.9 Deep root systems of two perennial grasses in a dry savanna climate.

Figure 32.10 Desert margin acacia thorn scrub. Notice the lack of grass species or other ground cover. These acacias have especially vicious thorns to help ward off herbivores.

longer into the dry season, although the common *Acacia* sheds its leaves soon after the rains have passed. Others, like the strange baobab tree (Fig 32.11), conserve moisture within their thick pithy bark. In the driest areas only the Mopane tree survives. Savanna trees are not tall – a mere 6 m or so high – often with thorny branches (Figs. 32.12 & 13) and nearly always with flattened crowns.

Rainfall exercises the main control over savannas, yet within any climatic zone, soils too can be very important. For example, on sandy soils with a rainfall between 75 and 250 mm/year, thin, wiry grasses such as *Aristida* spp. and *Cenchrus* spp. (Fig. 32.9) are the dominant grasses, joined by the sand-binding *Panicum* on the least stable soils. By contrast, on clay soils *Cymbopogon* spp. replace *Cenchrus* as a partner to *Aristida*. As rainfall increases to 600 mm all the wiry

Figure 32.12 Acacia 'parkland' is a typical form of savanna in better-watered regions. This photograph was taken mid-way through the dry season when many of the grasses had been eaten by cattle and wildlife.

Figure 32.11 Baobab, the 'upside-down tree', is so named because its branches look like roots. It stores moisture in its pithy bark.

Figure 32.13 The defensive thorns of the acacia tree.

species disappear, broad-leaved grass species become more important, and the frequency of *Acacia* increases.

Of all the world's regions, the savanna is the one where insects appear most obviously dominant. The savanna ecosystem is particularly influenced by two types of insects, the **termite** and the **locust**. In the drier savannas the normal range of larger soil organisms, such as earthworms, is absent. Their role of churning the soil and redistributing any dead organic matter that falls on the surface is taken over by the harvester termite. Termites are gregarious insects, constructing large nests, some just below the surface, others as great castles of soil whose turrets rise several metres above ground level (Fig. 32.14). The constant nest building causes soil material to be reworked and organic and mineral matter to be mixed together. At their maximum density termite colonies can number over a hundred per hectare, covering the ground with mounds only a few metres apart and needing so much vegetable matter that they begin to eat large quantities of living grasses as well as dead materials.

The effect of termites on the ecosystem is purely as local decomposers, whereas the locust, a migratory insect, can have a regionally devastating result. The locust can occur in large numbers over most of Africa north of the Equator, but its effect is so drastic because it can eat its own weight of food a day and a large swarm can weigh 50 million kilograms and travel up to 3000 km in a season. Locusts occur as solitary insects in most savanna ecosystems; only when they breed to a level at which overcrowding occurs do they gather into a migratory swarm – migration is simply a response to local ecosystem imbalance.

Figure 32.14 This giant termitarium has been built by hundreds of thousands of termites, each no more than a few millimetres in length.

THE IMPORTANCE TO SAVANNA OF FIRE

Fire is one of the most important natural phenomena in the tinder-dry savanna (Fig. 32.15). Often started by lightning during a period of thunderstorm activity, fires may sweep over hundreds of square kilometres of savanna before burning themselves out. Under purely natural conditions, any area of the savanna is set on fire every three to five years on average. Because fire is so common, the natural vegetation is closely adapted to survive repeated burning – either it has fire-resistant bark, as in the case of trees and shrubs, or it can reshoot from the base of a charred trunk (Fig. 32.16). Perennial herbs normally propagate by rhyzomes that remain protected below the soil surface. The seed-cases of annuals are often especially resistant to fire and allow the seed to survive and germinate when the next rains come.

Fire has both advantages and disadvantages for a savanna ecosystem. Burning releases nutrients such as phosphorus that are bound up in the woody vegetation, thus making them available to help further plant growth. However, some leaf litter that would otherwise have been converted to humus is lost; and the burned vegetation is less able to protect the soil from rainsplash, causing clay-textured soils to become crusted over, promoting overland flow and thereby reducing even further the amount of water that reaches the soil. Nevertheless, in areas where fires are frequent, the soil-plant system comes into a balance quite quickly after each fire. It is in areas where fires are less frequent and where substantial leaf litter has accumulated that fires can be particularly destructive, burning with great ferocity and destroying not just the vegetation tops but many of the roots as well. Grasses are especially severely hit by these really hot fires.

Strange as it may at first seem, plants become tolerant to fire and later cannot survive properly without it. The grass *Thermeda*, for example, only thrives and grows vigorous new shoots if it is burned back every 3-5 years; without such fires the dead shoots and stems from earlier years begin to choke the ground and prevent the new leaves from developing fully. Fire is also beneficial to the survival of species like *Thermeda* in that it restricts colonisation by species from other floristic regions. In particular,

Figure 32.15 Fire is not destructive in the savanna. Notice that the tree on the left of the fire has remained unscathed and looks little different from the unscorched tree on the right.

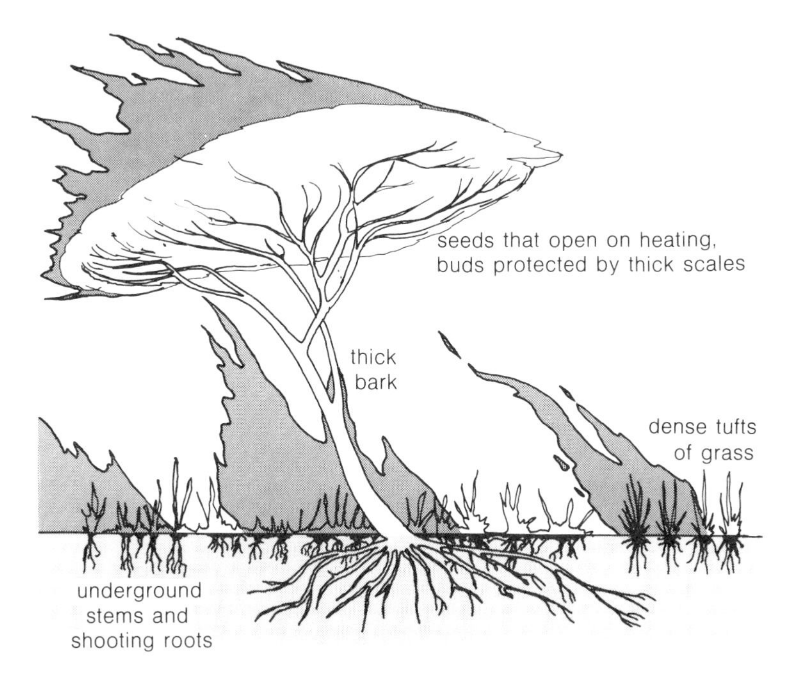

Figure 32.16 Fire-resistant plant features.

regular burning has probably been responsible for keeping down the number of woody shrubs and trees and thereby maintaining the openness of the savanna.

32.3 Biomes of the temperate mid-latitudes

In contrast to the tropical and subtropical lands, the mid-latitude biomes are primarily controlled by seasonal changes in temperatures (Fig. 32.17). Many of the mid-latitudes experience summer temperatures at least as high as those of near tropical lands, but in winter temperatures often plummet to below freezing. Vegetation is therefore primarily adapted to withstand a long period of cold conditions, although in continental and rainshadow locations a combination of both low temperature and drought resistance is needed.

In areas where there is a moisture surplus for much of the year **forests** are widespread: deciduous in the Northern Hemisphere; evergreen in the Southern Hemisphere. Where moisture deficits exist for periods in excess of four months , forests begin to give away to the great continental **grasslands** of the North American **prairies**, South American **pampas** and Russian **steppe**.

Grassland

Temperate grasslands cover large areas of the semi-arid mid-latitudes. Summers may be as hot as in the tropics, but winters are cold, frost occurs for upwards of 100 days a year, and winter precipitation falls as snow. Precipitation is between 250 and 750 mm/year, apparently making many of these areas too dry for tree growth. A gradual transition to forest occurs only near the wetter margins. At present there are virtually no trees over most of the grasslands and in this they are quite distinct from the savannas. However, as with the savanna, grasses have at least as much of their biomass below the soil surface (in the form of a deeply matted extensive root system) as they do above it. Characteristic grasses in North America, for example, are blue stem, needle grass, and the tussock grasses including buffalo grass and blue gamma (Fig. 32.18).

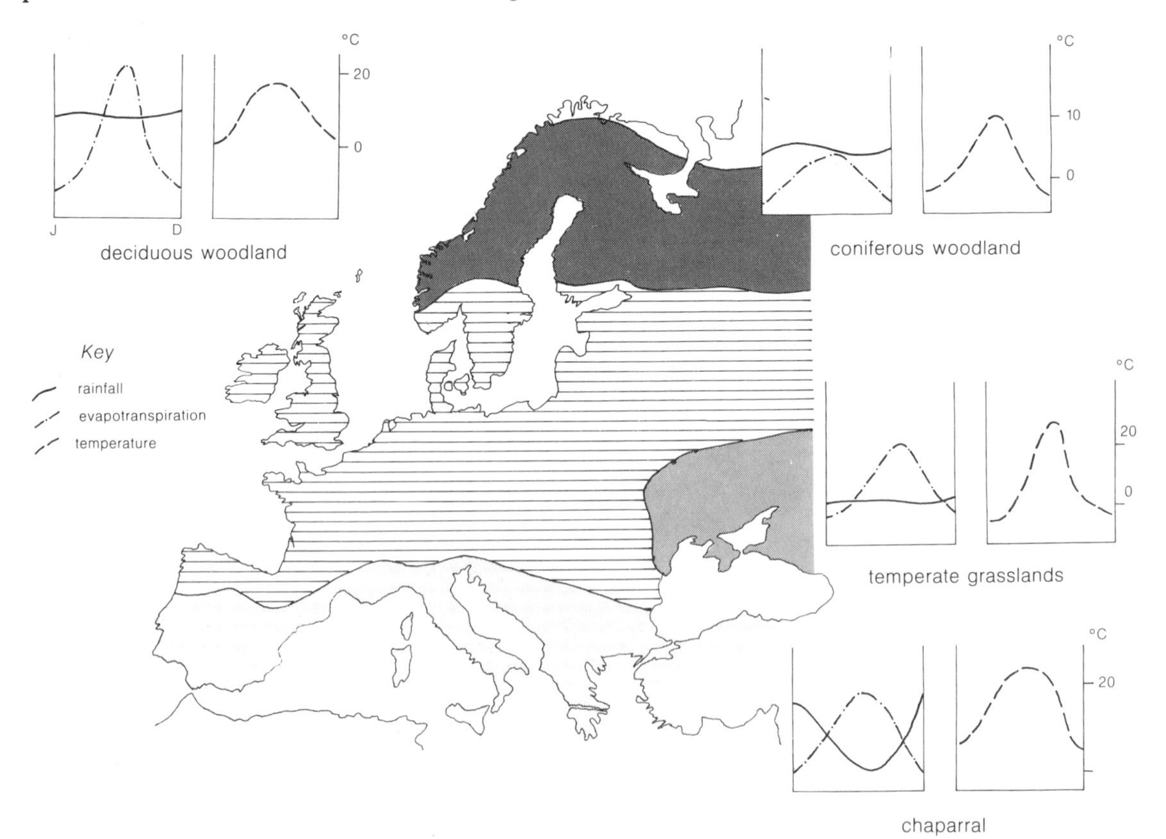

Figure 32.17 The environmental setting for mid-latitude/high latitude biomes.

Figure 32.18 Undisturbed prairie with a rich variety of grass species, herbivores such as the bison, and rodents such as this prairie dog.

Because the soils are frozen near the surface in winter and dust-dry in summer, grasses either need to ripen and set seed quickly during the spring when winter snows melt (as does blue stem) or they need deep roots and a resistance to wilting which enable them to carry on growing slowly into the autumn (as does buffalo grass).

Grasses of different species grow throughout the summer, so herbivores always have sufficient food, providing they forage widely. The great herds of buffalo, so famous a part of US history, migrated seasonally across the grasslands of the Great Plains and Prairies in a pattern that helped conserve the forage. The grassland ecosystem also contains many small burrowing rodents such as the gophers and prairie dogs, and these act as a sourcre of food for wolves, coyotes and snakes.

Through extensive foraging, the temperate grassland ecosystem, probably more than any other in the world, has been maintained, if not formed, by its natural fauna. Temperate grasslands may be a degenerate form of open forest, like the savanna. If this is the case then fires must have burned the trees away and saplings prevented from regenerating by the extensive grazing of animals.

Woodland and scrub

Warm temperate woodlands belong to the margins of the Mediterranean Sea, Chile and California, south-east Australia and southern Africa, where rainfall typically occurs during a mild winter season and summers are hot and dry. Plants do not have to adapt to withstand frost, but they must be able to withstand a long drought. Eucalyptus trees are the dominant species in Australia, their tall, graceful branches hung with small grey-green leaves adapted to lose little water by transpiration. So successful are the eucalyptus at adapting to this environment that they have become dominant forms in the vegetation of many other areas, spreading quite naturally after introduction. They have become an important element of the flora in the Mediterranean, in California and even in the East African Highlands, where they are prized as a fast-growing source of firewood.

Despite the incursion of eucalyptus into areas outside Australia, elsewhere the vegetation is still mostly dominated by various species of evergreen conifer such as the redwood and juniper. However, in many areas the vegetation rises to no more than scrub height and in these places small trees grow, each with

a canopy height of between 3 and 4 m. These are also normally evergreens, including the leathery grey-green-leafed olive or the deep-green-leafed cork oak. Many forms of aromatic pine are also common. This scrub vegetation is called **maquis** and **guarrige** along the Mediterranean coast (Fig. 32.19), and **chaparral** in California.

With a long, dry summer, fire acts as an important influence on vegetation forms, and the vegetation has many protective adaptations. Fire also helps to decompose the tough leaves and bark of dead plants, releasing minerals such as phosphorus for further use by the growing vegetation.

Figure 32.19 Woody, small trees and shrubs form the characteristic vegetation of this region in southern France.

Figure 32.20 Contrasts between summer (left) and winter (right) in a British beechwood. Note the relative absence of ground cover.

Forests

In the humid lands of the mid-latitudes, lack of moisture is rarely a limit to growth. The main influence on the biome is seasonal variation in sunlight. In contrast to the seasonal tropical forests, where the deciduous habit is related to the dry season, in the Northern Hemisphere leaves are shed at the end of the (warm) summer and grow again at the end of the winter (Fig. 32.20). In this respect there is a dramatic contrast between Northern and Southern Hemispheres, for in the Southern Hemisphere the equivalent forests are all of evergreens.

The major differences between the forms of trees in the tropics and of those in the mid-latitudes are related to their energy and moisture needs. In the tropics deciduous trees have narrow, leathery leaves and produce soft, pulpy fruits; in temperate regions the need to gather all available energy has resulted in deciduous trees having broad, thin leaves, while the seed is characteristically a hard nut. Differences in seed character reflect a totally different means of propagation: in the tropics trees rely on animals to carry their seed, and the seeds have to be *attractive*; in temperate lands only some trees (such as beech and oak) have their seeds distributed by animals; other seeds are largely carried by the wind (as with winged sycamore) and must survive a harsh winter probably exposed on the ground – hence *resilience* is the keynote.

The number of tree species in deciduous forests is much smaller than in the tropics. There are so few dominant trees, they can readily be listed: oak, beech, elm, lime, chestnut, maple, hickory, alder, poplar, yew, birch, tulip tree, and hazel. But even these few species rarely occur together; rather they occur in groups (**stands**) of two or three, dominating a forest for large distances.

The deciduous forest belt is really a region in transition; to the north lie the cold lands of the boreal forest; to the south the seasonal tropics, both with well-defined characteristics.

The balance of trees in any area is influenced by local factors of soil, slope and drainage. In Britain, for example, willow and alder occur in especially wet sites; acid soils are dominated by sessile oak and beech, whereas on loams the lowland oak, elm and lime predominate.

With less energy to be shared by the flora there is a sparse understorey in a temperate forest. Birches and holly are typical understorey trees, but otherwise the ground level is confined to plants having one of two specific adaptations: the group that flowers in spring and that set seed before the tree conopy can close in over them (primrose, anemone and bluebell) and the group that can tolerate shade (ivy and arum lily).

32.4 Biomes in harsh environments

High-latitude forest

Although at first sight the monotonous stands of conifers of the boreal forest (**taiga**) zone may appear to have little in common with the luxuriant and diverse trees of the tropical rainforest, they both share a problem of survival in heavily leached soils. Conifers tend to be shallow-rooted, not only to ensure that they capture the nutrients released by weathering but often also to keep above the permafrost that underlies much of the zone. Extreme waterlogging is another environmental feature that accompanies fine-textured, poorly draining soils. Many soils are formed from glacial till which has a low potential fertility.

Under such adverse environmental conditions the vegetation of the high-latitude forests still manages a net primary productivity similar to that of the deciduous forests despite a short summer season. Unlike deciduous trees, conifers do not have to spend part of the summer developing new leaves.

Figure 32.21 Coniferous trees are naturally quite widely spaced in a boreal forest, and a dense ground cover is possible away from direct shade.

In their natural state, boreal forest trees are quite widely spaced and much light reaches the forest floor, allowing a dense ground cover of bilberry, heather, mosses and lichens to form (Fig. 32.21).

Within the high-latitude coniferous zone, trees have to adapt to cope not only with an adverse climate, but also with a variety of moisture regimes. Spruce frequently dominates on the wetter sites, whereas pine is more prevalent on drier slopes and fluvio-glacial gravels. The more nutrient-rich valley-bottom sites colonised by spruce enable them to achieve a higher net primary productivity than pines.

Conifers are adapted in several ways to low temperature, low insolation and high snowfall. Their evergreen habit enables them to gain the maximum insolation from a sun that remains low-angled throughout the year. In addition needles resist transpiration while the tree shape and downward sloping branches foster the shedding of snow and prevent the branches from breaking.

Conifers face the same problem in the high latitudes as in the Mediterranean: a lack of available water in one season. In high latitudes the ground is seasonally frozen and in effect the trees have to survive in a seasonal desert. They have thick, waxy leaves with stomata hidden in rows on the underside of each needle.

Conifer nutrient cycles differ considerably from those of the deciduous forests. Needles are renewed only once in several years and they are both nutrient-poor and coated with waxes and resins which make their decomposition on the forest floor very slow. The rate of needle decay is further slowed by the cool conditions experienced in boreal regions, a regime that discourages soil fauna and reduces the rate of decay and incorporation. The low rate of decay and the poor release of cations prevents humus formation and so allows all cations to be leached away. In these conditions trees capture sufficient nutrients only by direct transfer from rotting vegetation to tree root via fungi (as in a rainforest).

Arid lands

True **deserts** are confined to those areas where evapotranspiration exceeds rainfall in every month, where there is no seasonal pattern of rainfall and where there are sequences of rain-free years. Yet despite the harshness of the environment some forms of vegetation manage to survive, helped by a wide range of adaptations. Indeed, because adaptation takes on many forms, desert flora are perhaps the most diverse in form of any biome (Figs. 32.22 & 23).

Adaptation follows several distinct lines:

(a) survival as *seeds*, followed by rapid germination, growth, flowering and setting seed – the life cycle of such annuals may be over within days, the seed dormant for several years;

(b) survival through very *deep roots* – the mesquite bush has a tap root that may sink 15 m through the soil and rock in search of water;

c) survival through *water retention* within the biomass due to hard, leathery leaves and a high-bulk stem – cacti are the most obvious example.

Figure 32.22 Plants such as the mesquite and creostose bush have roots that may be more than 10 m deep.

Figure 32.23 Cacti are the best known features of desert vegetation. Their soft, water-filled stems are protected by thorns and a leathery surface.

As with other vegetation forms that survive harsh conditions, much of the biomass (except in annuals) is retained below the parched surface and away from dessicating winds and sun. In some cases less than one per cent of the biomass may occur above the surface.

High-latitude and high-altitude vegetation

Beyond the limits tolerated by forests lie the great open lands of the arctic and sub-antarctic tundra and high-altitude mountains. In these harsh regions shelter is the keynote, vegetation adapting a mosaic pattern as it responds to changes in local climate and soils. More than anywhere else local habit is the dominant feature influencing species distribution (Figs. 32.24 & 25).

It is tempting to classify the low-growing vegetation of all cold, harsh regions together. However, the *diurnal* fluctuations in energy that occur at high altitudes in the tropics compared with the *seasonal* fluctuations in the arctic and antarctic make such comparisons unrealistic. It is only convergent evolution that makes the vegetation patterns of arctic and sub-antarctic flora look similar, for their species composition is quite different.

The biomass of the high-altitude **tundra** is about a tenth that of the boreal forests, and even this figure may at first seem remarkably high. However, with other 'desert' flora, plants of the arctic tundra ensure their survival by keeping up to 80 per cent of their biomass below the surface. Throughout the tundra, soils experience a water surplus; waterlogging is severe. As a consequence plant decay is extremely slow and partially decaying organic matter may make a higher proportion of the soil than mineral matter. By contrast, high-altitude sites have very low precipitation and are virtual deserts. Despite these differences, both regions have a unique pattern of nutrient recycling, being completely dominated by nitrogen. This is achieved because rodents (such as lemmings and rock hyrax) consume large quantities of the biomass and return it in a decomposed and highly nitrogenous form via their faeces. They are the chief means of recycling, taking on the role played by earthworms in mid-latitudes.

Figure 32.24 The typical tundra plant is very low growing, making use of the shelter of rocks.

Figure 32.25 High-altitude plant species are very diverse. Giant lobelia and groundsels dominate the highest vegetation of mountains in tropical Africa. This photograph was taken at over 4500 m.

Farming systems

33.1 The nature of farming

The nourishment of animals depends on organic substances – carbohydrates, fats, vitamins and proteins. Animals cannot themselves manufacture these substances directly from inorganic minerals; all nutriment must be obtained directly or indirectly from plants. Animals obtain energy by breaking down the energy-rich products of photosynthesis – plant products – into simpler, lower-energy molecules. For example, when glucose (a carbohydrate) is eaten, it is combined with oxygen, releasing energy, carbon dioxide and water. This is the process of **respiration**, the exact opposite of photosynthesis.

Much of the energy used by plants is employed in making cellulose. Most plant tissues are made from cellulose, which is therefore an enormous energy resource for animals. Unfortunately, humans cannot decompose cellulose so it has no direct nutritive value to people. They can only digest seeds, fruits or tubers, which contain mostly carbohydrate and protein. Otherwise people must exploit animals which *can* convert the cellulose into a form of energy usable by humans. Ruminating animals, such as cattle, have many bacteria living in their digestive tracts: some of these bacteria can decompose cellulose.

At each stage in a food chain, only about 10 per cent of the consumed energy is passed on. Direct consumption of plant material that *can* be digested by humans is at least ten times more efficient as a method of obtaining energy from plants than is obtaining it indirectly in animal meat. This is why bread, rice and the other cereal grains are a staple diet and a dominant feature of societies where high-protein food is scarce; animal meat is not an essential part of any human diet and is only found in those areas in which food is readily available or in which cereal crops cannot be grown. However, in the half of the world where cereals will not grow, people must turn to pastoral activities because eating meat is the only way of providing food.

The world pattern of agriculture

Cereal grains need a certain amount of energy to ripen – to complete the conversion of nutrients into carbohydrates and protein. This energy must come directly from the Sun. Regions in the high latitudes receive far less insolation than those nearer the Equator; the Sun's rays fall obliquely and provide less energy per unit area. Although areas in the Arctic receive continuous daylight through the high summer, cereal crops grown in such areas would not ripen.

Plant growth depends not only on insolation but on the surrounding environmental (**ambient**) air temperature. Usually an ambient temperature of at least 6 °C is needed if plant metabolism is to rise to a level at which growth can occur. Even areas such as tropical mountains that have high insolation are unable to ripen cereal crops if they also have low ambient temperatures. Plant growth is controlled also by water supply, for two reasons. First, plants receive most of their nutrients by sucking soil water through root hairs. The rate at which water is taken up by plants depends on transpiration from the leaves; this in turn is controlled by insolation, ambient temperature and relative humidity. Hot, dry, sunny conditions induce rapid transpiration and hence increased uptake by the roots of water and nutrients. A cool, humid environment causes little transpiration, making it hard for plants to bring in the nutrients needed for growth. This is another reason why plants grow faster in warm environments. If the soil does not contain enough moisture (due perhaps to a long dry season or to a sandy soil), the rate of transpiration cannot be balanced by a supply of water via the roots. At first this causes a reduced rate of growth, but in severe conditions it causes irreversible plant wilting.

Too little water can limit crops; too much water can have an equally undesirable result. Few plants can survive for long in totally waterlogged conditions, for root hairs are used to gather not only nutrients but also oxygen; they cannot extract oxygen directly from water. Certain crops, such as rice, have adaptations which enable them to survive waterlogging; many crops will not survive even damp conditions – moist air encourages the growth of fungi which can damage both leaves and seeds.

Choice in farming systems

Farming is a skilled process of soil husbandry. A farmer needs a very clear understanding of the environmental controls on his crops if he is to have any long-term chance of producing sufficient to provide him with either subsistence or an economic living.

Environmental factors play a major role even in developed countries. In the US, for example, corn (maize) will produce an economic yield only in those areas having a warm season above 10 °C for over 120 days together with ample moisture well spread over the growing season (see Figs. 24.1 & 27.11). These conditions are best met in central Iowa and Illinois – the heart of the 'corn belt'. In Minnesota corn becomes increasingly limited by low temperatures; in Nebraska low rainfall is the limiting factor. Southward, into the southern states, corn production is limited by infertile soils. In climatically similar areas of the world, corn is a major crop, in developed and in developing countries – Mexico, Argentina, Indonesia, Thailand, the Balkans and eastern Europe.

Wheat is more tolerant of droughty conditions than is maize; it is found widely throughout the semi-arid world – in Africa, in India and in southwestern USA, for example. However, wheat will also tolerate a cool spring and can therefore grow in western Europe, and from Oklahoma and Kansas into Canada.

By contrast, the low-lying lands of Denmark and the Netherlands which are mostly based on heavy clay soils, are often too cold and wet for cultivated crops. Consequently, wide areas are seeded to perennial, shallow-rooted ryegrass, whose growth extends from early spring to late autumn. Silage derived from the grass is then supplemented with grain grown on the more favourable land.

Various environmental factors, such as sunlight, temperature, effective precipitation, topography and soil, together determine whether it is feasible to introduce a certain crop into a given area (in just the same way as there is a range for each wild species). The actual choice of crops is also related to the economic and cultural environment of the society doing the farming, as is the manner and intensity of cultivation. Factors such as the density of the population, the distance to the market, the level of technology, and the society's cultural heritage all play as large a role as do natural forces (Fig. 33.1)

The great emphasis throughout the world on cereals (which provide over 50 per cent of the world's protein and energy needs) reflects not only their food value, but also their relative ease of culture, harvest and transport, and their wide range of climatic adaptation. The dominance of certain crops in a region therefore occurs not only because these are well adapted to the environment and give high yields, but

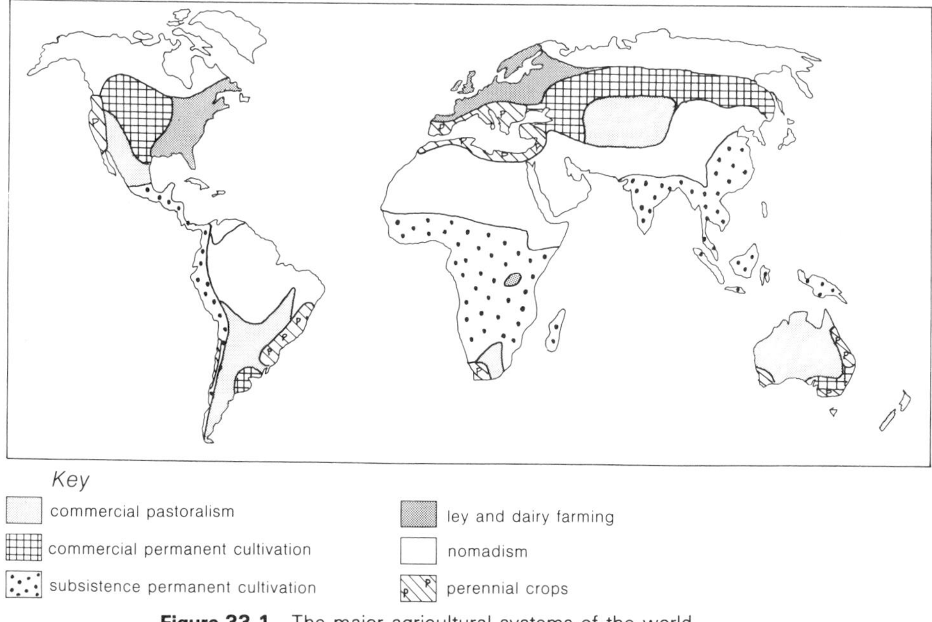

Figure 33.1 The major agricultural systems of the world.

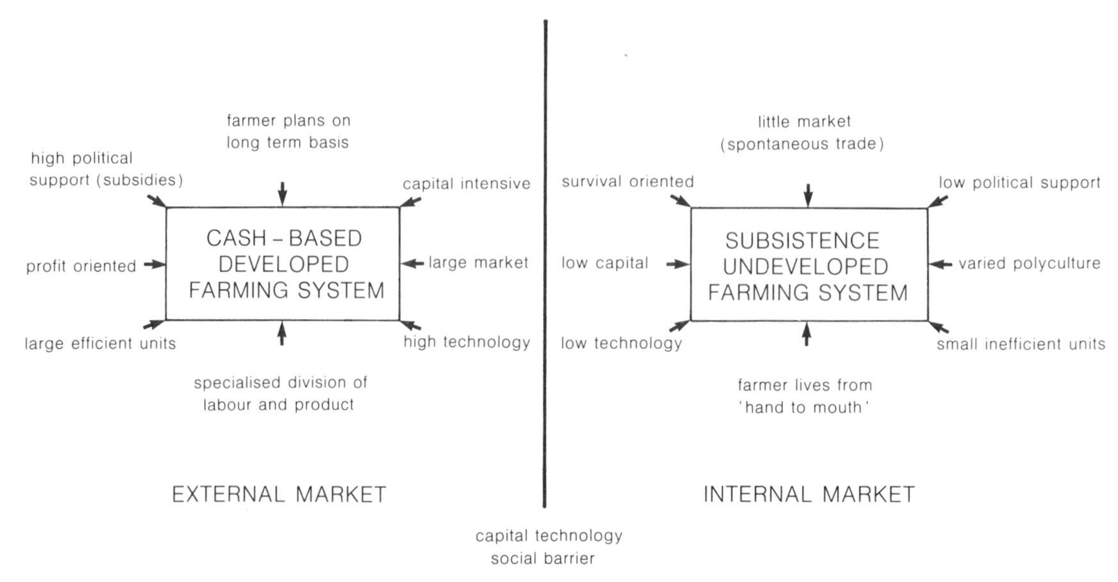

Figure 33.2 Developed and developing farming systems: the two extreme positions.

also because they fit well into a plan of farm management, have a low risk of failure and have a market (which allows these crops to provide an income for the farmer). This does not mean that other crops could not be grown there – for instance, potatoes and sugar beet could be grown successfully in the Corn Belt – but such other crops would, under present conditions of demand and revenue, yield a lower economic return. The assessment of risk and return has a great deal to do with agricultural systems.

In developed countries where machinery is intensively used and labour costs are high, farmers tend to specialise. Each can then develop considerable expertise in overcoming the vagaries of the weather, crop pests and market fluctuations. But these 'simplified' farms do not work well where subsistence is the rule – here diversity of food is desirable (Fig. 33.2). Farmers are nearly always extremely rational in organising their farming systems to meet their own objectives, but these objectives may often not be economic ones.

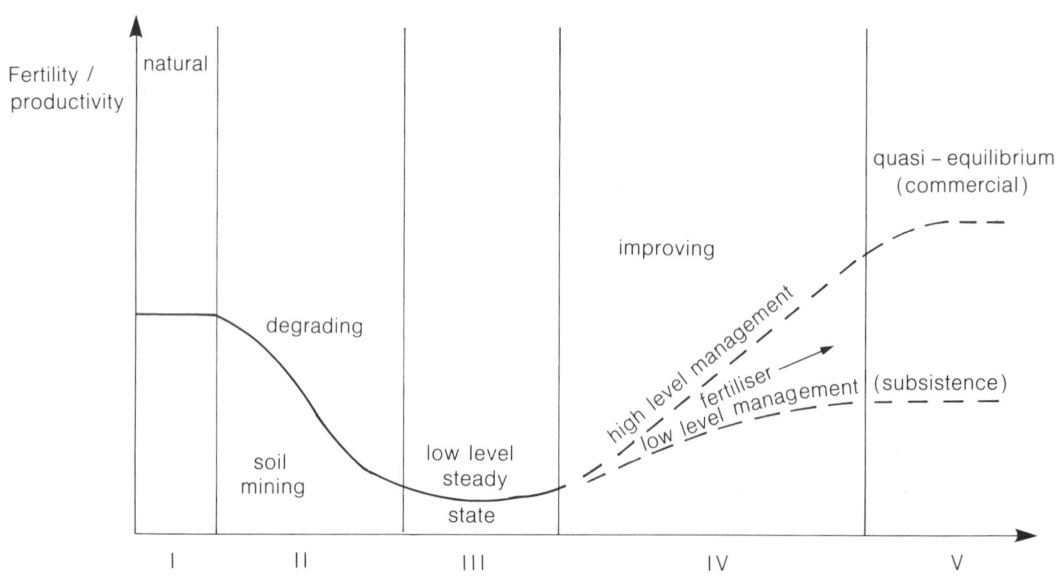

Figure 33.3 Soil fertility/productivity transition model.

33.2 Classification of farming systems

As we have seen, a farming system is a group of closely related activities determined by the nature of the land, climate, labour, capital, risk and style of management. H.Ruthenberg suggests that farming systems can be classified into:

(a) shifting cultivation;
(b) fallow;
(c) ley and dairy (mixed farming);
(d) permanent cultivation;
(e) arable irrigation;
(f) perennial crops;
(g) grazing.

Furthermore, each system may be either stable or unstable depending on the nature of its inputs and outputs. Several categories are possible.

(a) Stable (steady-state) systems may be
  (i) at a *high* level of soil fertility and output (e.g. wet rice growing in Thailand);
  (ii) at a *low* level of soil fertility and output (e.g. permanent upland farming in India).
(b) Unstable systems may be
  (i) in quasi-equilibrium, whereby soil fertility is maintained artificially (e.g.North America, northern Europe, Japan);
  (ii) improving, due to drainage, irrigation, and the like;
  (iii) degrading, due to inputs being fewer than outputs (e.g. dry farming on slopes in tropical Africa) – in this case the land is being *mined* and will move either into a man-made desert or to a low-level steady state (i.e. (a) (ii)).

Observation suggests that an area may pass sequentially through several of these states as perception of the problem and level of management change (Fig. 33.3).

Shifting cultivation

Shifting cultivation is the term for agricultural systems that involve an alternation between a few years of cropping selected and cleared plots and a lengthy period when this soil is rested. Cultivation *shifts* within an area that is otherwise covered with natural vegetation. Shifting cultivation is basically a pioneering cropping system which was once used in temperate areas; it is now found almost exclusively in the humid and semi-humid tropics (in New Guinea, Borneo, Burma, Zaire, and the Amazon rainforest, for example) and in the savannas. However, in the savanna the forest does not regenerate and is replaced by a *sub*-climax vegetation, dominated by tussock grass.

Shifting cultivation is still the main way in which the Amerindians of the upper Orinoco Basin in South America obtain their staple foods. A village of about 200 cultivate perhaps a dozen small gardens each of one hectare or less, growing mainly cassava – a plant with edible tubers – or bananas (Fig. 33.4). These gardens are neither exclusively devoted to the main

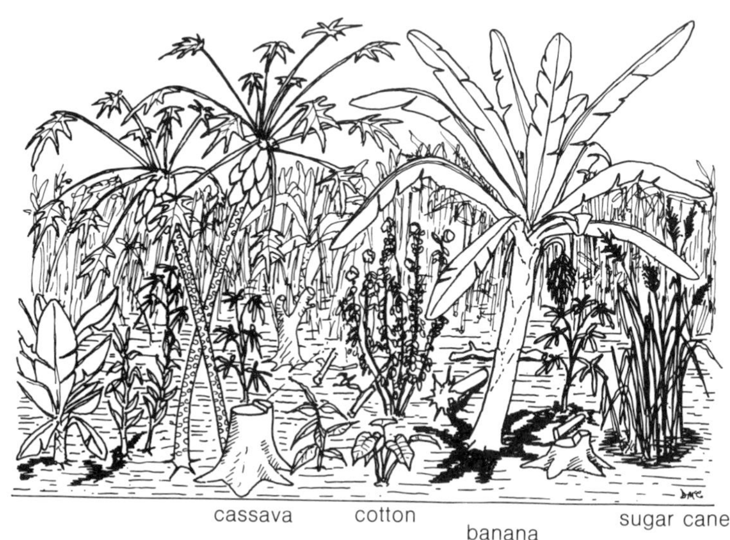

Figure 33.4 Schematic profile of a polycultural 'garden', based on photographs of a Waika headman's plot near Ocamo, Venezuela.

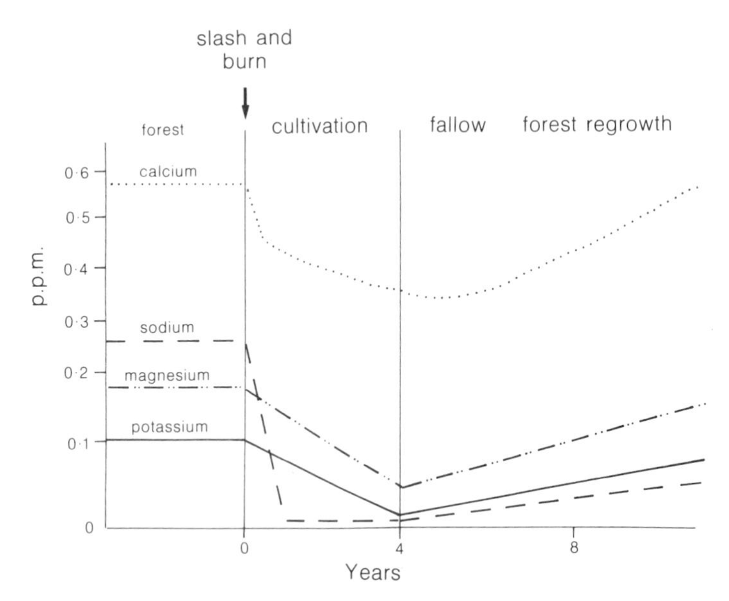

Figure 33.5 The change in soil cations (nutrients) during and after shifting cultivation.

crops nor planted systematically. The pattern of planting is controlled by the need to avoid the tree stumps that remain after burning, and in between the main crops a great variety of other plants such as cotton are allowed to grow. There is considerable practical value in this apparently haphazard system; not only does it increase the variety of crops, it leaves little bare soil exposed to rainsplash or sheetwash. Such a **polyculture** system exploits the land to the full, much as happens in the nearby forest, relying on the ability of species with different growth habits and root systems to thrive in one another's company. However, despite the 'natural' pattern of agriculture, the toll on soil nutrients is still high (Fig. 33.5) and a period well in excess of 12 years is needed for natural recovery after even a short period of cultivation.

A system of shifting agriculture is also able to cope with the problem of pest control: burning sterilises the fields before cultivation, and the duration of crop growing is too short to allow a major build-up of 'pests'.

Fallow systems

Shifting cultivation can only be maintained where there is little population pressure. It has been retreating fast in developing countries with high population growth because the time that can be allowed to rest the land becomes increasingly shorter and the farming system more stationary. Eventually more than 33 per cent of the arable and temporarily used land is cultivated annually, and farming then works on a **fallow system** (Fig. 33.6). A fallow system does not allow sufficient time for the natural vegetation to regenerate and scrub (bush) vegetation

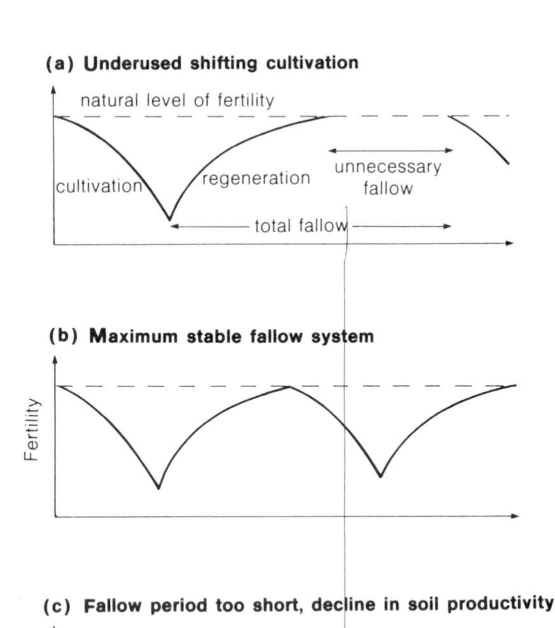

Figure 33.6 If the farming system relies on natural revegetation of the land there is a critical fallow period which must not be shortened, or soil productivity will fall and soil mining will result.

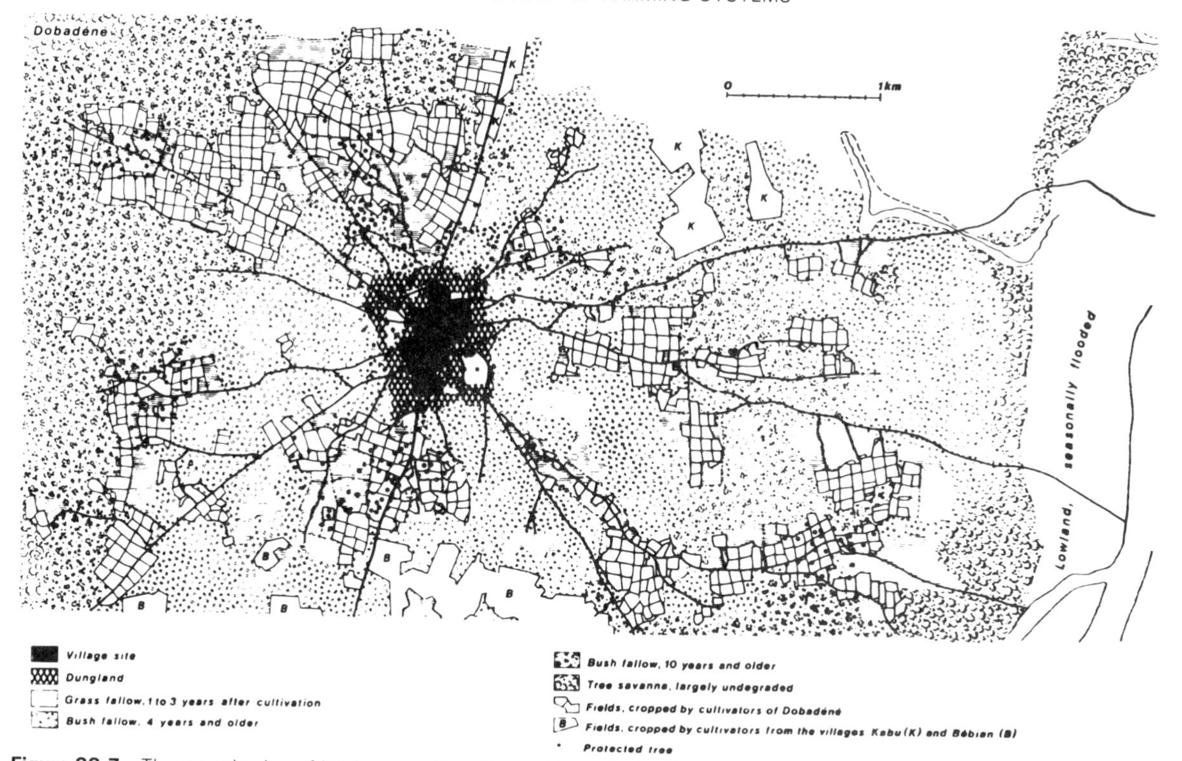

Figure 33.7 The organisation of land around the village of Dobadéné, Chad, showing a variety of fallow systems typical of the Sahelian zone.

dominates with even this being kept in check by grazing sheep, cattle and goats.

With fallow systems farm holdings become more clearly defined and field boundaries are established; housing becomes stationary and cultivation is by hoe rather than by slash and burn (Fig. 33.7). Fallow systems are far more important than shifting cultivation, accounting for the majority of non-commercial farm output in tropical regions. Because insufficient time is allowed for soil regeneration, there is a gradual reduction in soil fertility, although for many years this has been more than compensated for by the larger area under cultivation. However, in time the fallow system declines to a low-level steady-state productivity.

Animals are kept on fallow ground and form a vital part of the food supply for farmers of this system. However, with no special feedstuffs and grazing low quality the animals have quite low nutrient intakes (Fig. 33.8). There is no specialised breeding, and stock tend to be small and with low fat and lean content. Thus, with no control on the grazing or breeding, animals tend to fatten in the rainy season but become increasingly thin towards the end of the dry season, at just the time they are most likely to be needed as a source of food or to help cultivate the ground.

Some of the developments and problems of a fallow system are shown clearly in South America where, in countries like Ecuador, new roads have recently opened up to new settlers large tracts of rainforest. Here the government has given fifty hectares

Figure 33.8 These small cattle are an important part of tropical farming systems, but they do not provide a high milk or meat yield.

to each farmer willing to clear the rainforest. Fifty hectares may seem a large plot, but in the inherently infertile rainforest soils very long periods of fallow are required to allow satisfactory regeneration without fertiliser applications. As farmers cleared the forest to plant coffee and cacao for cash, and yucca, plantains, maize and rice for food, they found it impossible to leave sufficient ground for effective fallow. As a result of over-use the ground in many of these holdings has now become exhausted and many would-be farmers have had to abandon their land completely, some moving to new ground but many others leaving the countryside and seeking work in the ever-growing cities.

Ley and dairy (mixed farming) systems

The term **ley** is used to describe the use of several years of grass and legumes as fodder for livestock after several years of arable cropping. It is more efficient than a fallow system because it maintains a higher level of fertility, can support far more animals, and at the same time provides a better use of farm labour. However, it is an advanced system in which a high level of management and large capital investment are needed to maintain fertility in a state of quasi-equilibrium. Ley and dairy systems are the typical **mixed farming** systems of the developed world, with crop rotation, control of grazing, fenced fields and the production of hay or silage for winter feed of animals. Cereal production is quite often used in part as animal feedstuff to encourage fattening, while considerable care is taken with diets to ensure maximum growth rates of livestock. Understandably, these systems are heavily mechanised with an emphasis on low labour cost per unit output. Labour is used continuously throughout the year rather than seasonally as with less developed systems.

Ley and dairy systems are most common in temperate and subtropical regions. These areas have higher capital resources, and rapid growth of palatable grasses rarely occurs in the tropics. Furthermore, animal health risks (from the tsetse fly, for example) are far higher, and there is a shortage of trained farm personnel to run a high-technology system. Ley and dairy farming is found only in the upland areas of the tropics, for example in the Eastern Highlands of Kenya. Even in this cool humid area the proportion of ley and dairy is small: the high level of production of perishable commodities requires a highly organised marketing and retail system, and this has still not been developed in the tropics.

Many farms in the UK operate ley or dairy systems and grass is the country's most important crop, accounting for 60 per cent of the area under cultivation. Even a fall in grassland area (due to greater arable production and urban expansion) has not reduced livestock numbers, because the productivity of grassland has been more than raised in proportion.

Traditionally, UK farms have contained a mixture of livestock and/or arable farming. The reasons are these:

(a) the land quality varies rapidly in fertility, drainage and steepness;
(b) rotations help maintain soil fertility and reduce the risk of diseases building up in the soil;
(c) rotations even out the work load over the year, keeping full employment of men and machines;
(d) a wide variety of crops spreads the risk to income caused by fluctuations in yields and market prices.

More recently, however, farms have specialised because:

(a) there are considerable economies of scale available (for instance, in 1973 the average cereal crop was 30 ha/farm whereas the optimum size is over 150 ha/farm; dairy herds were 36 cattle/farm whereas economies of scale are most effective at over 70 cattle/farm);
(b) technological advances have reduced the risk of crop failure;
(c) EEC policy has helped to stabilise prices and make long-term planning on just one crop more practicable.

Because Europeans are able to choose to eat meat instead of a diet dominated by carbohydrate (as would be the case in the developing world), a large part of the cereal yield is consumed by livestock; the UK is actually a net importer of *food* grain. Together, milk and cattle meat make up 40 per cent of the output of British agriculture, although there are considerable regional variations according to climate and land quality.

Some of the characteristics of a ley system are shown in Devon where ley and dairy systems have traditionally been the mainstay of agricultural practice. There is a long growing season of over eight months and frost is rare. Nevertheless, proximity to the moisture-bearing air masses of the Atlantic gives Devon a considerable water surplus in most months and restricts the possibility of large-scale cereal cultivation. In addition, many slopes are too steep for mechanised farming, and the soils, developed mainly on mudstone rocks, yield a fine texture that can readily become waterlogged.

The area near Honiton, to the east of Exeter, is typical. Here there is a long tradition of livestock farming, using supplementary arable cultivation for winter feed (Fig. 33.9). As autumn sets in and evapotranspiration rates decline many of the fields become very prone to 'poaching' by cattle unless they are drained. Despite the long growing season it is

Figure 33.9 Typical high-density animal husbandry, associated with mixed farming.

therefore important to keep the stock off the land and in specialised buildings during winter. Where the soils do prove to be sufficiently well drained to allow arable cultivation, the yields of between 3 and 4.5 t/ha provide a much greater fodder store than the 1 t/ha of hay that is taken from permanent pasture. In consequence there is pressure for the proportion of arable land to be increased. Field drainage is installed wherever it is economic, and as land management and crop breeding improve it is likely that many of the ley and dairy systems will gradually be transformed into permanent cultivations.

Permanent cultivation

Fallow and ley systems typically have between one and two thirds of the cultivated land under arable. It is possible to push the proportion of arable land far higher, even when livestock production is the main aim. Denmark, for example, is world-famous for a permanent cultivation system that keeps livestock in stalls, feeding them on crops grown in the fields (Fig. 33.10). Yet a system of such high productivity is inherently unstable and is very fragile. Farmers are able to get away with the system only by regular application of fertilisers and organic matter, and by heavy doses of biocides. With fields bare for a large part of the year, soil erosion is also a continual hazard. Permanent cultivation is easiest to manage in humid temperate regions where soils do not dry out and become liable to wind erosion, and where rainfall intensities are sufficiently low to make water erosion small as a problem. Nevertheless, the cloudy conditions of these regions are not the most favourable for high crop yields, and high levels of humidity present many problems with disease and rot.

Permanent cultivation is extremely troublesome in the tropics where rainfall intensities are high and soil

Figure 33.10 Danish farming is renowned for its intensive nature. Co-operative systems and the use of high technology are important. Barley and oats are growing in this field to be used as animal fodder.

erosion and leaching become formidable opponents. Some form of **terracing** or other conservation system is always required on sloping land (Fig. 33.11). As a result, most permanent cultivation is found in sub-humid and semi-arid areas, where leaching and erosion may be less severe and where fodder crops do

Figure 33.11 On sloping ground, terracing becomes a vital part of any permanent cultivation system. Here, bananas and cassava are being grown.

not make satisfactory progress. Large parts of North America, Central Asia, Africa and India are cropped on a permanent basis, although the levels of applied technology and yields vary dramatically. But these areas are not without their problems, for they have the highest risk of drought and crop failure of all. All arable farmers have to adopt strategies to minimise their risk of loss; the variety in cropping patterns stems from the many ways in which farmers choose their strategies.

The risk of failure drives permanent cropping systems in two distinct directions – towards a commercial system or towards a smallholder system. Commercial farmers try to reduce risk

(a) by using high levels of fertiliser to compensate for a season with low water levels;
(b) by management systems that conserve soil water, such as cultivating every other year;
(c) by choosing rapidly maturing crops, to make the best use of available moisture;
(d) by accumulating a capital reserve from the sale of large yields in years of good rainfall, to carry them through the drought years.

In a smallholder system farmers try to reduce risk

(a) by delaying planting until the rains have clearly come to stay (this strategy carries with it the penalty that a large proportion of the minerals weathered out in the dry season are leached before the plants grow);
(b) by preferring drought-resistant crops such as millet to less tolerant but higher-yielding crops such as maize;
(c) by not planting at times which would maximise yields but instead phasing the planting throughout the rainy season in order to distribute the risks.

COMMERCIAL SYSTEMS IN BRITAIN

The Vale of York in eastern England reflects many of the problems and opportunities involved in permanent cultivation of humid temperate regions. Here soils have been formed on a mixture of glacial sands and clays whose monotonously flat landscape is well suited to extensive mechanisation. Although the flat land was once an ill-drained marsh, extensive river dyking and soil drainage have enabled the natural fertility to be exploited. And the effort has proved worth while, because this is a favourable region for permanent cultivation, having one of the driest climates in Britain (annual rainfall about 670 mm) with warm sunny summers (the July average temperature is 17°C). The glacial heritage gives a wide range of parent materials, with sands and clays occurring on almost all farms. The variety is used to advantage: sands can be ploughed early in the spring and, because the soils are 'warm' and crops ripen quickly, the crop can be harvested early; the water-holding clays, on the other hand, need to be ploughed and seeded in autumn when they are at their driest. Because the wet soils take a long time to warm up in spring (they are 'cold' soils), harvests are late. It is a contrast in timing which allows a more even distribution of labour and machinery across the year. Soil variety is also helpful when harvesting root crops such as sugar beet and potatoes: crops planted on the clays can be harvested on dry days, and during periods of wet weather the sands remain workable.

Farms in this area are classified as 'cash roots', particularly where sands occur. Here more than 40 per cent of the output comes from sugar beet and potatoes and a further 30 per cent from cereals. Sugar-beet areas have increased over the last decade, spurred on by the EEC policy of becoming self-sufficient in sugar. The change to sugar has also helped to offset pest problems that had begun to develop serious proportions with continuous growing of potatoes on the same land. Potatoes and cereals suffer from cyst eelworm, sugar beet from free-living eelworms; their control is the main reason for the rotation of crops. Potatoes are now grown on the same land only one year in five.

Farmers now find direct sales from a farm shop a profitable outlet for some of their produce, and with vegetables, fruit ('pick-your-own') and eggs forming the largest demand this has considerably influenced the pattern of land use. Labour-intensive forms of agriculture, such as dairying, have seen the most dramatic decline.

Under careful management, large parts of the Vale of York are highly productive for cropping. Yields of 50 t/ha of both sugar beet and potatoes are not uncommon, but with livestock numbers small there must be heavy expenditure on fertiliser both to maintain nutrient status and retain strong soil structure. Furthermore, large fields with sandy-textured soils are very prone to wind erosion, especially in March, April and May, after ploughing and before the crop has grown sufficiently to protect the ground. In an attempt to reduce the problem, soils are often rolled directly after ploughing. A seed bed is then roughed up on the surface of the compacted soil. In the past, when labour was very much cheaper, many fields were improved by marling. Nowadays the expense and scarcity of agricultural labour has meant that harvesting of sugar beet and potatoes is highly mechanised even on small farms.

COMMERCIAL SYSTEMS IN NORTH AMERICA

By far the largest areas of permanent cultivation occur in areas where leys would not be a viable commercial alternative; these are the world's natural grassland zones.

Both of the greatest plainland areas – the Great Plains/Prairies and the Steppe – have experienced revolutions in their farming systems during the last century, and now form some of the most important grain-growing and beef-rearing areas in the world.

In the USA the core region of the grassland is the Corn Belt (Fig.33.12), a land where the native grasses have long ago been replaced by specialised strains of maize (corn) and soybeans in rotation. Here, with the experience of the 'Dust Bowl', even gently rolling land is contour-ploughed and kept under a protective vegetation cover for as much of the year as possible. Substantial cultivations of humus-forming crops such as alfalfa also form a prominent part of the cultivation practice. Everywhere the emphasis is on highly capital-intensive, high-energy farming (Fig. 33.13). In areas where rainfall is insufficiently reliable to allow corn to be grown with a high yield, it is replaced by the more drought-tolerant wheat, especially those strains of winter wheat which can make the most use of the moisture provided in early spring by melting snow. Even the Dust Bowl regions of the Great Plains can now be planted to crops, again using special strains of early-maturing wheat. In the driest area, where there is insufficient moisture to allow continuous cropping, the land is cropped only in alternate years. Between crops the land is left fallow and kept free of vegetation for a year to allow soil moisture to be restored to a level that will support the next wheat crop.

Figure 33.13 Permanent harvesting on a grand scale in the Corn Belt. Teams of combine harvesters are hired to gather the crop quickly and efficiently. Hiring the harvesting service also saves the farmer from extra capital outlay.

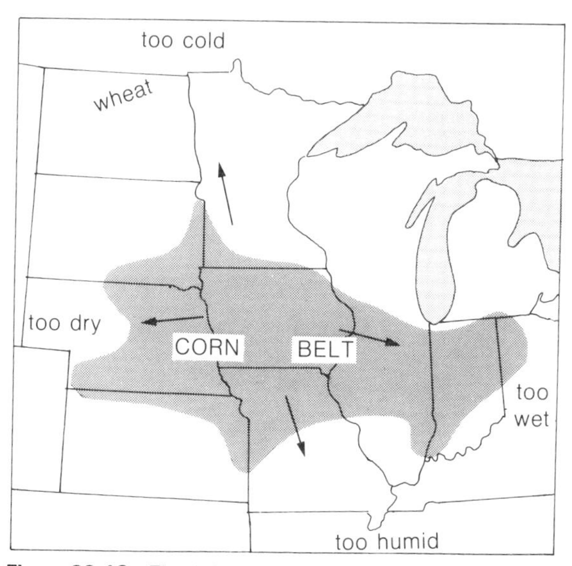

Figure 33.12 The US grasslands. The region that grows corn (maize) has specific climatic requirements. The northern grasslands specialise in wheat.

SMALLHOLDER SYSTEMS OF THE DEVELOPING WORLD

Although the greatest production is derived from the plainlands of the developed world, by far the largest population depending directly on permanent cultivation lies within the sub-humid tropics, especially in Africa and the Indian subcontinent. Here the balance between crop cultivation and pastoralism is largely governed by the available moisture supply, which in turn is related to the length of the dry season. Towards the equatorial margin of the savanna, dry seasons become shorter and water more plentiful, and there is an increasing tendency to rely more heavily on cultivation. This is not to say that animals form no part of the farming system – often they provide not only motive power but essential protein. Nevertheless, animals become relegated to areas beneath trees where crops will not grow, to roadside verges and so forth.

Where rainfall quantities permit, the basic element of the diet changes from dry land millet to the more water-demanding maize, cow peas and groundnuts.

Each farmer attempts to grow crops to provide the needs of his family, together with a small surplus or **cash crop** to buy essential clothing and hardware, to pay taxes and so on (Fig. 33.14). There is little road infrastructure to allow the sale of perishable foodstuffs, and the main cash crops are therefore the less perishable commodities such as tea, coffee beans, cotton, tobacco, coconuts and cashews.

The exact nature of the farming pattern depends on the distribution of rainfall. In East Africa, where the movement of the sun brings two rainy seasons (one in May-June, the other in November-December), it is possible to grow two crops; in other areas, with only one rainy season, there is a restriction to one crop. Yet despite the increased availability of rainfall, these areas are no less subject to upward population pressure than those in drier areas. Whereas once it would have been practicable to keep large areas fallow and to adopt a form of shifting cultivation, this is no longer possible; at best land is only left fallow for short periods. In all other cases the land has to be kept in more or less full use. Under continuous cultivation animal manure is still vital; otherwise expensive fertilisers must be bought. Settled farming at a subsistence level can increase the risk of famine. There is no hard-and-fast rule about survival in the savanna; as in other parts of the world, there are great variations in soil quality and rainfall reliability.

These variations, and the adaptations forced on the people, are shown very clearly in south-east Kenya.

How great a risk each community in this land can afford to take (concerning failure of crops or decimation of herds during a drought) depends on their approach. They have three options: they can

(a) tighten their belts;
(b) develop a surplus to help them out over difficult years;
(c) raid neighbouring territory.

In the hill lands the chances of obtaining adequate moisture for crop growth are good, the carrying capacity of the land is high and population densities reach 600 /km^2. People are therefore stable and largely cultivators (Fig. 33.15). In the mid-altitude plateau region the *chances* of adequate moisture are reduced to 30 per cent even though the *average* rainfall is 760 mm/year. Problems of food supply each year have therefore kept the population density down to 50-100/km^2 and cultivation includes a mixture of water-demanding maize and beans that will provide

Figure 33.14 Subsistence permanent cultivation, Rabai near Mombasa, Kenya. Coconut palms provide little shade to hamper the growth of maize in the smallholding or *shamba*. Small amounts of animal manure are used to help maintain soil fertility.

Figure 33.15 With reliable rainfall, high populations can be supported on small areas of land. Each thatched mud dwelling in this photograph is the home of a family of upwards of five people.

good yields when rains are plentiful, but also bullrush millet and cassava which will survive a partial drought. In general yields are lower and the people are therefore partly dependent on livestock, even though they may have to travel long distances to find adequate grass. Locally lands have become badly overgrazed and gullied (Fig. 33.16). Because it is marginally cultivable land, this is the area most prone to chronic crop failure and maize commonly has to be brought in.

In the low-level plains rainfall is very unreliable, and the chances of sufficient rain for agriculture are sometimes as low as 10 per cent. Population densities remain below 10 Km^2. To try to grow even some crops people use huge quantities of land on shifting-cultivation systems that often involve fields 16 km apart, with each field containing a multitude of drought-resistant, rapidly-maturing crops sown in what appears to have been a fit of temporary insanity. But the system is really geared to reduce the risk of crop failure, for rains are very patchy and with widely spaced fields the people stand the best chance of catching some rainfall. Nevertheless the tradition of hunter-gatherer remains – as evidenced by the importance of honey and of ivory poaching in the local economy.

Systems with arable irrigation

In some parts of the world it is possible to find water either from rivers or groundwater sources to supplement the natural rainfall and help provide a more

Figure 33.16 The area around this water-hole has been severely overgrazed and soil erosion is now a desperate problem. However, the livestock do not belong to pastoralists; they belong to cultivators who need to keep livestock in case of crop failure.

reliable supply for plant growth. These **irrigated systems** vary greatly in both form and area, ranging from primitive **bunds** (earth walls) built around a field to retain annual floodwaters, to those using the most advanced sprinkler and drip-feed techniques. However, flooding basins or furrows and sprinklers are the most commonly used systems. Basin flooding is the simplest because furrow techniques require extensive field levelling and the use of machines. Large furrow-fed irrigation systems are largely found associated with commercial operations such as the Gezira scheme of Sudan (Fig. 33.17) where cotton is the main crop, or in the Rio Grande, USA, where vegetables are grown.

If irrigation water comes from natural runoff it may still only be available seasonally, and this will control the pattern of land use. As a result, a dry-season crop may be alternated with a wet-season, irrigated crop. It is common, for example, to find irrigated rice being alternated with fallow in south-east Asia, and legumes with irrigated rice in Sri Lanka.

Some form of wet-rice system has always been the most important form of land use in south-east Asia (Fig. 33.18), but even in Africa and Latin America wet rice cultivation is increasing rapidly. No other agricultural product in the tropics and subtropics has as great an economic importance as wet rice, even though cotton (in India and Sudan), sugar cane (in India and Pakistan), groundnuts (in West Africa), wheat (in India and Pakistan) and vegetables are also irrigated extensively.

In temperate regions cereal crops are frequently irrigated either by rolling-sprinkler or centre-pivot methods, based mainly on pumped groundwater supplies (Fig. 33.19). Here it is applied only when moisture stress builds, thereby conserving water and allowing the unrestrained growth of the crop. However, temperature remains the major control and cold winters mean that only one crop per year is possible. With the year-long growing season of the tropics, irrigation can bring far higher improvements in yields. For example, in the Philippines it is possible to have continuous cropping and get three rice harvests each year. In the extremely high levels of management operated by the vegetable growers of Singapore, *eight* harvests each year are commonplace.

Irrigation brings not only benefits but also penalties. There are costs involved in distributing water, and large fertiliser applications are required because of the heavy demand on soil nutrients. In many cases these costs can outweigh the advantage of greater production. This is why most tropical irrigation farming (which, because it is based on subsistence cannot afford to pay for high-cost inputs) is

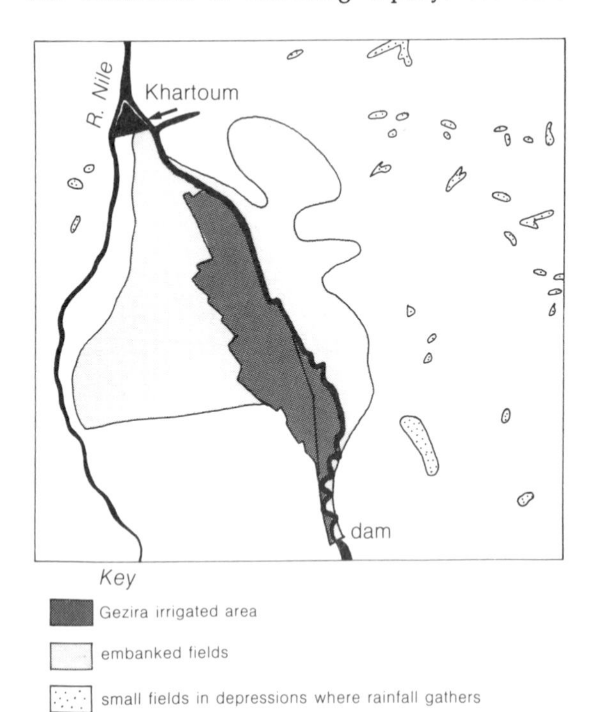

Figure 33.17 The Gezira irrigated area is literally a man-made oasis in the Sudanese desert. It is surrounded by thorn scrub; the only other reliable cropping being from embankments designed to catch water from the annual Nile flood.

Figure 33.18 Irrigated permanent cultivation is the lifeblood of Asian agriculture. The main crop is paddy rice, yet even this high level of intensive cultivation is unable to provide all the food requirement of some countries.

(a)

(b)

Figure 33.19 Some of the most impressive forms of pumped irrigation schemes can be seen in the areas near the 'mile high' city of Denver, Colorado. Here, in the shadow of the Rockies, both rivers and aquifers are used to provide water for a wide variety of cereal crops. In this area many fields are circular (the shape of the area swept out by the half kilometre-long booms that rotate about a pumping point), a system known as central pivot irrigation (a). Elsewhere, sprinkler booms on wheels trundle across the fields driven by their own motors (b). This places tremendous stress on the water supply.

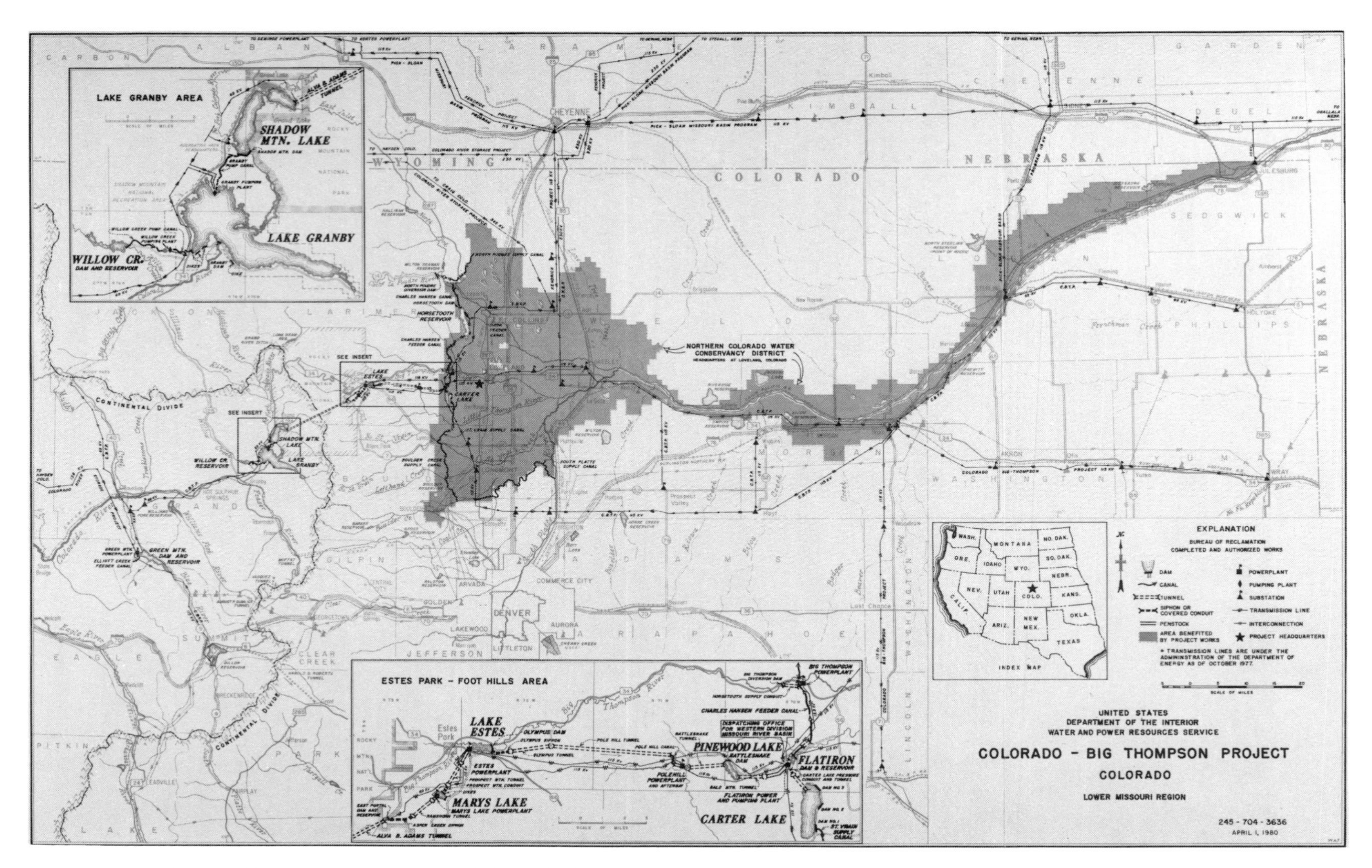

Figure 33.19 cont. (c) Colorado's Great Plains region is now irrigated by water that would once have flowed out to the Pacific Ocean!

confined to natural wetlands where water is easy to provide and where rivers bring an annual replenishment of nutrients. Such conditions allow continuous heavy cropping without artificial inputs and without soil deterioration (Fig. 33.20). Yields of rice in Java can support 2000 people per square kilometre even though the same land has been cropped for thousands of years.

The importance of irrigation is most easily demonstrated in places which have low levels of management. For example, in the vicinity of the Brahmaputra floodplain of Bangladesh, a monsoon climate produces extensive natural flooding each summer that leaves some low-lying fields over 1 m deep in nutrient-rich water. In these areas rice yields are over 3 t/ha. By contrast, the adjacent flood-free slope soils have only the moisture stored from rainfall and they only produce a harvest if sown with quick-growing 'dry' rice. Even so, yields are mainly below 1 t/ha because the soil now lacks fertility, and growth is restricted by the moisture stress that soon builds up as the dry season begins.

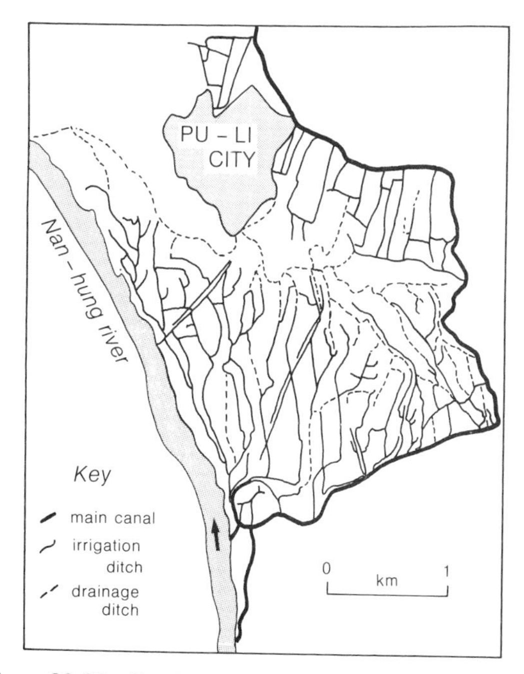

Figure 33.20 Simple gravity-fed irrigation scheme for 600 ha of floodplain in the Pu-li basin of Taiwan. Notice that, just as the human body needs arteries to supply fresh blood, and veins to remove it after use, so irrigation schemes need both irrigation *and* drainage ditch networks. Irrigation from the Nan-hung river allows two crops of wet rice to be grown each year; without irrigation only one crop would be possible. The complex pattern of inlet gates (called headings) is designed and operated to give extremely close control of the water and thus ensure maximum crop growth.

The tragedy that can befal! both permanent 'dry' (non-irrigated) cultivation – and even irrigated cultivation in tropical lands – is shown most vividly in regions with high population pressure, such as the Sahel and India. For example, large parts of India lie in the belt of seasonal and irregular rains that characterises the sub-tropics. India has 700 million people to feed and is increasingly affected by the variability of rainfall. In 1982 the monsoon failed. It was the worst drought since 1966, although four other severe droughts occurred in the intervening years.

India has two crops: **kharif** (mainly paddy rice) and **rabi** (mainly wheat). The failure of the rains hit the kharif crop which provides India with 60 per cent of its food grain.

In Bihar State half the average rainfall reduced the yield by a quarter; in West Bengal 13 of the 16 districts suffered drought and they were unable to provide water to 200 000 ha of paddy; in Uttar Pradesh, India's largest state, the yield was a quarter down because rain did not fall until the third week in July and the growing season was therefore curtailed; in Madhya Pradesh the erratic course of the rains caused a 50 per cent shortfall of the paddy. The Punjab – the wheat-growing granary of India – cannot make up the losses incurred by shortfalls of this magnitude suffered by the larger States. In 1981-82 India was obliged to import 5 million tonnes of grain, adding a further drain on her foreign-exchange reserves.

Systems with perennial crops

Most perennial crops come from shrubs and trees which have growth cycles of several decades. These include temperate crops (such as apples and pears), but the majority of perennial crops are grown in the tropics and subtropics, either on estates or on smallholdings. The main reasons for this concentration are that other crops do not thrive in warm humid regions, and that the perennial crops are mostly of tropical origin and cannot be grown elsewhere.

Because of the high levels of soil leaching, erosion and disease, farmers in the humid tropics tend to move towards perennial rather than annual crops wherever they can. For example, bananas produce as much energy per hectare as a root crop such as cassava, with less labour per unit of starch. Banana groves offer an almost complete soil cover and, because so little is removed as product, they help maintain soil fertility or even improve it.

Tropical and subtropical crops include not only bananas but sugar cane, pineapples, tea, coffee, coconuts, oil palm, rubber, cotton and cacao. By convention, any land that is planted with perennial crops is called a **plantation**. If it is large and under commercial management and employs a substantial number

Figure 33.21 Sugar-cane plantation and refinery, New Yarmouth, Jamaica.

of paid workers, the plantation is called an **estate** (Fig. 33.21). In general, estates develop directly from new land, whereas smallholdings follow from the consolidation of land as population pressure forces peasants to abandon shifting cultivation for fallow farming.

There are three main types of crops, each controlled by different factors:

(a) perennial field crops, such as sugar cane, pineapple, sisal and bananas, owe their distribution to market access and good transport;
(b) shrub crops such as tea, coffee and cotton, are not perishable but are closely controlled by climate, soil and the availability of large amounts of manual labour (Fig. 33.22);
(c) tree crops, such as cacao, rubber, coconut and oil palms, need only low amounts of labour and are not perishable, and consequently have the widest distribution.

In the past, estates dominated the perennial crop market; today there is a large and increasingly important contribution from smallholders, either in collaboration with estates and using their processing facilities, or with the government-owned processing stations. Coffee in Indonesia is grown almost entirely by smallholders, while cashew-nut production in Kenya is organised around a government-owned processing station. Smallholders cannot afford to have all of their land devoted to one crop, so the importance of perennial crops in a region dominated by

(a)

(b)

Figure 33.22 (a) A tea estate on the Mau escarpment of Kenya. The tea bushes are about waist high and their leaves have to be picked by hand. (b) Estate workers housing near the central processing plant.

smallholders may not at first be apparent. With the perennial crop there will be a variety of subsistence crops such as yam, rice, maize or beans.

In areas where the plantations have been established for a hundred years or more, they will invariably be found on the most level land, often near a river and certainly in a position of good communications. This means that they will be sited either on alluvial soils (which receive nutrients leached from upslope), or at worst on soils of the lower hillslopes. The only exceptions are crops such as tea and vines, with which slopes are needed for drainage purposes. These plants thrive on leached soils. A further reason for the apparent success of tropical plantations, especially this century, has been the widespread and recurrent use of fertilisers, fungicides and pesticides, as well as the benefits of large-scale scientific research into the most effective crops. Here, then, is a prime example of technology being applied to a system to override natural ecological balances.

CASE STUDY: PLANTATION CROPS IN MALAYSIA

Malaysia is a developing country still economically dependent on its agriculture. It lies within the tropical-rainforest zone and still has, especially in Borneo, large areas of undeveloped forest land. The government and large private companies want to exploit some of the remaining forest and develop plantations that will generate additional foreign earnings. The concern for foreign currency largely overrides the rights and wrongs of exploiting the forest at all; the main concern is to ensure that there will be an adequate and lasting return from any developed land. In Borneo, for example, annual rainfall averages just under 3000 mm, falling on about 200 days each year; while there are about 2200 hours of sunshine available to help ripen crops.

Weather statistics are essential when choosing the type of crop to be planted, as when framing any programme of agricultural work, such as felling, burning and planting. Detailed understanding of soils and slopes is needed to assess the capability of the land. In North Borneo there are relatively few areas of steep slopes and the main problem is draining some of the most gently sloping land. Soils are acid (pH 4.3) and dominantly clay-textured red earth; all have a low nutrient content and will need heavy fertiliser applications to keep them productive. These soils are a failure under peasant subsistence systems.

There are four potential tree crops that could be grown in this area: cacao, coconuts, rubber and oil palms. Oil-palm yields are not high in Borneo, probably because of a lack of suitable pollinating insects; rubber is very susceptible to fungus disease in high-rainfall areas, and tapping is made more difficult.

In addition to purely ecological considerations, major economic restraints are imposed by the value of each crop. At present cacao yields the highest return per hectare, followed by coconuts, rubber and oil palm. All of these products fetch widely varying prices each year depending on the world supply; this in turn depends on climatic fluctuations. However, recently cacao has been the most stable, yet another reason for preferring this crop. Future plantations in Borneo are, for the ecological and economic reasons outlined above, liable to concentrate on cacao as a single crop or to mix cacao and coconuts whenever it is thought prudent to lessen the risks of growing just one product.

Grazing systems

Pastoral activities may be undertaken either by choice, in addition to cultivation, or of necessity, in areas where cultivation is impracticable. In both tropical and temperate regions most of the livestock are kept by cultivators at very high stock densities. Although there are vast areas of natural grassland (dry savanna, steppe, pampas, chapparal, hill and mountain pastures), the feed quality is poor and quantity low so that only extensive grazing systems of low stock densities are possible.

There are several varieties of grazing practice:

(a) **total nomadism**, whereby people and animals move to the available grazing and are based on no permanent centre – usually found where rainfall is below 200 mm/year;
(b) **semi-nomadism** or **transhumance**, whereby the farmers have a permanent home to which they return, and near which they may cultivate land on an opportunist basis (they are still forced to travel long distances in search of adequate grazing) – this system is mainly found where rainfall is between 200 mm/year and 1000 mm/year
(c) **ranching**, whereby pastoral activities are organised on a commercial basis and in which animals are kept on a fixed area of land;
(d) **upland grazing**.

Because the success of ranching is dependent on markets, sheep rearing is most common where distances are great (for instance, in South-West Africa, Australia and South Argentina). Sheep can also survive in the harsher conditions where rainfall is as little as 150 mm/year or where frost and snow are common. Cattle require a great degree of husbandry, they cannot survive on rangelands with rainfall below 400 mm/year or where winters are harsh, and they are less easily transported. In consequence, they are found nearer to good transport routes (for example, in North Argentina, South-West USA and East Africa).

Ranching may be a high-technology system but it is often not a stable form of farming: as population pressures increase, so arable cutivation gradually replaces it, especially in dry lands where irrigation becomes available. Indeed, it is the semi-arid rangelands that will have to be used to support the future demands from the world's ever-growing population.

The causes and problems of grazing systems are nowhere better illustrated than in the African savanna, where primitive nomads and highly sophisticated ranchers often live side by side.

Farming is never an easy occupation, even in areas with fairly reliable rainfall. Faced with extreme variability from year to year and the great contrasts between wet and dry seasons, semi-arid pastoralism on the Tropics is one of the most fragile of all agricultural systems. Most natural vegetation adapts to the dry season by becoming dormant, shedding its leaves or developing thick leaves that reduce transpiration. To help eke out the scarce water resources, plants are spaced widely and have extensive networks of roots which extract moisture from a large volume of soil. Any form of reliable crop cultivation, on the other hand, is only successful if there is sufficient rainfall to allow rapid and sustained growth.

Thus, faced with crop uncertainty, it has been natural for the peoples with a long dry season to adopt some form of pastoral agriculture.

THE NOMADIC PASTORALISTS

Pastoral farming is the most reliable form of agriculture available in these fragile lands, but animals require feed thoughout the year so herdsmen cannot remain for long in an area where the vegetation is shrivelling and the water holes are drying up. To make the best use of forage, pastoralists must therefore be nomadic, following the seasonal movement of rainfall much as the wildlife has done for millions of years. When the rains have been good it may not be necessary to travel long distances in search of sufficient food but when the rains fail long and arduous journeys are undertaken.

Nomadic pastoral farmers obtain their food directly from their animals, as milk, meat and sometimes as blood. To maintain a sufficient food supply, four or so animals must be kept per person. The type of livestock kept by any group varies greatly with local preferences, with the tolerance of animals to heat and drought, with the nature of available forage and with animals' resistance to disease. As a result not only cattle, but sheep, goats, donkeys and camels are all widely herded.

Under this 'natural' system there is very little livestock management, except to guard against theft and carnivores. Pastoral tribes therefore usually set up temporary stockades (**manyattas**) in the area of grazing and herd their animals into these each night. But apart from this elementary protection there is no attempt at improving the stock by selective breeding or supplementary feeding. The herd size is allowed to be self-regulating, increasing in years when there is plentiful forage following plentiful rains, and reducing drastically by starvation when droughty years result in poor plant growth. Under these conditions the variation in herd numbers depends on the hardiness of the animals: in a recent period of drought a Masai tribe in Kenya lost over half its cattle, while their goats (which can eat even thorn scrub) were affected hardly at all.

Figure 33.23 A dominantly pastoral agriculture in East Africa. No cultivated 'gardens' can be seen in this photograph, merely clusters of huts surrounded by overgrazed savanna. This is a steadily decreasing form of agriculture, and one that initiates desertification.

With uncontrolled farming methods, severe overgrazing with consequent soil exposure and erosion are almost certain, especially in droughty times when animals can denude a landscape of every blade of grass (Fig. 33.23). As population numbers increase and tribes attempt to keep correspondingly more animals on the same amount of land, pressure on grazing and the degree of land degradation can only get worse. The pastoralists are faced with a 'catch 22' situation: with too little food and no proper breeding, meat output is very low; yet increasing animal numbers to try to provide more food results in overgrazing which destroys the very resources on which both animals and people depend.

SEMI-PASTORALISTS

Where there is sufficient moisture for some form of crop production and where pressure on grazing land is very high, many pastoralists have turned to primitive cropping in order to supplement their food supplies. However, now tied more closely to a definite area by their cultivated gardens, these farmers arrange for their annual migrations (**transhumance**) to coincide with a simple pattern of cultivation. And because they return to the same tribal lands each time, in areas of high population density there is a massive tendency to overstock and therefore overgraze. Here, where crops play a vital rôle in providing livestock forage, herds of animals are even more likely to die of starvation in times of prolonged drought because the stock density always exceeds the available grassland supply. Again the downward spiral is evident, with poorly fed animals not only less able to produce meat, but also more prone to disease. Recent investigations in Botswana, for example, have shown that the weight of an 18-month old calf reared under traditional methods is on average less than half that of a calf kept in the same area but under a modern commercial ranch system.

Semi-pastoralists do have the ability to supplement their food, and that of the animals, by growing crops. But the variability of the rainfall makes this a risky business and the style of farming is not helpful in obtaining the best yields. Furthermore, without applications of fertiliser a downward spiral of crop yields accompanies the downward spiral of livestock quality mentioned earlier. It is the combination of these two distressing features that has caused much of the semi-arid zone to be prone to famine, especially in the drier, desert marginal areas called the Sahel.

The traditional semi-pastoral system is well illustrated by the cattle-owning Baggara tribe of the Sudan. These people have their homeland in the broad-leaved savanna to the south of the semi-arid thorn savanna and scrub. Their annual cycle begins in May with cattle grazing in the tribal lands until the first summer rains allow the crops to be planted. When the rains really set in and the mud and flies make life unpleasant, they migrate north to the thorn savanna, where they remain until the rains end. By October all animals are brought back to their tribal areas where they stay while the crops (if any) are harvested. However, by November the grazing and supplies of water from the seasonal pools are finished and the animals are herded farther south. Even in these more favoured areas shallow wells eventually have to be dug to provide a water supply until the first rains occur in April. At this time the Baggara herd their animals north to follow the first flush of vegetation back to their homeland, to complete their annual transhumance cycle.

This pattern has recently been altered in several ways. The digging of new deep water holes which provide water throughout the year has allowed people to remain in the same areas. However, this has caused widespread overgrazing. In turn, lack of feedstuffs has meant that cattle production is increasingly more marginal and more and more emphasis is being placed on cultivation.

In West Africa the pattern of semi-pastoralism is similar to that in the Sudan, transhumance occurring within the limits imposed by the desert to the north and the regions of tsetse-infested savanna to the south. However, in areas where cattle are privately owned but grazing is free, there is a tendency to stock to the maximum, but there is no incentive to improve the quality of the grazing. The seasonal use of the natural environment by nomads or transhumance systems is well suited to obtaining the maxium yield for the least effort and with the minimum of damage, providing population densities are kept within the carrying capacities of the range. Governments tend to disapprove of these systems which cause great difficulties in administration, tax collecting, education, health and so on. To try to reduce migration, more

Figure 33.24 Not a blade of grass is to be seen in the area near Gotheye in Sahelian Niger. Few trees survive because all the wood is being taken for fuel. Unless the pressure on land can be relieved only disaster waits.

watering holes have been provided; but they have always resulted in herdsmen staying too long in one place and the land being overgrazed (Fig. 33.24). At the same time more permanent settlements have increased pressures on arable lands, causing a decline of the traditional shifting cultivation. There has been little attempt to maintain fertility by means of organic manures. The shortening of the fallow period has meant a decrease, not only in soil fertility but in the amount of 'bush' (young trees) which has traditionally been so important for timber, fuel and shade.

COMMERCIAL RANCHING

There is a tremendous gap between traditional herding methods and those of the commercial ranch, whether it be owned by government or privately. Ranches are characterised by investment in specially bred stock, fenced land, pumped water supplies, and the prevention of disease. In contrast to the nomad, who has to put up with the land as he finds it, ranches can invest in bush control by thinning or clearing unwanted woody growth. But commercial ranching is only viable if output can be increased to a level above expenditure. For example, it has not been found worthwhile to use fertilisers on the savanna, despite the increase in plant growth that this brings, simply because the increases in yields do not match the costs.

In all cases the main interest is in beef cattle and measures have been taken to eliminate both the herbivorous wildlife that would compete for food and the predators that would kill the cattle. Cattle are fed supplementary protein-rich concentrates in the dry season to maintain a regular growth rate throughout the year (Fig. 33.25). But above all this ranches practise a form of rotational grazing, while stock numbers are kept down to a level which prevents overuse of the grass (Fig. 33.26). Water is provided from boreholes spread evenly over the land to ensure a uniform pattern of grazing, while land is carefully burned over to ensure the maximum crop of grass and the minimum of woody growth.

Ranching is a way of improving the economic basis of the drier savanna lands, but as with all other types of advanced farming, it is capital-intensive and out of the reach of the tribal herdsmen. To create the ranches, large tracts of land previously grazed by nomadic herds are fenced off, thereby making the nomads' plight worse in times of scarce forage. At present there remains a basic conflict of interest which will take a lot of perseverence to resolve. Unlike the humid Tropics, there is no longer enough land for everyone.

Ranching is also a major activity of the drier temperate grasslands. Here indigenous Indians have long ago been displaced and farming is exclusively commercial. As with commercial range farming in the Tropics, temperate grassland ranching in South-West USA, in the Argentinian pampas, in Australia and elsewhere is a well organised industry, highly capitalised and technologically aware. Since the 1930s, when overgrazing was rife, stock densities have been reduced, ranges improved by seeding, 'weeds' controlled by spraying with herbicides and by using a rotational-grazing system whereby the grass is rested for 4-12 months on a long rotation.

Figure 33.25 Ranchland cattle in the savanna (notice the fencing). Compare the size and vigour of these cattle with those in Figure 33.8.

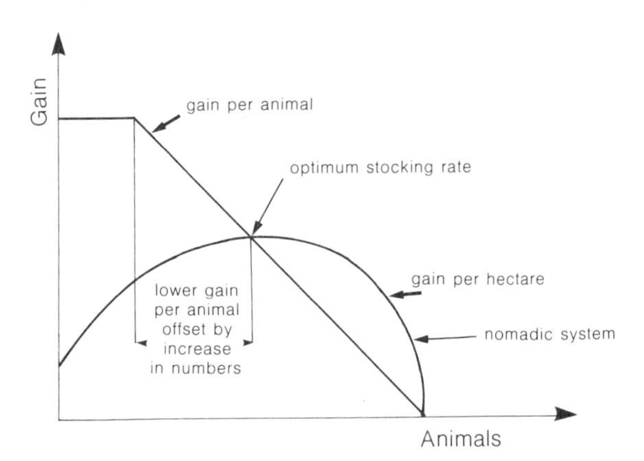

Figure 33.26 How a ranch determines its optimum stocking potential to get the highest yields and prevent overgrazing.

UPLAND GRAZING

It is hard to think of a greater contrast than the camel-herding nomads of North Africa and the pastoral farmers of the British uplands, or those in Norway, the Massif Central of France, or the Swiss Alps. But these upland farmers too are operating grazing systems, though they are constrained by temperature rather than by low rainfall. For example in upland Scotland rainfall exceeds evapotranspiration by a good margin every month of the year, soils are heavily leached and always acid. The problem is compounded by short growing season and low summer temperatures. This combination of environmental factors severely reduces not only the rate of growth but also the variety of plants that will grow. Cultivation is usually out of the question, but even pastoral activities are restricted because the upland grasses, themselves growing on nutrient-deficient soils, are usually tough, slow-growing, and of limited nutrient value for animals. Climate also limits farmers to rearing only the most hardy animals. Farming is therefore forced to be extensive in character.

Animal-rearing for meat is an economic proposition only on lowland good-quality grasses, unless it is subsidised by government. Upland farmers must therefore specialise in something else – wool production, lambs or bullocks for sale to lowland farmers for fattening, or breeding stock. All these elements usually play an important role in the uplands, sheep-rearing based on pure-bred Blackface and Cheviot flocks being the most common activity. However, because pure-bred flocks are generally self-replacing, there is little local demand for ewe lambs for breeding. Also, the grazing lands place strict limits on the total size of flock kept for wool. As a result there are extensive surpluses which are sold at Highland markets and bought by lowland farmers either for breeding stock or for fattening. Because lowlands and uplands are separated by large distances, sheep trading involves considerable transport. Lowland farmers do not travel directly to highland markets, but rather lowland dealers buy from highland markets and sell to farmers in their home districts.

In the high-cost environment of the developed world, pastoralism really only survives with a price structure maintained by government subsidy at an artificially high level.

33.3 Developments in farming

Farmers are popularly believed to be among the most conservative members of the population, yet the innovations that have been adopted within the last thirty years have been dramatic. There are three main developments.

Change from pastoral to arable farming

There has been a progressive shift from pastoral to arable farming in both the developed and developing world (Fig. 33.27).

In the developed world the shift has been promoted by economic factors: cereal and fodder crops grown in the field, harvested, and given to stall-fed animals have proved to be more cost-effective than allowing animals access to uncontrolled grazing. This improvement in productivity has permitted farmers to increase output on a land area that steadily declines due to urban expansion.

In the developing world, change from pastoral to arable activities is primarily a result of increasing population pressure. In a subsistence economy there are large numbers of farmers; as the population grows, so (in proportion) does the number of farmers. Their need for land can only be satisfied by a reduction in extensive farming systems.

Figure 33.27 Sheep farming in the Lake District.

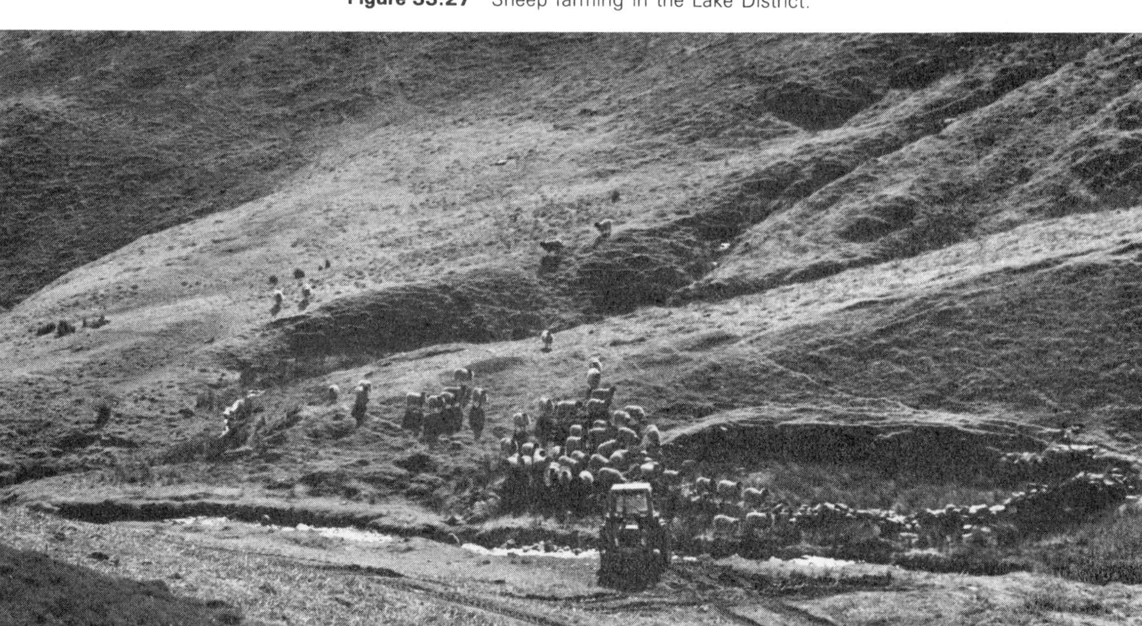

New varieties

New varieties of crop and better breeds of animal have gradually been developed. In particular the development of hybrid varieties of cereals such as wheat and rice has allowed increased yields (up to three times the earlier yields) in some parts of the developing world, particularly in Asia and Latin America. Unfortunately the **green revolution**, as the use of these new hybrids has been called, has had little impact on Africa, where climatic, soil and disease problems have so far made the introduction of most hybrids impractical.

Irrigation and fertilisers

The use of irrigation and fertiliser has increased. There are still large areas of the world that could be used for cultivation, and there is also scope for intensifying the use of land already under cultivation. For example it is estimated that the Sahelian region of Africa has a potential for 12 million ha of irrigated land, yet at present only 80,00 ha have been irrigated.

Land productivity can also be vastly increased by fertiliser application. For example, in the Chibi district of Zimbabwe the yield of maize on old, unfertilised land is only a couple of bags per hectare. The application even of low levels of animal manure has increased yields to about 1 tonne/ha.

In 1972 Indonesia produced two rice crops on 10 million ha of paddy, yielding 21 million tonnes of grain, the equivalent of 160 kg/head. It provided 70 per cent of the energy and 100 per cent of the protein required by the country's 130 million people, but the rice yield was only 20 per cent of that typical of Japan. The difference in yield is due almost entirely to the difference in plant nutrition. For example, about 40 kg/ha of nitrogen is taken from the fields with the harvest each year. Rainfall naturally returns about 10 kg/ha, but if the nitrogen applied were equal to the 200 kg/ha applied in Japan, then a fivefold increase in food supply would be possible from the same land.

Indonesian farmers do not add fertiliser because they cannot afford it. One reason is the policy of the government to keep down agricultural prices: this in turn is made possible because the food markets of the world have been depressed for 100 years by the huge surplus production of American farmers.

The cost of feeding the world

When we think of the resources needed to supply food to humans we need to consider not only the resources used by the farmer, but also those used in storing, transporting, distributing and cooking the food. With a rapidly expanding population we need to consider what resources remain underexploited and where they are on the Earth's surface.

Although the oceans occupy over two thirds of the Earth's surface, and fishing provides about 10 per cent of the protein available at present, the potential for increasing this proportion is low, and such a change would be expensive in energy. It is probably impossible to maintain a rate of catch more than twice the present one. Feeding the world's growing population therefore depends on improved agricultural yields. Whether it will be possible to maintain a rapid rise in production depends not only on the natural resources available, but on the people, their willingness to work effectively with the land they own and in technological advances (Fig. 33.28).

Some of the natural resources needed for agriculture are already limited: choices will have to be made about the use of each resource. Soon we may realistically have to ask such fundamental questions as:

(a) Should people continue to be allowed large gardens which are mostly unproductive, or should this land be allocated to farmers and people asked to live in high-density flats?
(b) Should water in semi-arid areas be used to fill swimming pools or irrigate crops, and should irrigation water be subsidised to encourage its wider use and therefore greater food production?

Yields can be improved drastically by the application of fertilisers. Nitrogen, in particular, is usually in short supply, but to produce a nitrogenous fertiliser requires the use of large supplies of natural gas. Perhaps by the end of the century 160 million tonnes of nitrogenous fertiliser will be applied to fields worldwide each year; this would require nearly 300 million tonnes of fossil fuel.

When you add the enormous use of fossil fuel for fertiliser to that required to run tractors and other machinery you discover that at present the energy used to produce eggs, milk and meat in the machine-intensive, developed areas of the world is often several times greater than the energy *in* the food! In the developing world the ratio is reversed. But what would happen if the agricultural systems of the developed world were adopted worldwide? How long then would the supply of fossil fuel last?

The energy used in preparing food by farmers, enormous as it is, is small compared with the energy used to transport and cook it, even in developing countries. Curiously, both in India and the US, for example, about twice as much energy is used to transport and cook the food as is present in the food itself. Finding the energy to move food from producer to consumer and then to cook it could become a greater problem than finding the energy to operate the farms.

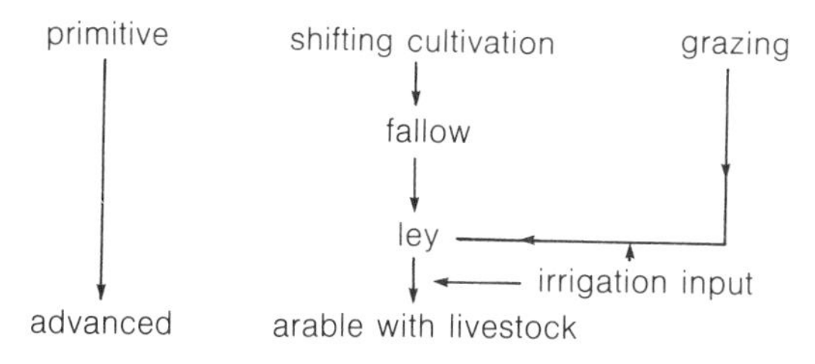

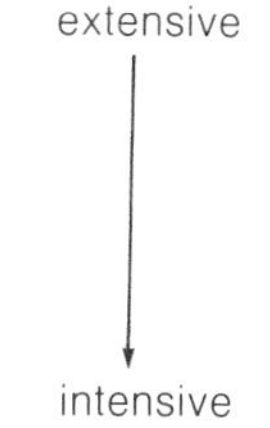

Figure 33.28 The evolution of farming systems.

As more and more land is brought into cultivation, so an increasing proportion will be of marginal quality. There will be a greater need for adequate drainage in some areas and irrigation in others; some will need massive doses of fertiliser, others will need stringent measures to prevent erosion; yet others will need levelling or terracing. All will need roads to get their produce to market and all will use machines and energy at every stage of production. Just to provide the necessary pesticides, for example, will need vast capital investment and many dozens of new factories.

The world's resources could not cope with these requirements for long. Yet the biggest resource remains relatively unexploited, even though it has been the kingpin of the Green Revolution. This resource is the plant itself. There will be ever more reliance on genetic research; plants must be made more efficient at using the resources that they are given. At present a crop of wheat transpires 99 per cent of its water just keeping cool and gaining nutrients; only legumes, of all the plants in the world, are able to fix their own nitrogen. Plants that used water less indiscriminately and that produced their own nitrogen would clearly be of enormous value in helping to produce more food. Such plants are not very far off.

There is great potential for improvement. At present a hectare of corn grown in the US mid-west produces about 60,000 kilocalories of food energy – enough to sustain 24 people all year. At a similar level of production the present world population could be fed from 0.17 billion ha. Already 1.4 billion ha are under cultivation – eight times the minimum, even without advances from genetic engineering. One of the principle differences between *potential* and *actual* yields from farmland occurs because most farmers cannot afford to use fertilisers and must leave large tracts of land fallow instead. About 10 per cent of farmland is used to grow non-food crops, such as cotton and rubber; 20 per cent is destroyed by pests; and the remainder used to rear livestock. But the fact that the mid-west US farmer gets 6 tonnes of corn per ha whereas a farmer in Pakistan gets no more than 1 tonne per ha is also due to soil variations. Differences in soil quality between regions will always mean that it will never be possible to achieve the same level of efficiency of production everywhere. These variations show up particularly strikingly in Africa, a country with a quarter of the world's land (Fig. 33.29). Here, naturally fertile chestnut and alluvial soils cover only 14 per cent of the land. The largest area of land (about a third) is covered with tropical red soils which are naturally infertile and which need massive fertiliser inputs to produce acceptable long-term yields. A further fifth of all the land needs irrigation. Therefore half of all potentially arable land can only produce crops if it is aided. Such aid uses large amounts of energy, either in producing and applying fertiliser or in pumping water.

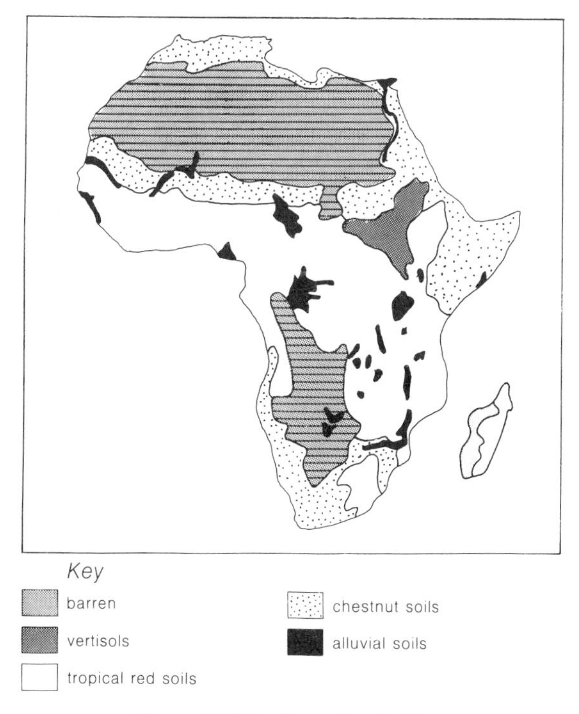

Figure 33.29 The soils of Africa.

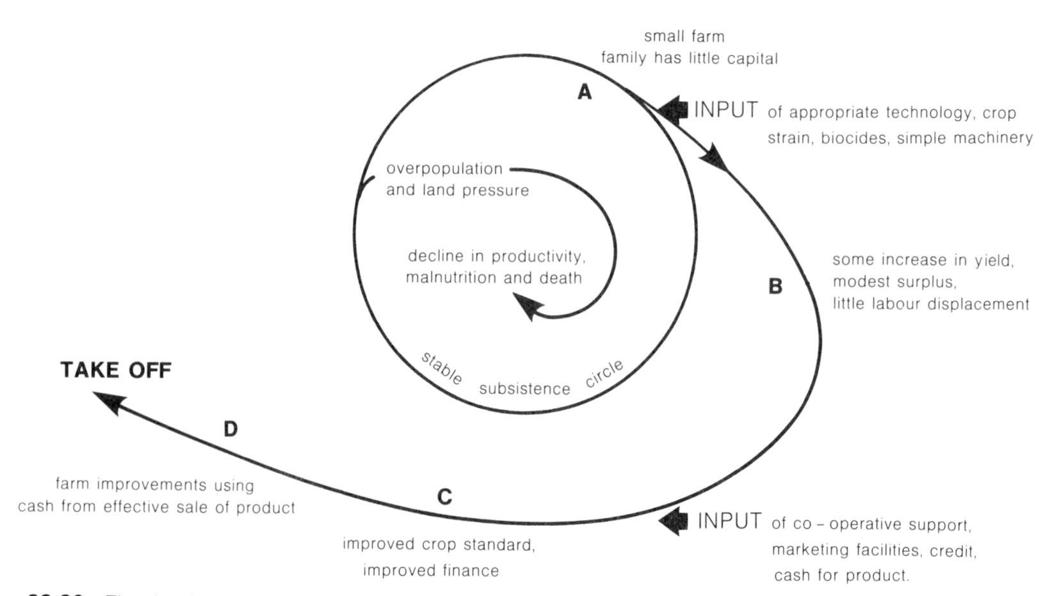

Figure 33.30 The development spiral helped by appropriate government aid; and the vicious downward spiral without it.

The state of farming

Some of the broad controls on farming systems have been mentioned above. However, some of the worst effects of environment can be combatted by skilful use of the basic biological principle of natural selection. Higher-yielding strains of all cereal crops have been bred, and also strains with shorter stalks which are less likely to fall over in storms, a problem especially just before harvest. New strains have greatly improved resistance to specific diseases. The technological innovations in the last century, both for increasing plant and animal yields, have been both rapid and dramatic, leading especially to the 'green revolution'. But many of the techniques for increasing yields are *energy-intensive,* and the energy comes from costly fossil fuels. Nutrient-deficiency in fields can be corrected using artificial fertilisers, made in a factory – at a price. Water can be pumped from underground and sprayed over fields to correct a moisture deficit – at a price: the cost of the machinery and the fuel to run it. The green revolution is thus primarily geared to a capital-intensive agricultural system – it can sometimes seem very inappropriate to the subsistence farmer in the developing world. In many poor countries livestock and man are in competition for the available grain, which is why grain is fed only to animals that are efficient converters of calories to tissue (chicken and pig) and animals that are needed for work (oxen). And although milk would be of great dietary value the world over, great productivity can only be achieved by a high-nutrient diet, which means a large input of money, so milk remains scarce in developing countries.

Agricultural location

34.1 Factors affecting land use

Farmers are influenced in what they produce by their natural environment, by the level of available technology, by market demand and by competition from other users of land. Near to cities industry competes for both land and labour, leaving only intensive (industrial) forms of farming such as market gardening. Other land is abandoned and allowed to degrade to wasteland or is used for recreation.

The pattern of farming away from a city does not necessarily follow a standard system. In the USA, for example, the southern states have deep, fertile soils and a warm humid climate which allows intensive farming. However, the area also suffers suppressed industrial employment, and this keeps a large number of people on the land both close to the cities and farther away. By contrast, the cool, humid northeast is the USA's most industrial region: agriculture is less productive, and the agricultural workforce is small. In these areas the rural population density is high – not because of a thriving farming workforce, but as the result of counter-urbanisation. In the south-west, industrial competition is limited and the main restriction on farming is lack of water. Here the change to low-intensity dry-land cultivation or ranching occurs at the city boundaries.

Despite the regional differences shown by the examples above, there is some justification for suggesting that farming types and employment change away from local centres of demand. In countries where the labour is less mobile than in the US or Britain, there is also good reason to suppose that land use changes with distance from the village homes of the farmers. This can be seen, for example, in Sicily (Fig. 34.1). Sicilian farmers live communally in villages, and must journey daily to their small plots scattered within the surrounding fields. Under these conditions the locations of different crops depend on the amount of attention required. Vines, for example, need extensive care (about 90 man-days/ha/year) and are therefore concentrated within 1.5 km of the village; less-demanding olive groves (needing 45 man-days/ha/year) begin to supplant the vine beyond 2 km. With little irrigation potential, wheat requires relatively little attention for much of the year (35 man-days/ha/year) and completely dominates the outer fields.

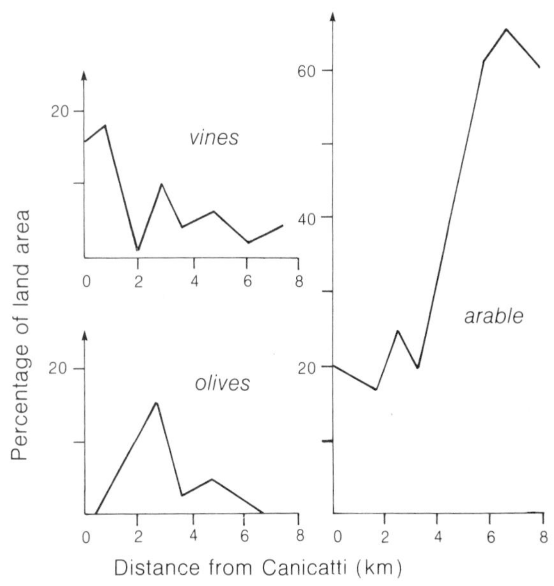

Figure 34.1 Land use near the village of Canicatti in Sicily is arranged in concentric zones according to the intensity of attention required by each crop.

34.2 Location in the commercial world

One of the first people to try to analyse farming patterns was A. von Thünen, who lived near the Baltic city of Rostock. Transport in the 19th century was so poor in this area that farmers were obliged to trade only with their local town. Each town 'catchment' was in effect an independent unit, isolated from its neighbours. Von Thünen argued that farmers engaged in commercial production would organise their use of agricultural land primarily to allow the most profit from each crop, bearing in mind the cost of transport to the local market. Thus the pattern of crops and livestock in an area would result from competition between the various types of production, and each field would be used for the type of farming that yields the highest net return of money per hectare of land (the **economic rent**). The most suitable land use for this simple case can be calculated by assuming that the climate and soils are uniform for the whole region and that the cost of cultivation is the same for all farmers. The main factor to consider is therefore the cost of transporting the product from the field to the market and hence the net return of each product per hectare.

Suppose the local town needs supplies of milk and wheat, both of which could be supplied from any of the surrounding farmland. The townspeople are prepared to pay more for their milk than they are for bread. A farmer can obtain a higher profit for each hectare by keeping dairy cows on grazing land than by using the land to grow wheat. However, milk is much more costly to transport than wheat, so farmers a few kilometres beyond the town boundary find more of their profit being consumed by transport costs than do those farmers nearer the town. At some distance from the town, therefore, transport costs exceed the profit and milk production becomes uneconomic (Fig. 34.2). Wheat cultivation is more attractive to farmers at some distance from the market because transport costs are lower. Not only is there a distance from the town at which milk becomes unprofitable overall; there is a *smaller* distance at which milk becomes *less* profitable than wheat. Beyond a certain distance farmers will therefore choose cereal rather than milk production in order to maximise their profit.

If all farmers were to adopt this method of calculating the most profitable land use, there would be a concentric pattern of land use, centred on the local market town. Close to the town there might be a ring in which dairying was most important; beyond this a ring in which potatoes were significant; and further out again a ring in which wheat cultivation would dominate. It is this **ring pattern** that forms the basis of von Thünen's analysis.

Von Thünen's analysis also provides a way of examining the effects of different *qualities* of land. At any given distance from the town transport costs are constant. But the yield of a crop from some land will be higher than that of the same crop grown on less fertile land (at the same distance from the town). So although it may be sensible to grow the crop on the more fertile land, it may not be economic to grow it on the poorer land, where yields are lower but transport and labour costs are the same. A farmer may prefer to use the poorer land for some other, more viable kind of farming (unless, that is, increasing demand from the town raises prices and so makes a profit possible even from the poor land). Because both transport costs and land quality are liable to vary, in practice no one ring will be kept to only one product; rather within each ring there will be a preponderance towards a small number of land uses, each of which is viable under the local circumstances of yield and distance to the market.

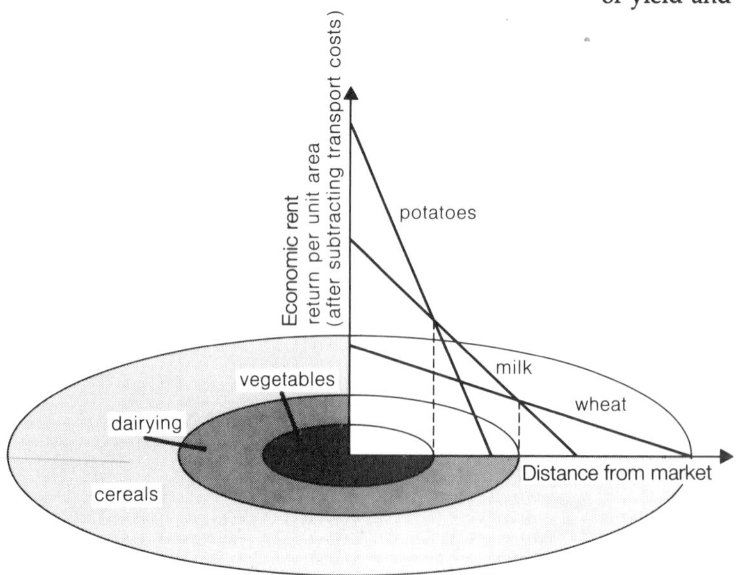

Figure 34.2 The concentric ring pattern of land use that would emerge from von Thünen's analysis applied to a single market centre.

We have assumed so far that every farmer wants to maximise his profits, and is therefore sensitive to the relative costs of his produce. But farmers are also interested in a satisfactory quality of life, not simply in wringing every last penny from their land. As a result, some land is not farmed at its greatest potential, but may still yield a satisfactory return for its owner. This adds an unpredictable factor to the analysis. In practice some farmers (**satisficers**) only change to a new land use when their present method of farming becomes positively disadvantageous; others (**maximisers**) will change more quickly in order to maximise profits. This satisficer/maximiser element, when added to variation in soil quality, gives a random aspect to land use. It is also the reason for the patchwork variety to fields which makes them such an attractive part of the landscape (Fig. 34.3).

Today, with rapid transport, it is unlikely that the produce of any one farm will be sent only to any one town. Products such as milk are now bought at standard rates by government agencies like the British Milk Marketing Board. The cost of transport is thus evened out over all the producers and plays no part in locational decision-taking. As a result the calculations of net return and of the most suitable land use become extremely complex. Indeed, von Thünen's analysis has been criticised because no one town can command all the produce from the surrounding farms. In Britain towns are so close together that a simple land-use pattern based on circular zones is inappropriate. With so many possible deviations from the simple model originally proposed by von Thünen, it may seem surprising that his ideas have not been abandoned decades ago. However, the basic model still serves as a basis against which to judge the complex patterns of the real world. Some forms of identifiable ring structure do exist in many areas. For example, the 'inner' ring of market gardening often survives, even in densely populated countries. Farmers in the outer rings begin to treat the combined areas of high population density as single market centres (Fig. 34.4). Maps of agricultural intensity in

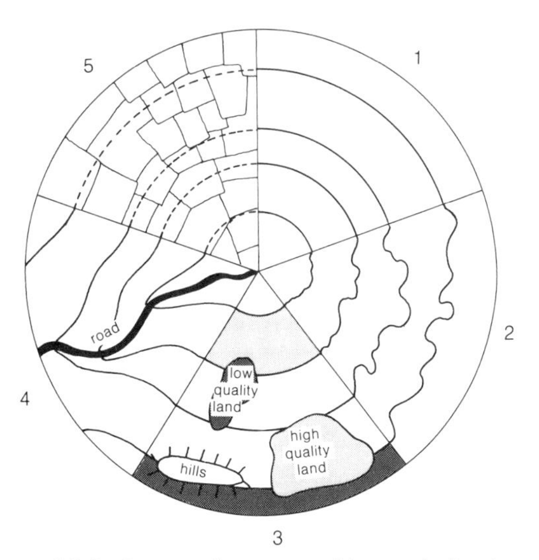

Figure 34.3 Factors that may add complexity to von Thünen's ring pattern of land use.

1 All farmers try to obtain maximum profit, land is of uniform quality and transport costs are the same.
2 If some farmers change their land use to obtain maximum profits (maximisers) while others only do so when their land use becomes uneconomic (satisficers).
3 If the land varies in quality due to soil conditions or relief.
4 If there are major transport arteries to reduce transport costs.
5 The effects of farmers trying to spread their efforts over more than one activity so they do not have all their eggs in one basket.

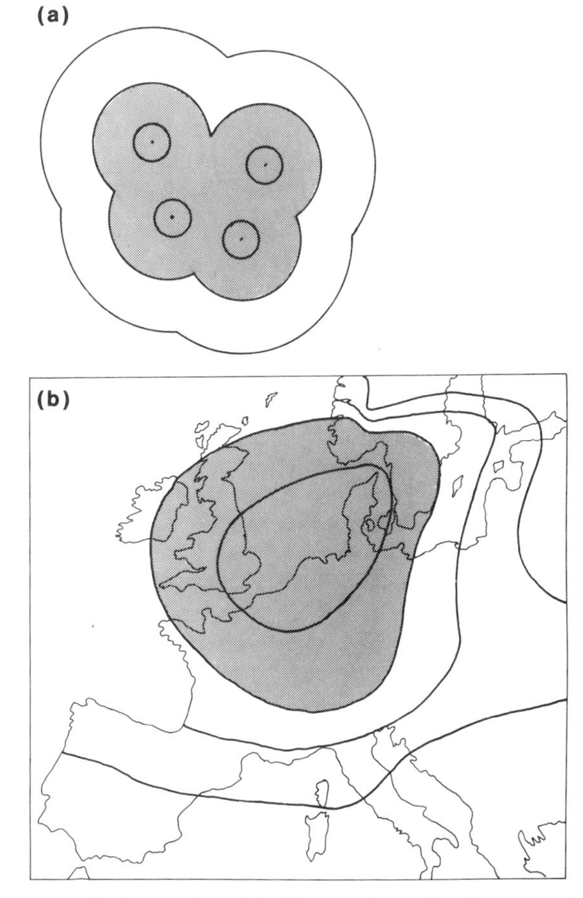

Figure 34.4 (a) A model of the way inner rings fuse around population centres. (b) The decreasing intensity of land use away from the central cluster of population in Europe.

Europe show a decline in farming intensity away from the main urban areas (Fig. 34.4b); the concentration of sheep farming in Australia and New Zealand could even be regarded as an example of a ring pattern on a global scale, Oceanian being on the periphery of the world markets.

34.3 Location in the subsistence world

Von Thünen's model has as its basic premise the assumption that farmers produce a surplus which they wish to sell in a market for a profit. However, at the time von Thünen was making his analysis, many of his neighbouring farmers would still have operated on subsistence levels; they would not have been strongly influenced by market forces. The same is true today. Much of the developing world has a dual agricultural system, one based on commercial principles and amenable to analysis based on transport cost, the other – and the major – sector based on subsistence, with only a minor amount of produce to sell in a market. In the latter case, the main aim is to grow a variety of crops that will yield a reliable food supply throughout the year, so everyone in a given climate grows much the same range of crops: there is no pattern of segregation. Crops grown for cash are not likely to be for sale at a local market, but to be bought for fixed prices by the government at official purchasing stations. People in these circumstances do not decide which cash crop to grow according to the costs of its transport, but according to:

(a) what will be bought by the government;
(b) what is easy and reliable to grow;
(c) what is not perishable.

34.4 Government influence on land use

A government planning use of the nation's land does not begin by asking: 'What are the physical conditions that will influence land use?' It begins instead from the number of people to be fed. These people, many of whom are voters who can keep the government in power, can be fed either by maximising home produce or by importing food. Imports must be paid for by exports in some other sector of the economy. In general terms, it is tactically advantageous to keep down imports (and therefore lessen any balance-of-payments problem) and strategically desirable to be as self-sufficient as possible, in case friendly relations with supplying countries turn sour.

This is the basic policy that a government has to try to implement. Only after the basic policy has been defined is attention turned to the farmers, to see what they are producing. Of course, each farmer strives to maintain a satisfactory a standard of living for himself and his family, by producing to suit market demands, whether at home or abroad. Within the limits imposed by climate and soil, natural market forces control farm produce. But what if the production that suits the individual farmer is less than his potential maximum, or if the variety of agricultural goods produced by the nation is unbalanced? In such circumstances, the government would to seek to influence farming patterns.

Some government policies are drastic and have substantial effects on the pattern of land use. In centrally controlled economies, such as those in many communist countries, the state attempts to control exactly both the type and amount of each product, to fit in with some previously defined plan. In practice this system does not prove to yield the greatest effort on the part of the workforce, and actual production is always well below the target potential. In most democracies, farmers and farm workers are represented by a substantial political lobby: a government that wants to stay in power has an interest in balancing national need against the desire to keep this sector of the population satisfied.

Two of the most powerful tools at the command of government are the **subsidy** and the **quota**. If the government decides that the country would be better off as a result of increasing production of a particular crop, as for example beet sugar, in order to reduce imports of cane sugar, it can award a guaranteed subsidy in beet production. This encourages farmers to cultivate beet. At the same time the application of import quotas suppresses competition. Governments often act for social as well as purely economic reasons. For example, sheep farming in highlands is heavily subsidised, not primarily to reduce imports, but to prevent the socially undesirable depopulation of highland regions. Governments pay for subsidies out of taxes on other sectors of the economy. Thus, by artificially controlling the pricing structure of agriculture, government can and does play a major role in overriding normal market forces and land-use patterns.

The EEC – a study in intervention

The European Economic Community (EEC) is a prime example of a 'national' government trying to control agricultural produce. Some 70 per cent of all funds contributed to the EEC budget by member states is used to subsidise agriculture and therefore to influence land-use patterns. The large expenditure involved reflects the EEC's aim to become agriculturally self-sufficient in all basic commodities, and to maintain the economic status of farmers.

The EEC operates its subsidies by means of import quotas and by guaranteeing farmers at least a minimum price for their produce, no matter what the world price may be. The minimum price at which government intervenes in the market is called the **intervention price**. The intervention price is designed to give farmers a 'fair' return for the effort expended and to keep their standard of living comparable with that of the industrial workers. The guaranteed price is also designed to provide stable market conditions in which farmers can have the confidence to plan ahead. In this way they should be able to maximise their efficiency.

Intervention pricing is a system of considerable vision and power, but it has a major drawback. Prices are set at a level that will allow the small, relatively inefficient farmers of southern Europe to maintain a reasonable standard of living. This gives very large profits to the most efficient farmers and creates surpluses. The surpluses must be dealt with:

(a) by selling them at a loss in the world market;
(b) by giving them to developing countries as food aid;
(c) by storing them as 'butter mountains' and 'wine lakes'.

This aspect is very worrying because it does not encourage the most efficient farm size, it allows what would otherwise be uneconomically small units to survive, and it does not allow for quick adjustments to changes in supply and demand within the market. Looked at another way, the EEC spends 70 per cent of its total budget trying to distort von Thünen's rings.

The way farmers respond to financial incentives provided by the CAP is illustrated by the expansion of the wheat area in the UK following our joining the EEC (Table 34.1) and the expansion in area sown to oil seed rape, following a decision to produce more vegetable oil and therefore offer better incentive prices (Table 34.2).

Table 34.1 Wheat yields and production in the UK.

Year	Yield (t/ha)	Production (million t)	Area grown (million ha)	UK population (millions)
1937	1.0	0.9	0.9	45
1958	2.6	2.4	0.9	52
1964	3.7	3.4	0.9	53
1977	4.9	5.2	1.5	55
1983	5.7	10.8	1.9	56

Table 34.2 Rape production in the UK (area, 000's of ha).

1980	1981	1982	1983	1984
92	125	174	222	269

34.5 Environment and farming: a case study

There are many difficulties in assessing the influence of the physical environment on land use in areas in which farmers employ high technology. However, in Dorset there are major contrasts between upland and lowland which allow the role of the physical environment to be assessed with some accuracy. Dorset has a 'backbone' made from a range of chalk hills whose trend is north-east to south-west. Separate lower-lying areas are predominantly clays. In the south-east there is an extensive area of heathland which has developed sandy-textured, acid soils.

In 1952 it was possible to divide the county of Dorset into eight land-use regions that were physically separate units. Such a striking correspondence appeared between these land-use units and the natural land types that one commentator remarked: 'the physical environment in Dorset has continued to exercise the strongest influence in fashioning the general pattern of the agricultural landscape' (Tavener, 1952). There was a clear pattern, for example, of arable/sheep farming on the chalk uplands and intensive dairying in the clay vales.

In the 25 years to 1977 there were considerable advances in crop and animal breeding, in machinery, and in the balance of pricing structures of differing forms of agriculture. Some or all of these factors seem to have affected the land-use decisions made by individual farmers in the period: while arable and sheep remain most important on the chalk uplands, and pastoral activity is mainly focused on the clay lands, there is now more variety within each region. The large urban area near Bournemouth seems to override other factors influencing land use. For example, the quantities of dairy produce, potatoes and market-garden products have increased near the town, showing how technology and demand/price structures can to a considerable degree override the natural landscape variety.

Over the whole county the broad simple relationship between geology and land use is breaking down. The pattern of farming has now become quite complicated (Fig. 34.5). For example, the mixed-farming areas no longer form a definite 'region' associated with clay soils. Although arable dominates on the chalk hills, there are significant areas of arable outside the region of chalk-derived soils. Farmers have been increasingly able to exercise choice in how they wish to use their land; this choice was not possible with the lower levels of technology of the past. Such results are a reminder not to use landscape labels to describe farming patterns in the simple way that might have been appropriate in the past. Market-pricing structures, rather than environment, are often the most influential features now within any climatically uniform region of the developed world.

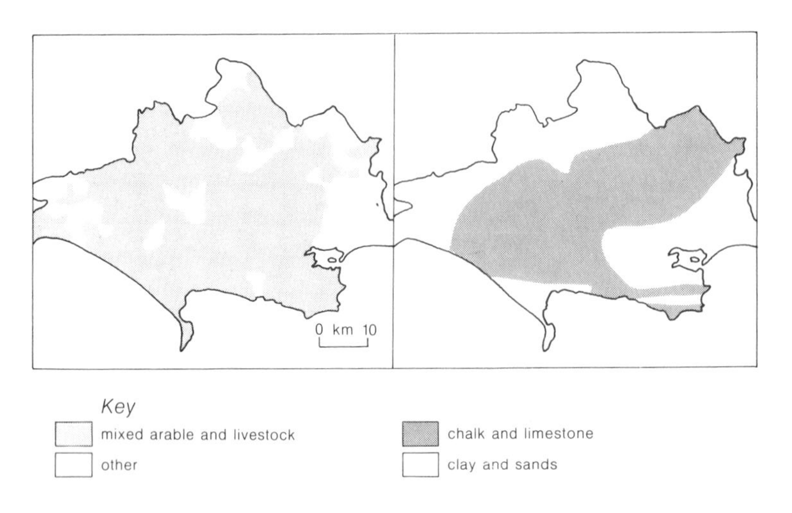

Figure 34.5 Farming patterns within Dorset are no longer as clearly related to physical conditions as they were only a quarter of a century ago.

34.6 The green revolution

In the last few years attention has focused on the introduction of new **high-yielding varieties (HYV)** of cereals, especially of wheat and rice. These have been heralded as the solution to the world's food problem. The introduction of these HYVs was hailed as 'the **green revolution**'. One author wrote of IR.8, the so-called 'miracle rice': 'The development of IR.8 and its dessemination throughout Asia is [...] literally helping to fill hundreds of millions of ricebowls once only half full'.

Such a view is simplistic in the extreme, and induces unwarranted optimism. It suggests that the countries to which the rice was introduced were transformed overnight by replacing their primitive strains of rice with the new sophisticated variety. This is both to overestimate IR.8 (and in particular its ability to grow in all conditions), and to undervalue the painstaking development of other rice strains by both farmers and scientists in the local rice-growing environments.

It is true that spectacular increases in yields of both wheat and rice have been recorded in Pakistan and northern India. However, the 'miracle' rice requires very closely controlled conditions; it does not yield well in areas of uncontrolled flooding, such as Bangladesh, nor in dry lands without extensive irrigation.

The green revolution was developed as a *complete* farming system, a 'package' of ideas and developments that had to be adapted *in full* for the benefits of increased yields to be realised. The 'package' included new seed varieties, chemical fertilisers, pesticides and weed killers, improvements to irrigation and the employment of new agricultural practices. In reality, many local farmers find only *parts* of the package acceptable or practicable in their circumstances. For instance, many farmers use the seed but do not use fertilisers: under these conditions the HYVs sometimes yield less well even than traditional varieties. Effective husbandry may require mechanical pumping from tube wells and the use of tractors; often one is adopted, but not the other. Farmers do not neglect parts of the package from ignorance, so much as from lack of capital or lack of suitability to their local conditions.

The green revolution has had other, equally dramatic social implications. The new HYVs are best grown under mechanised systems. These require relatively large fields and farm units if they are to be successful. In consequence, farmers who own land but used to rent part of it to others because they themselves could not manage it all by hand now use machines and need all their land to offset the cost of machinery. This consolidation of farms has deprived a large number of landless peasants of any opportunity to work the land. In this way the green revolution is actually depriving people of a means of providing for their livelihood. 'Extruded from the bottom of the pile, forced in desperation to leave their villages these "least enviable of men" [...] will swell the numbers of urban migrants and rural transients whose lot will be more terrible for being so often unseen and so easy to avoid seeing' (Farmer 1977, p. 418).

Part I

Population

35

Population distribution

35.1 The unevenness of distribution

The world is populated in a very uneven way (Fig. 35.1). The most densely populated areas lie in eastern Asia, in Europe and north-east America. There are also large concentrations of people in 'pockets' along the coast of South America, Africa and Oceania. By contrast, some areas, mainly deserts and mountains but also the great belt of tropical rainforest, support very few people. Finally there are areas with intermediate population densities, such as the interiors of America, Africa and Asia.

The distribution of the world's population may at first seem random, yet it is the result of a number of driving forces. The effects of these can, to some extent, be isolated and examined separately. This is the task of **population geography (demography)**.

35.2 Factors affecting distributions

One of the more important influences on human activity is the preference to live within a region of perceived comfort, where temperature and humidity are within tolerable levels. For example, many people prefer to live in climates offering temperatures within the range of 10 °C to 30 °C, and low to medium humidity. High humidity and high temperature is a combination that few can tolerate (Fig. 35.2). However, people can survive in climates beyond these 'natural' limits with the help of special clothing, houses, heating and air conditioning.

The more developed the culture, the more able are people to protect themselves from the undesirable features of their environment.

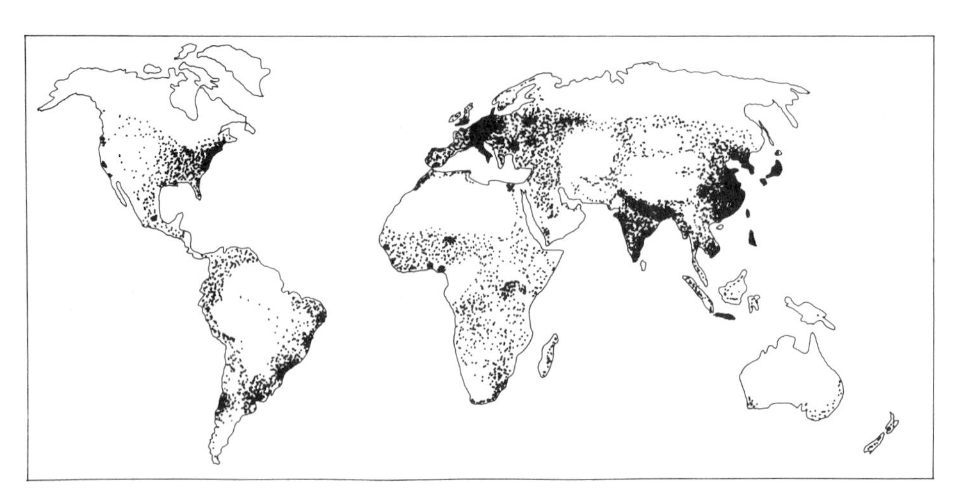

Figure 35.1 World population distribution (one dot represents 100 000 people).

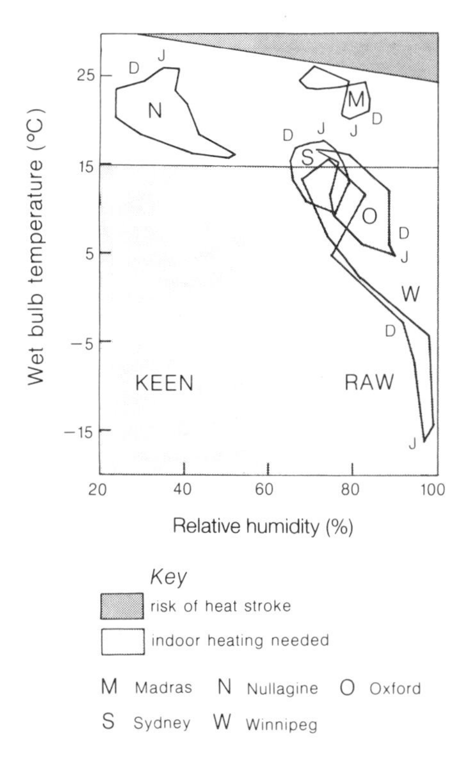

Figure 35.2 The human body is cooled by radiation and evaporation: both cause a sensation of cold. Thus temperature below −20°C can be tolerated in still air, but 0°C in a gale is intolerable. Similarly, 30°C in a region of high humidity is less tolerable than 40°C in the dry conditions of the desert. Each combination of factors produces its own physiological results: wet climates produce nervous depression and lethargy; dry climates create nervous energy, excitability and sleeplessness; wet-bulb temperatures over 30°C increase the chances of heat stroke; and continuous wet-bulb temperatures above 21°C have limited European colonisation by making sustained manual labour and even office occupations incompatible with continued health and comfort.

For this reason, and for the additional reason that differing cultures perceive different norms, it is difficult to find close relationships between people and environment. It is only possible to say that people *tend* to avoid harsh environments and *tend* to prefer those which are naturally more favourable. There are numerous exceptions even to this generalisation. For example there are few people in the harsh environment of Greenland or the Himalayas, but many in the harsh semi-arid lands bordering the Sahara; there are many people in the rich lands of Indonesia and lowland Mexico, but also high densities in the far poorer lands of southern Italy and Denmark.

Economic activity

People cannot be closely related to particular environments for another reason – they are largely driven by economic forces such as the need to earn money to buy food. Economic factors vary both with space and time; the water that once drove the wheels of industry in the Pennines still flows down the valleys, but technology has made direct water power redundant and the hills are now only sparsely populated. The positions of springs and fords that were of vital economic interest when settlements were founded in Britain in the Dark Ages play no part in the distribution of the great urban areas of today; the location of these areas is determined almost entirely by external economic factors.

The most one can say about the relationship of people and environment is that *people are opportunists*: they will find ways of using or modifying the environment to their best advantage as necessary, but once the need has passed they will then abandon the environment. In effect, therefore, the population distribution can most closely be explained by reference to an economic rather than to a physical base. Even then the relationship is not simple, for although the population density *tends* to increase with the level of economic activity (hunter-gatherer, cultivator, manufacturer), the exceptions – such as the corn belt in the USA or the plantations in Kenya, with their advanced economies but low population densities – show that this generalisation also has major limits. We do not have to look far beyond our own doorstep to see varying patterns of population even with the same economies. Thus France has a population density of 97 persons/km^2 whereas Belgium has a population density of 386 persons/km^2.

Culture

A large proportion of the activities of mankind are driven not only by an awareness of the environment and a need to survive, but by cultural factors (Fig. 35.3). Survival can be achieved in numerous ways, and each differing method leads to the development of a specific culture. For example, each culture has its own norms for mating, marrying and childbearing. In some southern Asian countries marriages take place at a very early age (sometimes younger than 10 years), and childbearing begins at puberty. In other countries (such as China) marriage is delayed until the woman is many years into her period of fertility. Contrasts such as these are fundamental to the structure of population, for early marriage and childbearing may easily lead to a very rapidly expanding population. Similarly, there are differing perceived

Figure 35.3 The human race is culturally and technologically very diverse. The European with his camera greets pygmies in the African rainforest.

ideals in family size. In Britain two children per family may be the norm today, but a century ago the norm was much higher. In some cultures the number of male children is still considered a measure of the father's status.

Most of the cultural practices have continued not for any survival need but because stress on the population has been insufficient to modify them. So far, man has been unique among the animal species in reproducing at a rate below his physical maximum. Many animals in the wild breed at the fastest possible rate until they exceed the carrying capacity of the land, when they die back from malnutrition and starvation. By contrast each time man has reached the apparent carrying capacity of the land, he has found ways of increasing that capacity to meet his needs – globally if not regionally. In the case of man, stress has also been reduced by migration to new lands – an escape mechanism largely denied to other animals.

Within the pattern of migration cultural influences have often determined destinations. Thus the northern Europeans migrated to northern America not only because these areas had been colonised by their ancestors, but also because the climate was similar and their accustomed life-style could be maintained. Spaniards and Portuguese migrated to their warm central and southern colonies in America. And though the Spaniards were well able to settle and survive in Mexico, they never gained control of the wetter regions of Colombia nor the colder regions of Argentina. Similarly, Britons settled easily in the cooler Commonwealth countries (Australia, Canada, New Zealand and South Africa), but settled less extensively in the hotter and more humid regions of the Tropics (eastern Africa, western Africa and India). Such cultural backgrounds are particularly striking in the colonisation of Kenya, where European settlement was quickly established in the cooler high plateau (Central Highlands region) rather than in the extensive surrounding plains. Not only was it physiologically more comfortable, but at that stage the Europeans had no cultural solution to plains farming.

Patterns of population are also influenced by attempts to maintain cultural identity – or perhaps to impose a cultural identity on others. Differences in race, religion, language, ethnic affiliation, social class and political beliefs have all led to tension and sometimes warfare. Such conflicts are still commonplace in both the developed and developing worlds, at a local level (as instanced by the Catholic and Protestant sectors of Belfast, Northern Ireland), at a regional level (French and Flemish speakers in Belgium, for instance), and at a continental level (as between the NATO and Warsaw Pact alliances). Animosity between groups is a powerful means of retaining cultural identity and therefore psychological security.

Disasters

Environmental, economic and cultural influences are the major factors that determine the distribution of population over long periods. However, disasters, both natural and those brought about by man, happen quickly and cause more rapid population adjustments. Their consequences are unpredictable and have to be seen as random changes in population composition and location.

Natural disasters are the least important to man: droughts, volcanoes, floods and earthquakes may kill many people, but they do not usually produce more than a temporary hiccup in the long-term population trend within a region. Social disasters – such as the extermination of many Jews in World War II, the expulsion of Asians from Uganda in the 1970s, and the two centuries of slave-trading across the Atlantic – have much larger and longer-lasting effects. In general the survivors from social disasters tend to migrate to new lands.

Political decisions

Today, more than ever, governments attempt to play a major part in maintaining cultural identity and in controlling the population size and settlement pattern. They use such devices as restricting immigration (as in Britain), imposing or encouraging various levels of fertility (as in China), and providing economic incentives to occupy one part of a territory rather than another (as in Italy). They do this not only within their own boundaries, but also, via their economic influence, in other countries. These external influences are seen at their crudest when, for example, the developed world markets reduce the price for staple goods produced by the developing world. Reductions in the demand or cost of rubber, cacao, sugar, copper and many other commodities have each had a major impact on the growth of the producer countries. So far, political decisions, of all the factors, have had the least effect on population movements. However, decisions such as that recently made by the Chinese government – to restrict each family to one child for the next hundred years, and thereby halve the total population – may have progressively more important effects on populations.

35.3 Population structures

The distribution of people throughout the world is a result of the effects of a dynamic and rapidly changing human population on a slowly changing physical environment. The number of permutations of interacting forces is infinite, but the structures of all communities are affected by certain basic biological factors:

(a) women reach childbearing age at about 12, go through a period of fertility until about 40, and then become infertile;
(b) men are fertile for longer than women;
(c) men have a shorter life expectancy than women;
(d) both sexes are especially liable to die from disease when they are very young and as they become old.

Within this broad framework there are slight natural differences in sex ratios. For example, in the US there are 95 male births to each 100 female births; in Japan the figure is 94 and in West Germany 91, whereas in China there are 106 and in India 107.

Many of the effects of these biological factors may be locally modified by cultural and technological variations. For example, urban populations are very different from rural populations in age and sex structure; developed technological societies differ from societies in the developing world. In urban societies there are more young adults than any other group; more males than females, and a higher life expectancy than for rural populations. The structure of a population both as a whole and for selected parts can be expressed by the equation:

$$PI = B - D \pm NM$$

in which PI = population increase, B = crude birth rate, D = crude death rate, and NM = net migration (the difference between the numbers of emigrants from an area and immigrants into it).

In a population that does not experience transfers of people with other areas, the structure of the society, displayed in terms of age and sex, would be of a pyramidal form (Fig. 35.4). Within this general form, cultural, social and economic factors will cause distortions according to local circumstances. For example, a population with a high birth rate but high mortality and low life expectancy (Fig. 35.4a) will show a wide-based but steeply inclined pyramid; whereas a population with a low birth rate but a low death rate and a long life expectancy will be narrowly based and decline with age much more slowly (Fig. 35.4b). Only primitive societies have both high birth and death rates (Fig. 35.4c). In consequence these countries have rapidly increasing populations. Most countries of the developed world have both low birth and death rates and a long life expectancy. Some have reduced their birth rates below a level needed to sustain the population at a stable level (**zero population growth** or **ZPG**). These countries have a population pyramid with a narrow base and an adult 'bulge' (Fig. 35.4d). They will have steadily ageing populations.

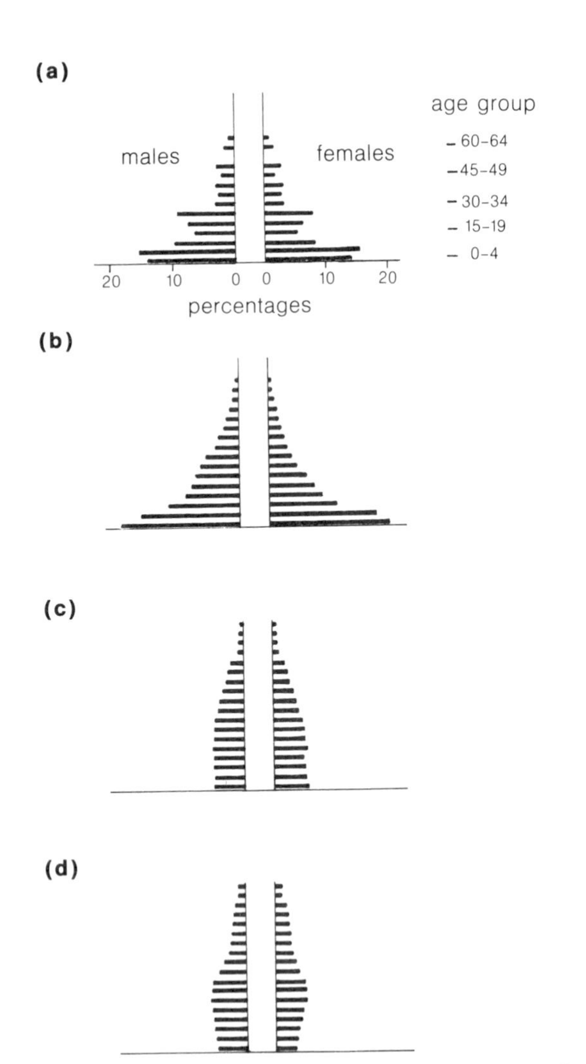

Figure 35.4 Population pyramids: (a) stage 1; (b) stage 2; (c) stage 3; (d) stage 4.

35.4 Population-resource regions

It is possible to identify 'population-resource' regions, where the cultural aspects of population combine with the physical and economic environment. Ackerman has suggested that there are five types of population-resource regions (Fig. 35.5):

(a) technology source areas of low population: resource ratio (e.g. the USA);
(b) technology source areas of high population: resource ratio (e.g. Britain);
(c) technology-deficient areas of low population : resource ratio (e.g. Brazil);
(d) technology-deficient areas of high population : resource ratio (e.g. Egypt);
(e) technology-deficient areas of low food-production potential (e.g deserts).

The type (a) regions are regarded as having the most advantageous population-resource combination; there is still much room for population expansion and a plentiful resource base dependence on any other region may be reduced. Although type (b) is no less technologically advanced and the ability to support the population remains good, more intensive use of resources is needed and a careful control must be kept on renewable resources. (For example, countries such as Britain could never have survived extensive soil erosion with as little economic penalty as was the case for the USA.) Furthermore, the close balance between population and resources encourages an outward trading approach, particularly with developing world countries that have surplus raw materials and labour for sale. Many less-developed regions with expanding populations cannot hope to achieve an (a) type of population-resource combination, and may evolve towards the (b) type.

The (c) type can be regarded as a favoured resource area which is still considerably underdeveloped. It will be able to support a much larger population on its known resources and therefore has a good prospect of making the transition to the (b) type. By contrast, the (d) type represents a population-resource environment in which there are already high levels of population but limited resources, resulting in a severe stress on the environment and little long-term chance of improvement. The (e) type is distinct from all others because the land is virtually uninhabitable and populations have always been low.

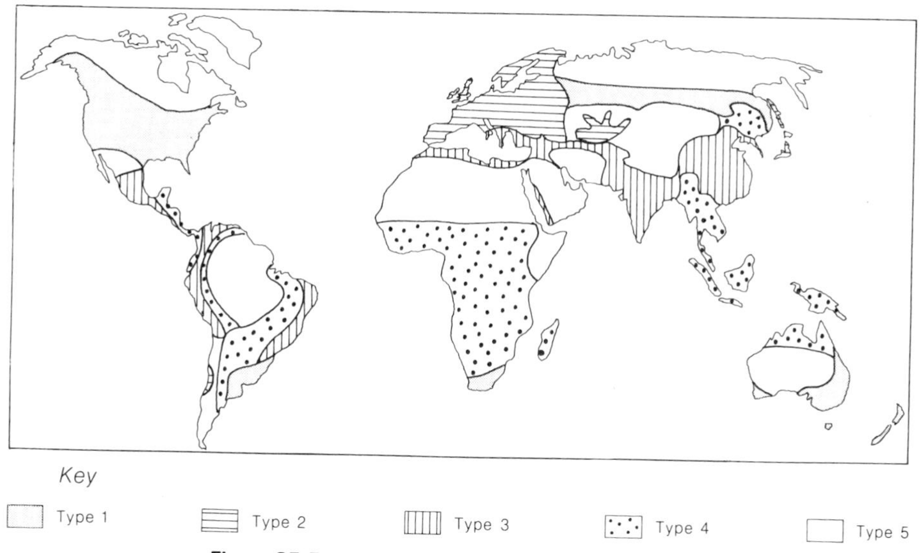

Figure 35.5 Generalized population/resource regions.

Population change

36.1 The importance of demography

In any study of world population it is necessary to know not only the present distribution of people, but the manner of their change in the past and the likely pattern of change in the future. This information, when combined with a knowledge of available resources, can then be used to help understand the way people live today. Information about change also serves a predictive purpose; it helps to demonstrate some of the future consequences of population change.

When many people think of population change, they think of growth. Population growth is a phenomenon that has dominated world history, but, as is the case today, the world very often contains both regions of growth and regions of decline.

36.2 The population explosion

It is now nearly two centuries since Thomas Robert Malthus wrote his *First essay on the principle of population*. In it he suggested that populations of all species grow ever faster until they reach the carrying capacity of the ecosystem in which they have developed. In this sense human expansion is no different from that of any other species: the human population doubles itself during an interval equal to 69/(annual rate of increase). Thus Kenya, a country with an annual rate of population increase of 4 per cent, will double from its present 14 million to 28 million in 69/4 = 17 years; China, with a current rate of increase of 1.2 per cent, will double from its present 1000 million to 2000 million in 57 years; and the UK, with a current increase of 0.15 per cent, will double from the present 56 million to 112 million in 460 years.

The growth of the world's population to the present 4 billion has been achieved by a doubling only once every 30 000 years (Fig. 36.1). However, most of the doublings have occurred within the last 200 years and the time for doubling is becoming ever shorter in many areas. The crisis brought on by this population explosion becomes apparent primarily as a shortage of food. Malthus, speaking from a low-technology 18th-century perspective, suggested that the accelerating growth of population was quite different from the expansion in available food supply, which would follow only a steady growth (Fig. 36.2). It thus seemed *inevitable* to Malthus that at some point the population would exceed the food resources of the world and mass starvation and death would occur. Malthus' prediction has not yet come true because technological improvements have kept the world food supply slightly ahead of the world population's needs (see Ch. 33).

36.3 The demographic transition

Populations are controlled by the balance between the rates of birth and of death, usually referred to as the **rate of natural increase**. The world is presently (1985) adding about 85 million people – the population of Nigeria, Africa's most populous country – to its total each year, because the annual rate of births (29/1000 members of the population) exceeds deaths (11/1000) by 18/1000. At this rate the world's population will have doubled to 8 billion people in only 37 years.

The figure for the average global world growth rate hides a wide degree of variation between the nations of the world. Many developed countries now experience **zero growth rate (ZGR)** and in some cases are even in decline (negative growth). On the other hand most countries of the developing world have very high rates of increase: the highest, Kenya, at present is over 4 per cent per annum (1985).

THREE VIEWS OF POPULATION CHANGE

These three charts provide three different perspectives on past and present population growth. Although each looks different, all are based on the same facts.

The top chart shows the change in the absolute size of world population from about 9000 BC (the beginning of the agricultural age) to the end of the present century. The middle chart shows population growth rates from AD 1750 to 2000. The bottom chart, like the top one, shows change in the size of the human population, but over a longer period (back to 1 million BC) and on a logarithmic scale.

The three charts suggest strikingly different impressions of population growth. The top one conveys the impression of an enormous population explosion beginning sometime after 1750 (the beginning of the industrial age). It points upwards at the year 2000 with no apparent limit. The middle chart indicates that these recent dramatic increases have been produced by relatively small, though accelerating growth rates. Annual growth was about 0.4 per cent between 1750 and 1800, crept up steadily until it reached 0.8 per cent in 1900–50, and then rose sharply to 1.7 per cent between 1950 and 1975. The line for growth rates points upward, but shows a recent dip, and is therefore not as dramatically threatening as that in the top chart.

The bottom chart shows that population has grown from somewhat less than 10 million on the eve of the agricultural age. In the first rise in the curve, population expanded gradually toward the limit supportable by hunting and gathering. With the adoption of farming and animal husbandry, a second burst of population growth began. Eventually – though much more quickly than in the first case – the technological limits were again reached, and population stabilised at around 300 million in the first millenium AD. From the late 18th century the industrial age triggered a third burst of population growth. It began from a much higher base and has covered a much shorter period, but on a logarithmic scale it appears no more rapid or unusual than earlier growth spurts.

Which chart best portrays the past and the prospects for the future? The top one emphasises the special character of recent population growth, setting current experience apart from thousands of years of earlier history. It conveys a sense of crisis. The middle chart highlights the substantial acceleration in growth rates, especially in the past quarter century, that produced this expansion, and the current downward trend in those rates. It suggests that managing population growth is possible. The bottom figure underlines the likelihood of an eventual equilibrium between population and resources, achieved either by a decline in birth rates or an unwelcome rise in death rates. It calls attention to the need to achieve equilibrium by a decline in birth rates.

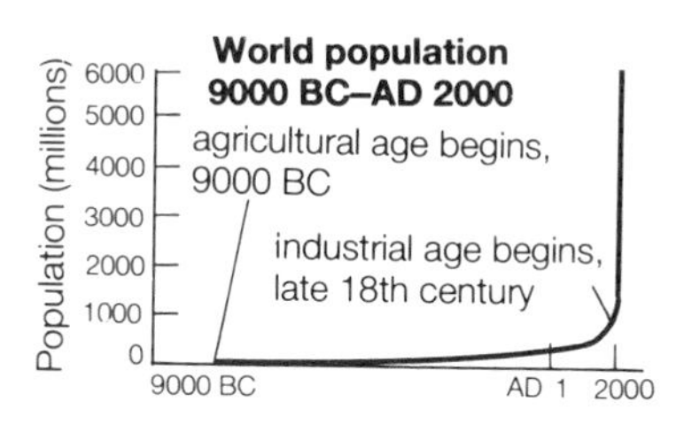

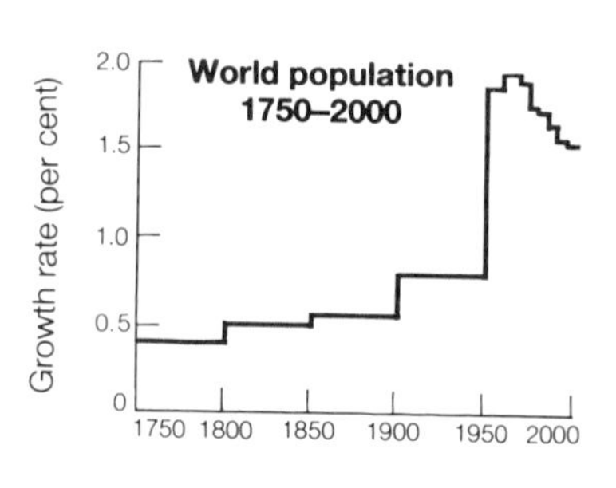

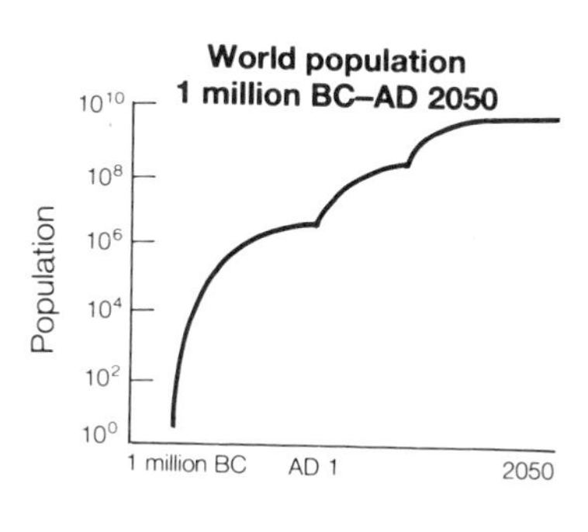

Figure 36.1 World population growth since 1000 BC.

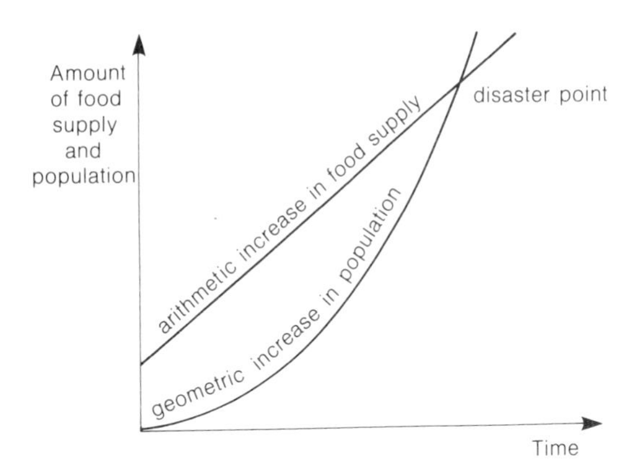

Figure 36.2 The Malthusian prophesy: in the long term, population growth will outstrip food supply.

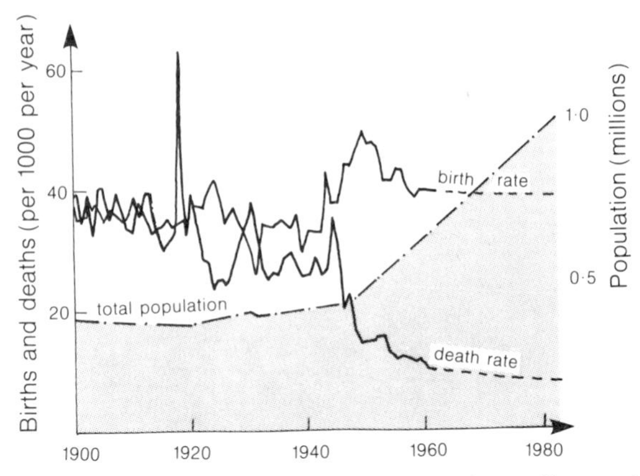

Figure 36.3 Developing world population change illustrated by Mauritius. Related changes in social and economic conditions and natural hazards are indicated.

Within the groupings 'developed countries' and 'developing countries' there are equally large contrasts. Thus, for example in Latin America the crude birth rate is 31/1000 and the death rate is 8/1000, giving a growth of 2.3 per cent; in Africa both birth and death rates are higher: 46/1000 and 16/1000 respectively, giving a growth of 3.0 per cent. Asia is in an intermediate position with a birth rate of 30/1000 and a death rate of 11/1000 giving a growth rate of 1.9 per cent. Similar comparisons are possible within the developed world, and even between adjacent European countries. Thus the USA has a birth rate of 16/1000 and a death rate of 9/1000, giving a growth rate of 0.7 per cent; the UK has a birth rate of 13.5/1000 and a death rate of 12/1000, giving a growth rate of 0.15 per cent; and France has a birth rate of 15/1000, a death rate of 10/1000 and a growth rate of 0.5 per cent.

These differing growth rates have only existed for a relatively short time. Two centuries ago the birth and death rates throughout the world were much the same, both between 40/000 and 45/1000. Half a century ago all the countries of the developing world had birth and death rates close to 40/1000 and a growth rate of no more than 1 per cent (Fig. 36.3). So what has happened to cause the world's growth rates to diverge so widely?

Changes in the growth rates can be produced by:

(a) an increase in the birth rate with the death rate remaining constant;
(b) a decrease in the death rate with a constant birth rate;
(c) an increase in the birth rate with a declining death rate.

Certainly the figures given above suggest that population growth has occurred mostly because of a dramatic decline in death rates; there is no evidence that it results from soaring birth rates. This change from high fertility and high mortality, first to high fertility and low mortality, and finally to low fertility and low mortality, is called the **demographic transition** (Fig. 36.4).

At present population growth is fastest in the developing world, but developed countries such as Britain experienced rapid growth rates in their recent past (Fig. 36.5). Before the agricultural and industrial revolutions, the population of Britain was held in check by disease rather than by malnutrition; the gradual improvement in health, brought about by rising living standards, changes in diet, better sanitation

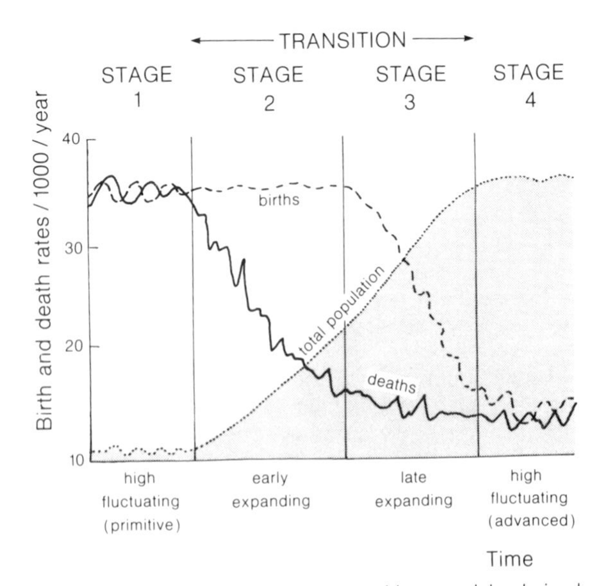

Figure 36.4 The demographic transition model, derived from studies of European population patterns.

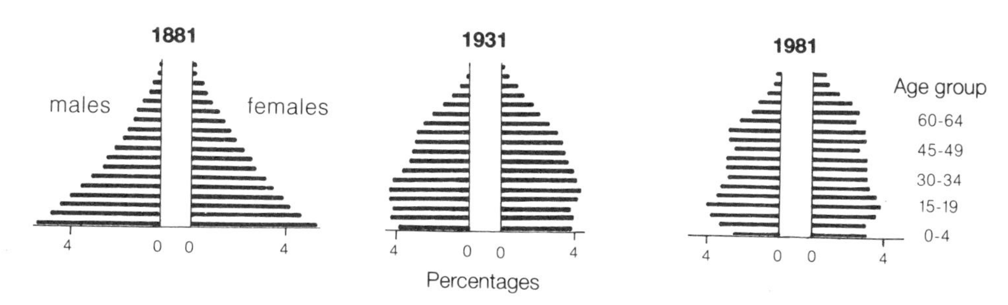

Figure 36.5 The demographic transition illustrated by the population of England and Wales. Notice how the widest part of the pyramid for stage 3 has moved up in stage 4.

and increased medical knowledge, *gradually* caused a reduction in the death rate. However, because of the traditionally high death rate, high fertility was an established part of the culture: although fewer children died, women continued to produce children and there was a significant increase in family size. It is for this reason that early and middle 19th century families in Britain were characteristically large. It was some decades before improvements in economic status and tendency for women to take full-time employment brought about a changing attitude to family size; fertility rates then began to fall. Gradually, however, as parents became more assured of the survival of their children, keen to improve their living standards by working rather than raising a large family, as children were, by law, no longer allowed to work to help support the family, and with the state taking over the role of support in old age that had been the traditional task of the extended family, there was a decline in fertility rates and a return to levels of low increase. But two vital changes occurred during this period, which were to result in major changes in the structure of society and in population growth. First, as mortality declined but high fertility continued, a large lag developed, allowing an enormous expansion of population (Fig. 36.6). Furthermore, as medical care, nutrition and living standards improved, so people suffered less stress at work and began to live much longer. This compounded the effects of high fertility and produced an even greater population increase.

Today the future of the population in countries such as Britain is related to birth rates, because most people live to beyond childbearing age. Reductions in infant mortality would have little impact because women are already able to choose the size of their families. And because – apart from natural disasters – it is only the fertile portion of the population that can cause a population change, prolonging the lives of the older people will no longer significantly affect the growth rate of the developed world populations.

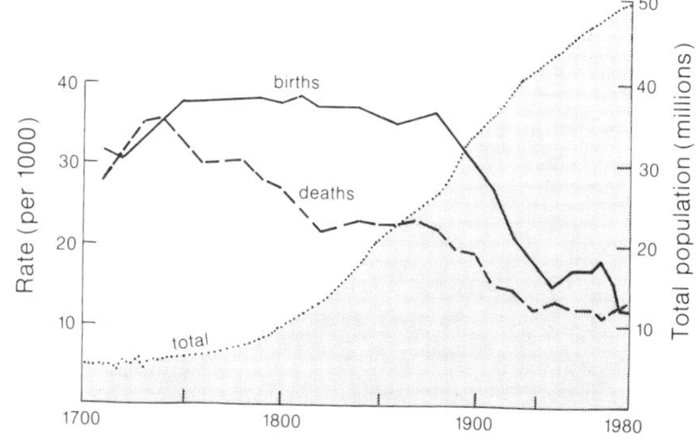

Figure 36.6 The pattern of population growth in England and Wales, 1710–1981.

A demographic model

The careful examination of the growth of European populations has led some people to propose a general model for population development both in the developed and developing world. This model has four stages.

STAGE 1: HIGHLY FLUCTUATING SOCIETIES

'Primitive' societies have a high birth rate and a high death rate, and consequently a low increase in population. Probably no country now falls into this category. The most isolated countries, which have been little influenced by the developed world (such as Afghanistan, Nepal and Zaire), still have high death rates, but even so mortality rates are consistently falling below fertility rates and the populations are beginning to grow (in Afghanistan, a death rate of 23/1000 and a natural increase of 2.5 per cent; in Nepal 21/1000 and 2.3 per cent; in Zaire 18/1000 and 2.8 per cent). These countries are therefore really at the take-off points for stage 2.

STAGE 2: EARLY EXPANDING SOCIETIES
These have high birth rates, falling death rates, and consequently a rapidly increasing population. There are many of these countries, concentrated mainly in the tropical regions of South America, Africa, central Asia and the East Indies. In these cases death rates have been lowered mainly by improved living standards, nutrition, medical care and sanitation. Birth rates remain very high because there has not been time for cultural change to promote a decrease in fertility, and because better living standards allow a higher and more effective fertility.

Many of these countries have such rapidly expanding populations that they can feed their people only with the help of massive food aid from the developed world. In general their population growth rates are greater than their increase in gross national product (GNP), so they are becoming poorer. Without improvements in economic conditions it is hard for them to create the cultural change needed to reduce fertility, although massive propaganda campaigns are beginning to show some effects.

STAGE 3: LATE EXPANDING SOCIETIES
These have low death rates, falling birth rates, and declining population growth rates, but the total is still rising because transition has not yet worked its way through. Countries in Latin America (such as Brazil) and India and China are the major examples of this stage. Although growth rates in Latin America and India are mostly still between 2 and 3 per cent, social pressures to reduce fertility have begun, chiefly instigated by government action. For example China has a concentrated programme of education, contraception and communal pressure. People are now fined for producing more than one child, and social benefits for the children are withdrawn. Not all programmes have been easy to maintain. India provides an illustration of how difficult it is to promote cultural change. In the 1970s a policy of male sterilisation in exchange for material rewards eventually led to the collapse of the government, providing a lesson that cultures have considerable inertia and must be altered gradually and with care.

Some developed world countries, such as Spain (growth rate 0.6 per cent), Portugal (0.6 per cent), Canada (0.8 per cent), New Zealand (0.8 per cent) and the USSR (0.5 per cent) have just completed the late part of stage 3. Their lack of overcrowding had until recently resulted in personal and government policies of procreation.

STAGE 4: HIGH FLUCTUATING SOCIETIES
'Advanced' societies have low death and birth rates, high populations and low growth rates. This category contains most European countries and those settled by Europeans. In many, demographic transition has only just been completed, however, and populations are still rising, as the large young population grows up to childbearing age. However, in the more established countries, such as the UK, the transition was completed as early as 1939.

Consequences of the demographic transition

There are two quite separate consequences of the demographic transition, one a problem for those countries that have completed the transition and the other, by far the more important, for those still undergoing the transition.

Completion of the transition has created an imbalance in the population structures of developed world countries, with older people making a progressively large part of the community. In Britain, for example, there may be 30 per cent of the population over 65 by AD 2030. This is putting considerable strain on both the social services and the economically active people, who have to support the elderly. A regressive population could lead to a position in which there are simply too few wage earners to meet the demands of the large numbers of elderly – a situation that would lead to a lower standard of living for all (Fig. 36.7).

In the UK there are already more than 11 million people over 60. The over-75s are increasing rapidly and will reach 3.5 million by the mid-1990s. Many live alone and are highly concentrated in 'retirement' areas such as North Wales and South Coast towns. In some areas they comprise 30 per cent of the population today.Many others are left in decaying inner-city areas.

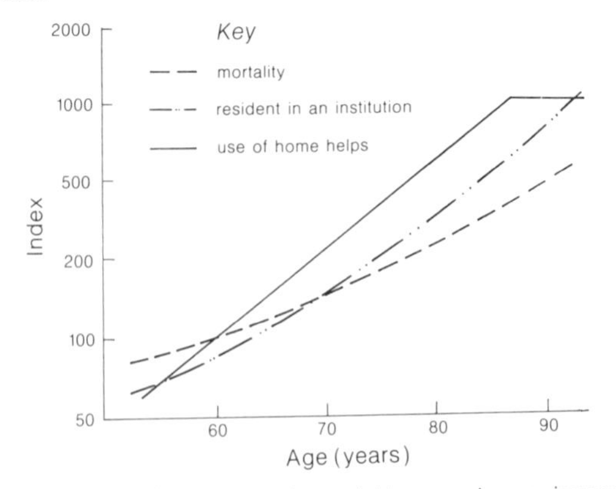

Figure 36.7 The increase in social/economic requirements with age, England and Wales.

The other major problem is produced by the fluctuation in fertility. For example, after World War II there was a substantial increase in the birth rates which later required an expansion of school, university and employment places. There was a drop in birth rates through the late 1950s and early 1960s, but then a recovery and a further substantial increase in birth rates in the late 1960s. Each of these fluctuations has made for great difficulties in planning. The most recent increase is now working its way through the school system, resulting in falling school rolls and school closures. A similar drop in university places is following.

The problems created by countries undergoing transition is of far greater significance to the world because they contain the majority of the population. The population growth during transition within the developing world is occurring at a much faster rate than that experienced by Europe during *its* transition phase. High technology, medical and food aid are now available to make *rapid* inroads on death rates. Consequently the decline in the rate of mortality in the developing world has been many times faster than that experienced in Europe. However, it has in general not proved possible to make equally rapid changes to the cultural habits of the people, with the result that the world is experiencing phenomenal population growth (Fig. 36.8). Indeed, there is much doubt about whether the developing nations, whose growth rates currently exceed 3 per cent per year, can achieve the demographic transition and reduce their populations even to the level of replacement before natural control arrives in the form of mass starvation, increased deaths and a return to the state of high fertility and high mortality from which they have so recently emerged.

The problem of population growth is even more serious than that occasioned by sheer numbers. Considerable political instability results in a world containing on the one hand a group of developed countries with stable populations and a high living standard, and on the other underdeveloped countries with high growth rates and low living standards. This instability is seen in the so-called 'North–South debate'. At present Asia, with a landmass smaller than that of Africa, is home to 55 per cent of the whole world's population and to 75 per cent of the *developing* world's people, a concentration which causes stress. However, in Africa, where only 14 per cent of the world's population live, there is similar pressure on the land because the food resources are very limited. Only the 11 per cent of the developing world's population that live in South America have quite high potential resources, though these are, as yet, not effectively exploited for the benefit of the population as a whole.

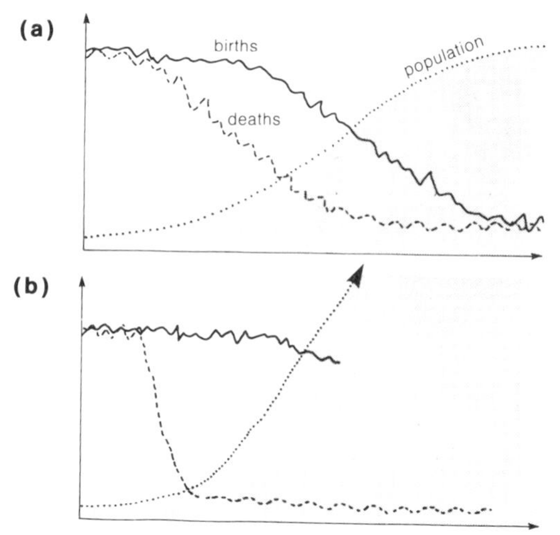

Figure 36.8 (a) Demographic transition where death rate declines slowly, e.g. Europe in 18th and 19th centuries; (b) demographic transition where death rate declines rapidly, e.g. Africa at present.

36.4 Population and the problem of resources

As the world population grows there is an increasing degree of competition for available resources. Food is the most vital resource. Although supply and demand have kept more or less in step through the centuries, there has been increasing evidence recently of a significant shortfall in supply. This scarcity was only temporarily alleviated by the fortuitous discovery of new strains of wheat, rice, and other grains, which enabled production to rise, initiating the 'green revolution'. Today food supply and demand are once more precariously balanced. As yet we have not seen major failures of world harvests in *both* maize and rice; shortfalls in one crop can be made good by surpluses in another. But the major producer of surplus grain, the USA, is a heavy consumer of energy in food production; as a result, its food is an expensive item for developing countries, which are already short of money.

All the efforts to increase the food supply by importing grain, fertilisers or energy are tied to total national production (GNP; Fig. 36.9). If the GNP can be increased faster than the population, there is a chance of more adequate food supply. Unfortunately the aid programmes from the developing world that in many countries make up the shortfall between GNP and demand are slowing down due to the world recession and shortage of resources. The lack of skilled labour for self-help and the crushing burden of a rapidly expanding population which doubles every twenty or so years will make the problem of preventing mass starvation that much harder.

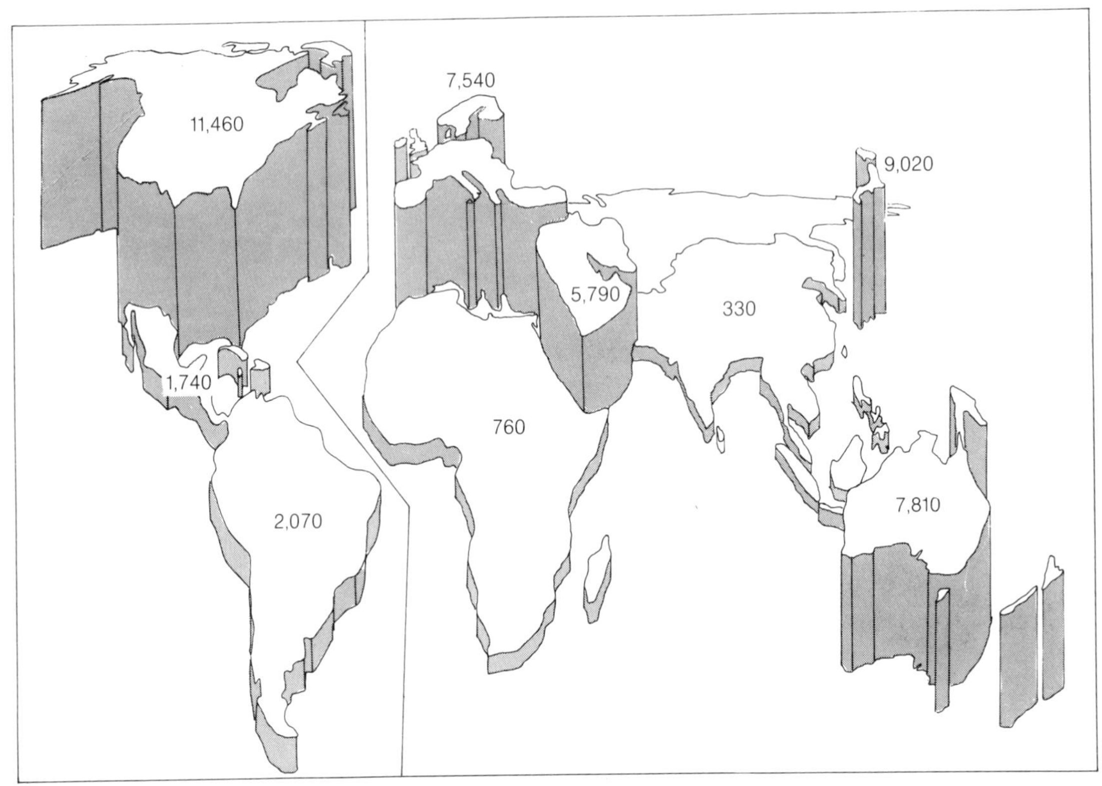

Figure 36.9 GNP *per capita* for the world in 1980.

37

Migration

37.1 The nature of migration

Migration involves the movement of people from one country or region to another on a permanent or semi-permanent basis. It is quite distinct from short-term movements of people, such as the daily journey to work, or movement to another country or region as tourists.

There are two types of migratory movements:

(a) The unrestricted drift of people as they expand into new areas gradually. These may be both short-distance movements (such as the migration of lowland people into uplands) or long-distance movements (such as the colonisation of America). Such movements are a form of diffusion.

(b) The compulsory movements of people as a result of forced expulsion, involving sudden, short-lived movement of certain segments of a population.

37.2 Models of migration

Every person who decides to migrate does so for a unique combination of reasons related to his environment and social pressures. Nevertheless, people in particular economic, ethnic, social and religious groups have often been subject to broadly the same pressures, and have therefore tended to move in distinctive patterns (Fig. 37.1). Towards the end of the 19th century Ravenstein proposed several 'laws of migration' based on his observation of migration patterns (Fig. 37.2). These included the following.

(a) Most people migrate only short distances (e.g. from the suburbs to a nearby city).

(b) Migration occurs in stages, a short movement of one group of people from an area leaving a vacuum which is filled by another short migration of a different group of people. Migration therefore occurs in 'waves' making a chain of **stepwise**

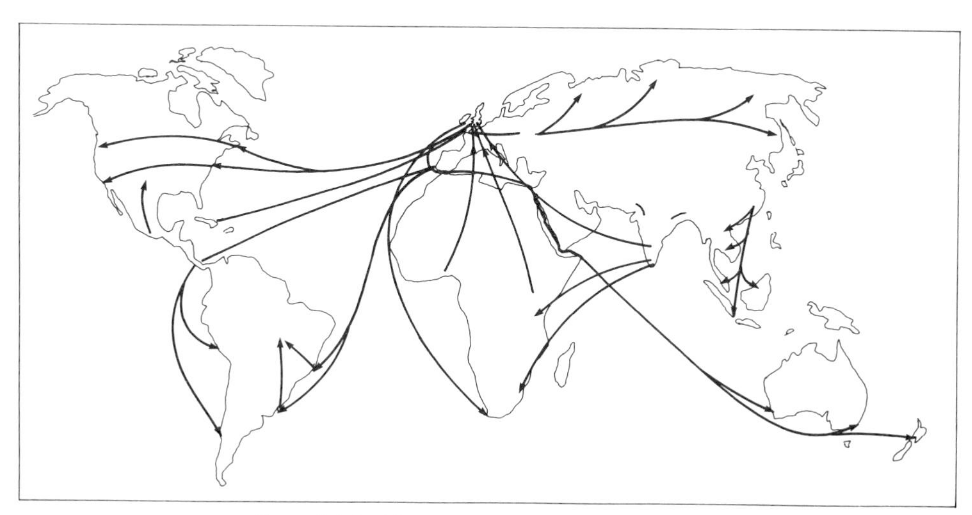

Figure 37.1 World migrations of the 20th century.

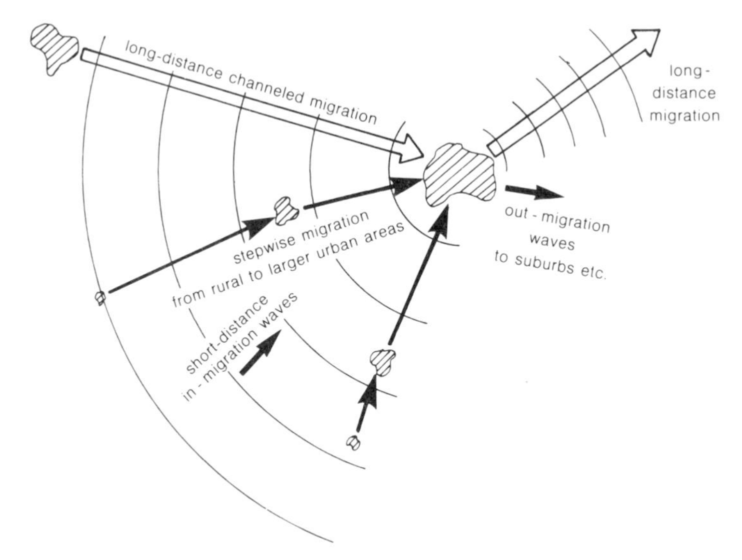

Figure 37.2 Patterns of migration.

migration (e.g. one group from suburbs to country, another from inner city to suburbs).

(c) Where migration takes place over a long distance the goal tends to be a large urban area; small urban areas are unknown beyond their immediate neighbourhood (e.g. British migration to Sydney, Australia).

(d) Urban dwellers tend to migrate more than rural dwellers: they have a greater awareness of distant environments, and urban community ties are weaker than those in rural communities.

(e) Men tend to migrate long distances whereas women tend to migrate much shorter distances. (This may no longer be true.)

In the 1930s Zipf tried to express the relationship between migration distance and numbers migrating in terms of an equation. He suggested that the volume of migration is inversely proportional to the distance travelled by the migrants.

$$N_{ij} \propto \frac{1}{D_{ij}}$$

where N_{ij} = number of migrants from towns i and j and D_{ij} = distance between these two towns.

This distance-decay model does not take into account the third law listed above. As a result some social scientists have suggested using a 'gravity model' in which the attractiveness of a city is also made proportional to its population.

$$N_{ij} \propto \frac{(P_i \times P_j)}{D_{ij}}$$

where P_i and P_j are the populations of towns i and j. There are many practical difficulties in measuring the effective populations, especially if migration distances are small and if migration is from diffuse urban zones. 'Distance' may also not be a matter of kilometres but of time or cost of migration, and of retaining links with friends and relatives. A migrant's view of remoteness and distance may well be a gross distortion of reality. For example, he may not be able to differentiate between regions all of which are at a great distance – he may lack experience and information (and so hold general opinions such as 'All tropical countries are the same'). Similarly because travel costs rise less quickly as distance increases great distances may not be a major restraint on the distance of migration. Migrants also tend to foreshorten large distances.

There will, for example, be little point in migrating to an urban area in a region known for its unemployment. Although some people might migrate to a large urban area without having been informed of the employment prospects or housing facilities, today many people might be expected to migrate to an area because they perceive it to have good prospects. Thus it would be reasonable to argue that the volume of migration over a given distance is in proportion to the number of opportunities thought to be at the destination (in comparison to those offered locally), and that it is also inversely proportional to the number of intervening areas that offer similar prospects. In other words, migration over large distances is likely only if there are few comparable opportunities nearer to the origin.

37.3 International migration today

Migration is often closely allied to the problems imposed by an expanding population. As the population grows so the environment has to be shared by an increasing number of people, and this eventually leads to a decrease in the quality of life. In extreme cases it produces a situation of 'shared poverty' (as with the shortage of food in the developing world, and of jobs in the developed one). Small populations do not suffer the stress that builds up in densely populated regions, a stress which invariably leads to recriminations against one part of society or another. It is often the rejected part of a society that migrates, as do those ambitious enough to want to seek a better life elsewhere.

Although unemployment in a welfare state is not as strong an incentive to move as is starvation in a poor country (as illustrated by the lack of migration in the UK in the early 1980s during the recession), the more enterprising are more liable to leave in both cases, a situation that deprives the source area of its most able people and may lead to further decline.

Not all people have the same goals when they migrate. Those leaving developing-world countries without already having skills will not be in competition with those highly qualified people moving between the developed countries or to newly developing regions. As a result one general international migration model applicable to present conditions is this: a nation gains migrants from countries less developed than itself and loses migrants to countries that are

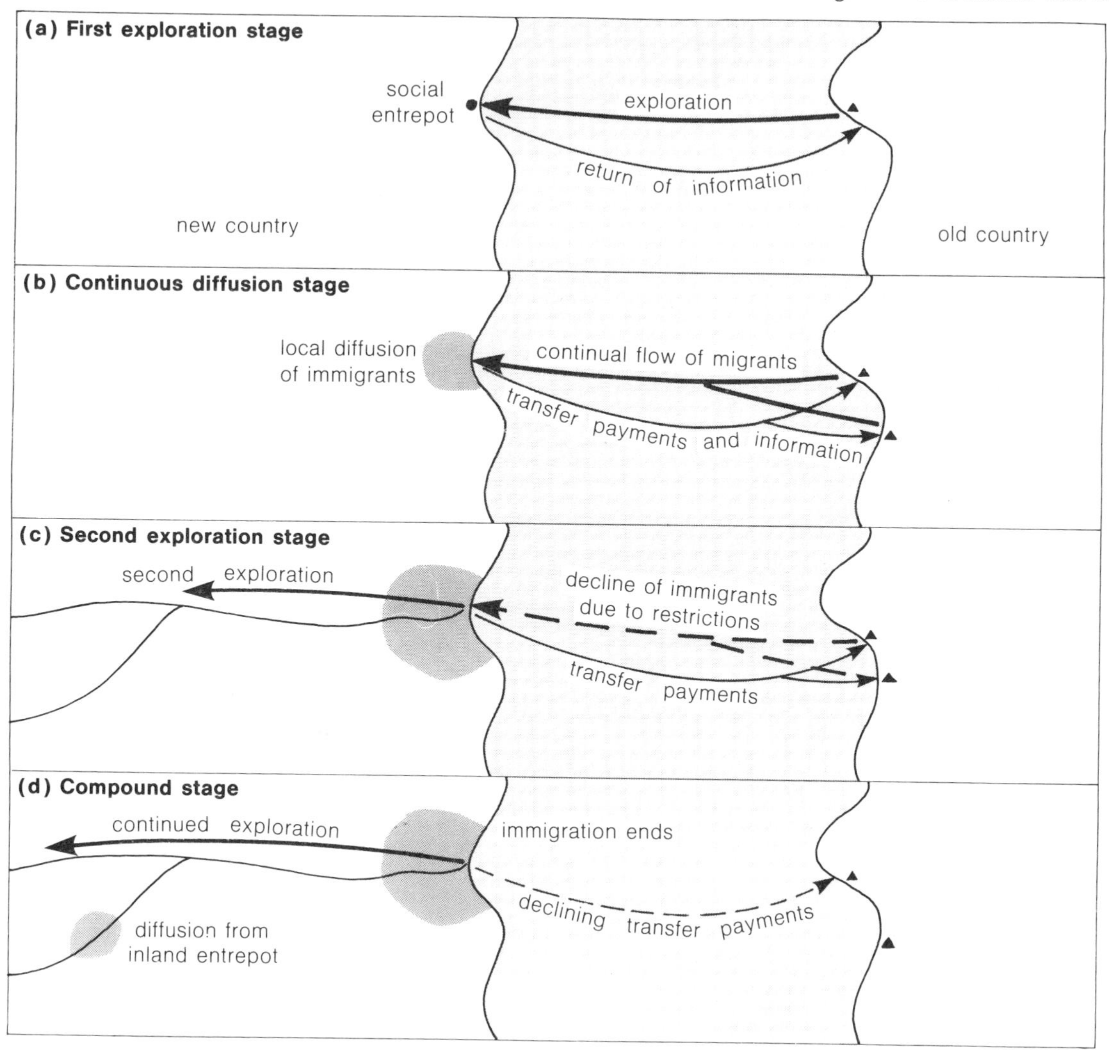

Figure 37.3 The social mercantile model.

more developed. The result is a reduction in the quality of skills in the source areas and an improvement in the quality of skills in the recipient ones.

In places where developed and developing countries are close together (e.g. Mexico and the United States; France and Algeria) or have special ties (e.g. the United Kingdom and the Commonwealth; Egypt and Saudi Arabia) the migratory pressure is very strong.

In the special case of the UK, whose Commonwealth contains both nations that are richer and nations that are poorer than itself, there is a large population turnover due to migration, but no large *net* increase in migrants. The UK has taken in a large number of untrained people at the bottom of the social hierarchy and sent out trained people at the top. This effect of Britain as a 'migration conveyor' has produced a large, often poorly educated, foreign population and caused a loss of highly educated people. 97 per cent of the UK exiles live in Australia, Canada, New Zealand, South Africa or the USA (Table 37.1).

Table 37.1 Net migration for the UK 1979–81 average (source: Census).

Emigrants ($\times 10^3$)	Country of origin	Immigrants ($\times 10^3$)
160	UK	60
20	New Commonwealth & Pakistan	35
45	alien	55
20	Old Commonwealth	30

The only advantage to the developing world of being at the bottom of the population conveyor is that migration is a partial stop-gap solution to population pressure, at least in small populations such as that of the Caribbean islands; but it quite obviously cannot be a cure. For the major developed countries to receive as migrants the excess population of the developing ones, they would have to receive at least all the future world population growth – a figure that would double the developed-world population in under 10 years!

37.4 The arrival of immigrants

It is of concern to know not only the source and destination areas of migration but also what happens to immigrants when they arrive at their destination. Clearly much depends on the social status of the migrant and whether he or she arrived as a result of choice or coercion. However, Vance suggests that many long-distance migration patterns can be represented by a two-stage model (Fig. 37.3). The *first* stage consists of **exploration** by a small number of people (usually to coastal sites, or **social entrepôts**) who return home and inform others about the new country or region. They retain close ties with their homeland and, for example, send money home. For a time these new migrants collect in the social entrepôt city until there is some reason for further migration, such as work opportunities or a better perception of the new environment. From this time a *second* stage of exploration will create new entrepôts of migration at some distance from the first. This in turn leads to a secondary pattern of **diffusion** into the surrounding countryside and a gradual adjustment to the new home and to the local culture (the **acculturation** stage). With immigrants now fully settled, links begin to be severed with the source country and payments to the homeland decline.

Examples of long-distance migration

CASE STUDY: THE USA

Slavery was introduced into the colonies in 1619, and in the following two centuries there was a large growth of the black population on tobacco and cotton estates in Virginia, Maryland and coastal Carolina (Fig. 37.4). Gradually the black population spread with the estate system across Carolina and Georgia. There were also small groups of free blacks in New York and in the New England ports, in service trades.

When slave-trading ceased in 1865 black field workers and their families tended to move from employment on plantations to share-cropping on nearby agricultural land. Small numbers also migrated to northern cities where there were already nuclear black populations. Very few went on into the northern countryside despite the fact that their southern background was rural. During World War I, recruiting officers went to the South to get workers

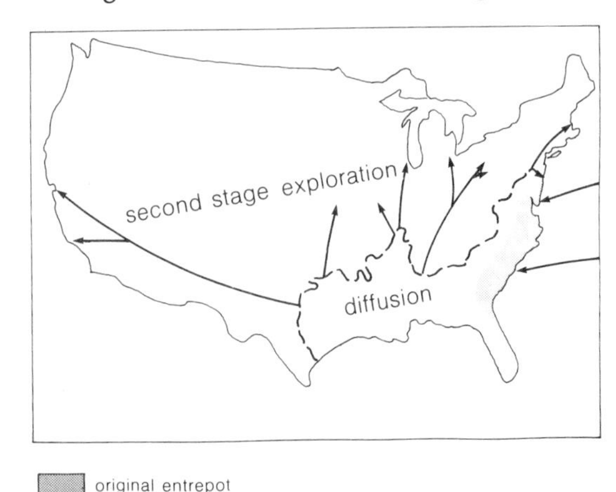

Figure 37.4 Movement of blacks in the USA.

for the munitions plants and shipyards of the North, with the result that the number of blacks in the northern cities increased sharply. Major centres of black residence formed in cities such as Cleveland, Detroit, Chicago, Baltimore, Philadelphia and New York.

Another phase of recruitment during World War II encouraged migration to the West Coast cities. By the 1960s most northern towns had a black community, although there were still virtually no blacks in the northern countryside. However, at no stage have these migrants fitted in closely with the indigenous (white) population and have remained in narrowly defined social ghettos, dispersed about the nation.

Mexican migration followed a similar pattern to that of the blacks. However initial migration was voluntary, growing out of failure in their home rural areas over the last 50 years due to uncontrolled population growth and a stagnant economy in Mexico. The small Mexican communities that lived in the four border states – Texas, California, New Mexico and Arizona – acted as a focus for waves of migration: 100 000 in 1850–1900, 200 000 in 1900–10, 400 000 in 1910–20, and 1 200 000 in 1920–30. As with the blacks, expansion of Mexicans to the rest of the US was instigated by labour recruiters. Large numbers of casual labourers were needed for farming sugar beet and cotton, for fruit picking and so on. Partly because of linguistic difficulties, these people have not become integrated into the US population, but remain isolated, retaining their culture and their language. Many are still illegal immigrants and denied any rights in the US.

CASE STUDY: THE UNITED KINGDOM

In Britain population migration has also been by initial colonisation followed by a second stage of expansion. Irish, Asian or West Indian groups are found in all conurbations. Of the New Commonwealth and Pakistan, Indians are most common in all except south-east Lancashire and West Yorkshire; Jamaicans are most common in Sheffield, Nottingham and Bristol, whereas Pakistanis show a clustering in the northern textile towns, immigrants from Italy are concentrated in the market-gardening and brick-making areas north of London. Poles, far fewer in number than New Commonwealth immigrants, appear to cluster in the coal-mining areas; while the Americans are highest in NATO defence-base and oil-prospecting areas of East Anglia, north-east Scotland and the Scottish Highlands. Irish are found in all cities but predominate in London, the North-West and Scotland. The non-Irish Europeans are concentrated in provincial Britain, whereas the Irish, Asian and Caribbean people are most common in metropolitan Britain. As in the USA, so far many of these people retain a strong, segregated urban bias (Fig. 37.5).

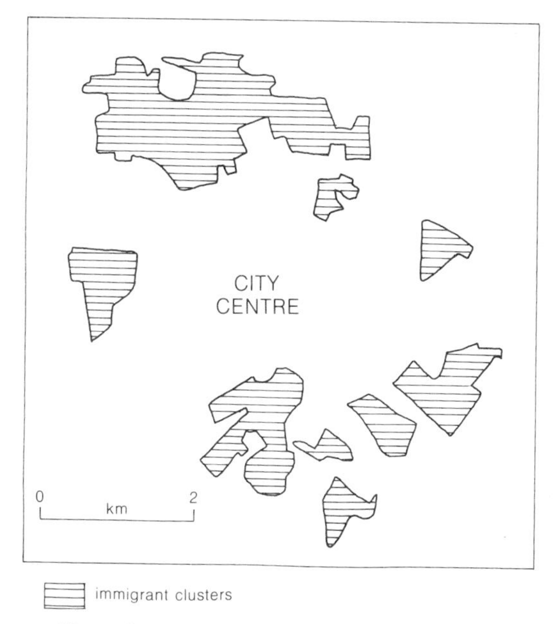

Figure 37.5 Immigrant clusters in Birmingham.

CASE STUDY: GUEST WORKERS IN EUROPE

One problem of the high, fluctuating population of advanced economies is the imbalance that can develop between natural wealth and national population. The rapidly increasing wealth of Europe in the 1950s and 1960s was not matched by a population increase, and a gradually widening gap developed between available jobs and the workforce. This directly allowed the migrant flow of labour from southern Europe (Greece, Italy, Spain and Turkey) to Germany (the 'gastarbeiter'); from former colonies (Algeria, Senegal and others) to France; and from the New Commonwealth to Britain.

Unlike Britain, Germany and France mostly took in **guest workers** on short-term contacts, yet despite the banning of recruitment during the recession which began in 1973 permanent or semi-permanent guest workers now number 4.5 million (7.5 per cent) of West Germany's population, and 1.8 million (4 per cent) of the population of France, as compared with 3.4 million (6 per cent) of the total population of the UK. In all cases, however, these people are largely unskilled and are liable to suffer heavy unemployment in times of economic adversity. They also tend to have larger families than the indigenous population, and their descendants will thus form a progressively larger part of the national population.

CASE STUDY: THE DEVELOPING WORLD

Migrations are not confined simply to the developed world, they have also affected many nations in the developing world. In these cases, however, migration has sometimes been closer to a form of diffusion (Fig. 37.6). Kenya, for example, is a main focus of cultural contact in Africa. There are four main ethnic groups: the largest group, the Bantu-speaking peoples (64 per cent of the African population), have spread eastwards from the Niger Basin in west Africa and now occur mainly in the south of the country; the Nilotic peoples (14 per cent of the African population) are also mostly dispersed in the south, having spread eastwards from their base region around Lake Victoria. Both of these groups are primarily cultivators. The third group, the Nilo-hamitic peoples, which include the famous Masai warrior tribe, have spread down from the north and now occupy the western part of the country; whereas the fourth group, the Hamitic people, have spread down from Somalia and still occupy the north-east. Both of these groups are nomadic pastoralists by tradition. Today their distribution through the rural areas of Kenya still shows clearly the origin of their movement. The cultural differences between each group have produced many tensions within the country, as shown so dramatically between the attitude of the (Bantu) ruling Kikuyu people with their tendency to adopt

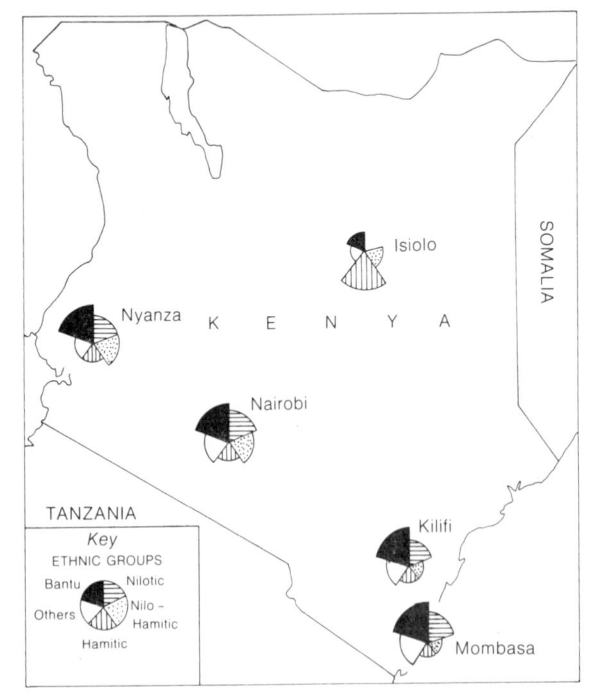

Figure 37.6 The distribution of ethnic groups in Kenya.

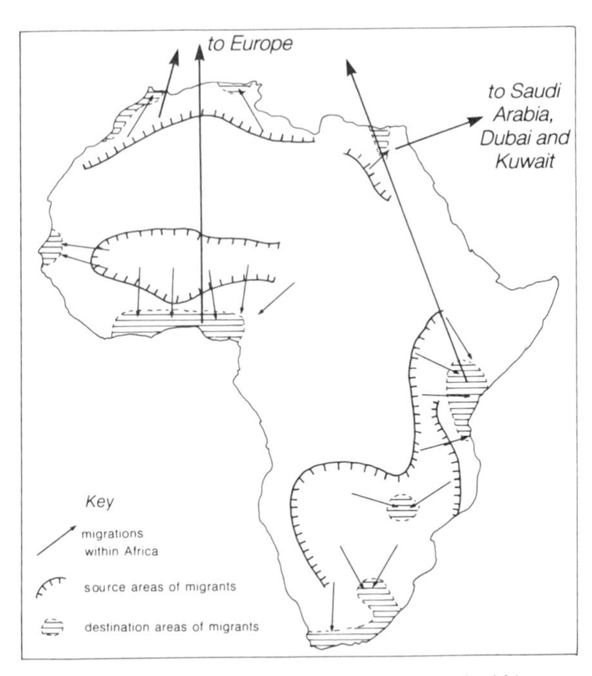

Figure 37.7 Major migrational movements in Africa.

modern ways, and the Masai who cling to their traditional ways and resist many of the attempts at advancement.

In contrast to these long-term changes, there have been major short-term population movements between those with low resource-population regions, such as Egypt and Ghana, to those of relatively high resource-population ratios, such as Dubai and Nigeria (Fig. 37.7). These are relatively short-distance international movements, mainly encouraged by the resource-poor governments. Thus Dubai (with an indigenous population of 30 000) has 180 000 immigrant workers and Nigeria (with an indigenous population of 86 million) had nearly 2 million Ghanaians until it expelled them in a dramatic purge in 1982.

The effect of migration on population structure

Because the majority of migrants are young adults, some countries that have experienced considerable immigration such as Canada have a relatively young population. Population pyramids in these countries look more like those for countries in stage 3 of the demographic transition. This pattern is most marked in urban, rather than rural, populations.

By contrast, those countries that have experienced prolonged emigration (such as Norway) tend to be deficient in young adults, thereby accentuating the ageing character of the population. Again, because members of urban populations are more mobile than those of rural ones, the pattern is most marked when urban populations are considered alone.

37.5 Internal migration

Internal migration, the movement of people within the boundaries of their own country, has traditionally been from rural to urban areas and from areas of perceived deprivation to areas of perceived success.

Internal migration to cities is a long-term feature of the developed world and is becoming an increasingly acute problem for those in the developing world. However the process of migration has been different in each case.

The developed-world model

Migration to the cities of western Europe varied with each region and with each stage of the Industrial Revolution. The first stage was the seasonal movement of young single men to the large cities. The beginnings of rural to urban movement were thus based on a specific sector of the population who did little to alter the structure of the urban population. The second stage was by migration of entire families in search of employment, a change in pattern that helped create the large urban-population expansion in the 19th century. Both *push factors*, such as lack of land, population pressure and underemployment in rural areas, and *pull factors*, such as dissatisfaction with the traditional environment and the prospect of better wages and conditions, were important in sustaining the long phase of rural–urban migration that led to the present form of cities in the developed world.

Much later a new phase of migration occurred as improvements in communications led to the growth of suburbia and the commuter-enlarged **dormitory towns** for the more wealthy, and finally to the **new towns** when urban renewal of the city centres began. This phase is leading to outmigration (**counter-urbanisation**) and creating many problems for inner cities.

The developing-world model

The push factors that influenced migration in the developed world of the 19th and early 20th centuries can never be repeated exactly in the developing world of today. Communications now allow much easier access to urban areas, while overpopulation in rural areas is increasingly rapid. Mainly as a result of better environmental perception, peasants want to be something other than peasants. It is now the pull factors that become most important: the prospect of better wages and conditions and, of equal importance, a job leading to some form of authority and status. But a more important difference lies in the speed of rural–urban migration. People have flocked to the cities in such numbers over such short periods that they have given many urban areas a distinctly rural influence which was never characteristic of cities in the developed world. Indeed, growth in the cities of the developing world is by a major early phase of suburban (shanty town) expansion, not a late outward spread as has been the experience of the developed world.

At present much migration is still initially by young men, and male population structures of most cities in the developing world are weighted heavily in favour of young adults (Fig. 37.8). Once settled, however, the men go back to their home villages – where the chances of finding a wife are better – to get married and then return with their wives to the city. This results in a large rise in the number of children (Fig. 37.8). Unfortunately,the immigrants' perception of the city is not always very close to reality; many people find themselves moving from a situation of rural poverty into one of urban poverty (but see p. 491).

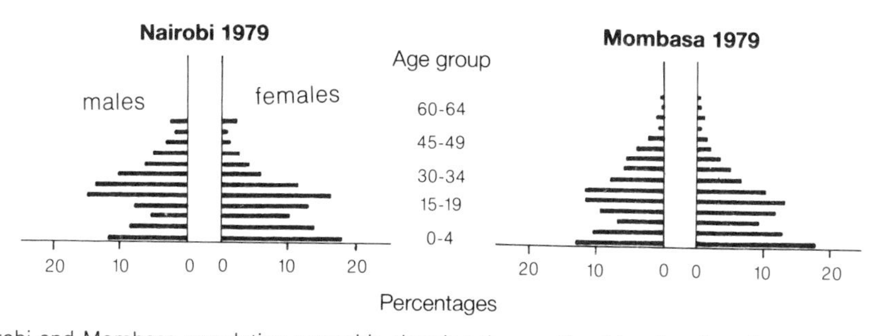

Figure 37.8 Nairobi and Mombasa population pyramids showing the results of immigration of young adults to the city. Men come in search of work, whereas women and girls are brought in from the country to act as 'nannies' for working mothers and to help with domestic chores. Other women come to the city as the wives of men who have recently found a city job.

Short-distance migration

People are able to move much more easily within than across their country's borders. The majority of these short-distance migrations have been from rural to urban areas, although there have been inter-urban movements in the developed world and even urban-to-rural movements (counter-urbanisation) near cities. Because the overwhelming proportion of migrants move towards cities there has been great stress on the fabric of urban societies.

Governments have been faced with rural-to-urban migrations throughout the world, and various strategies have been attempted to solve each unique set of circumstances. Strategies include:

(a) *regional development*, e.g. in Greece, Italy and Finland;
(b) *decentralisation* of government activity, e.g. in the Netherlands (Randstad) and in France (métropoles d'équilibre);
(c) *relocation* of the capital, e.g. in Brazil and Tanzania (in progress);
(d) support for *new towns*, e.g. in Britain and Japan;
(e) *dampening* of urban/rural wage differentials, e.g. in Zambia;
(f) *reorientation of education* towards agricutural interests, e.g. in Indonesia and Tanzania;
(g) *subsidies* for industrial relocation, e.g. in France and Sweden;
(h) *rural land reclamation*, e.g. in the Netherlands and Kenya;
(i) a *citizenship tax* on living in the city, e.g. in Seoul and South Korea.

Rural–urban migrations have had dramatic effects on societies. Rural depopulation has long accompanied the gradual shedding of labour, accentuated in years of agricultural crisis such as in France between 1875 and 1881 when competition from cheap imported American wheat and the destruction of vineyards by disease resulted in the '*émigration de la misère*' that displaced 840 000 people. But the changing balance can go too far, as typified by the expression '*Paris et le désert Français*', an allusion to the overwhelming preponderance of a capital which can parasitically sap the life of the provinces. Today Paris, with 8 million people, has a sixth of the country's population on 2 per cent of its area.

Occasionally migration can be very dramatic. Milan's post-war industrial boom saw its population rise by 1.5 million in just five years, as people flocked from Italy's rural south (the Mezzogiorno) to try to gain jobs in the prospering north.

Rural–urban migrations have changed the face of western European countries. Between 1950 and 1970 the number of cities with populations over 100 000 grew from 200 to 300 and the proportion of the total from 27 per cent to 35 per cent. Furthermore, the balance of both population and wealth is tipping more and more towards the centre of western Europe (which, because of the free movement of trade and jobs as part of the EEC policy, may now be regarded as equivalent to a large nation) and away from the periphery,creating a favoured central *core* Europe and a second-class *peripheral* region (p. 430). Indeed, some regions (such as the highlands of Scotland) have become so underpopulate that it is increasingly difficult to provide adequated modern services for those who remain. It is this problem, perhaps just as much as city overcrowding, that is taxing the minds of politicians, keen to prevent further migration before the situation become irrecoverable.

CASE STUDY: ENGLAND AND WALES

The widespread effects of internal migration on the welfare of a country are well documented for Britain. In 1981 the census revealed that all regions except the South-West had a net loss of people to the South-East (Fig. 37.9). The South-West received a net flow of

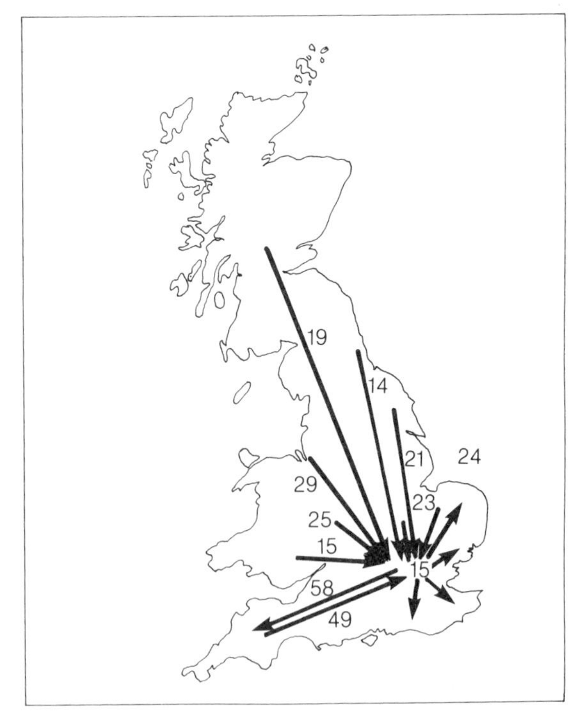

Figure 37.9 Inter-regional migration (1981) for standard regions. Arrows indicate the direction and magnitude of primary migration. Secondary movements can be obtained from Table 37.2. Notice the 'drift to the South' and the net gain of people to the rural South West. Inner London is losing people to the South East, at a rate that roughly matches the immigration from other regions.

people, but this was almost entirely from people seeking a home for retirement (Fig. 37.10).

The migrations can be classified as:

(a) a *north–south urban-to-urban* drift of young single adults seeking work;
(b) an *urban core–periphery* migration of those families seeking relief from inner-city congestion (counter-urbanisation);
(c) an *urban–rural* (and particularly coastal) migration of older people after retirement;
(d) a *rural–urban* migration from the remote hill lands of the periphery.

On top of these there were also the international migrations described above, whose effect was to provide input to the inner-city areas.

The small-distance urban-rural migration of families to dormitory villages has not taken place only around the large conurbations; it is a widespread feature of the country. This is illustrated by the pattern near Lincoln (Table 37.3). Its effect is to produce local segregation of the population, with the better-off moving to the peripheral locations and leaving the less advantaged in the city.

Figure 37.10 The increasing population imbalance due to migration of retired people (in Great Britain). The shaded areas indicate regions where the ratio of retired to working-age population was more than 20 per cent above the GB average in 1981.

Table 37.2 Inter-regional movements* (in thousands) in Great Britain in 1981 (Sources: Office of Population Censuses and Surveys: General Register Office for Scotland).

	Region of origin												
	Great Britain	North	Yorkshire & Humberside	East Midlands	East Anglia	South East			South-West	West Midlands	North-West	Wales	Scotland
						Total	Greater London	Rest					
Region of destination													
Great Britain	—	47	73	72	43	209	186	211	88	78	93	42	47
North	39	—	8	3	1	10	4	6	2	3	7	1	5
Yorkshire and Humberside	68	9	—	11	3	17	7	11	4	6	12	2	4
East Midlands	76	3	13	—	5	25	9	16	5	11	8	2	3
East Anglia	53	2	3	6	—	31	11	20	3	3	3	1	2
South East	218	14	21	24	23	—	115	73	49	25	29	15	19
Greater London	154	6	9	9	8	73	—	73	15	9	12	6	8
rest of South East	252	8	12	15	15	115	115	—	34	16	17	9	11
South West	107	3	5	6	4	58	17	41	—	12	9	6	4
West Midlands	66	3	5	9	2	20	7	13	9	—	9	6	3
North West	73	6	11	7	2	19	7	12	5	9	—	7	6
Wales	44	1	2	2	1	14	5	9	6	6	10	—	1
Scotland	45	6	5	3	2	15	6	10	4	3	6	1	—

* Figures are derived from re-registrations recorded at the National Health Service Registers and are lagged by three months to make allowance for the time between actual move and re-registrations. Movements within each area are excluded.

Table 37.3 Census returns for selected parishes south of Lincoln during the earlier periods of industrialisation, people left the villages and moved to the towns. Between 1821 and 1851 there is considerable growth, while between 1851 and 1901 there is stagnation or decline despite the rapid rise in national population. Migration has only been reversed since 1951 with a growth of dormitary settlement in the larger villages near to Lincoln (e.g. Waddington) and a stabilisation of agricultural employment. (Source: Census of England and Wales, HMSO).

Parts of Kesteven R. D.	1821	1851	1901	1921	1951	1961	1971	% change 1821–1971
Branston	700	1330	1210	1275	1963	2040	2440	248
Waddington	700	960	770	864	2676	4190	5370	667
Harmston	330	410	330	301	672	760	650	96
Coleby	320	420	400	352	443	390	390	22
Boothby Graffoe	160	210	170	188	249	180	150	− 6
Navenby	630	1060	780	824	851	810	940	49
Wellingore	730	910	590	538	660	610	610	− 16
Potter Hanworth	370	460	480	466	462	540	460	24
Nocton	370	510	480	553	1030	990	680	83
Dunston	410	590	570	532	561	500	480	17
Metheringham	630	1520	1520	1447	1816	1680	2140	239
Blakney	500	600	580	600	587	420	350	− 30
Martin	590	890	720	667	1063	760	710	20
Scopwick	230	410	430	423	1170	780	480	108
Timberland	500	640	550	430	527	510	440	− 12
Rowston	120	230	200	202	160	200	160	33
Walcot	510	610	130	472	499	470	470	− 9
Totals	7430	11760	9980	10070	18650	15830	16920	+ 128

In detail, it is clear that rural depopulation is at its most severe in the remoter highlands and islands of Scotland. Barra, in the Hebrides (Fig. 37.11), does not have any substantial industry and children of secondary-school age must board on the mainland. Many parents need to seek and move to work on the mainland, a situation which often results in grandparents looking after children and which disrupts family life. Only on retirement can most of the people afford to become residents once more.

Short-distance migration removes the wealth from inner-city areas; it makes the self-financing and provision of adequate social services all the more difficult. This is a problem shared by those regions chosen to be the focus for migration of the elderly. The pyramid the coastal town of Worthing in Sussex (Fig. 37.1) shows a large elderly population. In these areas taxes derived from the economically active population are often insufficient to meet the demands of social services needed by the elderly; the preponderance of older people in an area is driving the younger people away in search of a more lively environment; and the lack of a substantial wage-earning local market makes the area less attractive to potential employers. The imbalance between the revenue from taxes and the demand for services places great strain on local resources, and many seaside towns are making strenuous efforts to encourage a better balance of population.

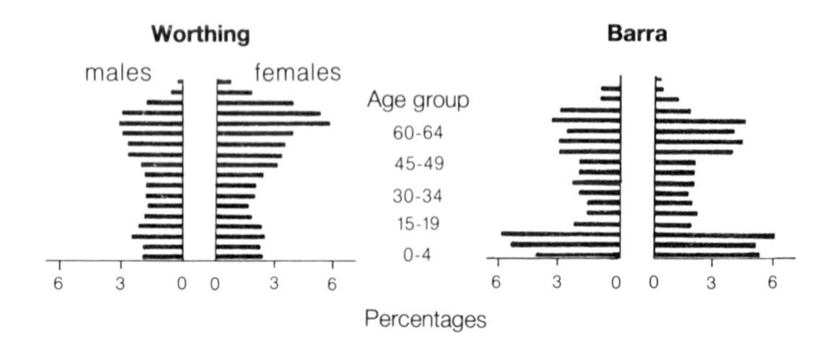

Figure 37.11 The effects of a peripheral region of constant disadvantage are reflected in the distorted pyramid for Barra. The inflow of retired people is the most noteworthy feature of Worthing.

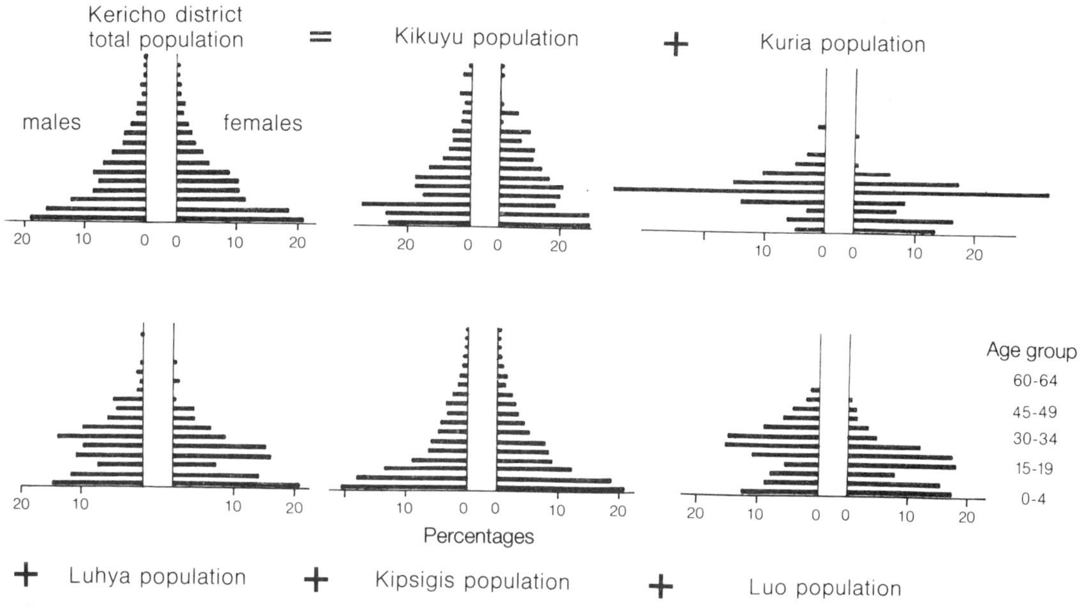

Figure 37.12 The Kericho District of Kenya may appear to have a progressive population structure, but it really contains many tribal groups who have migrated to the area, some within extremely restricted age bands (only the most important tribes are shown here).

CASE STUDY: EAST AFRICA

Internal migration is also selective within the developing world. In the rural Kericho region of Kenya, for example, there are representatives of over a dozen tribal groups, each with a distinctive cultural heritage (Fig. 37.12). The Kipsigis tribe have a 'normal' age–sex pyramid, showing them to be the indigenous population. They are, in common with most rural communities in the developing world, burdened with a large proportion of young dependants, as shown by the wide base to the pyramid. The shortage of young males is due to migration to nearby Nairobi. However, there is a large migratory element from other ethnic provinces, particularly the Kuria of Nyanza province, a nearby region on the Rift Valley floor with much less favourable land than that in Kericho. This is a recent migration and thus shows a heavy concentration in the young-adult age group. The Luo and Luhya contribute a much wider age range of people and have clearly been established long enough to raise families. The Kikuyu are the migrants who have been established for the greatest time and have a much more evenly structured population.

The reason for migration in this region is far different from that in Britain. Here the major resource is still the land, and when independence in 1963 made available the land once owned by Europeans, there was a great influx of migrants to occupy the area in small independent farmsteads. Nevertheless, dissatisfaction with opportunities in rural areas is reflected in the outmigration of young males and the departure of young females to help' in urban households. In Kenya, both city and country are expanding.

Migration policy: Italy's Mezzogiorno

Between 1951 and 1971 the population of the Mezzogiorno (the lower part of the 'boot' of Italy) grew by 900 000 – the difference between a natural increase of 5 million and a migratory loss of 4.1 million. Most of this loss was from towns with populations of less than 100 000; in effect it was a flight from the precarious economic conditions of the small rural centres. But the nearby cities have few jobs to offer, and migration to the cities simply results in a deterioration of urban living conditions. Overall the income of the principal cities declines, as does the percentage of non-agricultural activities.

In contrast with the industrial North, southern Italy is mainly an area of traditional subsistence agriculture. This disparity was partly caused by government attention being focused on improvements to the competitiveness of industry. Thus the South became predominantly a source of cheap labour and an outlet for goods from the North.

Unfortunately, attempts to stimulate industry in the South have led to the development of highly capital-intensive industries which use only small amounts of local labour and which require skilled workers from the North. Furthermore plant has been located at coastal sites, depopulating inland zones and causing overcrowding on the coast.

37.6 Migration – good or bad?

Migration can be useful if it relieves the employment pressure at home. Many labour-exporting countries believe this process is working well, and actively seek foreign employment for their citizens. For example, Pakistan's construction sector was helped by considerable migration to the Near East: in 1973, 800 000 workers were in construction and 224 000 were underemployed; by 1978 emigration had made underemployment in construction insignificant, and allowed wage levels in Pakistan to double. However, problems arise when outmigration passes a critical level and produces actual labour-shortages in some sectors. For example, because so many nationals have migrated, the plantation managers of the Ivory Coast must now import workers from Upper Volta.

Countries that export labour are extremely vulnerable, financially and socially, to manpower decisions taken elsewhere. For example, Morocco sent nearly 30 000 workers to Europe in 1973. By 1977 recession had caused the number to be cut to 1300, the employment 'safety valve' was abruptly cut off, and the money that could be sent home drastically reduced. In 1982, Nigeria expelled 2 million foreign workers back to neighbouring countries when it got into severe economic difficulties. This had catastrophic effects on the local labour markets of countries such as Ghana and Chad.

In the final 20 years of this century, the global labour force may grow by 600–700 million. The Food and Agriculture Organisation (FAO) estimates that 200 million jobs can be created in agriculture; the rest – possibly 500 million – must be found jobs in urban areas. Although many of the future population will be born in the cities where they must find work there will continue to be much rural–urban migration in the future.

Part J

The basis of urban activity

38

Natural resources and energy

38.1 The importance of natural resources

For much of history, people have been influenced by the world distribution of resources in a quite fundamental way. These resources are of two kinds: those provided from the **inorganic realm** (water to drink, for power, for irrigation, and for transport) and minerals (salts, rocks); and those from the **organic realm** (fish, forest, gas, and coal). Sometimes, as with soil, a given resource has both inorganic and organic components. Whatever their origin, the pattern of resources is basic to a study of human geography.

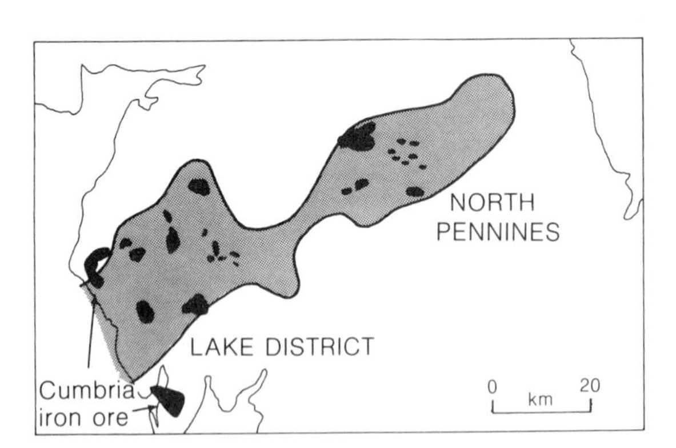

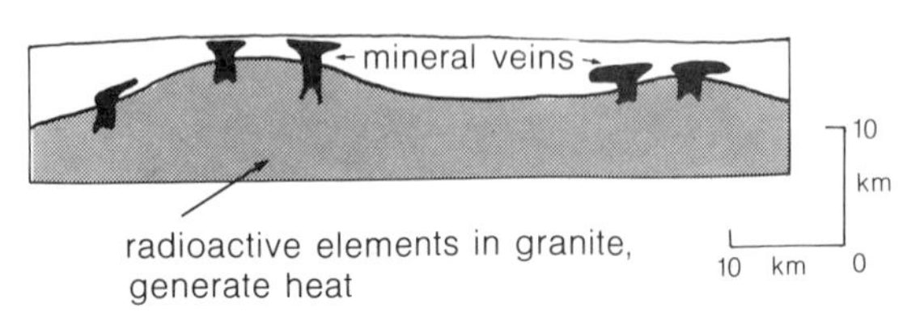

Figure 38.1 The location of vein deposits (above) in relation to the granites of the Lake District and North Pennines. The section below shows conditions at the time of mineralisation.

38.2 Inorganic resources

Minerals are scattered unevenly around the planet, often in zones associated with aspects of plate tectonics (Ch. 1). New rocks and minerals are provided only from sub-crustal magma and are therefore primarily located along plate boundaries. Destructive plate margins result in the emplacement of magma which later consolidates as batholiths and from which molten materials rise. These **volatiles** contain many **minerals** of considerable economic importance – gold, silver, tin, copper, lead and so on. As the volatiles rise through lines of weakness in the overlying rocks they solidify, according to their melting points, at different distances from the magma. This both separates the minerals and concentrates them into veins. Much mining is therefore directed along lines where batholiths have been uncovered by erosion or where they lie near to the surface (Fig. 38.1). For this reason the major mountain ranges have yielded many mineral discoveries. **Lodes** (veins containing metal ores) tend to be orientated according to the pattern of weakness in the rocks. For example the zone of mining in Devon and Cornwall extends east-west: this is the direction of the batholith emplacement and the line of weakness lies in this direction. For a similar reason, lodes in the Lake District are aligned south-west–north-east.

Destructive plate margins have been the scene not only of mineralisation but also of the production of **igneous** and **metamorphic rocks**, both of which are tough and therefore valuable as building materials. Granite is used wherever exceptional resilience is needed; for kerbstones, for facings on buildings and, among other things, for the foundations of the Eddystone Lighthouse. Metamorphic rocks such as slate break into sheets readily; and as they are impermeable, they make good roofing materials. Even decomposed granite (china clay) is an important material: it is used in making porcelain, as the inert base for pills, and to give paper its smooth and dense texture.

Although they are tough, igneous and metamorphic rocks are eventually eroded and the sediment is carried away by rivers. However, the minerals they contain, being more dense than most rock materials, quickly settle out in estuaries and on the continental shelf. These are called **placer deposits.** Here they form a reserve of ore whose exploitation by dredging is becoming increasingly important (Fig. 38.2).

Constructive plate margins offer quite different resources. Here basalts are produced and the crust is renewed. In this region metal-rich solutions, leached from the hot rocks by seawater as it circulates through the fractured upper levels of the crust, precipitate highly concentrated metal ores as **nodules**. One small region of the Red Sea, for example, contains 50 million tonnes of readily accessible ore; and the Pacific Ocean floor is littered with an estimated 100 billion tonnes of ore nodules, so thick on the ground they sometimes cover more than half the surface. These nodules will become extremely important in the near future; for example, three times as much copper is locked up in nodules as in the total land-based copper reserves (which are presently estimated as 350 million tonnes).

Sedimentary basins are a further source of specific minerals. These occur either on land or on continental shelves, mainly in areas that have been developed (or which are developing) as ocean trenches. Particularly important are the deposits of evaporites (**salts** such as calcium sulphate and sodium chloride) where there were once desert salt flats or lagoons; **coal**, which was deposited in coastal delta swamps; **oil**, the liquid remains of decomposed marine life; and **gas**.

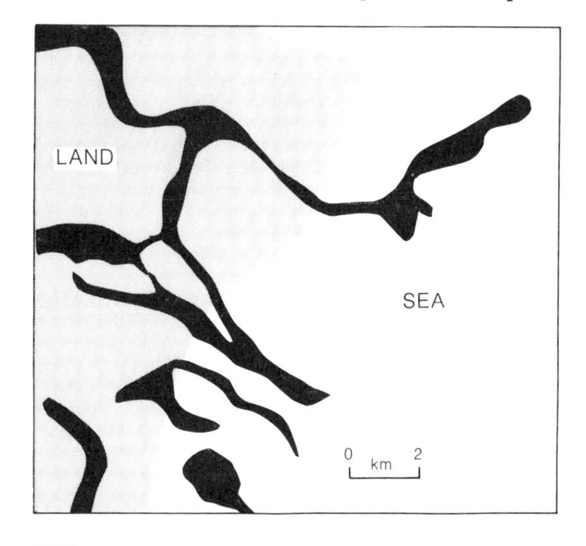

Figure 38.2 Tin ore is now recovered both onshore and offshore by dredging the placer deposits on the coastal shelf of Indonesia.

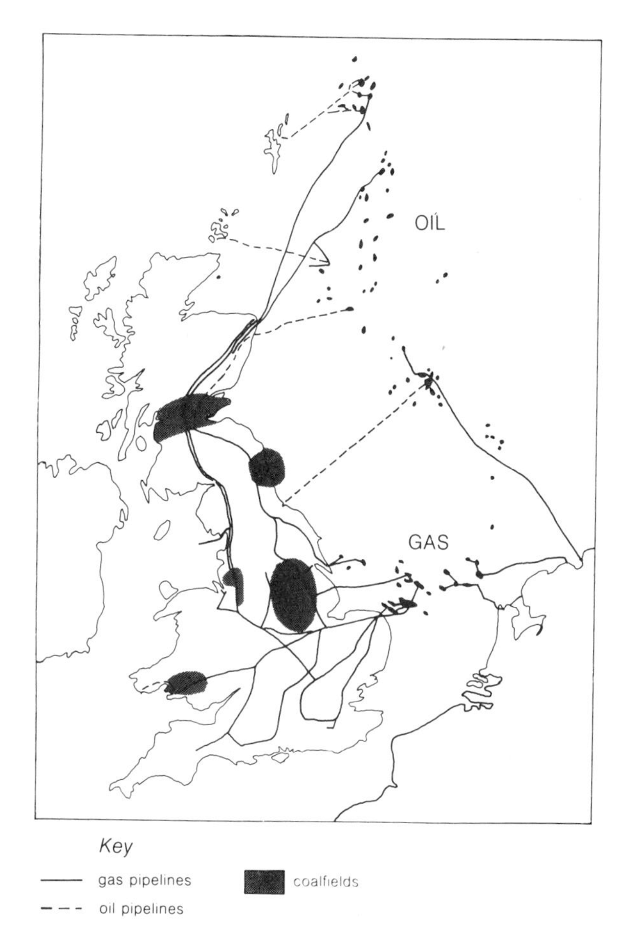

Figure 38.3 The major resources of Britain and the North Sea.

Gas and oil are often found with salt, for salt can act as impermeable cap-rock, preventing the gas from escaping during the formation of either coal or oil. The North Sea, with its rocks formed into basins and domes, provides a range of environments in which oil and gas may be trapped (Fig. 38.3). Gas is found in two separate locations, one where it has derived from coal, the other where it has formed from oil. Although gas and oil are found widely dispersed in many rocks, they are only economically exploitable when concentrated in domes. Exploration for these resources is therefore a 'hit-and-miss' affair, and any one dome ('field') usually contains quite limited reserves.

Oil contained in the shales of many countries

represents a very large potential reserve, although at present the cost of extraction is uneconomic. The mining of shales represents a major potential environmental problem because large volumes of shale rock would need to be extracted.

Coal is found in widespread sheets, formed at a time when much of the world's land surface was close to sea level. One set of related fields, for example, extends from Poland across western Europe to England and Wales. The same kind of coal is found in the Appalachian region of the USA, now far from the European deposits because of continental drift. Coal is not of uniform thickness, continuity or quality; like oil, it is of varying chemical composition and value. Some coal, such as the deposits in South Wales, is highly folded and faulted, and characterised by thin, discontinuous seams that are expensive to work. Other coal, such as that at Selby in Yorkshire, is thicker and less geologically disturbed. Coal in the USA and in Australia may be thicker still (more than 7 m thick in places), and so close to the surface that it can be open-cast mined, a process considerably cheaper than deep-shaft mining,. It is these variations in thickness and accessibility of coal reserves that make some coal-producing areas much more competitive than others.

Shallow-water sediments are sometimes rich in ores, especially iron. Both sandstone and limestone are ore-bearing rocks, but the metal content is often less than 30 per cent, in striking contrast to lodes, which are usually over 60 per cent metal. Sedimentary deposits, such as those in Northamptonshire in England or Lorraine in France, are worked because they are accessible and indigenous, but their high percentage of waste makes transport expensive. Iron-ore fields have had an important influence on the location of iron and steel works such as those at Scunthorpe and Nancy.

Possibly the most versatile of all sedimentary rocks is **limestone**; it is not only a tough building material and base for fertilisers, but good as a flux in iron and steel making. Formed in warm shallow seas as a precipitate from sea water, as the remains of plankton, or as coral reefs, limestone is the world's most widely distributed resource. **Phosphate** is another chemical precipitate of the sea bed, occurring as crusts, nodules or pellets in many tropical coastal waters. There appear to be at least enough reserves on the sea bed to last 100 years at present rates of consumption. While land-based reserves continue plentiful, though, the sea-bed ones will not be exploited.

Sands and **gravels** are an important resource for construction. Land-based deposits are either associated with **river beds** (as with the buried gravels of the River Thames floodplain) or with **glacial outwash** (as in Shropshire). However, these are dwindling reserves and most such aggregates will eventually have to be obtained from the **continental shelf.** In Britain the proportion of aggregate taken from the sea bed is already approaching a fifth of all production. This makes it all the more necessary to take care not to disrupt the coastal sediment budgets.

Relatively few economically useful minerals are formed in the eroding environment of continents (although some formed in earlier times are revealed). One important exception is **bauxite**, an ore of aluminium, often produced as a soil layer in laterites. Laterites are sometimes worth mining in tropical countries for both their iron and aluminium contents.

The world distribution of inorganic resources

NORTH AMERICA

Resources occur mainly along the east and west margins. There is a major zone of mineralisation the length of the Rockies (but not in Appalachia). Coal is found in Appalachia; less contorted beds lie in the sedimentary basin of central USA. Oil is found in the Gulf of Mexico, Mexico, Texas, Alaska and northern Canada.

SOUTH AMERICA

The zone of mineralisation provides resources in the Andes. Sedimentary iron occurs in Brazil. Oil appears widespread in the Amazon and in the Orinoco basins.

AFRICA

Africa is an ancient continent that has been stable for thousands of millions of years: it is therefore rather poor in minerals. Oil occurs in Nigeria in sediments of the Niger delta. Gold and diamonds are both very localised in South Africa. Copper in Zambia is one of the major reserves. Iron ore and phosphate occur in northern Africa together with oil in Libya.

ASIA

The zone of mineralisation associated with the Himalayas probably contains large reserves but so far they have only been exploited at the USSR boundary. An extension of the Himalayan zone forms the backbone of Malaysia and this has produced large tin reserves. The main oil reserves lie in the Middle East; only modest reserves have been found in other regions. India and China are particularly poor in exploitable resources and Japan is quite resource-deficient.

AUSTRALASIA

The main exploited reserves are from sediments, and the potential of inland sedimentary basins is high. Both iron ore and coal are extensively exploited. There are also reserves of oil.

EUROPE
This small continent has a high proportion of natural minerals, particularly coal. They have been exploited to a greater degree than elsewhere and Europe is now deficient in mineral resources. Western Siberia contains large oil and gas reserves, while exploration for oil and gas continue in the North Sea and other parts of the continental shelf.

38.3 Organic resources

Fish resources

The richness of the marine ecosystems depends on a continuing nutrient supply to enable the **plankton** (which form the base of the marine food pyramid) to flourish. Nutrient supplies derive from rivers, and therefore the regions of sea at the outlets of estuaries are often important for fishing. However, most nutrients are recycled from dead organic matter which decomposes on the sea bed. Shallow seas, where winds can induce turbulence which reaches the sea bed, are places where plankton thrive. In the North Sea winter storms cause nutrients to be stirred from the sea bed and, with the warmth of spring, this causes a plankton 'bloom'. Fish stocks are high in the North Sea and fishing is therefore an important industry. The shallow waters of the mid-Atlantic Ridge near Iceland provide a similar environment.

Regions where ocean currents well up to the surface (such as West Africa, where there is a permanent offshore wind) bring nutrients to the surface and are associated with plankton blooms. Many such areas exist in the Tropics, but the recovery of fish in the open waters there requires a high level of fishing technology, greater than that currently available to Third World countries.

It is believed that fish catches could be doubled before over-exploitation occurs. However many of the present reserves lie in the Tropics and include fish that, at present, are in little demand. The traditional Northern Hemisphere fishing grounds of the North Sea and Labrador are already over-fished and have recently become subject to quota limitations.

Timber

Timber is a vital resource to all countries. In the developed world it is now used little for fuel, but is extremely important for construction, chemicals and

Figure 38.4 Timber is a vital fuel supply in the tropics. Its over-exploitation and widespread burning for farming has, however, had disastrous consequences for soil erosion. This is the Rondonia area of western Brazil.

paper. In the developing world wood is used primarily for fuel, and next for construction.

Timber is analogous to any harvestable crop. Its production is best thought of as a form of farming. As with all forms of natural vegetation, the growth of timber depends on moisture, insolation and temperature conditions. The greatest gross primary productivity (GPP) is therefore achieved in the equatorial region. Areas of great moisture stress, such as savanna, steppe and desert, are either devoid of trees or have slow-growing, small tree species which have little commercial value. The GPP of deciduous and coniferous forests is not dissimilar, but far below that of the Tropics.

Tree growth and environmental conditions are important influences on the type of wood produced. Hardwoods (oak, beech, ash and elm), especially important for furniture, grow very slowly in temperate regions. Tropical hardwood (mahogany, teak, ebony and luan) grow faster and are more extensive. Softwoods (for construction, pulp, paper and chemicals) are provided mainly by conifers growing in cool temperate regions, but extensive stands of conifers have recently been planted in the Tropics.

Timber has to compete with agriculture for valuable land resources. It will grow in conditions of greater adversity than will many crops and is therefore found predominantly either on steep upland terrain in places of climatic harshness from temperature (such as Norway, Sweden, Finland and Canada), or in places where large-scale agriculture has not yet taken hold (such as parts of Amazonia and Malaysia).

Many countries have allowed their timber reserves to dwindle to hazardously low levels and are now dependent on imports. In Britain the Forestry Commission was instigated to help correct a situation of considerable deficit which had developed by the end of the 19th century. However, despite a vigorous planting campaign, the area devoted to forest in Britain is still too low to meet demand and considerable imports are needed.

Deforestation has also proceeded to crisis levels in many developing countries, due to widespread shifting cultivation, to increased pressure on the land for crop use (Fig. 38.4) and to the continued use of wood for fuel (Fig. 33.24). Wood for fuel is now extremely scarce in parts of Africa and India where a process of cutting without replanting has proceeded for many years. To become self-sufficient in wood, Kenya, for example, would now need to devote 9 per cent of the land area to forest – a difficult proposition for a country with a rapidly expanding population and an economy tied to agricultural produce. In such countries there is now the hope of growing trees and crops together – a system called **agroforestry**.

38.4 Energy resources

The availability of energy supplies is of prime concern to all countries, but the importance of energy differs between countries. In Ethiopia, for example, imported energy is necessary for the country's survival. In North America imported energy may make a difference between economic growth and recession (resulting in a change in lifestyle). Oil-exporting developing countries depend on the oil resource to build their economies.

The use of energy depends on the demands of technology and the economic development of each country (Fig. 38.5). In turn these are to some extent influenced by the relative pricing structure of each energy source. Coal, for example, was supplanted by

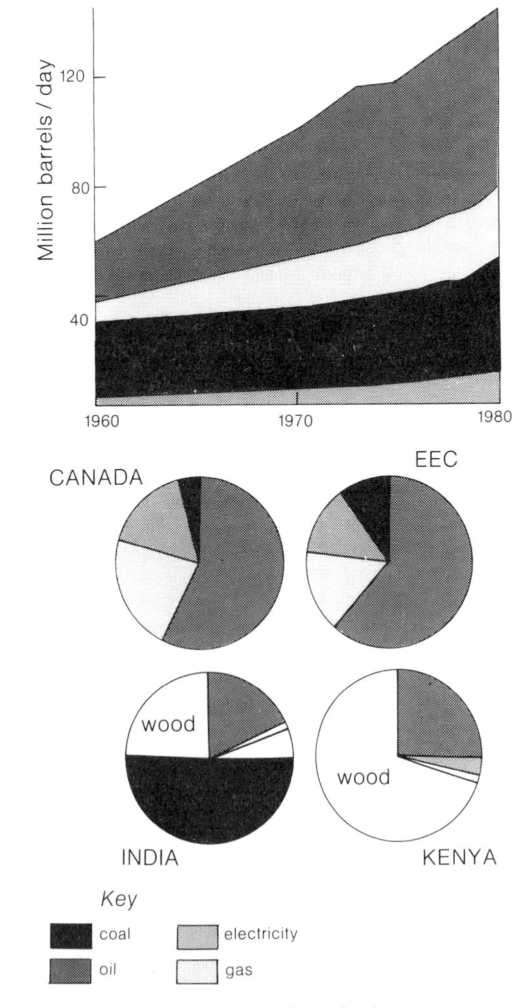

Figure 38.5 World consumption of primary energy and its use by the developed and developing world.

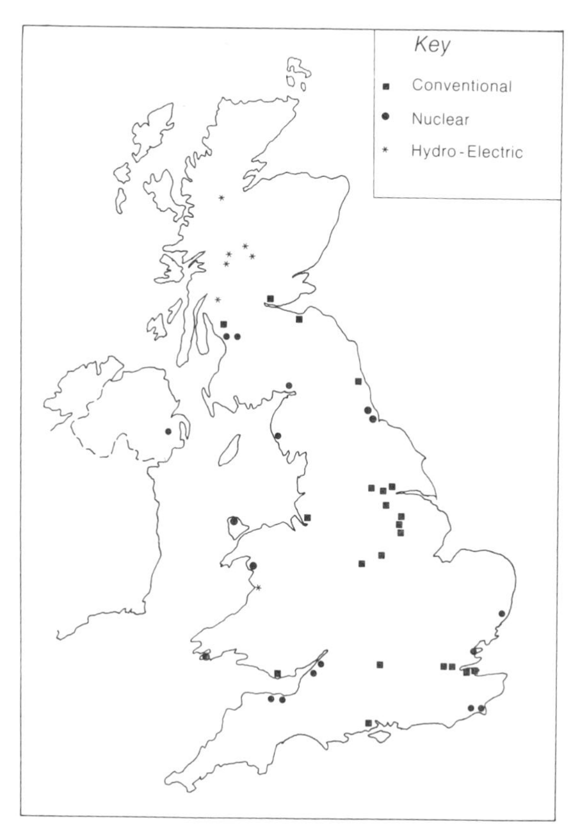

Figure 38.6 For many purposes, electricity is the most convenient form of energy. In the UK, as in most of the developed world, it is produced from a variety of sources. Nuclear stations are remote and coastal; coal stations are either on coalfields or on major rivers such as the Trent; most oil-powered stations are coastal or on rivers near large urban areas; HEP schemes are in the Welsh and Scottish Highlands.

cheap oil as the main direct energy source in the UK between 1945 and 1965, because the oil was plentiful and more flexible. Similarly, the use of electricity has grown markedly as new technology has been developed (Fig. 38.6).

Bridging the energy gap

Very few countries have an energy surplus: most need to import. This is often a substantial burden on the developed world; it can be crippling to developing countries. India, for example, has been involved in an upward spiral of energy costs: in 1950 oil imports used 9 per cent of revenues from exports; by the early 1980s this had risen to 80 per cent, although this figure has now been reduced. The only way to prevent a balance of payments deficit is to cut back on imports, but this could lead to economic growth below the rate of population increase.

However, it is not only the developing world that has an energy gap. Japan needs to spend half its export revenues on providing energy, and France and Germany are also both severely deficient in energy supplies.

Unreliable supplies and high-cost oil encourages development of nuclear power, but environmental concern has recently slowed the building programme for nuclear-power stations.

France at present plans to derive 30 per cent of its total energy needs from nuclear power, at the same time reducing its coal use to 17 per cent, its oil to 32 per cent and gas to 13 per cent. Even maintaining gas at this level poses political problems: imports must come from politically sensitive countries such as Algeria (as liquid) and the USSR (via a new pipeline).

The energy-resource base and its implications

Man's energy comes

(a) *directly* from the Sun (solar energy), the Moon (lunar energy), and the Earth (nuclear and geothermal energy);
(b) *indirectly* via stored energy in fossil fuels, or via wind power, water power and biological processes.

Direct sources yield replaceable forms of energy; only fossil fuels are non-renewable.

The total resource base is enormous. For example, direct **solar energy** received (insolation) is about 9000 times as much as the energy currently being produced by man (2×10^{10} KW). However, solar energy is highly dispersed and variable in amount, both seasonally and daily. At present its capture remains largely uneconomic. Similarly the effects of the Moon on the Earth are to create **tides** whose total energy is a hundred times that currently produced by man. But engineering techniques only allow tides greater than 2 m to be harnessed. The sites where this occurs are few, and although widely publicised, tidal generation of energy will remain of little significance.

The heat lost by the Earth's core is lost by conduction through crustal rocks, by the ejection of magma and by heating fluids in permeable rocks which then transfer the heat by convection. Convecting fluids form the basis of **geothermal power**.

Water moving through the hydrological cycle transfers as much energy to low-grade heat by friction as is produced in total by man. Only when it is concentrated into rivers can it be used (as hydroelectric power), and then only in restricted locations.

Wind power has a potential as great as that of tides, but again harnessing it provides major problems. Wind strengths are notoriously unpredictable.

The resource base of **nuclear energy** is immense, with uranium occurring at about 3 ppm in the top kilometre of the Earth's crust alone. Easily workable

Figure 38.7 Nuclear power plants in France, 1982. Notice the large proportion of inland sites, by contrast with the UK (Fig. 38.6). Further expansion will largely be decided on environmental grounds.

quantities of uranium are rather scarce but the new generation of fast-breeder reactors could make nuclear energy almost a renewable resource (Fig. 38.7).

Fossil fuels are quite different from all the other energy resources, mainly because they are finite. Coal is far and away the greatest single resource, amounting to 10^{13} tonnes. On the other hand oil and gas reserves are quite low, amounting to only 20–40 years' supply at present consumption rates.

OIL

At present oil is the overwhelmingly most important source of energy, greater than all the other energy sources combined. For this reason it is (in value) by far the largest single primary commodity in world trade.

The supply, distribution and use of oil is a preoccupation throughout the world, because of:

(a) its finite nature and the need to conserve its use, and to develop alternative energy sources;
(b) its impact on the balance of payments, international financial flows, inflation and economic growth;
(c) the dangers of supply disruption;
(d) its importance as the prime form of commercial energy used in developed and developing countries alike.

World production in oil grew by 8 per cent per annum in the 1950s and 1960s as developed countries made a massive switch from coal to oil, and as developing countries began their industrialisation using oil as a base. Although today oil production has slowed down (Table 38.1), even at present demand it will soon be exhausted. There are now only a few countries where production is large and still clearly expanding: Argentina, China, Egypt, Iraq, Mexico, Norway and the UK. Together these countries provide 17 per cent of the world's oil production. However, in none of them is future production likely to exceed the national demand. Several producing regions have now passed their peak of supply: the US (whose huge exploration effort is now demonstrating the law of diminishing returns), Algeria, Indonesia, Nigeria, Qatar and Venezuela. These countries provide 15 per cent of the world total. (The US *uses* 28 per cent of the world total.) Four of the most important members of OPEC (the Organisation of Petroleum-Exporting Countries) are dedicated oil conservers: Iran, Kuwait, Saudi Arabia,

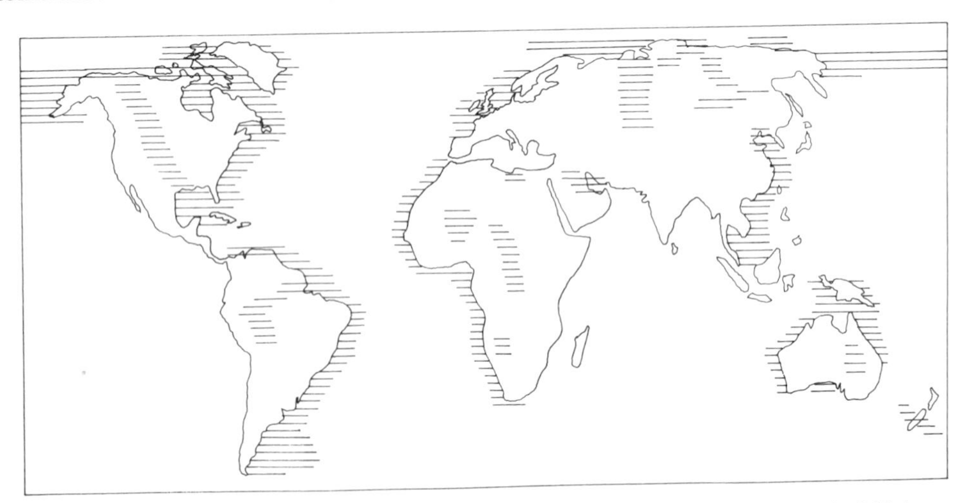

Figure 38.8 The world's oil reserves are concentrated in a relatively few sedimentary basins; only 240 have recoverable reserves; just 30 contain 90 per cent of the world's known recoverable oil and two of these (Arabia–Iranian and western Siberia) contain over half.

Table 38.1 Consumption of oil in the UK (million tonnes) (source: Road Transport Federation).

	1970	%	1975	%	1982	%
Fuel uses						
agriculture	1.33	1	1.39	1	0.98	1
iron and steel	5.64	6	3.27	4	1.22	2
other industry	16.92	17	13.69	16	11.60	16
road transport	19.27	20	21.54	25	24.98	34
other transport	6.09	6	6.34	7	6.34	9
households	3.05	3	3.27	4	2.15	3
fuel producers	24.20	25	20.67	23	12.68	17
other fuel users	10.54	11	9.26	11	5.29	7
non-fuel uses	9.82	10	8.28	9	7.60	10
total	96.86	100	87.71	100	72.84	100

and the United Arab Emirates (UAE). They provide 24 per cent of the world's production.

Despite wide-ranging exploration, the world's known reserves still appear to be highly concentrated in relatively few countries. The ten largest, with 20 billion barrels or more each, contain 80 per cent of the total reserve. In order, they are: Saudi Arabia, the USSR, Kuwait, Iran, the US, Iraq, the UAE, Mexico, Libya and China. Of this the middle East OPEC members have 55 per cent and Saudi Arabia alone has 25 per cent (Fig. 38.8).

COAL

The greatest reserves of coal lie in the USSR, China and the US; the remainder mostly in Australia, Canada, western Europe, South Africa, eastern Europe, India, Korea, Botswana, Colombia, Mozambique, Vietnam and Zimbabwe.

Coal consumption declined between 1945 and 1973 because it is less easy to use than oil, and its extraction is more costly. Nevertheless, the recent rise in oil prices has made coal competitive again. There are some near-surface coal seams over 10 m thick in the US and Australia, and these can be open-cast mined very economically. The US, Australia and Canada are currently the largest exporters; Japan, Italy and France the chief importers.

In the UK the demand for coal is at present about 100 mt/year, 66 per cent of which is used in power stations. There are reserves of 190 000 mt which will last for several hundred years at the predicted rates of consumption.

NATURAL GAS

Gas is likely to make an increasingly major contribution to world energy supplies, but it is expensive to transport and distribute. Nevertheless, it is likely to become the primary choice in developing countries (as portable cylinders of **liquid natural gas, LNG**) especially for heating, as traditional firewood sources run out. The existing LNG exporters are Algeria, Indonesia, Libya, the UAE, Brunei and the US (Alaska). The USSR is piping natural gas directly to western Europe.

TRADITIONAL FUELS

Wood and charcoal still meet 80 per cent of rural needs for lighting and cooking in the developing world. The vast quantities of timber used is causing rapid deforestation and a major fuelwood crisis has already developed. Indeed, because the majority of the world's population (60 per cent, or 2.5 billion) are rural, the situation will deteriorate. By the end of the century, when there will probably be over 6 billion people, 3 billion will still be rural, and even many in urban areas will continue to use either firewood or **dung**, because fossil fuels will continue to be too expensive (Fig. 38.9).

Figure 38.9 (a) Wood burned on an open fire is still the prime energy source for the majority of the world's population. (b) With firewood unavailable, many rural communities have turned to dung as a fuel. Not only is this an inefficient means of heating, but it also deprives the soil of much-needed organic fertiliser.

(a)

(b)

One way out of the fuel crisis, at least for heat and light, will be to produce **bio-gas** from animal dung by a process that also leaves the dung as a valuable fertiliser. There are already 7 million bio-gas plants in China using underground fermentation pits; more fuel is available in rural areas without transport being a problem. At present India has 90 000 bio-gas plants using a more expensive drum system.

HYDROELECTRIC POWER

The oldest form of power available to man comes from water. Yet its development from small-scale uses, such as to power flour or textile mills, to modern industrial use has not been without problems.

Direct water power suffers from many limitations. It severely restricts industry, by forcing it to have a riverside location; by keeping it to be dispersed, as factories have to be separated by sufficient distance for each to have a useful head of water (perhaps several kilometres); and by limiting the size of factory to the power locally available. These limitations all run counter to the requirements of efficient industrial production. Water power is therefore of value primarily when used to generate electricity.

Hydroelectric power derives directly from the conversion of the potential energy of water stored behind a dam into the kinetic energy of its movement, which can drive the turbines. Because kinetic energy relies on both the mass and the velocity of water, generation can be successful either with a large water mass with a much higher velocity (as on the Rhine and the Rhône), or with a smaller mass with a much higher velocity (as with most generating stations in the mountains of Norway, Wales and Scotland). Naturally the largest power plants – those at Boulder Dam, Arizona; at Kariba Dam, Zimbabwe; and at Itaipu Dam, Paraguay (Fig. 38.10) – have both large mass and high velocity. Hydroelectric generation is limited by relief and by water supply, and is therefore chiefly found in areas away from major centres of population. It can, however, be exported by means of transmission lines and is of paramount importance to many economies.

Figure 38.10 One of the largest hydro-electric schemes in the world is on the River Paraná at Itaipu, between Paraguay and Brazil. The total generating capacity is 12 600 megawatts, but because it involves using a high-volume, low head system is has been necessary to create a vast reservoir stretching back 150 km. Indeed, it is the environmental impact created by drowning large areas that, more than any other factor, will restrict the growth of such schemes in the future, especially in the more densely populated and developed countries.

Case study: Sweden's HEP system At first sight Sweden is not an ideal country for hydroelectric power generation. Sweden lies in the drier 'rain-shadow' area of the Scandinavian mountains, the land does not slope particularly steeply, there are few natural sites for large reservoirs, and the climate is harsh with temperatures below freezing point throughout the winter. However, an examination of Sweden's natural power resources reveals that the country has severe energy problems; there is no natural gas or oil and virtually no coal (Fig. 38.11). There is thus a large dependence on outside supplies and the need for the continuing goodwill of trading partners such as the unstable Middle Eastern countries. For strategic as well as economic reasons, government policy has been to try to generate as much energy within Sweden as possible.

Most northern rivers have a very variable flow, with a spring maximum and winter minimum. Demand for electricity has a winter maximum and a summer minimum. Supply and demand are matched by storing water and releasing it in a regulated manner through turbines. The large number of HEP stations is necessary because the landscape does not provide suitable sites for large reservoirs. At present Sweden is utilising 75 per cent of its potential HEP capacity.

GEOTHERMAL ENERGY

The Earth's crust acts as a very effective thermal lid, keeping much of the core heat inside. There is a substantial thermal gradient down through the crustal rocks – on average, 30 °C/km. If water is able to circulate deeply through rocks it will heat up and may reach temperatures far above boiling point (e.g. 150 °C at 5000 m). In most cases however, deep circulation is also slow circulation: rising water loses most of its heat to the cooler rocks that it passes on its way to the surface. Only in those cases where there is a relatively fast route to the surface – such as along a belt of fault-shattered rocks – will water temperatures remain high enough to provide **geothermal power**.

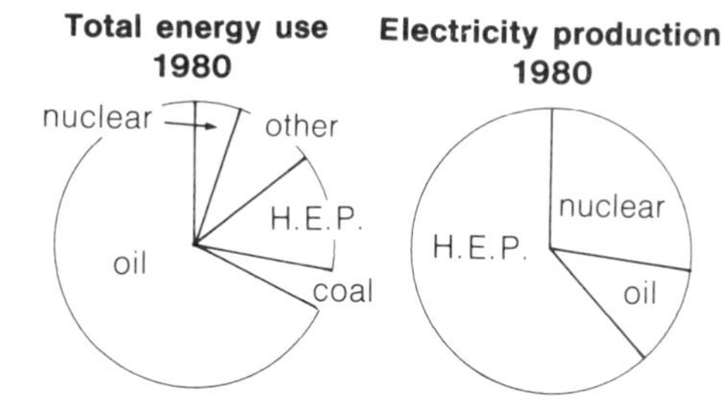

Figure 38.11 Sweden and HEP. Controlling the resource: (a) Variable natural river energy without control; (b) demand; (c) matching HEP output to demand using reservoir storage.

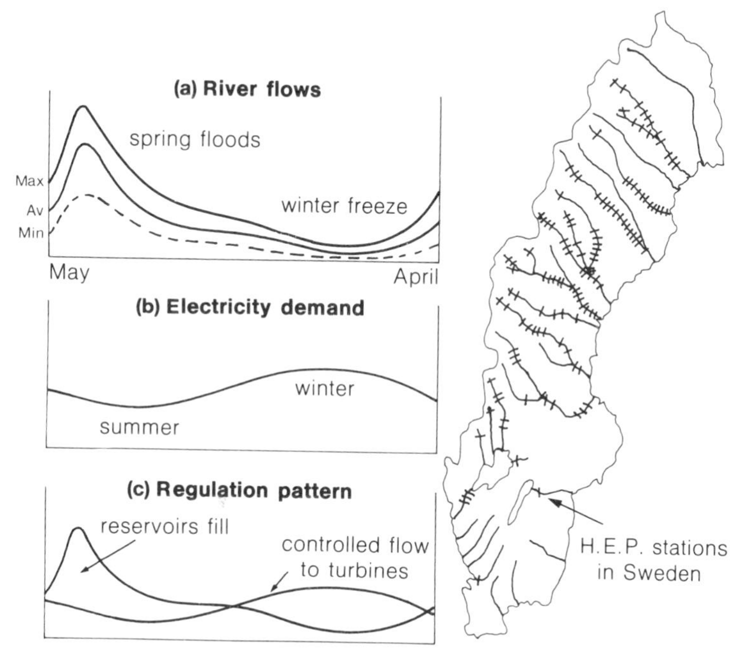

Most geothermal-power stations are no more than steam turbine generators. Naturally heated water must be above boiling point to be usable. Superheated water is normally obtained by drilling deep wells that tap the rising water source at depth, providing a direct surface access. As the superheated water rises up such a well, it boils and becomes steam which is fed directly to the turbines.

The basic requirements of a geothermal-power station are therefore:

(a) a source of permeable rocks that allow water circulation;
(b) rocks in which there is a high geothermal gradient;
(c) a shatter belt to allow rapid return of water to the surface.

Lower-temperature geothermal regions may be used to provide domestic heating systems – an application that has wider potential, especially in Britain where water temperatures are too low for use of geothermal power.

The principal areas of useful geothermal power are those with active volcanic activity, such as Iceland, Italy, New Zealand and Western USA where magma is actively rising to the surface. Near the Klamath Falls in Oregon, for example, seven thermal artesian springs and 800 thermal wells are used to provide substantial space heating. Here faulting is associated with the Basin and Range Province (Fig. 38.12). Water from adjacent mountain streams percolates through the valley alluvium, is heated from a deep source of magma, then rises quickly along fault lines emerging on the surface as hot springs.

Geothermal power is also potentially of great advantage to developing countries without energy resources. Kenya, for example, has been attempting to maximise its use of geothermal power and the 30 MW produced near Lake Naivasha in the Rift Valley provides 14 per cent of her electricity needs. At

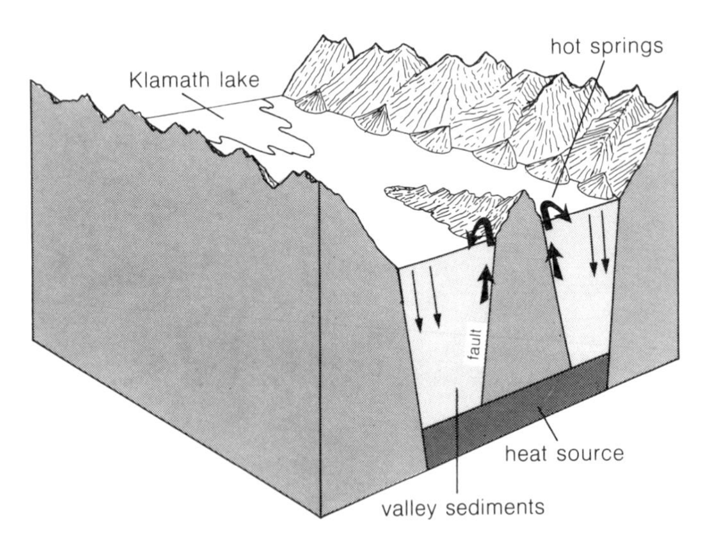

Figure 38.12 Geothermal power in the American Basin and Range.

present Kenya's imported oil takes almost all the revenue generated from coffee, the main export crop. As geothermal power is substituted for oil the export earnings saved can be directed to importing capital goods for industrialisation.

Italy, USA and the Philippines presently generate three quarters of the world's geothermal power, but by the end of the century nearly 40 per cent will be generated in the developing world, especially Mexico. However, future progress will involve pumping water into hot dry rocks. This will make geothermal power possible for every country (Fig. 38.13). It has been estimated that the world's *potential* geothermal power is 10^5 MW, and it will one day have a significant part to play in the world, not just a marginal one as at present.

Figure 38.13 The crustal heat flow in Britain, showing regions where it might be used for space heating.

39

Communications and trade

39.1 Communication

An important part of human geography concerns the study of the ways in which people move, transport their goods and communicate with one another. Many of these forms of communication occur in some form of channel – along a road, pipeline or telephone wire. When we study communications we are really dealing with **networks** made from **links** (routes) that connect **nodes** (centres such as cities, towns and villages).

Networks vary considerably with the form of communication involved. For example, pipeline networks are usually very simple because pipes are only valuable as a means of transport when high, relatively constant volumes of materials need to be moved. These give a **coarse mesh** network. On the other hand, networks of public footpaths are extremely complicated: they result from many years of contact on foot between settlements, and show a **fine mesh**.

Because transport systems play a large part in determining network geometry, it is first necessary to establish the constraints imposed on each means of transport and the environments in which they developed.

39.2 Transport systems

By foot

Economies that are primarily subsistence-orientated have little surplus wealth and engage only in limited trading. Sophisticated means of transport are therefore largely unavailable; distance travelled is mainly restricted to journeys that can be undertaken within a day. Most communication is between adjacent settlements, with people taking the shortest possible routes across the landscape. Repeated use of these routes leads to the formation of tracks.Because the tracks are in use throughout the year and in varying weather conditions, a network of good-weather tracks is often complemented by a network of bad-weather tracks, many of which are diversions to avoid wet or high ground.

Routes established in this way are 'user-orientated', often criss-crossing in a complex mesh. They still provide the chief links between settlements in many rural areas of the developing world, and their imprint on the British landscape is seen both in the system of public footpaths and in the little country lanes that form such an interesting part of the countryside.

Foot transport, with or without the use of pack animals such as mules or camels, or, latterly, of bicycles, is still the most common form of transport in the world (Fig. 39.1). Living in the developed world, it is easy to forget the much larger population of the developing world whose use of mechanised transport is severely limited.

The main advantages of human or animal transport are:

(a) it is very cheap;
(b) it is extremely flexible.

The disadvantages are:

(a) it is slow for any but the shortest distances;
(b) it cannot cope with large quantities of goods;
(c) it cannot cope with large sizes of objects.

Foot transport and animal transport cannot provide the basis for large-scale regional trade. Today, without some form of mechanised assistance, transition to a commercial market economy is not possible.

Figure 39.1 Until recently, rural transport was all powered by muscle: human or animal. This camel-cart is carrying bales of cotton.

By water

BARGES

With the exception of the direct use of pack animals, the oldest form of transporting goods is by water. Sometimes, as with logs, a simple raft can be built that allows goods to be moved without even loading them into a specially built vessel. However, most goods must be carried in boats, and this places strict limits on the waterways that can be used.

The depth of a boat below the water line is called the **draft**. Boats that are very wide need only a shallow draft to remain stable; although such craft could not well withstand gales on the open sea, they have considerable advantages in shallow rivers. Wide, shallow-draft river vessels built for carrying goods (**barges**) have been in use for many centuries.

The advantage of carrying goods by water is that very little effort is needed to move them: the water supports the boat and the cargo. The engines of large river barges, which can propel several thousand tons of cargo along canals, are no bigger than those of lorries.

Barges helped start the Industrial Revolution, by transporting bulk cargoes. Barges either used sails (as with the famous sailing barges of the Thames) or were pulled along either by men or horses on tow paths.

Although barges are now motor-powered barge transport is still cheap compared with all other mechanised means of moving goods (Fig. 39.2). However, it also has several important *disadvantages:*

(a) it is limited to the routes taken by rivers and canals, and these routes may not correspond with either the sources of the goods or their destinations;
(b) even with power-driven barges, movement is slow, making it quite unsuitable as a means of carrying perishable cargo;
(c) considerable time is needed to load and unload, and special facilities such as cranes and elevators have to be provided (although to some extent this problem has been overcome by the use of containers – see below);
(d) transport can be disrupted if the waterway freezes over in a harsh winter of if water levels fall in a dry summer.

The first canal ever built for industrial purposes was opened in 1761. Its use so dramatically reduced the cost of coal transport that there was an immediate canal 'boom' to link rivers, towns and coalfields. So important were canals that factories were built beside them, and distribution centres grew rapidly wherever roads crossed them.

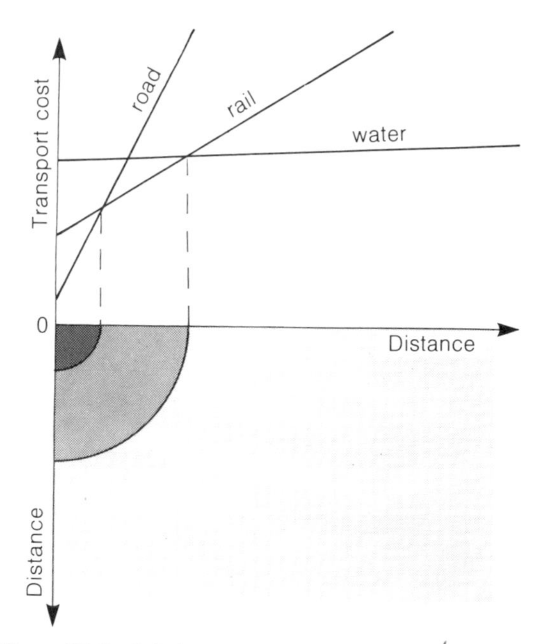

Figure 39.2 Relative costs for road, rail, water and air transport. This explains why, in general, short-haul traffic is by road, medium-haul mainly goes by rail, and long-haul by water, assuming equal availability of all transport systems.

The canal age did not last long in Britain: the small-capacity 'narrow boat' was no match for the railway and average hauls were short. By contrast,canal transport in other parts of the world prospered. Few countries are islands, fewer still are sufficiently small to take advantage of bulk movements by coastal ship. Barges therefore provide the only means of carrying bulk cargoes cheaply. In Europe, for example, rivers and canals are larger than in Britain and this enables much larger, and therefore more economic, loads to be carried. The success of waterways lies in their cost advantage: they can offer transport at as little as a fortieth the price of road transport. Thus , for example, the import trade of Switzerland is largely funnelled through Basel, whose Rhine port can receive the national requirements for bulk raw materials at the lowest transport cost. Basel's port is noticeably less prominent in the export trade for which manufactures are sent to a wide variety of destinations.

SEA-GOING SHIPS

No country is completely self-contained, wanting no goods from its neighbours and wishing to sell nothing that it produces. Raw materials are not distributed evenly over the Earth and rarely are they near to the places where they are needed. In many cases there is no alternative but to transport goods and raw materials from one continent to another by ship (Fig. 39.3).

Sea-going vessels come in many shapes and sizes:

(a) small ships are used to ply the coast and to deliver cargoes to small ports that the larger vessels cannot enter;
(b) larger ships are used primarily on long journeys, usually to carry bulky cargoes such as metal ores, oil and grain – they can only enter very large harbours and are usually designed to carry one specific cargo;
(c) **container ships** are both large and small. They have been introduced in the last twenty years to overcome the previously slow and expensive process of loading and unloading general and fragile cargoes at the dockside – a process which uses cranes and much dockside labour. Container vessels are loaded and unloaded by special cranes and the containers are of standard sizes to enable them to be stacked efficiently in and on ships, railway wagons, trucks and river barges.

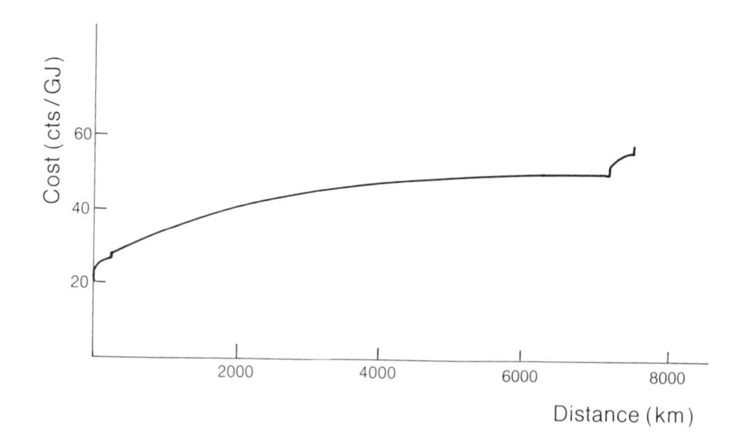

Figure 39.3 The cost of moving coal from Blair Atholl in Queensland, Australia, to Japan illustrates the effects of comparative costs and distances with different transport systems. The mine price in 1980 was 21 cents/GJ (gigajoule); overland transport in Australia and Japan (300 km) added 10.5 cts/GJ; ocean transport (7000 km) added 21 cts/GJ; loading and unloading were 5.2 cts making a total of 58 cts/GJ for 7300 km of transport. Notice how shipping makes up only a third of total costs.

By rail

Canals have been built in many countries of the world to supplement natural rivers and to bring water-borne goods closer to their destinations. However, as trade becomes more intricate and goods have to be moved between many locations the use of canals is often impractical. Because animals cannot pull heavy cargoes or travel long distances, they were never a match for canals. It was only when steam-powered engines became available in the early 19th century that there was an expansion of the railway system.

Railways have many advantages:

(a) they can reach places that would be impractical for water-borne transport;
(b) they provide a fast means of transport and are therefore suitable for carrying perishable foodstuffs;
(c) they can carry goods in bulk such as coal and iron ore,yet they are able also to transport fragile goods;
(d) they can transport large numbers of people efficiently from one centre to another.

Nevertheless the railways do still have disadvantages:

(a) railways need special track and are restricted by gradient, so that often expensive cuttings, tunnels and embankments are needed;
(b) trains must be spaced out along the track and have therefore to be strictly timetabled;
(c) it is difficult and expensive to provide flexibility in the number of trains, so they are most suited to regular transport of a constant volume of goods;
(d) stations need to be provided for loading and unloading, thus restricting the flexibility of delivery.

By road

TRUCK

Despite the enormous capital outlay needed to provide an intricate railway system, there is no doubt that railways have allowed countries to develop their industries efficiently, and they continue to provide a valuable long-distance transport system. Yet there is always the difficulty of loading and offloading at stations, in addition to that which occurs at the factory and at the destination. It is primarily to provide real door-to-door service that trucking has developed. Trucks have many advantages for goods transport:

(a) they offer completely flexible transport, literally door to door;
(b) they are flexible in time – it is easy to use more or fewer trucks, depending on the requirements;
(c) they yield reduced loading and unloading costs;
(d) they allow a reduced transport time over short distances for which loading and unloading time is an important part of the total transport time;
(e) they permit the entrepreneur access to the transport business with relatively little capital outlay (the cost of one vehicle).

It is these advantages that have given road transport such superiority in recent years both in the developed and developing worlds (Table 39.1). Even so, there are disadvantages to road transport:

(a) it is the costliest means of transport for each kilometre-tonne of journey, and although

URBAN TRANSPORT PROBLEMS

Segregating long-distance and local traffic is one of the major transport problems facing all societies. At present, many major through route systems still suffer from lack of upgrading to meet present demands. In the UK during the 1960s and early 70s, many urban motorway systems were begun, but they were stopped, partly through lack of finance, but also from increasing environmental concern. As a result of the termination of these schemes many major roads remain unimproved and others poorly connected to the local road network.

London's South Circular Road is a classic case of an unimproved major road that is permanently congested. It was to relieve such congestion that the M25 orbital road was constructed, encircling the capital. To get a measure of the problems that have to be tackled, it is helpful to know that, since 1950, traffic in London has risen by 250%. In central London the traffic density is now 50 times the national average. About three quarters of a million tonnes of freight are moved daily – mostly by road. Two and a half million deliveries a day are made within the capital. Indeed, half of London's daily freight tonnage consists of movements within the conurbation, rather than freight moving into or out of it. The orbital road will help reduce the impact of lorries, but it will do little to relieve private vehicle movements and the consequences: a reduction in average speed within the city of 10% between 1971 and 1980; the use of residential roads and shopping streets by through vehicles because the main roads are congested; a saturation level of onstreet parking; and an increase in traffic hazards.

transport costs can be more than offset by the reduced loading costs on short journeys, on long journeys and with large volumes of cargo going to a single destination, rail would be cheaper;

(b) it causes the most stress to the community as heavy trucks thunder along roads near to houses and through otherwise quiet and peaceful countryside;

(c) it causes more pollution than any other transport system.

Table 39.1 Freight transport in the UK (thousand million tonne kilometres) (source: Road Transport Federation).

	1970	1975	1979	1980	1981	1982
road	85.0	91.8	104.6	95.9	97.1	100.0
rail	26.8	20.9	19.9	17.6	17.5	15.9
total road and rail	111.8	112.7	124.5	113.5	114.6	115.9
road proportion	76.0%	81.5%	84.0%	84.5%	84.7%	86.3%
water	23.3	21.1	38.0	41.8	41.9	44.4
pipelines	3.0	5.9	10.3	10.1	9.3	9.3
total	138.2	139.7	172.8	165.4	165.8	169.6

Going nowhere fast on the South Circular

By LEE RODWELL

THERE are 36 pedestrian crossings and 47 signal controlled crossings along the South Circular. We seemed to stop at every one.

Not to mention the hold-ups when we let traffic filter in from the 320 streets with direct access to the A205, or the delays when buses blocked the only lane by stopping at some of the 144 bus-stops on the route.

EDGED

We joined the South Circular at the Kew end and stopped and started and edged our way over Kew Bridge and down Kew Road. The traffic flowed more freely down Mortlake Road until we hit the first of Andrew Warren's "black spots"—East Sheen.

He said: "The big problem here is one that is repeated along the South Circular — vans and trucks have to unload in front of the shops. Queues also build up through people waiting to use the two car parks behind Waitrose."

It was not far to the next Warren black-spot, Putney. We queued up to the approach of the traffic lights where the Upper Richmond Road (i.e. the South Circular) crosses Putney High Street.

Mr. Warren said: "Government statistics prove that a High Street is ten times more dangerous than something like a motorway which has limited access."

We press on, through the maze of lanes and signs that make up the Wandsworth one-way system. "Even the signs are confusing for the stranger," says Mr. Warren. "If you are heading for Woolwich, how do you know there which way to go?"

He was looking at two signs, one above the other. The top one said: "Sth. Circular Rd., The City, Brighton, Folkestone." The bottom one said: "Sth. Circular Rd., the North, The West, Oxford."

On to Clapham Common, down The Avenue and across the A24 at Balham Hill. As we approached the main Brighton Road, the A23, at Streatham Hill, Mr. Warren said: "This is the first bit of dual carriageway so far.

"That's what is so bad with the South Circular. So much of it is single-lane traffic, sometimes it goes to two lanes each way, but compared to the North Circular there is very little dual carriageway let alone three-lane traffic in each direction.

"The South Circular is just a series of high streets and residential roads linked by signposts. To suggest it acts as a viable inner by-pass for London, as is officially claimed, is a very sick joke."

HAMPERED

Across the A23 the three lanes almost immediately became a single lane. From there the South Circular winds across Dulwich Common and along to Forest Hill—another black spot. Vans unloading in London Road hampered progress and bus stops and traffic lights meant more queues.

Through Catford, the shops give way to houses and in Brownhill Road the front doors are just a car's length from the road.

Mr. Warren said: "Think how miserable that must be for the people living there—and how dangerous. A dog or a child could dash right out of the door and straight into the traffic."

Past Hither Green the road widens to a dual carriageway and we reach the legal speed limit of 40mph. "This is the best part of the South Circular," said Mr. Warren, "but it also has its dangers.

"After all the frustrations of the past miles you're tempted to put your foot down and take silly chances."

We negotiate the Well Hall roundabout — another notorious rush-hour trouble spot—without hold-ups and the rest of the journey to Woolwich was problem free.

Journey's end and Mr. Warren summed up: "The GLC is talking about doing something about odd junctions but that's just tinkering."

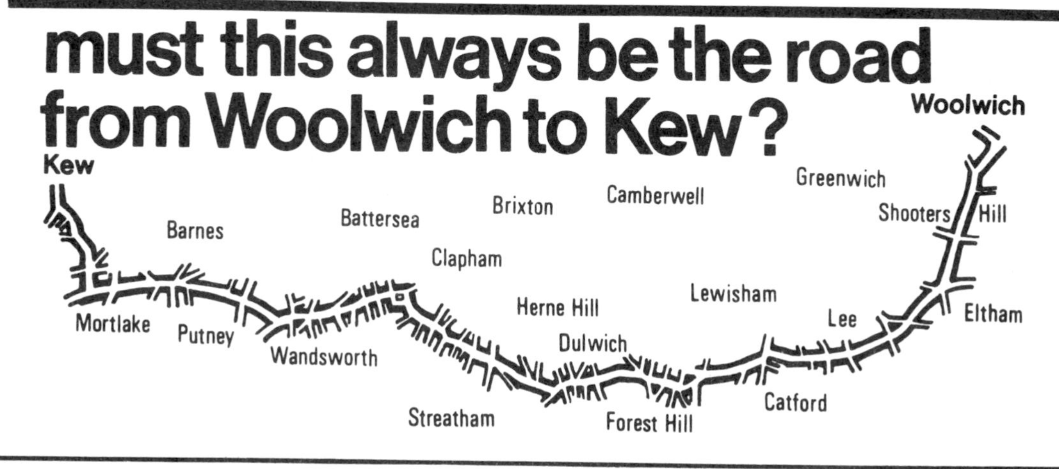

COACHES AND BUSES

The **bus** was first introduced to enable people to travel between places that could not be reached by railway. Nevertheless it soon grew to be a strong competitor with the railway. Coaches are the long-distance version of buses, but with more specialised networks they more nearly resemble a railway.

Buses and coaches are exclusively for carrying *people*. They have the following advantages:

(a) they are extremely flexible and can offer a door-to-door service;
(b) because they can run smaller units than the railway they can be cheaper both over short and long distances;
(c) they remove congestion from the roads by carrying more people than do motor cars;
(d) routes can easily be altered to match changing demands of the customers.

CARS

Motor cars achieve a high degree of flexible personal transport. However, they are often of greater convenience value than economic value (Table 39.2). Only some motor cars are used for commercial purposes. They have several major advantages:

(a) they provide a real door-to-door means of transport;
(b) they are ready for use when needed – people do not have to organise their journeys to fit in with public-service timetables;
(c) they are used for many domestic, social and pleasure purposes – to get the shopping, to go to work, to go on holiday and so on.

Many people recognise that they do have many disadvantages:

(a) they cause widespread congestion especially in city centres;
(b) large areas have to be provided for car parking;
(c) they are a major source of air and noise pollution;
(d) they are inefficient and expensive fuel-users.

By air

Nowadays aeroplanes seem commonplace, but they are newcomers to the transport scene. Their advantages are:

(a) they provide the fastest means of transport over long distances and (using helicopters) even over short distances – they are therefore suitable for rapid transport of people and urgent material such as letters and medicines;
(b) on very long journeys they are the cheapest means of transporting people.

Table 39.2 Passenger travel in the UK (thousand million passenger kilometres) (source: Road Transport Federation).

	1970	1975	1979	1980	1981	1982
road						
private transport	309	352	407	423	420	452
public transport	53	55	48	45	42	40
pedal cycles	5	4	4	5	5	5
total	367	411	459	473	467	497
rail	36	35	36	36	35	31
total road and rail	403	446	495	509	502	528
road proportion	91.1%	92.2%	92.7%	92.9%	93.0%	94.1%
air	2	2	3	3	3	3
total	405	448	498	511	505	531

There are, however, strict limitations on the use of aeroplanes, especially for short hauls:

(a) the majority of the fuel costs are used in taking off and landing: it is therefore nearly as expensive to use an aeroplane for a short journey as a long one;
(b) aeroplanes (and to some extent even helicopters) need large areas of open land for **airports** – often these cannot be provided near city centres, so the rapid journey time between airports may be more than offset by the subsequent journey from the airport to the centre;
(c) aeroplanes cannot carry large bulk cargoes economically and **air freight** remains limited to rather special goods. (Nevertheless, about 12 per cent of the UK's total trade [£13 billion] was transferred via Heathrow airport in 1982 – a reflection of the progressive change to low-bulk, high-technology production.)

Pipelines and cables

Some industries, such as petrochemicals, produce large quantities of liquids and gases that need to be delivered to a small number of places (Fig. 39.4). In these cases the cheapest means of transport, despite the high initial costs,is to use a **pipeline**. There are now many pipelines connecting, for example, the natural-gas fields under the North Sea with the land-based distribution points, or connecting oil terminals at the coast with chemical works inland. The use of pipelines has allowed many industries greater flexibility in choosing sites for their factories. This flexibility has been even more important in the case of electricity, where high-tension **cables** can transfer this vital power in the most flexible and economic way.

Telecommunications

In the past few years we have become so used to communication by telephone, television and radio that we seldom give more than a passing thought to this as a major means of transport.

When the Industrial Revolution began instructions had to be taken by hand between customer and manufacturer. This process became so complicated that eventually the **postal system** was introduced simply to transport all the letters and small parcels that people wanted to send. The postal service has grown tremendously since its foundation and is now a major user of rail, road and air transport. But it was the development of the **telegraph** and the **telephone** that revolutionised the way people were able to communicate. They provided instant contact between people, freeing them from personal visits or the time needed between sending a letter and receiving a reply. In the same way **radio** and **television** have provided instant communication to mass audiences and have tended to replace **newspapers**.

It is more difficult to assess the rôle of telecommunications – the transport of ideas and information – than that of rail or road in carrying goods. Nevertheless it is clear that telecommunications have allowed offices and factories to be located in different parts of the country, and offices to be sited outside the centres of cities.

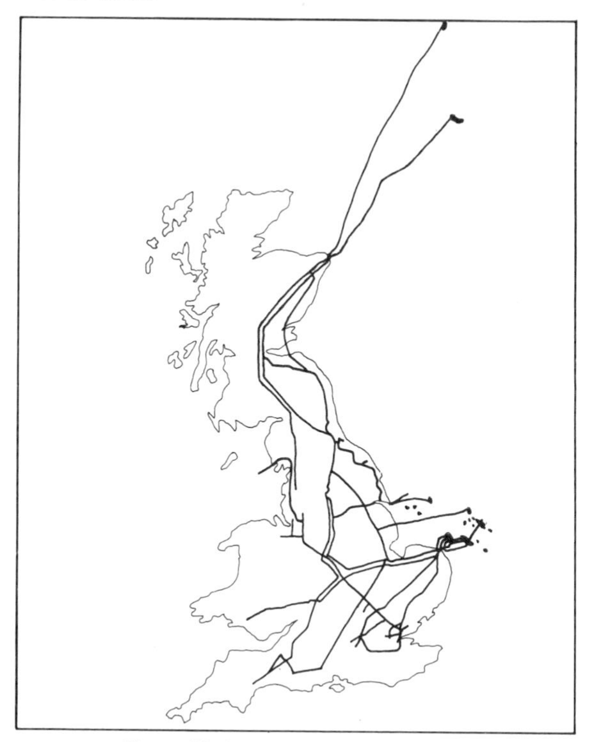

Figure 39.4 Gas pipelines in Britain.

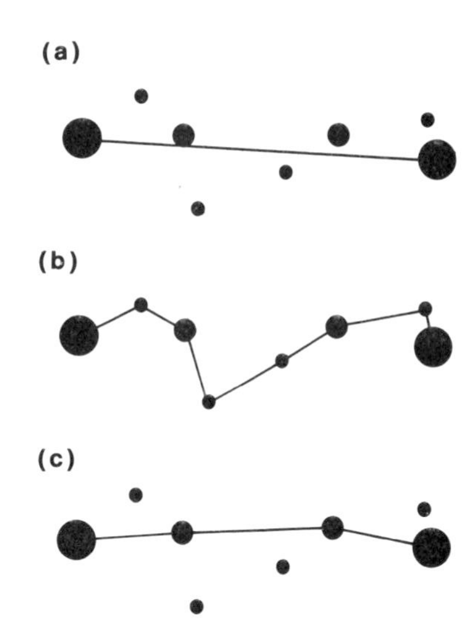

Figure 39.5 Location of routes: (a) shortest route; (b) maximise traffic; (c) optimise length and traffic.

39.3 Settlements and networks

The effect of settlements on networks

Transport systems vary considerably in their flexibility, and this partly explains the observed variations in route patterns. However, for each type of transport system the exact route constructed depends on its primary purpose. For example, routes designed to join two centres may ignore any settlements in between because these have little importance for the type of trade envisaged – this is true of airways and motorways (Fig. 39.5a). This produces the cheapest transport costs between major centres. At the other extreme, routes may be designed to connect all intermediate points either to provide a service (as with minor roads) or to maximise the traffic (as with bus services; Fig 39.5b). This extends the journey length and therefore transport costs are higher. Naturally there are many compromise route systems, such as those designed both to minimise journey lengths between major centres and to maximise traffic (Fig. 39.5c). This is achieved by connecting major centres with a route that sweeps as near as possible to all minor ones. Many railway networks have been designed on this principle. This simple idea can be taken one stage further by considering the networks that will connect a number of nodes together (Fig. 39.6), each network best suited to a different purpose. Thus the type of network suggests the reasons that underlie its construction. In general, in regions of clustered cities (Fig. 39.6) the pattern most closely approaches a system of least cost to the user; whereas in areas where settlement is sparse (Fig. 39.6a), it tends towards least cost to the builder.

PLANNING A ROAD SYSTEM

A road network for a residential area needs to provide for an orderly movement of traffic. It must serve each part of the area equally, thereby allowing maximum access and potential development.

The requirements of a road network are matched very closely by a natural river system: a network of tributaries provide an even drainage density that focuses water into a larger trunk stream. 'River flow' in a residential area consists of two daily 'flood' peaks when people go to and return from work. Through the day there is a much smaller and steadier 'groundwater flow' as people go on shopping and service trips.

An efficient road network should be amenable to drainage-basin analysis. What are the implications if it is not?

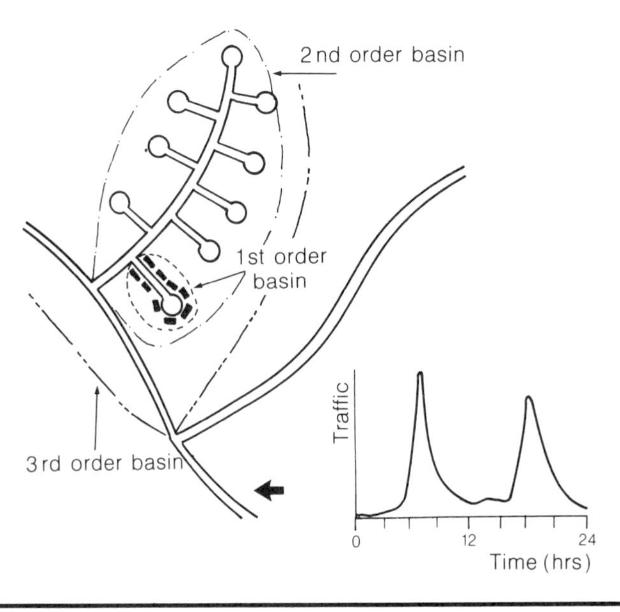

Physical limitations on networks

The position of a route can be affected by changes between two transport systems. For example, a cargo of electronics from the EEC via Rotterdam to Birmingham must be carried first by ship and then by rail or road. The choice of which East Coast port to use is based on the need to achieve a minimum *total* cost of transport. Because sea transport is cheaper than road transport, the cheapest route in this case is not a straight line but a dog-leg. Lösch has shown that the port will be located as near as possible to the point where

$$f_1 \sin x = f_2 \sin y$$

the same law of refraction as used in physics (Fig. 39.7). [In practice, however, there are other economic and political elements involved which means the sine law rarely works very well. The use of a port can even depend on the attitude of trades unions.]

In many areas relief imposes a constraint on route construction. A route built to connect two points on either side of a range of hills which simply took the shortest path across the hills would involve considerable building costs, because of the need for cuttings, embankments and tunnels. This solution is only adopted when the volume of traffic, and therefore the return on road building, is high, as in the cases of the M62 connecting Leeds with Manchester across the Pennines; and the Grand St Bernard tunnel in the Alps connecting Italy with Switzerland.

An alternative stategy is to avoid the hills altogether, choosing a lowland route of least cost per kilometre. Such a route will have great length and few benefits to the user. Most minor roads in upland areas are of this kind. For many major links a compromise is adopted, balancing the higher construction costs of a short hill section against the savings of a shorter journey (Fig. 39.8). The proportion of high-cost hill route and low-cost plains route can be calculated by using the refraction equation.

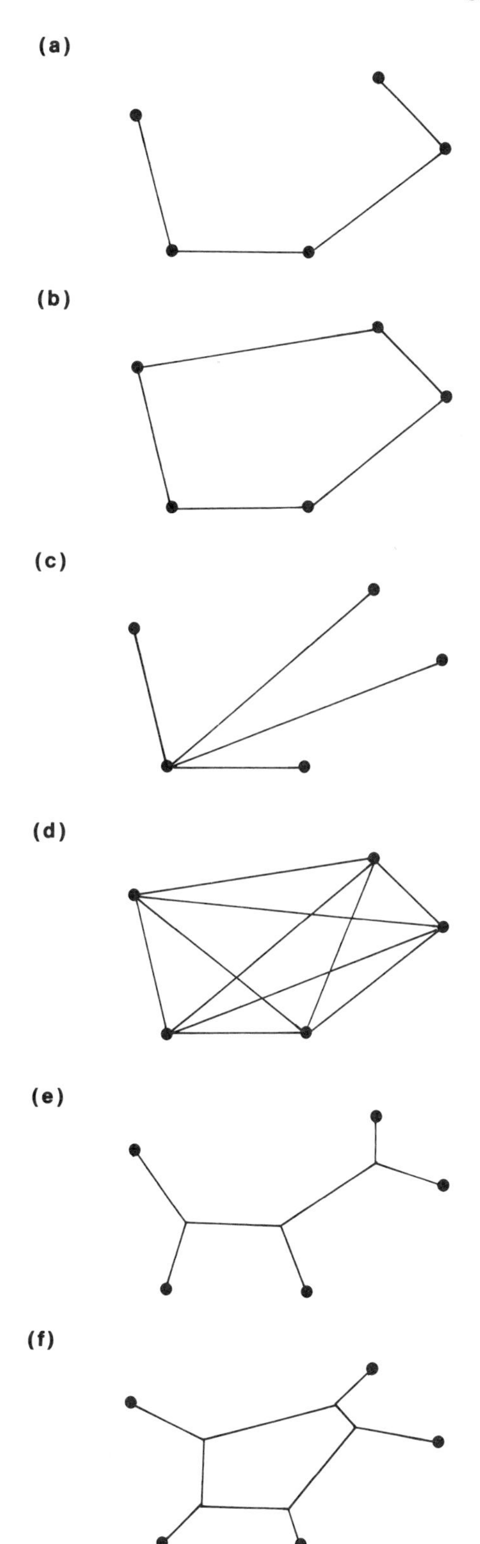

Figure 39.6 Networks to suit various purposes: (a) minimum distance network for visiting all points (e.g. bus route); (b) minimum distance to visit all points and return to the start (e.g. milk round); (c) connecting one point to all the others (e.g. the French road network radiating from Paris); (d) connecting any point to all the others (e.g. local road network; fits $K = 3$ Central Place Model); (e) the shortest route pattern to connect all places (i.e. a more efficient form of (d) for the road builder, but more time-consuming for the user); (f) the general solution of (d) and (e), a compromise between builder and user, employing a ring road with access spurs (e.g. new housing estates).

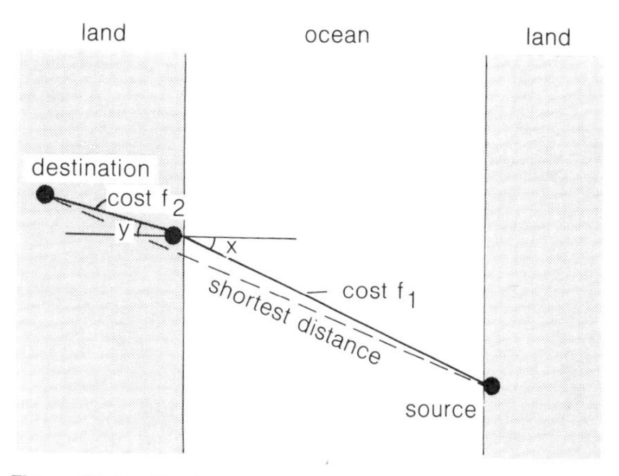

Figure 39.7 The least cost route determines the choice of transhipment (break of bulk) port after Lösch.

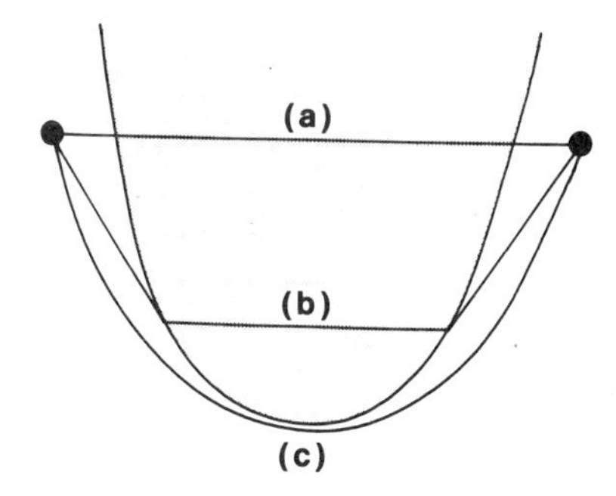

Figure 39.8 The problem of a hill range: an alternative solution of the route refraction analysis (after Lösch).

The effects of networks on settlements

In many areas the settlements pre-date the route network. In some instances, though, route networks are established first: these then determine the subsequent settlement pattern. The route pattern developed in the 'Old World' was influenced by the settlement pattern, whereas colonisation of the 'New World' occurred after route networks had been established (Fig. 39.9). As a result the Old World networks tend to be complex and more hexagonal in pattern, whereas the New World networks are more rectangular, with diagonal connections to important places (**long-distance desire lines**).

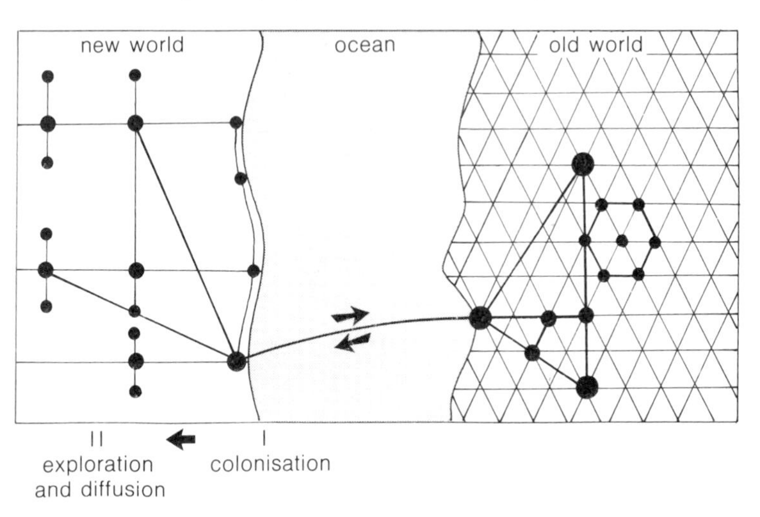

Figure 39.9 The historical factor in network development. The new world pattern on the left contains leading routes of exploration and secondary feeders. The success of one coastal port leads to cross route long range **desire lines** and an increase in network connectivity. The old world settlement pattern on the right is based on central place theory, with superimposed long-range desire lines and distortions due to the development of manufacturing centres.

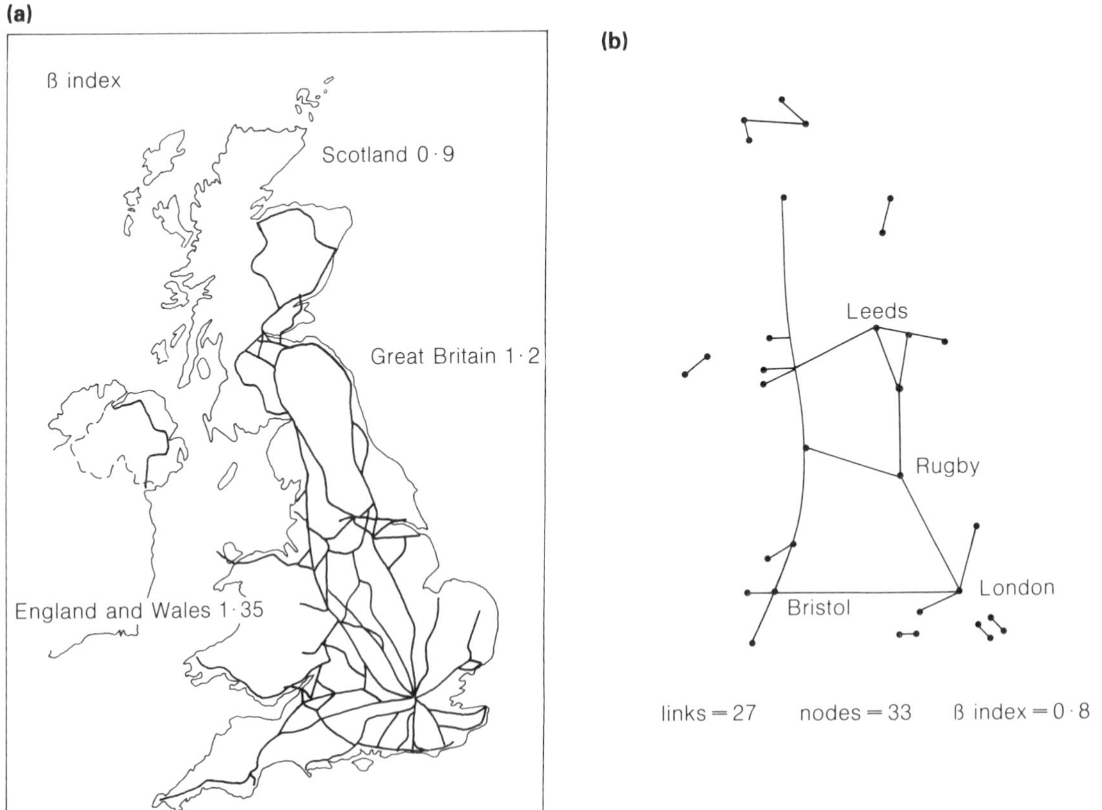

Figure 39.10 The simplest index to calculate is the β index. Compare the sophisticated rail network (Inter-City) (a) with its equivalent primitive motorway network (b).

Network connectivity

Networks of links and nodes show many variations. From a trading standpoint those places which have the greatest accessibility to all others will be in the most advantageous positions for the location of distribution industries, and for quaternary industry (particularly the headquarters of multi-plant firms and government administration). Because of the intricate nature of industrial and personal communications, developed countries with substantial internal trade each have networks with a high degree of connectivity; whereas in a developing country the connectivity will be poor.

To assess the success and function of a transport network it is therefore helpful to have a quantitative measure of the **connectivity** of networks and nodes within networks. Several indices have been developed to express the degree of connectivity.

(a) The **beta index** is calculated by dividing the total number of links by the total number of nodes. This provides a measure of connectivity for a network, and it is useful for assessing the actual or potential development of a region or country. As a rule of thumb, networks with a beta index above 1.0 tend to belong to advanced economies; those with an index below 1.0 to developing economies (Fig. 39.10).

(b) The **König number** uses as an index of connectivity the maximum number of links by the shortest path from each node to the other nodes in the network. In this case the lower the number the more central the place in the transport system (Table 39.3; Fig. 39.11).

(c) The **Shimbel–Katz index** is calculated as the sum of the links by the shortest path from a node to the other nodes in the network. It provides a measure of the centrality of any node: the smaller the number, the greater the centrality (Table 39.3; Fig. 39.12a).

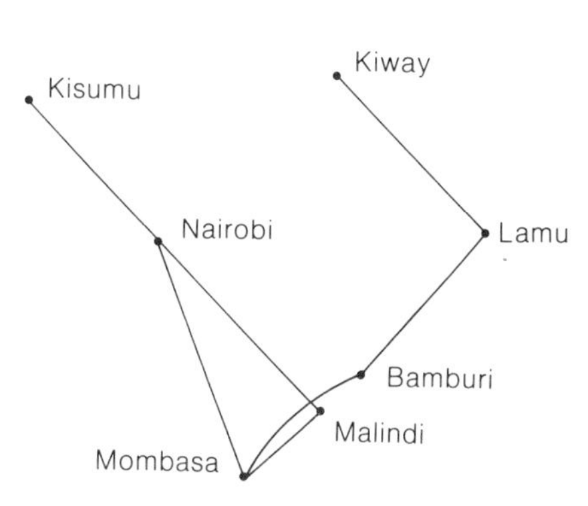

Figure 39.11 Construction of a network matrix. The Shimbel–Katz number shows that Mombasa is the best connected city for domestic flights in Kenya.

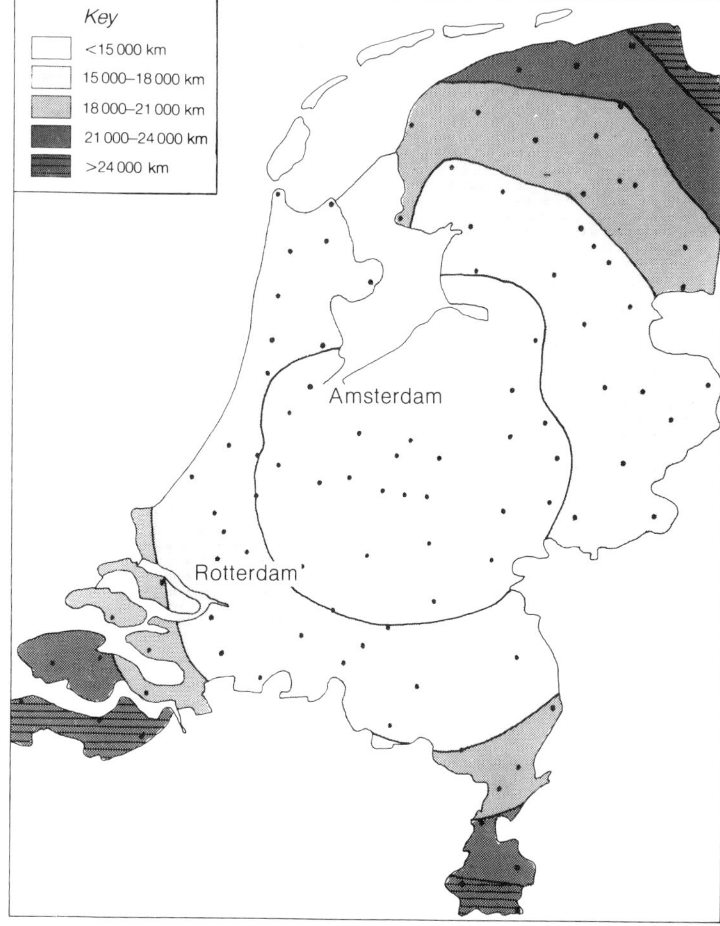

Figure 39.12 Isolines of accessibility by road in the Netherlands. This identifies areas of remoteness from the national centre of gravity near Amsterdam. Notice that Rotterdam has a low order of national accessibility.

Table 39.3 Measures of connectivity.

	Ku	Ni	Ma	Mi	Bi	Lu	Ky	König number	Shimbel number
Kisumu (Ki)	—	1	2	2	3	4	5	5	18
Nairobi (Ni)	1	—	1	1	2	3	4	4	12
Mombasa (Ma)	2	1	—	1	1	2	3	3	10
Malindi (Mi)	2	1	1	—	2	3	4	4	13
Bamburi (Bi)	3	2	1	2	—	1	2	3	11
Lamu (Lu)	4	3	2	3	1	—	1	4	14
Kiway (Ky)	5	4	3	4	2	1	—	5	19

(d) Isolines of accessibility can be drawn to connect the points where the sum of the distances to all the places in the area of study (measured along the shortest route) is the same (Fig. 39.12).

The impact of networks on industrial location

The influence of raw material and market on industrial location is strongly influenced by the route network connecting the two (see Weber's analysis, p. 394). But in the case of a ring network, where there are several sources of raw materials and the market is evenly spread around the route system, there is no unique optimum plant location (Fig. 39.13). In practice the manufacturing plant is often located at a node that corresponds with one of the raw-material sources. Remember that routes are normally established before sources of raw materials are discovered.

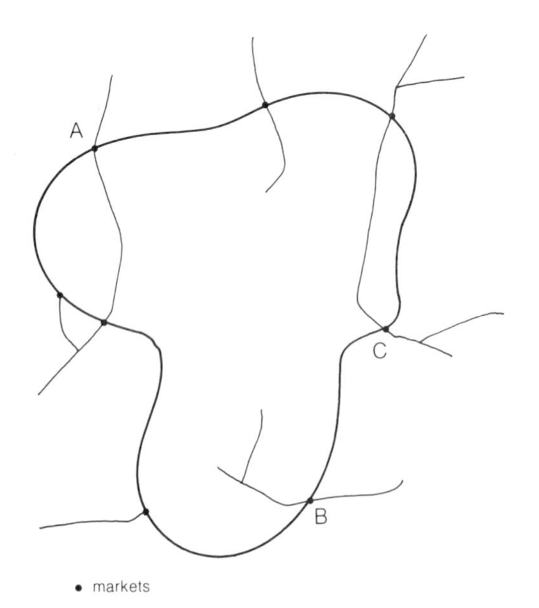

Figure 39.13 A route problem for locating a factory. Raw materials are located at A, B and C, and the market is evenly spread at all nodes.

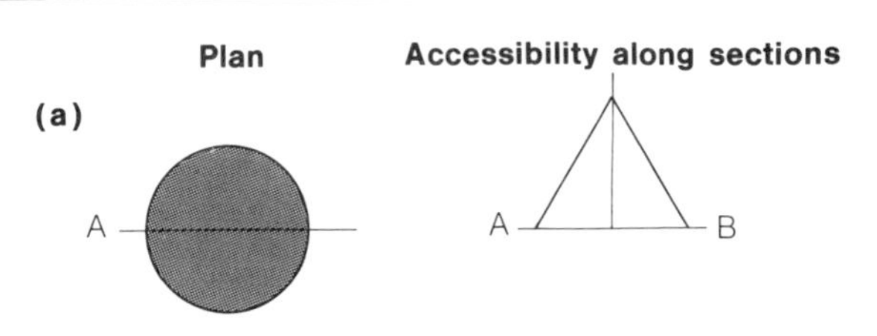

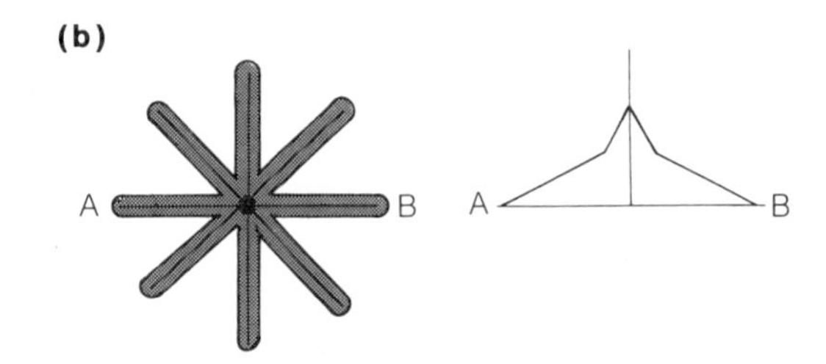

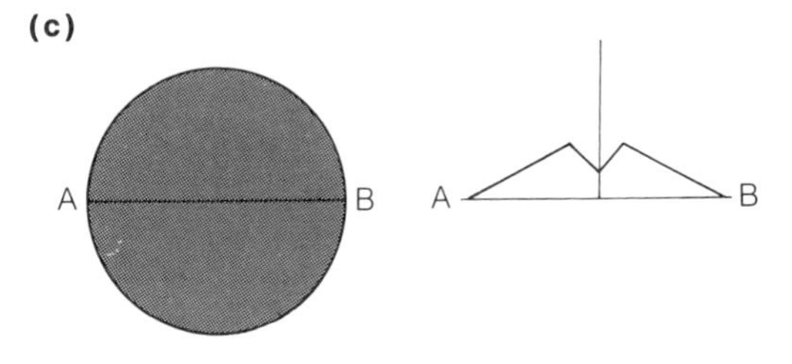

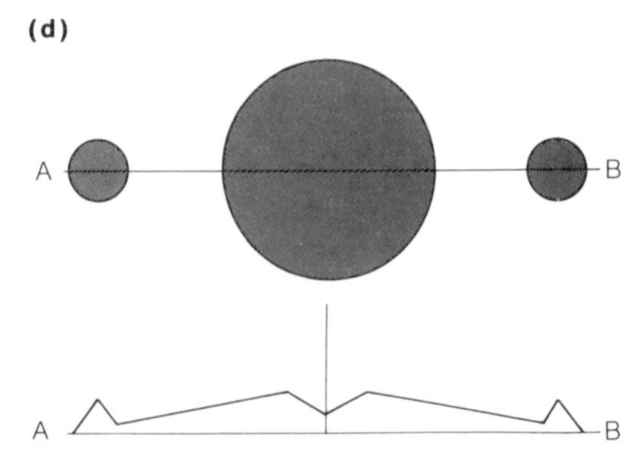

Figure 39.14 (a) The walking city; (b) the public transport city; (c) the 'rubber' city; (d) the 'motorway' city.

Transport system and city shape

The shape and size of an urban area reflects not only its functions, but also the transport systems available to the community. Thus urban areas where walking is still the chief means of transport (Fig. 39.14a) are small, and have high densities (narrow streets and tall buildings) in order to reduce the distance and time of travel within the area. Cities with good public transport occupy larger areas, but their built-up zones closely correspond to the radial transport networks (Fig 39.14b). With the development of private transport there is no longer a need for development to remain closely tied to public-transport systems, and the undeveloped wedges *between* the 'spokes' are filled in by low-density suburbs (Fig. 39.14c). Finally, with increasing dominance of private transport, accessibility to the centre declines due to congestion, and many people prefer to live in the surrounding towns, making the journey to the city by motorway (Fig. 39.14d). Thus access to the centre via motorways becomes easier than access from the suburbs.

Manchester's growth illustrates each of these city structures (Fig. 39.15). In 1845 the city was still based largely on walking distances, with only embryo transport corridors. By 1891 horse-drawn buses had extended the radial development. By 1905 the pattern was further accentuated by more effective public transport, using train as well as tram. However, by 1950 the rise of private motor cars had allowed much infill. Infill has continued, coupled with out-of-city commuting via the motorway and main-road network.

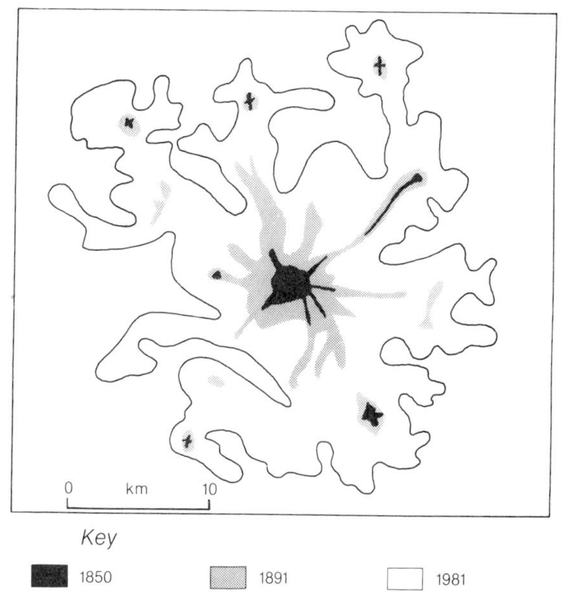

Figure 39.15 The growth of Manchester, 1845–1980.

39.4 Reasons for trade

Originally **trade** comprised a direct exchange of goods and services. However, such a **barter** system is inflexible, and trade now occurs primarily through the intermediary of **money**.

Trade does not necessarily occur because people produce different goods. In many cases even regions or countries that are able to produce the same goods choose to specialise in one part of their potential product range and trade for the other parts. Specialisation and trade creates greater benefits than self-sufficiency (subsistence), a property which underlies all economic growth. Economic growth can only be achieved through trade, a point worthy of note by people who suggest establishing trade barriers.

The principle that trade leads to economic growth partly results from the **law of comparative advantage**. Suppose there are two countries, *A* and *B*, each capable of making the same goods – say, bicycles and cars. They both have the same number of workers, but one set of workers is more productive than the other; country *A* has an absolute advantage over country *B* in producing both kinds of goods.

The production figures are

	A	*B*
Bicycles (output/man/day)	5	3
Cars (output/man/day)	2	1

If they each have 2000 workers and divide their work force equally between making bicycles and cars the daily totals will be:

	A	*B*	Total
Bicycles	5000	3000	8000
Cars	2000	1000	3000

But if they specialise to some degree – say, country *A* decides to employ 1200 workers in making cars and country *B* decides to employ 1300 in making bicycles – then the output will be:

	A	*B*	Total
Bicycles	800 x 5 = 4000	1300 x 3 = 4200	8200
Cars	1200 x 2 = 2400	800 x 1 = 800	3200

It can be seen that the total output is greater if each country specialises in producing the goods for which it has the greatest relative or comparative advantage. It is therefore of greatest advantage to both countries to adopt a degree of specialisation and trade with one another. Of course, the relative value of the two products will determine the amount of actual trade that occurs.

The results of specialisation and trade abound, although mostly the reasons are more complex than the simple example outlined above. The law of comparative advantage can be seen to operate in the motor-car industry, where electrical components are a speciality of the UK and gearboxes of Germany. Multinational companies use their plants as though they were trading with each other; within Europe, for example, Ford and GM make gearboxes in one country, engines in another and assemble the cars in a third.

The law of comparative advantage is important: it helps to explain why developed countries trade with one another when they are all able to make the same goods. On a world level, there is also considerable trade in commodities that are not available within each country. These are mostly raw materials for industry, fuels and food, in the developed countries; and food, fuel and manufactured goods in the developing countries.

Case study: Britain as a trading nation

The pattern of British trade shows how trading develops and changes continually. Because Britain was the first country to undergo an industrial revolution it was able to manufacture goods when others were not. Britain began to concentrate on the manufacture of goods *even though* her agriculture was more efficient than other neighbouring countries (because of the preceding agricultural revolution; enclosures; better seed and animal breeds, and so on). Britain began to trade manufactured goods for agricultural products because of the law of comparative advantage: there was more profit to be made from industrial goods than agricultural ones. In turn this was to the advantage of the countries with which Britain traded because they also were able to produce more. As a result cotton came from India to Britain and textiles were sent from Britain to India.

Imports of cheap food were not entirely beneficial. Indeed, the import of cheap food led directly to British farmers having to shed labour and become even more efficient just to compete. Growing success in one sector was therefore offset to some extent by considerable problems in an older sector.

The same pattern of change was continued through time. As the developing world began to produce basic manufactures with cheap labour, so Britain's comparative advantage in those fields waned, while at the same time its advantage in more specialised manufactures, services and quaternary activities strengthened.

It is interesting to observe what long-term effects trade has had, particularly in agricultural production. Trade has allowed Britain's population to grow beyond the level where it can be sustained by natural production at the standard to which people have become accustomed. There is a point, therefore, when trade moves beyond a level where it is an *advantage* to a level where it becomes a *necessity* if living standards are not drastically to fall.

39.5 Examples of trade

Agricultural goods

British agriculture produces only 55 per cent of British food supplies. It is at a comparative advantage in dairy products and livestock, and these form the major component of national production. It is at a disadvantage in wheat production, so much of this is imported. Britain cannot grow tea, coffee, citrus fruits and so on; these foods also are imported. However, fruits are not staple items but luxury foods: their consumption, and therefore the trade, changes rapidly with price fluctuations.

Developed world economies tend to dominate world agricultural trade mainly because developing countries have large indigenous populations to feed and their systems of production are inefficient. Countries such as the USA, Canada and Australia have large land areas and comparatively small populations. Consequently they produce most of the world's non-tropical food-trade items (Table 39.4) as well as most of the manufactures.

Agricultural exports are vital to developing world economies (Table 39.5). They are extremely liable to fluctuation due to climatic and demand factors, and they are subject to intense, often near-fatal competition. For example, Indonesia has 10 million coffee growers. In 1976, when coffee prices almost doubled, Indonesian farmers rushed to take advantage of the price.

Indonesia is now the world's largest producer of *Robusta* coffee. Coffee forms 12 per cent of the country's export earnings and provides a cash crop to the millions of peasant farmers who grow 93 per cent of Indonesia's coffee. But the change to coffee has not been advantageous because of subsequent changes in

Table 39.4 Total value of world agricultural exports, 1980 (US $).

	Developed	Developing	Total
forest products	45	9	54
cereals	35	5	40
livestock	31	3	34
tropical products	17	28	45
oil seeds	7	9	16
raw materials	9	8	17
fish	9	6	15
others	9	2	11
totals	208	82	290

world market prices. Exports of *Robusta* account for 90 per cent of Uganda's foreign exchange, and Indonesia is therefore in direct competition. The giant producer of *Arabica* coffee, Brazil, also has a major stake in world production. The introduction of Indonesian coffee has caused supply to outstrip demand, leading to depressed world prices and loss of income to all farmers.

Trade also varies with contact and awareness. For example, improved contact with western societies has had a major impact on food trade in Nigeria. Wheat and rice used to be foods eaten by the middle classes on special occasions; today they are near-staple parts of the diet. Between 1976 and 1982 rice and wheat imports each rose from insignificant quantities to 1 mt. Sugar has become another major import (600 000 t 1982) due to increases in brewing, in soft drinks and in food processing.

Table 39.5 Agricultural exports in comparison with manufactures (source: Lloyds Bank economic report).

Kenya	*% share in 1980*
coffee	22.2
tea	11.9
fruit and vegetables	5.2
hides and skins	2.0
sisal, pyrethrum	3.7
agriculture (total)	45.0
manufactures	13.0
services	42.0
overall total	100.0
Ecuador	*% share in 1981*
bananas	8.9
fish	7.1
coffee	4.1
cocoa	5.8
agriculture (total)	25.9
petroleum	60.8
manufactures	13.3
overall total	100.0
Bangladesh	*% share in 1981*
raw jute	16.8
leather	7.7
tea	6.5
fish	4.6
agriculture (total)	35.6
jute manufactures	60.4
other manufactures	4.0
overall total	100.0

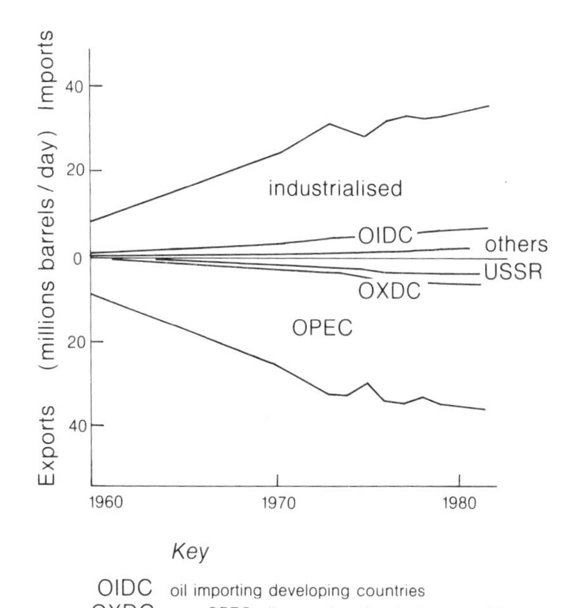

Figure 39.16 World oil trade, 1960–83 (million barrels/day).

Oil

Energy supplies (mostly oil) are the most important primary materials in world trade. Many countries have negligible energy resources and must import all the energy they need. By contrast there are few countries with surplus energy, and of these most are in the Middle East. Trade in oil increased spectacularly in the 1950s and 1960s, but demand is very dependent on cost, and the large price increases in 1973 and 1979 caused a virtual stagnation in demand. Nevertheless trade in oil is still of the order of 35 million barrels per day, mainly from OPEC (the Organisation of Petroleum-Exporting Countries; Fig 39.16). Despite their large oil production the US and USSR play little part in oil exports; the US is already becoming a large importer and the USSR is expected to become an importer within the decade.

Tourism

In many countries **tourism** is one of the largest, if not *the* largest, industries and sources of foreign exchange. A country or region uses its resources (landscape, sea, beaches, sunny weather and warm climate or other features of interest) to attract people. By paying for holidays people enter into trade.

Countries such as the UK and West Germany offer a wide range of entertainment facilities. They are economic core regions though, so more people take holidays beyond their national borders than arrive from abroad. For example in 1981 West Germany received US $ 7 billion from foreign tourists, while German nationals spent US $ 20 billion abroad, in effect generating 14 per cent of German *imports*. By contrast, Greece, a peripheral country within the EEC, received only US $ 2 billion in 1981 but, in comparison with the rest of its economy, the tourist trade provided one fifth of total of goods and services. Kenya, a developing country with the resources of coral sands and safari parks, earned US $ 168 million in 1981 making it the third largest foreign-exchange earner.

Quaternary activity

When primary materials or manufactured goods move from one country to another or when services such as airlines carry people on holiday, the trading pattern is easy to follow. However, the industrialised countries in particular now enter into a less tangible form of trade that involves the provision of office services such as insurance and banking. This is the so-called **invisible-trade sector**, and is of vital importance to the UK, for example, in maintaining its balance of payments. The growth of invisible trade is mostly reflected in the increase of office activities and the Central Business District.

39.6 Trading systems

In the same way that large producers and customers can command better discounts and prices when dealing with wholesalers,so large economic units are often in especially favourable bargaining positions. Many of the relatively powerless oil-exporting countries grouped together into OPEC in order to increase their bargaining position with respect to the industrialised countries and multinational concerns. This does not much affect the pattern of trade because the oil resource must be carried from producer to consumer in the normal way, but, by influencing the price structure, it has tended to control the *volume* of trade in recent years.

Trading systems can also be established to promote trade within an area. Thus the USA is a free-trade area made up of its member states. The USA is the most advanced form of trading bloc: it has progressed from trade, through political unity, to become one country. The European Economic Community (EEC, or 'Common Market') is a much more recent trading bloc established soon after World War II and still increasing in size. Here the purpose is to provide a secure and stable enlarged market whose trade can, to some extent, be isolated from the fluctuations of world conditions. The Common Market has been particularly effective in influencing the pattern of trade in agriculture. For example, when the UK joined the EEC in 1971 Denmark was more or less obliged to join because its agricultural exports were so closely tied to supplying the UK market. The EEC has long-term political as well as trading objectives, but its effect at present is to promote trade within the market. Thus the UK has turned away from existing trading partners in the rest of the world, and more to Europe in its trade since joining the EEC. **Trade barriers**, such as those established by the EEC, go against the pattern of free trade and growth based on comparative advantage; for this reason, trading blocs need to be large to be successful.

Manufacturing industry in advanced market economies

40.1 Categories of industry

Industrial activity forms the economic base on which the living standard of many countries depends. Although relatively simple in the early days of the Industrial Revolution, industrial activity today takes on such a wide variety of forms that it is often impossible to refer to it as a whole. Because of its variety, industry is normally separated into four basic categories in such a way as to reflect the position of each industry in a chain leading from resource to market.

(a) **Primary (extractive) industry** is the term used to denote those activities that derive materials directly from the Earth's surface without changing them in any way. These vary from quarrying and coal mining to agriculture, fishing and forestry.

(b) **Secondary (manufacturing) industry** is the term used to denote those activities that process raw materials into products whose character is distinctly different from the original resources. They are typified by the engineering and chemical industries.

(c) **Tertiary (service) industry** concerns the distribution of the products of either primary or secondary industry: it involves commercial transport, wholesaling and retailing, as well as personal services. It is often the link activity between the extractive and manufacturing industries and their customers. It can also be a productive industry in its own right, as in the case of tourism.

(d) **Quaternary (office) industry** is concerned with information, knowledge, and the exchange and generation of ideas. This is a difficult category to isolate because office functions occur in each of the three other types of industry. However, some activities are entirely office-based: they include the professions (such as law and architecture), financial institutions (banks), administration (local government), and research institutions, universities and schools.

Although a manufacturing base is still vital to most countries in the developed world, successful sales increasingly depend on managerial expertise and marketing. These are office-type activities. At the same time industries which are entirely office-based – banking, insurance and the like – and those that are chiefly service-based, such as tourism, make important contributions to national earnings. Because of these changes the number of office workers has increased dramatically. In the 1950s alone, total employment in Britain rose by 7 per cent, but office employment rose by 40 per cent. Although the rate of office expansion has now slowed down the proportion of people working in offices is now greater than the proportion in manufacturing, a phenomenon that has earned the current development of developed countries the title of **post-industrial phase** (Table 40.1).

Table 40.1 UK employment and GNP (source: Central Statistical Office).

	Employment				
	1961		1984		
Category	millions	%	millions	%	GNP (%)
primary	1.5	6.2	0.7	3.3	2
secondary	10.0	46.7	7.2	35.4	28
tertiary	10.8	47.1	13.3	61.3	70
total	22.3	100.0	21.2	100.0	100
unemployment		1.5		4.0	

40.2 Models for industrial location

Manufacturing industry consists in processing a variety of raw materials into new products that can be sold either to other manufacturers or consumers. The sole purpose is to obtain as high a profit as possible, a purpose which controls both the location and scale of the factory, and the product made.

There are two basic approaches a factory owner can adopt when examining the best way to make a profit: he can look either at his costs or at his potential trading area.

If, for a moment, we assume that the factory is of such a size as to achieve the greatest economies of production, the only variable costs are those incurred in transporting the raw materials to the factory and in distributing the finished product to its market. Clearly it would be desirable to locate the factory at such a point that these combined transport costs (of materials and product) are kept to a minimum.

In the simplest case the raw material will be located at a single point, say a dock, and the market will be at another point, say a motor-car works. Assuming that there are no difficulties in obtaining a workforce

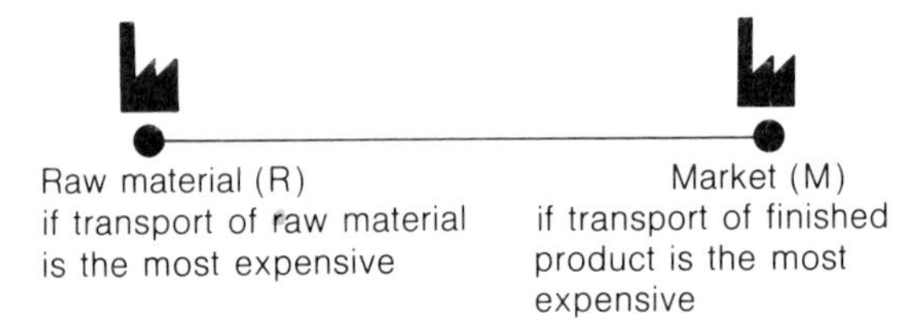

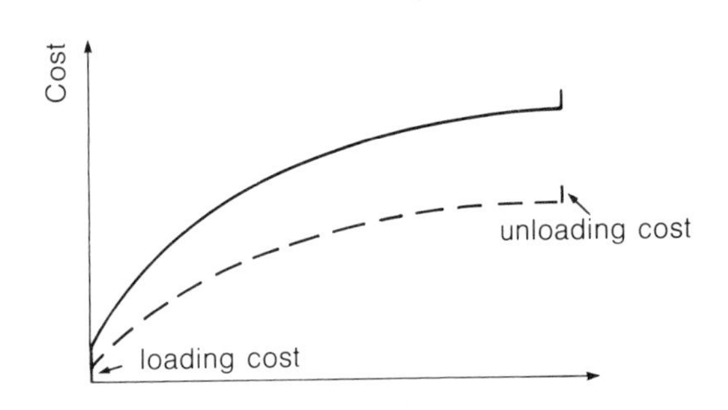

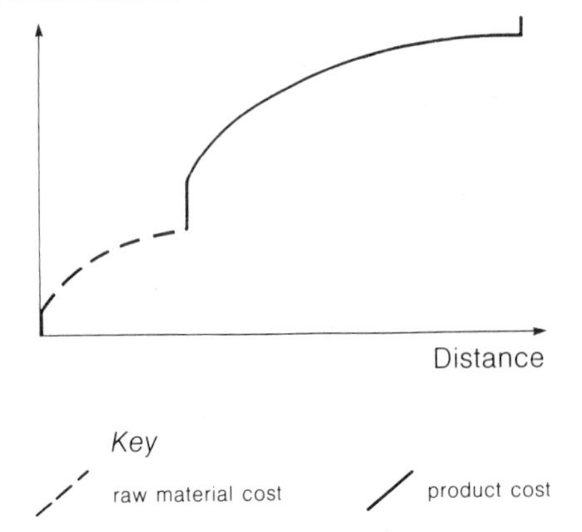

Figure 40.1 (a) Locations based simply on transport costs. (b) An intermediate location not only incurs additional loading and unloading costs, but uses transport systems in the most expensive way. This could only be justified at a transhipment point such as a port.

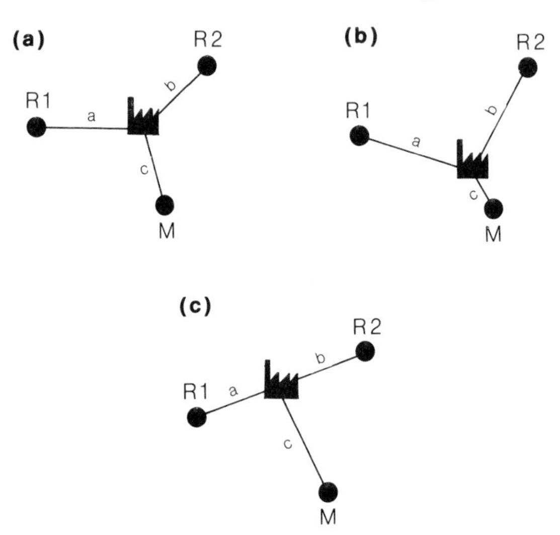

Figure 40.2 The Weber industrial location model, based on minimum transport costs. (a) If the cost of transporting both raw materials is equal, and the cost of transporting the finished product is the same as that of all raw materials, then the least cost location for a factory will be where distances $a = b = c$, that is, the factory will locate midway between raw materials and market. (b) If the raw materials are cheaper to transport than the finished product, probably because the product is bulkier (e.g. polystyrene foam) or more fragile (glass) the factory will locate nearer the market ($c < a = b$). (c) If the raw materials are dearer to transport than the finished product, probably because much of the raw material is consumed (e.g. power stations) the factory will locate near to the raw materials ($c > a = b$).

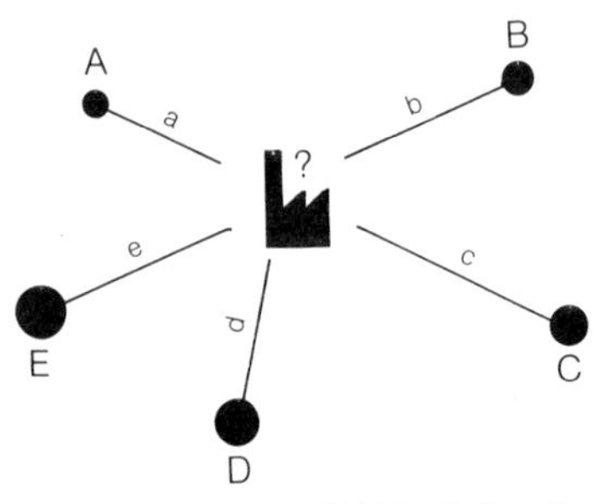

Figure 40.3 The general form of Weber's location problem.

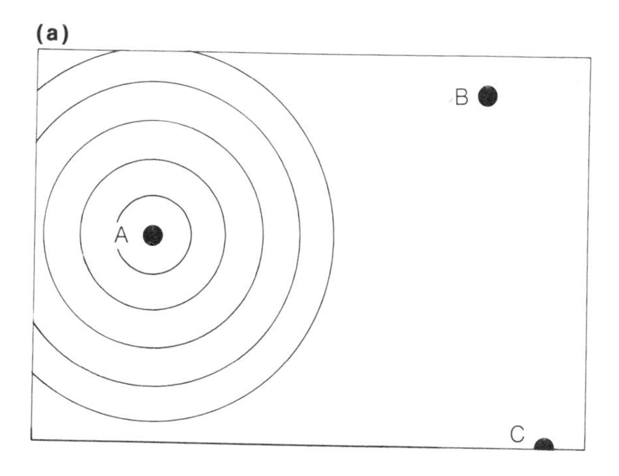

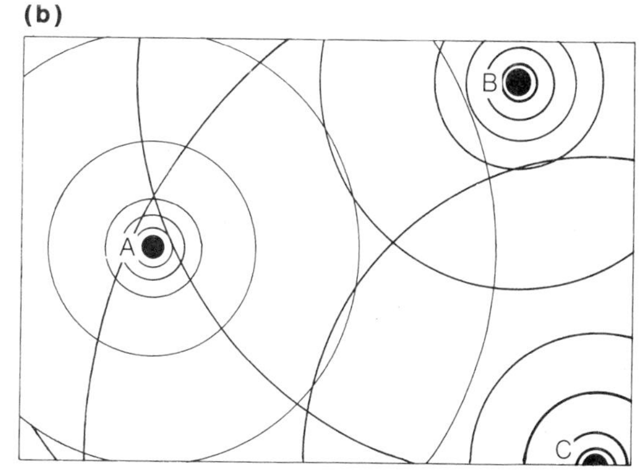

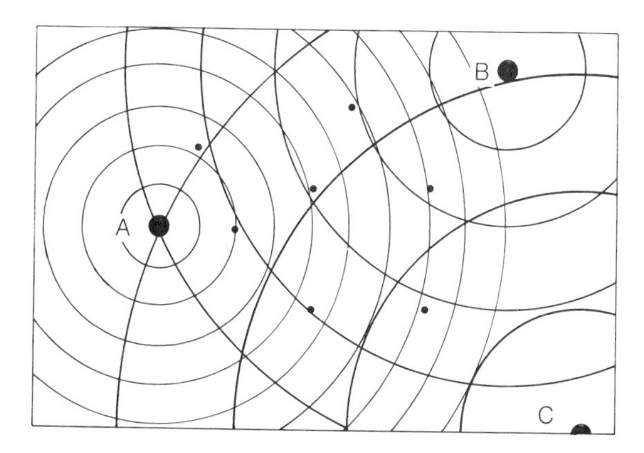

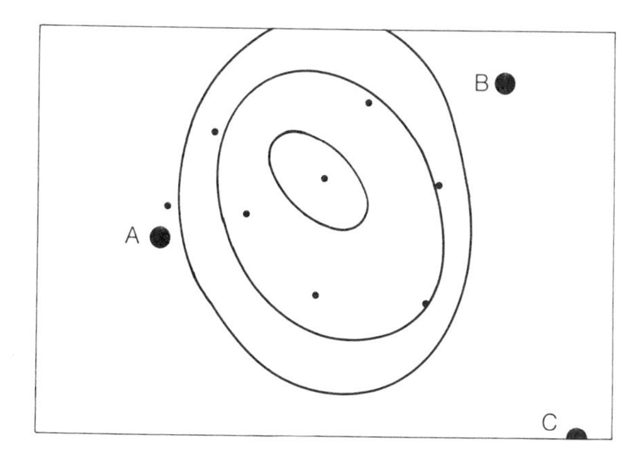

Figure 40.4 (a) The graphical solution to Weber's location problem for two raw materials and one market. (1) Draw lines of equal cost (isotims) outwards from each centre. (2) For a number of locations x_1, x_2, x_3, etc., chosen to cover the region within A, B and D, add up the isotim values e.g. $x_1 = 10 + 4 + 15 = 29$. (3) Use these cumulated costs to construct a 'contour' map of total costs. The region within the lowest value isodapane (contour) has the least cost. (b) Isotims modified to show the tapering effect of transport costs

wherever the factory is located, the best position is the dockside if it is the raw materials that are the more expensive to transport, or beside the car works if it is the finished product that is more expensive to transport (Fig. 40.1).

In general, however, factories obtain their raw materials from more than one source, while the labour pool will be available only near towns (and must therefore be considered as another raw material). In this case the location for least transport costs will be less easy to define (Fig. 40.2).

The Weber approach

Weber attempted to find the minimum-cost location for a factory whose materials were at *A*,*B* and *C* and markets at *D* and *E*, when the costs of transporting both materials and product (*a*,*b*,*c*,*d* and *e*) are all different (Fig. 40.3). The general equation is:

$$\sum_{1}^{n} mt = m_1t_1 + m_2t_2 + m_3t_3 + \ldots$$

to be a minimum

where *m* is the cost of the material or product; *t* is the cost of the transport; and *n* is the number of transport movements.

The easiest way to find a solution to the problem is by a graphical technique (Fig. 40.4). On the assumption that conditions are the same everywhere (that it is an **isotropic surface**) concentric circles representing, say the transport cost of materials *A* and *B* and of the finished product *D* can be drawn outwards from each location. These circles of transport cost/unit weight are called **isotims**, and their spacing reflects the cost of each form of transport. Isotims from *A* are more closely spaced than from *B* because material *A* is more expensive to transport, or because a greater volume is needed to make the product.

The pattern of isotims can now be used to develop lines of equal transport cost (called **isodapanes**). To do this, total cost is calculated at a number of points and an isodapane map drawn. Thus, at point *X*, the cost of transporting *A* is 30 units, of *B*, 34 units, and of *D*, 14 units. Therefore the total cost is $30+34+14=78$ units and the 80-unit isodapane is drawn to encompass it. The solution is found within the isodapane that has the *lowest* total.

Notice that this model makes several assumptions – that there is no preferred transport route, that there is no physical hindrance to communications, and that transport costs vary in simple proportion to distance. In practice none of these assumptions is valid. Not only are physical restrictions real, but transport costs 'taper' with distance (Fig. 40.1), and there are important additional loading and unloading costs if more than one transport system is used. Furthermore, the whole pattern may be influenced by pricing structurres enforced by government. All these factors greatly complicate the application of Weber's analysis to real problems.

The Lösch approach

In Weber's model the market is assumed to be concentrated at specific sites. This may be an appropriate assumption for some industries, but it is certainly not the case for many, especially those that sell goods bought directly by consumers from retail outlets. In this case it may be more appropriate to argue that the greatest income derives from locating as near as possible to the largest potential market.

Consider the transport costs from any factory to its customers. Costs of delivery increase with distance from the factory gate, but, because the manufacturer has to cover his costs, he may charge for delivery, thus putting up the effective price of his product to distant markets. The higher price will make the product less attractive; sales may decline with distance from the factory until at some distance transport, and therefore selling cost, is so high that sales will cease (Fig. 40.5a). Furthermore, there may be competition from similar factories nearer to the market and this may restrict the trade area to an even greater degree. Thus, if markets were uniformly spread, competition would ensure a regularly distributed pattern of factories, all of which had interlocking hexagonal trade areas (Fig. 40.5c). (For explanation of the derivation of a hexagon pattern, see p. 454.)

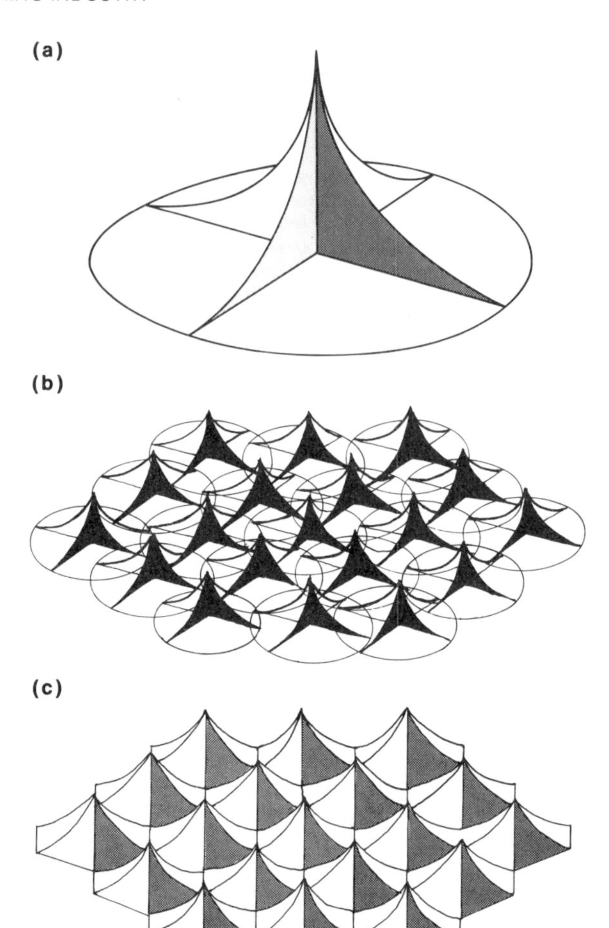

Figure 40.5 The Lösch industrial location model based on greatest revenue: (a) the form of a single trade area; (b) competing trade areas; (c) stable, non-overlapping hexagonal trade areas.

Weber versus Lösch

These two models of industrial location put emphasis on different aspects of the manufacturing process. Both, in their own ways, contain attractive ideas, though we might suspect that a combination of the two would be most appropriate. Both make assumptions about the distribution of raw materials and markets that are unlikely to be true of the real world, but they do focus attention on the possible reasons for industrial location. However, before the location of industry can finally be assessed, additional and sometimes overriding factors must be taken into consideration. These are

(a) economies that might be derived from changing the scale of production and thereby reducing the number of plants;

(b) attractions of clustering;
(c) government policy;
(d) labour costs.

40.3 Clustering of industries

Large units

There are a number of factors that encourage factories to be in large units. In essence, the factory system is designed to produce standardised units by mechanical means. One of the main costs is therefore **machinery**. Any piece of machinery is most economically run at a certain size, and this may determine the factory size. In general, large-capacity plant works at lower cost, so factories tend to become larger and fewer.

The workforce in a factory is at its greatest productivity if the scale of operation is such that manufacture can be separated into a number of narrowly defined, specialised tasks. As each worker becomes skilled in his task his rate of production increases beyond the level reached if there were several tasks to perform. The production-line systems employed in motor-car manufacture illustrate such **labour specialisation** very clearly.

Furthermore, the volume of output from one manufacturing stage may influence the stage that follows, for each part of the manufacturing process has to be closely integrated. Integration tends to increase the size of the factory. Consider the motor-car factory once more. Suppose the engine shop is able most economically to produce 10 units/hour and the gearbox division 15 units/hour. The body shop will receive an uneven supply of its main components unless the supplies of engines and gearboxes can be matched. It would be undesirable to reduce gearbox production to 10 units an hour because this would lower productivity and make each gearbox dearer. The most economic strategy would be to increase both engine and gearbox production lines until output reached the lowest common denominator, in this case 30 units/hour. A larger body shop is now needed and the result is greater car production. If the market is limited it is therefore sensible to have a few relatively large and more economic plants than several smaller ones whose production lines operate at lower efficiency.

As plants become very large, the market area widens to a size at which the economy gained by enlargement is offset by the increased average transport costs. Very large plants may actually suffer a *decrease* in productivity. In practice, therefore, there is in each area of manufacturing a certain most economic size. In steel production the most economic unit may be about 3 mt/year. Applied to Britain, where output is only about 10 mt/year, this means that only four plants are needed. But, as a counter to plant economies, greater concentration leads to increasing distribution costs. It may be more economic to run more steel works slightly *below* capacity than to have to pay higher transport costs (Fig. 40.6).

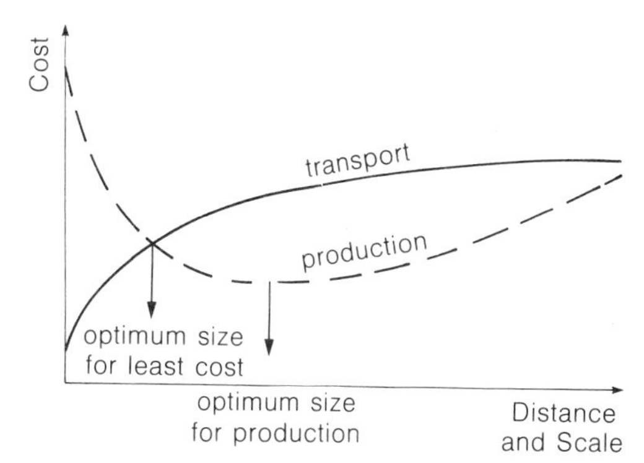

Figure 40.6 There is a trade-off between economies of scale and cost of transport that often results in a factory size below optimum production size.

Many large firms have more than one factory. In such cases, the firm needs to attain the greatest productivity from all the plants taken together. Often this can be achieved by making each factory specialise in just one part of the manufacture. Faced with reduced demand, a multi-plant firm, which has a number of production sites, is liable to close one of its factories and concentrate production in the remaining units. On the other hand, a single-factory firm will resist shutting completely. Multi-plant firms may therefore destabilise a regional economy (Table 40.2). Nevertheless, the large size of multi-plant firms

Table 40.2 Changing employment in the major manufacturing firms of the North-West region of England, 1975–82.

Examples	Employment ($\times 10^3$) 1975	1982
(a) Foreign-based multinationals		
Philips	45.7	25.0
Heinz	10.0	8.5
(b) UK multinationals – controlled outside the North-West		
British Leyland	164.0	80.6
ICI	132.0	74.7
British Steel	223.0	103.7
GEC	171.0	145.0
Courtaulds	123.7	62.6
(c) UK multinationals – controlled within the North-West		
Pilkington	23.0	18.3
Turner and Newall	20.8	16.5

means they can afford to spend more on research and development and on advertising, and are therefore more likely to keep on top of the market. For this reason they are often the first to grow when economies are buoyant.

Concentration within an industry

The number of firms that can survive in an industry depends on the economies that can be gained from increasing in size. Handicrafts, for example, rely on individual skills and can rarely benefit from economies of scale and concentration. On the other hand, high-volume chemical production may only be profitable at high levels of output. Each industry therefore has a 'natural' level of concentration. A large number of takeovers in an industry may indicate that many firms are still too small, and stability has not been reached. However, at some level of concentration no significant increase in profit will result from buying up more firms (except for companies wishing to diversify their product range) and stability will have been achieved.

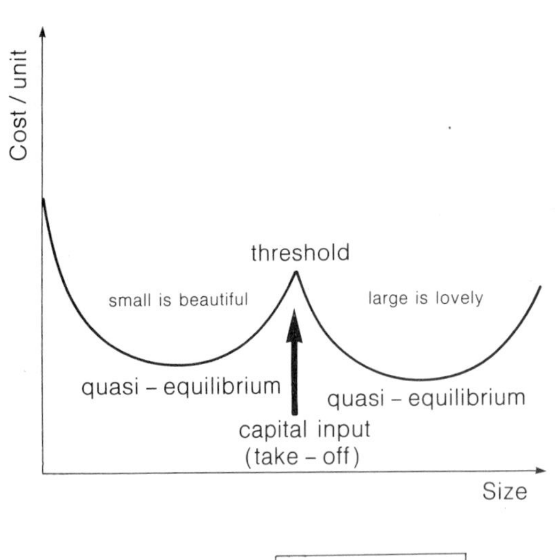

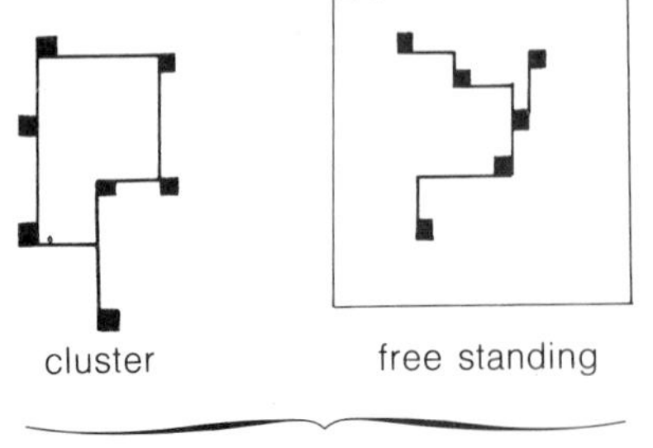

Figure 40.7 Industrial unit thresholds.

There may be more than one stable size of unit (Fig. 40.7). For example, it is often possible for both large and small units to coexist and yet have similar product lines. The difference is primarily a matter of capital investment. Whereas a group of small firms can link together to manufacture a product, each employing only small amounts of capital, a substantial capital input would enable large-scale takeover followed by regrouping into a smaller number of large units. The degree to which such restructurings have occurred depends on the investment policy of those with the capital; and this in turn depends on the level of expected profit. As a result, industries with small profit margins often have small industrial units sited alongside larger ones. Both types of structure are illustrated below.

Single-site complexes

A number of industries such as primary metals and chemicals can only operate cost-effectively on a very large scale and if complete control can be achieved over the entire manufacturing process. To do this they must invest at a level far higher than other industries and must occupy large, freestanding sites. For these industries access to a large labour force is often a prerequisite as are good communications for raw materials and products and an adequate power supply (often about the same as that required by a small town).

When the Bethlehem Steel Corporation decided to build a new steel works in the Mid-West of the USA, they had certain basic site requirements: they needed a site *long* enough to run a continuous 5 mt/year operation that included raw materials storage, blast furnaces, steel furnaces and rolling mills; a site *large* enough to allow for any changes in technology that might require new plant; *access* to a deep-water harbour for import of 4 mt/year of iron ore; *access* to a road and rail network primarily for shipment of product; a *central position* within the perceived market area; and a nearby urban development that could provide a *workforce* of 7000. The final choice of site, on the shore of Lake Michigan at Burns Harbor in Indiana, was made because this could offer all of the site objectives and save capital by using the nearby State harbour.

The British Steel Corporation (BSC) (see Fig. 40.8) were faced with a very different problem. They wished both to expand their steel-making capacity and to modernise their equipment. Teesside was one favoured area because the BSC already owned land of suitable size and shape as the result of nationalisation of the once-private Dorman-Long works (Fig. 40.9). There were already favourable harbour facilities and a trained labour force, so they were mainly concerned with replacement, a process that was well established on

Figure 40.8 An aerial view of British Steel Corporation's Redcar works shows the size of the growing steel complex located on the banks of the River Tees.

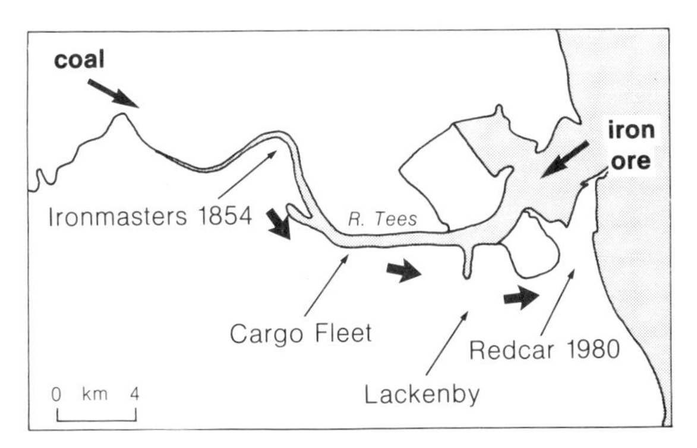

Figure 40.9 The changing site of steel-making on Teesside reflects continued faith in the region's basic locational strength.

Teesside from the days of the earliest iron works in the old Ironmasters site. Nevertheless, the continued selection of Teesside for the largest new steelworks in Europe was the result of exactly the same site selection procedures that the Bethlehem Steel Corporation had undertaken.

Regional swarming

Industries with the highest degree of geographical localisation are not the large plants, but the small ones. This is because a large plant is developed to be freestanding, and its economies are *internal* (as in the case of the steelworks), whereas the small-sized plant finds its economies *externally*, say in the production of components. These small firms are linked together

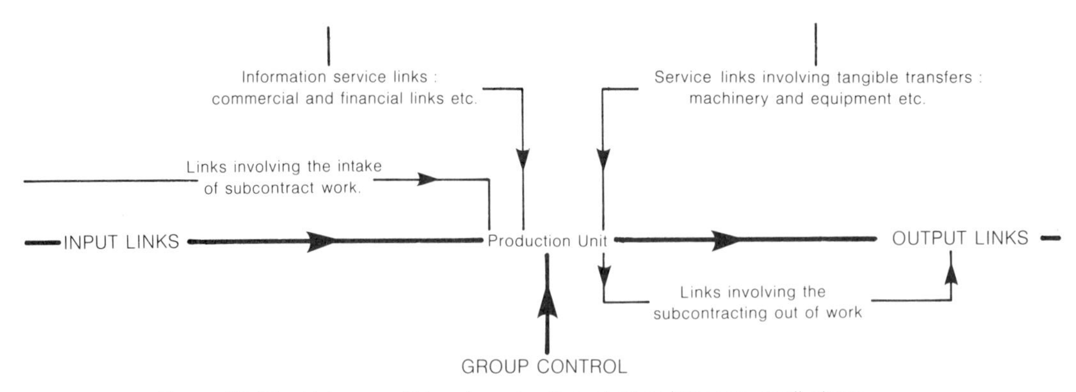

Figure 40.10 Linkages within a free-standing plant or between small plants.

through operational contacts such as flows of materials and exchange of information (Fig. 40.10). For example, the West Yorkshire textile towns have many factories that supply machines, dyes and so on to the mills. Similarly the Black Country, to the west of Birmingham, is dominated by metal-working firms each sharing a high degree of linkage.

In each of the three typical Black Country industrial sectors (iron founding, drop forging, and lock and latch making), linkages are strong, although they vary with individual circumstances. Large plants are relatively freestanding, make little use of local subcontracting services and find their main customers outside the conurbation. Nevertheless, they tend to purchase materials and equipment from local sources. On the other hand, small plants are heavily dependent on the local industrial economy for buying in material, subcontracting and selling, because they are technically unsophisticated and cannot afford to maintain specialist skills. Lock and latch making is different from the other sectors: it supplies, via wholesalers, a nationwide market, but the small size of the plant and the single product line requires few subcontractors.

In some ways the factory swarm behaves much as though it were a giant freestanding factory. There may be several seeming competitors, each working on the most efficient scale, supplying a market that is big enough to take their combined product. They occupy a position in the manufacturing process equivalent to individual machines within a factory.

40.4 Factors affecting industrial location

Raw materials

Raw materials become an effective locational pull only if there is a large loss of bulk on processing. But even in these cases, and especially if the source of raw materials is overseas, the manufacturer may, for security reasons, prefer to instal his expensive processing equipment within his own country. Thus in effect it is the ports that are the raw-material locations. With this qualification, the location of oil refineries and petrochemical plant can be regarded as raw-material-orientated; they are, in general, located at ports. Both divisions of the chemical industry at Teesside, for example, show raw-material orientation. The ICI works at Billingham, largely engaged in producing fertilisers, uses salt pumped from a deposit below the factory; and the petrochemicals plant, at Wilton, which uses oil as a raw material, is as close to the coast as practicable (fig. 40.11). Even so, the location of these two factory complexes is not dominated purely by considerations of raw materials: they also benefit from direct interlinking, sending byproducts by pipeline under the River Tees in order to extend their product range. Thus there is a significant element of aggregation in locational decisions, a factor highlighted by the presence of Shell, Philips, Monsant and other chemical combines in the same area.

In the past raw materials played a much more significant rôle, especially where coal was used directly to power factory machinery. Much industrial location is therefore influenced by historical inertia,

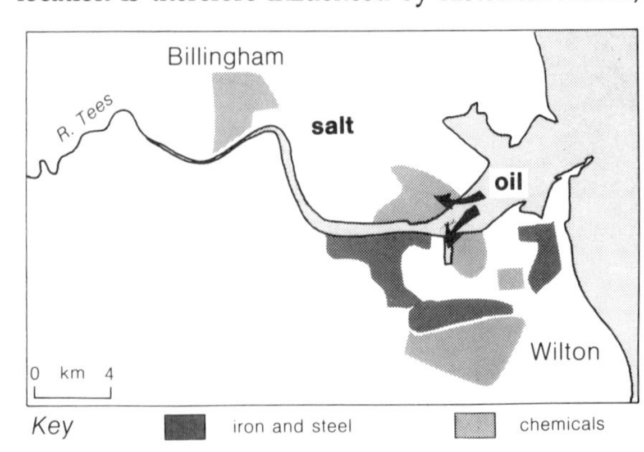

Figure 40.11 Raw material orientation on Teesside.

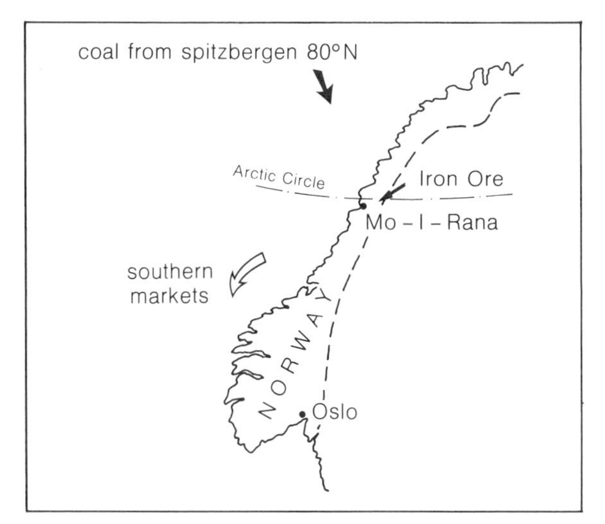

Figure 40.12 The location of the Mo-I-Rana steelworks.

the stress imposed by an old site not yet being an overwhelming disadvantage. The iron works at Scunthorpe in Lincolnshire is a classic example of industrial location on an iron-ore field using ore of low metal content (less than 30 per cent). Nevertheless its continued survival at this site has been seriously questioned and it now relies perhaps as heavily on government policy as on advantageous location. Indeed, an apparently obvious location

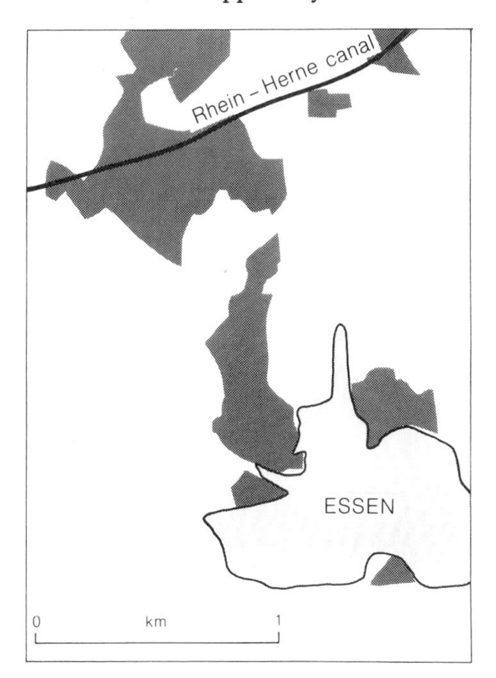

Figure 40.13 Transport orientation near Essen, West Germany.

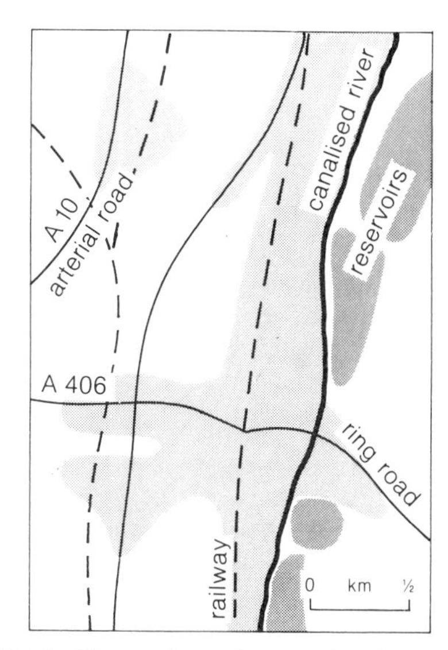

Figure 40.14 The earliest phases of industrialisation in London were concentrated near to navigable waterways such as the River Lea. Railway networks reinforced this early pattern. However, when road transport improved during the 1930s, there was a possibility of greater locational flexibility and the newer industries were attracted to the arterial and ring roads.

beside a raw material can be misleading. A striking example is the iron and steel works at Mo-I-Rana in Norway, situated almost astride the Arctic Circle on an iron-ore deposit (Fig. 40.12). In this case it is not simply the raw material that attracted the plant to this remote location, nor that keeps it there – it was a political decision to provide employment for people in an otherwise disadvantaged region.

Today there are few industrial plants that are completely tied to raw-material locations. Perhaps those that remain most closely tied are some food-producing plants. If peas really are to be frozen within a couple of hours of picking, the processing plant must literally be among the fields.

Transport costs

The Weber model focuses attention on transport costs, although these are not always the dominant factor in location. In general, however, the costs of carriage increase with distance and it is an advantage to hold these costs to a minimum. Costs, measured in say £/tonne/km, increase progressively slowly with distance. They may be of less importance than many other factors, and can often be reduced further by increasing the size of the transporting vehicle. The classic example of vehicle size being increased to minimise freight costs has been the development of supertankers to carry oil and metal ore across many thousands of kilometres of ocean. The main influence of transport on industrial location is therefore through the cost of transshipment from one form of transport to another. (Places at which such transfers occur are called **break-of-bulk points**.) The relocation of the iron-making division of Krupp from Essen, in the Ruhr heartland, to Rheinhausen, beside the Rhine (Fig. 40.13), was primarily to take advantage of direct unloading of iron ore. The concentration of industry around the River Lea in North London was primarily to take advantage of both a break-of-bulk point and a nodal road-and-rail position (Fig. 40.14).

Markets

As Lösch explained, industrial location is often not simply a matter of minimising transport costs to a single location but for a widely scattered market. For products that have a great added bulk (such as soft drinks and beer) the distribution costs are all important. The decision of Carlsberg, the brewery giant in Copenhagen, to set up a major brewery plant in Northampton reflected a need to be located within the area of maximum market demand; in this case just south of the M1–M6 junction, between the main sales outlets of London, the Midlands and the North.

Today most factory locations have a market orientation, though at first sight this may not be easy to see. Japan has a central market location for its motor-car exports simple because it has a world-wide distribution of sales. But it is not only the market access of the single factory that has to be considered; most plants today are parts of larger groups and here it is the *group* costs and revenues that have to be maximised, a feature that may not directly be evident in the location of any individual component plant.

Power

In the past it would have been difficult to separate industry from its source of power because power was the least mobile of all resources. In the earliest phases of the Industrial Revolution, textile mills had to locate in the upper reaches of valleys in order to use water power. Indeed, it was the need for direct water power that kept early industry strung out along swift-flowing rivers.

Today, however, there is much less direct dependence on local sources of power except where power transport is uneconomic, as is the case for HEP generation in the fjordlands of Norway. The iron-and-steel industry, for example, which twice changed its location in the search for energy supplies (first adopting a forest location for charcoal supplies, later a coalfield location for coke), has improved its use of coal so greatly that it now regards the supply of iron ore and the distance to markets as the major factors

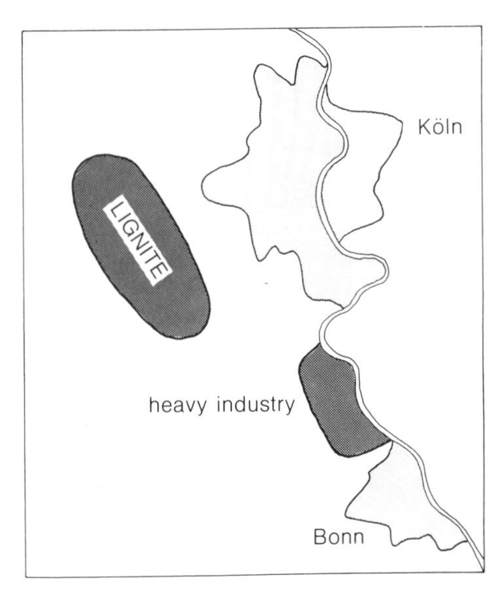

Figure 40.15 Heavy industrial concentration near Die Ville lignite fields, Cologne, West Germany.

influencing location. Most steelworks now have coastal locations, and only the Sheffield steelworks is near to coal. Industry nowadays usually locates beside an energy supply only if this is convenient for other reasons. Thus the petrochemicals plants and the aluminium refinery near Cologne in West Germany use electricity generated from the nearby lignite deposit, but they are primarily market-orientated, sited as close as practicable to the Ruhr conurbation (Fig. 40.15).

Labour

Labour is an extremely important factor in most locational decisions. Although it may be of relatively small concern to the manager of a capital-intensive plant such as a petrochemical works (in respect of which other factors such as raw-materials location may weigh more heavily when choosing a site), for more labour-intensive forms of manufacturing and for all service and office activities, the supply, quality and skills of the workforce are of paramount concern. Indeed, despite the increased use of machines in automatic production and of computers in reducing the number of clerical staff, the higher wages that have to be paid to the remaining, more highly qualified workforce means that the total wage bill still remains a major factor of production cost. In the developed world labour costs still represent virtually half the total production cost. These high costs may persuade a manufacturer to move to a region of lower labour cost, where savings in labour cost may more than offset an increase in net transport costs.

Industry is more than ever concerned not just with locating near a supply of labour, but also with finding people with the right skills. Many industries owe their success to a relatively small number of key managers, production specialists and innovators. Location may be strongly influenced by the need to choose an environment that is convenient and pleasant for these key workers. (This contrasts with the reduced influence of production workers, whose jobs can easily be filled from the general workforce in any area). It is the reliability rather than the availability of the production workforce that is of increasing importance. Given two areas equally advantageous from all other points of view, a manager is more likely to locate in the area where strikes, unrest and absenteeism are less common. Thus the perceived militancy of Merseyside workers has been suggested as a major reason why management may be reluctant to establish new factories in the area.

For similar reasons, managers look for productivity and willingness to adapt to new methods of working as a major factor in choosing a location. Indeed, variations in productivity can more than offset differences in wage levels between regions, and may even encourage managers to choose a region of higher wages. In those industries where national wage agreements have been negotiated by the Trades Unions, there is no regional difference in wage levels; regional variations in adaptability, reliability and productivity then take on even greater significance.

Capital

The supply of capital, usually in the form of a loan from the government, from banks or from other financial institutions, is the lifeblood of all industry. The uses of capital are twofold:

(a) to acquire buildings and plant;
(b) to provide money with which to buy raw materials, to pay the wage bill and to pay other recurrent costs.

This capital is repaid from the eventual sale of the goods or services.

In the past a considerable proportion of the total capital was tied up in buying the land and buildings, and this caused industry to be relatively immobile. Once established, industrial location became subject to a great deal of **inertia**. Recently the growing tendency for industry to relinquish ownership of land and simply to lease factory space has removed one of the major restrictions on relocation. Without extensive capital tied up in the site, industry can afford to be much more flexible in its location, and it is then liable to move to where it can obtain the greatest marginal returns from the changing patterns in its labour, materials and market.

Because the supply of capital involves an investment risk, there is a tendency for such capital to be loaned for smaller businesses to people in the proximity of the loaning institution, where the financial institution can monitor progress. The location of the loaning institution may therefore have an important locational pull on the industry. The same may also be true with government (as in the classic case of the growth of Paris), but government may also be able to counter the tendency to agglomerate near capital sources by using regional development policies and by giving money to industry in peripheral locations.

Technology

With many industries depending on new product lines for their continued survival, **research and development** (R&D) is becoming an increasingly important part of any industry. Although research and development may involve relatively few people, it can use a substantial amount of capital. Centres for the advancement of technology, such as universities and government research stations, can provide important input to industry.

Because innovation is largely concerned with the sharing of information, there is a tendency for private research and development to locate near to these established centres. For multi-plant firms there is often considerable advantage to be gained from separating research, development and trial production from the mass production of established products.

As a corollary, areas deprived of access to innovation become less favoured areas for growth. This applies equally to the peripheral areas of the developed world and to the developing world as a whole.

Product life cycles

It is usually unrealistic to think of industrial production as a static activity, comprising a single plant producing a single product over a long period of time. In general individual products go through a sequence of evolutionary change from initial **innovation**, through a **growth** stage and lastly to a stage of **maturity**. As this happens they may be produced in a variety of factories under a range of conditions.

In stage 1 a newly discovered (innovated) product is most likely to be produced near to the corporate headquarters of a firm where a close watch can be kept on its progress. Headquarters tend to be situated in the most prosperous areas of advanced economies where the customers can afford the relatively high initial price associated with trial production, and where potential consumers will be most receptive to new products.

Initially, the product will be able to maintain a high price because of its monopoly, but eventually competition from other manufacturers will lead to a decline in market share for the innovator and force a change of policy. This will involve an attempt to expand the market and reduce costs by a higher volume of output. Stage 2, therefore, sees a sales drive to reach export (non-local) markets. These, too, will later suffer competition and, if overseas, may also suffer import restrictions. Nevertheless, in many cases the foreign markets are sufficiently large to encourage the innovator to set up a new plant within the export market This is stage 3.

In time, competition from other manufacturers who may have newer plant and possibly lower production costs, will threaten both domestic and export markets. In response the innovator may seek to lower his own costs by setting up production in developing countries (which are themselves not significant markets), using their cheap labour for production. This is stage 4. These factories are used to supply not only the other export markets, but also the original domestic market. At this stage, therefore, the original plant in the home country may cease production of the product.

There are many European electronics groups that have gone from innovation to maturity very quickly. Originally producing under their trade name in their own country, they have subsequently developed factories in other parts of Europe and then in the developing world. For example, Grundig started in West Germany; opened a second factory in Northern Ireland; and then a third in the Philippines.

40.5 Evolution of industrial location

Multinational companies

The product life cycle for manufactures is concerned with the changing location of factories and products in response to a changing pattern of costs. In this respect it illustrates Weber's location analysis in a dynamic form.

A quarter of the capitalist world's manufacturing is in the hands of **multinational companies**. These have plants situated in many countries, and the opening and closing of factories and shifts in product locations are invariably made on least-cost decisions. However, these costs are not as easy to define as those for a single-plant firm: the multinational has to maximise its *group* profits and minimise its *group* costs. As the product-life-cycle theory shows, factories are liable to be relocated first to market sites abroad (stage 3) and then in developing countries where labour costs are lower (stage 4; Fig. 40.16). For example, even by the early 1970s a chief executive of Agfa-Gevaert had remarked about the manufacture of cameras: 'Very labour-intensive products can no longer be made economically in Western Europe and North America. As labour costs account for a large proportion (up to 50 per cent) of the production cost of cameras and other products, the importance of low-wage regions in the long-term planning of the enterprises is increasing'. In 1983,while closing its Liverpool tyre plant and shedding many of its tyre interests to the Japanese, Dunlop was expanding its plant in Malaysia. Thus Dunlop are now operating a policy of diversifying into new activities and using its UK base to pilot-test production prior to licensing manufacturing overseas.

The location of plant is further complicated by the increasing insistence of countries on having a production base. Thus major UK firms have found it necessary to buy a North American manufacturing base in order to obtain access to new technology and a launch pad from which to internationalise their operation. ICL, for example, acquired the Singer Business Machines Corporation in the USA, specifically for this purpose.

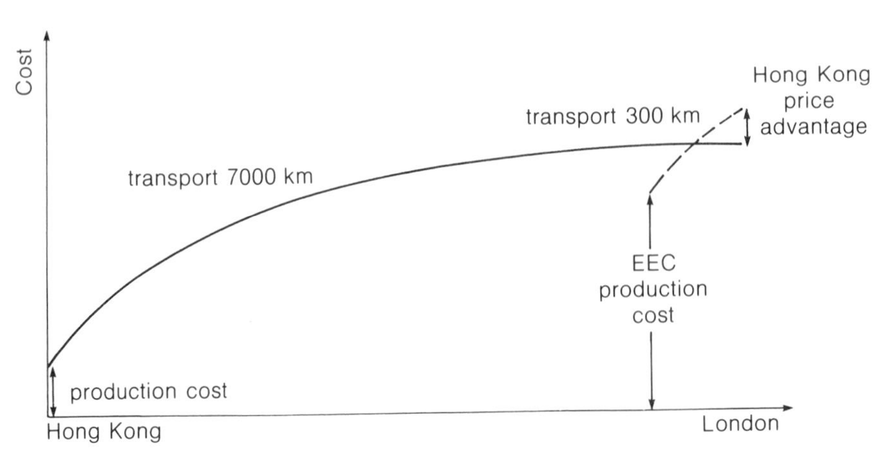

Figure 40.16 The role of labour as a major factor in production costs is illustrated by the final costs of making transistor radios.

Urban areas of the developed world

The industrial environment is constantly changing with the needs of society. This is true not only for the least-cost location of a plant, but also for the site and environment in which it is established. Through this century the old central manufacturing districts of cities have given way to the modern suburban industrial park (see also p. 429). Much of the innovation in the use of industrial space has occurred within **industrial parks**; these offer large areas that can be planned on an integrated basis, and although most industry still lies outside these designated areas, government planning policies are liable to encourage further use of such sites.

The keynote of industrial activity is the need to be profitable in the face of competition. However, people are discovering that it is as important to make the working environment attractive as it is to make it efficient; there never was any need for the 'dark satanic mills'.

There are certain key trends that will influence industrial development in the future, some social, others strictly economic. In summary they are as follows.

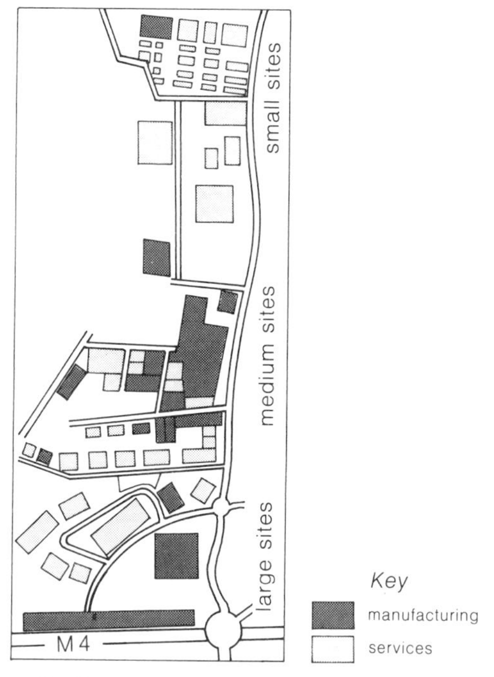

Figure 40.18 The Basingstoke Road industrial estate, Reading, showing segregation in structure.

(a) *Environment:* There is a growing concern to preserve the environment and controls on development are becoming ever tighter (see also Ch. 45).

(b) *Energy costs:* It is increasingly clear that the cost and supply of energy will be a major consideration. Sometimes the need to conserve energy will help the environment, but environmental controls often reduce the efficiency of energy use. Higher energy costs can be expected to drive some industry to peripheral sites where compensatory savings can be made through lower labour costs.

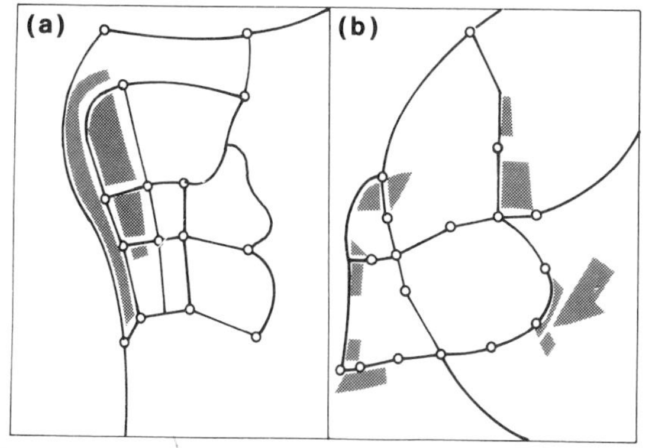

Figure 40.17 Two of Britain's New Towns. (a) Stevenage, one of the earliest towns, had segregated industrial and residential areas, giving long journeys to work. (b) Washington, one of the later towns, has industrial areas scattered through the residential network, thereby reducing the journey to work.

(c) *Integration of land uses:* In the 1950s, 1960s and 1970s local government often tried to segregate industry and housing, but in some places there is now a trend towards neighbourhood units that include elements of all land uses (Fig. 40.17).

(d) *Specialised industrial clusters:* The advantages of plant clustering are considerable. Each cluster group has its own special problems and operates best if the site is designed for this purpose. Typical groups are research/industrial parks, and warehousing and distribution centres (Fig. 40.18).

(e) *The social needs of employees:* There will be a trend not just to maximise the efficiency of industrial premises and locations, but to provide as pleasing and social an environment for employees as is possible, on the grounds that productivity is related to job satisfaction. It is therefore important to have access to commercial, retail and recreation services (Fig. 40.19).

(f) *Technological change:* This will be the most far-reaching force in the future, especially in the field of communications technology. But innovations in processing systems and transport methods will clearly have major impacts in industry, often causing changes in factory layout and locational shifts to motorway and airport sites.

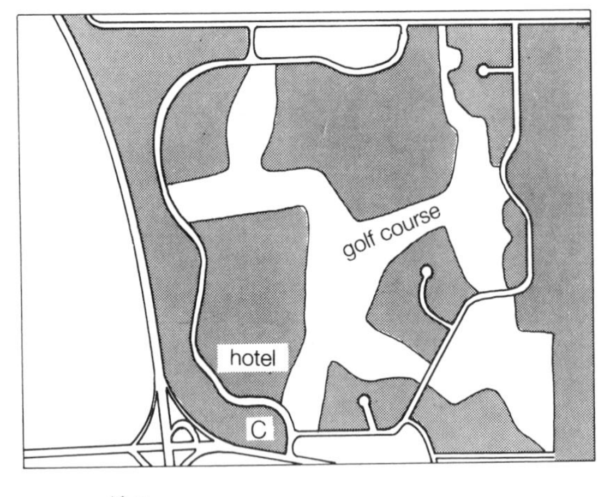

Key

office / light assembly

C commercial centre

Figure 40.19 The Inverness industrial park near Denver, Colorado, is a multi-use project with over a third of the space devoted to employee use. The site includes an 18-hole golf course. The developer clearly felt that the value of the remaining property was sufficiently enhanced to make the large recreation area economically justifiable. Notice that the golf course has been sited on flood-prone land that would, in any case, have been expensive to develop. Using the river area also provides a 'linear' golf course and gives a maximum number of commercial sites with a golf-course frontage.

Figure 40.20 The distribution centres for MFI plc show how the plan of buildings is now being organised to help computer-controlled handling. The system features an 'in line' handling system and out-of-town retail outlets.

(g) *The shifting pattern of location:* There will be a continuing shift of employment and production from old coalfield sites toward the rest of the country (north to south in Britain; north to south in Germany; north-east to west and south in the USA; and so on). There will continue to be a strong shift to non-metropolitan areas, which in turn will cause a major shift in markets. This 'natural' trend is frequently being strongly countered by a government policy of redirection back to the metropolitan heartlands (p. 425). If redirection is ever to be achieved, the main developments will have to be on old railway or owned land (the London Docklands, for instance), simply because these are large sites. It is only such large sites that can be successful because new handling methods demand single-storey accessibility units (see also Fig. 40.20).

Two case studies

This chapter has demonstrated that the location and evolution of manufacturing industry is the result of a complex interplay between many factors. Although these factors can be isolated in order to illustrate principles of location, in practice changes in location involve decisions which attempt to balance all the factors in such a way as to maximise the advantages of each.

To show this clearly, this chapter concludes with a description of the locational decisions made by two large multi-plant companies, one in the North of England, the other in the south. These case studies also show the widely differing individual circumstances that influence decisions made by each firm, and the role of government as a major locating force. The examples emphasise the dynamic nature of industrial products and location: no longer is it possible to assume that factories will remain static in their choice of location for many years, especially as there is now usually very little capital tied up in the building. Most buildings are acquired on leasehold and are owned by insurance companies, pension funds and the like; manufacturers have only to finance the equipment.

THE METAL BOX COMPANY

This is a British-based packaging concern that has traditionally specialised in the manufacture of metal containers. However, with the advent of new plastics technology, Metal Box is thrusting forward with the production of a wide range of plastic containers, partly to suit the expanding range of household consumer products, and more recently to compete with the traditional glass containers in the soft drinks and beer markets.

The filling locations for soft drinks and beers are primarily in the Midlands, north-west England and Scotland, rather than in the South of England, filled bottles being transported to southern markets by road. The raw material for making plastic bottles is obtained from a new ICI plant established in the Wilton Complex on Teesside. Bottle production also places a considerable demand on electricity supplies.

Metal Box was seeking a location which was within reach of the markets and of raw materials, a location in which there were a skilled labour force, good industrial relations and a favourable attitude by the local authority (to ease establishment). A location in a Special Development Area (for which large capital grants are available) was a further important incentive.

The new factory was located in 1982 at Wrexham, in the North Wales SDA, where redundant skilled steelworkers could be retrained. The local authority provided a purpose-built factory on an old coal-mine site and arranged for leasing at attractive rates. Wrexham is near to the Lancashire motorway complex and provides good access to the company's markets, while also being close to the residentially desirable areas of West Cheshire – an important factor in attracting a good management team. A location in nearby Liverpool (the traditional home of Metal Box activities in the North-West) was ruled out on the grounds of its being an unsuitable site for a factory, and its having a less desirable residential environment for the management.

The new factory is close to another bottle-making plant belonging to Metal Box, where an established management team and a network of skilled subcontracting manufacturers and services has been built up since the earlier plant was established in 1969. Thus it is possible to use some of the management structure to run both factories, and workforce training can be accomplished at the existing plant, allowing a more rapid commissioning of the new factory.

A. JOHNSON AND CO.

This company is part of a Europe-wide engineering organisation whose headquarters are in Sweden. Johnson makes stainless-steel food containers and processing equipment.

It started in Sweden as a shipping line, carrying Newcastle coal to Sweden and bringing stainless steel from Sweden to Tilbury. In 1934 it began to manufacture dairy equipment in Britain, choosing a site at Acton that was near to the imported steel and within the rapidly expanding manufacturing zone of West London. It was also close to another Swedish-based company, perhaps for psychological support.

However, the product range gradually changed and widened to include large tanks for the brewing industry and it became clear that the Acton site was unsuitable, being too small already and with no room for expansion. These constraints led to the decision in the late 1950s to move out of London to a 'green field' site, at Wokingham in Berkshire. Wokingham was near to Heathrow and therefore attractive to management who regularly flew to and from the parent firm in Sweden. Furthermore, an increasingly large proportion of the products was for export and Wokingham is well placed with respect to both Tilbury and Southampton Docks. Wokingham Council could also offer council housing accommodation for the skilled workforce and, on moving, 75 per cent of the workforce moved with the company.

In the 1960s the product range changed again, this time moving towards the design and commissioning of high-technology food-processing plant and away from direct manufacture of equipment. The manufacture was to be accomplished by specialist steel firms who were scattered throughout southern England. These changes prompted a further move to a building that contained a larger proportion of office and less production space. With emphasis on highly qualified and skilled staff such as graduate engineers, the overriding need was to retain the workforce; a short move only was possible unless they were to impose unacceptable disruption on the staff. In contrast to the aggressive marketing of the Clywd Authority in Wrexham, the local councils remained inactive in this relocation decision; unemployment is not such a pressing problem near Wokingham and no special financial incentives could be offered. Johnson has now moved to a factory in the industrial estate of South Reading, next to the M4 junction and within 10 km of their old factory.

41

Service and office industries

41.1 The post-industrial society

Until perhaps thirty years ago it was widely assumed that manufacturing was the essential foundation for development throughout the world, and that expansion and economic growth rode on the back of the manufacturing sector. It was believed that there was a natural progression as a society developed, beginning with agriculture or mining, then advancing into manufacturing and finally to services. In this belief many developing countries sacrificed their agricultural sectors and concentrated on the development of manufacturing. The acquisition of a large-scale manufacturing plant became a symbol of economic maturity, a matter of national prestige and a measure of the standing of the country in the eyes of the world at large. South Korea became dedicated to steelmaking, as did India and Nigeria.

Yet it was already becoming clear that there was no simple single path to growth and even that growth did not arise simply as a result of establishing a system of manufacturing. Nowhere was this more clearly demonstrated than in Britain. Despite being the first industrial nation, Britain had a rate of growth lower than many more recently industrialised competitors. Furthermore, a chain of events through the 1960s and 1970s confirmed that it was possible for other forms of activity, especially services, to grow and become major, often independent, even dominant forms of employment and providers of the national wealth (Table 41.1; Fig. 41.1).

Table 41.1 Changes in the UK GNP, 1976–81 (percentages). The remaining change in GNP was primarily due to changes in oil revenues from the North Sea (source: Central Statistical Office).

	1976	1981	Change
manufacturing	28.4	23.0	−5.4
services	54.6	56.6	2.0

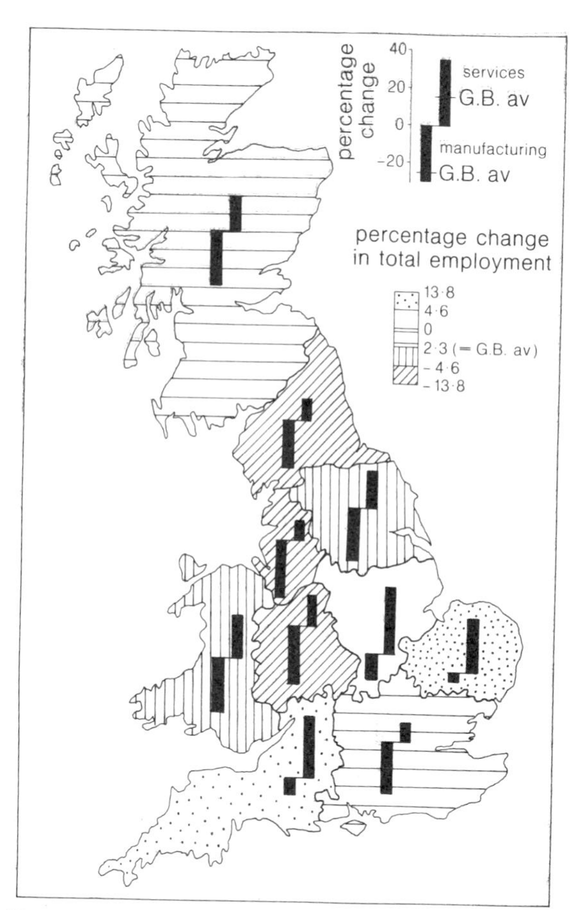

Figure 41.1 Regional employment change in Britain, 1971–81.

In many instances developing countries were, in any case, simply not able to develop a strong manufacturing base: they were deficient in natural resources, both of energy and raw materials; they had unskilled workforces and were under-capitalised. Furthermore, the headlong rush into more and more manufacturing was creating considerable overcapacity – countries began to rely on continued world growth in demand just to consume the products.

The world rise in oil prices following the 1973 Arab – Israeli war burst the manufacturing 'bubble', caused world recession, and prompted many countries to take a cool, hard look at the structure and pattern of world employment. But even before this catastrophic event, the role of manufacturing as an employer in the industrialised world was fast diminishing, as automation and technological change made unskilled and semi-skilled jobs increasingly redundant. In effect the world was becoming a **post-industrial society** (Fig. 41.2).

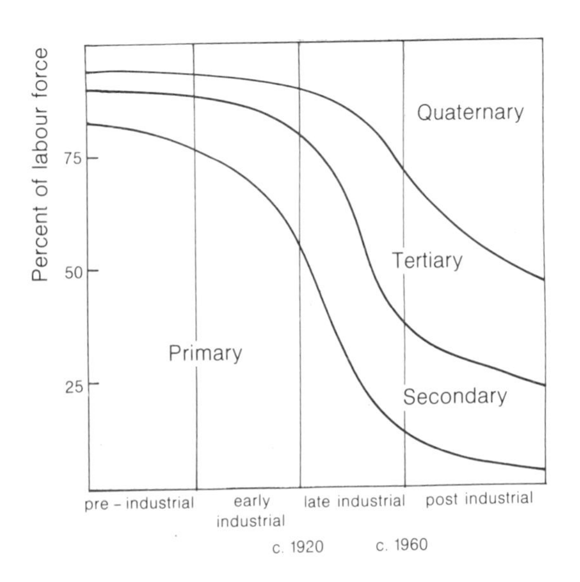

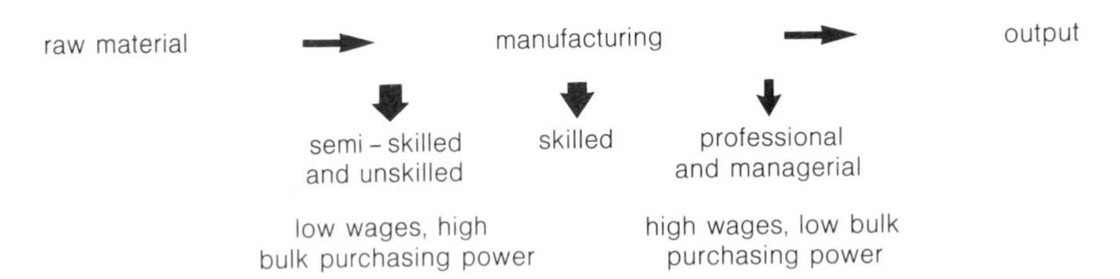

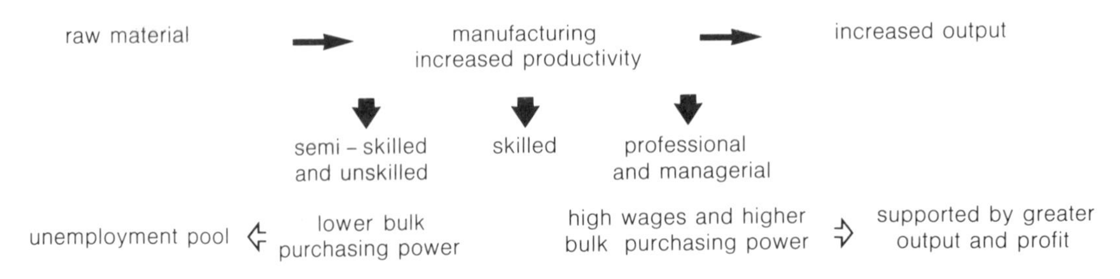

Figure 41.2 The evolution of the industrial workforce. In 1982 the UK had 62.3 per cent of its labour force engaged in Tertiary and Quaternary activities, 33.8 per cent in manufacturing, and 3.9 per cent in the primary sector.

Employment changes

The changes that have overtaken much of the developed world can be illustrated by the example of London. Until the 1950s London had forged ahead, increasing its share of the manufacturing sector in Britain until this was larger than any other. Electricity to power machines, road transport for flexible delivery, and the rise of consumer products were all in London's favour. Yet between 1961 and 1981 London lost three-quarters of a million jobs in manufacturing industry (1961: 1430000; 1981:680000) and today it has almost lost its manufacturing heart.

These jobs have not been lost simply because manufacturing industry has relocated in more attractive areas. Of the total job losses 7 per cent have occurred because of factory relocation in new towns; 9 per cent have relocated in Special Development

Areas and 11 per cent to areas with no special incentives. But this only accounts for 27 per cent of the job losses; over 29 per cent of losses have occurred because changes in technology have caused employers to cut the workforce, and a remarkable 44 per cent, by far the largest proportion, were simply caused by firms going out of business, their skills no longer required in a rapidly changing world. Many of these losses were in old, labour-intensive jobs, and their loss turned previously thriving central urban areas into wastelands. The cities of the developed world were experiencing **deindustrialisation**.

The situation was very different in the developing world where there was no tradition of manufacturing. To encourage manufacturers, developing-world governments increasingly turned to protectionist measures and required foreign-based companies to transfer at least some of their processes from already industrialised countries. The rate of change of location varied according to the temperament and the level of skills available. In Africa, where regular industrial employment is new, and where people have only been prepared to work for as long as was necessary to meet immediate needs, there has been relatively little progress. In Asia, where the temperament is more in tune with regular work, there has been an increasing level of industrialisation, particularly of work initiated by Japan. Hong Kong and Singapore, the Philippines and Taiwan were even singled out by multinational companies because of their low-cost, hard-work capability.

But although manufacturing can be expected to increase in all developing countries, this is primarily to meet their domestic needs; manufacturing will not be a major growth sector of employment either in the developed or the developing world. In fact the reverse will be true, for the growth of the urban population in developing-world cities, and therefore the demand for jobs, is so much greater than in industrialised countries.

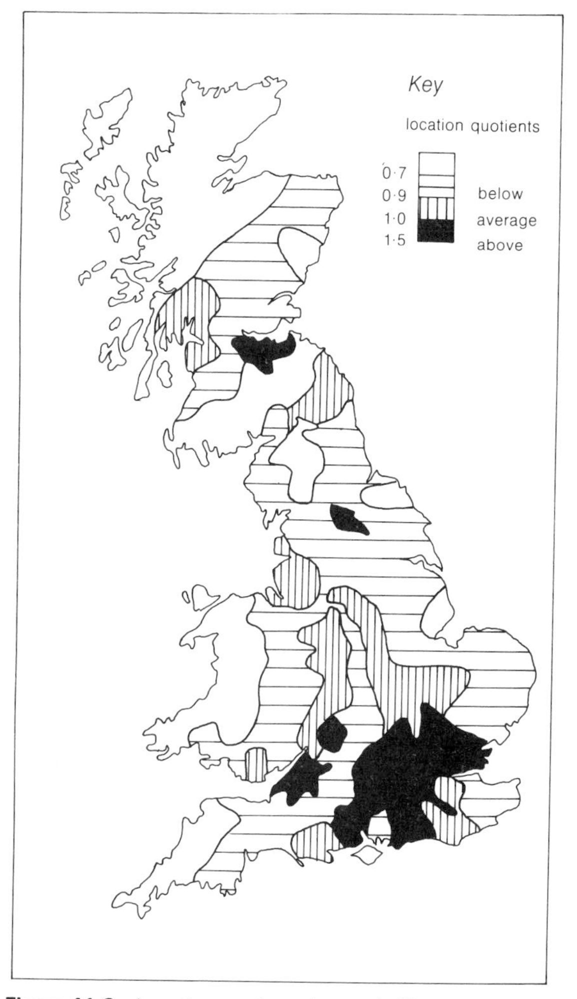

Figure 41.3 Location quotients for total office employment in Britain, 1971. Notice the strong localisation in the south-east of England and that many of the major conurbations other than London, Edinburgh, Bristol and Leeds have no more than the national average in office employment.

41.2 Kinds of services

The growth of services

In most large manufacturing centres of the industrialised world, there is a large office-employment sector. In the 1960s and 1970s, when employment in manufacturing was falling, the growth of the office sector was thus able to cushion job losses in some areas (Fig. 41.3). Nevertheless, the skill requirements of manufacturing and office activities are very different, and only occasionally allow job switching. What happened was that the perceptions of school children changed: an increasing proportion planned to go into some form of office or other service activity, rather than to follow traditional apprenticeships in manufacturing. As older people left employment and retired, young people moved into a different sector of employment. Whereas in 1960 only a third of London's working population had some form of office employment, by the beginning of the1980s the proportion had risen to nearly half, with other forms of service making up a substantial proportion of the remainder.

Changes in technology have not left office employment and location untouched, however; the reverse has become true as computer technology has gathered momentum. Word processors and computers begin to replace the less skilled office jobs. Indeed, since 1975 there has been an absolute decline in the number of office jobs in London; and the growth of office-sector employment throughout the developed world has consistently been slowing down.

Government administration has traditionally been a major employer of office staff. In Britain, for example, it employs over two million people directly. Tasks varying from tax collection to collating census returns are performed to allow the country to be administered effectively. Without significant manufacturing employment, the office sector has also been the major employment sector for many developing countries. Indeed, many people arriving in the city move directly from agriculture to some form of service activity. If they have a relation in the administration they can usually be found a job somewhere, though the result may be to extend the chain of command and make the administration ever less effective.

Office activity is only one form of service. But because it is often highly concentrated into the central business district and in purpose-built premises, it is often very conspicuous.

Services cover a wide range of activities, including hairdressing, shops, sewage, water maintenance, refuse collection, education and welfare (Table 41.2). Although each of these activities has an element of office work, they rely primarily on a large staff giving personal and direct attention to the needs of the community. The great diversity of services and the fact that, in their very nature, services are necessarily labour-intensive allows them to support a large amount of employment.

Table 41.2 The UK Health Service, 1981. An example of a service industry, indicating the proportion of each type of service (source: DHSS).

Activity	Number of staff ($\times 10^3$)
medical and dental staff in hospitals	47.5
nurses/midwives	491.6
professional/technical	80.1
works and maintenance	34.2
administrative/clerical	133.3
ambulance personnel	22.0
ancillary	220.0
general practitioners	30.1
dentists	15.1
other professionals	9.1
total	1083.0

The main contrast between the industrialised and the developing worlds lies in the rate of change from manufacturing or agriculture into services. In the industrialised world there has been a long period in which manufacturing was the dominant employer, and services have taken over gradually; the developing world must establish service sectors that effectively generate wealth and make services a viable leading sector of their economies. Simply adding more and more people to the government administration cannot help in this task.

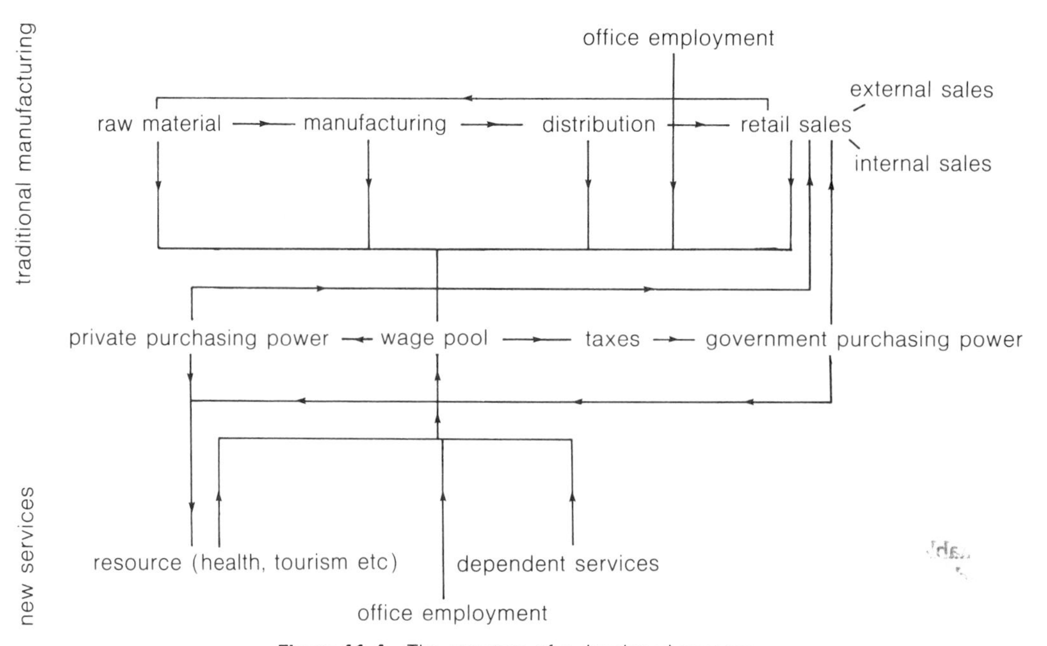

Figure 41.4 The structure of a developed economy.

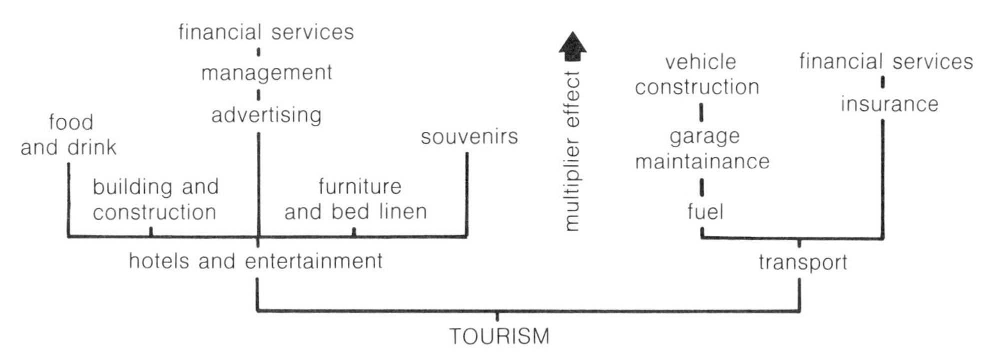

Figure 41.5 Some aspects of how the multiplier effect works in the tourist industry. The wealth generated indirectly may be two or three times the direct revenue from tourism.

Dependent and self-supporting services

Not all services are 'parasitic' on manufacturing in some way. Whereas social services such as the health service (which, in Britain, employs 1 million people; Table 41.2) can only be maintained by continual government funding, others, such as tourism, are totally freestanding and self-supporting, themselves acting as leading sectors and generating wealth for other services (Fig. 41.4).

Tourism can be an export sector of the economy, using various attributes of the environment as a resource base (as raw material). In the case of London, the eight million foreign visitors that arrive each year expect to find a combination of history, culture and entertainment. To provide for the needs of tourists, 250 000 people in London work in hotels or restaurants, drive taxis or coaches, act as couriers, and so on. Indeed, direct tourism services provide a larger economic contribution to Britain than the financial and professional services (the 'invisible earnings') generated by the office services of the City of London. Tourism is also an important service in so far as it offers much unskilled and semi-skilled employment, in contrast to many other service industries. Furthermore the multiplier effect works in the service sector just as it does in manufacturing (Fig. 41.5).

Tourism is a major source of employment in areas of Britain where there is no manufacturing. The 'golden mile' of Blackpool is perhaps the best example of one vast service industry. However, the importance of tourism as a leading sector is most dramatically demonstrated in the developing world, especially among the smaller nations (Table 41.3).

The small nations of the Caribbean can offer sun, sea and fine beaches on paradise islands whose coral sands are fringed with palms. Moreover, their best season for reliable sunshine corresponds with winter in Europe and North America, making them ideally suited to offer winter-break holidays. In the past most of the island economies have been almost totally dependent on a narrow range of agricultural products such as sugar cane and bananas. However the value of these products varies greatly with world demand and the quality of the harvest. By contrast tourism offers a chance to diversify economies into a more stable field. By concentrating on the development of luxury hotels and the high level of personal service thus required, the islands have made tourism especially labour-intensive. There is commonly a labour force equivalent to at least one employed person per hotel room.

The effect of tourism is vital for the industries directly involved – hotels, building, airlines, taxis, car hire and charter yachting – and also, through the multiplier effect, as a source of income for at least twice as many people.

Table 41.3 Employment in Tunisia (source: ONTT study).

Jobs created directly by tourism in Tunisia		
hotels	0.38	person years of
international transport	0.02	employment per
domestic transport	0.04	hotel bed per
restaurants	0.01	year (56%
government	0.02	occupancy rate)
production and sale of handicrafts	0.65	
total	1.12	
Jobs created indirectly by tourism in Tunisia		
agriculture	0.34	
other	0.44	
total	0.78	
Grand total	1.90	
Jobs created by investment in tourism in Tunisia		
hotel building	1.6	person years of
hotel renovation	0.4	employment per
infrastructure	0.7	hotel bed
total	2.7	

MANAGEMENT OF THE TOURIST INDUSTRY

In order to develop local resources for tourism, entrepreneurs must:

(a) *publicise the resource*, because if people do not know about the resource they will not visit it;
(b) *make sure the resource is accessible*, because difficulties with accessibility will make people unwilling to take the trouble to visit;
(c) *develop facilities* that will enable tourists to enjoy the resource fully while, at the same time, enabling the tourist industry to tap the money the tourists have to spend.

Tourism can be thought of as short-term migration, but its effects are long-term and dynamic, involving a sequence of evolutionary stages both for the tourists and the people who live in a potential tourist area. Consider, for example, an island that has not yet been developed for tourism (Fig. 41.7). Until a pioneer resort is established, potential tourists will have little interest in or information about the island. However, after a resort has been established, feedback of information by pioneer tourists gives a better perception of the resource for those who may follow later. As pioneer tourism develops into

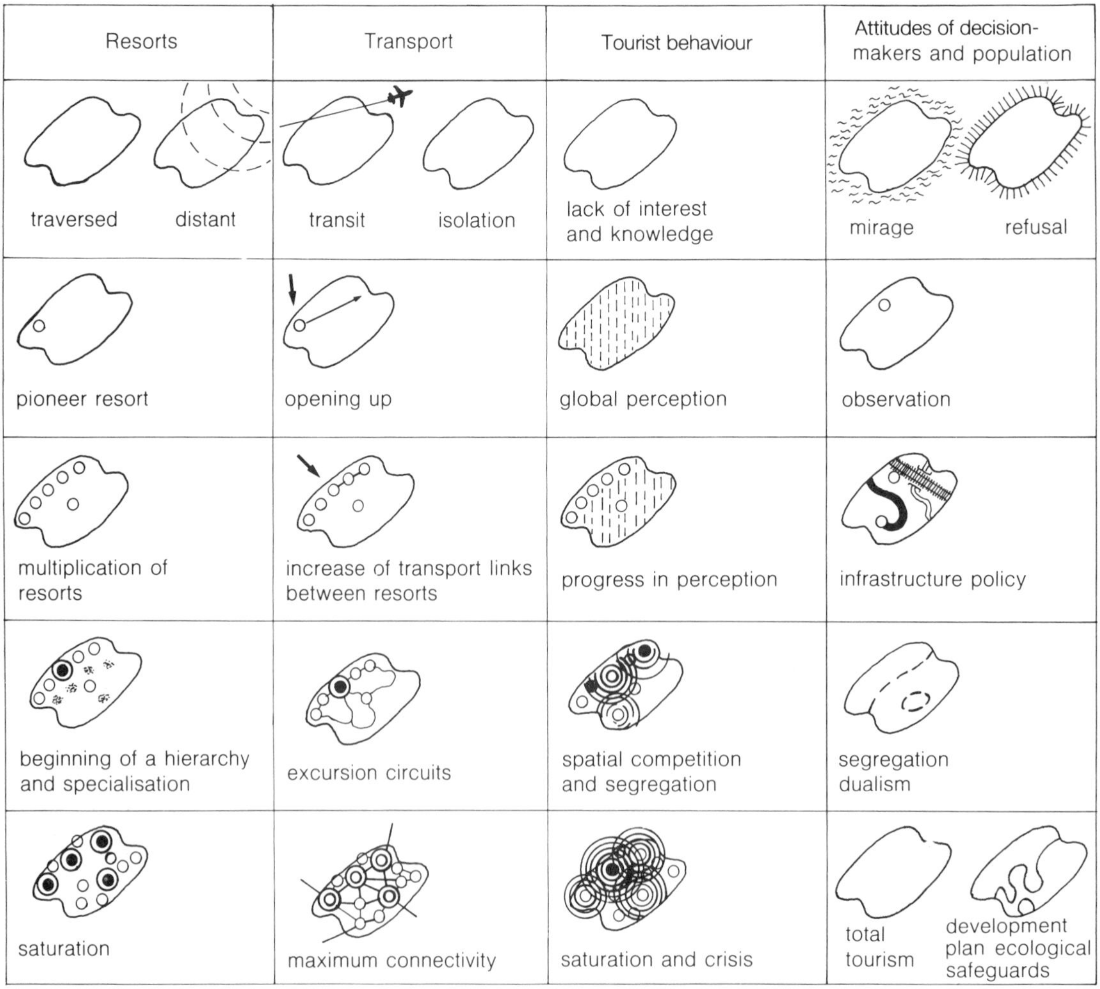

Resorts	Transport	Tourist behaviour	Attitudes of decision-makers and population
traversed distant	transit isolation	lack of interest and knowledge	mirage refusal
pioneer resort	opening up	global perception	observation
multiplication of resorts	increase of transport links between resorts	progress in perception	infrastructure policy
beginning of a hierarchy and specialisation	excursion circuits	spatial competition and segregation	segregation dualism
saturation	maximum connectivity	saturation and crisis	total tourism development plan ecological safeguards

Figure 41.7 The evolution of a tourist industry.

mass tourism so more resorts are opened up, links with the outside world increase, and the tourist potential of the rest of the island is more fully investigated. This, in turn, leads to developments in the road infrastructure, service industry, and the spread of tourism away from the original pioneer area.

Further development sees competition from the resorts, each attempting to produce a distinctive identity and to cater for a specific range of tourists. However, while the tourist industry develops there may be increasing resistance to the change in lifestyle from those living on the island who are not directly involved in tourism. Furthermore, the increased intensity of tourism may drive away the original type of tourists. At this final stage the islanders have to decide whether they will allow themselves to be overwhelmed by tourism or whether they will try to develop their island for many activities, zoning part for tourism and attempting to preserve the remainder.

Management of the tourist resource in large countries is much more complex than for a small island. But for small and large alike tourism is often a major source of foreign earnings. In North America and Europe tourism constitutes one of many service industries. However, it may be the major export earner of small developing countries and thus become the leading sector for economic growth, generating employment and stimulating other domestic industries. Nevertheless, unless a country has the organisational capacity to manage tourism it may find the industry taken over by foreign-based companies. For example, in the Caribbean islands up to 60 per cent of the goods and services required by tourists are imported because they cannot be produced locally. And because the major hotels are owned by foreign-based companies a large proportion of the profits also go abroad.

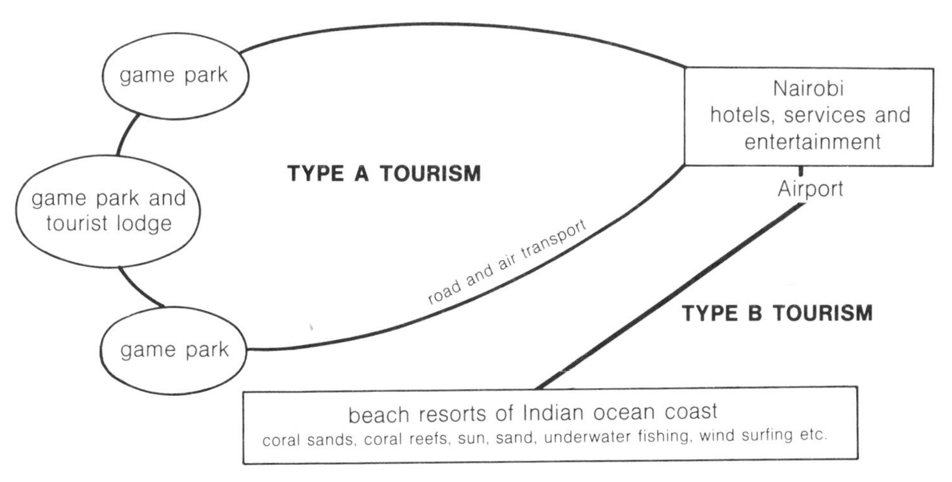

Figure 41.6 Part of the structure of Kenya's tourist industry. Here the resources are widely spaced and a chain of tourist lodges and minibuses feed tourists around recognised circuits (Type A tourism). Such tours are expensive and can only cater for up-market demand. To capture part of the mass market, Type B tourism provides beach-side hotels at lower costs. These holidays are more static but cheaper.

Kenya is another example of a developing country in which tourism is important. Kenya is the most industrialised economy in East Africa, yet its manufacturing industries only account for 13 per cent of its gross domestic product. It is a country with a good record of political stability, a vital prerequisite for a country wishing to attract tourists. Furthermore it has large natural resources of wildlife in its game parks, and mile after mile of coral sand beaches astride the Equator (Fig. 41.6). It can also capitalise on the cultural heritage of its people, especially the striking Masai warriors and the crafts of its many tribal groups. Although the number of people visiting the country is far smaller than the number who arrive in London each year, Kenya is proportionately more dependent on tourist revenues. Tourism is at present the third largest foreign-exchange earner (after tea and coffee) and has the highest rate of growth for any one sector.

Social and personal services

Social and personal services depend mainly on a supply of wealth from those employed in primary or in secondary industry, or in resource-based services like tourism. They are either paid for indirectly through taxation (as in the cases of the health service, education, refuse collection, the armed forces and the police) or directly (as with plumbers, hairdressers and garage machanics). Because of these varying sources of finance the growth of this sector of the service industry depends heavily on the wealth of the country and the political approach adopted by government. Thus there is only moderate potential for these services in the developing world. In countries such as Britain, the high level of personal taxation enables a wide range of services to be offered. For example the number of jobs in British welfare services has expanded by 1.5 million since 1951.

41.3 The location of services

Services are closely tied to their markets. Transport and industrial services are mostly found occupying premises alongside those they services, thus enhancing agglomeration. Tourism services are linked with their resources and may therefore be in rural as well as urban locations. By contrast personal services will be more closely associated with residential areas, or be in highly accessible central places. Many small-service businesses operate from their homes, using light vans as mobile workshops. Other, larger services operate from central depots where stores are kept and vehicles are serviced and garaged. These bigger firms need larger sites and are more often found in industrial estates where ground rents are cheaper and where materials are readily available from wholesalers (Fig. 41.8).

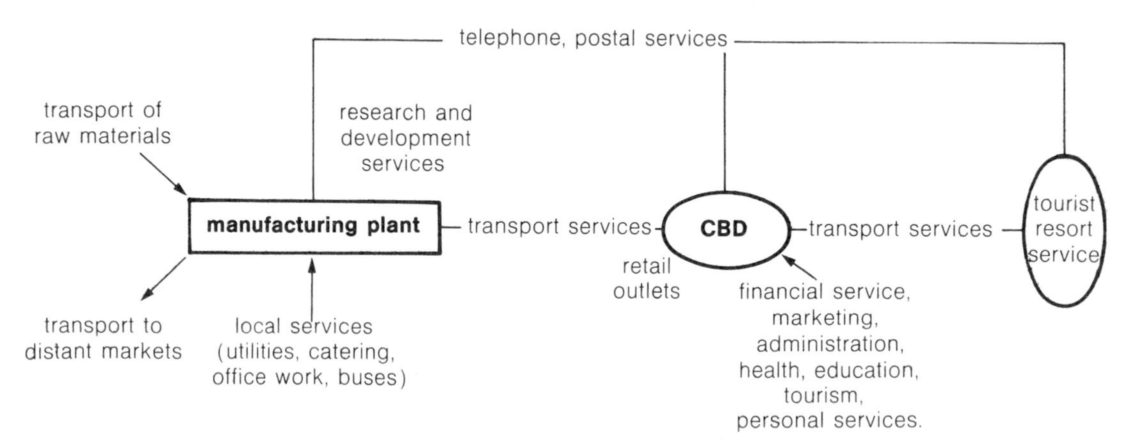

Figure 41.8 The relationship between a manufacturing plant at a low-cost peripheral site and the services offered. Some services are highly concentrated in the CBD; others, such as research and development, may occupy rural sites; yet others are clustered near to the plant.

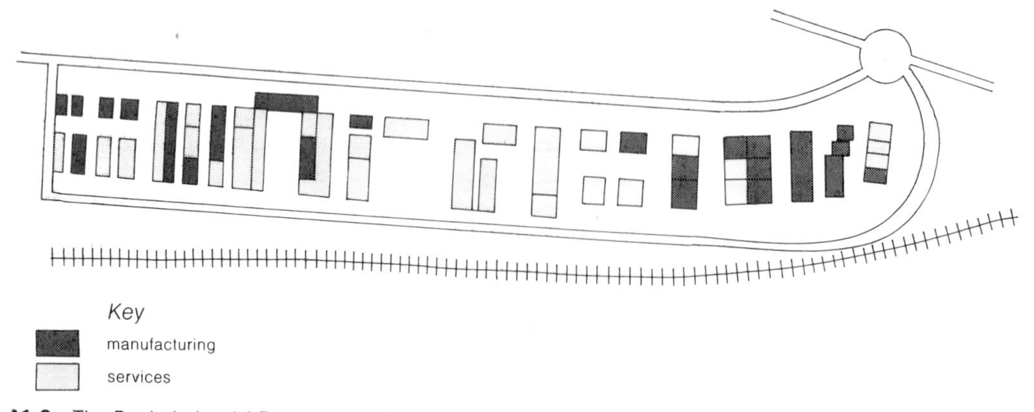

Figure 41.9 The Battle Industrial Estate, a smaller inner urban industrial area with a high proportion of service activities.

Health, social and educational services are located in relation to the population. Many hospitals and schools were built within residential zones in the 19th century, a practice of integration that has continued with the growth of the urban zone. The trend to larger, more cost-efficient units such as comprehensive schools and regional hospitals has required large sites, and these can often only be provided at peripheral urban locations or by large-scale urban renewal.

The locational requirements of the armed forces are different from previous examples. The choice of location depends on whether internal or external security is the main concern. In many developing countries in which there is a problem of severe political instability, barracks are located within the urban areas, while heavier equipment and stores are kept at sites peripheral to the urban areas and near national borders. In the developed world, problems of *national* defence are often more important, and locations at military ports or at communications nodes are often chosen. Major depots and training areas are normally located on land which has a low agricultural potential, typified by the British Army's base on the infertile sandstones near Aldershot in Surrey and by the US Army's weapons-storage and testing sites in the Utah and Nevadan deserts.

41.4 Strategies in providing services

The service multinational

There has been a clear trend towards diversification within the service industries, just as there has been within manufacturing. For example, banks have moved into other financial services, such as insurance, and have expanded their operations overseas. Similarly, the hotel and leisure industry has diversified to include car hire, shipping and many communications industries. Grand Metropolitan Hotels, for instance, began as the side interest of a property developer; with the post-war rise in the leisure industry, hotels became a major growth area and hotels were purchased overseas. Later there was a further expansion into country hotels, motels and inns; then into the retail and distribution sector, with the acquisition of Express Dairies. Later still, the Berni Inns chain was added to the company, and the entertainment group Mecca, with its bingo halls and betting shops. Finally GMH bought up several breweries.

The purpose of this strategy was to integrate GMH's functions in an effort to stabilise its environment, providing both satisfactory returns and maximum growth. In times of recession, though, service multinationals can be every bit as destabilising to the local economy as can multinational manufacturers.

Government and services

Many service industries have come under government control to allow economies of scale (as have electricity and gas), to provide the financial backing to those industries that need to offer a social and a welfare, as well as a commercial, service (as have health, the post office and the railways), and often to ensure military or national security (as with the police, the fire service, and the armed forces). However, many decisions to nationalise an industry have been taken on political (ideological) grounds, rather than on commercial ones.

41.5 Contrasts between industrial sectors

Service industries do not have a 'manufacturing' operation to provide a central focus for their activity; their major assets are dispersed to the selling or distribution points. Whereas in manufacturing the sales force is both the only dispersed part of the organisation and a small part of the total activity, in services the situation is reversed, with most of the assets at the salesroom. This system of operations requires a much higher degree of managerial supervision and co-ordination than does manufacturing, and the 'head office' function is relatively less important. This has allowed dispersal of services from central business districts.

If we adapt fast enough the jobs will still be there

Christopher Huhne

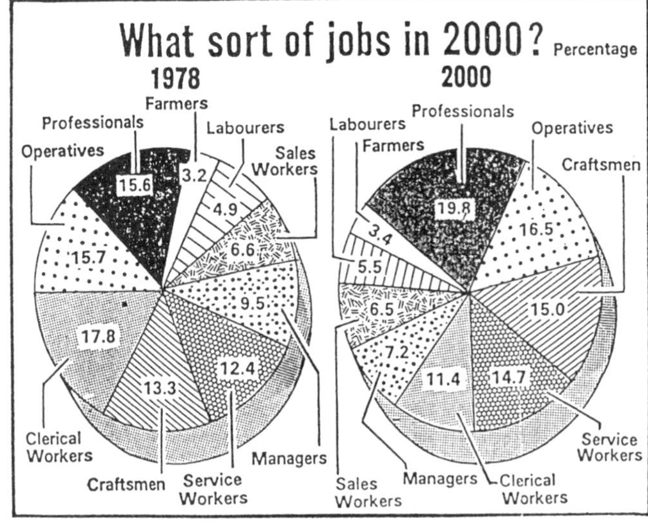

WHAT SORT of future will our children really inherit? The most popular economic scenarios tend either to project an imminent paradise in which an abundance of human ingenuity provides for every unimagined want, or alternatively that society will be characterised by a growing gulf between ever richer haves and ever more deprived have nots.

Far from it being likely that machines will displace workers, it is more likely that there will not be enough workers to operate all the machines we will want.

Equally, the changing structure of the labour force will not be dominated by a decline in production workers, who will probably increase their share of jobs. But there will be a dramatic fall in office workers and a rise in the number of professionals.

"The Impacts of Automation on Employment 1963-2000" relies on Input-Output Analysis, a method clearly different from the more usual attempts to build computer models of the economy. It constructs a detailed picture of each sector of the economy, rather than just looking at the big aggregates like Gross Domestic Product which are all that is necessary for short term forecasting.

This approach makes sense because most of the impact of new technology comes through its gradual adoption as one old process is phased out, and the new one phased in. The structure of the economy is like the proverbial oil tanker which takes 10 miles to change course.

Several important social implications stand out. Production workers — the blue collar working class — do lose many of their existing jobs, but they are rehired in growing industries required to produce new equipment for the technological revolution.

The people who are really hammered are clerical workers whose share in the labour force declines from 17.8 per cent to 11.4 per cent as office automation gathers pace. There is simply enormous scope for improving office techniques.

At present, the average American office worker uses only $2,000 compared with $25,000 of capital equipment for a factory worker.

As Dr Duchin, a woman, points out, the implications of this change for economies like Britain and the United States where there are a lot of women in the labour force — and a disproportionate number in clerical work — are that the trend towards more women working could be rapidly thrown into reverse **unless woman can adapt themselves to expanding occupations.**

These calculations about the changing occupational structure are the ones about which the Leontief team feel most confident, but they also shed some light on the saloon bar wisdom that new technology is going to destroy jobs in aggregate (as opposed to destroying specific jobs).

When the researchers projected the growth of demand for various categories of goods and services as compiled by the US Bureau of Labour Statistics their favoured technological scenario simply generated too many jobs for what is likely to be the size of the American labour force in 2000. So they had to develop an alternative scenario which assumes a lower rate of technical change to create fewer jobs.

In other words, the study strongly implies that the likely technological changes up to 2000 are going to increase the demand for labour rather than reduce it — exactly as past technological changes have done. If unemployment does stay high, it will be for other reasons such as the deflationary response of governments to wage bargainers' price-raising pay claims. It won't be due to new technology.

Other evidence shows that such structural unemployment is small now, but it is clearly still a possiblity on the Leontief scenarios, which is why the researchers underline the importance of training, particularly at this stage, of the teachers. They point to a successful schools computer programme in France which spent 70 per cent of its budget allocation on teacher training and only 30 per cent on the actual "courseware" — in contrast with American experiences of going for the gadgets before the school teachers know how to use them.

What the Leontief study underlines heavily, though, is that all those who talk about technological change destroying jobs do so at their — and our — peril. The real moral is that the faster we adapt ourselves to the skills and the opportunities of technical change, the more likely we are to be able to enjoy lower unemployment and higher living standards.

Economic growth and regional development

42.1 Stages of growth

Economic growth, on global, national and regional scales, has attracted much study. A model widely used to describe the gradual evolution of an economy, from subsistence to an advanced stage, has been proposed by Rostow. Looking initially at national economies, Rostow suggested that five stages of growth can be recognised:

(a) a traditional *subsistence society*, dominated by agriculture;
(b) an environment with the *preconditions for take-off*, in which there is an expansion of trade and the introduction of more advanced technology;
(c) a *take-off* stage, at which traditional methods have largely been replaced by modern ones, an increase in investment is occurring and growth is becoming self-sustaining;
(d) a *drive to maturity*, during which a more complex pattern of industrialisation emerges, urban society dominates over rural society and a transport infrastructure develops;
(e) a *mass-production, mass-consumption* stage, where service industries and social welfare become important parts of the economy.

Most nations and regions have not developed to stage 5. The reasons for disparity in advancement are the subject of this chapter.

42.2 The problem of unequal resources

In an ideal world, the resources needed for agriculture and industry, for welfare and education, for amenity and for employment would all be evenly distributed, the level of technological development everywhere would be the same, and there would be no differences between the standards of living in one region and in another. In fact, the distribution of the world's resources is far from even; moreover, at each location they can only be exploited for a limited time, before they become exhausted. Climate, for example, ensures that the intensity of agriculture varies from area to area, and accidents of history have allowed technology to prosper in some areas while remaining poorly developed in others. There can be few people who want to maintain so uneven a distribution of the global wealth, yet there are no easy means for sharing the benefits of success, neither between countries nor even between regions in the same country. Regional differences therefore occur because of the distribution of resources and because of the impossibility of sharing the benefits of economic development evenly. Indeed, the development of one region often occurs at the expense of another.

The problem of uneven development would be lessened if both industry and labour were perfectly mobile: whenever a resource ran out or the location of industry had to be altered, because of changes in markets or technology, labour would move with it. But such mobility is impossible in the real world: people have homes to be paid for; they have children at school whose education they do not want to disrupt; and they have friends and relatives nearby. Industrial structures cost a lot of money to build and equip and they cannot easily be dismantled and moved; and a workforce takes time to train and to become an efficient integrated unit. For all these reasons neither industry nor people are willing to move unless there is some overriding reason.

This **inertia**, or tendency to remain in one place even when the going gets bad, is readily seen in all countries. In Britain, for example, northern regions have been disadvantaged with respect to southern ones for most of this century, yet although there has been substantial migration, the great majority of people and industries remain in the locations of their birth and/or growth.

Economic growth and regional development focuses on change in time. Although it may be possible to explain the pattern of settlement or the location of an industry at a moment in time, this only reveals part of the picture. As in the natural environment, industry and people are in a dynamic equilibrium, a relationship that is continually altering, sometimes apparently catastrophically and at other times almost imperceptibly, yet in the long run maintaining a balance.

City and region

Cities lie at the heart of any growing region. They are predominantly the creation of economic forces, because economic growth within a country becomes impossible beyond a certain level without the development of major centres to focus activity.

People can be spread evenly over a region only if they rely entirely on the land as a resource. Even rural regions desire a range of services, and these can be provided economically only at certain central places. Furthermore, the goods that people want to buy mostly have to be manufactured in a small number of places, in order to achieve economies of scale. Consequently, some of the service centres develop industrial functions. At this stage they often enter upon a self-generating, self-sustaining cycle of urban and industrial growth that is at the heart of regional differences.

Because governments of towns and cities want their communities to grow and prosper, they try to encourage the growth of industry that may lead to increased employment. Three basic urban functions relate to growth:

(a) production;
(b) managerial functions;
(c) distribution of goods and services.

Initially economic development is determined by export-related manufacturing growth sectors, new industries that exploit some technological innovation (Fig. 42.1). Which industry this proves to be will depend on the local circumstances, for each growth industry has its own locational requirements. However, although manufacturing industry is of great significance in the growth of the small or medium-sized town, it is often of less importance in respect of large towns, in which services and office industries are predominant. Indeed medium-sized towns often seek to continue their growth by increasing their proportion of managerial functions.

Thus it is possible to observe a sequence of stages for urban growth:

(a) a *subsistence-urbanisation* stage at which the town only caters for the service demands of its immediate region (its **hinterland**);
(b) a *take-off* stage at which small cities begin to attract manufacturing industries: take-off is produced by a **leading sector**;
(c) a *tertiary growth* stage as service industries are attracted by the presence of the manufacturing sector, leading to a sharp rise in employment;

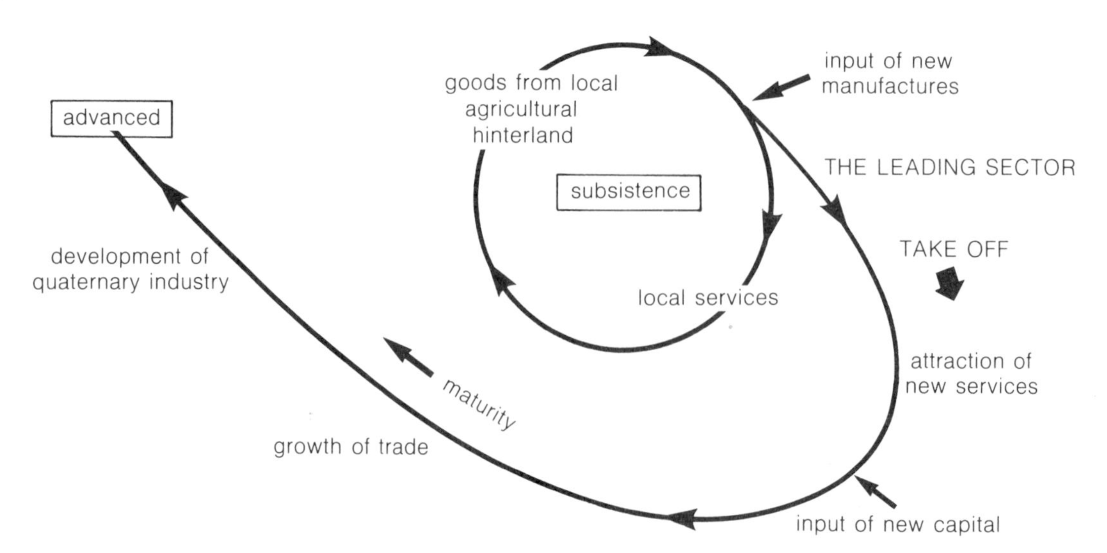

Figure 42.1 The economic development spiral helped by appropriate technology.

(d) an *office growth* stage, at which the service and office functions begin to take over from manufacturing and cause sustained growth, to a large degree independent of a change in the welfare of manufactures.

Cities in stage (a) are likely to be smaller than 50 000 people (e.g. market towns such as Saffron Walden), those in stage (b) tend to have populations between 50 000 and 100 000 (e.g. Cambridge); those between 100 000 and 250 000 (e.g. Reading) have a balance between secondary and tertiary industries and are early in stage (c), while those up to 500 000 (e.g. Nottingham) have a dominance of tertiary industries. The million-plus cities (e.g. Birmingham) are led by quaternary functions and are in stage (d).

The dominance of quaternary industries in million cities has been demonstrated recently by the continued low rate of unemployment in London, despite the drastic reduction in its inner-city manufacturing sector (but see p.428). However, there are a limited number of very large cities whose growth can be sustained by quaternary industries. Regions are therefore mainly concerned with the success of medium-sized cities and thus with the continued prosperity of manufactures. Here the problem is that products tend to have a limited life, and that technologies change and are subject to new attractive forces. Unless a region can continue to attract the expanding industries, to compensate for the decline of its older counterparts, it will be at a disadvantage relative to others.

This principle can be illustrated by a simple example (Fig. 42.2). The central region has developed a manufacturing sector (a **leading sector**), producing the conditions for **economic take-off**. Areas nearby remain at a subsistence-urbanisation level and are

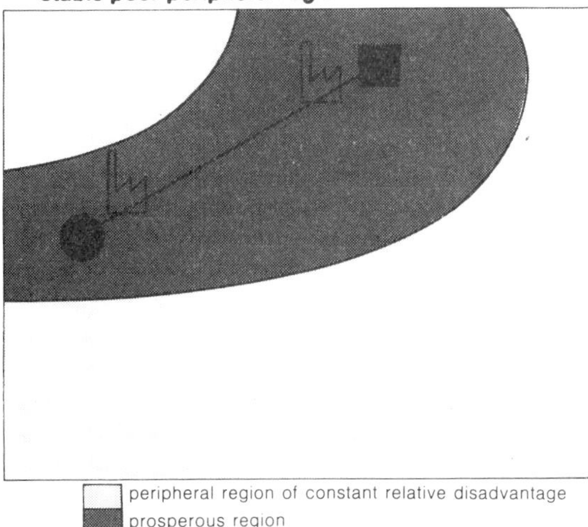

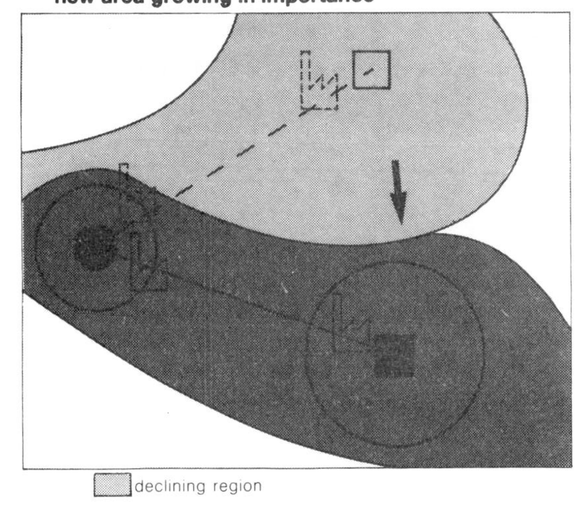

Figure 42.2 Economic advancement or decline when manufacturing activities dominate the structure of wealth production.

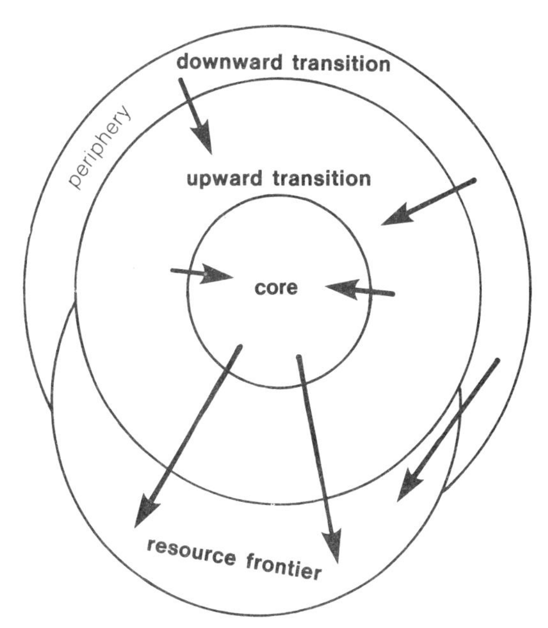

Figure 42.3 Friedmann's centre–periphery model of economic development suggests that four zones exist: (a) growth core region; (b) a region peripheral to the core but pulled along by it (upward transition); (c) a peripheral remote rural region or area containing exhausted raw materials and/or old industries undergoing downward transition; and (d) a resource frontier region on the verge of new expansion.

therefore economically at a constant relative disadvantage. As a result the area is divided into a prosperous **core region** and a disadvantaged **peripheral region**.

The exploitation of raw materials may lead to industrial development both at the source and the market (see Fig. 40.1); both therefore share in the prosperity. However, raw materials are finite and in time will become exhausted or be replaced by raw materials located elsewhere. This does not affect the market, but causes a decline in the prosperity of the raw-material source region; the region moves into a phase of **downward transition** towards a lower economic status. At the same time the new source region (the **resource frontier**) begins to prosper and attract new industry.

The core region is the most stable, assured of the longest period of growth, and may well develop quaternary and tertiary industries that make it even less vulnerable to changes in raw-material location. Areas near to the core will be pulled along by the demand from and prosperity of the core, benefiting from the need for manufactures. In this sense the zone surrounding the core shows an **upward transition** when compared with the peripheral region which, lying beyond the economic influence of the core, remains **constantly disadvantaged**.

These concepts of regional division based upon economic performance were described by Friedmann (Fig. 42.3). The pattern of growth and decline has also been examined in other ways. Myrdal, for example, considered the problem of sharing finite resources. He maintained that the success of one city or region would have to occur at the expense of others (Fig. 42.4). This is because whatever the growth sector may be, the chances are that it will be concentrated

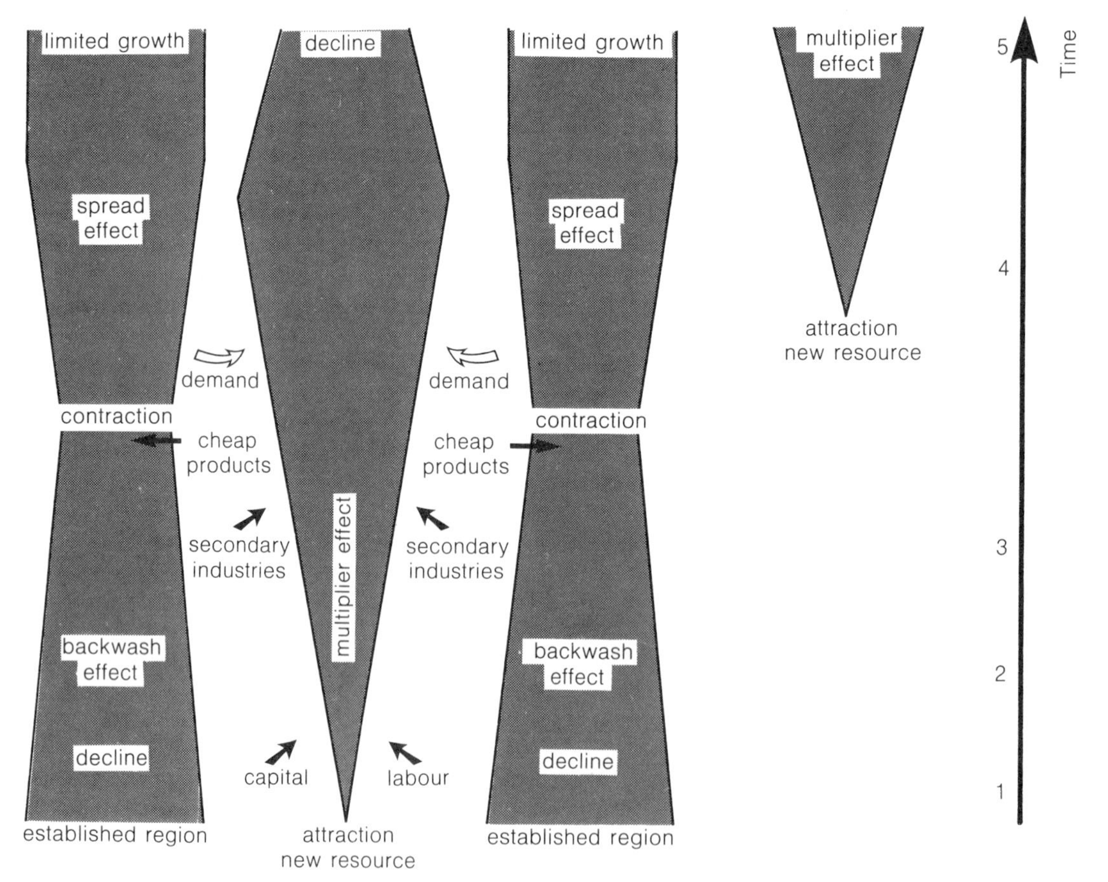

Figure 42.4 Myrdal's model of cumulative causation suggests that (1) as one region expands others must decline as the finite capital and labour go to the growing region. Growth has a multiplier effect (2), success breeding on success and attracting new industry and services. At the same time, declining areas suffer a *backwash effect* (3) as they cannot compete with the new goods and cheaper prices of the growth area. Eventually, demand from the growth area requires additional supplies from the other regions and a *spread effect* (4) leads to limited growth instead of decline. Then a new resource will begin to grow elsewhere and the process will be repeated (5).

into a small number of places and it is these that will prosper. The converse is that unless regions keep being chosen as locations of growth industries, increasingly they will become dominated by industries that are fading away or at best have finished expanding. The regions that contain these 'traditional' industries will be at a positive disadvantage.

'Knock-on effects' (**cumulative causation**) work in two ways. On the one hand growth may be self-reinforcing (the **multiplier effect**) because it brings in new capital, allowing other industries to expand or attracting new industries to the region. Growing industries generate a need for more labour, they tend to pay higher wages and this allows the workforce to buy more goods and services. A wealthier population can afford higher rates in order to provide new public facilities and a better living environment altogether. On the other hand, the knock-on effects of declining industries are also dramatic (the **backwash effect**). Declining industries result in higher unemployment, a lower ability to contribute rates, and therefore declining finance for local government. Because people have less money, they are inclined to refurbish their houses less often or less well, thereby creating less demand in dependent manufacturing and service industries.

Local government likewise will have less money to maintain public facilities. Decreased shopping turnover leads to a downturn in shopping quality. In extreme cases the region takes on a 'down-at-heel' appearance which acts positively to discourage growth industries from choosing this region as a location. Only in time may the success of a region lead to increase in wealth of disadvantaged regions (the **spread effect**).

42.3 Contrasts between regions

Regional contrasts as an aspect of development

Before manufacturing becomes important in a country the only significant contrasts between regions will be due to the relative productivity of the land in each. However, in the early phases of manufacturing, the areas of growth gain in prosperity rapidly whereas those that remain tied to agriculture lag further and further behind. Thus industrialisation increases the contrast between regions as is clearly seen, for example, between the industrialised coastal cities in Brazil and its primitive hinterland, and between the Niarobi-Mombasa industrialising 'corridor' in Kenya and the tribal agricultural regions that characterise the areas to both north and south (Fig. 42.5). As industrialisation proceeds more and more regions become involved, dragged onwards by the increasing demand of the leading region. The relative contrasts become smaller once more. Thus, although people in Britain are acutely aware of their regional contrasts, the difference is in no way as great as that between the village peasant and the urban steelworker in, say, India.

Regional character

Regions have two parallel economies: one supplying the local market and the other, the 'export' economy, supplying goods to other regions. It is the export sector that holds the key to economic success, for it along generates new wealth on which local goods

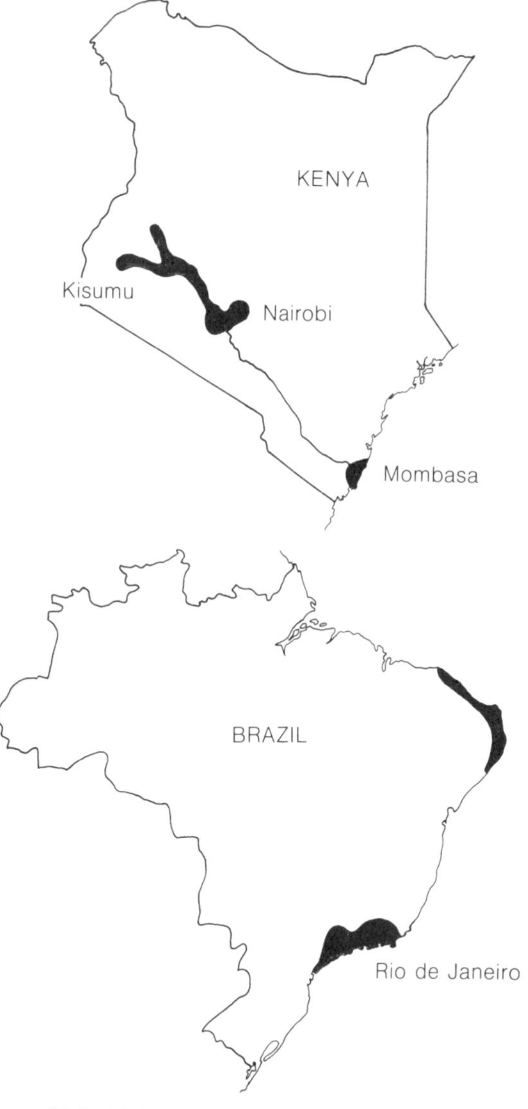

Figure 42.5 Major regional contrasts in developing countries.

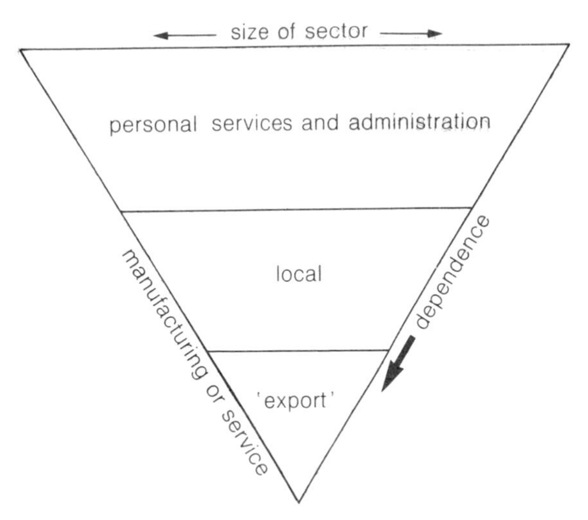

Figure 42.6 Economic sectors and their relationship. The basis of a region's growth is 'export' (bottom sector), which provides money to pay the workforce. In turn, they are able to buy their goods locally (middle sector) or commission personal services or administration (top sector). The middle and top sectors depend on the bottom sector for their prosperity.

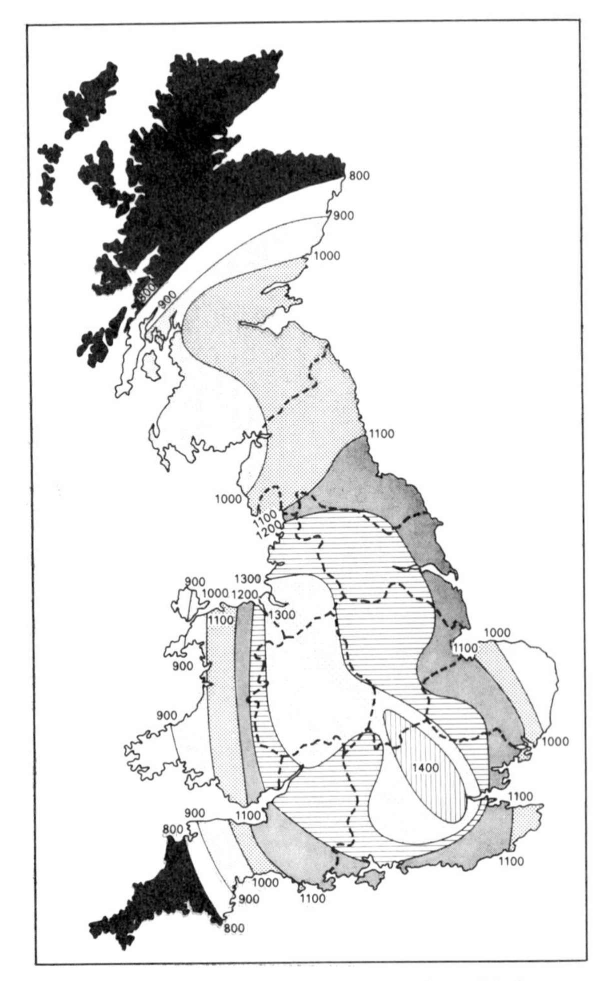

Figure 42.7 Economic potential in Great Britain.

and services depend (Fig. 42.6). Because this dependence takes the form of an inverted triangle, if the growth of the manufacturing base fails the dependent economy will also fail. In general, therefore, it is important that a region maintains the growth of its export sector in money terms, at least as fast as the national average growth. Regions that become very specialised may find this difficult if demand for their speciality wanes (as has happened in the Tyne-and-Wear region, which specialised in shipbuilding and marine engineering). The most stable regions will most probably be those of diversified structure, following the maxim of not keeping all one's eggs in one basket. The greater breadth of industry has been one of the strengths of the Midlands and of the London region.

Regional position

Unless a region has an extremely vital resource on which all others depend, there is great advantage from a marketing point of view in having a central location. With higher transport costs, regions that are peripheral to the main centre of economic activity tend to be poorer. For a peripheral region to compete in cost terms with one having a central position, the higher transport costs will have to be offset by a reduction in other costs. Because there is little opportunity to cut costs either of raw materials or of energy, the main burden of economy has to fall on wages. Thus people in peripheral areas tend to earn less and, as discussed earlier, this produces a knock-on effect. Any industries that are attracted to peripheral locations tend, in consequence, to be those offering low wage levels, a feature that further suppresses the purchasing power of the region. Moreover, there is a tendency for large firms to split their operations, keeping their offices at a central location and their manufacturing plant in the cheaper peripheral region. Such action produces further contrasts between occupations and wages of central and peripheral locations.

The 'centrality' of a region can be assessed by plotting the accessibility of each region to the total potential market (Fig. 42.7). In Britain regions often look 'inward' for their markets because the sea creates substantial barriers to trade. Yet even in continental regions, national boundaries and consequent tariff barriers restrain international trade. The removal of tariff barriers, therefore, can play a vital role in changing the fortunes of a region. For example, when the Benelux countries joined the EEC in 1957 they found themselves at the centre of an enormously expanded market, and in consequence their prosperity increased greatly.

Regional migration

There is often a strong resistance to movement from a 'home' region to another region whose people and customs are less familiar. Nevertheless, many people do move each year for a variety of reasons. The most mobile sector of the population is people in their twenties and thirties. They are the least likely to have established 'roots' , and they often need to move in order to gain promotion or to get a job at all. In general, however, the people who move tend to be those of better education and of the 'middle class', and they tend to be single men rather than single women. At the older end of the scale are people who are not seeking a job but somewhere of low economic activity, such as a peaceful seaside town.

The consequences of these differential migration patterns are important, for the younger people with greatest abilities tend to move to central regions, reinforcing the economic potential of the centre, whereas retired people move to the peripheral regions, lowering the productivity of these regions yet further. Selective migration patterns sometimes play a substantial role in creating contrasts between regions.

Figure 42.8 The Assisted Areas (areas for expansion) in Britain, as of 1 August 1982.

Ethnic and religious regions

In many countries regions can be created purely because of differences in ethnic or religious character. This happens particularly where there are strong contrasts between peoples: Czechs and Slovaks in Czechoslovakia; European and Asiatic peoples of the USSR; Flemings and Walloons in Belgium; Protestants and Catholics in Ireland; tribal regions in Africa are just some examples. In each case the origin of the discrimination is different and is largely historical in character, rather than economic.

Government reaction to regional contrasts

Governments in democratic countries are elected primarily to care for the well-being of all the people. In consequence, one of their principal aims is to try to influence the pattern of wealth created by market forces, by encouraging industry or people to move to selected, less fortunate regions. In general it is difficult to get *people* to move; in any case, the result would be to put stress on the housing stock of the area of immigration while causing the virtual depopulation of emigration areas and a breakdown of social and cultural cohesion there. Industrial enterprises are fewer than people, and more susceptible to financial considerations, so they can be more readily influenced in their choice of site.

Governments have two main ways of influencing industrial location: the 'stick' and the 'carrot'. The stick can take the form of the refusal of permission to all new industrial development in a given region (as happened to Inner London in the 1960s), or the imposition of special tariffs that make location in high-tariff areas unprofitable. For example, the 'Pittsburgh-plus' tariff system for steel in the USA was designed to protect the Pittsburgh region by charging for *all* steel freight from factory to consumer as though the steel had originated in Pittsburg. The carrot takes the form of financial incentives, such as a rebate on factory or office rents, or a subsidy for each employee taken on within a specified area; and the provision of new purpose-built factories or offices. In Britain the regions of greatest disadvantage (Fig. 42.8) are designated **Assisted Areas**. Within eleven of them lie

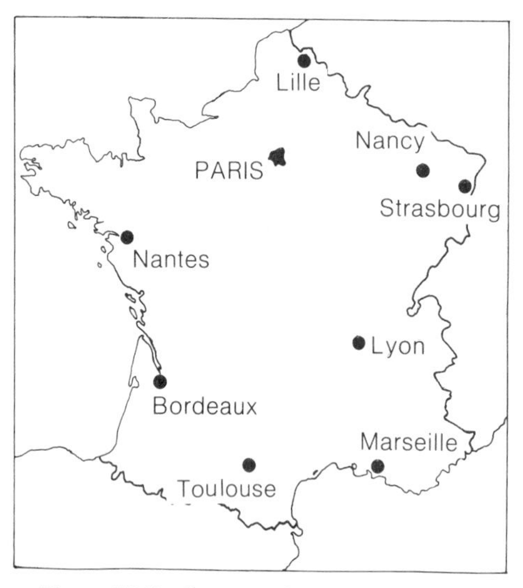

Figure 42.9 France: métropoles d'équilibre.

further advantageous areas called **Free-Enterprise Zones**. Special concessions include:

(a) 10-year exemption from rates;
(b) larger tax allowances;
(c) exemption from development land tax;
(d) priority treatment for customs warehousing and relief from customs charges for goods imported to be processed and then exported;
(e) abolition of requirements for Industrial Development Certificates.

Thus, for example, the derelict 190 ha London Docklands at the Isle of Dogs has been declared an Enterprise Zone to help attract industry to a particularly hard-hit section of London's inner city in which the workforce in the docks has dropped from 26 000 in 1967 to zero today, and in which nearby manufacturing employment has fallen by about 30 per cent.

Other countries have somewhat different arrangements designed to suit their specific circumstances. In France, where overcentralisation is the main problem and all regions beyond the Paris Basin are peripheral regions, attempts have been made to stimulate growth in peripheral cities (*métropoles d'équilibre*; Fig. 42.9). However, in all cases the cost of government measures aimed at evening out the regional differences is high. In Europe, for example, it is often between one and two per cent of the GNP.

42.4 Regional change

Regional decline

NORTHERN BRITAIN

Movements of resource-frontier regions have been critical to the regional development of Britain. In medieval times the chief source of textile manufacture was East Anglia (Fig. 42.10). At the start of the Industrial Revolution the fast-flowing streams of the Pennines became the new dominant resource, the valley-head region of the Pennines became a resource frontier and East Anglia experienced downward transition. Later, when steam replaced water power, the nearby coalfields, together with those in the Midlands, South Wales and Scotland developed as resource frontiers and the Pennine valleys suffered downward transition. Throughout this period London remained an expanding market and core region.

When direct steam power was replaced by electricity and industry was able to diversify its location, much new technological industry developed at the major markets and thereby reinforced the success of the core. At the same time competition from abroad led to the decline in overseas markets (especially the Commonwealth) and to the reorientation of the major

Table 42.1 Shift-share analysis for employment in the north-west region of England, compared with Great Britain 1971–7 (source: Lloyd & Reeve).

Type of area	Total shift number	(%)	Proportional shift number	(%)	Differential shift number	(%)
(a) inner regions of conurbations	−84 000	−76	10 000	76	−94 000	−93
(b) large towns dominated by manufacturing	−8 000	−7	−27 000	−71	20 000	31
(c) large towns dominated by services	35 000	75	3 000	22	31 000	51
(d) small towns dominated by manufacturing	−19 000	−17	−11 000	−30	−7 000	−7
(e) small towns dominated by services	12 000	25	400	3	11 000	18

Total shift reflects deviations from national trends; they are negative for inner conurbations and manufacturing centres and are positive for those places with a developed services sector.
Differential shift reflects performance regardless of structure of employment.
Proportional shift reflects changes in structure; the positive inner city results reflect the importance of the CBD functions.

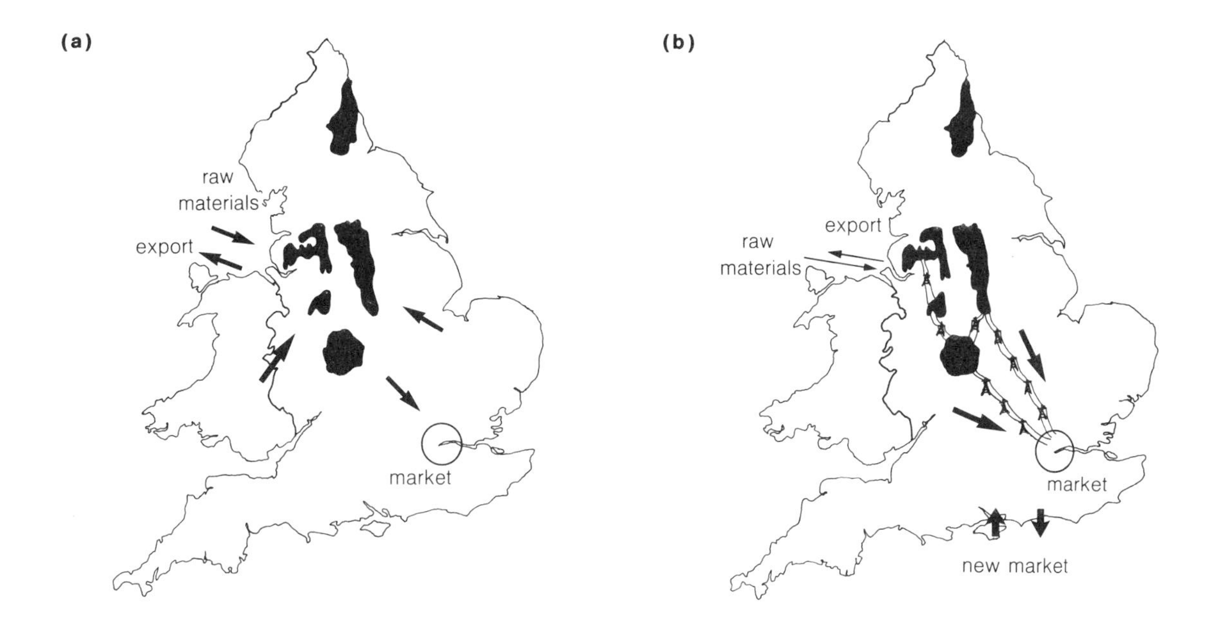

(c)

Figure 42.10 (a) The introduction of industrialisation: the decline of traditional textile areas and concentration on coalfields. Constantly disadvantaged peripheral regions, Wales, south-west England; downward transition region, East Anglia; resource frontier regions, Lancashire and West Yorkshire. (b) Changing raw material base: industry less coal-dependent after the advent of electricity, and raw materials increasingly imported from abroad. Trade focuses to Europe rather than the whole world. Constantly disadvantaged peripheral regions, Wales, south-west England; downward transition regions, Lancashire, West Yorkshire; resource frontier region, south-east England. (c) Dismantling a steel rolling mill, West Midlands: the sad face of downward transition.

market towards Europe. This led to urban areas on the coalfields experiencing downward transition (see Table 42.1). At the same time the 'Home Counties', previously at a constant relative disadvantage along with the remainder of rural Britain, began to develop as an upward transition region, forming a service and manufacturing base using electric power and the increasing flexibility of motor transport to distribute goods. Multiplier effects subsequently led to further growth in the South-East, with manufactures being sent back to the North and creating a backwash effect.

Within the past ten years the situation has changed once more. Growth is no longer concentrated in inner cities but in non-metropolitan regions. The Home Counties are no longer upward-transition but resource-frontier regions, gaining high-technology industry and using a pleasant environment and good international communications as a resource. The spread effect is also pushing northwards, for although much research and development takes place in the resource-frontier region, much mass-production manufacture is transferred to the downward-transition regions of the North where labour costs are lower. In its wake has moved the outward wave of office employment, at present undergoing a major locational change from CBD to peripheral parts of the core.

NORTHERN NORWAY

Norway's regional development is critically determined by its physical geography. The core region, Oslo, is located several hundred kilometres from the

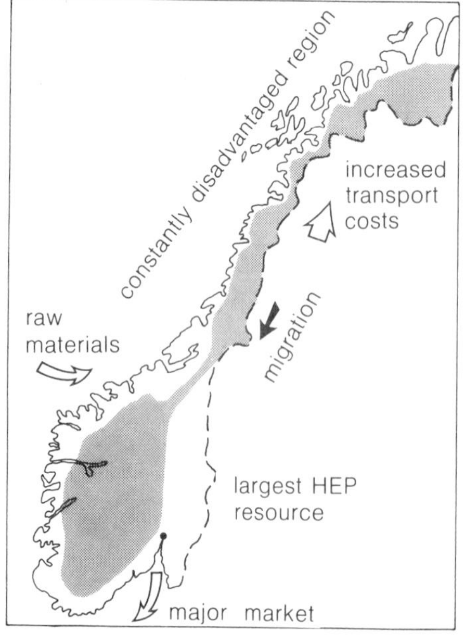

Figure 42.11 Regional disadvantage in Norway.

upward-transition regions of Bergen and Trondheim, linked to them only by the umbilical cords of rail and road (Fig. 42.11). Resource-frontier regions are dotted along the coast, using HEP as a basis for aluminium smelting or similar basic manufactures, or relying on the impact of North Sea oil.

The main markets for Norwegian goods lie to the south, in the EEC. But even before industrialisation and exploitation of external markets there were regional differences between North and South, because the agricultural base of the North is so limited by climate. This has simply been reinforced by the development of HEP, whose potential is far greater in the South than the North. The North is therefore a peripheral region of constant (or even increasing) disadvantage.

Higher transport costs will work against the development of northern Norway. They are especially acute, however, because of the difficult terrain. The only way this area can compete is by compensating for its high transport costs by offering low wages. Much of the younger, most mobile, section of the population have migrated to the South in search of higher wages and more varied social life. This leaves the North further disadvantaged: it removes part of the workforce and makes the area even less attractive to prospective industry.

The government has tried to balance out some of the disparity – for example, it has set up a subsidised steelworks at Mo-I-Rana on the Arctic Circle which uses local iron ore there instead of shipping it to a southern steelworks. However, steelworking is a capital-intensive industry and therefore not a good regional investment in areas with high local unemployment. Circumstances work against the growth of tertiary activities and even associated manufacturing industry because of the peripheral location. For example, aluminium, even after refining, is a 'raw material'. Goods made from aluminium (such as drink containers) are usually bulky and costly to transport. They usually therefore need to have a market location.

The only resource the North can offer is its distinctive environment and particularly its 'midnight sun'. To help tourism make the North a successful resource-frontier region, the Norwegian government has been actively improving accessibility, replacing the fjordland ferries with bridges wherever practicable and reducing the isolation of the Arctic regions.

Success in particular regions

CASE STUDY: THE THAMES VALLEY

Whereas many other parts of Britain have been suffering decline, the area between Slough and Reading

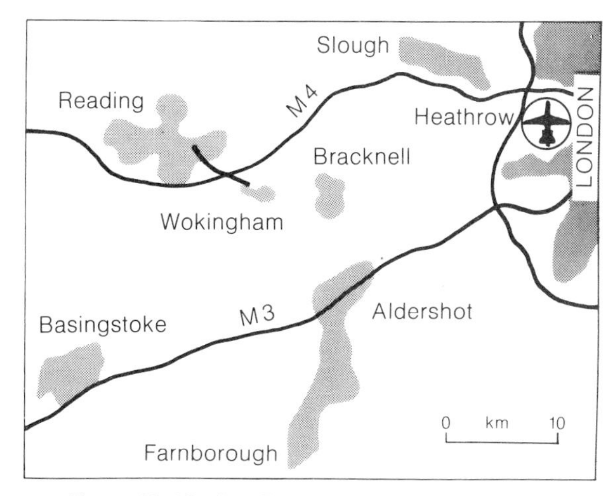

Figure 42.12 The Thames Valley growth region.

to the west of London has been experiencing spectacular growth (Fig. 42.12). Here growth has been led by the high-technology electronics-based industries, the high level of accessibility afforded by rail and road systems, and proximity to a major international airport. It is particularly interesting because expansion is occurring at a time when nearby London is experiencing considerable difficulties in attracting new industry. Indeed, Reading, the regional capital of Britain's 'silicon valley', has one of the fastest growing needs for office space in the country. One of the major resources that the Thames Valley has to sell is its environment – pleasant countryside and lack of major-city congestion; yet it is close enough to the city to make use of central functions.

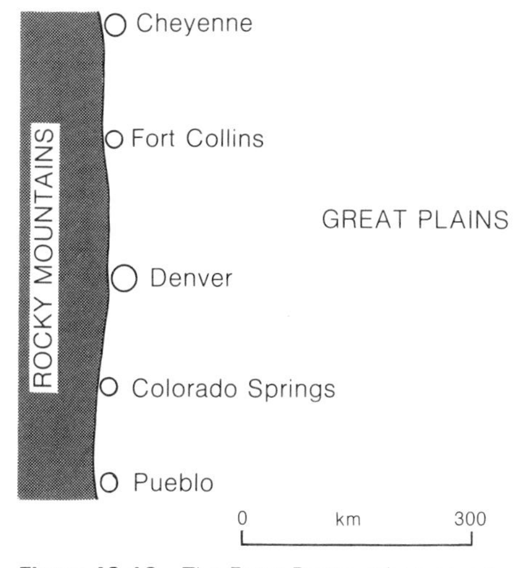

Figure 42.13 The Front Range urban corridor.

CASE STUDY: THE FRONT-RANGE URBAN CORRIDOR

The Front Range Urban Corridor is an area of central USA, immediately to the east of the Front Range of the Rocky Mountains (Fig. 42.13). Here, in a semi-desert environment, a string of isolated towns centred on Denver, Colorado, is experiencing rapid growth far from any established urban centre. Like the Thames Valley in England, the Front Range has developed electronics and information-technology industries. Midway between the major urban centres of the east and west coasts of the USA, the area is proving attractive; not so much to manufacturing as to research-and-development activities of large multi-plant firms. Again, one of the resources that has attracted firms has been the environment for its skilled personnel, with clear, invigorating air, and the possibility of recreation in the mountains nearby.

CASE STUDY: MARSEILLE-FOS

The region around Fos-Lavera just to the west of Marseille on the mediterranean coast of France was nicknamed 'Europort of the South' because its rate of growth was three times the European average in the early 1970s. It has developed, like Reading and Denver, in the post-industrial era when the location of industry, instead of being controlled by the location of raw materials, is primarily influenced by the tastes of the workforce – and who would not choose the French Riviera to live if offered the choice?

Fos contains a natural deep-water harbour on a coast that has no tidal problems; supertankers can arrive at any time, not only at high tide as is the case with North Sea ports. However, the area does not have a naturally developed service hinterland and so, when the port was begun in 1962, this was a priority for development.

Fos was chosen as a centre for heavy industry – a new, more convenient site at which to import the raw materials for steelmaking and petrochemicals at a time when industry was growing rapidly. This is quite different in concept from the growth at Reading or Denver. And Fos also has a more peripheral location: on two counts it is therefore a much higher risk than the others. The decision was also partly political, for Fos was planned to help reduce the depopulation of the South. However, since the oil crisis of 1973 things have not gone well; with established plant having more capacity than the world needs, the world steel and chemical industry does not need new plant. Fos has been an industrial development in a post-industrial society.

42.5 Core–periphery relationships

Europe

Mapping Europe using a number of criteria that might be regarded as indicators of economic progress (agricultural intensity; economic potential; major urban centres) provides a consistent pattern: an egg-shaped core of high activity surrounded by the peripheral regions of southern and eastern Europe and by the mountainous west and north (Fig. 42.14).

Within the core, national centres of industrial activity have no directional orientation, yet in the peripheral regions the national centres consistently show an orientation towards the core. This points to a long-term stability of the core-periphery relationship and suggests that the concentrations of capital and political power in the core and of the low-skill labour supply in the periphery are self-perpetuating processes which will be difficult to reverse. Even EEC aid programmes are but a drop in the ocean in terms of redressing the imbalance.

Two of the main indicators of the core-periphery relationship are migratory movements and tourism. Migrants move towards the core in search of work. In doing this they help to feed the prosperity of the core for, unless the core is prospering, there will be no need for migrant workers. Similarly, tourists move from the crowded core to the more pleasant environments of the periphery. However, the tourist revenues on which the periphery depends are closely connected to the economic success of the core, for without surplus wealth there would be little possibility for tourism.

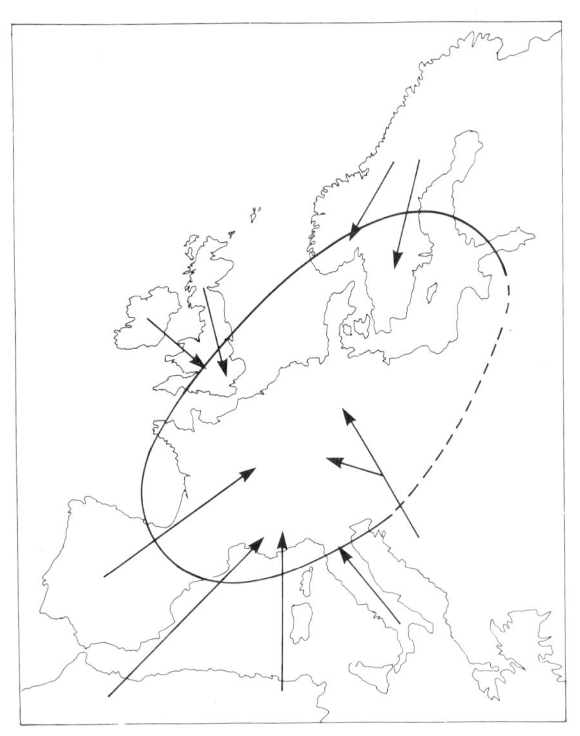

Figure 42.14 Core–periphery in Europe. Arrows indicate main post-war migration paths; the shaded area is the egg-shaped core of Europe.

The core-periphery relationship is further accentuated by the multi-national firms which are based in the core and which use the periphery as a region of low-wage production, thus tending to hold down the wealth of the periphery.

It has been suggested that peripheral regions

(a) lack control over their own resources;
(b) are deficient in innovation and can therefore not develop new industries independent of the core;
(c) have weak internal linkages and communications;
(d) are more vulnerable to economic change than the core, being the first place in which plant closes and wealth declines, and from which workers return home.

If all of these assertions are true, Europe's core-periphery pattern will be difficult to change.

The world as a whole

With the great improvement in communication systems, the world has effectively become a single system, consisting of two parts: one part – the core region – comprises the developed countries of Europe and North America, Japan, South Africa and Oceania; the other part – the peripheral region – comprises the 'Third World' of Latin America, most of Africa and most of Asia. Although such a simple two-part division glosses over large differences in wealth *within* each region, it is true that the peripheral region as a whole depends on continued trade with the core countries.

It has been argued that the specialisations in industrial goods in the core and in the production of primary commodities in the periphery will, through international trade, maximise production and consumption and benefit both core and periphery. In practice, instead of the benefits being shared equally, there has been a continued system of unequal exchange, with the peripheral countries having a rate of growth far below that of the core. The domination of the periphery by the core, on a global scale, is usually thought to be due to the lower bargaining power of the Third World countries. Today such dominance is largely focused, perhaps to too great a degree, on the multinational corporations, which are seen as a substitute for colonial trading companies. Under increasing pressure from the Third World countries, they have sharply increased their rate of investment in manufacturing in the periphery, rather than simply extracting raw materials there. Furthermore, the developing countries have begun to see that they cannot become developed while remaining excessively dependent on others. Global core-periphery relationships (sometimes called the **North-South debate**) form an extremely important but complex topic, and can only be introduced in this section. Aspects of the problem are discussed in detail in other chapters.

The developing world

The commercial development of southern Asia is often looked upon as the product of European imperialism. It is pointed out that many of the most important industrial and commercial regions of southern Asia (the core regions) have coastal locations because they are the entrepots for goods destined to be consumed in the 'primary' core region of Europe (Fig. 42.15). In this sense both the cores and the communications networks that linked the countries were designed both for easy export of agricultural surpluses and the import of manufactured products from Europe. Nevertheless, a considerable proportion of the commercial and industrial trade of these Asian cores is, and always has been, directed at internal rather than external markets. Indeed, southern Asia had a well developed and ancient network of urban centres before the first Europeans arrived. Much of the internal revenue provided an infrastructure that allowed the development of many large-scale industries, such as coal-mining, iron and steel, and transport engineering. On the other hand, imperialism largely preserved the traditional village-based rural structure. As a result, two types of core region grew, the one consisting primarily of commercial coastal centres (Calcutta, Madras, Bombay and so on), the other of largely administrative centres set in a rural landscape (Delhi, Hyderabad and the like).

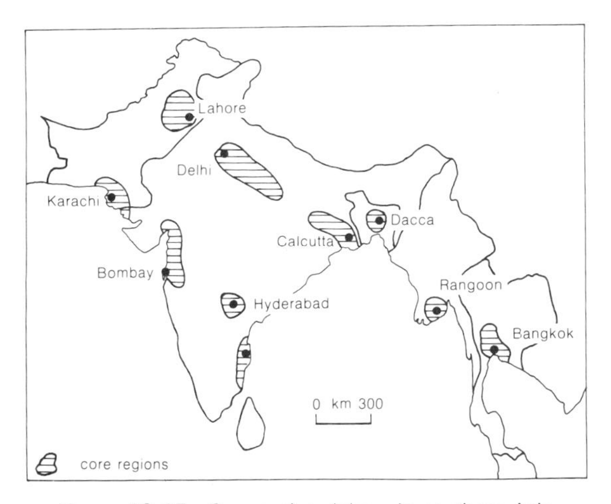

Figure 42.15 Core and periphery in southern Asia.

Post-independence has seen a division of southern Asia which left rural East Pakistan (now Bangladesh) divorced from its natural core of Calcutta, forcing it to establish and develop a new core region around Dacca.

All governments have attempted to redress the imbalance between core and periphery by instigating large development programmes, such as irrigation of the peripheral dry-land zones of the Indian and Pakistani Punjab, or the north-west of Sri Lanka, but in general the established core regions have forged ahead still faster. Some 70 per cent of the total industrial employment of India, for example, is concentrated in six states, and half the total is concentrated in the states of West Bengal (core, Calcutta) and Maharastra (core, Bombay). This suggests that the southern Asian countries are still in an early stage of economic growth, with core and peripheral areas at near maximum inequality. It will take a long time before the economies mature and regional disparities are significantly reduced.

Two contrasting examples

KUWAIT: A RESOURCE FRONTIER

A thousand years ago the Middle East was a major centre of learning and science. Yet by the 20th century world development had left much of the Arab world as little more than a backwater region of farmers and fishermen. For example, only a generation ago, Kuwait was a small seaport on the Arabian Gulf coast, surrounded by desert and best known for its pearl fisheries. Today Kuwait has moved from being in the periphery to being a resource frontier, becoming one of the world's major oil producers (Fig. 42.16). In the 1940s only 100 000 people lived in Kuwait; today there are 1.5 million. As a result they have had to telescope development into a very short time. However, Kuwaitis are protected by a lavish social system and by subsidised housing. The large revenues from oil have been invested abroad, and the interest received now helps to form a more stable, long-term base for development.

Kuwait is a place of opportunity. Palestinians, Egyptians, Lebanese and Iraqis; Indians, Philippinos and Koreans come to Kuwait as waiters, domestic servants and construction workers; Americans and Europeans come as architects and engineers. Kuwait has thus become a cosmopolitan centre where 60 per cent of the population are not Kuwaitis. The oil money has also allowed Kuwait to participate in experiments to improve its agriculture and manufacturing base. Kuwait is now trying to move from a

Figure 42.16 The landscape of a resource frontier: Kuwait.

purely resource-frontier situation dependent on only one resource, to a more broadly based economy, so that it can eventually join the stable core.

TANZANIA: A REGION OF CONSTANT DISADVANTAGE

In the developing world it is often not possible to identify small and specific disadvantaged areas within a country in the same way that is possible in core regions. This is because the vast majority of people live in rural areas, as subsistence farmers with a very low income. For example, in the sub-Saharan zone of Africa, over 90 per cent of the 310 million people could be classed as disadvantaged, most of them having an annual income of less than £150 each. Indeed, many developing-world countries have experienced negative growth rates in gross national product. Regional development here demands nothing less than planning how to transform the national situation.

Much of the problem stems from land fragmentation, which allows a farmer too little land and forces him to seek to rent more, often at extortionate prices. At the same time he is forced into a cash economy to pay taxes and to buy fertiliser, clothes and so on without the means to guarantee the cash. The result is extensive borrowing and rural indebtedness. Faced with a virtually hopeless rural situation, many young men turn to cities as a means of gaining a living, but this in turn makes rural labour both scarce and expensive. Under these circumstances even the growing population and increasing demand for food do not allow farmers to produce more, for they lack not only the money to invest but the workforce to help.

It has been common for governments faced with this problem to accept monetary aid from abroad to achieve 'integrated rural development' or the 'green revolution' in the hope that, by throwing money at the problem, it will go away. In this manner improvements to infrastructure (such as building rural roads, constructing irrigation dams, and establishing research stations and marketing organisations) take the place of the more fundamental need to restructure the rural society. Improvements to infrastructure are vital but they can only be of long-term benefit if social engineering precedes or accompanies them.

Rural development is not helped by the national bureaucracy, an important employer whose function emphasises repetition and reiteration rather than innovation. In particular, most governments are opposed to devolution to the regions because they see this as destabilising. Rural development therefore needs:

(a) *land reforms*, to change the land rights and to consolidate the land into plots that are economically viable;
(b) *agrarian reform*, to organise an efficient system for the delivery of new seeds, fertilisers, pesticides, machinery and technical aid, credit and to provide a stable market for agricultural produce;
(c) *infrastructure development*, to build new roads, add irrigation and drainage wells and boreholes and to replan and resettle the rural population in a way best adapted to the land capability and their needs;
(d) *social amenities*, such as schools and clinics;
(e) *institutional development*, to create integrated ministries that can provide finance, technical assistance and arrange marketing;
(f) *political participation*, to help the rural people to see the benefits of change and take a part in decision-making.

At present only a quarter to a third of most developing-world budgets go to rural development, despite the fact that the overwhelming proportion of the population live in the countryside. Many governments adopt a purely investment approach, and pay little heed to social engineering. One reason for this is that it enables them to blur over ethnic or social problems, especially as they may have no popular power base to attempt more wide-ranging change.

One of the few governments that has taken the bold step of attempting some form of social engineering in order to improve its society is Tanzania. Here the government decided that the only way to improve living standards would be to remove people from their dispersed settlements and to regroup them into villages (Fig. 43.9). The intention was that people would work collectively on the villages' communal lands and be paid according to their needs and their labour contribution, a truly communist system. This process of 'villagisation' (called **ujaama**) was designed to prevent irregularities in rural income, to provide service centres and trade, and to create foci for rural industries (in order to control the drift of labour into the incapacitated urban system). At the same time villages were to promote political and economic integration of the ethnically diverse population. A hierarchy of villages was therefore established whose pattern more closely matches Central Place Theory than rural-settlement patterns in the developed world.

When the plan started, Tanzania's rural population made up 94 per cent of the country's total, but the country possessed only 48 villages each with over 5000 people (0.4 per cent of the population). Today everyone lives in villages. Nevertheless, this has by no means solved the problems. For example, the Ilekanilo village in Mwanza province now owns 3325 ha of land. It needed 2412 ha to support the original population, leaving only sufficient spare land (at present used for fallow) to provide food for the population growth within eight years. The government has argued that to improve yields and prevent nutritional standards from falling a communal effort is needed. However, communal farming here, as in China, has not led to the greatest efficiency. At present each family needs 3.2 ha/yr to grow 1.25 ha of grain and 0.7 ha of roots (cassava and yams) for subsistence and 1.3 ha of cotton to provide cash flow. At the same time livestock continues to be an important element in the farming system but there is less and less land for it. Clearly further important decisions will have to be made quickly – only time will tell whether 'villagisation' has helped.

42.6 Assessing regional advantage

In order to assess the character of a region, some quantitative measures of its performance need to be established. It is especially vital to separate the internal and 'export' employment sectors, for the export sector is the only one that can bring new wealth into the area. The proportion of each employment sector can be obtained by calculating the **location quotient** (LQ) indicating the degree to which a given industrial group is of greater or lesser importance locally rather than nationally. The LQ is obtained by dividing the

Table 42.2 Location quotients for a hypothetical region.

	LQ 1970	LQ 1980	LQ change
mining	0.1	0.1	0.0
manufacturing including			
primary metals	0.9	1.0	+0.1
machinery	1.2	1.3	+0.1
electrical	1.7	1.8	+0.1
transport equipment	0.8	0.6	−0.2
food	1.0	1.1	+0.1
textiles	1.2	1.2	0.0
chemicals	2.7	2.6	−0.1
transport	0.9	1.1	+0.2
wholesale and retail	1.0	1.0	0.0
finance	0.9	0.9	0.0
services	1.0	1.0	0.0
government	0.7	0.7	0.0

percentage of local employment in a given industry by the percentage nationally in the same industry. It provides a good indication of those industries that underpin the regional economy: an LQ of about 1 is average, implying that the area is simply self-sufficient (subsistence); an LQ greater than 1 indicates an export; an LQ less than 1 indicates that the region must import this category. It is only those industries that have LQs well above 1 that bring wealth into the region. From this index regional strengths can be assessed (Table 42.2).

These figures show the region to be dominated by chemicals and electrical machinery and being relatively under-endowed with office-type jobs, a situation which has changed little over the ten years (1970-80).

Further evidence of trends can be obtained by **shift-share analysis**, whereby the local employment in each industrial sector is compared to the national norms over a number of years. Thus the **regional share** (RS) trend might be:

RS (1970–80) =

$$\left(\frac{\text{regional employment 1980}}{\text{regional employment 1970}}\right) - \left(\frac{\text{national employment 1980}}{\text{national employment 1970}}\right)$$

If this yields a positive figure, the region is growing; while a negative figure shows a decline (an upward- or downward-transition region). The **structural shift** is the change expected because of the type of industrial structure within the region; the **locational shift** is the change that cannot be explained by regional share or structural shift. Areas with positive values are growing, whereas those with negative values are declining.

Thus the North-West of England shows a heavy net loss of employment in manufacturing both in inner cities and in large and small towns that are dominated by manufacturing and which have old, relatively inaccessible sites (Table 42.1). However, inner cities (−93 per cent) fared worse than large or small towns (−31 per cent; −7 per cent). Only the free-standing towns without a dominant manufacturing sector have captured the little net growth available in this recession period (+51 per cent).

Part K

Settlement patterns

Rural settlement

43.1 Changing patterns

Inevitable changes will occur in the distribution of people as new forms of employment open up and old ones decline. In Europe nearly two centuries of industrial revolution has seen an enormous swing from country to city. Among the *push factors* that have driven people from the country have been:

(a) an absolute decline in the numbers required to farm the land;
(b) an absolute increase in the rural population due to lower mortality rates;
(c) a change from rural ('cottage') to urban ('factory') industry.

In addition there have been *pull factors* attracting people to cities, such as;

(a) increasing employment;
(b) better working conditions and pay.

In Europe these changes have occurred at different times: in England over 75 per cent of the population were already living in urban areas by 1900 in contrast to only 25 per cent in 1800; in France, however, agriculture was still the major employer until 1950.

Some shifts of population have been particularly dramatic. For example between 1951 and 1956 the population of Milan increased by an incredible 1.5 million, and of this 70 per cent came from rural southern Italy. Changes such as these make even greater the contrast between the thriving cities contained within the 'Golden Triangle' at the corners of which lie Birmingham, Dortmund and Milan, and the stagnating remote rural areas on the periphery of Europe. The outlying areas today are almost reduced to the level of population reservoirs for expanding core areas (Fig. 43.1). As a result, areas of severe depopulation such as northern Norway, western Ireland and northern Scotland already have difficulty in maintaining viable communities and services; they now have little chance of economic recovery because they no longer have a population base. Efforts are being made to slow down depopulation of some areas (such as Sicily) before they suffer the same fate. Because most of the people leaving the farms are young, those remaining tend to be old. Marriage and birth rates in the areas left remain low and the process of population decline accelerates. It is a trend which, if it continued, would render large areas of Europe virtually uninhabited. But there is another side to the coin: with increased leisure, more urban people want to visit the countryside for recreation. These are the people who buy holiday homes in remote villages and then leave them unoccupied for much of the year, adding further to local stress. Even in more centrally located areas, what were once farming villages have become dominated by commuters. In the long term the area available for agricultural land will decline and the area used for urban expansion, recreation and woodland will increase. It is a trend that will come hard to the more traditional sections of the rural community.

The trend for rural urban migration is also gathering momentum in the developing world, where, at present, there is still a preponderance of people in the countryside. Here the stress on land is becoming even greater than in the developed world, despite low levels of mechanisation. Better communications are opening up new horizons to rural dwellers, increasing their perception of the wider world and sharpening their dissatisfaction with their own circumstances. Better roads have also brought a new commercial structure to diversify the opportunities for buying and selling goods, although commercial and near-subsistence activities are still very different.

Although there is a tendency for urban people to treat population movements of the kind outlined above as though unusual, there have always been changes taking place in rural settlement. Populations in the countryside are not fossilised, but continually

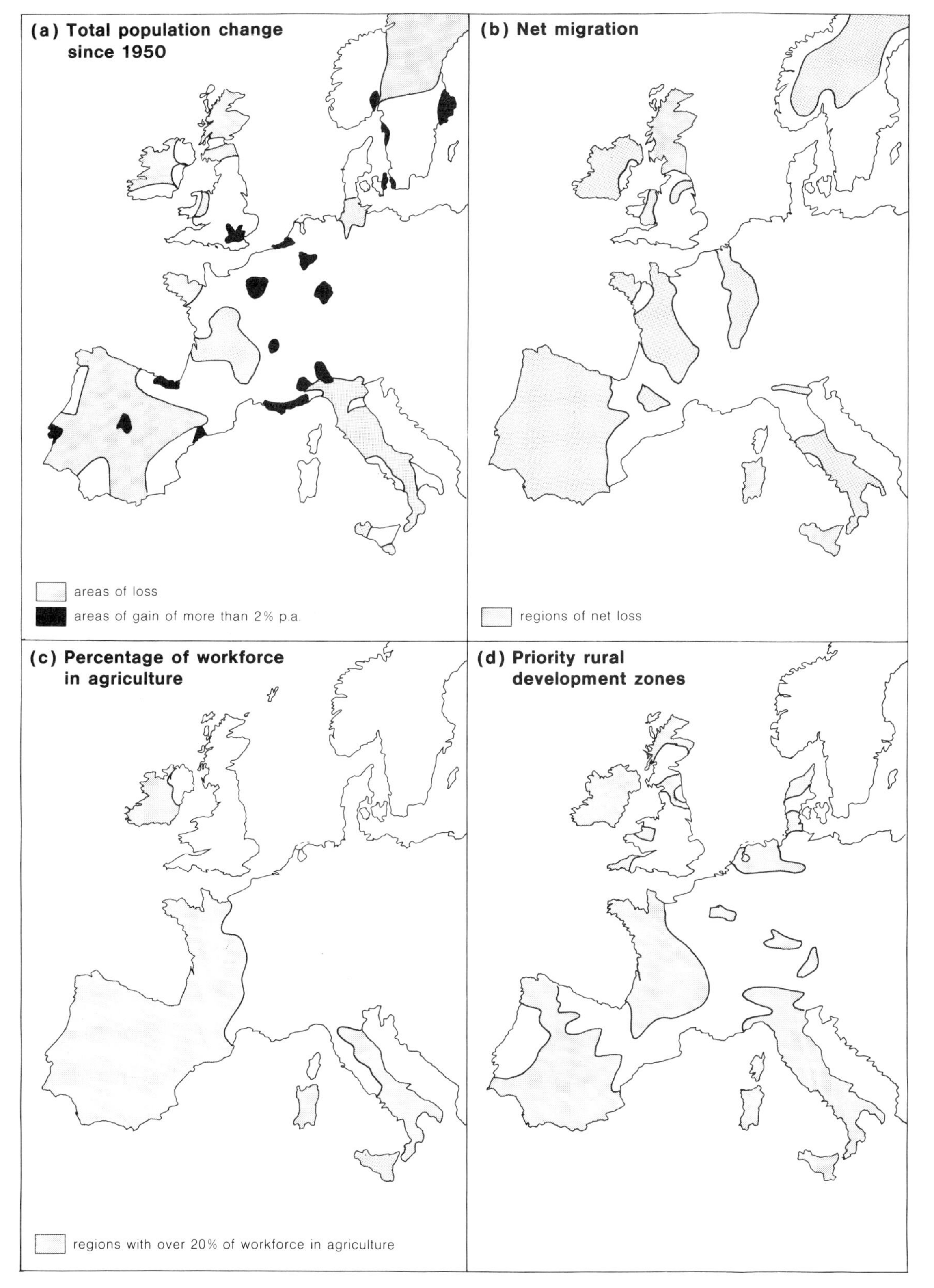

Figure 43.1 Problem regions of rural Europe can be identified by changes in population and degree of migration. Many such areas have been classified as priority rural development zones.

changing in response to internal and external stress. In the following sections some of the factors that have influenced the development of settlement patterns are outlined.

43.2 The basis of rural settlement

A rural **settlement** is a place where a group of people live more or less permanently, allowing them to share problems of shelter, defence, cultivation and services as well as to enjoy social activities.

The exact form taken by any settlement or pattern of settlements depends on the environmental and cultural factors and the level of technology available. For example, because of good communications (road, post and telephone), farmers in Britain and North America are able to live in farmhouses isolated from one another while still feeling part of, and being supported by, a form of 'scattered' or **dispersed** settlement (Fig. 43.2). On the other hand, where communications are poor and technology is primitive, people often have to live closer together. This is a principal cause of **nucleation**, whether it be as hamlet, village or town (Fig. 43.3).

All forms of rural settlement are concerned with the continuing need to provide food, drink, clothing and shelter from the environment. The ease with which these necessities can be obtained depends on three control factors:

(a) the level of population;
(b) the actual area exploited at any one time;
(c) the nature of the rural economy (farming techniques and the level of technology).

The interaction can be summarised as in Figure 43.4.

With low levels of communications and the absence of industrial influences, there is always a simple relationship based on local supply and demand. When a subsistence-orientated population evolves until it has used all of the available land, it reaches a threshold which it must cross if it is not to suffer an increasing death rate. This change involves both improvements in technology and a more commercial approach to farming, such as only growing a limited range of crops. However, the land can only be worked for the crop it will grow best if communications are well enough developed to allow the inter-regional transfer of products (trade; p. 000): the farmers in a given region will grow cabbages to the exclusion of other crops only if they can sell their crop to a known market and obtain other foodstuffs from distant

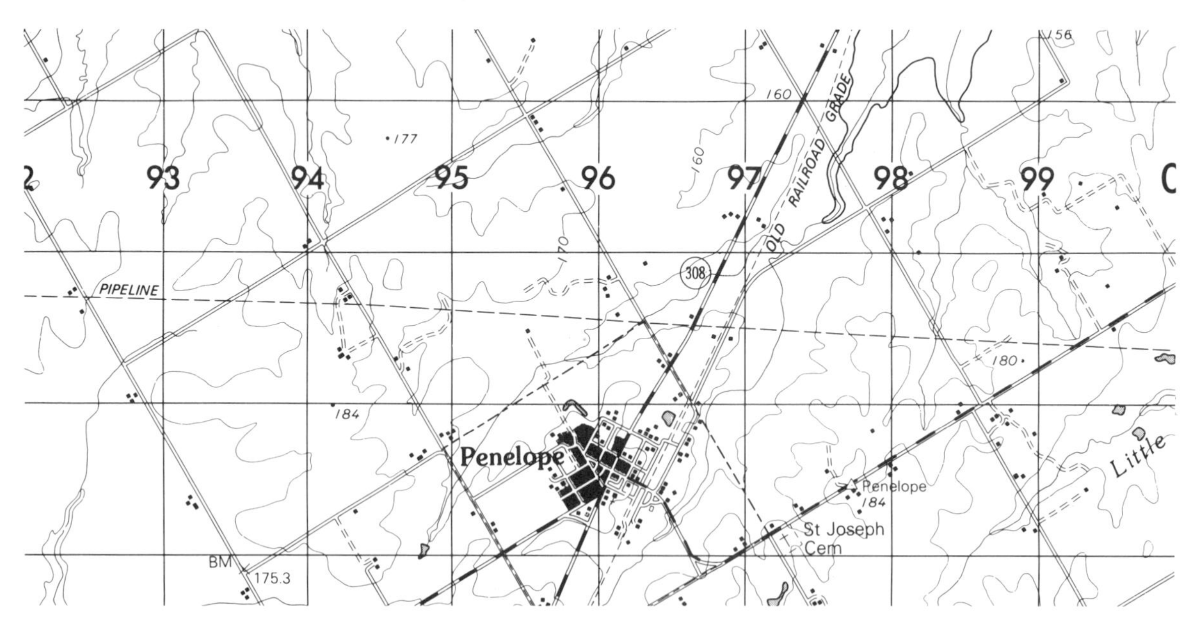

Figure 43.2 Penelope, near Hubbard, Texas, USA. The great sweep of the central plateau country that connects the Rocky Mountains to the Appalachians stretches down into Hubbard in Texas. This semi-arid region is under extensive ranching with some cropland, and was colonised at a relatively late stage. Most roads have been laid out on a grid-iron system, not only in the villages, but also across wide tracts of country; only the road that follows an old railroad track stands apart. Village separation is often over 20 km, a service network that would collapse without cheap gasoline and extensive use of the motor car. However, although roads have been laid out, there has been no attempt to plan the settlement pattern, villages and townships responding to natural, rather than planned forces. Outside the township, each isolated farm is shown as a small cluster of buildings spaced widely along the roads.

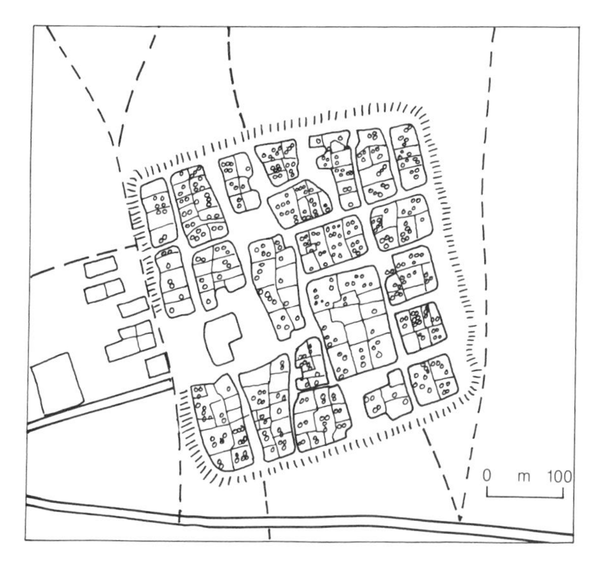

Figure 43.3 Ungogo, near Kano, northern Nigeria. Ungogo is a nucleated settlement with a population of about 1800. The surrounding land was, until recently, communally farmed and therefore a central location for all the community was a sensible choice. Nevertheless the overriding consideration in settlement type has been *defence*. Within the town wall the small space is divided into family compounds each with a frontage on a path. With the arrival of colonial stability, defence was no longer needed. The town wall is now a ruin, and expansion is occurring beyond it.

regions in turn. This is a fundamental distinction which causes major contrasts between many rural populations.

The settlement pattern near Yaoundé in the Camerouns of Equatorial Africa can be regarded as an example of settlement in the **pioneer stage** (Fig. 43.5). The equatorial rainforest provides one of the most difficult areas for both settlement and communications; in the Camerouns at least, it remains largely untouched. Here, even in this relatively accessible region near the capital city, Yaoundé, settlement comprises no more than clusters of huts spread along the dirt roads. Subsistence agriculture and low fertility of the land result in a small trading potential,

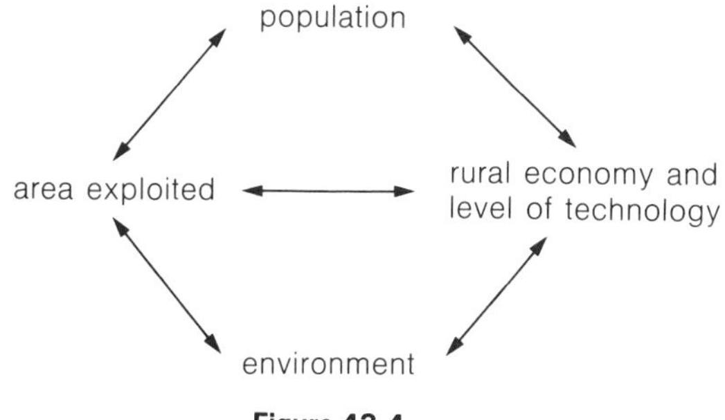

Figure 43.4

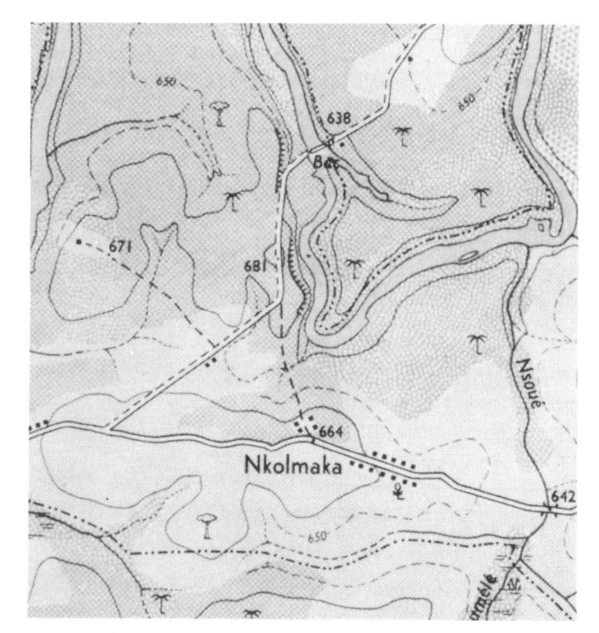

Figure 43.5 Yaoundé region, central Cameroon, Africa.

and none but the most rudimentary services are offered.

By contrast, Yala in the west of Kenya lies in the most fertile part of the East African savanna. It is therefore able to support a large rural population. However, the people still rely on subsistence cultivation from small plots, and they do not partake of significant external trade. On the map (Fig. 43.6) each

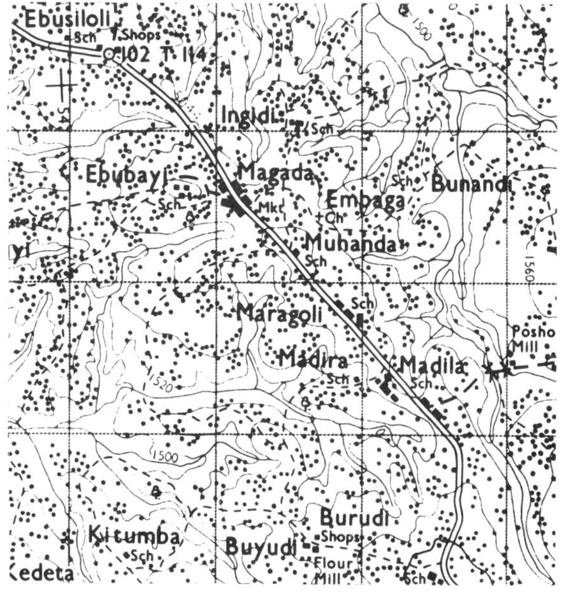

Figure 43.6 Yala area, north of Lake Victoria, Kenya.

dot represents a family hut. Settlement is dispersed because each family needs a plot of land to grow crops (Fig. 43.7). Population is no less dense several kilometres from the road, because rapid transport and the varied goods available at the government-established shops are not important in the day-to-day lifestyle of these people. Many of the schools and churches that lie scattered between the huts have been established and are still run by missionaries.

The pattern at Yala can be regarded as an **established** stage of rural development, even though it lacks a service structure. The land has all been cleared and now carries a capacity population, whereas in Yaoundé there is much room for expansion.

A pattern similar to that in Yala prevails in parts of Asia, especially on flat land subject to annual inundation. In Bangladesh, for example, where the more basic system of flood irrigation has not been replaced by systems based on the controlled flow of water in feeder canals, there has been little tendency to nucleate (Fig. 43.8), the changing rural density simply reflecting changes in the land quality.

The pattern of rural settlement that evolves, however, is just as much a reflection of the social and

Figure 43.7 High population density with subsistence arable farming near Yala, Kenya (see also Fig. 33.15).

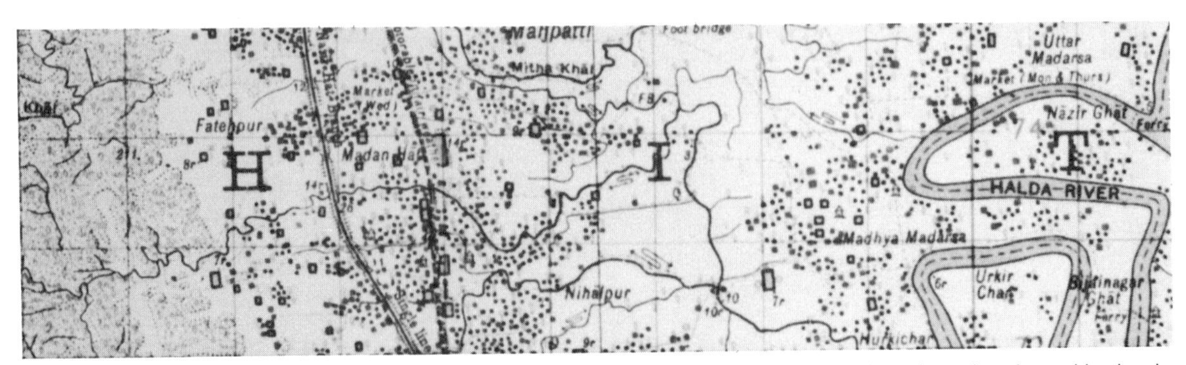

Figure 43.8 Bangladesh. Contrasts between the fertility of the Halda River floodplain, where intensive rice cultivation is possible, and the poor quality of the hillands are reflected dramatically in the density of settlement. Irrigation by annual flooding is the main technique used, although gravity-fed canals are also employed. Notice the absence of a settlement hierarchy. (The map is now quite old: present government policy is not to make new maps available to the public, for security reasons.)

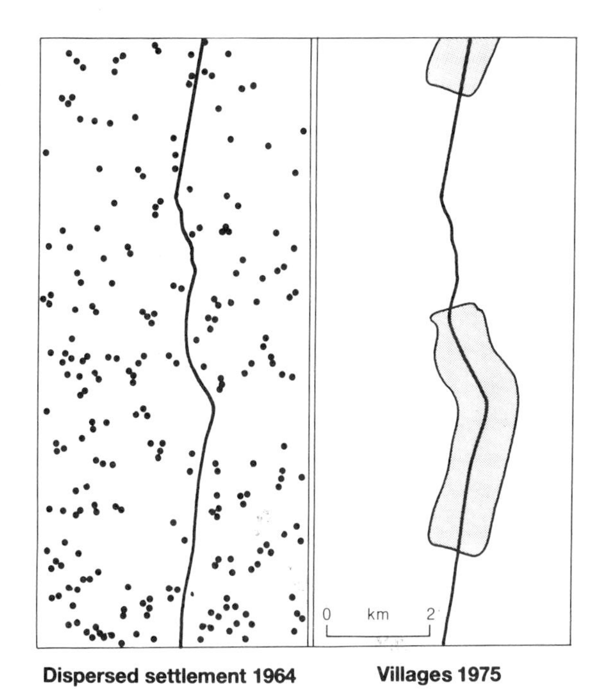

Figure 43.9 Tanzanian *ujaama* scheme.

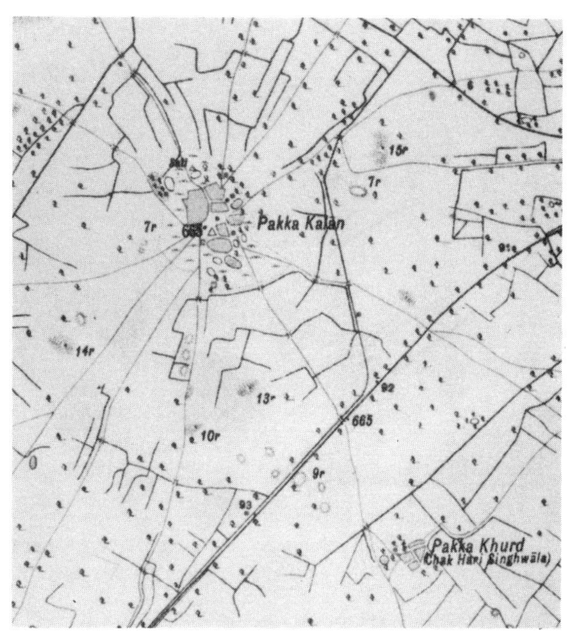

Figure 43.10 A nucleated settlement pattern within an area of extensive irrigation, Punjab, India. (The map used here is now quite old: present government policy is not to make new maps available to the public, for security reasons.)

political structure of societies as of the nature of the land. For example, land basically similar to that in Yala exists in Tanzania, but here, where the political philosophy is very different, the government has recently engaged in a determined campaign to change the rural-settlement pattern without any change in the rural economy. Between 1965 and 1975 virtually the whole of Tanzania's rural population were moved (many by coercion) from their dispersed hut settlements, and resettled in specially constructed villages (Fig. 43.9; see also p. 360). At the same time the previous system of private landholding was overturned in favour of a communal system of shared land and profits, with work parties going out to tend communal crops each day. The rural settlement pattern in Tanzania is thus very new and entirely a product of a political idealism.

Nucleated settlements are common in other parts of the developing world, but most have evolved over many hundreds of years in response to particular problems. For example, in parts of the Indian subcontinent and South-East Asia the need for communal attention to sophisticated irrigation systems has done much to encourage nucleation. In India a severe caste system of landowners and landless labourers has reinforced the tendency for settlement to occur in villages (Fig. 43.10), although such villages, being mainly based on low-income labourers, again do not provide a demand for more than rudimentary services.

Although the original pattern of rural settlement in Western Europe may have had much in common with that presently seen in the developing world, the rise of commercial activities and the need for ready access to markets has caused many differences, particularly in the plans and functions of the individual settlements.

One of the most enduring settlement patterns of Europe occurs in the Beauce region of the Paris Basin of France (fig. 43.11). Here the loess overlying chalk is eminently suited to arable cultivation, and in the Middle Ages a pattern of open fields was established. Although land is no longer divided into smaller units when it is inherited, farmers have still been reluctant to build farmhouses away from the villages. This is because the restructuring of land holdings into consolidated units is much more recent, having taken place at a time when motor transport provided easy access even to distant fields. Furthermore, a purely arable farming system does not require daily attention to each field. For these reasons the villages have simply developed into clusters of farmhouses, while the accessibility of market towns has suppressed the development of local shops. A further stimulus to continued nucleation is the considerable expense and difficulty of obtaining water supplies in chalk regions.

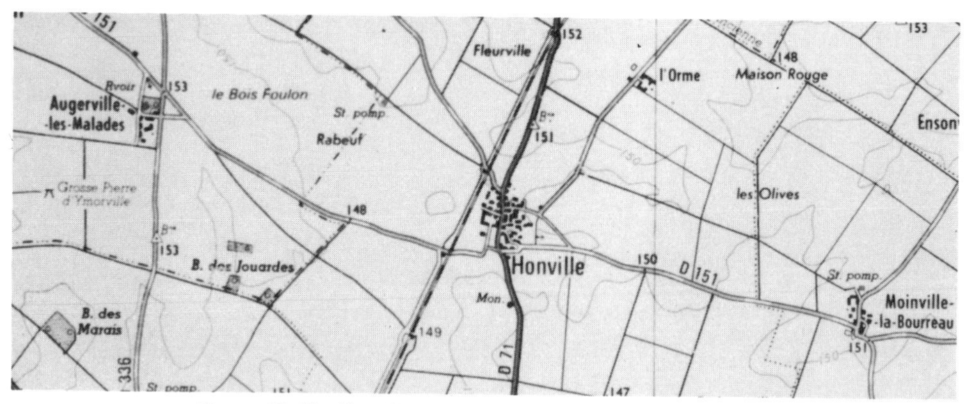

Figure 43.11 Honville, in the south-west Paris Basin, France.

Most water is drawn from deep underground supplies.

By contrast, much of the land of the Netherlands had to be won from the sea. Here the pattern of settlement goes hand in hand with reclamation systems and thus reflects the level of technology prevailing at the time the engineering works were begun (Fig. 43.12). The area near to Loppersum was reclaimed at a very early stage, perhaps as early as the 13th century. As people were then only able to recover small areas (called **polders**) at a time, land was won in small units, each new field being a little more distant from the village site. Not being sure of keeping the water out, the settlers raised the villages on soil mounds as a protection against flooding. This type of reclaimed land remains relatively poorly drained and few farmers have built their houses beyond the raised village sites.

De Beemster polder was drained in the 17th century, the whole area being encircled within a single ring dyke. In this highly structured landscape, farms, roads and settlements have been laid out carefully on a grid: farmhouses are evenly spaced along the roads; and each farm has its shortest boundary along the road – this not only keeps roads to a minimum but creates a degree of social cohesion by providing areas of relatively dense housing. The roads have almost become linear villages, although the service and other central functions are still provided by villages such as Westbeemster (Fig. 43.12b).

Most recent of all are the villages that have been established on the South Flevoland Polder (Fig. 43.12c). Just a few years old, the pattern of rural settlement here really does reflect rapid transport: farms are isolated in their fields and villages planned at distances far greater than in the past.

Denmark's rural settlement pattern shows yet further variation, reflecting a radical alteration of the land use within the last 150 years. In the 19th century feudal estates were divided up and the land was reallocated in small units. This has produced a high-density but dispersed pattern of farmhouses, the majority sited along the roads. The remnants of the large feudal villages in which the estate workers once lived now act as service centres (Fig. 43.13).

(a)

(b)

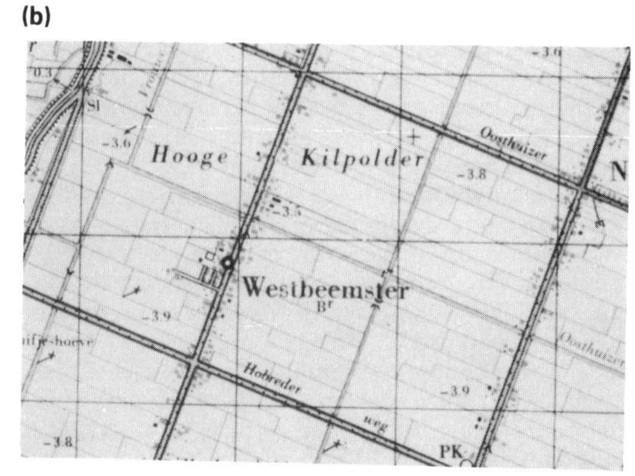

Figure 43.12 (a) Leermens, Netherlands; (b) Westbeemster, De Beemster Polder, Netherlands.

Figure 43.12 (c) South Flevoland Polder, 1976.

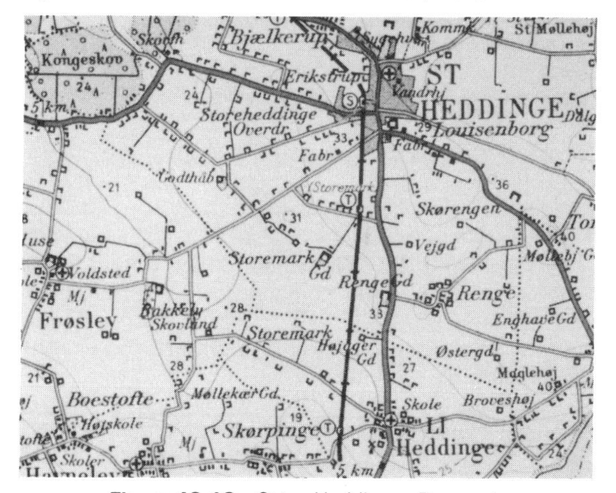

Figure 43.13 Støre Heddinge, Denmark.

43.3 Britain: a case study

The form of settlements is the result of a long heritage of responses to varying environmental, social and economic changes as has been demonstrated briefly above. Nowhere is this more evident than in Britain, where the majority of settlements are over 1000 years old. By looking in some detail at rural-settlement patterns in Britain one can better understand the complex problem of settlement evolution in general.

Before looking at the special historical events involved it is important to consider the environmental basis on which the choice of many settlement sites must have rested.

The environmental basis of site selection

In cool temperate regions there is a considerable advantage in gaining the most heat possible from the environment (Fig. 43.14). This is particularly important in winter when a slope having a southerly aspect will receive direct insolation for a much greater part of each day. For this reason, in areas with pronounced relief there is a tendency for settlements to be sited more frequently on the northern sides of valleys. This is promoted by two other factors. At night, radiation from the upper valley slopes causes cool air to drain into the valley bottom with an associated increased risk of frost. Second, because floodplains in valleys are prone to inundation, soils there remain wet and difficult to work for some while each spring, so better drained sites such as river terraces would be preferred locations.

However, there are other environmental factors to be considered: a point bar is a region of sediment accumulation and is less prone than the outside bend of a river to erosion during flood, and the inside of a meander bend can be most easily defended against attack; however the outer bank of a meander provides deep water suitable for boats, whereas a

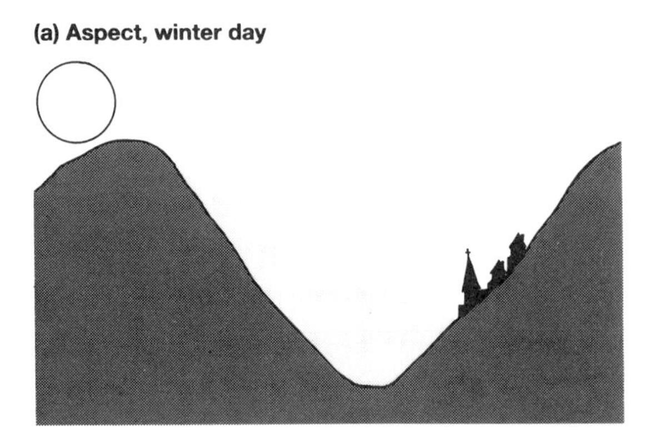

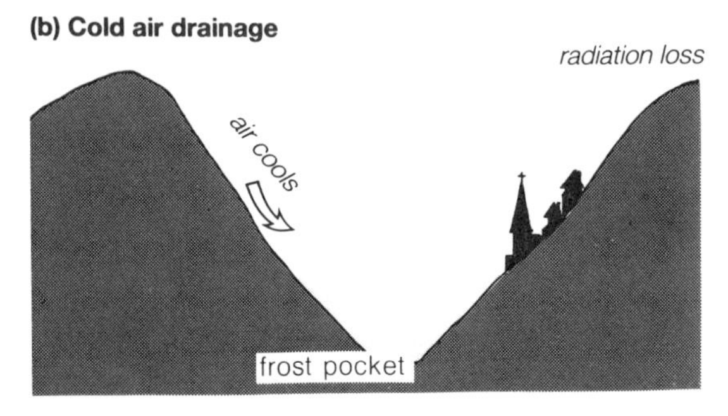

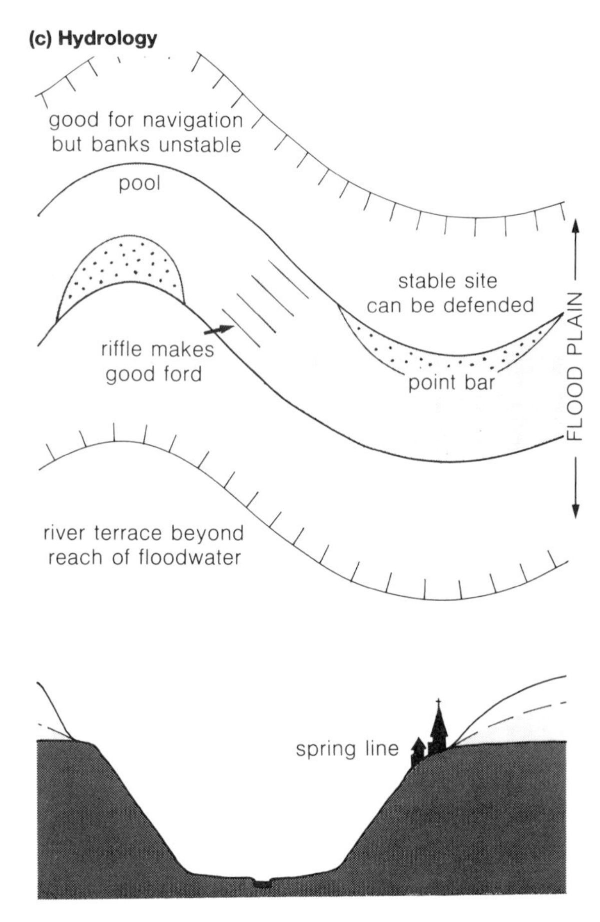

Figure 43.14 The environmental basis of site location.

streambed riffle provides an evenly shallow region which permits the river to be crossed. Clearly the sites chosen result from consideration of all such factors.

Early settlement

Although there are many environmental factors that influenced the selection of sites for settlement, by the time the Romans arrived in AD 43 there were few parts of Britain that had not been settled or used by prehistoric peoples, and by the later Iron Age many lowland areas had already lost their forest cover. One of the reasons for the Roman conquest was to acquire the rich resources of this country.

Settlements ranged from remote hill farms, peasant villages, vast prosperous mansions and fortress towns, to thriving regional market centres and religious shrines, all linked by a network of main and local roads and protected by well garrisoned forts. Grain, cattle, iron, lead, silver, timber and cloth were produced to support both local populations and a standing army. But with the collapse of the Roman Empire the Anglo-Saxons were able to take over in the south and east of Britain. They were far less sophisticated than the Romans, and in those regions under their control many roads fell into disuse, market towns began to decay, and there was a return to subsistence agriculture.

In detail, the pattern that evolved was extraordinarily complex; in general, small hamlets based upon a form of communal agriculture became the norm. In each case the patterns of clearance and of land use reflected both the history of the particular settlement and the variety of land available. By the 12th century the characteristic English village had emerged in many lowland areas with good soil. It comprised a church, a manor house (or lord's residence) and associated peasant houses, the whole being surrounded by arable lands worked in common and divided into strips. Such nucleations served many purposes: they made communal co-operation easy, they allowed the 'journey to work' to be minimised and they offered protection; but above all else they placed the peasantry where they could work the lord's farm, be controlled and taxed by his bailiff, and be near to the church they supported by their tithes. The geometric regularity of many hundreds of village plans shows that this aggregation was not haphazard. Only in the woodlands and on the hills, away from the richer plains and valley lands, were scattered farmsteads and irregular hamlets still to be found.

The developing pattern

By the Middle Ages most villages were already 'old', and most, together with their lands, had been incorporated into feudal estates. These estates might contain several types of land:

(a) good land, with few physical limitations to agricultural use;
(b) intermediate land, with moderate limitations to use because of soils, relief or climate;
(c) poor lowlands, with severe limitations such as heavy- or very light-textured soils;
(d) poor uplands, with severe limitations of altitude, climate or soils.

Clearly the better-quality land can support more people than can poorer land, and the pattern of settlement is therefore largely related to land quality (Fig. 43.15). The good land is occupied by numerous villages and hamlets, each surrounded by their own fields, their lands fitting together in a roughly polygonal pattern (the 'ideal' shape would be a hexagon: see Christaller's model, p. 454). By contrast, the poor upland is suitable only for rough grazing and is outside the limits of cultivation. Hamlets rather than villages prevail, despite the larger land-holdings, a reflection of the lower yield from the land. Hamlets are sited as close to the better land as possible, and often in a sheltered site – rarely are they placed centrally within their parishes.

The intermediate land is transitional: local custom or choice can dictate the direction of emphasis. As a result, a more complicated pattern of land use is characteristic. Because of the limitations imposed by soil and climate settlements are still widely spread and the land is generally less populous.

The shape of villages

Village shapes have had a considerable resistence and have survived through all the changes in function since they were established. Such survival may seem surprising; after all, a village is not a very large entity. But, as with many patterns established by man, village shapes have survived because they have not presented any considerable disadvantage or obstacle to progress. Nevertheless most villages contain very few signs of the houses that were built even a couple of hundred years ago, let alone the mud-and-wattle

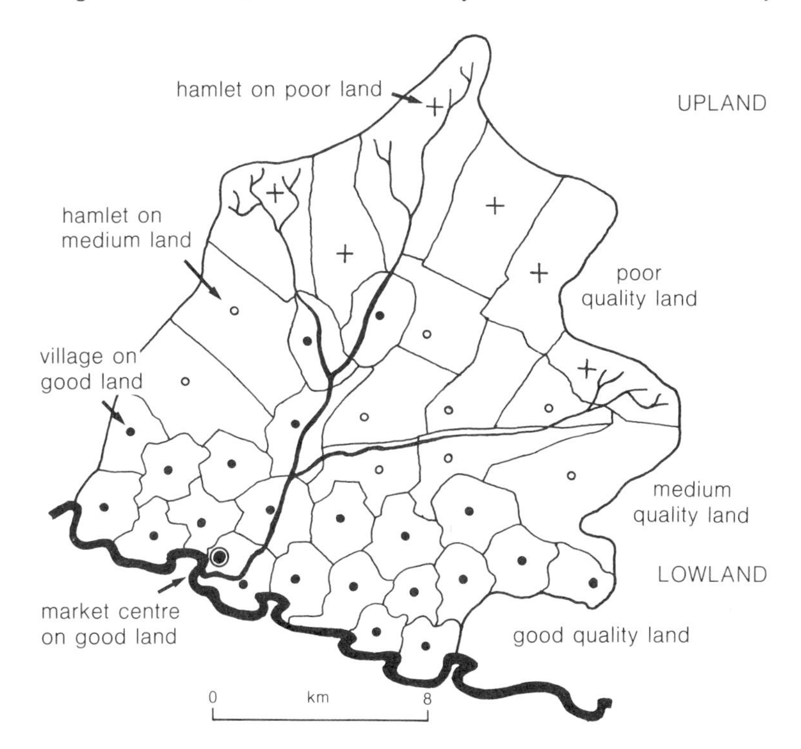

Figure 43.15 By the Middle Ages, agricultural land was fully used, the farming system and pattern varying according to the quality of land use.

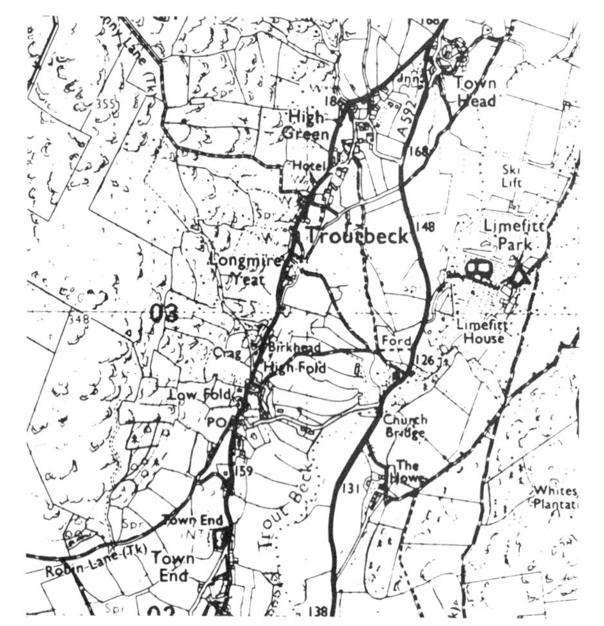

Figure 43.16 Troutbeck, Cumbria occupies 2 km of valley-side, just above the potentially flooded and agriculturally valuable bottom land. Above the village lie fells too steep for building. The exact site is partly controlled by a spring line and therefore water availability. Troutbeck is an example of a linear village shape, typically associated with regions of pronounced relief or major communication lines.

structures that were the homes of the original inhabitants. The shapes remain largely because houses were demolished and rebuilt in a piecemeal fashion, under which circumstances it is rarely possible to alter the basic pattern without considerably disrupting village life. Such disruption would have been resisted because roads often form property boundaries.

Village shapes were closely related to their environments. Thus a valley floor, valley-side trench or ridge top could contain only a linear settlement unless considerable effort was made to build on the hillslopes (Fig. 43.16). Both locations are common because valleys and ridges also provide good communications, and a variety of land units can be reached from each. Where the landscape influences shape less strictly, however, villages are often simply compact groups of dwellings. Sometimes the houses spread around a pond if water could not be obtained from a stream; others cluster near a river access point, usually a riffle (wide and shallow) where animals could reach the water and cross it most easily. Yet others are on small knolls of better-drained land (Fig. 43.17). On the other hand, in areas where pastoral activities dominated, it was often essential to keep the animals within safe surroundings overnight, a requirement that often led to a **green village** in which the peripheral houses could be joined by wicket fences to make a stockade.

(a)

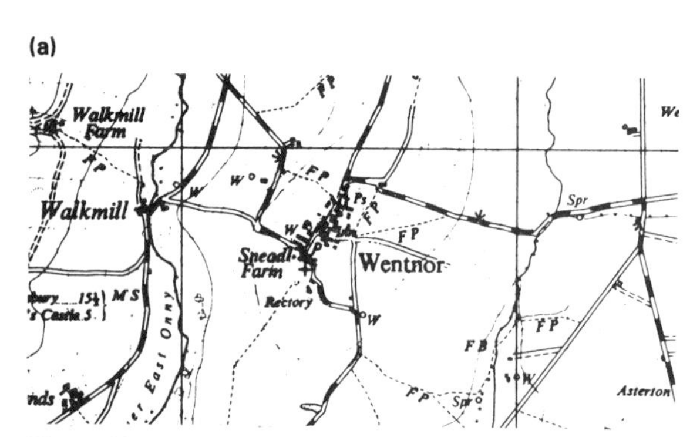

(b)

Figure 43.17 (a) Wentnor (Herefordshire) and (b) Great Edstone (North Yorkshire) share many features in common. Both are on slightly elevated ground near to a river; they are approached by a 'ridge' road following the drainage divide. Wentnor also acts as a focus for a cross-valley route. A small number of central functions (church, inn, post office, primary/junior school) dominate the villages, which today are surrounded by enclosed fields and isolated farmhouses, features added to the rural landscape as late as the 18th century. Notice that the road pattern was largely established after the settlements.

(a)

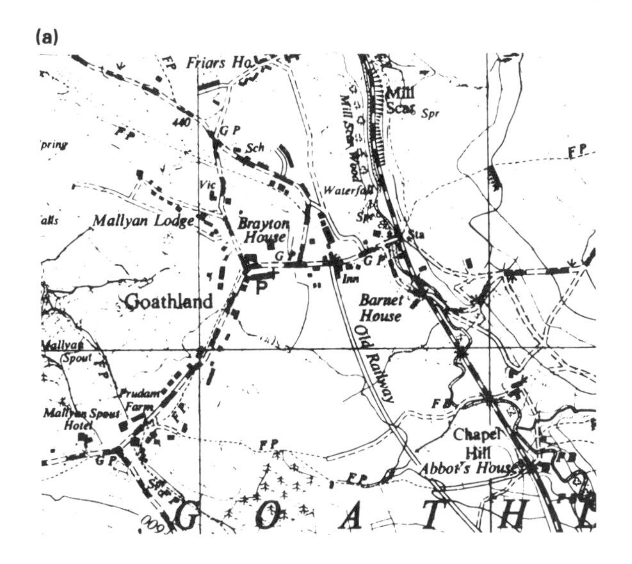

Table 43.1 Agricultural employment change (source: EEC).

Country	% on land 1910	% on land 1982
UK	9	2.7
France	71	8.4
West Germany	37	5.5
Spain	56	18.3

Figure 43.18 Open (green) villages are often associated with pastoralism. (a) Goathland, North Yorkshire, is a settlement of widely spaced dwellings; (b) Mortimer, Berkshire, has a large square green (notice the 'daughter' settlement of Stratfield Mortimer by the stream, and the recent expansion of Mortimer to the west).

(b)

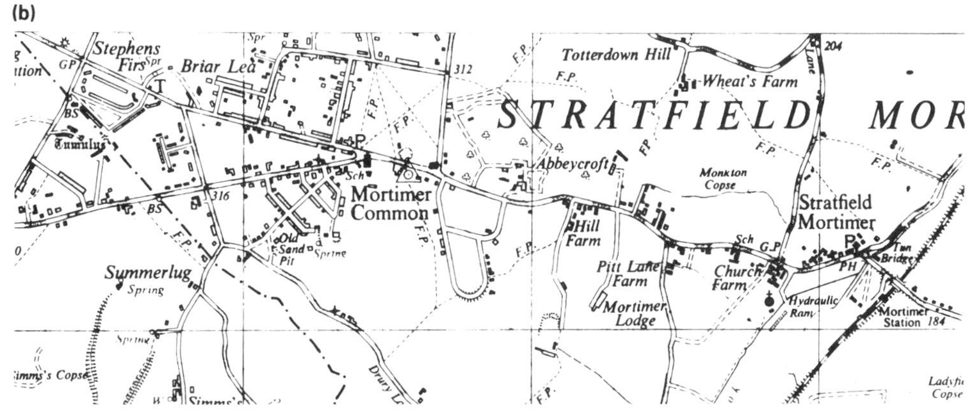

Rural response to stress

Village plans have often endured, at least in part, but the pattern of rural settlement may have changed dramatically. The pattern seen today is a result of a long and complex history in which many villages experienced stability, expansion, contraction, site moves and even total destruction. For example, the Black Death of 1348-50 reduced the rural population by a third, made many villages inviable units and led to widespread abandonment. The gradual rise in national population in the 17th and 18th centuries resulted in the expansion of nearly all rural settlements. By the 18th century, agricultural innovations were gathering momentum and the traditional fallow system using large communal open fields was replaced by ley and dairy systems characterised by enclosures. In turn enlosures placed considerable stress on rural communities by depriving them of many jobs at a time of high populations. This was a two-way squeeze only relieved by the mass exodus to towns and cities.

While the farm labourers were seeing their employment opportunities decline, farmers were moving from the villages to new houses built at the centres of the newly compacted and enclosed land. Thus the wealthier section of the community left the villages, thereby further accelerating the villages' decline. Through the 18th and 19th centuries, therefore, and after three centuries of expansion, the villages and hamlets contracted. The rural settlement was not redundant: some farm labour was still required and there was a growing tide of service functions that could be focused on the villages. Nevertheless, there was severe rationalisation at this time and many hamlets today are mere shadows of villages that once housed many farmworkers and their families. The grouping of a pub, the general store and a handful of

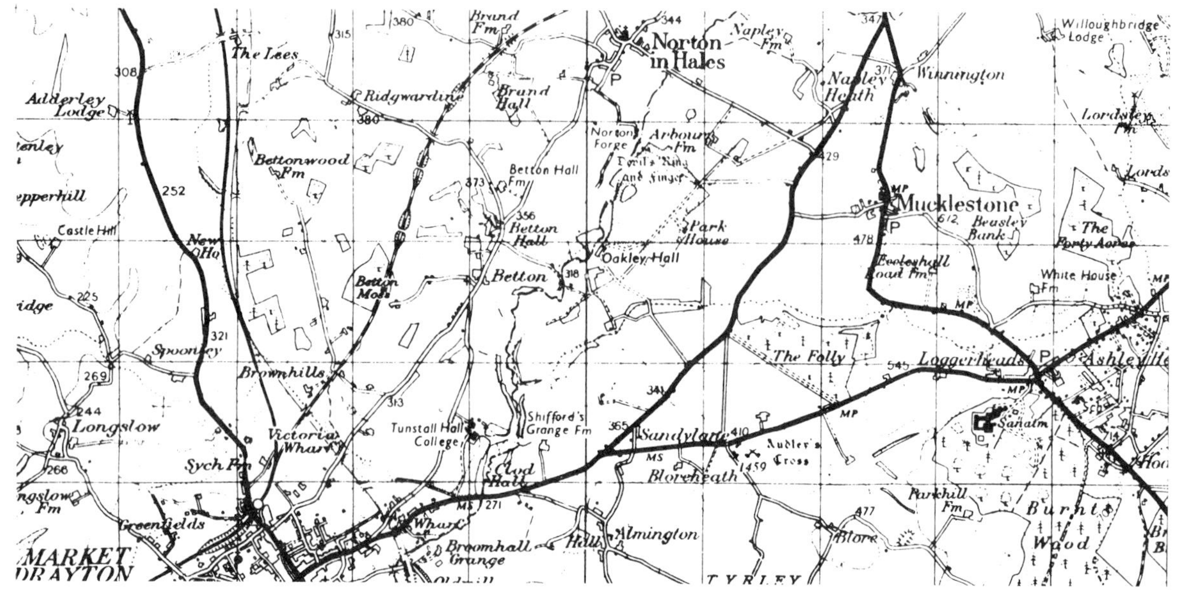

Figure 43.19 The Market Drayton area, Shropshire. In the rainshadow of the Welsh mountains and with only gently undulating land, both arable and pastoral activities are possible. Farms have a substantial arable component and are large; the agricultural population is small. Farms are organised into large estates and are now run as highly mechanised 'industrial' units. Villages and hamlets exist only as minor service centres, their inns catering now primarily for the businessman's lunch and the urban dweller's evening out. This is not a particularly popular area for second homes, and once-substantial villages have dwindled to hamlets.

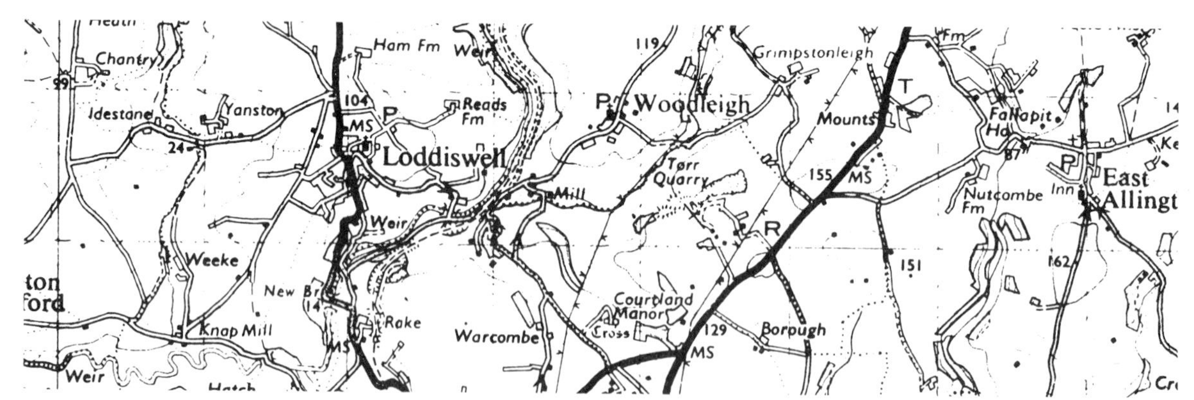

Figure 43.20 Loddiswell, South Devon. The mild climate of South Devon offers the opportunity for both arable and pastoral farming, but heavy clay soils and a long growing season favour ley and dairy systems. The plateau is deeply dissected, placing great restrictions on communications, and roads mainly follow either ridges or valleys. Many settlements have developed where ridge and cross-valley routes meet. Although the villages are now smaller than for 200 years, the period of contraction seems to be over, as the climate, scenery and proximity to the sea attract those seeking a place to retire or a second home. The permanently resident population is thus becoming increasingly old, placing stress on the social services. At the same time, the total population fluctuates widely both from weekday to weekend and from summer to winter. Overall, there is too little consistent demand for retail services and the number of functions offered within the rural environment continues to decline.

houses, represents a temporary level of stability (Fig. 43.19).

Since the advent of motor transport some rural settlements have declined even further. By contrast others have been revitalized. An increasing number of people now choose to live in the country even though they must earn their living in the city. This has caused the growth of those rural areas which can be reached relatively easily from the large urban areas, swelling the villages with landscaped housing estates (a process called counter-urbanisation). Scenically attractive but remote rural regions have also experienced a partial revival as city dwellers buy second homes for use at weekends and during holidays (Fig. 43.20). However, for the permanent residents of such settlements, second-home owners are a mixed blessing. They may prevent the hamlet or village from fading away completely, but they introduce an outside element that does little to enhance the social fabric of the community or to provide a consistent demand for services.

Urbanisation

44.1 From rural to urban settlement

No society can remain static if it is to survive. Not only is it important to understand the rural context in which settlements were founded: one must understand also the dynamic element that propelled some settlements from being hamlets to being market towns or major cities.

Taunton, in Somerset, provides an example with which to investigate the process of urbanisation (Fig. 44.1). Founded by the Saxons as a small defensive settlement on a dry terrace site beside the otherwise marshy banks of the River Tone, it gradually grew into a medieval market town, helped by the local resource of sheep wool for textile making and the additional wealth of a priory.

Subsequent growth was limited by poor communications with areas beyond the local hinterland. The arrival of a canal and later a railway provided some stimulus to trade, and there has been a substantial addition of commercial and industrial functions. Thus, in a series of phases, over more than 1200 years, Taunton has grown into the medium-sized manufacturing, market and administrative town of today.

Taunton's growth has clearly been influenced by the changing external environment, by competition from neighbouring settlements, and especially by its degree of isolation. Taunton has its own set of environmental and historical influences, and these give the town its unique character. But underlying the individual details of Taunton's development there is, as with all other towns and cities, evidence of a general pattern of evolution. The discovery of such patterns forms the subject of this chapter.

44.2 Influences on urbanisation

Historical factors

In the 12th and 13th centuries something quite remarkable happened all over Europe. Out of the inward-looking feudal system, with its oppressive controls on production by the Church and by landowners and its consequent low productivity, there blossomed a whole range of small settlements whose primary function was not to house the workforce of the surrounding fields, but to make things and to trade goods brought from afar. As currency, the

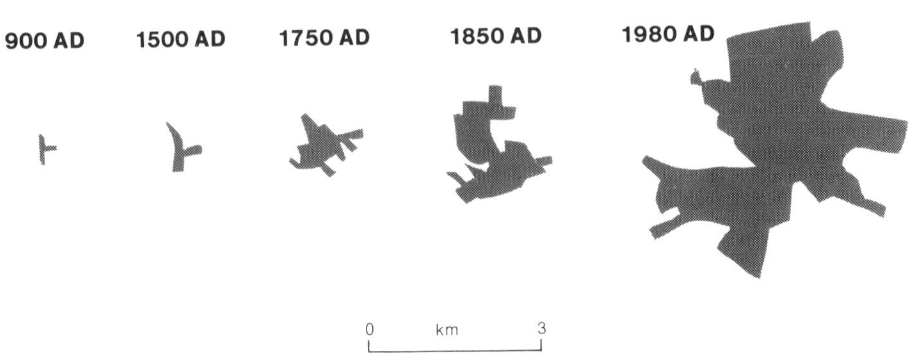

Figure 44.1 The evolution of Taunton, Somerset.

villagers used whatever tiny surplus of food they had saved. As the population grew and the cultivated land expanded, so the total volume of surplus grew also; in time, it provided a means whereby those with more specialist skills could leave the land and concentrate on their manufactures. These artisans and traders – the potter, the smith, the farrier, the mercer and others – could not survive simply on the business created in their local village; they needed to seek a more accessible **central place** which a larger population could also reach. In seeking such a central place they came together to form the embryo **market towns**.

The location of these centres was not so much a matter of sound geographic judgement as of speculation on the part of the local feudal lords, who recognised in the rise of trade a possible means of increasing their own revenues. In these two centuries 2500 speculative sites were chosen as potential trading centres (**charter towns**) in England alone. The evolution of market towns was a two-step process. The first step was the formation of the 2500 charter towns, each subtly different as regarded its location. These variations were the result of the random whims and landholdings of each lord and were unrelated to the overall needs of a successful trading network. However, the second step – selection through survival in the struggle for existence – produced order in the settlement pattern. Like the thousands of seeds scattered by a seed pod only a few survived and grew: the very few 'towns' that had the most appropriate combination of characteristics – a sufficient sphere of influence, ease of access, and a dry flat site. Of the 2500 starters many were in poor positions for communications or too close to their competitors. Today most have disappeared completely; others remain merely as hamlets. Yet even most of those that survived to become market towns have grown no further. For example, the West Country towns of Bodmin and Okehampton, and to some extent even Taunton, still have only the small-market-town status they enjoyed in the Middle Ages (Fig. 44.2). Only the most successful have evolved into cities.

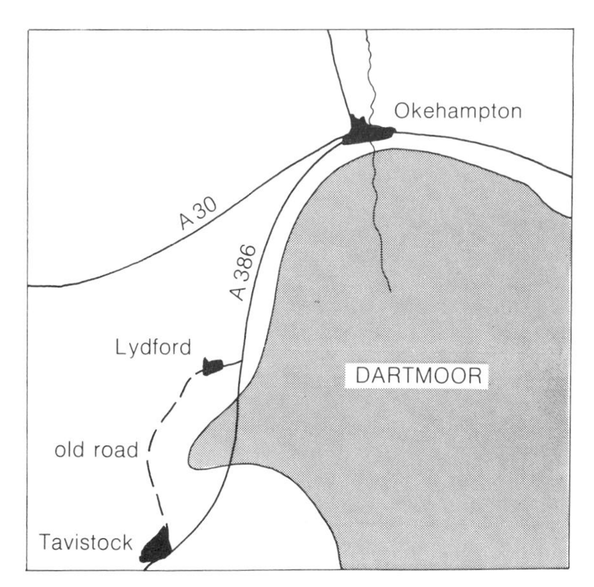

Figure 44.2 Okehampton, Lydford and Tavistock were all charter towns on the edge of Dartmoor in Devon. Lydford's situation is inferior to that of the other centres and it has remained a small village. Even the new trunk road bypasses it. Nevertheless, neither Okehampton nor Tavistock can offer more than average quality access in an area where there is a low population density. Without an adequate hinterland they have never developed beyond small market town status.

Trade

Charter towns largely aimed to cater for people living in the immediate hinterland, but another element was also present: the migrating fair, which brought with it a wider variety of goods than would normally be available. The goods exchanged at fairs – salt, grain, fish, wool and metals – may seem mundane to us, but they formed the embryo of what was later to become a vital part of the urban function: **inter-regional trade**. Even today the functions of a town can still be clearly classified into two groups: local trade and inter-regional trade.

Trade in these early towns was largely concerned with agricultural products. In 1600 the chief occupation in Leicester, for example, was still that of grazier (graziers went to the fields each day), with only embryo forms of industrial employment provided by weavers, tanners and glovers. Services such as those provided by grocers, butchers and carpenters were also represented. Nevertheless, the range of products that could be made in 1600 was so limited that it did little to encourage town growth. Iron-making was still a rural activity that needed large forest reserves of charcoal, and most woollen and leather goods were made by cottagers as a part-time activity. As long as communication remained poor the many small towns formed during the 16th and 17th centuries were relatively secure. This lack of communication ensured their monopoly of the custom of the local area and made difficult the growth of large-scale enterprise. However, changes in the traditional pattern of manufacture were slowly forced into being

(a) by the demand from an increasing population for further textiles;
(b) by the greater availability of iron and its manufactures;

(c) by the wars that put large demands on replacing equipment;
(d) by improvements in communications.

These changes allowed the rise of large provincial centres (such as Canterbury and Reading) with much larger hinterlands, and therefore they automatically caused a corresponding demise in many charter towns. By the 1600s there were less than a third of the market towns of the 14th century. In the 16th century too there were important voyages of discovery to the Far East and America, and the goods brought back (such as tobacco, from America in 1566, and sugar, from the West Indies in 1590) added diversity to the products sold from a town. By the 16th century, therefore, many market towns such as Taunton, confined within their defensive walls since the 12th century, either had knocked these down or were bursting at the seams. Not that any of them occupied very large areas, because size was largely controlled by the difficulty of communication on foot.
Instead of being dispersed, houses crowded in on one another, which led to worsening sanitary conditions and the increasing likelihood of fires. Many towns had their overcrowding problem solved by damaging fires, as did London in 1666 (the Great Fire). Nevertheless the narrow streets and crowded forms of many town centres remain even now as a legacy of this early phase of urban development – York's Shambles area and Ledbury in Hereford are good examples.

44.3 Market towns

Market towns reflect both their heritage and the cultural environment in which they have developed. Small market towns in Britain today have retained much of their structure and function from the pre-industrial period, despite the influx of people seeking retirement, second homes and so on. Many retain the plan of the early centuries and some of the 16th- and 17th-century timber-framed houses: Ludlow, Helmsley (Fig. 44.3) and Ledbury are among many examples. Although many saw a phase of cottage industrialisation (such as the wool industry in the South-West and in East Anglia), this function has largely been captured by large manufacturing centres, and they are now left as service centres for the surrounding community, much as they were four centuries ago. Weekly markets (successors to the 'fairs') still function, and the main streets are still lined with shops selling a basic range of goods. Any 'industry' remains subordinate to the service role, concentrating on repair work or small manufactures.

The pattern and function of small market towns in Europe follows along much the same lines as those in

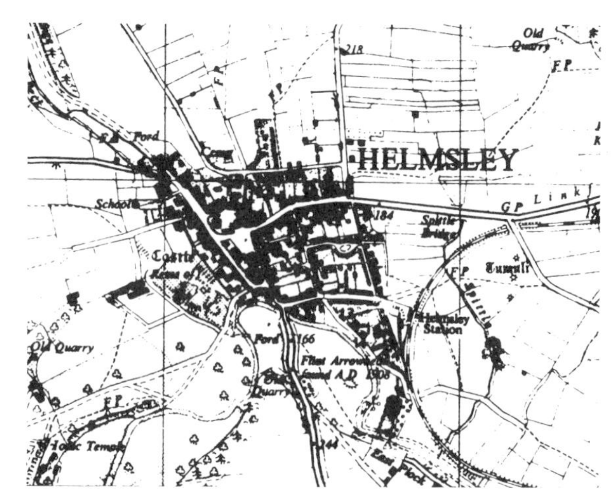

Figure 44.3 Helmsley is an ancient route centre on the edge of the North Yorkshire Moors, developed around a market square.

Britain. However, in much of the US, market towns have no heritage of poor communications, and towns are laid out on a regular grid pattern with considerably more internal space. Services are also related to a hinterland whose populations move by car rather than on foot. As a result shops all have large car parks and are spaced further apart. With rapid transport it is no longer necessary to concentrate all the service functions together and – as in

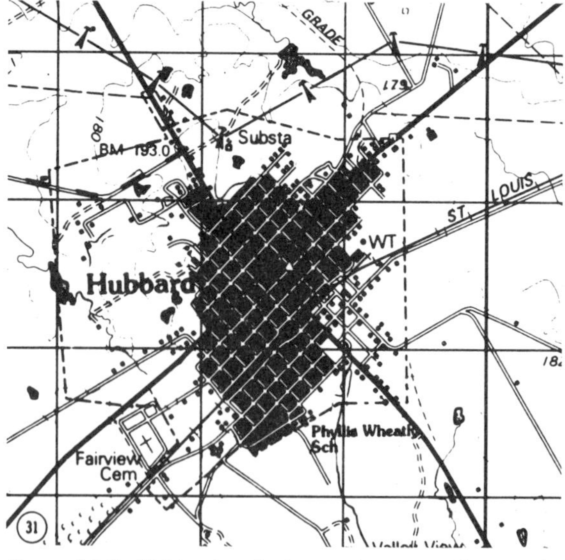

Figure 44.4 Hubbard reflects modern growth on the northern Texas rangelands, with services scattered evenly along the access roads. There is no market centre; the closest thing is a shopping mall on the *outskirts*. Hubbard's size is determined by the needs of its local community.

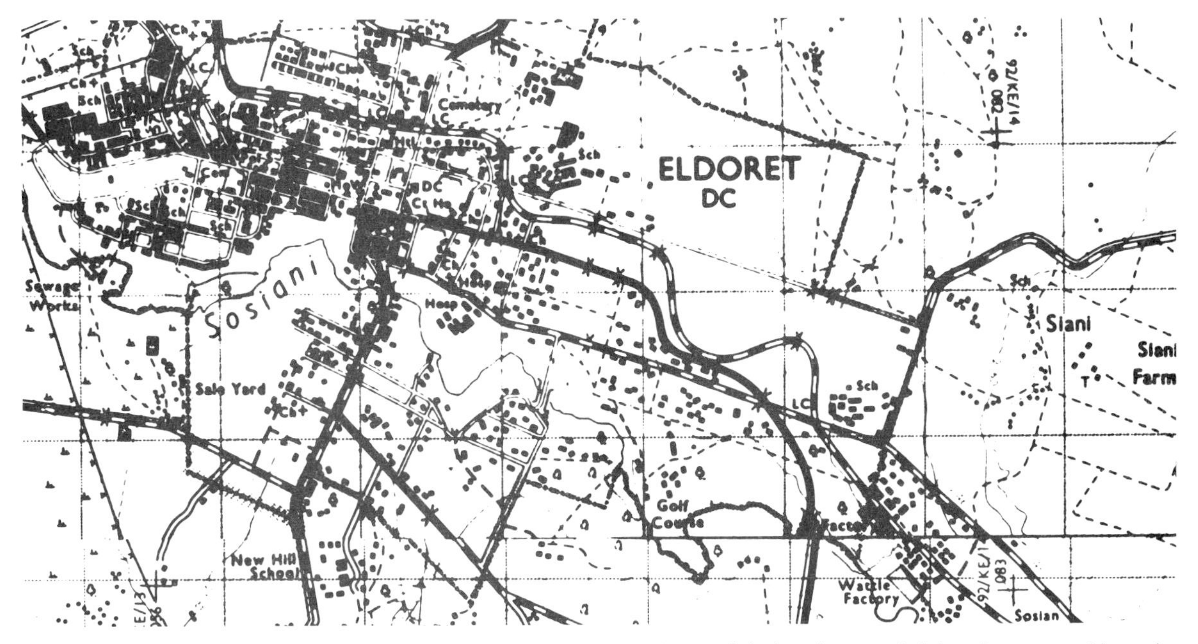

Figure 44.5 Eldoret, Kenya, is a market town situated among plantations and designed as an administrative centre with a subsidiary service role. Within the town, the main street contains nearly all the shops, but they do not all have specialist functions; rather they are mainly general stores to supply such essential goods as candles, aluminium cooking pots or maize. A further contrast is the lack of a residential area; most people still travel in from surrounding farmland each day or live in their shops.

Hubbard, Texas, for example (Fig. 44.4) – these are scattered more evenly along the main access highways.

Helmsley and Hubbard, although different in plan, offer the full range of services demanded by a developed society. Eldoret, in Kenya (Fig. 44.5), on the other hand, is a market town that has been 'imposed' on a totally agrarian subsistence society whose needs do not really stretch much beyond those that can be provided by a travelling market. These travelling markets still fill a prominent role in Kenyan rural society and this must mean that the nature of the market towns is substantially different.

The rural population visit Kitale perhaps once a month to buy specialist items such as candles, paraffin and aluminium cooking utensils (Fig. 44.6). Truly specialist shops would be a nonsense in such circumstances: the main street of Kitale consists of a collection of general stores, each offering the same range of goods. And although the town has government offices, police station, hospital and veterinary office, these are not closely integrated with the other town functions; rather they represent an attempt by central government to superimpose some of the features of towns in the developed world.

Pucarani, a small market town near La Paz in Bolivia (Fig. 44.7), provides another contrast, both in form and function. The plaza (square) is the focus of the town's activities, being in effect the 'CBD' – the location of many of the shops, and the location of the weekly market. As in Eldoret, the shops in Pucarani sell a wide range of inexpensive goods. However, each shop tends to cater for the needs of a particular rural community, not only selling goods but also buying produce from those customers who prefer to trade or barter instead of using cash. These retailers subsist on a multitude of very small purchases made during the week, and especially on market day. There are often close ties between people in towns and those in the countryside in the form of co-parenthood (godparents) and these reinforce the link between shop and community.

In contrast with towns of the developed world, the retailers of Pucarani also act as wholesalers, collecting local produce in their shops during the week and on market day, and then taking this for sale in La Paz or in nearby weekly markets. In both Eldoret and Pucarani the concept of aggressive marketing is completely alien to local attitudes and therefore shops do not have an acute need for a central location.

The weekly market specialises not only in food, brought by the rural farmers to sell, but also in items such as bicycles and radios, and clothes such as sweaters; items which tie up too much capital for the local shopkeeper. Because of low levels of inter-urban communication, people do not go to the higher-order settlement to buy their specialist items; the higher-order functions come to them.

Figure 44.6 Although there is a covered shopping area in Eldoret, there is still much demand for an open-air market where producer–sellers can trade the surplus from their farms, or where travelling traders can ply essential wares such as plastic buckets and jerry cans. Thus there is a stable core of retail shops supplemented by an open market; a situation closely parallel in concept to that of many provincial towns in the developed world.

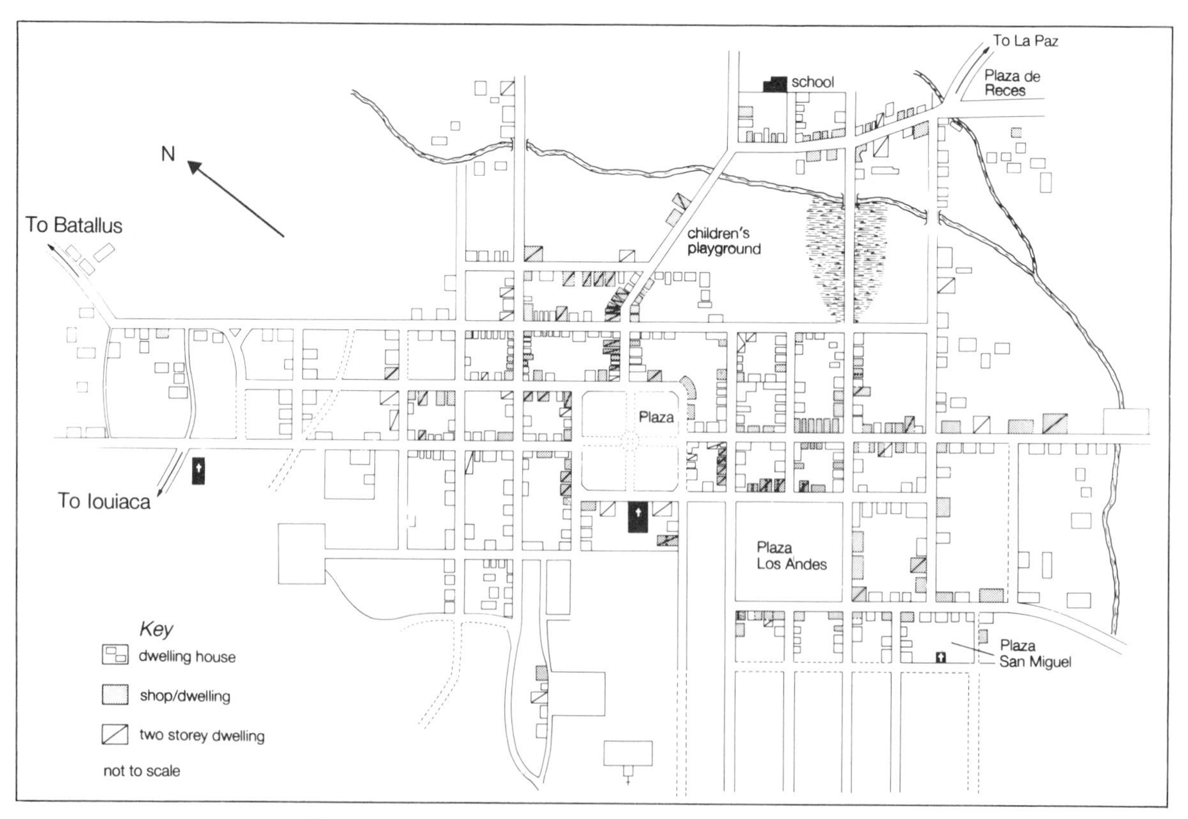

Figure 44.7 The South American market town of Pucarani.

44.4 Urban growth

In Britain between 1700 and 1800 (the latter being the date of the first true census) the population began to show the first signs of rapid increase. At the start of the 18th century there were about 5.5 million people: by 1750 there were 6.5 million, and by 1807 almost 8 million. This magnitude of increase could not but affect the nature of the landscape. Fortunately the 18th century also saw major steps forward in the improvement of agriculture, including innovations which removed the need for fallow land, provided better forage (and therefore permitted larger stocks of animals), improved implements, and raised the quality of stock and seed. Indeed it was the agricultural revolution, with its new breeding, cultivation and enclosure systems, that improved productivity sufficiently to cause people to leave the land and live in the towns. The results on the landscape were dramatic – the open, windswept and treeless arable fields, divided into strips, disappeared; new roads were laid out and rectangular fields with hedge boundaries were established. Land was used to its best advantage for the first time, use being dictated primarily by soil and drainage.

But the increasing food supply and greater production had just as dramatic an effect on the urban population. Fewer people were needed on the land: the remainder either left to seek employment elsewhere, or lived in poverty, relying on a dimishing supply of part-time jobs. Fortunately, advances in technology that were to result in the Industrial Revolution produced machines that needed workers. Many people left destitute in the countryside drifted to towns to seek such employment as there was, for even though labour in towns proved to be arduous and poorly paid, conditions in the countryside at this time could be far worse.

The distribution of urban centres

Rural settlements (hamlets and villages of less than about 2000 people) were originally designed to provide homes for people who worked on the surrounding land, although they later took on a limited role in providing the basic services needed on a day-today basis by their local communities. By contrast *urban* settlements have, from their inception, been places designed primarily to offer specialised services and goods; they provide homes for people engaged in services or manufacture.

Most urban centres have grown out of existing rural settlements simply because, by accident, they happened to occupy favoured sites that enabled people from the surrounding trade areas to reach them easily. Each urban centre interacts with its neighbours to create an interlocking pattern of trade

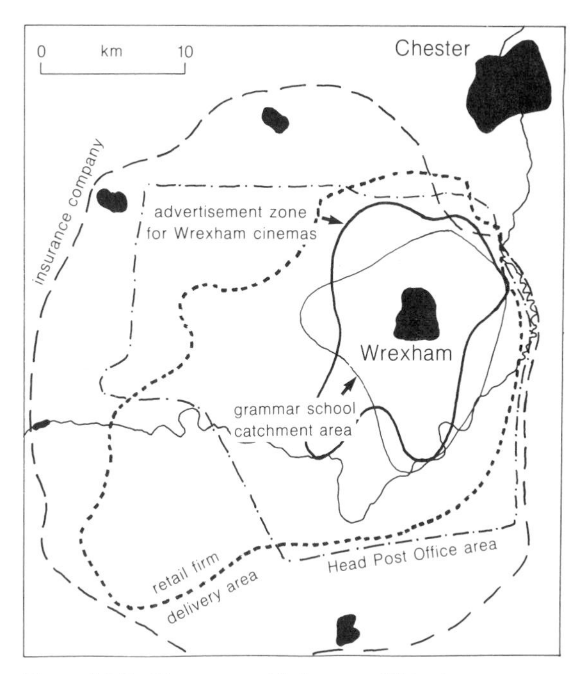

Figure 44.8 The zones of influence of Wrexham.

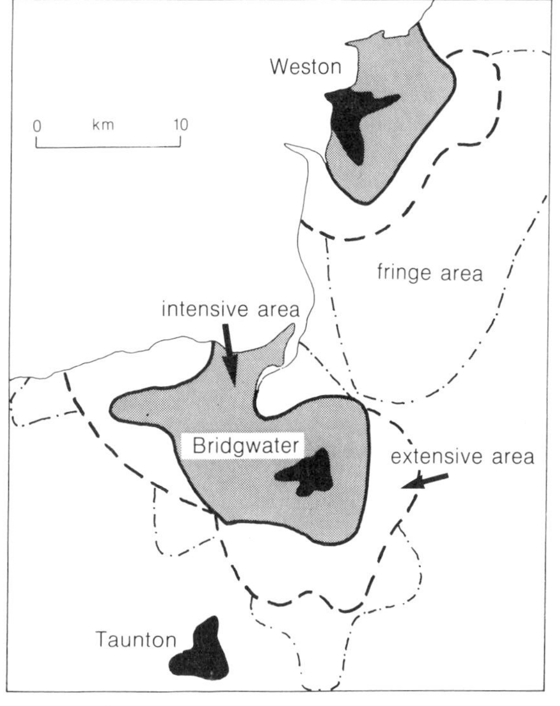

Figure 44.9 Intensive, extensive and fringe areas of Weston-super-Mare and Bridgwater.

areas, their sizes and spacing determined by competition. Thus villages and their parishes developed to occupy just those areas of land sufficient for the survival of each community, whereas market towns grew out of the village pattern to provide more central places for specialised functions.

The **sphere of influence** of any urban centre depends upon the type of service offered: people might be prepared to travel farther to buy an unusual suite of furniture than they might to buy an ordinary pair of shoes; they might go farther to visit a special rock concert than they might to see a film at a cinema. Observations of this kind have been substantiated by a number of careful studies based on services offered (as with Wrexham; Fig. 44.8) or the frequency with which people use a particular centre (as with Bridgwater, Fig. 44.9). In every case, however, it is clear that there is a complicated interplay between the urban centre and its trade area, an interplay that determines the character of the region and that influences the size and spacing of service centres.

Time also plays a critical role in the development of pattern. The pattern in Britain can be regarded as 'mature', having evolved over the better part of 1500 years. Rather like the colonisation of bare ground by vegetation, however, several growth stages were involved:

(a) *primary colonisation* by many settlements of equal status (a small range of species);
(b) *complete cover* of the ground to give maximum density but little variety;
(c) the *gradual emergence* of slow-developing but larger settlements (species) that rise above and eventually overshadow the primary colonisers, causing many to die back but leading to a pattern of settlement with much greater variety of function and size;
(d) *final stabilisation after competition* to a level less dense than that at the early stages, still with much variety, but with each settlement occupying an 'ecological' niche and distributed with more regularity over an area.

The result of this pattern of 'natural selection' of settlements is that each will probably evolve to offer a range of distinct functions appropriate to its location within the **urban fields** of influence of its neighbours. As Haggett put it, looking at a 'mature' settlement pattern is like 'looking up at the night sky. We can at once distinguish great galaxies and constellations, made of clusters of population of vastly different sizes. The few great centres of metropolitan population stand out clearly, while at the other extreme the myriad of small rural communities lies at the limit of our powers of [...] discrimination.

However, the wide range of settlement sizes that at first seem merely a random jumble begins to make more sense as it is studied systematically. For example, when scientists investigated the urban centres of many countries in the developed world they discovered that in some of those countries settlements were related by size in a remarkably straightforward way. This relationship is called the rank-size rule. The **rank-size rule** simply says that if you arrange all the urban centres of any region in descending order by population, then there is a regular ratio between the rank order of each and its size compared with that of the largest centre. For example, the second city contains half as many people as the largest; the third city a third as many; and so on (Fig. 44.10).

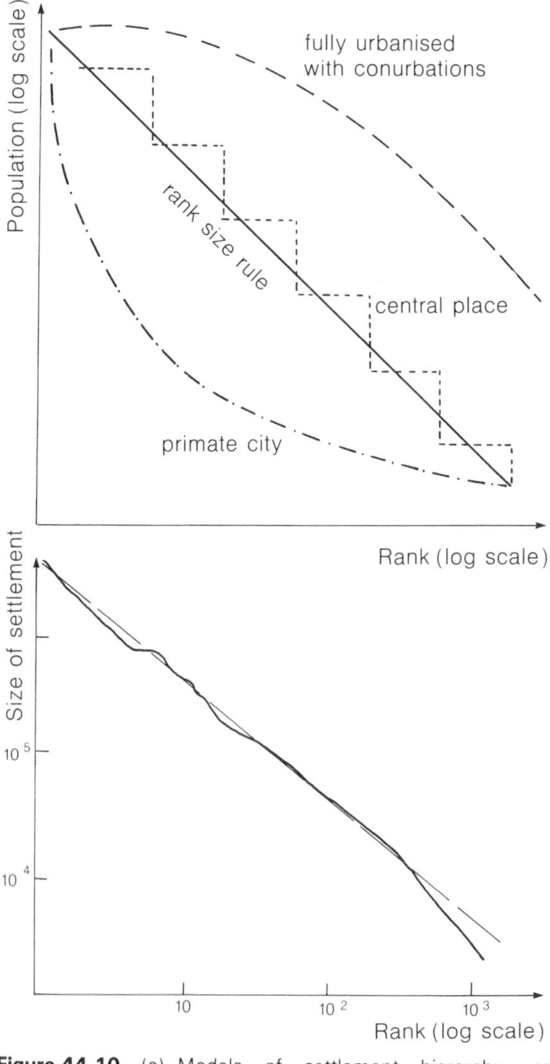

Figure 44.10 (a) Models of settlement hierarchy, as discussed in the text. (b) The rank size distribution for the USA in the 1930s. Since then it has tended to look more like the conurbation pattern.

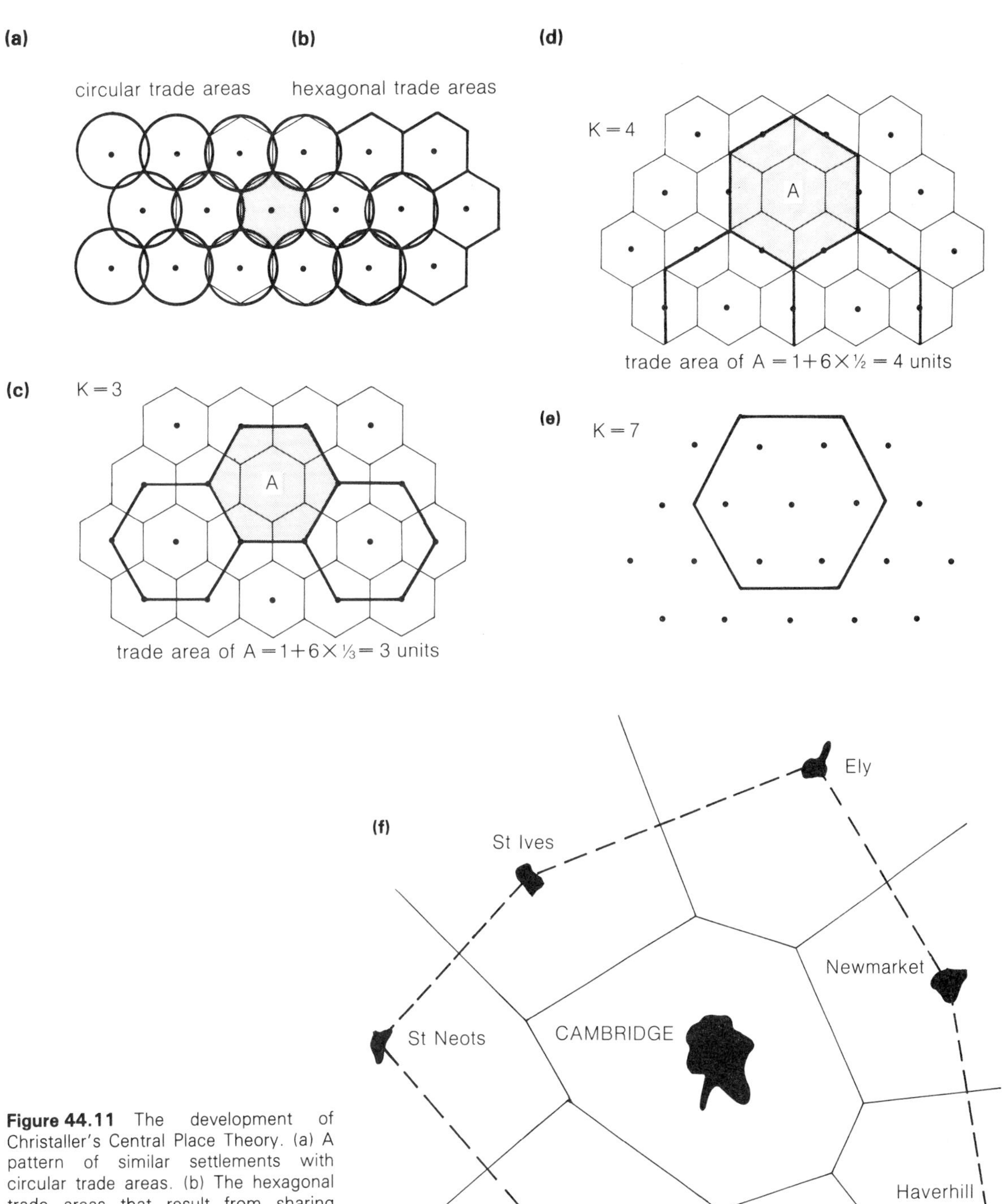

Figure 44.11 The development of Christaller's Central Place Theory. (a) A pattern of similar settlements with circular trade areas. (b) The hexagonal trade areas that result from sharing regions of trading overlap. (c) The hierarchy on $K = 3$ principle. (d) The hierarchy on $K = 4$ principle. (e) The hierarchy on $K = 7$ principle. (f) The area surrounding Cambridge shows a close resemblance to the classic Christaller pattern.

Christaller's central place model

The rank-size rule suggests that the spacing of settlements is governed largely by size (population). Because large settlements are few they are widely spaced, whereas the larger number of smaller centres have a closer spacing. The problem is to produce a model that will satisfactorily explain this complex distribution.

It is possible to begin to build a model of settlement spacing from a local level. To do this easily a number of simplifying assumptions about the character of a region must be made. These are as follows:

(a) Throughout the region there is an even distribution of purchasing power; that is, there are no especially wealthy or poor areas.
(b) People go to the nearest centre to buy all their goods or to obtain their services.
(c) The trade areas of the urban centres completely fill the region, neither overlapping nor leaving regions without services.
(d) There is a certain trade area needed to support a town.
(e) The bigger the town, the larger the trade area needed.

As the Wrexham and Bridgwater studies revealed, trade areas tend to be variable in their shape, but this is largely due to special local conditions. In general, trade areas will be roughly circular. The first problem in constructing a settlement pattern model is therefore to find a means of packing as many circular trade areas as possible into a region while keeping to condition (c) above. If all possible regions are served, circular trade areas must overlap. But on the assumption that everyone will go to his or her nearest centre, these areas of overlap can be bisected, one half being allocated to each centre (Fig. 44.11). This process yields polygonal trade areas.

By packing similarly sized trade circles together evenly (so as to provide services to every part of a region) an even pattern of settlement is created in which trade areas are *hexagonal* – a pattern that looks rather like a honeycomb.

This arrangement represents the distribution of *villages* and their trade areas. A pattern of larger centres (market towns), containing more specialised functions, can be superimposed on this pattern by assuming that every other village develops into a market centre.

There are, however, several ways in which these larger trade areas can be arranged and still fit the geometric pattern. Each arrangement is related to a specific economic consideration. For example, if direct access from each village to the centre is the most important factor, then the villages may be expected to lie on the corners of the trade area (Fig. 44.11c). With this arrangement, however, most villages will have a choice of three market towns, all equally near. Only the people within the small area in which the market town lies will owe their total allegiance to it. Nevertheless, on average, it is probable that a third of the people from each peripheral village trade area will visit each nearby market, thereby providing a trading population three times that of the basic 'village' trade area; an arrangement described as '$K=3$'. In a similar way larger towns, with more specialised functions, may be expected to build on these market centres and have trading populations three times greater in turn (that is $3 \times 3 = 9$ times greater than that of the basic village trade area). With this system the number of settlements of each level of specialisation will fit into a geometric sequence: 1, 3, 9, 27, 81, ...

As an alternative, suppose the villages found great convenience in being sited as close as possible to main roads linking larger urban centres. In this case the trade areas would be arranged as in Figure 44.11d. Now the peripheral 'village' trade will be shared equally between the two nearest centres, an arrangement conveniently described as '$K=4$'. It gives a pattern of settlements of 1, 4, 16, 64, ...

As yet a further system, suppose that administrative control was the paramount consideration. In this case each village would owe its allegiance to only one larger centre, a system that can be produced by a slightly larger basic hexagon (Fig. 44.11e). Each larger centre would now have the allegiance of seven basic village population units – a '$K=7$' pattern with a hierarchy containing 1, 7, 49, 343, ..., settlements.

This simple way of modelling the pattern of settlements was proposed by W. Christaller in the 1930s. He had observed that, in his home region of southern Germany, where there were few industrial resources and where settlements functioned primarily as service centres, there appeared to be five great regional capitals: Frankfurt, Munich, Nuremburg, Strasbourg and Stuttgart, each separated from the other by an average of 178 km. Similarly there were a large number of provincial capitals whose average separation was 108 km, country seats only 21 km apart and villages spaced, on average, by a mere 7 km. Thus Christaller saw the service settlements (**central places**) as being ordered into a sort of tiered system or hierarchy, where each size of settlement and its spacing across the landscape related to its function and status within the region. Such a landscape can still be seen where services, rather than manufactures, dominate a regional economy (Fig. 44.11f).

The Löschian landscape

The hexagonal model Christaller developed was a major step forward because it allowed a pattern of central places to be evaluated in terms of their most

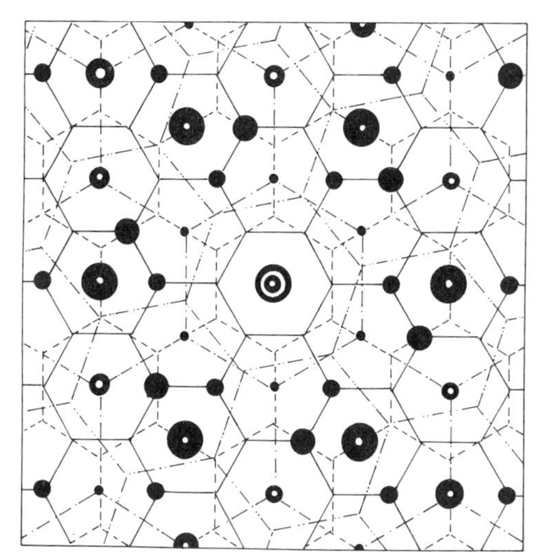

Figure 44.12 Hexagonal service areas are again assumed, with distinctive-sized zones of influence for different types of services. Here, however, there is no constant ratio between the various sizes of hexagonal service areas, so that they do not 'nest' in the manner characteristic of Christaller's system.

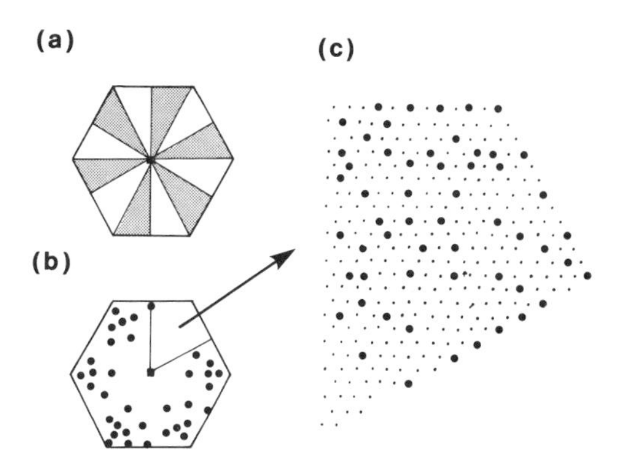

Figure 44.13 A Löschian landscape with (a) alternating city-rich (shaded) and city-poor sectors, (b) distribution of all centres within one sector, and (c) distribution of large cities.

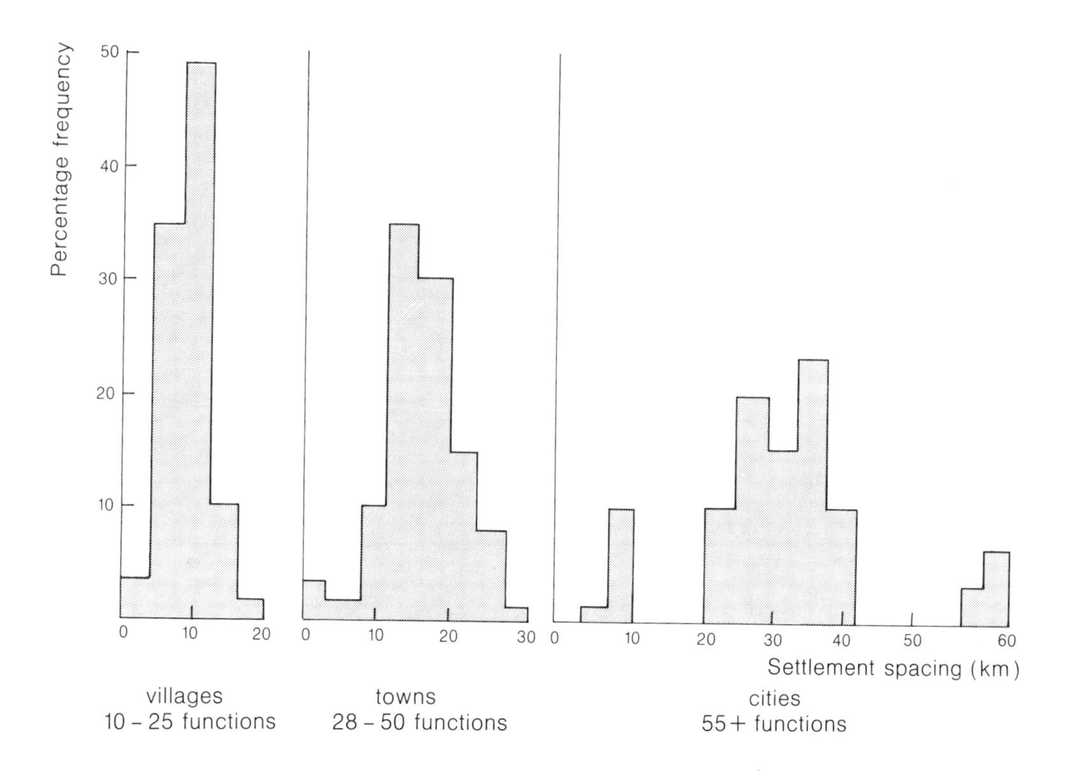

Figure 44.14 Histograms of overland distances separating towns in Iowa.

important cause. Thus, for example, settlements could be seen to be influenced most by access to *markets* ($K = 3$ system) or by major lines of *communication* ($K = 4$ system). However, although valuable as an initial stage in the modelling of real patterns, the geometric pattern is too rigid to allow its general use. In addition, if all central places fitted a strict hierarchy, pronounced steps should be apparent in the rank-size rule graph, each step related to a particular level of central place (Fig. 44.10). In fact, the continuity of the rank-size rule suggests that real settlement patterns are arranged in a more complex way and that settlements grow to a whole variety of sizes.

It was A. Lösch who helped bridge the gap between the simple step models of Christaller and the continuity of the real world. Lösch saw Christaller's system as a special case of the real world. He argued that most settlements are not influencd by just *one* economic pull (market *or* communications *or* administration) but probably by a *combination of many*. To illustrate the results of this, Lösch tried to superimpose many of Christaller's simple networks, centring each network on the most important city (Fig. 44.12). To make the greatest number of central places of each network coincide, he rotated the nets about their centre.

The pattern of settlements that results from Lösch's work looks almost random. However, it is no more random than the pattern of waves that appear at sea as a result of the interaction of a small number of simple wave trains. Of prime importance, however, is that this system permits a nearly continuous sequence of settlement sizes and allows settlements of the same size to have different functions. Similarly, large places do not necessarily have to have all the functions of the smaller central places.

One final interesting feature of the Lösch network is that it yields zones that are 'city-rich' and others that are 'city-poor' (Fig. 44.13). These zones of intensified activity are characteristic of areas bordering the communications lines that radiate from major cities.

The strength of Christaller's model is in its simplicity. It is especially useful to show the importance of trade in agricultural regions. For example in Iowa in the 1930s there were indeed characteristic distances separating towns of similar status, and trade areas of towns had a distinctly hierarchical form (Fig. 44.14). However, so many other events concerned with the use of industry have overtaken the majority of regions that Christaller's model or even that of Lösch has only limited application.

The influence of trade and industry

Traditionally, urban areas have grown out of rural settlements; yet urban areas that act purely as service and administrative areas, and that contain only a relatively small amount of industry, cannot grow to be very large. Large urban areas cannot be explained satisfactorily by Central Place Theory alone (Fig. 44.15) because they no longer have total dependence on the nearby hinterland, but are related to distant markets through inter-regional trade. There is thus a change from the market town whose function is primarily that of providing local trade and whose resource – the people on the surrounding land – was evenly distributed to one whose function is primarily to engage in services and manufacturing, and whose resources are often highly concentrated. This change completely alters the pattern of settlement.

Two new factors begin to take on an increasingly major role in explaining the pattern of these larger settlements:

(a) the **gateway** concept of an external input or resource that leads to basic processing industries and transport services;
(b) the concept of **agglomeration** whereby interlinked and dependent manufacturing and service industries cluster together to achieve economies of scale and to reduce transport costs.

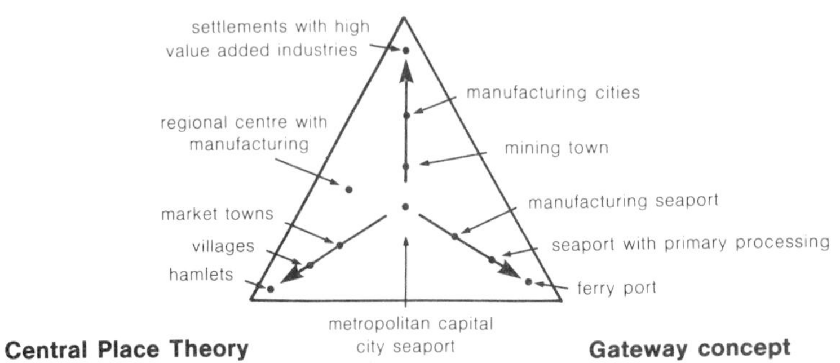

Figure 44.15 The difficulty of explaining the pattern of settlements by any one simple concept shows clearly on a triangular graph. Here, settlements have been arranged to show the degree to which each of the three basic concepts (central place, gateway, agglomeration) can offer a satisfactory explanation. Notice how the metropolitan capital city seaport, being the most complex form of settlement, draws equally on all three.

One major gateway was produced when explorers travelled out from Europe and discovered a whole range of new commodities. Sugar and tobacco were discovered in the 16th century, cotton in the 18th century, and so on. Each new commodity found a ready market in Europe and thus stimulated trade. Trade with areas abroad automatically required ports (gateways) in both the exporting and importing countries. The goods were mainly of large bulk with some desired ingredient that was relatively small. Cane sugar, for example, contains mostly waste fibre and a little glucose. For the importing country there was considerable sense in reducing the bulk near to the port, but neither port nor processing location is related to the network of local trading centres on which they are superimposed. In effect, therefore, trade and processing industries distort the Christaller or Lösch landscapes by adding centres (cities) whose roles are not primarily connected to those of their immediate neighbourhoods (right-hand side of Fig. 44.16). These gateway cities are not the end-product of the growth of the agricultural region; rather they are places of further growth, attracting other industries and people.

The effect of trade on the 'new worlds' discovered by European explorations was rather different. Colonisation imposed an advanced culture and city pattern, which had been developed over centuries in

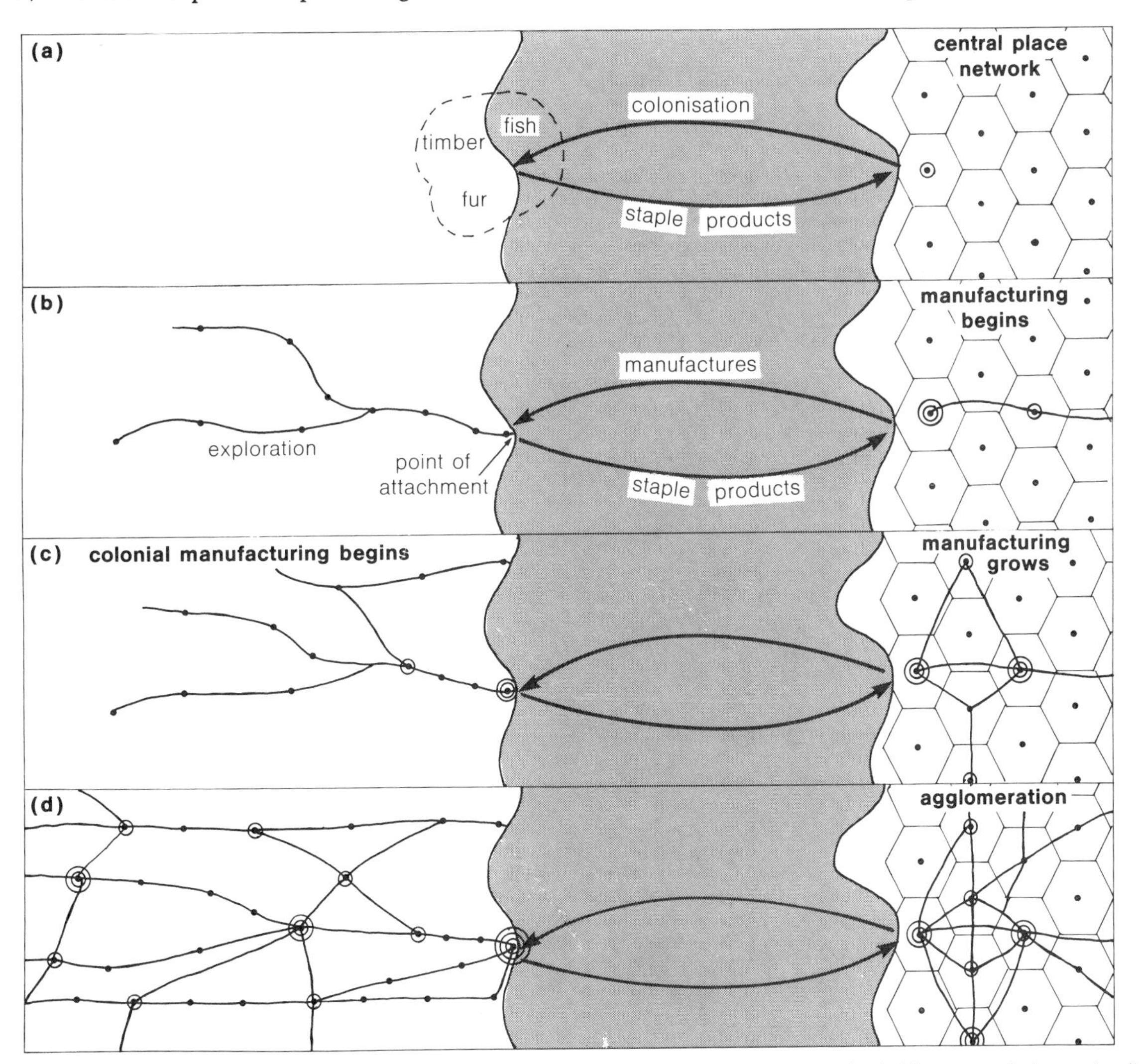

Figure 44.16 The growth of settlements based on (a) opening gateways, (b) development of subsidiary central places, (c, d) subsequent agglomeration. Notice that pioneer gateways are opening inland all the time (entrepôts) as well as extending along the coast.

Europe, on to countries which themselves had developed *no* hierarchical structure. In these places the city port immediately took on the dominant role regardless of the nature of the hinterland. New York for the American colonies in the 17th century, Calcutta for India, and Mombasa for Kenya are good examples. From this gateway or **point of attachment**, lines of communication were built into the hinterland. Inland, new centres (such as Nairobi in Kenya) were established to feed goods back to the port. Later the lines of communication became more intricate as the areas were more fully colonised and exploited (left-hand side of Fig. 44.16).

Patterns of settlement based on external trade (in commodities) therefore have little to do with a hierarchy of central places. Similarly, those settlements based on the local availability of a resource such as coal will be but little influenced by the pattern of trading settlements in which they are growing. Indeed, the simple fact that they attract large populations who wish to exploit the resource automatically makes them important centres for services, whether near an established service centre or not. This is the agglomeration effect. As a result their zones of influence may be very small compared with the number of services offered (Fig. 44.17).

Multiplier effects

Trade and industry place a directly disruptive effect on a central-place model. However, in many cases, the location of a resource dictates that they develop at entirely new sites. As industry grows it can distort the central-place model even further, because it constantly attempts to minimise costs by close linkage and by economies of scale. In many cases interdependent manufacturing units can minimise costs by agglomeration. Because industry needs a workforce for its operation, concentration of industry will cause a corresponding concentration of population.

The growth of population in these industrial centres may be many times greater than that caused directly by the agglomerating industries. This is because a **multiplier effect** operates. Suppose, for example, a chemical factory is built which requires a workforce of 5000. Each worker will probably have a family which will need not only housing but services – groceries, hardware, utilities, transport and the like. Thus the initial 5000 may generate several times as many jobs, most of them in service industries.

Some industries, such as motor-car manufacturing, have a high degree of linkage between the assembly plant and the component makers. Service activities are also commonly clustered, adding to the size of the urban area. Many firms offer only a partial service (such as office cleaning or office-machinery servicing or consultancies). They therefore need to operate in a 'swarm'. Agglomerations often grow to the point where a small-scale service cannot afford to be elsewhere.

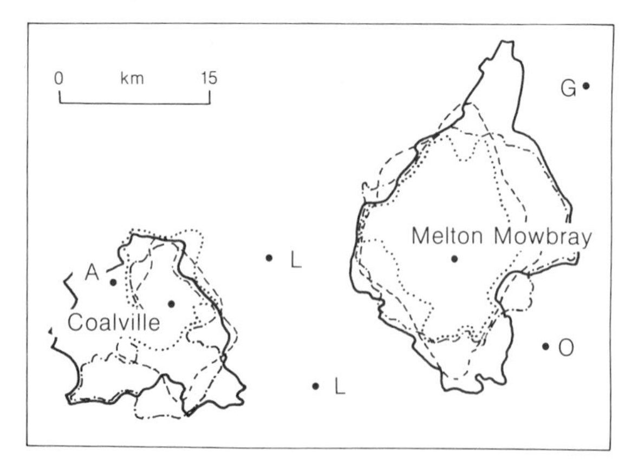

Figure 44.17 The contrast between the outward-looking trading town of Melton Mowbray and the introspective town of Coalville, whose growth and employment largely centre on coal mining and have little to do with the surrounding rural area.

Conurbations

The resources for industry may be located at a point (at a port, for instance), or within a small specific region. This is the case with coal-mining: many mines can tap the same coalfield, all of the mines being within a few miles of each other. In the early days of the Industrial Revolution coal was required in such large amounts to power steam-driven machinery or to make iron, steel and gas, that in an age of difficult transport it came to dominate location. However, once sited on the coalfield and within easy reach of a mine, much industry could have a choice of locations. In practice the iron-and-steel industry grew up in the Black Country area of the West Midlands as a collection of small independent firms, each focusing on one agricultural village on the coalfield (Fig. 44.18). Thus the pattern of early industry fitted well within a central-place model. (The same was true of the Ruhr in Germany; Fig.44.19). With time, however, the agglomerating and multiplier effects caused each village to grow into a manufacturing centre whose expansion was like so many fingers reaching out in a grimy embrace. And as the urban fields interlocked, so the Black Country became a **conurbation**, finally destroying the last vestiges of the central-place pattern.

Not all conurbations have grown on a specific resource like coal. Major metropolitan centres such as London and Manchester have simply grown by

Pre-industrialisation

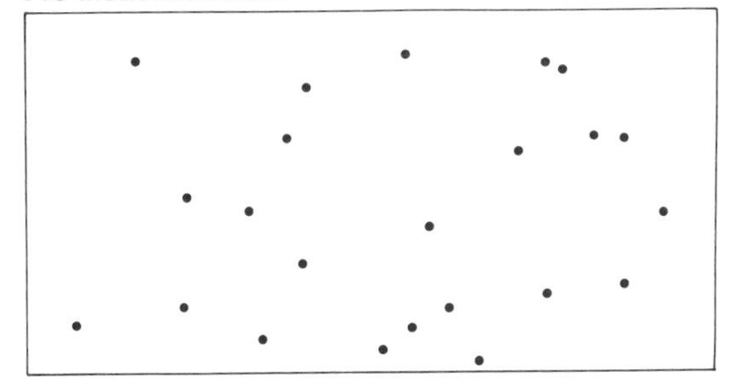

Post-industrialisation

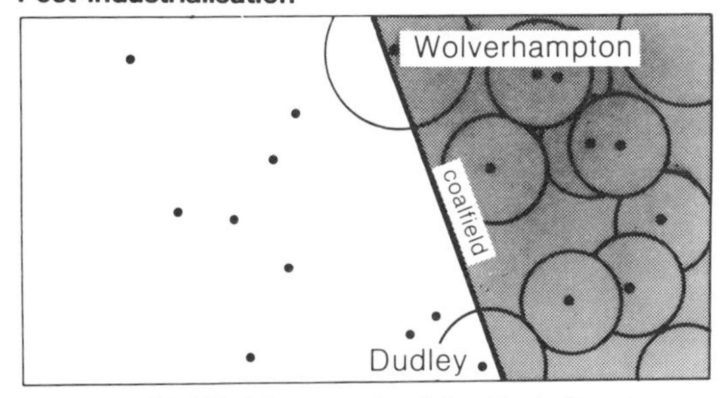

Figure 44.18 The growth of the Black Country.

agglomeration from a central point, slowly but surely enmeshing what were once peripheral urban centres. In these cases, in particular, a hierarchy of sorts remains; a hierarchy evidenced largely in the distribution and size of neighbourhood shopping centres and in the structure of local government.

Variations from the rank-size rule

Having now looked at settlement patterns varying from those totally based on agriculture to those dominated by conurbations, it is time to ask the question: 'Does the rank-size rule represent a fully developed urban pattern?'

This question can be answered by looking at those countries where there is considerable deviation from the rule. Most are developing-world countries in which one city completely dominates the others. Examples include Sofia in Bulgaria, Khartoum in the Sudan, and Kuala Lumpur in Malaysia. This is possibly because, in countries based on a simple subsistence-level agricultural economy, one city – the **primate city** – can provide all the urban functions necessary for the welfare of the people (Fig. 44.10, lowest line). London was a primate city for Britain before industrialisation. The rank-size rule, however, suggests that a mature pattern has the urban functions divided to allow many regional capitals to develop. Such **decentralisation** can only become important as a country becomes more complex and diversified in its activities.

From the initial evidence, therefore, it is tempting to suggest that many urban-settlement patterns evolve from a primate city pattern towards a pattern with a closely structured and balanced network of cities.

But is this the end of the line? A balanced pattern was certainly true of the USA in the 1930s and even of Sweden in the 1950s (Fig.44.10). However, since this time conurbations have dominated the settlement patterns of many countries. As a result, countries of the developed world are beginning to diverge from strict adherence to the rank-size rule again, this time in favour of a pattern that is dominated by several fairly equally sized conurbations, a so-called **binary pattern**. In Britain, Greater London may still dominate with a population of 7 million, but the West Midlands (2.8 million), Greater Manchester (2.7 million), West Yorkshire (2.1 million) and Clydeside (2.0 million) have all risen to dominate the pattern as well. The same is true of West Germany where the Ruhr has 5.5 million, West Berlin 3 million; Rhein-Main (Frankfurt) 2.2 million, Munich 1.8 million, and

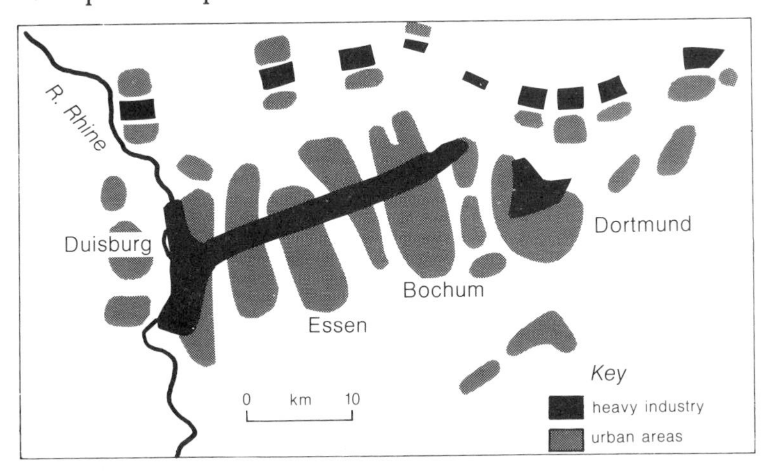

Figure 44.19 The Ruhr conurbation.

Hamburg 1.8 million. It is also true of the USA and many others. In other words there is a tendency for a number of similarly sized major centres to dominate the upper end of the settlement pattern. It seems, therefore, that a situation where the majority of the population live in conurbations is the next stage of settlement evolution and that the rank-size rule represents only a special case in a continuous process of evolution.

45

Cities of the developed world

45.1 The evolution of cities

The most characteristic, yet enigmatic, urban form is the city. There have been many studies of the internal structure of cities, both in the developed and developing parts of the world. The most fundamental result has been the realisation that no one standard model will represent all cities. In Europe, for example, cities grew first in the Middle Ages when technology was but little developed, the main period of their expansion coinciding with the Industrial Revolution. These cities were therefore based on manufacturing, and often closely tied to resources such as coal. In contrast, cities in the developing world are now evolving at a time of advanced technology and frequently possess only a minor manufacturing sector.

The pre-industrial city

The pre-industrial city was small (less than 100 000 population) and had an internal structure dominated completely by the centre. Here government and religious activity took precedence over any form of trade. Surrounding this core were houses of the élite, with the poor people and outcasts relegated to the periphery. Within the city, ethnic separation was at a maximum; the class structure was very marked; there was hardly any opportunity for social advancement; there was little division of functions into sectors and there was almost no specialisation. Trade guilds placed restrictive practices on all crafts and thereby held down the level of economic activity. In this type of city there was little to attract people from the land.

The 'Western' city

As far back as the Middle Ages in Europe dramatic changes were occurring that were to alter the size and form of the pre-industrial city beyond recognition. The important factors were as follows:

(a) *The growth of a system based on producing a surplus for sale* (as opposed to simply producing that needed for subsistence). This change was directly responsible for the rise in commercial activity, for the search for new products to offer for sale, and for development of new techniques of manufacturing. This led to the Industrial Revolution.
(b) *The concentration of political power* into one centre for each country. With power focused and administration more centralised, there was also a focusing of wealth and the centre began to attract migrants. This caused a rapid expansion of the centre, which became the **primate city**. The growth of the primate city automatically led, at least for a time, to stagnation elsewhere, such that the gap between city sizes grew.
(c) *The rise of the new idea of orderliness,* of planned and gracious city centres and of improving buildings. This caused a flurry of rebuilding, which destroyed the jumbled form of many pre-industrial cities forever.

The small pre-industrial city could remain dominant only as long as the majority of people used the land as a *resource*. Agriculture automatically keeps the population dispersed, and changes in the demand for agriculture-based products occur only slowly. On the other hand, manufacturing, commerce and services use land only as a *site;* to minimise transport costs and to share a potential labour pool, they tend to

Figure 45.1 A panorama across the provincial town of Reading in 1974, looking east. At this time the centre was being rebuilt. A new shopping, office and local government complex was being constructed (foreground), bypassed by an inner distribution road. The CBD shows clearly, although its office blocks are even more prominent today. On the left, the railway forms an expansion barrier. The university parkland lies within the suburban housing on the right. Low-status terraced housing along the arterial road contrasts with better-status dwellings on higher land to the right.

group together. Manufacturing is also a much more flexible economic sector: it is able to create new products, to stimulate demand and to respond quickly to increases in that demand.

Slowly at first, the growth of the modern city of today gathered momentum. Cities grew which contained both order and disorder, the whole bound together by the great link of convenience (Fig. 45.1). At its cornerstone lay the two major thrusts of industrialisation and commercial competition whose influence still dominate the city structure.

The industrial city

So long as the main function of the urban area is to provide services for a primarily agricultural population spread evenly over the landscape, it is possible to maintain a hierarchical pattern of settlement much like that described by Christaller. Industrial growth completely disrupts this orderly structure by concentrating demands at selective growth poles. As a result the strict hierarchy dissolves and is replaced by a more random distribution of urban centres.

Through the process of industrialisation, entirely new centres will develop close to specific resources. Trading cities that already have embryo commercial structures are most likely to be able to benefit from, and give support to, less rigidly tied industries, providing finance, good distribution and a substantial local market. Such commercial towns often sprint ahead of others as industrialisation proceeds, eventually growing in an upward spiralling, self-reinforcing manner (Fig. 45.2).

45.2 Factors affecting land use in cities

Once a city begins to direct its endeavours towards manufacturing as well as towards providing services, a natural conflict arises between potential users for city sites. This competition is accentuated by the need to provide housing for the workforce at a reasonable distance from their place of employment. As a result

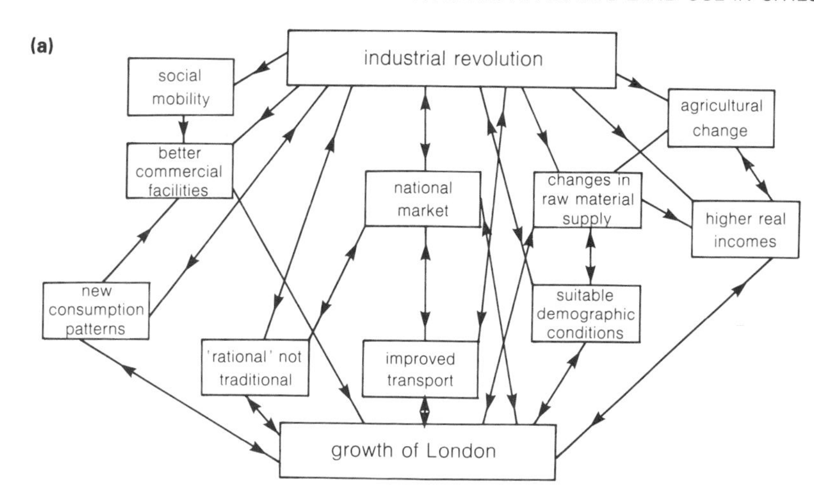

Figure 45.2 (a) The explosive growth of many established commercial centres such as London was due to the two-way flow of information and innovation, with city growth and industry reinforcing each other in a cumulative causation sequence (b).

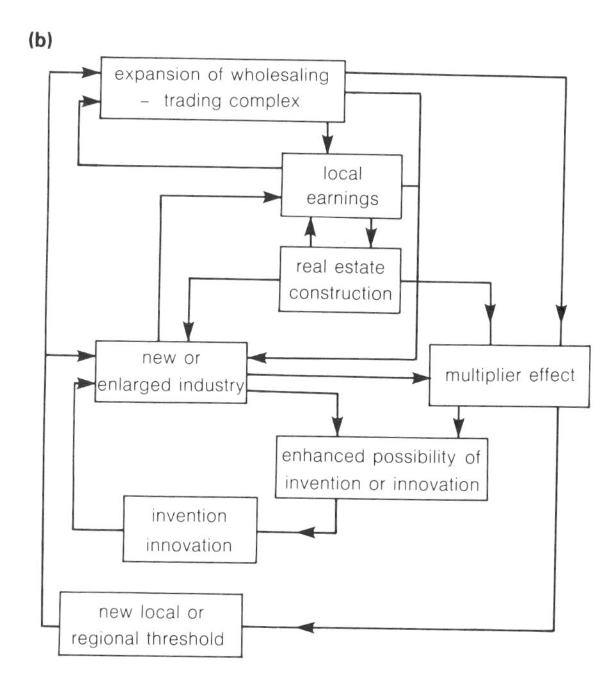

the structure tends to develop towards a number of zones, each with a particular function. Exactly what pattern these zones make depends in part on the character of the competing functions and in part on the transport available: if people must walk to work, they need to live near workplaces, and segregation is restricted; but if they can travel by train, car, tram or bus, their degree of freedom increases and segregation becomes much more pronounced (Fig. 45.3). These factors, and the unique city structures they have produced, are discussed in the sections that follow.

Bid-rent analysis

People live and work in a city for a variety of reasons. However, in a free market economy, success results in an increase of wealth. In many instances the location of the workplace within the urban structure is of vital importance in ensuring its success.

A shop, to take a familiar example, needs a position such that as many people as possible walk past it. The size of the potential market depends on the type of goods being offered for sale: a small food shop would gather sufficient custom simply by having a frontage on a main road or street corner; a supermarket needs a much larger turnover to compensate for its cut prices, and may therefore need to be at the junction of two main roads in order to gather sufficient custom. Even this more central location may be insufficient if there are no other shops nearby, for many people try to do all their shopping in one trip. It is therefore advantageous for shops to cluster. For the most specialised shops of all, perhaps only the most accessible (and therefore the most central) location within the city will provide sufficient custom. But specialisation and high accessibility do not encourage low prices; on the contrary, as central sites are in demand by other types of employment, and as the owners of these sites can charge a higher rent than site owners elsewhere, the profit margins will need to be substantial.

In the past before road communications allowed greater flexibility of site, industry also needed a central location simply to be able to gather raw materials and to distribute its products by canal or rail. Today it is those functions that require not only high accessibility within the city and beyond, but also which need to cluster for ease of personal contact and exchange of information, which can gain the greatest advantage from a central location. International

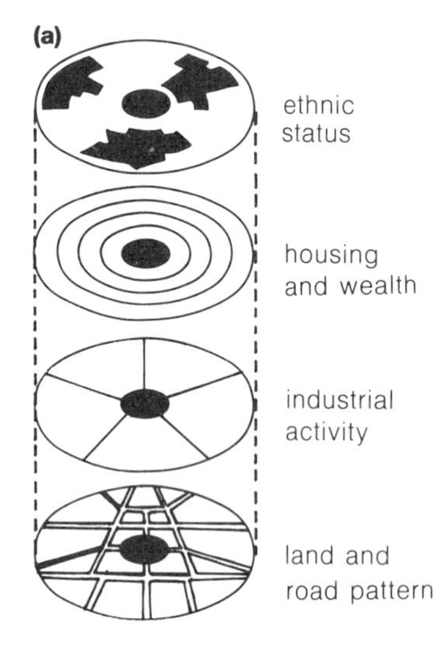

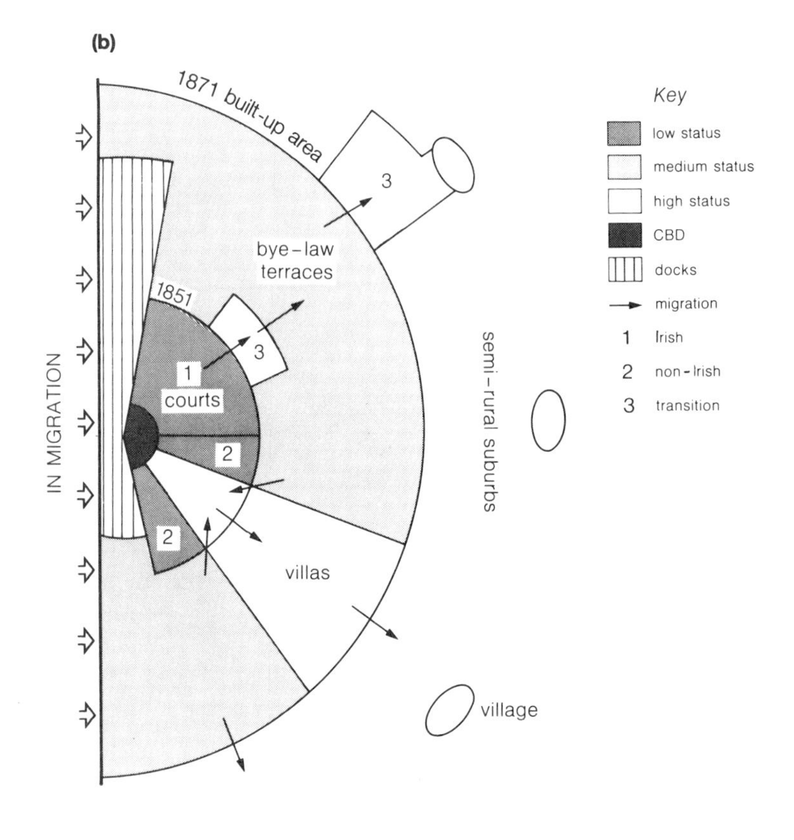

Figure 45.3 (a) A city is made from many parts, each of which has its own distinctive pattern. Thus industrial activity may be distributed sectorally, housing patterns may be distributed concentrically, and ethnic status may occur as groups. Taken together, an almost random 'cellular' pattern is produced. (b) The structure of Liverpool at a time of changing transport technology in 1871, showing the main areas of land use and direction of migration along railway lines. Notice the dominantly concentric residential pattern with ethnic grouping. It was observation of Chicago's similar development that led Burgess to propose a concentric model of city structure in the 1920s.

business, government administration and financial institutions have, in the past, found a central site particularly advantageous, and they have been able to pay the highest land prices. Today they cluster to form the zone of the city called the **central business district (CBD)**.

If market forces alone influence the location of businesses, the pattern of land use in a city will result from competition between the various potential users: at any given location, we might expect the land to be used by whichever competitor can pay the highest price. In effect, the site is auctioned – all potential users *bid*, and the one who can afford to pay most rent acquires the site.

For each urban function there will be a location at which it could make most profit. In the example of specialised shops, the most profit will be gained by being in a prominent position within the main shopping area of the city centre. If the shop is in a side-street location, fewer people will go past and the profit will consequently be lower. The shop owner may therefore decide to pay a higher rent for a central site rather than a lower rent for a peripheral one, gambling that a greater profit will more than compensate for the greater rent. At a certain distance from the centre, so few people would pass the shop that the profit obtained would be insufficient to pay the rent (Fig. 45.4a). The **bid-rent curve** shows the rent affordable at varying distances from the centre. In the case of speciality shops, such as those selling exclusive jewellery, this curve will be very steep, for a peripheral location will spell disaster. By contrast, a firm of wholesalers, which would find a central location advantageous both for receiving its goods and distributing them all over the city, cannot simply put up its prices to match its location: it cannot afford to make a higher bid. Indeed, the amount of extra rent it could afford to pay for a central location will only be the amount gained by having lower transport costs. The bid-rent curve for wholesaling may well be

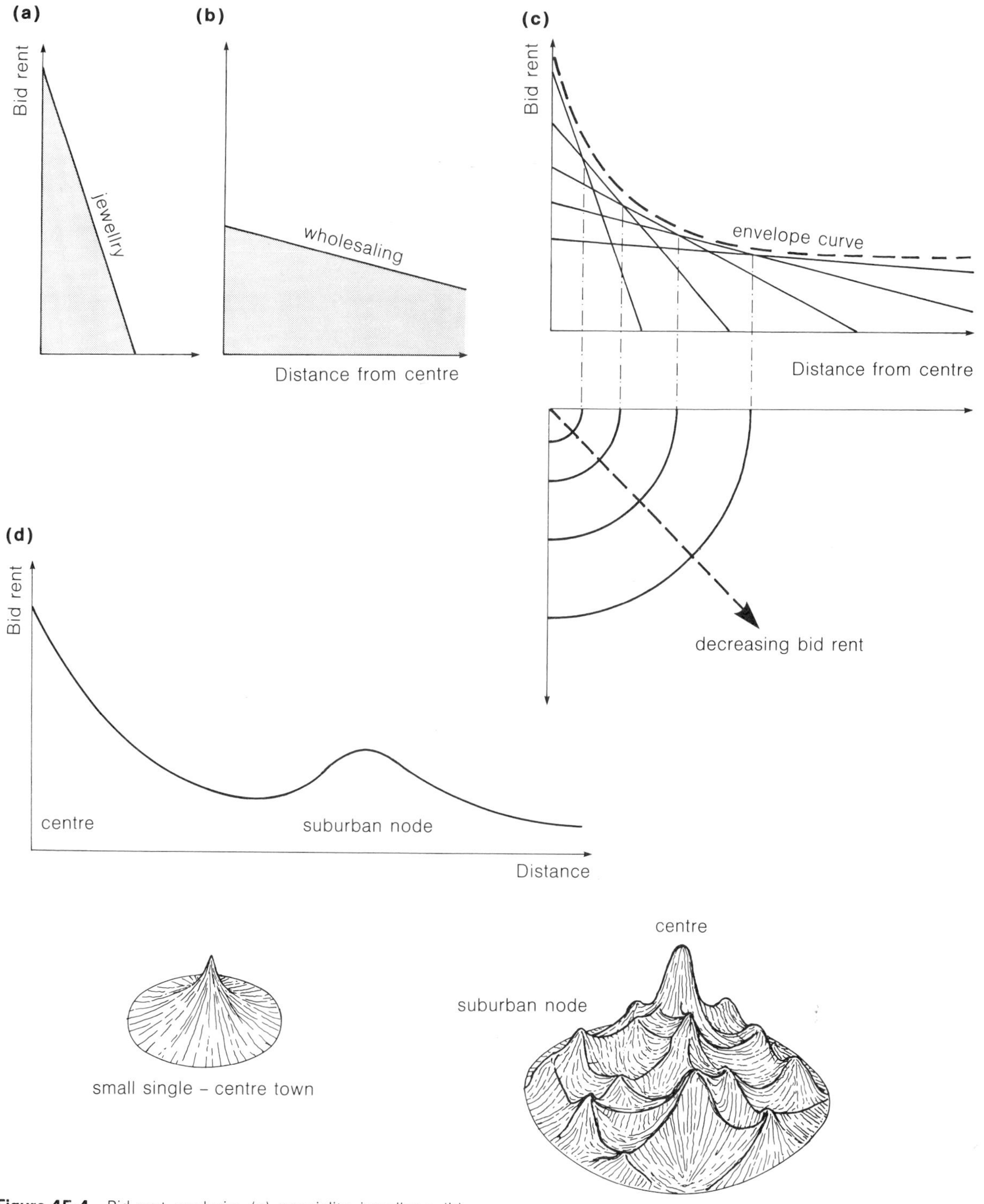

Figure 45.4 Bid-rent analysis: (a) speciality jewellery; (b) wholesaling; (c) the pattern of commercial land use based purely on competition; (d) bid-rent cones for small and large towns.

as in Figure 45.4b. Because the speciality shop can afford to bid (and pay) more for a central location than can the wholesaler, it will occupy the place of greatest accessibility; the wholesaler will occupy a more peripheral location.

In this example we have looked only at specialised jewellers and wholesalers. In reality there will be many types of activity, each with its own bid-rent curve. In general it is found that if all the bid-rent curves are drawn on the same graph they will define an 'envelope curve' such as that in Figure 45.4c. This reveals the highest rent the competing users are prepared to pay at each distance from the centre. Thus the envelope curve is in a sense a cross-section of the city; shown in three dimensions it is seen to be a cone (Fig. 45.4d).

Competition for a suitable location will occur outward from more than one centre – not only outward from the centre of a large urban area, but also outward from major crossroads (such as where ring and arterial roads meet). These too have good accessibility; indeed, with congestion occurring in the city centre, these junctions may have even greater accessibility for many purposes. Such desirable locations will generate subsidiary bid-rent cones (Fig. 45.4d). Further competition for other desirable

Turn again, Dick Whittington

So you think it costs the earth to rent office space in the City of London. A new "rent contour" map drawn up by the estate agents Hillier Parker May & Rowden invites you to shed your misconceptions. What you pay in the square mile depends on where you want to be. Roughly 85% of the lettable space costs less than the peak prices charged for premises hard by the Bank of England. In recent months, some rents have risen above £30 per square foot.

Only a small minority of office users, however, actually pay these astronomical rents. Even tenants with rent reviews every five years generally pay less than the open market rent between reviews. Also, over quite short distances, there are startling differences in rents. For example, over the 460 yards between 99 and 246 Bishopsgate, rents fall from £32 to £14 per square foot. Put another way, every yard north costs you 4p per square foot less. Another steep gradient lies between Lloyd's in Lime Street and Mansell Street on the eastern boundary of the City. Between those points, rents drop at the rate of £2.50 per 100 yards.

The lowest rents (about £10 per square foot) are to be found in the administrative area of the City. If you want to pay less than that, make sure you get your bearings right as you walk out of the City. North of Liverpool Street or south of London Bridge costs well below £10. Walk 40 miles west to Reading, though, and you will still be paying a tenner a foot.

What it costs to sit in the City

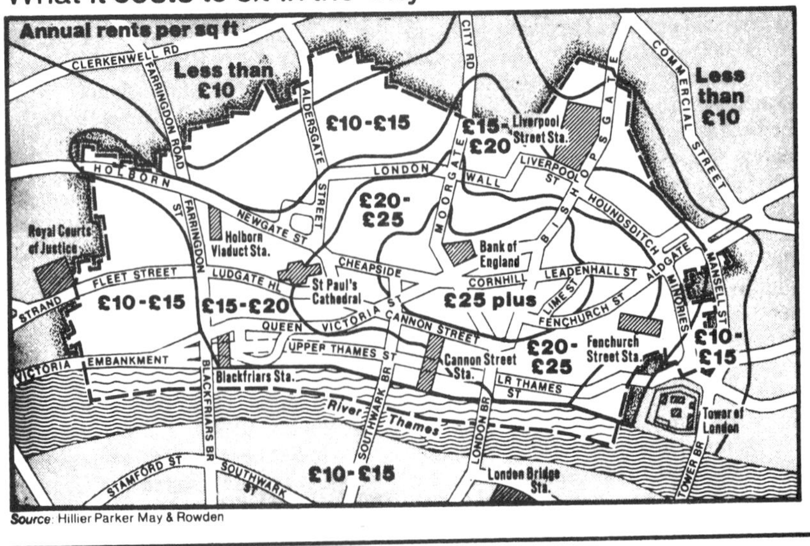

Figure 45.5 Bid rent at work in central London. Here the shape of the CBD is mapped by the rental costs of office space in 1983. Notice how the CBD declines more sharply towards the east and north than towards the west.

locations will result in a mosaic of varying bid-rents within a real urban area, although the principle behind the resulting pattern remains simple (Fig. 45.5).

BID-RENT UPWARDS

The city is a three-dimensional structure and bid-rent operates to a large extent upwards as well as outwards from the centre. Thus it is common to find retail outlets occupying the ground floor of a CBD site, while the floors above are occupied by offices and other functions. By the nature of their activities, industrial premises are often restricted to ground-floor sites, but services, entertainments, offices, residences and some warehousing can all function from upper stories (Fig. 45.6).

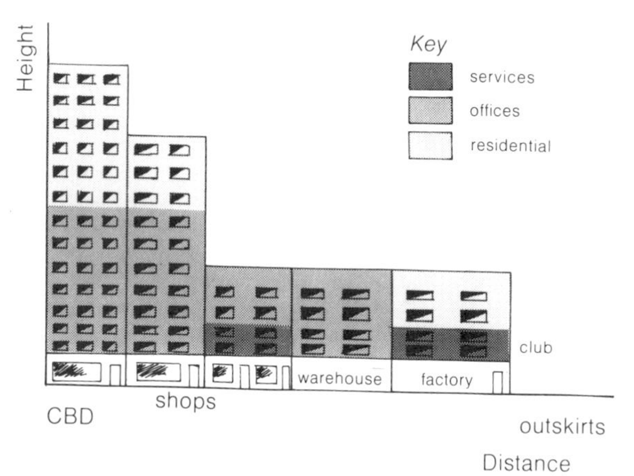

Figure 45.6 The vertical structure of land use in a city.

BID-RENT THROUGH TIME

The argument for segregation of functions because of varying ability to pay rent has so far taken no account of time. In practice time becomes very important for several reasons. First, the only way a city can grow is to expand outwards, which means that much of the newer development of a city will be at the periphery. Each new growth takes place at a different stage of technological development, with new ideas being put into effect along the way. Thus some buildings, which may have been very suitable in design for their original purpose, gradually become less valuable for more modern uses. In addition, as buildings age, they become increasingly costly to repair, a factor that may cause their owners to sell them in favour of more modern buildings elsewhere. For example, the once-prestigious houses built by wealthy Victorians who had large families and servants are mostly too large for the families of today. The more affluent sector of society therefore chooses mainly to live in smaller, more modern houses. In turn, this means that the only way the old, large buildings can be managed profitably is by dividing them up into flats for use by a less affluent group. In a similar way, old warehouses and factories are often subdivided and re-let to smaller firms which, in general, are not individually able to pay as high a bid-rent as could their predecessors. In this way there is a continual change of building use, and therefore of bid rents, as they are occupied by people or organizations of varying socio-economic status.

Forces in equilibrium

A city is in a state of dynamic equilibrium, its structure and shape the results of contrasting influences. These influences can be separated into those that cause functions to want to move outwards to the newly developing regions (the **centrifugal forces**), and those that tend to draw them towards the centre (the **centripetal forces**). The *centrifugal* forces include:

(a) lack of space to expand;
(b) lack of a suitable site for changing needs;
(c) high bid-rents;
(d) congestion, reducing efficiency of movement to and from the site;
(e) government attraction to out-of-town sites or restrictions on development at central locations;
(f) more spacious living environments.

The *centripetal* forces include:

(a) prestige derived from a particular central address;
(b) high accessibility;
(c) the ability of certain types of function (such as business premises, specialised shops, and entertainments) to cluster;
(d) living environments nearer to central entertainments.

45.3 Models of land use in cities

All the factors outlined above influence people's choice of a location within the urban area, whether for work or as a home. Convenience, market forces, transport networks and the ability to pay for the desired site all influence the final decision. The result is a pattern of land use within the city which appears simple but is in practice difficult to summarise. Of the several attempts to form relatively simple models, the most important results are

(a) the concentric model;
(b) the sector model;
(c) the multiple-nuclei model.

The concentric model

The simplest model of a city is a development of the bid-rent analysis. Ability to pay rent exercises control over many commercial and industrial functions, segregating them into concentric zones. This principle can be extended to include residential development and even agriculture (Fig. 45.7). Clearly commerce will be able to pay the highest rents and will therefore occupy the CBD, with specialised shopping and entertainments. Only government administration is not controlled by the competition of the market place, and this opts for a central location for reasons of convenience.

Outside the CBD should lie a ring of commercial premises not quite able to afford the high central rents, together with any light industry that is particularly dependent on patterns of transport. Here, too, will be the service firms who deal primarily with businesses in the CBD.

Beyond this zone lies the first zone of residential development, again segregated according to the ability to pay for sites: first the houses with multiple occupancy, and beyond them the houses owned by

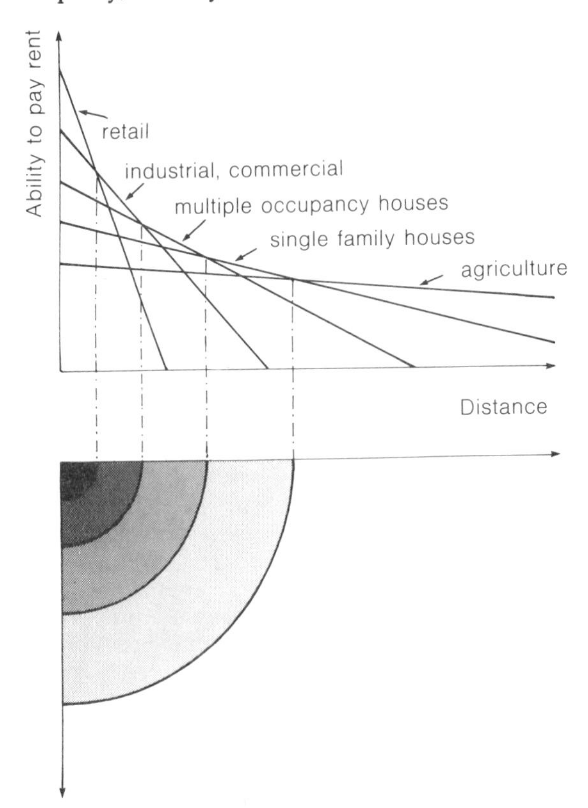

Figure 45.7 The pattern of occupancy derived from the bid-rent analysis that produces concentric zones of land use within a city.

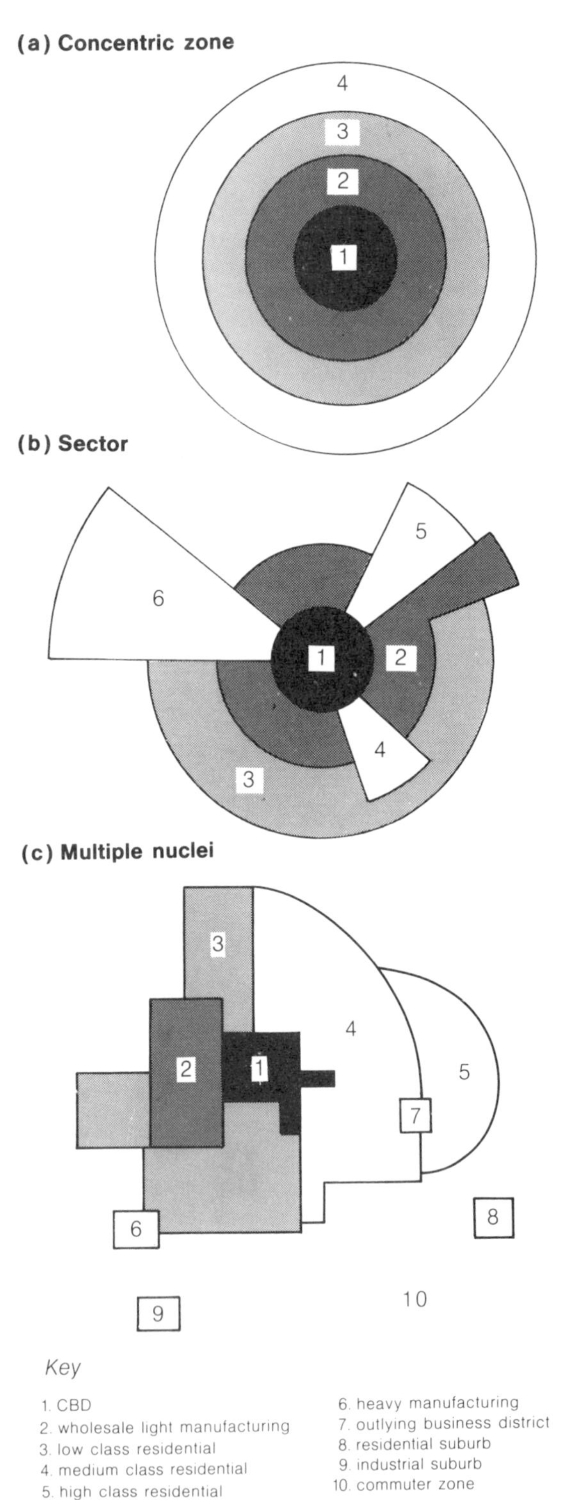

Figure 45.8 Models of the developed world city.

single families. Only where all urban competition has ceased do agricultural activities take over – they are such extensive users of land that they cannot afford a high rent per unit area.

A 'concentric city' model (Fig. 45.8a) concentrates on bid-rents for a homogeneous area and takes little account of the peculiarities of the city site or of changes that may take place through time. Thus it takes no account of the mixture of functions that result from the reluctance of people to move. Thus the zone beyond the CBD may contain the vestiges of industry that has not yet moved out to a less congested site, and large houses in multiple occupancy, remnants of the time when this zone was the home of the more affluent. In the *observed* pattern of occupation, therefore, each zone has a mixture of functions all in some way able to afford the same scale of rent. Thus industry may find itself cheek by jowl with the poorest (immigrant) sector of the community who, because they live in overcrowded, multiple-occupancy dwellings, together pay quite high rents. One particularly characteristic association of older, often decaying, industry and housing – which includes not only immigrants and those without families, but also less socially acceptable occupations such as prostitution – is often described by the terms **twilight zone** or **inner city**.

The pattern of multiple function in concentric zones outwards from the CBD was first described by Burgess, in a study of Chicago in the 1920s. Bid-rent analysis provides some theoretical support for his observations.

The sector model

The difficulty with the concentric model is in its rigid geometric pattern; in detail it becomes increasingly unsatisfactory as a model in describing real cities. For example, it does not recognize the effect of transport networks. People tend to decide on where to live in terms of *time* as well as *distance*.

Because it takes less time to travel by major through-routes than by side roads, zones tend to distort from circles into star shapes, with extensions occurring along major roads and railways.

But however important time of travel may be to the commuter, it has even greater importance for industry.

Factories are much more liable to locate near to main lines of communication than in the relatively inaccessible areas between. Inevitably, therefore, a pattern of **sectors** forms, with industry occupying positions near routeways. In such positions firms minimise the time taken in reaching both other parts of the same city and distant cities; the distance between factories is also at a minimum. This becomes especially important for those factories that supply others with components (ones that have a high degree of linkage). As industry tends to become clustered, so other functions likewise concentrate into sectors or wedges. For example, areas of more expensive housing and of industry are unlikely to occupy adjacent sites. Each sector can be expected to grow outwards by 'welding on' related functions. For example, once a housing sector has become established as high-class, sites adjacent to the established area will become expensive as builders of new houses attempt to capitalise on its reputation. The sector model of urban structure (Fig. 45.8b) is not static: as the city expands, so the sectors also will expand, outwards. There are thus some similarities between the concentric and sector models: although functions may segregate into sectors, each sector grows outward, so the pattern of age of the buildings is concentric (Fig. 45.9).

The multiple-centred city

Both the concentric and sector models describe some aspects of city functions. However many larger cities show a structure which is more complex than that indicated by either model. In addition, the pattern of development of many cities is further influenced by the relief, so cities could never conform strictly to any simple geometrical plan. Furthermore, some functions (such as docks and the shipbuilding industry) are closely tied to a specific site and will, of necessity, be independent of any formal structure.

Indeed, many cities are really too large and complex to be organised around any single centre. They are perhaps more like living organisms. Many have their functions arranged almost in a cellular form, with distinctive types of land use having developed around a number of growth points or nuclei (Fig. 45.8c). The CBD may still occupy a position of considerable accessibility but in a large city there may be no single most accessible point. For example, London has grown up around the twin centres of 'Westminster' and 'The City'.

From an industrial viewpoint a cell-like pattern may develop for two reasons. First, as mentioned above, industries that are interlinked can minimise costs by clustering, while they achieve further economies by using the same services. Second, manufacturing industry is unable to afford high ground-rents, largely because factories occupy large areas. As a result industry will be unlikely to concentrate near the city centre even along major routeways, but rather at some distance from it where rents are lower. Peripheral locations are also an advantage to the workforce, who cannot afford high rents for their homes. Similarly, shops and services providing a similar function frequently cluster so as to provide a

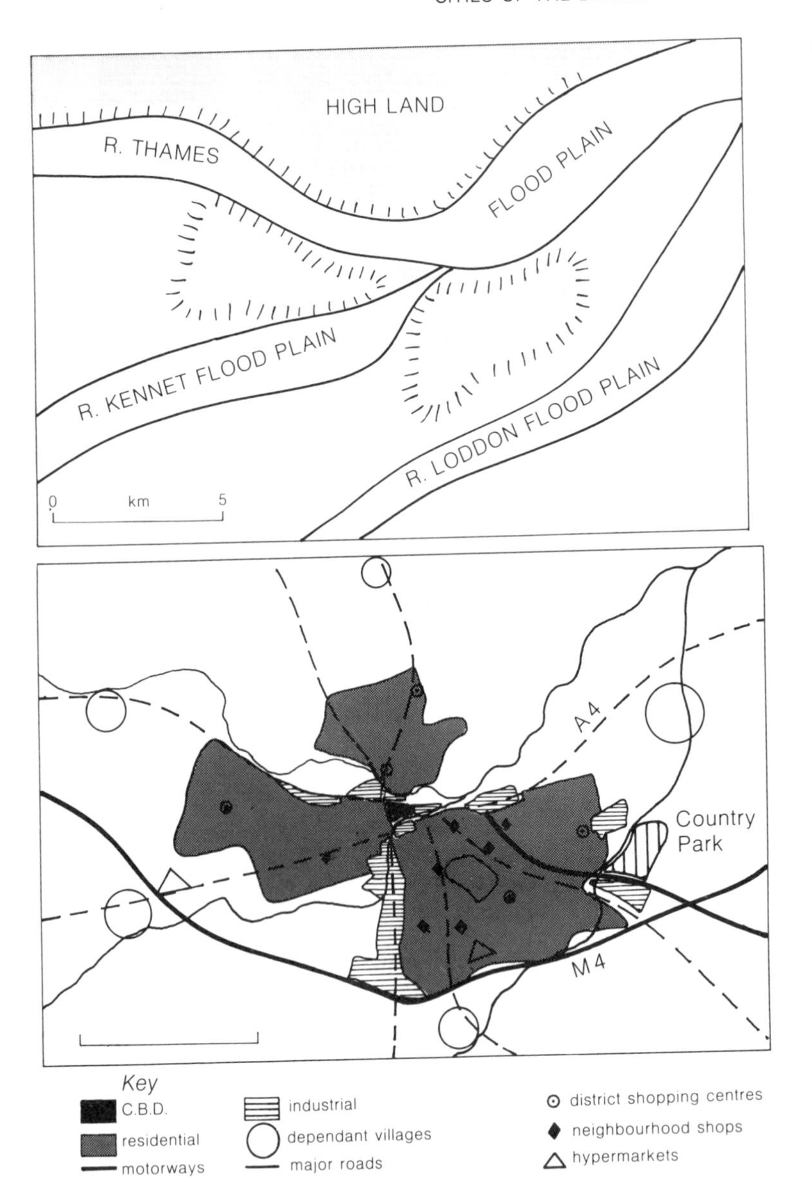

Figure 45.9 (a) The physical basis of Reading. (b) Reading's sectoral development.

recognisable area for customers and clients. Examples include building-society offices, clothing shops and solicitors.

In reality, therefore, a city, because of its large size, complicated functions, and the restrictions imposed by the physical features of its site, will show broad elements both of concentric rings and of sectors, but these patterns will, in detail, be broken up into distinct cells. Current trends in planning are based on segrating the functions of a city, so the multiple-nuclei model is almost bound to be the most relevant in the future.

45.4 Features of the urban landscape

Offices

The growth of offices in the centre of many large towns and cities has been the most dramatic development in the urban structure. The demand for office space in central areas has resulted not only in increasingly extravagant bid-rents but in the expansion of the CBD into surrounding zones. In consequence, much of the older housing in areas that were once peripheral to the central core has now been pulled down and replaced by office blocks, thereby enlarging the 'dead heart' of the city (Fig. 45.10).

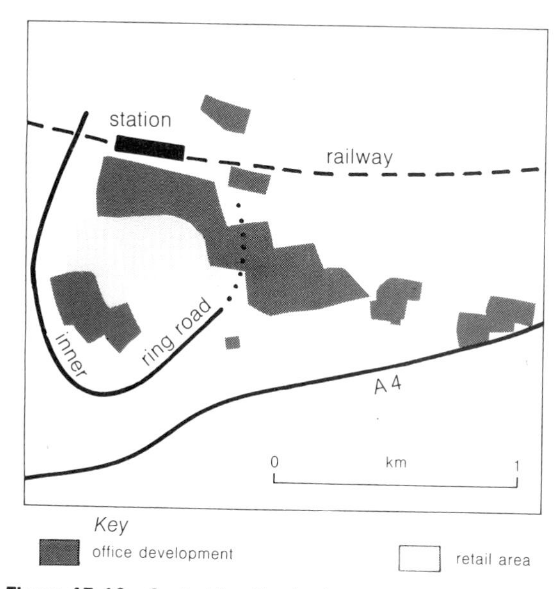

Figure 45.10 Central Reading land use. The preferred policy area for office development sends out an eastward salient towards London, acquiring land by urban renewal. The office area now stretches from the railway station towards the M4 motorway link, stressing the vital role of accessibility in the survival and growth of a CBD.

The growth of offices in the CBD has not been the same in every large city. For example, the South-East of England has a higher proportion of jobs in offices in research-and-development than anywhere else in Britain. Furthermore, the largest firms in the country tend to have their headquarters in Central London, with only firms engaged in some specialist activity located elsewhere. The dominance of office activity in the South-East and especially in Central London has been self-reinforcing: the best-qualified people in peripheral regions being forced to move if they wish to obtain the best-paid jobs.

Offices cluster in city centres because of personal contact, even in an age of advancing telecommunications. For example, because most research-and-development is located in the South-East, a firm located in a peripheral area could well find itself out of touch with the latest developments and fall behind in new products. There is thus a strong incentive to locate at least the decision-making part of the firm near to the research-and-development part. The CBD offers accessibility both to sources of new information and to branches scattered throughout the country.

Today offices locate in CBDs, and especially that of the capital, because it is information, not access to markets, that is the top priority. In addition most personal contacts, so vital for marketing, are more easily made in a large centre where there is a highly diversified range of skills in quaternary services such as advertising, financial investment advice, computer

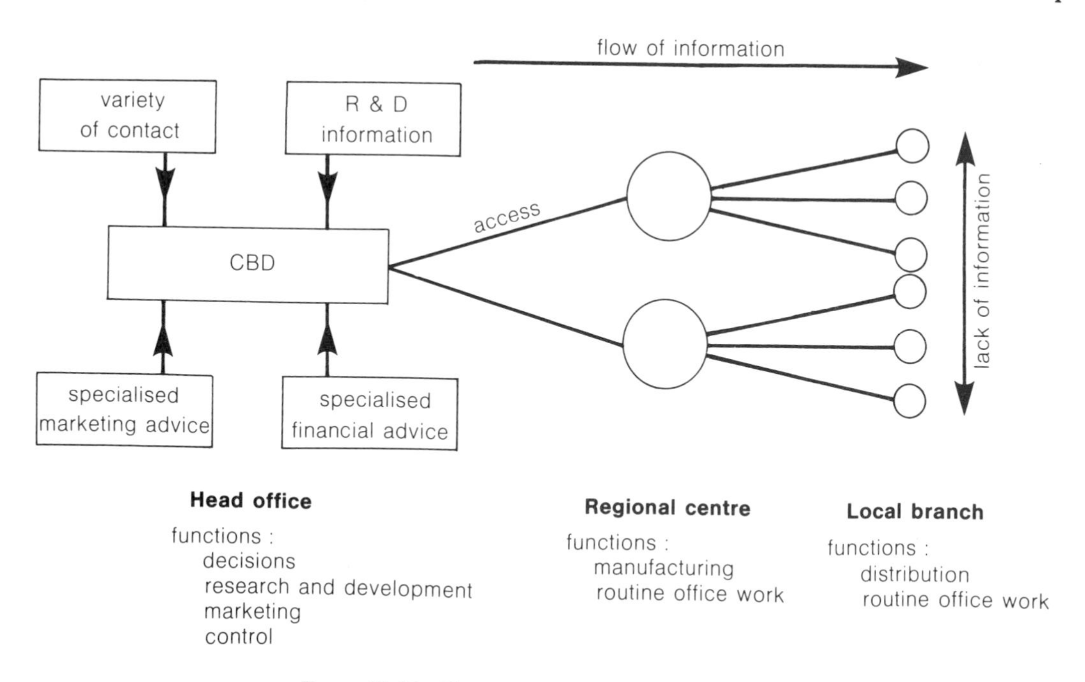

Figure 45.11 The structure of office development.

services and so on. A central location is therefore encouraged by the *quality* of the information as well as by its availability.

Most firms have more than one plant, and the CBD offers a location of greatest accessibility both to branches and to local information contacts (Fig. 45.11). Thus offices tend to cluster for the same reasons as manufacturing industry – only instead of exchanging materials they exchange information. Figure 45.12 shows that there is functional segregation within the CBD. The frequency not only of telephone but of personal contacts within particular groups is so great that firms prefer to cluster together.

Besides the large firms there are also a multitude of small businesses, each trying to grow by finding a niche. Many of these firms do not survive, so there is a large turnover in business accommodation. Those that do survive and grow often make many moves in search of the most appropriate site for their level of business. Businesses in the CBD tend to be extremely mobile.

When an office business grows it may be able to separate out those functions that need to be in the CBD and those of a more routine nature that require few exernal contacts and that can satisfactorily be undertaken in the suburbs or in the CBD of a regional centre where costs (bid-rents) are lower. There is, in consequence, a pattern of continuing growth and dispersal within the office sector of the CBD.

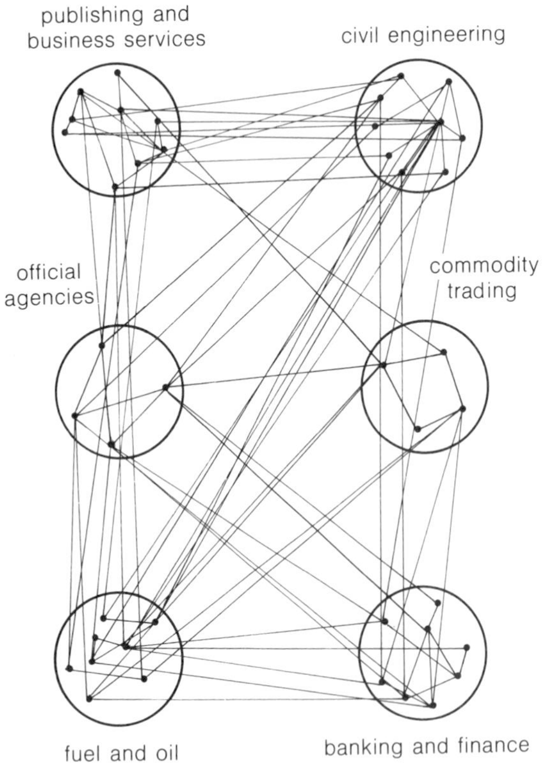

Figure 45.12 Telephone contacts within Central London illustrate how offices cluster to exchange information.

GOVERNMENT POLICY AND THE CBD

Governments first become aware of the importance of the CBD to offices as the flow of commuters imposes greater and greater strain on the urban transport system. Yet a policy of restricting office growth in central areas must be treated with caution: the prospect of increasing job opportunities is often an attractive one. A policy of dispersing offices can in any case be *too* successful, as many cities in the USA illustrate. In these cases CBDs have reformed near freeway intersections, leaving the central areas deprived of the employment they once enjoyed. This is also happening in London.

Despite the higher costs of location in the CBD, the advantages of clustering in a point of maximum accessibility still hold sway. But the location in a central cluster has long-term implications, for although dispersal to a cheaper peripheral site may provide short-term economies, 'like birds of a feather, it is only a large office centre that can offer the number and variety of local contact opportunites that can sustain an office organisation when external conditions change' (Goddard 1975).

Industry

Industry has been the cornerstone of growth in most developed-world towns and cities. Manufacturing industry generated the wealth on which the commercial and service sectors grew and still largely depend. Most industrial development began in the inner-city areas because, when urban areas were small and transport systems less flexible (canal or rail as opposed to road), a central location was the only one that could provide direct access to markets and to a labour force, and at the same time allow industries that depended upon one another to cluster together.

In fact it is not so much a central location that is vital as an adequate communications network which allows raw materials to be brought in and goods to be efficiently despatched. As road transport became more important, therefore, employers were able to consider alternative locations for their factories – along arterial roads between cities, for example, or on ring roads just outside the major built-up area. But the growth of industry away from the central zone was not just a matter of convenience; on the contrary, industries began to find that they were being penalised for remaining in a central position.

As urban areas grew, and particularly as cities developed into conurbations, it became harder and harder for large lorries to fight their way through the inner-city congestion. Competition for land from users who could make more profit from a central location forced rents up and left manufacturers either with increasing overheads (if they rented their premises) or with the temptation to sell (if they were freeholders). Furthermore, in line with local government policy of segregating industry from residential areas, new, outer-suburb **trading** or **industrial estates** were built near main roads. These planned estates were a major factor in causing many industries to move out from the centre, for there was the opportunity to move and yet remain clustered.

Once the shift of industry from the inner city had begun, the advantages of clustering in the centre were lost, and with them a major reason for the remaining factories to stay. Many remain today only because of inertia; they are not yet so disadvantaged by remaining that they are losing money. These factories cannot win the bid-rent battle in the long run, and their departure is hastened by government policy that discriminates against them. Urban-renewal schemes have tended to give precedence to the location of residential development in prime sites, leaving the least desirable and often least accessible sites for industry. Moreover, when urban renewal has included the demolition of factories, new purpose-built units have sometimes been offered back to the factory owners at rents so high that they cannot afford them.

Finally, the growth of technology has meant that some industries once tied to their location in some part of the city have now been released from their restricting sites. For example, industries that once needed to be near the docks because of the awkwardness of moving cargo on and off ships, can now use easily-moved containers, and thus be sited with other industries in the suburbs. One of Britain's biggest container handling areas is at Stratford, in North-East London, not beside the London docks; the container terminal is at Tilbury, 60 km away.

All these changes in industrial location have led to a reduction in the jobs available to inner-city people: this in turn has caused high unemployment, and a fall in relative earnings and real income. The changing location of industry has been a prime reason for the 'crisis of the inner city'.

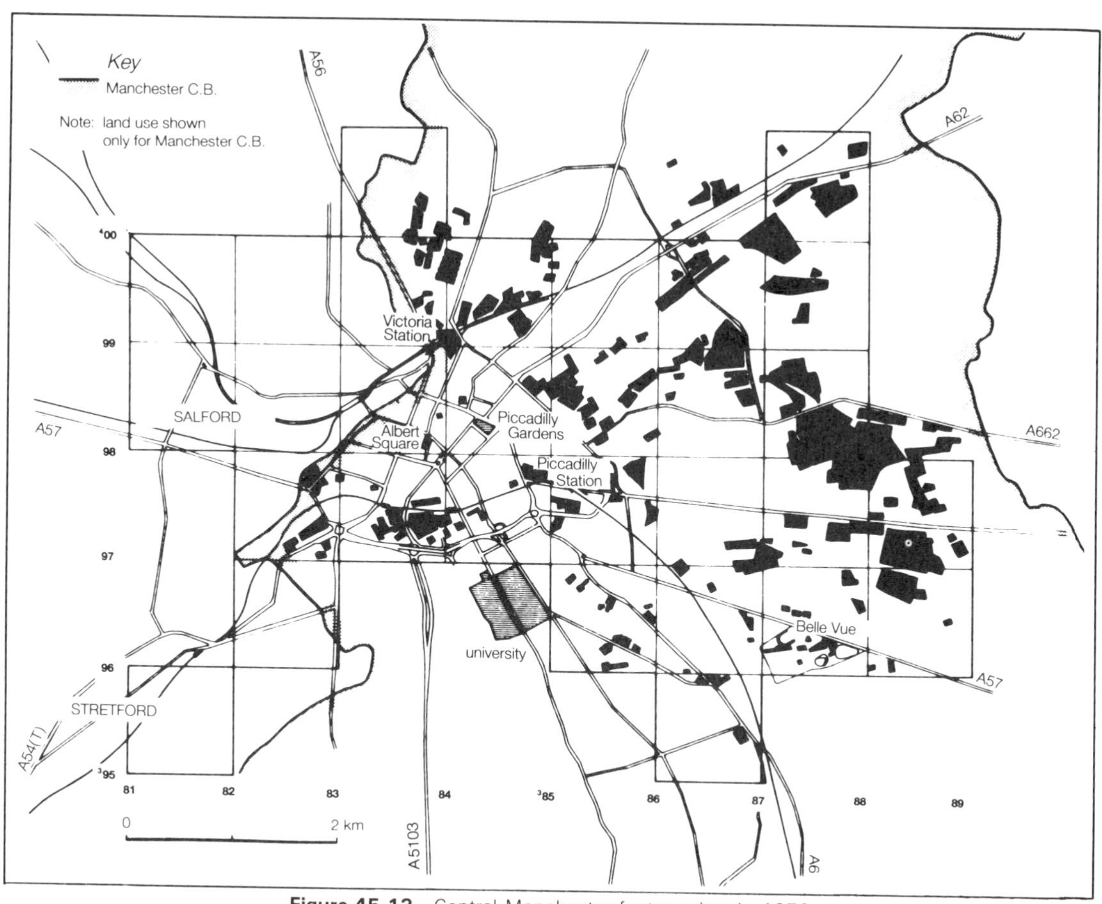

Figure 45.13 Central Manchester factory sites in 1972.

CASE STUDY: MANCHESTER

The growth of Manchester was entirely due to the factories and to the factory system associated with them. Manchester's early industry, focused on textiles, engineering and clothing, developed particularly close to the docks that handled most of the raw-material imports and the exports of finished goods (Fig. 45.13). In the spaces between the factories and warehouses, and spilling out to make an encompassing ring, were the closely packed terraced houses which were the homes of the workforce.

As roads replaced canals and railways as the most effective means of distributing goods, and as trade began to focus less on the export market and more on the home market, the industries began to exploit the new transport system. However, the narrow and congested inner-city roads proved to be a severe disadvantage and many firms moved to more peripheral sites. Because there was no more room in the inner-city region, new-technology industry was forced to seek locations elsewhere, mainly in the suburbs (Fig. 45.14). Now many of the old city-centre industries have moved or faded away, and their premises have either been demolished or subdivided for use by small firms. The result has been a dramatic loss of employment (Fig. 45.15). For example, between 1966 and 1972, in a period of national growth before the oil crisis, the inner core of Manchester suffered a net loss of 30 000 jobs in manufacturing, equal to a loss of a third of the total inner-city manufacturing employment; by 1975 a further 10 000 jobs had gone. But it was the closure of the large plants that was particularly severe because of the knock-on effect this had on the small firms that depended on them as their major market. Without drastic improvements in the access of the inner city and in the decaying environment it will be virtually impossible to get the businesses back.

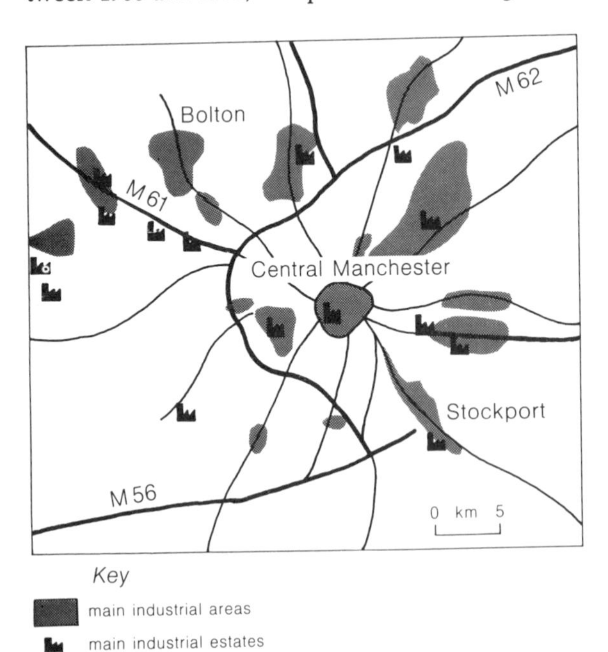

Figure 45.14 The Greater Manchester conurbation. This map clearly shows the close correspondence between industrial location and main arterial roads. Notice how the cheaper rents of 'satellite' towns such as Bolton have enabled them to retain much manufacturing.

Retail function

WITHIN TOWN

In many ways the retail functions of an urban area operate as though the city is composed of a collection of villages and market towns condensed into a small space.

The population of the city has everyday domestic needs that can support a local shopping area (the 'village' function), more specialised requirements that will support a neighbourhood shopping centre (the 'market town' function), and very specialised needs that can only be met by city-centre shops serving the whole community and a wide area beyond the urban limits. It would be reasonable to suppose, therefore, that the retail functions will be spread throughout the city in a sort of collapsed Christaller pattern, with 'local' shops occurring most frequently, and 'neighbourhood' shops at greater spacings, all ranged about a single city-core shopping zone. In

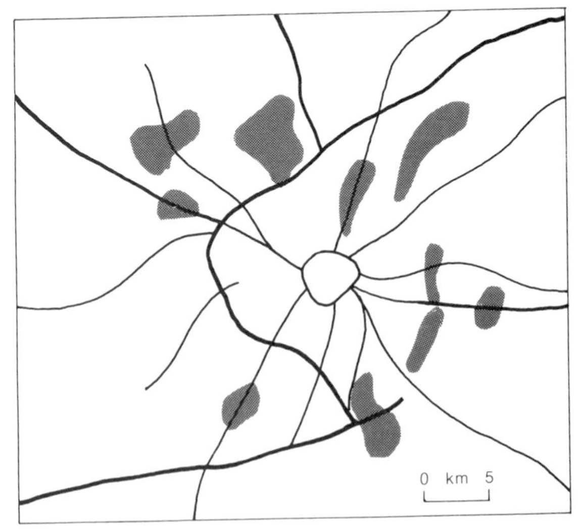

Figure 45.15 The areas of Greater Manchester to gain employment in manufacturing between 1967 and 1972. Notice that the areas are not necessarily traditional major manufacturing zones (compare with Fig. 45.13) and that the inner city has a net loss.

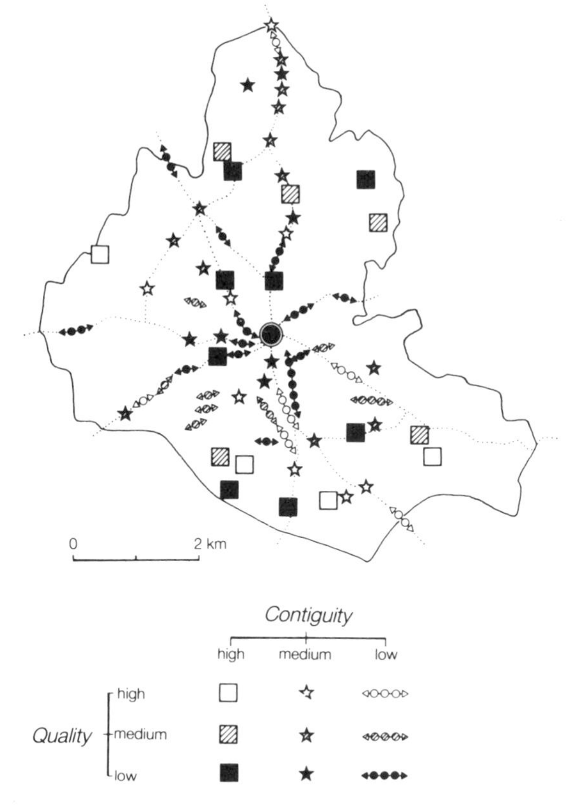

Figure 45.16 The pattern of shopping in Stockport. Inner-city areas are characterised by poor-quality linear shopping areas; those in the suburbs by high-quality compact shopping centres.

fact, several factors prevent such a pattern developing in any detail. First, the city does not have an even spread of population, nor an even distribution of wealth; there are industrial zones without high populations, rich-housing sectors and poor-housing sectors, and so forth. Some groups of the city population are more mobile than others. The result is a pattern that reflects these unevenly distributed qualities of each city.

In Stockport (population 140 000), near Manchester, for example, a study revealed 1562 shops grouped into 72 centres (Fig. 45.16). When these were graded for specialism and quality a wide variety was soon apparent. The majority of centres were of medium-to-low quality, with only 11 per cent in the top three of the five quality grades established. Clearly the majority of shopping areas were catering for the day-to-day needs of the medium-to-low-income mass market.

Further study revealed that the quality of shopping areas increased with the distance from the city centre: near the CBD shopping area quality was suppressed by the influence of the centre itself, whereas towards the suburbs competition from the city centre diminishes and the income of the residents increases. For similar reasons of competition, the quality of shops also increases with the size of the centre, with the major suburban neighbourhood shopping centres standing out as pinnacles of high quality, together with those situated in areas of high residential status. All of these lie within a zone 2–3.5 km from the centre of Stockport. Quality also improves in those centres away from the main arterial roads because such roads

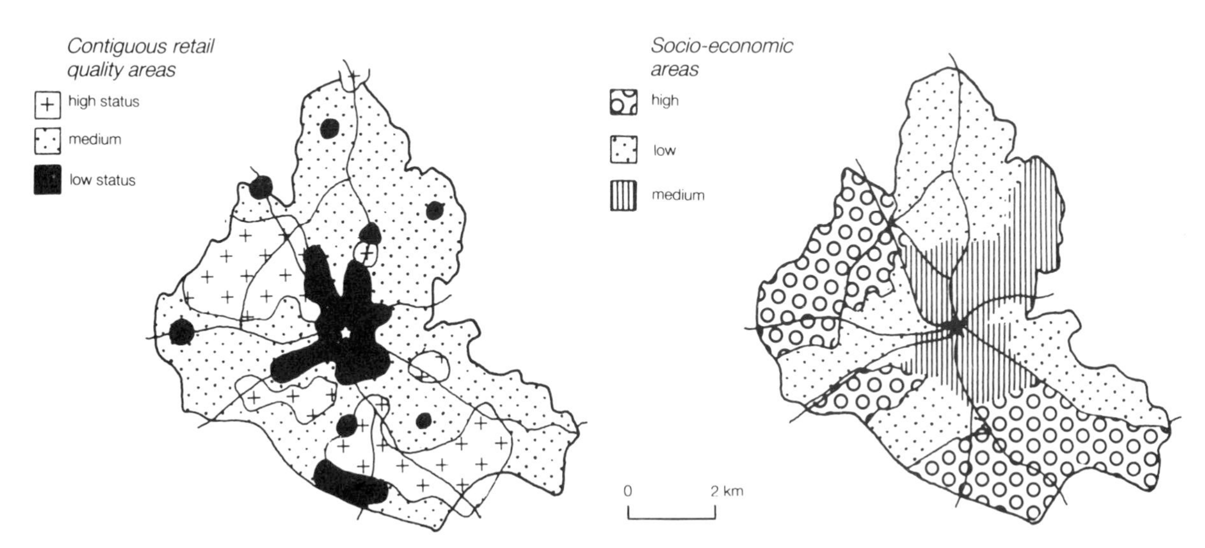

Figure 45.17 There is a close correspondence between the purchasing power (socio-economic group) of a neighbourhood, and its shopping centre quality.

are unattractive as housing environments. Indeed, the quality of shopping areas is closely related to the pattern of the city's wealth (Fig. 45.17).

The shape of each shopping centre is also closely related to its position and function within the city. For example, with the exception of the city-centre shopping core, shops of the inner city tend to be arranged in a line, poorly clustered as regards function (often containing a mixture of retail and other functions), of low quality, and sited along main roads. By contrast, the high-quality shopping centres in the suburbs are away from main roads, highly concentrated, and cater exclusively for multi-purpose domestic shopping trips. It is this latter group that fits most readily into a hierarchical system such as that proposed by Christaller (Fig. 45.18a).

The city centre contains quite different shops from the neighbourhood and district centres, as is clearly shown by the centre of Coventry (Fig. 45.18b). Here there is a large proportion of specialist shops clustered about the 'centre of gravity' of the pedestrian precinct, further subdivided into distinct zones of high quality in the core and low quality at the periphery.

OUT OF TOWN

The most recent development of shopping patterns has been catalysed by the increasing difficulty of access to central shopping areas. Out-of-town shopping centres, commonly called **hypermarkets**, were first a feature of American cities, in which such a large proportion of people have cars. In France hypermarket chains such as Monoprix and Mamouth quickly became household names, while in Britain the lead was given in northern England by the Asda chain.

The exact location and style of hypermarkets varies considerably, although all rely on peripheral locations with high accessibility and low bid-rent, which

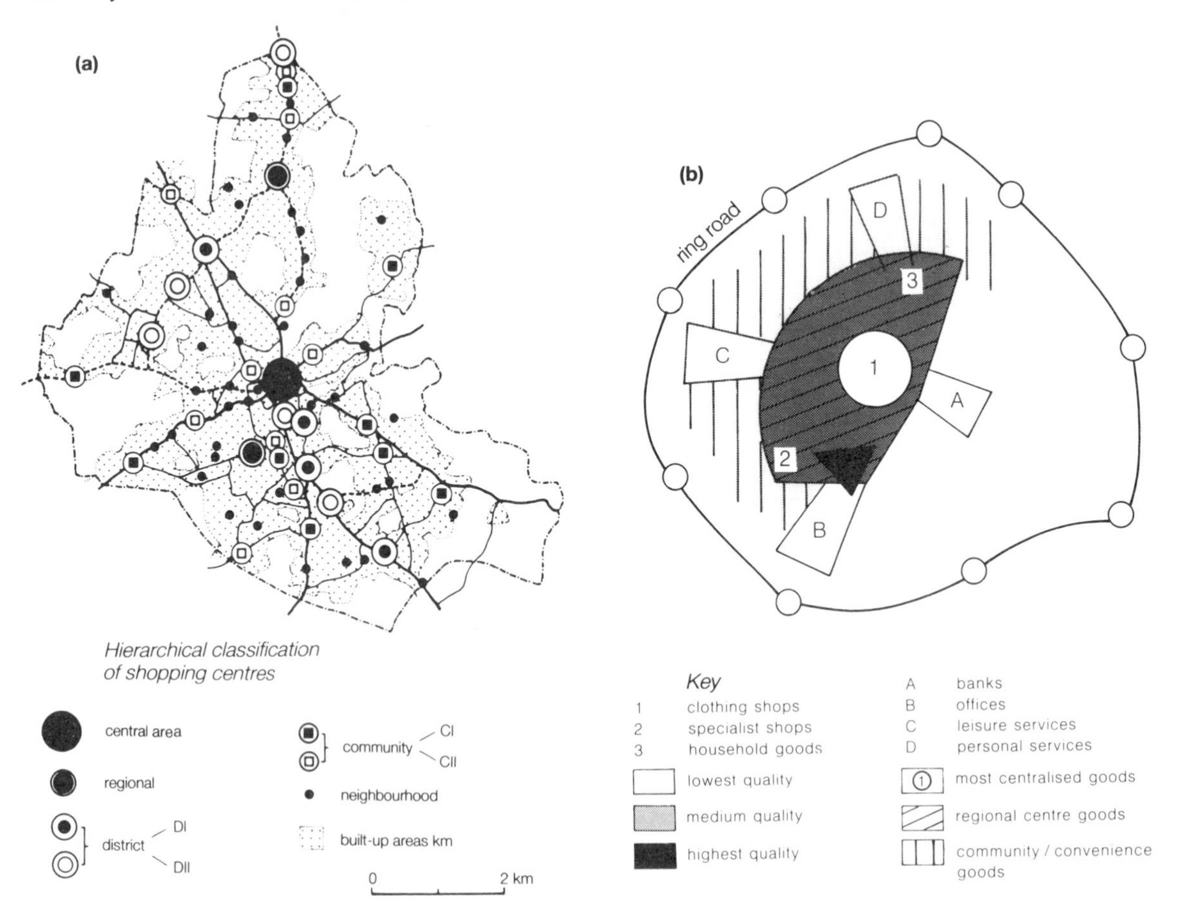

Figure 45.18 (a) The shopping areas of Stockport can be separated into six main categories: 1 central area, 2 regional areas, 9 district shopping areas, 19 community areas, 41 neighbourhood centres, and 247 isolated shops. These areas make up a form of Central Place hierarchy for the city. (b) The central shopping functions of Coventry.

allow them to provide units sufficiently large for low-cost, one-stop shopping (Fig. 45.19a). Some, such as Asda near Huddersfield or Savacentre near Reading (now Europe's largest hypermarket), are situated at specially developed motorway or trunk-road access points, to encourage people to shop from a wide rural as well as urban hinterland. Others, such as Asda in Chapeltown, several kilometres from the cramped shopping centre of Sheffield, have used old industrial sites (Fig. 45.19b). In this case, both the shops and an extensive car park were built on cleared ground. But, like the others, the Chapeltown hypermarket is not designed to cater only for the most mobile sector of the community. British Rail has improved its Sheffield-Chapeltown-Rotherham service, providing more frequent trains and advertising the services as a new shopping link. In addition, Asda provide free shoppers' buses that go to the outlying villages once a week. Clearly, the hypermarket is destined to be a major influence on urban shopping patterns.

Residential

THE INNER CITY

The zone between the suburbs and the CBD is called the **inner city**. It is the most complicated of all the regions in a city. Mainly the product of the building boom that accompanied the earlier phase of expansion during the Industrial Revolution, houses and workplaces are frequently intermixed.

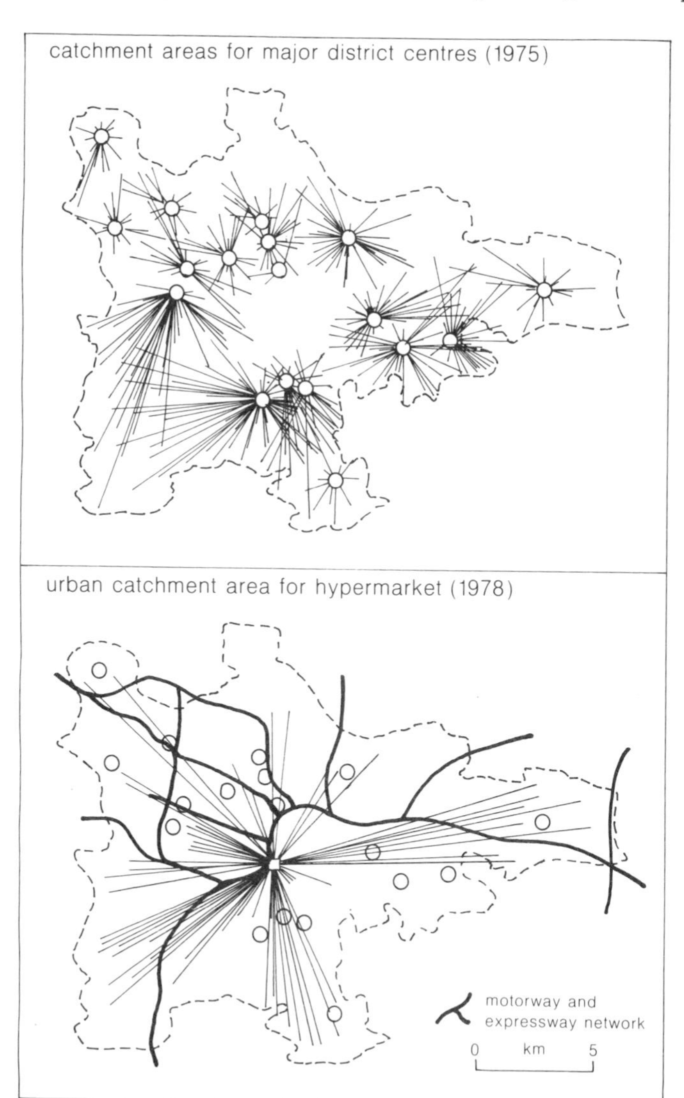

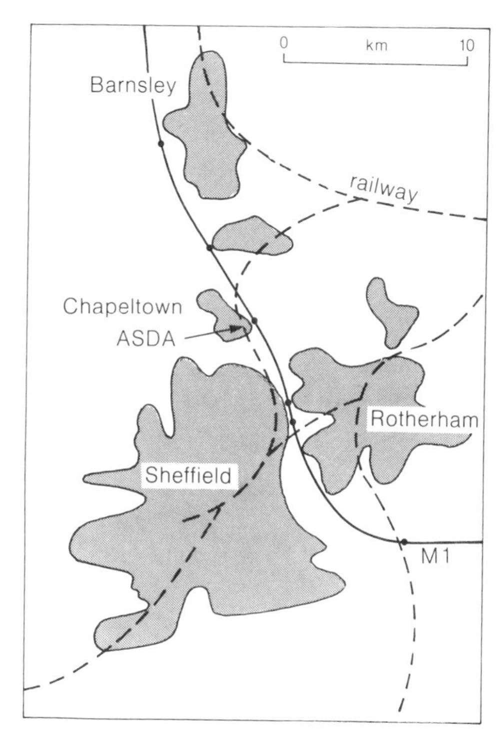

Figure 45.19 (a) Contrasts between the catchment areas of district shopping centre; and a hypermarket in Glasgow. (b) The location of Asda in Sheffield.

Most of the houses were built on a grid-iron pattern, with little aesthetic consideration and in an age when working people did not own any private transport. Some of these houses were built by factory owners especially for their workforce. This was a particularly characteristic form of development in the North of Britain, where banks of back-to-back houses were built in large numbers. However, in all areas, most of the housing stock was built for rent by speculators and many were built to very poor standards and with little or no internal sanitation. Even before the end of the 19th century many of these houses were falling into disrepair. Legislation eventually brought an improvement in mimimum levels of housing and the so-called 'bye-law' or 'tunnel-back' terraced house, with its kitchen and toilet built on to the rear of the property, became the standard form of building. The legacy of poor construction from the earlier period caused much of the housing stock to fall into a very dilapidated state even by the beginning of the 20th century. To try to alleviate the suffering that poor housing caused, voluntary non-profit-making trusts were formed to provide blocks of flats for rent. Although well intentioned, their efforts could not keep pace with demand; but they showed the way for government, which took over responsibility for housing the less fortunate. Thus it was that the zone belonging to the industrialising phase of city growth became a mixture of industry, privately rented terraced houses and municipally rented blocks of flats.

Much of the inner-city fabric remains old and built to a poor standard. Many of its industries are in cramped and old-fashioned buildings, its roads quite unsuited to the pace of today's progress. The inner city is a zone of low bid-rents, of the socially disadvantaged, of the unskilled and the immigrant. These are the people within a city least able to help themselves and most susceptible to manipulation by others. When the CBDs were rebuilt larger industries moved to new locations and those with sufficient money drifted to the suburbs. The inner cities of all Western countries fell further and further into decay throughout the first half of this century.

Simply because it consists of older, poorer housing intermixed with factories, the inner city epitomised for some people the way the city had gone wrong, even as far back as the turn of the century. These people began to ask: 'What is a city for?' In 1904 Garnier suggested that it served:

(a) as a place for living in;
(b) as a place for working in;
(c) as a place where people want to move about;
(d) as a place of recreation.

It seemed that what had gone wrong with the 19th-century city was that all these functions had got muddled up. And if this was the cause of the problem, then an entirely new approach was needed – a *planned* approach, with cities *zoned* according to their functions. But more than this, new ideas were introduced that began to shape the way that people thought about housing for the masses – ideas that are characterised by Le Corbusier's ominous phrase 'the house is a machine for living in'. From this developed the idea that the old cities, which had grown up piecemeal and chaotically, must be torn down and replaced with new buildings in a setting designed as a coherent plan. This plan should focus at the middle on a complex of office blocks, the *CBD*. Beyond the office zone, and surrounded by public open space, there should be a *residential zone* of municipally (publicly) owned blocks of flats in *housing estates* where the residents could be isolated from the noise and inconvenience of traffic. Beyond these should be the *industrial zone*, on the edge of the city, surrounded in turn by a great belt of woods and fields (a *green belt*) beyond which would be a ring of satellite 'garden cities' or *new towns*.

The concept of planning outlined above pays no attention to the realities of the market place, of bid-rents or of personal wishes. It is a totalitarian idea on a grand scale and basically quite alien to a democracy. And yet, because World War II wreaked so much destruction to the inner cities within Europe, cities had to be rebuilt quickly. To achieve this the state took a positive role in the housing market and passed legislation that allowed local authorities not only to decide what each area of land should be used for, but to purchase it compulsorily. Planners correctly saw once-proud Victorian inner cities as dowdy, obsolete and inefficient. They recognised that the inter-war years had allowed the suburbs to consume too much valuable agricultural land. Overall, their aim was the replacement of the old housing, but at a sufficiently high density not to require the use of more agricultural land. Unfortunately, the result was the wholesale demolition of areas of inner-city houses that remained standing and their replacement with monolithic blocks of high-rise flats and broad, soulless urban motorways (Fig. 45.20). This result did not improve the quality of life; far from it. The social cohesiveness of former communities was destroyed and replaced by a faceless 'concrete jungle'. Small factories were driven from the inner city, which took away the employment base of the community helped and led eventually to inner areas having the highest unemployment rates. It is a pity such good intentions instead created one of the great social disasters of the century.

URBAN PLANNING

In the first part of this century architects planned to create an abstract art form for cities, rather than a place that would be comfortable to live in.

(a) They chose high-rise blocks and segregated functions, divorcing people from their work and from one another.
(b) Slowly there has been a return to a mix of functions and low-rise development more in keeping with the pattern that developed naturally in older cities.
(c) However, the artistic experiment of the post-war period will remain a dominant feature of inner-city and new-town geography for a long time to come.

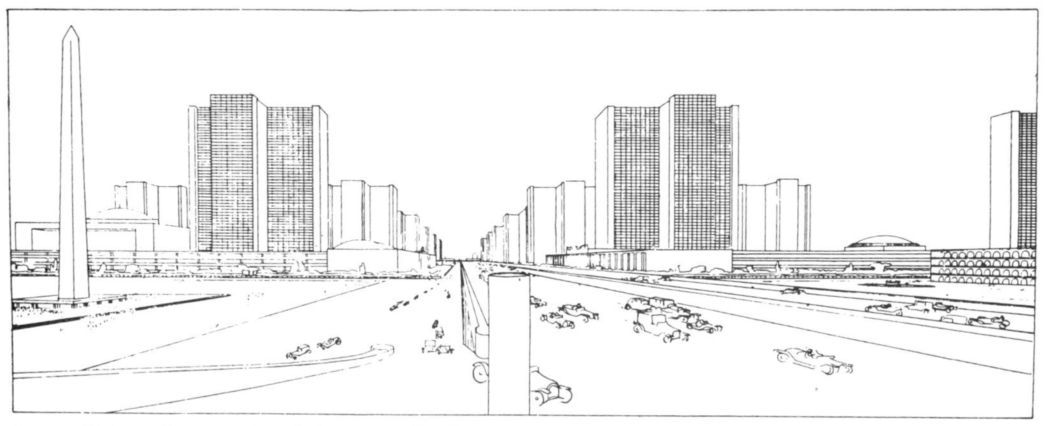

A superhighway forms a triumphal entryway leading directly into the centre of the Contemporary City (1922). From *Oeuvre complèt de 1910–1929*.

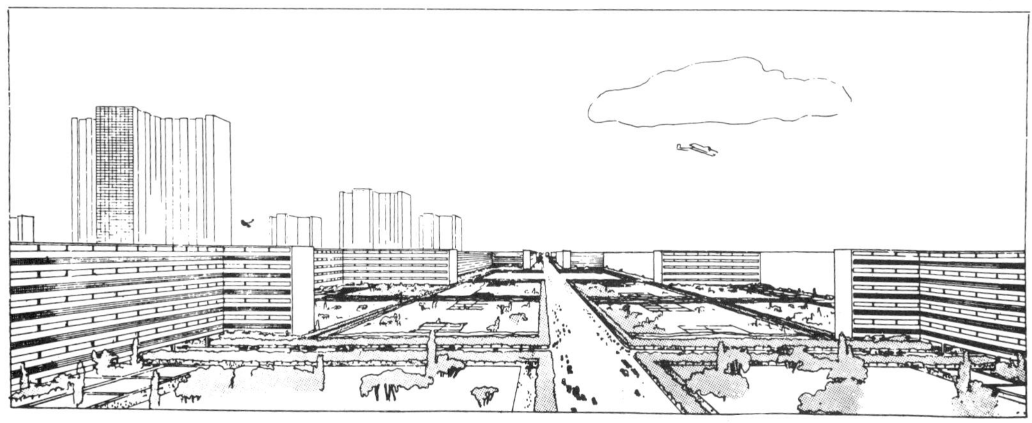

The vast apartment blocks where the elite of the Contemporary City live, surrounded by parks. From *Oeuvre complèt de 1910–1929*.

Figure 45.20 Inner-city urban renewal in Glasgow. A dream come true, or the reality of a nightmare?

THE SUBURBS

Suburbs, predominantly residential areas that surround cities and towns, form the largest component of the urban landscape, growing ever outwards from the core.

Before the invention of the railway there was a limit to the physical extent to which a city could grow, for without rapid transport people simply had to live close to their places of work. Thus 18th-century London society lived alongside the slum dwellers of Soho; and today's inner-city London boroughs were at that time no more than little villages in the country.

The railway altered everyone's concept of the city, although at first only the better-off were able to take advantage of it. The railway owners, rather than restrict travel to the wealthy few, actively sought to tap the lesser wealth of the many. Soon there were 'workers' specials' at rates low enough to make rail travel a real possibility for most people. As a result the small, select groups of large mansions which had earlier clustered around the suburban railway stations were joined by serried rows of terraced houses for the skilled working class. By the outbreak of World War I the edge of London was already seven miles from St Paul's Cathedral. In Leeds, more typical of the smaller cities, transport also allowed the growth of suburbs beyond the tightly packed urban core (Fig. 45.21). But even these dramatic expansions were overshadowed by the tremendous growth that took place between the wars when the urban area of many cities doubled.

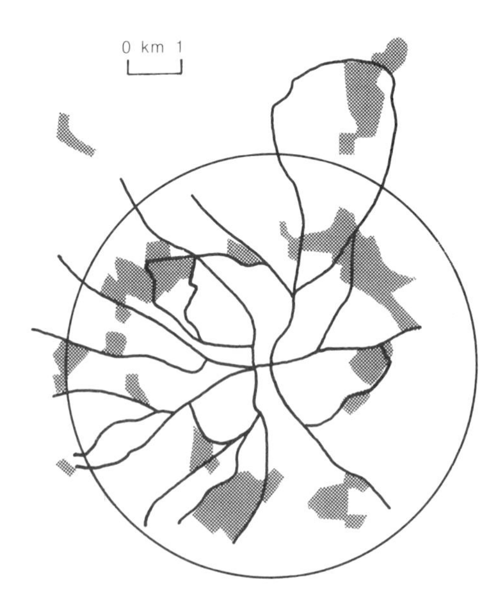

Figure 45.21 The close relationship between local transport and building development in Leeds, 1890–1919. The shaded areas have been called 'street-car suburbs' because their development was only possible following the introduction of the tram. The circle is 4 km from the city centre.

The growth of the inter-war suburbs was a phenomenon of a period without any planning co-ordination. It was an unstoppable onrush of the urban area, not from a simple need for somewhere to live, but as a result of the combined expression of three-quarters of a million imaginations to leave the pollution of the central city and find somewhere better to live. In the 1920s suburbs like Ilford to the east of London were simply villages in the country; by the 1950s they were completely integrated parts of the city sprawl (Fig. 45.22).

The growth of the suburbs was not a smooth development – suburbs were built around villages in the country, often following the lines of the expanding railway network. One newspaper advertisement announced:

> Notice is hereby given that the Southern Railway will open a new station at Purley: the electric train service will be in operation from March 25th (1928) direct to London Bridge.

Soon after, builders were quick to add to their advertisements: '*You* will want a house in Purley'. Indeed, one of the things the developers sold with their houses was the absence of other houses around.

The first phase of suburbia stretched tentacles out along the main railways and engulfed the rural villages. Then came the backfill, closing the gaps between the villages and the metropolis, making the villages a part of the city and turning them into suburbia.

In Britain there are considerable contrasts between suburban development in the North and the South. This is because the main phase of growth occurred just after World War I, when new technologies such as the widespread adoption of electricity had begun to free industry from the coalfields. The North was severely hit by depression and suburban growth was limited by lack of money. By contrast, in the area near London, where 80 per cent of all new factories were locating, there was an expansion of population and an increasing ability to pay for improved housing.

In southern Britain the growth of suburbia was mainly achieved by small builders, using small plots of land. Each builder constructed perhaps four or a dozen houses at a time; with the profits from this sale he built on a plot, often using a slightly different design. This small-scale building created variety within suburbia. It controlled the street plan too: progress was made by occupying farmland field by field, widening and enlarging the existing country lanes.

Today suburban development continues, especially on backfill sites and around the expanding provincial towns. However, the days of the small builder are past, and the development of new sites is in the hands of large building firms which are able to plan and build estates complete with shopping centres and other facilities (Fig. 45.23). New planning concepts have been introduced, with clusters of houses built around traffic-free open spaces, and there are often pedestrian-only routes to shops and schools. However, as before, these are largely for the private house-buyer; they do not cater for the lower-income groups, and the new suburbia remains as middle-class as suburbia has always been.

THE NEW TOWNS

The two most dramatic effects of urban planning lie isolated by the chaos of the suburbs. But while the inner city was in a state of decay and undergoing a massive upheaval, the growth of the city as a whole

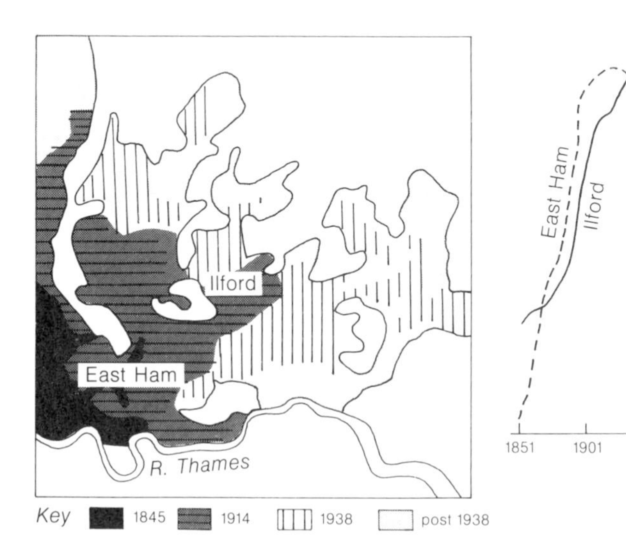

Figure 45.22 The change in the population structure and the size of the urban area were closely linked with mass transport. The expansion of first East Ham and then Ilford in north-east London closely follows the advance of the railway into Essex. Notice, however, that expansion was not even, rather it occurred in bursts, following extension of the railway, infilling not occurring until much later, when motor transport allowed greater flexibility of movement.

Figure 45.23 The new style of suburban development.

had been stopped by another planning concept: the **green belt**. The continued outward rush of the inter-war suburbs was divorcing inner city from country, while placing greater and greater strains on the transport systems needed to get people to work. In an attempt to stop the city growing too large and unwieldy, a wide belt of land surrounding the existing suburbs was designated 'green belt'. In it no new building could be started without special permission; and its width was intended to make urban growth outside it unattractive. Green-belt land was not an isolated concept designed solely to contain cities and towns, it was planned in conjunction with inner-city renewal and the development of **new towns** (Fig. 45.24). After all, the suburbs had only grown up because people had desired to be away from the dirt and congestion of the city core, so if the inner-city environment could be improved, planners argued, people should want to move back. Nevertheless, **urban renewal** on a grand scale would inevitably displace people, at least temporarily, and in a post-war environment of rapid population growth, new provision for them was urgently needed.

The new town was derived from the notion of the **garden city**, the first of which was started at Letchworth (Hertfordshire) in 1903. Two thousand hectares of countryside were developed into a pleasant environment for people to live while being close to their place of employment. The plan, by Ebenezer Howard, took houses from their stark regimented rows and gave people gardens and open space to enjoy. The whole scheme only needed the loss of three trees.

Letchworth was a private development – one man's dream of a new society. At the time the close tie of industry to resources severely limited its expansion. Welwyn Garden City was built after World War I as another experiment: it was based on light

(a)

Cumbernauld
Glenrothes
Glasgow
Edinburgh
Irvine
Livingston
East Kilbride
Newcastle
Washington
Peterlee
Newton Aycliffe
Middlesbrough
Preston – Leyland
Skelmersdale
Manchester
Runcorn
Warrington
Newtown
Telford
Birmingham
Redditch
Peterborough
Corby
Northampton
Milton Keynes
Hemel Hemstead
Cwmbran
Cardiff
Letchworth
Stevenage
Welwyn
Newbury
Harlow
Basingstoke
Hatfield
Bracknell
Basildon
Crawley

(b)

(c)

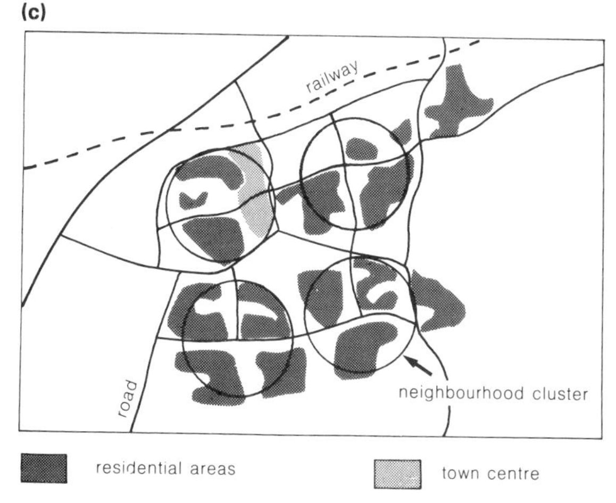

Figure 45.24 (a) New Towns in Britain. (b) Telford New Town shopping centre. This is an example of planned central shopping in an air-conditioned environment. (c) The structure of Harlow New Town.

industry, electricity, and improving communications. However, it was the New Towns Act of 1947 that stimulated the planning of most of Britain's new towns. Around London especially, self-contained new towns were planned at various places beyond the green belt. Harlow in Essex, for example (Fig. 45.2b), is 40 km from London: it was planned as a self-contained housing and industrial community of 90 000. Initially it was nicknamed 'Pram City' because of the large number of young families that first moved there.

In Harlow there are four clusters of neighbourhoods, with large open spaces between them and a town centre largely free from traffic. For the first time industry was zoned separately from housing. The whole town plan was controlled in structure by the form of the land, which is framed by the low boulder-clay Rye Hills and the valleys of the River Stort and its tributaries. The shallow valleys were preserved as open spaces and buildings were put on the lower slopes of the hills to add interest to the views.

In the early days of Harlow's development people were required to prove that they had a job before they could rent a house, thereby preserving the self-contained nature of the town. Harlow was built before many families had cars, yet although the town is well within the commuter belt for London, the substantial industrial presence in the town enables a large proportion of Harlow's residents to work locally.

In contrast to Harlow, Peterlee, in County Durham and midway between Hartlepool and Sunderland, was built to replace the scattered and depressed mining villages of the Pennine coalfield with a town big enough to support the people and to provide the social amenities that the old villages lacked. It was also designed to provide alternative employment to coal mining and increasing job opportunities for women. Built as the trend moved towards prefabricated construction with concrete, it is perhaps less sympathetically adapted to the landscape than is Harlow. Peterlee is also situated in the generally depressed area of the North-East, where continuing pit closures have left their mark.

Milton Keynes is one of the later generation of new towns. Just south of Northampton and beside the M1 motorway, Milton Keynes actually has a higher population density than many ordinary towns, but its careful design disguises this. It is intended to house 200 000 people eventually, which means this is really a **new city**. However, it is the site more than anything that has ensured the success of Milton Keynes. Centrally placed between the West Midlands conurbation and London, its advantages as a distribution centre have attracted light industry as diverse as Coca-Cola and Volkswagen, and also the Open University. But it is industry far different from that in older areas, sited in special areas with direct motorway access. Here factories are mostly single-storey because so much handling of goods is done by mechanical means with fork-lift trucks, conveyor belts and computerised transfer systems.

46

Cities of the developing world

46.1 Accelerating urbanisation

Traditionally the developing world has been an overwhelmingly agrarian society (Fig. 46.1). In much of Africa less than 10 per cent of the population live in towns; in Asia about 15 per cent and in South America about 30 per cent. By contrast, North America and Oceania have 50 per cent of their populations in large urban centres, while in Britain the proportion rises to 80 per cent.

Even within the developing world there are considerable contrasts and much rapid change (Fig. 46.2). For example, most North African and Middle Eastern countries have small agricultural resources and their people have always concentrated in urban centres, acquiring their food primarily by trade. By contrast, in countries south of the Sahara, subsistence agriculturalists have traditionally found little use for urbanisation and cities have largely been developed by colonial administrators.

Within the past few decades, however, there has been an increasing tendency for towns and cities in every part of the developing world to grow at the expense of the countryside: in 1800 only about 2.5 per cent of the world's population lived in towns of over 20 000, whereas by 1900 it was nearly 10 per cent and by 1950 just over 20 per cent. With the urban population growing ever more rapidly at least 50 per cent of the world's population will live in towns and cities by the year 2000. The UN estimates that by the year 2000 the developing world should account for 1500 million of the 2400 million living in towns and cities. Indeed, Asia's population alone is such that its urban population will rise from 265 million in 1960 to 1000 million by 2000 AD (Fig. 46.3).

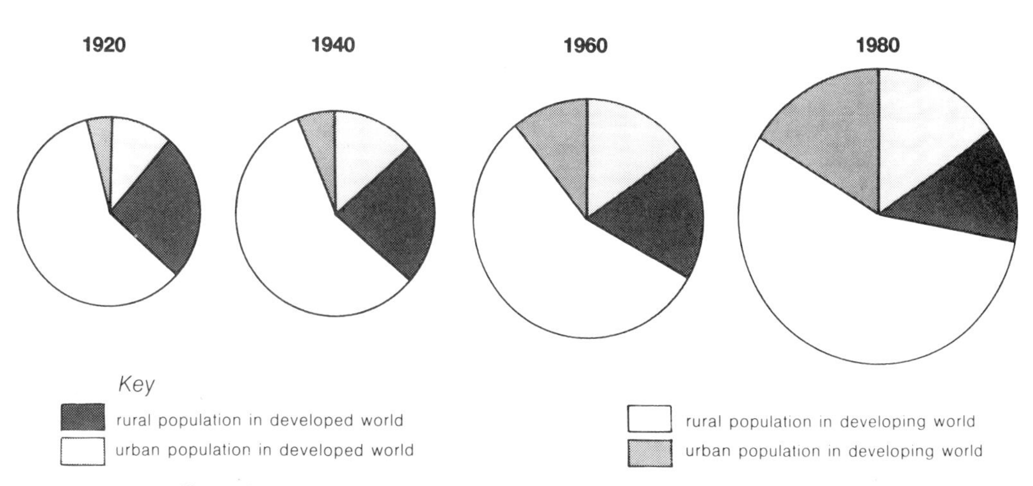

Figure 46.1 The size and composition of the world's population, 1920–80.

(a)

(b)

Figure 46.2 The high-rise flats of commercially thriving Hong Kong (a), contrast with the shanty town area of Delhi (b). Both demonstrate the impossibility of attempting to construct a single model of urban structure in the developing world.

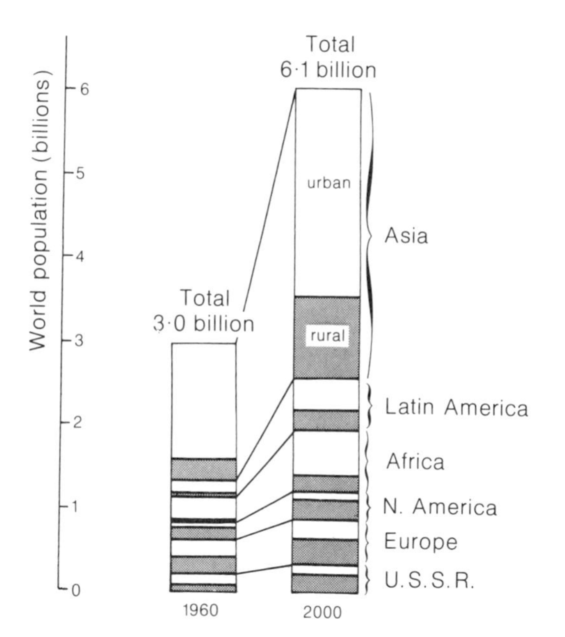

Figure 46.3 Estimated growth of world urban and rural populations.

46.2 Problems of developing-world cities

Rapidly growing cities

It is both the scale and the rate of urbanisation that cause severe stress in the developing-world cities. In countries with very limited financial resources there is an immense problem in providing homes, water and electricity, in collecting garbage, and in providing sanitation (Fig. 46.4) and public transport. In some cities transport is so difficult that many people organise their lives so that they can reach everything they need on foot. Difficulties in providing sufficient housing in a planned and orderly manner have led to the creation of many **spontaneous settlements** (popularly called **shanty towns**), whereby people build homes for themselves on any unused land within or marginal to the established city. The result is that developing-world cities often have a quite different *structure* from those in the developed world.

The *pattern* of growth in developing-world cities is also substantially different from that of the developed world. This is partly because of a lack of investment, and partly because growth is taking place at a different stage of technology. In the developed world there was a long period in which manufacturing was the dominant employer, with services taking over as the leading sector for employment only slowly. In the developing world urbanisation is occurring faster than the growth of manufacturing employment, and

(a)

(b)

Figure 46.4 (a) Water availability at a communal tap, Cartagena, Colombia. (b) The lack of even basic sanitation, Calcutta, India.

most people arriving in the city therefore move directly from agriculture into services. This has charged services with the role of the leading sector in the developing-world economy, and therefore with the responsibility of generating wealth. However, as greater wealth can only come from some types of service industry and must partly follow from improvements in the manufacturing base, developing-world cities frequently have a top-heavy employment structure, with too many people in dependent, non-productive services. As industrialisation becomes more capital-intensive this imbalance will be hard to rectify, and there will be an increase in underemployment and unemployment in developing-world cities into the foreseeable future.

Difficulties in employment have led to the survival of the traditional system (here called the **bazaar system**) whereby those in work help other members of their families to get a job. But because the amount of money available is limited, increasing employment in this system results in lower wages for all and the growth of **shared poverty**.

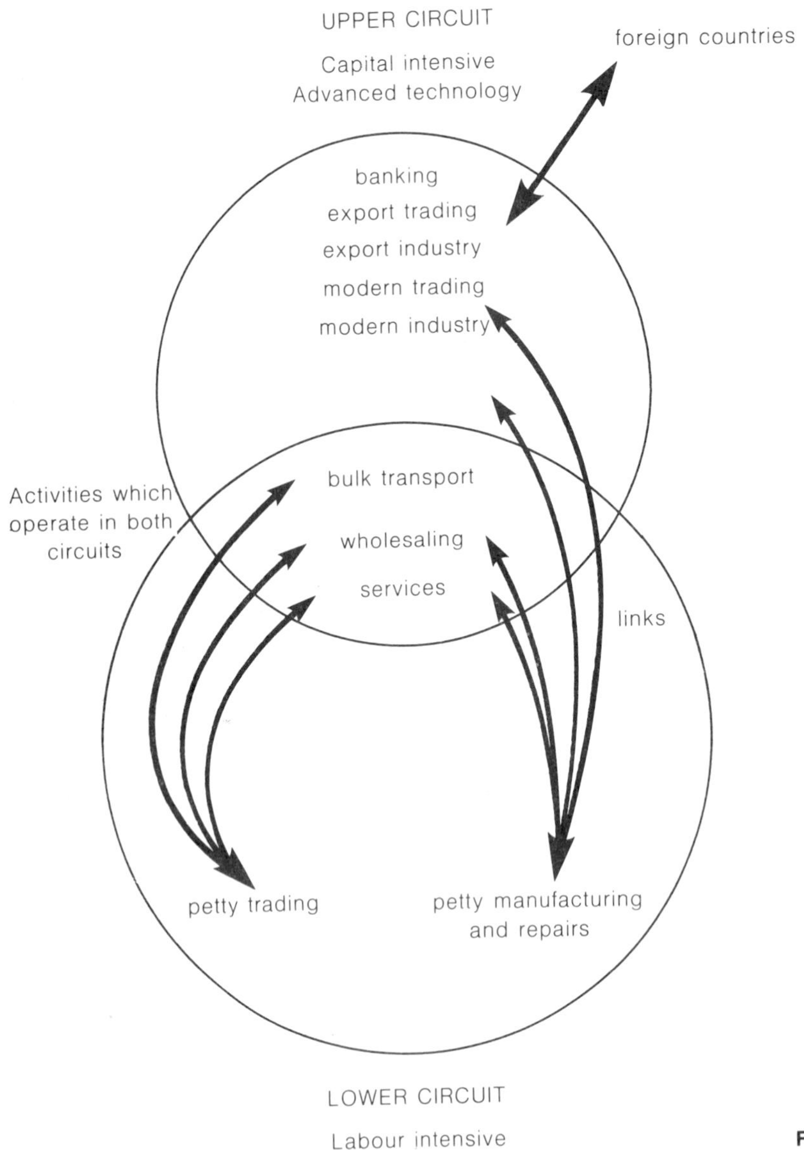

Figure 46.5 Dualism in the economy of the developing world cities.

In most developing countries the government administration is the largest employer of labour, and it too is run on the bazaar system. By continually adding more people into the chain that provides the goods or services, the government manages to absorb the growing population, even though it makes for ever-greater difficulty in obtaining those goods and services. Within the economy there is a contrast between the commercial capitalist system of modern-style manufacture in capital-intensive factories, working through a highly organised and specialised structure, and the old bazaar semi-subsistence economy, working through a system of shared labour and ad hoc exchanges of goods and services (Fig. 46.5).

Part of the reason for this dualism is the colonial heritage of many countries. During the colonial period many of the cities operated on strict hierarchical principles in which the European colonial administrators and businessmen separated themselves from the indigenous population, often allowing them only very limited entry to the ranks of the professional classes. Each community was encouraged to carry on its own independent way of life; the commercial shopping centre developing in the European organised sector; the bazaar semi-subsistence market area (Fig. 46.6) continuing in the native sector.

The firm-based and bazaar systems can run together in a stable way, providing some form of employment for everyone. However, if there is a substantial takeover of the bazaar sector by the firm-based sector, as has been the case in some small countries such as Jamaica, widespread unemployment results.

Ethnically mixed cities

When colonial governments began to exploit the resources of their territories, they found that the native population was often unable or unwilling to take part in manufacturing, in services or in commercial agriculture. In Africa, for example, the British had to import large numbers of Indian labourers to build the railway from the coast at Mombasa to Kampala in Uganda. Later, the Indians remained to become the hub of East Africa's commercial life. In Malaysia, the colonial government found it had to import large

Figure 46.6 The bazaar (permanent covered market) is one of the most distinctive features of Medinas. Its narrow streets remain a place of craft activity.

Figure 46.7 Problems of ethnic segregation in Nairobi: these barricaded and guarded shops belong to the Asian community.

numbers of Chinese to run the tin mines that were the foundation of the economy. Like the Indians, the Chinese adapted to a commercial urban environment very quickly. On the other hand the Malays remained ambivalent to commerce and tended to stay in the countryside. As a result the flourishing commercial life of Malaysia's capital, Kuala Lumpur, was built up almost entirely by the industrious efforts of the Chinese. Today Kuala Lumpur is essentially a Chinese city, the Chinese making up 60 per cent of the population. Even the Indians outnumber the Malays who, with only 15 per cent of the population, still largely live in enclaves within the city. The result is a **plural society** because under the protective umbrella of the colonial administration, the different groups were not forced to merge in any way. Since independence the Malays have taken over most of the top administrative jobs, although the commerce and manufacturing remain firmly in the hands of the Chinese and Indians.

The control of much commercial business by ('foreign') ethnic groups is a source of much social friction in countries throughout the developing world, and its harshness is often much greater than in countries of the developed world. During the period of the attempted coup in Kenya in 1982, for example, many Africans took the opportunity to ransack the shops of Indians, while leaving the shops of fellow Africans untouched (Fig. 46.7).

46.3 Residential patterns

In the developed world residential patterns are mainly the result of commercial competition and of government planning policies. The characteristic pattern of housing thus has high-cost housing in the core, then outward-spreading rings or sectors of lower-cost housing, the oldest and least expensive in the inner city, and the more expensive in the outer suburbs. By contrast, the colonial history of the developing world has often produced separate 'European' and 'native' areas within a city. There is frequently an inner-city élite region, often near the CBD, and also a number of higher-class residential suburbs within easy reach of the city centre. In ancient cities the European developments were often achieved by wholesale demolition of native housing, and these remain as enclaves within the dense housing pattern of the lower-class quarters (Fig. 46.10). However, in the majority of cases the overwhelming growth of urban areas has been due to the arrival of rural people seeking employment in the city.

Such is the rate of influx that the city authorities have little chance of providing sufficient housing to meet the demand and even if they could, the rents they would have to charge would be beyond the means of the immigrants. As a result, the vast majority of people are left to make their own arrangements. The native areas of the inner cities therefore contain many new arrivals who are seeking employment and who have not become sufficiently well established to have any permanent housing of their own. They may live in lodging houses, on roof tops, on verandahs or even on pavements. Eventually most will attempt to form some permanent dwelling, moving away from the centre to more peripheral areas where land is available. River valleys (Fig. 46.8) and steep hillsides are often colonised by these people. They begin by constructing their homes from the variety of materials that they can buy or pick up from waste heaps, but these cardboard and tin shacks are gradually rebuilt with more substantial materials as the people begin to earn money and can afford to buy bricks and other more permanent materials. The spontaneous settlements gradually mature into established suburbs. During this process it may be difficult to decide at a glance whether the area is being built up or demolished, for housing additions take place piecemeal, while roads are still unsurfaced and service provision is rudimentary (Fig. 46.9).

The gradual development of spontaneous settlements into suburbs takes place in a manner that is extremely varied from region to region and depends heavily on local circumstances.

In general, spontaneous developments are not simply random collections of shacks; mostly they have a well developed structure, for it is in everyone's interest to see that a network of thoroughfares develops with the new housing. Dwellings are often organised into groups that are equivalent to rural village units. The government helps as best it can by providing at least some water via standpipes. It is the existence of an internal structure that allows the spontaneous developments to become upgraded into suburbs.

Figure 46.8 Nairobi's growth has produced an increasing proportion of squatter settlements, mostly in the poorly drained valley bottoms.

Many of the major differences between the housing arrangements of the developed- and developing-world cities stem from the contrasting way in which house ownerships are transferred and new houses are built. In the *developed* world houses are bought and sold in such a way that people can gradually move to better housing as their circumstances improve and their savings increase. The release of money by house sale provides capital which allows further professional building to match demand. Furthermore, the whole system is underpinned by financial stability, a policy of low interest rates for house purchase, and a large middle class with the salaries to afford to buy property. In the *developing* world the national wealth is spread out much less evenly, and there are drastic differences between rich and poor. Most importantly, there is no large middle class and therefore no supply of money to finance the growth of the housing stock. Governments do try to build public housing, but their resources are very limited. As a compromise, some governments provide a concrete 'pad' as the base for a house and leave the construction of the fabric of the building to the prospective house owner.

Figure 46.9 The apparently ramshackle development contains a considerable element of structure. Each house is being built in stages as its owner can afford more materials; there is no money for street surfacing.

Eye sores as vote banks

By USHA RAI

NEW DELHI, March 19.

SOME 700 jhuggies have come up the nallah behind the Nehru Stadium in the past two months. The residents of the slum call it the Rajiv Gandhi camp and with the talk of election in the air are hopeful of getting alternative sites, if not now in a few years.

Most of the jhuggies here were demolished after the Asiad but a handful were allowed to stay to house workers for the unfinished construction work in the area. Six months ago there were 100 jhuggies in the area. Today their number has increased sevenfold and it is pertinent that not all are set up by construction workers.

The pradhan or head of this squatters colony is Udaybhan Singh, a Congress worker who was also the pradhan of the Govindpuri slum. Other influential people from the area are actively involved in the affairs of the squatters. It is with good reason that the jhuggies have been termed the "vote bank" of politicians. The name given to the camp and the involvement of political workers makes it look as if it has official sanction and this attracts more squatters.

A few residents claim they have each paid Rs. 200 to the pradhan for permission to put up a jhuggi. With another contribution of Rs. 20 each the squatters have installed four handpumps. Eight bathrooms (four each for men and women) provided earlier by the National Building Construction Corporation for their labour force, now serve the entire slum. These are wholly inadequate for such a large number of dwellers and the entire area stinks.

But the Rajiv Gandhi camp is only symbolic of the 500 odd clusters dotting the Delhi landscape — each having a pradhan with political connections as well as muscle power to legitimise their existence.

The Subhash camp in Dakshinpuri has 2500 jhuggies, Pulpehlad camp near Tughlakabad about 4000, the Okhla industrial area another 3000 and the janata camp at Seelampur 700. In the West Delhi, near Janakpuri, there are four camps each with 700 to 800 jhuggies. In Naraina and Prem Nagar there are more clusters of jhuggies. On the Old Yamuna bridge some 3000 jhuggies have come up. About 1000 jhuggies dwellers of this area were given plots in Seelampur and Jahangirpuri in 1975, they were all back six years later.

This is the first of a two-series report on Delhi's proliferating slums and shanty colonies.

The problem of jhuggies is, however, not new. During the emergency 1.50 lakh jhuggies were cleared and the owners moved to the resettlement colonies. On March 1, 1977, only 25,000 jhuggies remained in the Union territory. The survey in January 1978 revealed that the number had swelled to 80,000 in six months of the Janata regime. After the Asiad there were 1.13 lakh jhuggies and today their number has more than doubled. Every day more jhuggies are being added to the existing clusters.

One of the reasons for this rapid proliferation is the 50,000 labour force that came to Delhi for the Asiad projects and stayed on, adding to the filth, dirt and foul smell. This is despite the clause in the contract with the labour contractor categorically providing for the removal of all jhuggies and encroachments before the final payment is made to the work force.

In addition, there is a daily influx of labour from the neighbouring states. No one who enters the city wants to leave, for employment opportunities and wages are better in Delhi than elsewhere. Because of the slow pace of development of satellite towns even people working in Ghaziabad, Faridabad factories stay in jhuggies in Delhi.

A major attraction for squatters is the promise of 25 square yards of developed land by politicians. The government's scheme of providing shelter to the squatters while removing the eyesores from the city proper has, however, not worked out. Seventy-five per cent of the plots given during the emergency and 50 per cent during the post-emergency period have been sold at prices ranging from Rs. 3,000 to Rs. 5,000. The squatters have made quick money, moved back closer to their area of work and now are sitting pretty in anticipation of another plot, says an authoritative DDA source.

THE "GODFATHER"

Thus overnight what is a small group of jhuggies starts mushrooming. More and more people infiltrate into the existing clusters till it becomes a sprawling slum. In parks, land allotted to group housing societies, private plots which have not been constructed on and even in the available space in the resettlement colonies the jhuggies come up. It is difficult to identify the infiltrator and even more difficult to take action against him.

One of the big colonies of squatters is along the railway line at Ashok Vihar, near Subzimandi. Some 8000 jhuggies have come up there. This area had been totally cleared of jhuggies during the emergency. Under the Janata regime the first few appeared and their number has grown since. Most owners are labourers working in the Mandi and the Wazirpur industrial area. This massive colony has 20 or 30 pradhans, each extending his patronage to a cluster of 50 to 200 jhuggies. The Congress councillor, Mr Deepchand Bandhu, is the "godfather" of this slum.

At the Lok Nayak Jayaprakash Hospital, overlooking the operation theatres, is another cluster that the DDA has been battling to remove. During the emergency some 3000 jhuggies were cleared from the area and residents given land in Nand Nagari. In the Janata regime Mr Sikandar Bhakt allowed 450 of them to come back. Mr Arya, a former Congress councillor, allowed another 700. Mr Ramesh Dutta, the new councillor, is credited with bringing in another 1500 persons to establish their jhuggies.

46.4 Variety in developing-world cities

Because many cities do not have the overriding and centralising force of commercialism to draw their functions into a coherent geometry, the concentric, sector and multiple-nuclei models of the developed world have only marginal relevance to the developing world. The Western traveller visiting Cairo, Nairobi, Calcutta or Bogota will find great difficulty in discerning the overall pattern of city life. Because each city has grown in relative isolation, its form is strongly influenced by the local history and culture. It is therefore not possible to construct a single model of the developing-world city; rather it is necessary to work with a number of models, each reflecting the culture and history of its region.

The Middle East: Cairo

Egypt has always been governed from its cities and many of them are very old. Today Cairo contains 10 million of the country's 45 million people and is more dominant than ever. Because the people in the fertile agricultural regions of the Nile Delta perceive that power and wealth are inseparable, they have flocked to the city in search of better opportunities and wealth. Paradoxically, the more services the government provides in the rural villages to make the life of farmers more tolerable, the more the farmers realise the discrepancy between their life-style and that of the city and the more they migrate. Currently some 200 000 people leave the land each year, partly attracted by the supposed benefits of the city, partly driven from the land by the system of land inheritance which has fragmented landholdings to such a degree that even subsistence farming is becoming impossible for many. For those without land, jobs as labourers are even more uncertain, and activities such as brickmaking are extremely arduous.

Cairo's growth, structure and prospects must therefore be seen in the context of a long history coupled with a dramatic influx of farmers. However, its main streets and CBD stem from a quite different source: the desire of one of its rulers to turn Cairo into a European city. In the late 19th century the heart of the city was ripped out and new boulevards constructed; old native housing was demolished to make way for a European housing quarter, and the city walls raized to the ground (Fig. 46.10).

Today Cairo is a mixture of old and new jostling side by side (Fig. 46.11). But although the centre may look like a Western commercial city (Fig. 46.12), there is no real wealth to underpin the economy. Foreign aid, oil, cotton and remittances from Egyptians working abroad bolster the country's economy. Most people cannot earn enough to be able to afford to buy Cairo's imported Western goods, and they do not

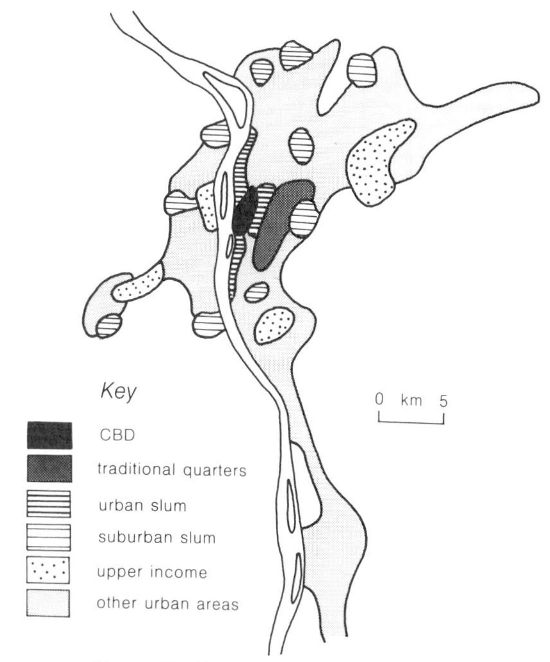

Figure 46.10 Land use in Cairo.

Figure 46.11 The crowded Medina areas of Cairo, split by 19th-century roads and now partly redeveloped by municipal housing projects.

Figure 46.12 The Nile front, Cairo. Here bid rents operate and the best sites are used by international hotels for western tourists.

have the skills or financial resources to build industries on a scale that would allow them to compete as manufacturers in a world market. Most people find employment in services, sharing out the small real wealth generated by export earners such as tourism. Most rural workers find employment as petty traders, drivers, clerks, domestic servants and in other services, and discover that life is hardly better in the city. Indeed, by living in the old quarters they experience a world that is economically, politically and culturally like an overcrowed village – save that, unlike the village, there are in the city only weak social relations and no support from an 'extended family'. The impossibility of earning a decent wage has driven many rural people on a two-step journey, first to Cairo and then abroad to countries such as Jordan, Saudi Arabia and Kuwait where they can earn four or five times as much money. At present 3 million people, one in eight of the workforce of Egypt, are employed abroad.

Industry has long found the severe congestion that plagues the city an intolerable burden and has moved to the periphery, although there is a multitude of small craft and service industries crammed into the old quarters. In these areas people live in incredibly cramped accommodation, with an estimated 500 000 living on rooftops and verandahs. Population densities often exceed 100 000/km^2 and this has compelled many people to build peripheral spontaneous settlements. Today the city sprawls out onto valuable agricultural land and reaches to the base of the pyramids at Giza. Pressure on space is so great that tombs near the pyramids, and other city cemeteries (including the famous City of the Dead), are now used by the poor as housing.

Africa south of the Sahara: Nairobi

Nairobi, the capital of Kenya, lies in the sub-Saharan zone of Africa, one of the least urbanised areas of the world. With a population of 1.3 million, Nairobi dominates the urban hierarchy not only of Kenya, but of much of East Africa (Fig. 46.13).

Traditionally, urban centres of sub-Saharan Africa have a coastal bias. Most were founded or developed by colonial powers, whether of Arab or European origin. Thus, Mombasa, in Kenya, occupies a defensive island site along an otherwise inaccessible coral-fringed coast. These coastal towns were intended to

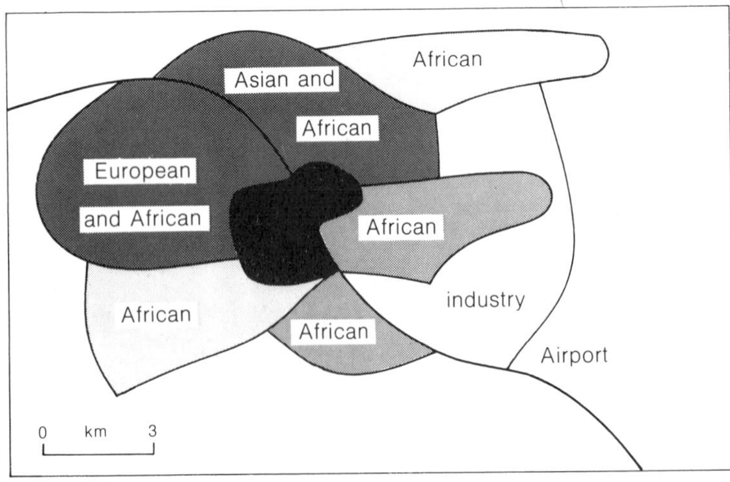

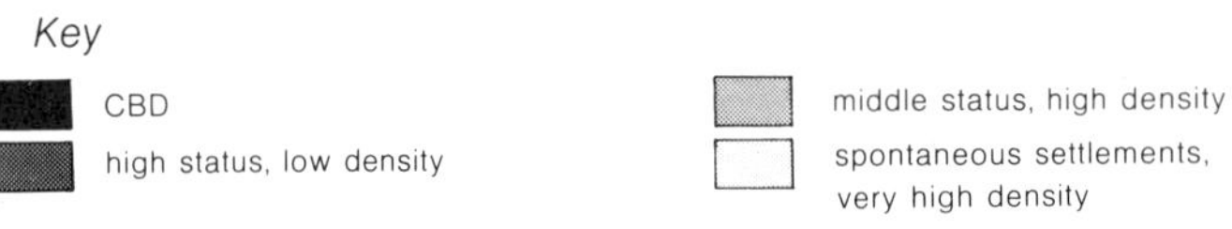

Figure 46.13 Land use in Nairobi.

be no more than entrepots for goods leaving the colony and destined for the Middle East or Europe. However, in the 19th century, exploration and colonisation were carried inland to enable the resources of the port hinterlands to be exploited. Nairobi grew up at the site of a railway construction camp in the Eastern Highlands region of Kenya. Here Europeans found an equable climate from which they could administer the country.

With the colonial administration based at Nairobi, lines of communication were built radially outwards, allowing the city to become a centre in which to gather agricultural goods from the productive uplands before sending them to Mombasa for export. The radial transport structure has also ensured Nairobi's development as a primate city, where much of the country's wealth and influence are concentrated. Such affluence as exists in Kenya is highlighted by Nairobi's imposing CBD (Fig. 46.14).

Nairobi is built on a lowland swamp and has major problems of building and sanitation. The most difficult areas to drain, the shallow Nairobi River and its tributary, have never been used for permanent buildings. Today these neglected regions are acting as the foci of the spontaneous settlements that are forming an ever-more extensive part of Nairobi's structure.

Nairobi is a city built for colonial Europeans by imported Asian construction workers and now ruled by Africans. Asians still perform most of the commercial functions in the city and this leads to considerable ethnic resentment from the disadvantaged African majority. As a result, Nairobi's residential districts are highly segregated, and a feeling of considerable social tension is a pervading element of city life.

Southern Asia: Calcutta

In India, the special form of colonial administration provided the basis of another unique city form (Fig. 46.15). As with other colonies, there was a paramount need for a port that could transfer agricultural goods from the hinterland to ships destined for Europe. The other main requirements were for administration and defence, and thus port, government offices and fort were built together. Furthermore, these European functions were separated from the native town by an open space which provided a defensive field of fire in times of attack. The planned European sector, with its large houses and wide streets, developed quite separately from the native sector, which grew in an *ad-hoc* manner and was characterised by narrow streets and high-density housing (the classic 'walking city', p. 454). With the

Figure 46.14 The Europeanised CBD of Nairobi, focused on the high-rise Kenyatta conference centre.

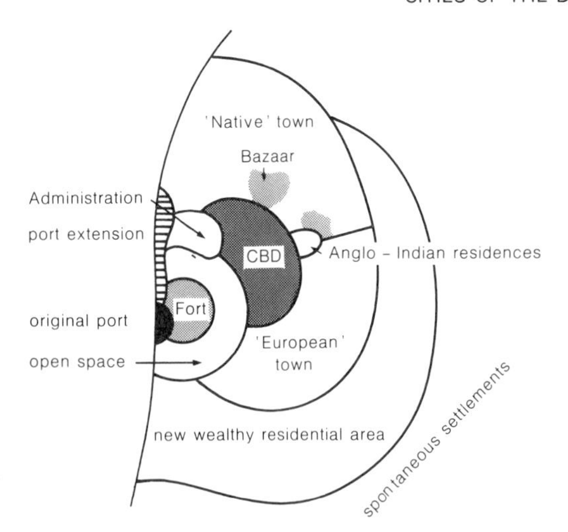

Figure 46.15 The structure of an Indian colonial city such as Calcutta.

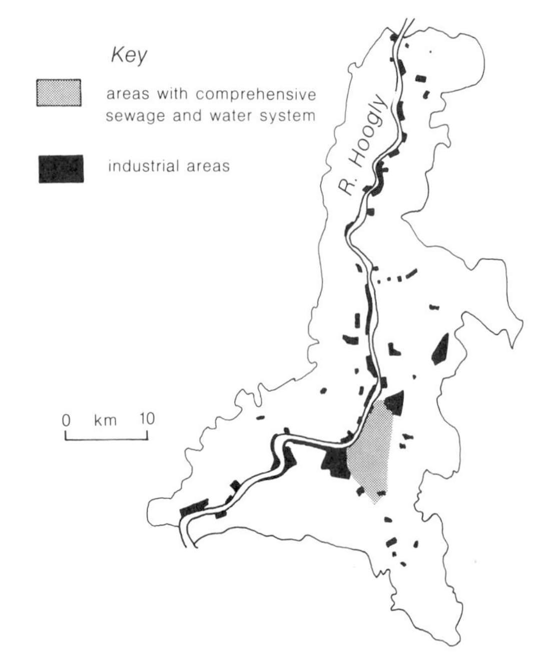

Figure 46.16 The site of metropolitan Calcutta.

rise in trade, there was a gradual development of both office and commercial functions and a CBD grew near the government offices. As the rural-urban drift gathered momentum, the outskirts of the city continued expanding with the growth of spontaneous settlements.

Calcutta has been called the worst city in the world. It shows some of the most extreme forms of urban decay and under-development. It is an example of the problems that can come from under-capitalised urban growth of an uncontrolled and largely spontaneous nature. Calcutta should never have been a city. It lies on the vast Ganges Delta 130 km from the Bay of Bengal, surrounded by marshes and salt lakes, with a water table nowhere more than a metre or two below the surface. Yet despite the physical difficulties of the site, by the early 19th century there were already 300 000 people crowded along the banks of the Hoogly river, with warehouses stretching along the waterfront on the remainder of the levée (Fig. 46.16). At the upper end of the commercial zone spread the spacious planned European settlement, but surrounding the whole complex on every piece of dry and not so dry land among the marshes was an ever-growing ring of native shanty huts, already mostly slum. When factory-style industry was introduced in 1854, jute processing mills were able to use some of these people, but the possibility of factory employment increased the reputation of Calcutta and drew even more people to its marshes. Because the mills needed a river frontage, Calcutta gradually expanded down stream. Soon local coal and iron ore were also part of the trade, and a steel works and engineering factories were constructed. Today the Calcutta conurbation stretches 65 km along the banks of the Hoogly River.

Although conditions in the city have always been squalid for the majority of its inhabitants, lack of farmland in the delta has driven ever more people in search of any work that might be available in the city. This, together with the natural city increase and an influx of refugees from Bangladesh, has swollen Calcutta's population to over 10 million. The problems faced by the city administration are of gigantic proportions, but they are made worse by the unfavourable nature of the swampland site which makes the provision of basic services such as water and sewerage so much more difficult. Calcutta has been described as a city of 10 million with the facilities for a quarter of a million. Poor sanitation and low levels of nutrition have also resulted in Calcutta being the worst city in the world for cholera. In Calcutta any sort of accommodation is in pitifully short supply, and despite 700 000 being housed in single-room 'bustee' lodging houses within the old town, over 100 000 people have to sleep in tents or on the pavements each night (Fig. 46.17).

Latin America: Bogota

Latin America was colonised mainly by the Spanish and Portuguese. Most Spanish cities grew in the same mould, dictated by the colonial administration, with a central plaza surrounded by a grid-iron pattern of roads. At the centre of the town the government offices were constructed, and this in turn acted as the

Figure 46.17 Basic problems of housing and sanitation are acute in the uncontrolled urban growth of Calcutta. Here a mother struggles to look after a family in a tent made from rags.

focus for the colonists and encouraged the wealthy élite to live close to the centre.

Latin American cities were slow to develop any form of industrialisation and they remained purely administrative and commercial centres well into the 20th century. Only Buenos Aires, Montevideo, São Paulo, Rio de Janeiro and Mexico City were restructured to take some advantage of manufacturing. In these places the central plaza was transformed into a CBD which operated along commercial lines, gradually extending along the main road in the form of a **development spine** with skyscraper office blocks, shops and hotels (Fig. 46.18).

More recently industrialisation and commercial activity has taken place in the other major Latin American cities. However, with a lack of financial resources to carry services into every part of the city, there has been a tendency to take services primarily to the region containing the wealthy. Because industry also requires services, factories have been attracted towards the development spine such that industry and élite housing are much closer to one another than would be the case in developed-world cities.

Latin American cities display the contrasts between rich and poor perhaps more overtly than cities in other parts of the developing world. They have also seen the most rapid growth in urbanisation, with growth rates often exceeding a 12 per cent increase in population per year for prolonged periods. Spontaneous settlements are widespread and a characteristic feature of all Latin American cities. In Brazil, for example, a quarter of the population of 90 million live in spontaneous settlements (called **favelas**); a million live in the favelas of Rio de Janeiro alone. The élite continue to occupy an amount of land grossly disproportionate in relation to their number.

Perhaps fewer than 5 per cent of the population of each city may occupy over 25 per cent of the land.

All of the characteristic development features of Latin American cities can be seen in Bogota, the capital city of Colombia. Here the spine began to develop in the early 1900s with the establishment of trolley-bus routes northwards from the central plaza (Fig.46.19). By the 1970s it had reached the El Chico district some 9 km away. The spine contains a mixture of office blocks, light industry, speciality shops, restaurants, embassies and services catering for the wealthy. To either side of the main road lies the élite residential area, consisting of detached houses, often built after European or American styles. A 'mock-Tudor' house may thus be seen adjacent to a 'Californian ranch' house in the centre of Bogota.

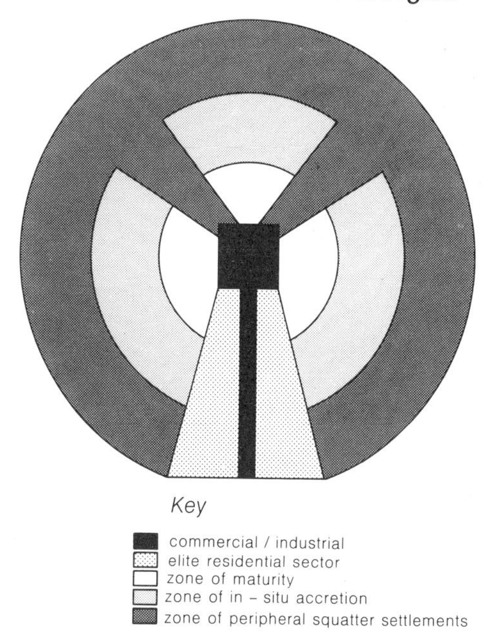

Figure 46.18 A model of a Latin American city.

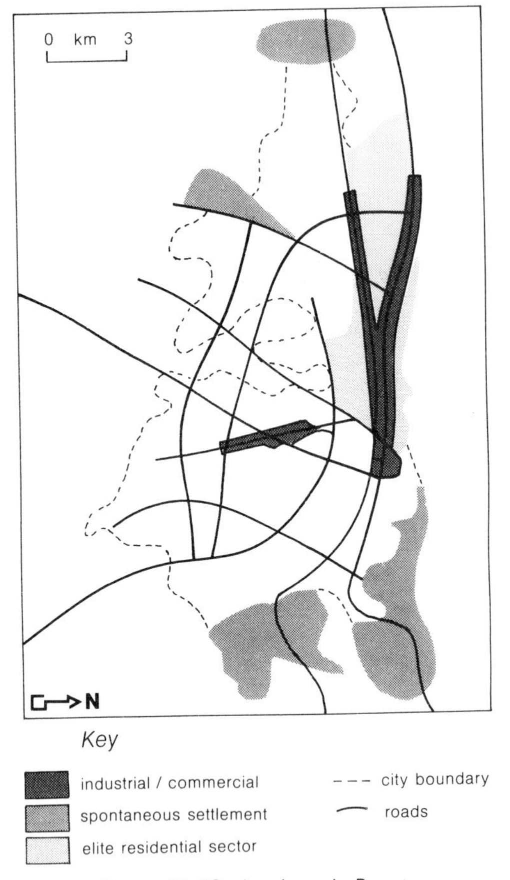

Figure 46.19 Land use in Bogota.

Figure 46.20 Carrying water in the peripheral shanty area of Bogota.

Surrounding the élite zone, and acting as a buffer between élite and new spontaneous settlements, lies the zone of maturity. This once held spontaneous settlements, but it has gradually become more established, and now most buildings are of brick or concrete. Beyond this again lies the newer spontaneous-settlement zone, now steadily climbing the surrounding steep hills or filling in the swampy river-valley bottoms. Electricity, water and sewerage are virtually absent from this zone (Fig 46.20), although there is a great deal of optimism from the inhabitants who aspire to make their areas as good as the zone of maturity. Meanwhile the spontaneous-settlement zone continues to grow.

Part L

Revision summaries

The line drawings in these revision summaries have been prepared to the standard expected of a student in an examination, rather than emulating professional book illustrations.

Part A Earth

Tectonic processes

1 The Earth has an inner *core* (which generates the magnetic field), an outer *mantle* containing convection currents, and a thin rigid *crust*.
2 Continental materials are *granitic*; ocean materials are *basaltic*.
3 The height of any section of the crust depends on *isostatic forces*. Most mountains have roots.
4 The concept of *plate tectonics* considers the Earth as a number of rigid plates with *subduction zones* where the plates meet, *ocean ridges* where they split apart, and *wrench faults* where they move past one another.

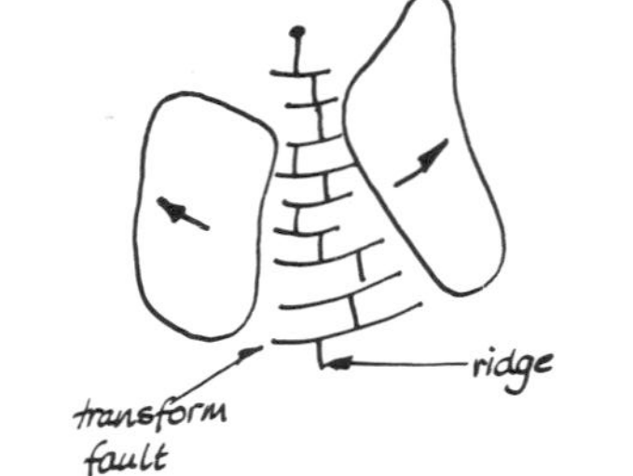

5 Subduction zones are marked by island arcs and trenches; *spreading centres* are marked by ocean ridges and rifting.

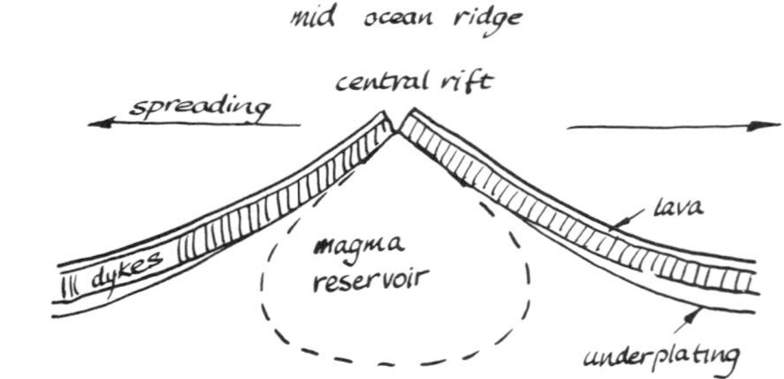

6 Plate tectonics is part of the long process of *continental drift* in which Pangea split first into Laurasia and Gondwanaland, then into smaller plates.

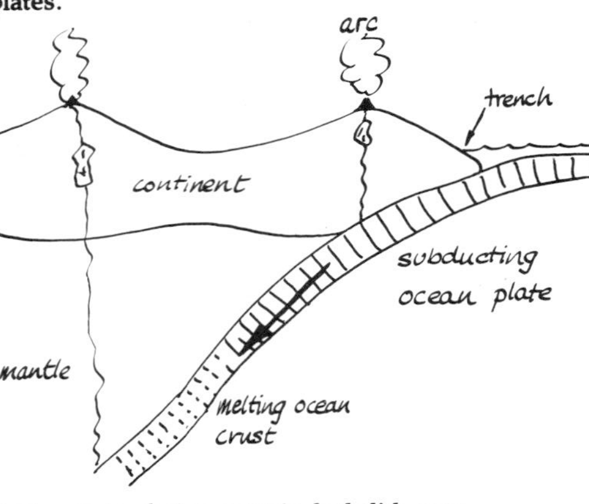

7 Mountain chains contain *batholith cores*, *metamorphic zones* of compression and heating, and marginal *sedimentary basins*.

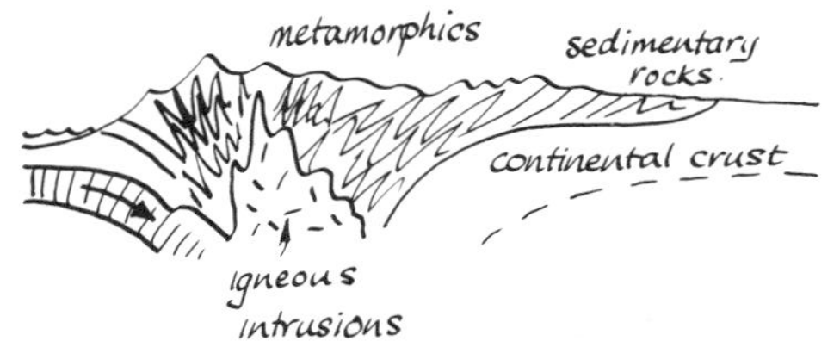

8 *Earthquakes* form when rocks fracture. Waves move out from the *focus*; tall buildings are damaged by low-frequency waves, low buildings are damaged by high-frequency waves.

9 Earthquakes associated with wrench faults occur spasmodically but maintain a long-term average rate of activity.

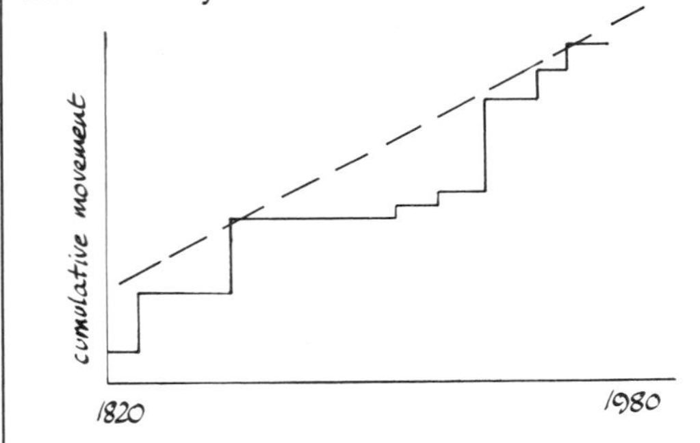

10 Earthquakes cause direct vibration damage, initiate landslides, cause ground liquefaction, rupture utilities, and damage dams and bridges.
11 *Volcanic eruptions* occur when *magma* rises up a profound fracture. Eruptions are often accompanied by earthquakes.
12 Volcanic products are lava, ash, steam and gases.
13 *Basaltic volcanoes* erupt quietly; *andesitic volcanoes* erupt violently.
14 Volcanic eruptions cause damage by ash fall, lava flows, mudflows and floods, and by initiating fires.
15 Volcanoes and earthquakes are most common at plate margins.

Part B Atmosphere

Weather and climate

1 Weather denotes the everyday character of the atmosphere; *climate* is the long-term average pattern of the weather.

General circulation

1 Areas equatorwards of 38° receive more heat than they lose; areas poleward of 38° lose more heat than they receive. The unequal distribution of heat drives the *atmospheric circulation*.
2 The global circulation consists of:

(a) a large *convection* (*Hadley*) *cell* in the Tropics where vertical movements of air are most important;
(b) a region of lateral mixing in mid-latitudes;
(c) small convection cells over the poles.

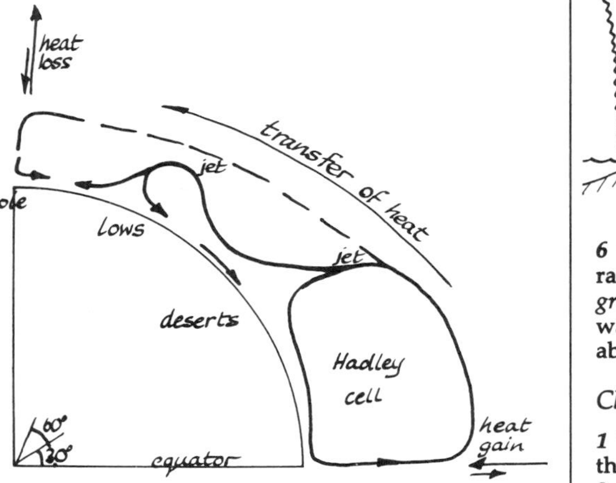

4 Rotation of the Earth is responsible for the pattern of mid-latitude mixing. Air moving polewards of 20° becomes so unstable it forms into large waves in the upper troposphere (at whose centres lie the *jet streams*). The shape of the waves is strongly influenced by mountain ranges and by temperature contrasts between oceans and continents.
5 The *troposphere* is transparent to solar short-wave radiation and is therefore heated from below by long-wave terrestrial radiation and by conduction and convection.

6 Clouds and carbon dioxide absorb long-wave radiation and radiate part of it back to Earth. This *greenhouse effect* keeps the lower troposphere warmer by 25 °C than if clouds and CO_2 were absent.

Cloud formation

1 The troposphere is heated from below and is therefore essentially unstable.
2 Air rising quickly obeys the gas laws, cooling at the *DALR* (10 °C/km) when dry and at the *SALR* (6 °C/km) when fed with latent heat from condensation of vapour.

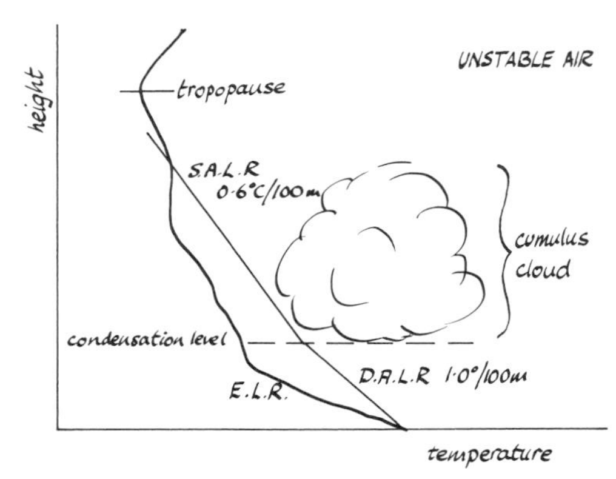

3 A parcel of air will only rise if it has buoyancy, i.e. if it is warmer and less dense than its surroundings.
4 The *temperature profile* of the troposphere is very important in controlling convection. On average the environment has a *lapse rate* of about 7 °C/km.
5 Air can be absolutely unstable (DALR and SALR less than ELR), in which case convection cloud is possible; conditionally unstable (DALR greater than ELR but less than SALR), in which case convection cloud is possible only after air has been forced beyond condensation level; and absolutely stable (DALR and SALR greater than ELR), in which case no convective cloud possible.
6 The condensation of water vapour usually occurs on *condensation nuclei*.
7 There are three basic *cloud forms*: *cumulus* (pillow cloud), caused by convection in unstable air; *stratus* (layer cloud) forming widespread sheets, caused by forced ascent in a stable atmosphere; and *cirrus* (wisp cloud), high-level layered cloud made from ice crystals.
8 Cloud types are not mutally exclusive and one type very often develops into another, e.g. stratus develops cumulus tops in conditionally unstable air.
9 There are two *precipitation-forming* processes: the *Bergeron process*, using ice crystals and water droplets in cold clouds: and the *coalescence process*, using sweeping updraughts of air in warm clouds.

Climates

1 Most *climatic classifications* rely on temperature and rainfall variations. Temperature regimes are strongly related to latitude; rainfall is related more to areas of atmospheric convergence (*ITCZ* and *polar front*) and the location of moisture-bearing winds (*trades*). The most commonly used classification is that devised by *Köppen*.

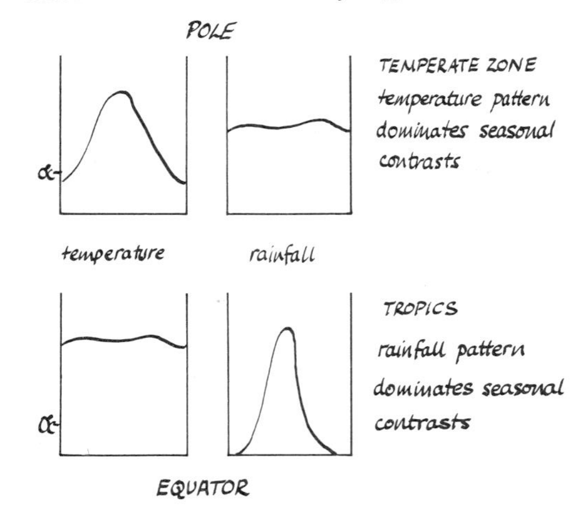

2 The *Thornthwaite* classification is based on variations in the annual *water budget*. This scheme has direct practical applications.
3 Regions near the equator lie within the *inter-tropical convergence zone* (*ITCZ*) with weather dominated by daily thunderstorm activity.
4 Tropical regions near the edges of the ITCZ have a climate with pronounced wet-and-dry seasons as they are influenced alternately by the ITCZ and the *sub-tropical high-pressure belt*.
5 *Deserts* remain dominated by dynamic subsidence of the sub-tropical high-pressure belt.
6 There are considerable contrasts between east- and west-coast climates in the Tropics: the trade winds and ocean currents make east coasts wet and keep west coasts dry.
7 Some sub-tropical and tropical regions experience *monsoon* conditions with an abrupt start to the wet season. They are often associated with rapid shifts in the sub-tropical jet stream.

8 *Hurricanes* are tropical storms that move into higher latitudes.
9 Mid-latitude weather patterns are characteristically variable. There are no distinct wet or dry seasons.
10 The variability of mid-latitude weather is caused by continual passage of *anticyclones* and *depressions*.

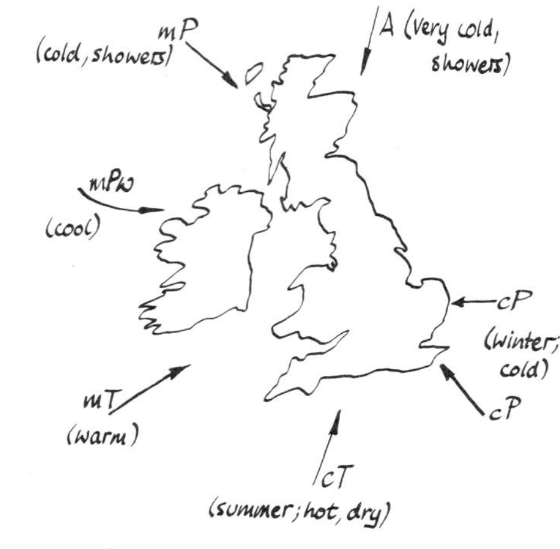

AIR MASSES

11 Depressions cause the interaction of air masses from polar and tropical source regions. These are mP, mPr, A, cP, cT, mT.
12 Polar air masses are cold and unstable; tropical air masses are warm and stable.

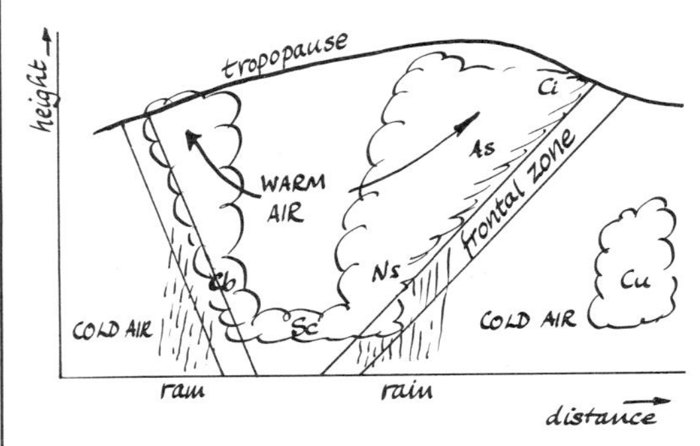

13 Depressions may contain warm-front, cold-front and occluded-front cloud belts. Some thin cloud may also form in the warm sector.
13 The *cold front* is a much more active feature than the *warm front*, finally catching up the warm front and causing an *occlusion*.

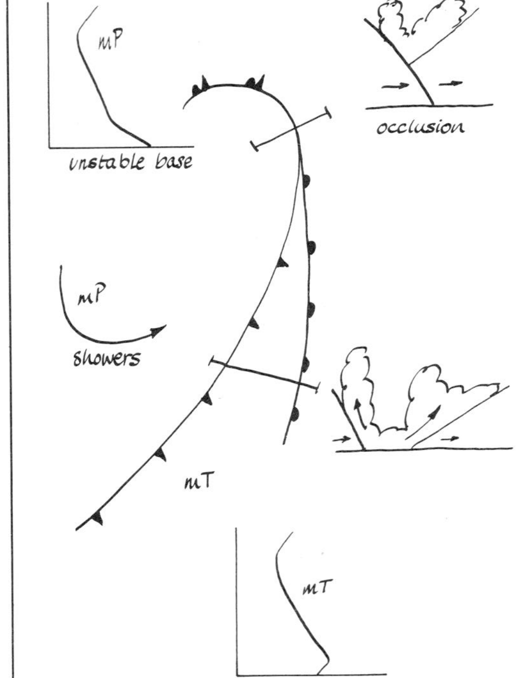

14 Temperature contrasts between land and ocean cause *onshore* and *offshore breezes*.
15 Air forced over a mountain range may give a *föhn effect*.

16 Daily temperature variations give *valley winds* in early morning, *thermals* by midday, and *mountain breezes* by night.

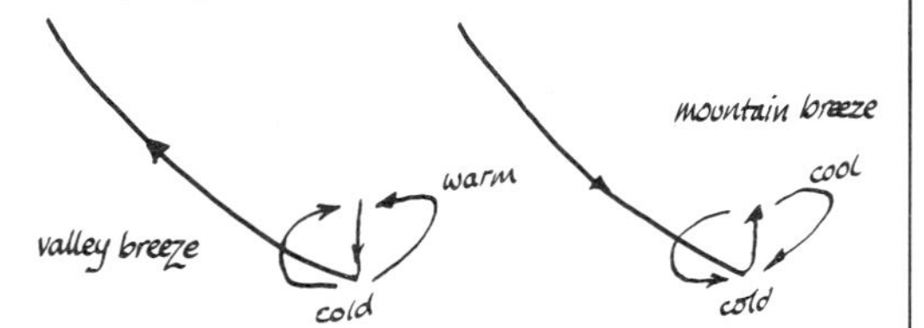

17 Urban areas cause *heat islands* in winter with calm weather.

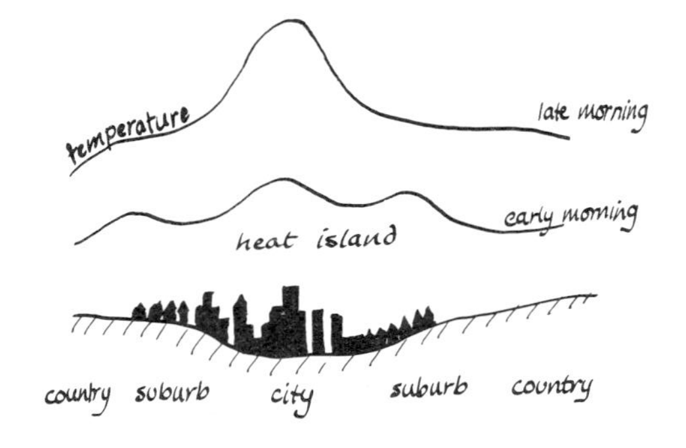

18 Local-climate effects can enhance the build-up of *pollution*; the dispersal of pollutants causes *acid rain*.
19 Emission of CO_2 may cause global warming and alter global climates, making some areas drier and others wetter.

Part C Water

The water cycle

1 The water cycle involves the continuous transfer of water between ocean, land and atomosphere.

the hydrological cycle

2 When rain falls on land there is *loss* due to :

(a) interception;
(b) evaporation;
(c) transpiration;
(d) replacement of soil moisture and ground water.

The remainder returns to the ocean.
3 *Infiltration* depends on rainfall intensity and *soil permeability*. Infiltration decreases with time towards a constant low value.
4 Water moves in the soil by vertical *percolation*, lateral *throughflow* and *groundwater flow*.
5 Streams are fed by

(a) throughflow converging in concavities;
(b) the build-up of a head of water in soils near streams or in aquifers;
(c) infiltration-excess overland flow (high rainfall intensity);
(d) saturation overland flow (long rainfall duration).

the hydrograph

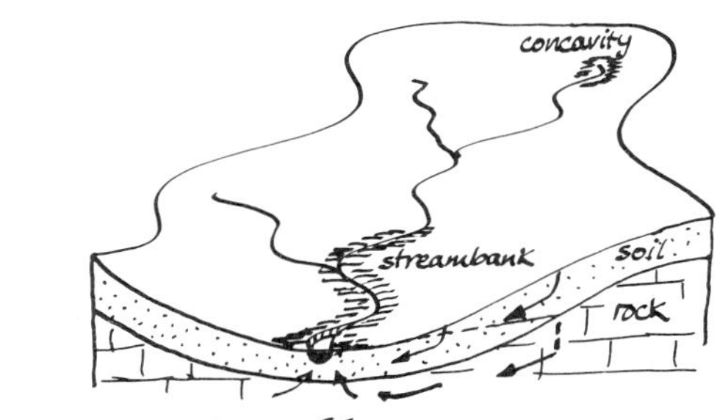

sources of runoff

6 A *water balance* is the difference between *precipitation* and *evapotranspiration*. In Britain *water deficits* are common in summer and surpluses common in winter.

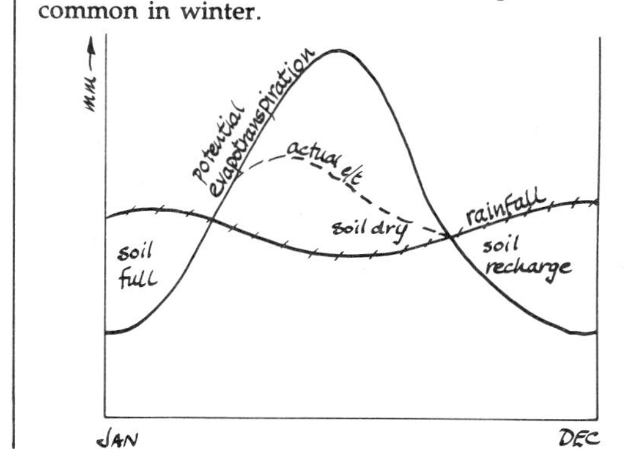

7 Lake storage reduces peak streamflow.

Drainage networks

1 Drainage networks develop from *sheetflow* to *rills*, to *gullies*, to *stream channels*.
2 There is a belt of no erosion near *basin divides*.
3 Streams are initiated considerable distances from divides, then grow headward.
4 Drainage networks can be divided into first-order, second-order, third-order (etc.) streams. There are regular relationships between stream *order* and the number of streams in each order, stream length and basin area.

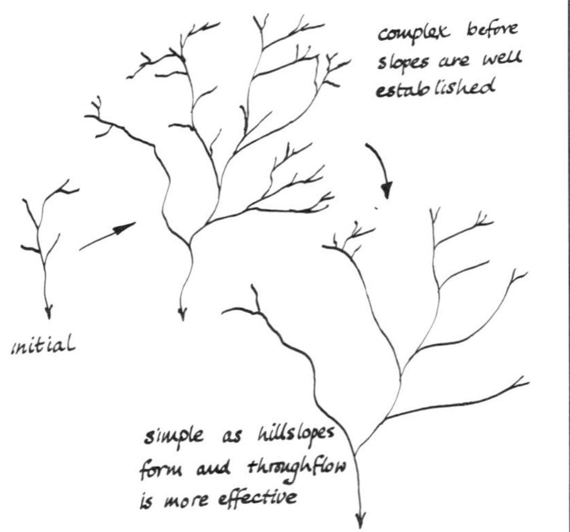

5 Drainage density is the ratio of overland flow to throughflow.
6 Streams evolve first into a dense network, then *rationalise*. Fingertip segments become fewer as hillslopes lengthen and transfer water more effectively.
7 River *long-profiles* are concave upwards but show many irregularities.
8 Long-profiles change repeatedly with time, sections flattening as sediment builds up, then steepening as trenching occurs.

Channel shape and process

1 There are three basic stable *channel shapes:*
(a) single meandering channel;
(b) braided channel;
(c) single straight channel.

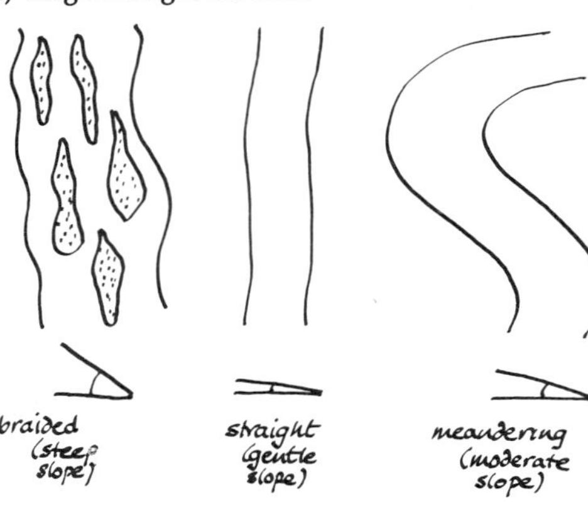

2 Straight channels occur on gentle slopes; meandering channels on moderate slopes; braided channels on steep slopes.
3 River *meander wavelength* is closely related to channel width and basin area.
4 Transport and erosion within a channel depend on available energy. Most energy is dissipated as *frictional drag* on bed and banks.
5 Rivers *flow* fastest in a central surface thread that zig-zags across straight channels; in curved channels centifugal force throws water to the concave bank.
6 Channel shape changes occur according to the *law of minimum variance*.
7 Control on speed in a channel is produced by

(a) roughness;
(b) channel shape;
(c) meandering;
(d) pools and riffles;
(e) ripples and dunes;
(f) braiding.

8 Channels in cohesive clays have steep banks; channels in non-cohesive sands and gravels are shallow and have gently sloping banks.

9 Meandering rivers evolve a distinctive pattern of *pools*, *riffles* and *point bars*.
10 Floodplains develop by meander-belt migration.
11 Meandering rivers evolve randomly but negative-feedback processes keep them in check, causing *cut-offs*.

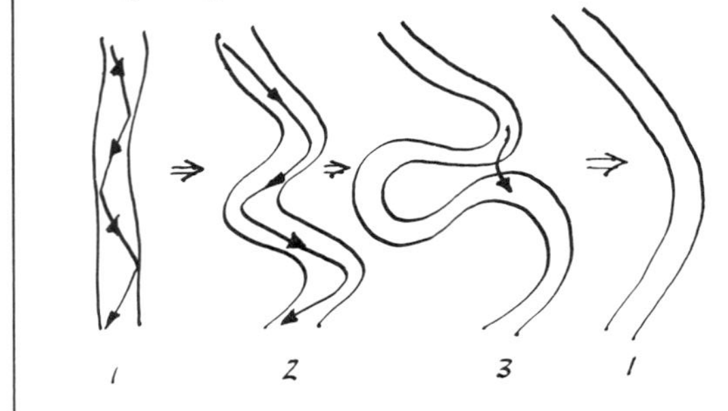

12 Each *sediment size* has a *threshold velocity* which will cause it to move.
13 Movement occurs by *solution*, *suspension*, *saltation* and *traction*.

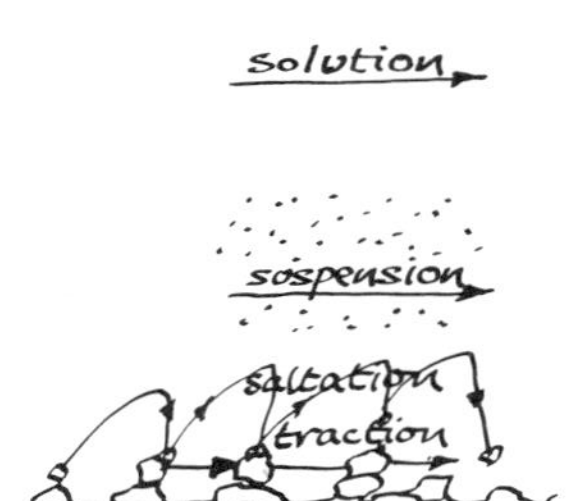

14 There are limits on movement imposed by each transport process and rivers rarely scour to bedrock.
15 Where cobbles form part of the sediment load they cause *bed armouring*.
16 River channels are primarily *alluvial*.
17 Peak sediment movement may not always coincide with peak river flow.

Water and man

1 Rivers develop slowly in *dynamic equilibrium* but achieve order through fluctuation about a long-term average; the natural tendency for rivers to meander and transport sediment must be understood before *river modifications* are attempted.
2 River straightening through meander cutoffs causes downstream deposition and upstream incision until a stable long-profile is regained.

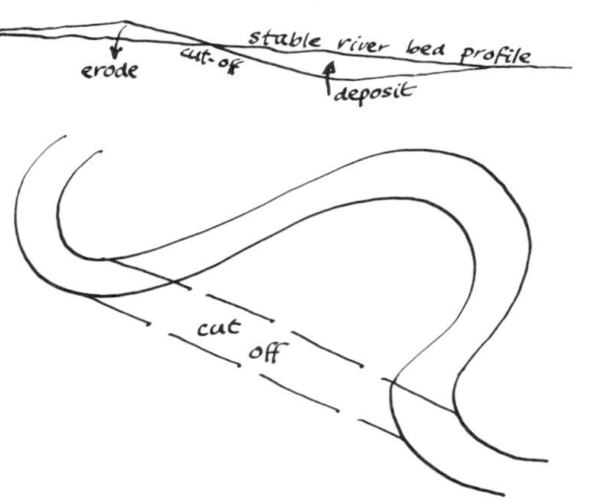

3 Straightened rivers would naturally regain a meandering pattern; they can only be held in place by *revetments* and *training walls*.
4 *Land drainage* achieved by cutting deeper river channels will result in channel-bank collapse, in widening, in upstream incision, and in downstream aggradation (a process similar to rejuvenation).
5 Reservoirs cause *siltation* and loss of storage. They also make rivers downstream more erosive.

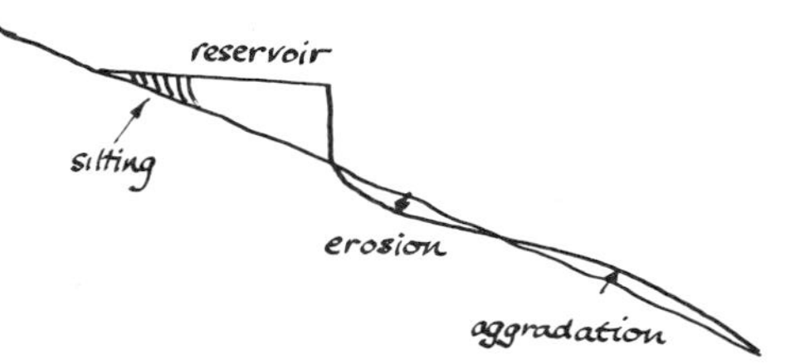

6 *Floods* are caused naturally on average once every two years in humid temperate regions; annually in humid tropical regions; and nearly every time rain falls in semi-arid and arid regions.
7 Floods are most likely:

(a) when a storm moves across a drainage basin in a downstream direction;
(b) when there is little vegetation;
(c) when storms become fixed;
(d) when storms are of long duration;
(e) when snow and ice melt;
(f) when there are exceptionally high tides;
(g) when farmland is artificially drained;
(h) when land is deforested.

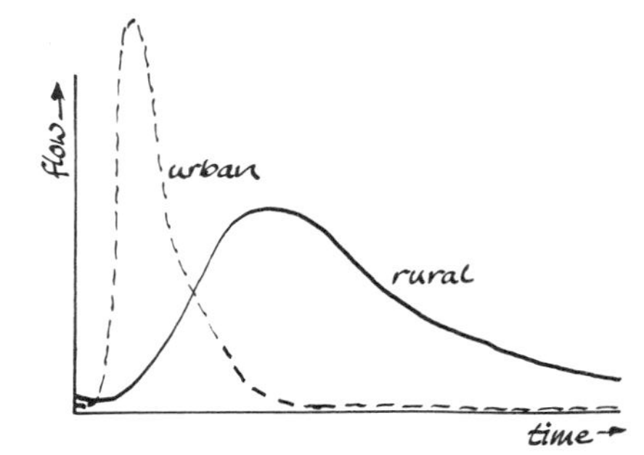

8 Choice of *flood control* depends on the cost of damage when it occurs, the *flood return period* and the cost of protection work.
9 Small-scale controls include *embankments*, *floodways*, *retention ponds*, specially designed car parks and roofs, recreation ponds, and bridges.
10 Large-scale controls involve reservoir construction.
11 *Droughts* are usually long periods without rainfall. They are serious when water demand exceeds the dry-weather flow.

Water supply

1 Unless some water is stored, river flow is too variable for intensive use.
2 Used water must be returned to the river for further use to prevent the need for large-reservoir storage.
3 *Water supply* can come from *reservoirs*, *aquifers* or by *desalination*.
4 The simplest supply scheme uses headwater reservoirs in areas of water surplus with transfers by river to downstream demand areas.
5 In lowland areas *bunded reservoirs* are needed.
6 *Aquifer storage* can act as a supplementary reservoir.
7 Inter-basin water transfers are a logical answer to the most effective use of water resources.

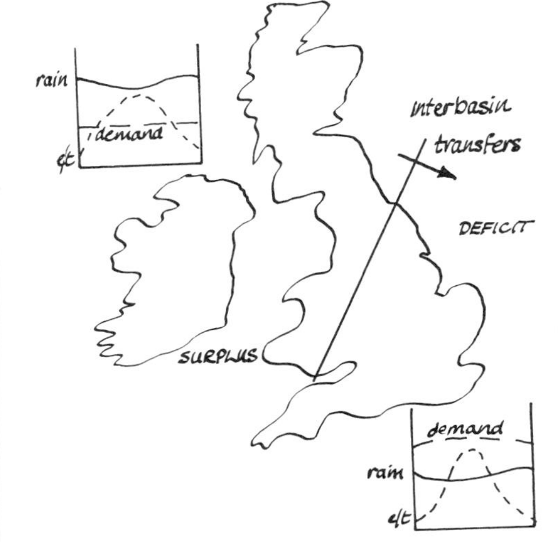

8 *Irrigation* is the world's largest single user of water.
9 Groundwater supplies used for irrigation are often severely overused.
10 Water supplies must be protected against *pollution* by allowing sufficient *dilution* and encouraging natural *self-purification*.

Part D Weathering

Weathering processes

1 Rock is often made more prone to weathering by the formation of joints and other cracks by dilation.
2 *Dilation* can have a direct effect by causing *sheeting* from cliff faces.
3 *Weathering* is the action of atmospheric processes on rock which reduces its size and which may possibly change its composition
4 *Physical weathering* by *frost shatter* or *salt-crystal growth* in pores (*exfoliation*) causes disintegration of bare rock.

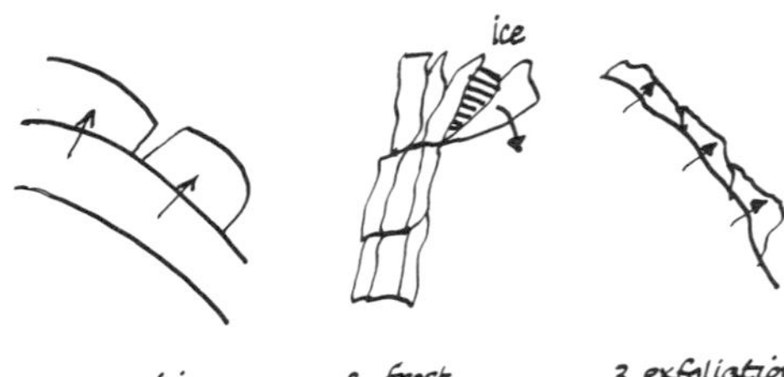

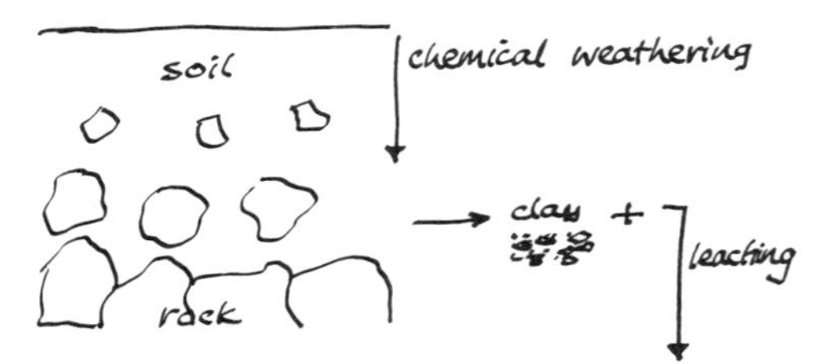

5 Physical weathering can only continue if the debris is carried away; the debris otherwise protects the underlying rock from further action. It is therefore not important on soil-covered slopes.
6 *Chemical weathering*, mostly by *hydrolysis* using acidified rainwater, causes rock to decompose and change chemically into *clay minerals* (the precipitate) and *soluble ions* (leached away).
7 Chemical weathering produces the medium for plant growth and leads to *soil formation*.
8 Chemical weathering is fastest in warm, moist conditions and is not hampered by a layer of weathered debris (soil) on the rock surface.
9 Because chemical weathering is a surface reaction it is most effective in rock that is highly fractured or has a high porosity.

Hillslope transport

1 Mechanically weathered coarse debris forms *screes*, and moves by *rock avalanches* after it exceeds its *threshold angle of stability*. It gives straight slopes.

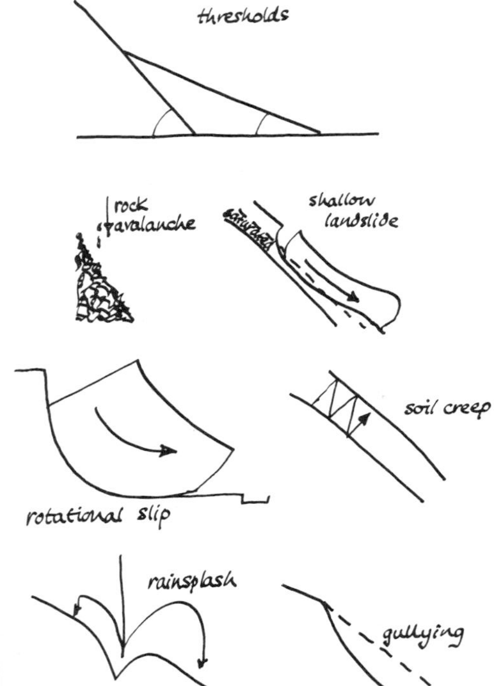

2 Screes lies at angles determined by static and dynamic *friction angles*.
3 Soil is *cohesive*; it is held together by electrical forces on clays, by organic gums, by iron oxides, by surface tension and by plant roots.
4 Soil tends to fail in wet conditions when *water tables* rise and give buoyancy. At such times *surface tension* effects are lost.
5 *Soil failure* results in sliding over bedrock (a *shallow slide*), leaving a circular scar.
6 Many slides develop into *mudflows* as movement vibrates the soil particles apart.
7 Sandy soils flow more readily than clay soils.
8 Landslides and flows occur above the threshold angle of stability for *saturated* soil. They lead to straight slopes.
9 A *terracette* is a form of landslide, in thin soil.
10 *Soil creep* occurs due to changes in soil volume brought about by changes in moisture. Creep is very slow but works on all slopes. It leads to convex slopes.
11 *Rainsplash* operates on bare soils with high-intensity rainfall and yields convex slopes.
12 *Sheetwash* and *gullying* occur on bare soils with high-intensity rainfall and soil-covered saturated soils. They produce concave slopes.

13 In unconsolidated material landslides are deep-seated *rotational slips*. They yield irregular straight slopes.

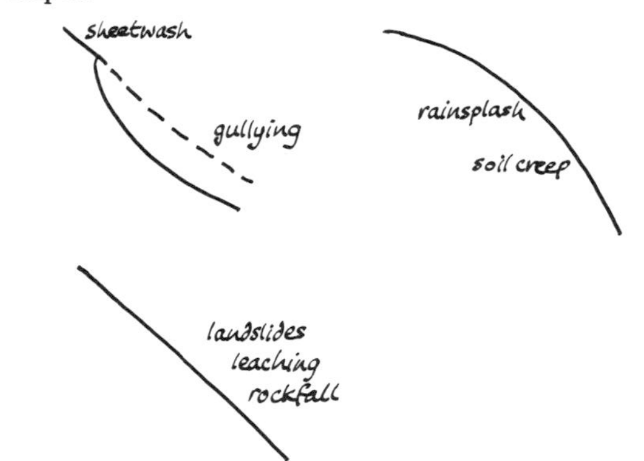

Hillslope shapes and formative processes

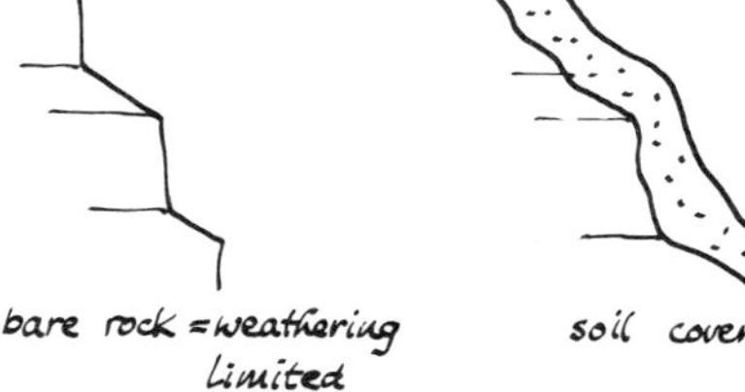

14 Concave–straight–concave hillslope profiles show three processes each dominating on a different part of the slope.
15 Loss by *solution* and *leaching* leads to concave slopes if it diminishes downhill (temperate areas), or to parallel retreat if it is uniform over all the slope (tropical areas).
16 All types of slope are parts of a continuous and overlapping series (that includes river channels).

Landscape evolution

1 Slopes dominated by *weathering-limited slopes* (in semi-arid, arid and periglacial regions, and on sea cliffs) show bare rock and sharp, angular junctions between slope elements. They tend to form by parallel retreat throughout all phases of landscape evolution.
2 Regions dominated by *transport-limited slopes* (humid regions) have soil and vegetation cover, and have smooth, gradually changing slope elements. They tend to evolve by decline, then by parallel retreat during a period of dynamic equilibrium, and finally by retreat again.
3 The *cycle of erosion* shows a sequence from incision and *increasing relief*, through a long dynamic-equilibrium stage of *maximum relief*, to a protracted stage of *decreasing relief*
4 The sequence of landscape development migrates upstream.
5 *Valley alluvium* is a natural occurrence in landscape evolution; it occurs when the upstream yield of eroded material temporarily exceeds the trunk-stream capacity to remove it.
6 *Landscapes evolve in a sequence of steps* due to *isostatic uplift* followed by *denudation*. This, together with climatic fluctuation, causes *river terraces* and a fluctuation in *sediment yield* in rivers.
7 Hillslopes do not show short-term effects of changes in base level because they form so slowly.

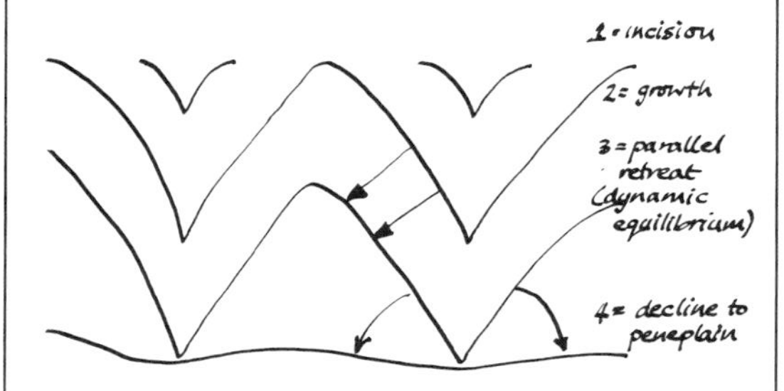

the 'cycle' of erosion

Part E Warm environments

Tropical landscapes

1 The humid Tropics have high temperatures and a water surplus in every month. Chemical weathering dominates and produces clay soils tens of metres thick.
2 Erosion is by hydrolysis and solution rather than by landslides or river abrasion; rivers have few stones for abrasion.
3 River profiles have rapids separated by very gentle stretches and are closely controlled by rock structure.

tropical river profiles

4 Differential weathering rates produces *inselbergs* separated by *etchplains*.

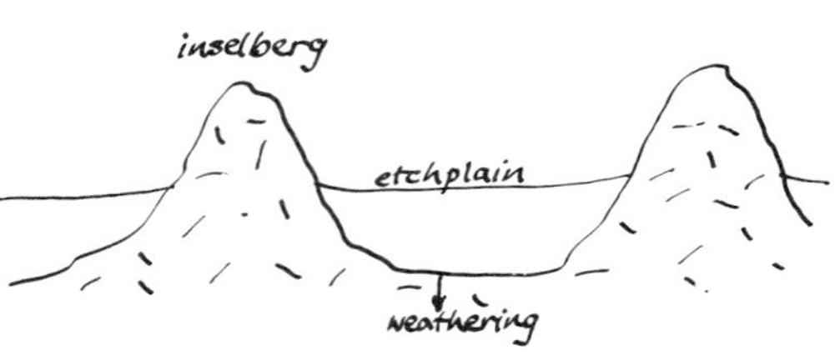

5 Tropical areas with a rainy season cause the development of *cuirasses* in soils. These become crust-like and resist erosion, leading to the formation of *tableland* landscapes.

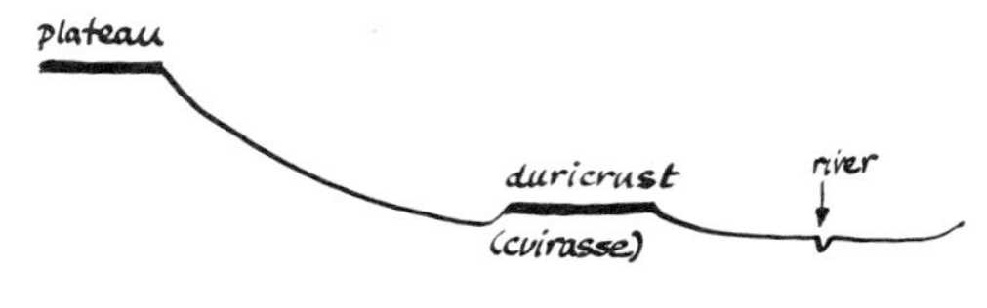

savanna tablelands

6 In semi-arid and arid landscapes mechanical weathering is more important than chemical weathering.
7 Overland flow is common; most steep slopes are bare and weathering is limited.
8 Large *alluvial fans* and *wadi fills* occur because storms are too short in duration to transport debris far.
9 Much weathered material is moved by traction, saltation and in suspension. Sand moves to form *dunes*, *draa* and *rhourds*.
10 Sand sometimes erodes rock pavements to leave *yardangs*, long grooves in rocks.
11 Many features of arid regions seem to be relics of a former dry period.
12 Tropical coasts are often fringed with low plains rather than cliffs, due to the reworking of river materials.
13 Much beach debris is fragmented coral.
14 Coral reefs fringe much of the coast and protect it from erosion.
15 In arid areas, moving sand dunes present the greatest hazard to permanent settlement.
16 *Flash floods* from wadis can be very destructive.

Part F Cold environments

Periglacial processes and landforms

1 Periglacial regions have mean annual air temperatures below 0 °C and permanently frozen ground, but there are many variations.
2 Periglacial environments occur on steep mountain slopes where ice cannot accumulate, in continental regions where the supply of snow is limited, or near to the edges of ice sheets.
3 *Frost shatter* is the dominant process on steep slopes.
4 Above 40° debris moves by *rock avalanche.*
5 On slopes between 20° and 40° *snow avalanches* transport frost-shattered debris.
6 On slopes less than 20° extensive *ground ice* forms, and in the seasonally thawed *active layer* movement is by *solifluction* and *frost creep.*
7 Solifluction forms *terraces* and *lobes* on slopes as gentle as 2°.

8 In areas away from mountains solifluction is the most widespread and most important periglacial process.
9 On flat or gently sloping *ground, frost heave* and *ice contraction* under intense cold cause *ice wedges* and *patterned ground* to develop.
10 On floodplains and other areas where water is near the surface, *pingos* may develop as advancing *permafrost* forces water to the surface.
11 The net result of periglacial conditions is to sharpen up the profiles of steep slopes and to transfer the debris to gentle slopes, mantling them and filling in valleys.
12 Frost shatter of valley-bottom rock due to advancing permafrost may break up the ground and make it easier for glaciers to quarry.

Glacial processes and landforms

1 Glacial processes depend only on the effects of moving ice-masses.
2 Glaciers usually follow and modify pre-existing valleys.
3 Snow is converted first to *névé* (firn), then to *ice* by *pressure* and *regelation.*
4 Ice is plastic and deforms slowly under continued pressure to flow in valleys.
5 Under low pressure, when ice is subject to tension, it breaks up to give *crevasses.*
6 A *warm-based glacier* moves by flow and also by sliding over its base – this kind is the most erosive.
7 A *cold-based glacier* moves only by flow and remains stuck to its rock floor – it erodes relatively little.

8 Ice erodes because rock debris *quarried* from one part of its bed *abrades* other areas.

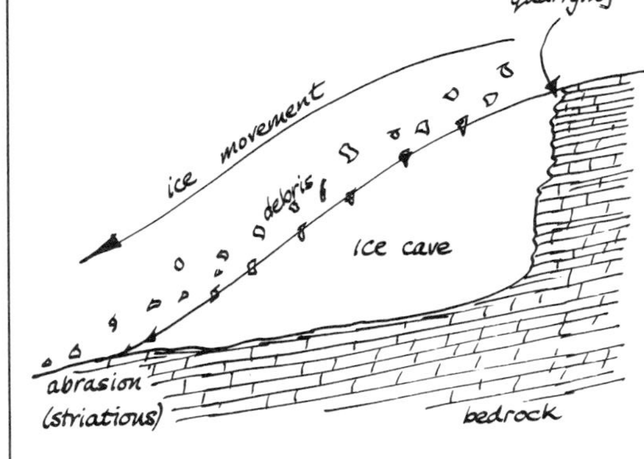

9 Quarrying occurs where ice pressure is low; abrasion occurs where ice pressure is high.
10 *Striations* result from abrasion; rough *plucked surfaces* result from quarrying.
11 *Cirques* are formed as small valley-side glaciers slip in their hollows with a rotating motion.

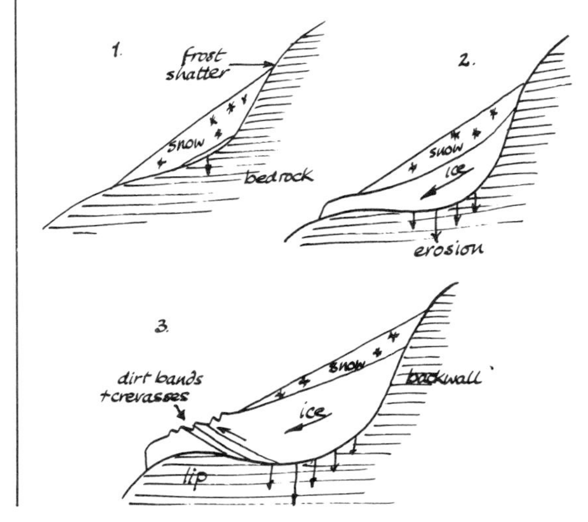

12 A *cirque lip* forms where rotation reduces ice pressure on the bedrock.
13 With increasing ice thickness, abrasion first increases to a maximum value and then declines again until, under very thick ice, it stops completely.

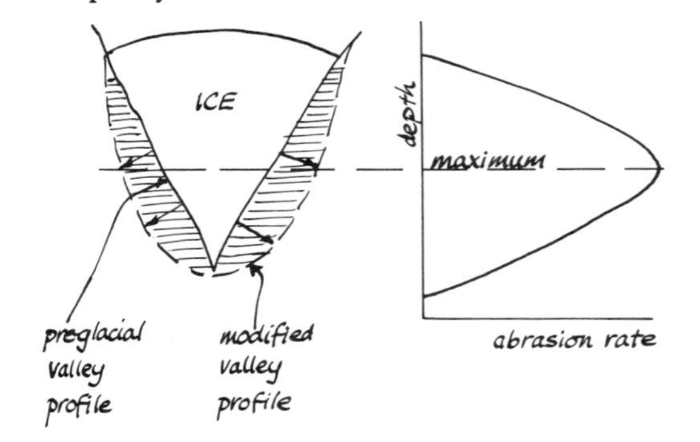

14 During an ice age the zone of maximum erosion moves from the valley floor to the divides.
15 When ice overwhelms a mountain region it spills across divides and forms *glacial diffluence troughs*.
16 Rugged mountain tops have never been overwhelmed by ice (e.g. the Alps); smooth mountain tops have been submerged beneath an ice sheet (e.g. the Lake District).
17 As debris is carried from its source area it becomes smaller and is crushed to *rock flour;* new debris is larger and more angular; the mixture, once called *boulder clay*, is now called *glacial till.*

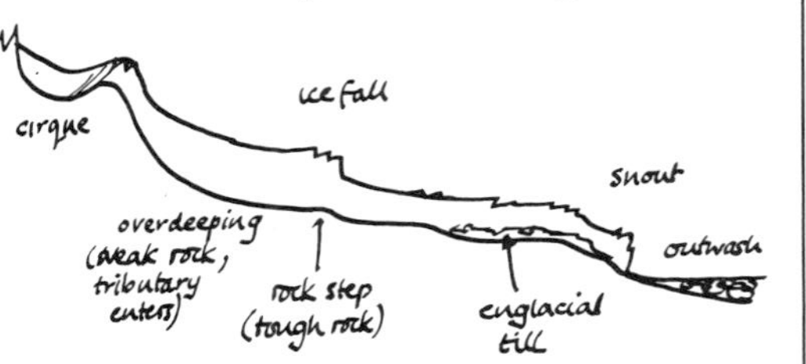

18 Ice sheets may erode as they advance, but most erosive features are hidden beneath later deposition.
19 Glaciers may deposit till, but much is later washed away by meltwater streams. *Lateral* and *medial moraines* produced by frost shatter and plastered over the valley sides rarely survive after glaciers have melted away.
20 If ice advances at the same rate as its margin melts, a *terminal moraine* will form. Terminal moraines are often dissected by meltwater streams.
21 Beyond the confines of the ice, meltwater washes across the landscape in braided streams creating an *outwash plain* or *sandur*.
22 *Kames* are crevasse infillings.
23 *Eskers* are the beds of subglacial streams.
24 *Drumlins* are mounds of till moulded by ice which continues to advance; *kettle holes* are depressions in till indicating the former position of buried ice blocks.

25 Depositional features are more likely to survive on plains where ice *stagnates* (downwastes) than near the ice margin or in valleys where ice *retreats* (backwastes) and meltwater streams can wash debris away.

Cold climates and man

1 Periglacial activity can cause many *hazards,* including disasters (such as avalanches crashing into homes). Deflecting walls and barriers must be built for protection.
2 Building on permafrost may cause either *settlement* (if permafrost melts), or *heave* (if permafrost grows).
3 Glacial till has provided much of the *soil* for lowlands in the mid-latitudes; lenses of outwash material provide useful *aquifers* for water supply.
4 Local variations in glacial deposits have often played a major part in *affecting the site of a settlement* and in the construction of cuttings for roads and railways.
5 Glacial diffluence troughs provide direct communication links and control settlement patterns in most mountain environments.

Part G Coasts

Wave processes

1 Waves have definite shapes and sizes described in terms of *wavelength*, *wave height* and *wave period*.
2 Waves occur in patterns called *wave trains*.
3 The form of a wave train depends on wind *duration*, wind *speed* and wind *fetch*.
4 A mature sea created by a strong wind is called a *fully arisen sea*.
5 Storm conditions produce small-wavelength, steep-sided *storm waves*.
6 Storm waves decay away from source regions and change to low, broad, large-wavelength *swell waves*.
7 *Wave energy* depends on both height and period; some swell waves contain more energy than some storm waves.
8 Waves approaching shallow water are retarded by friction with the sea bed.
9 An irregular sea bed causes wave refraction, concentrating energy into regions of shallow water.

Coastal processes and landforms

1 Wave energy is dissipated by

(a) heat produced from wave turbulence;
(b) transporting sediment;
(c) erosion of rock.

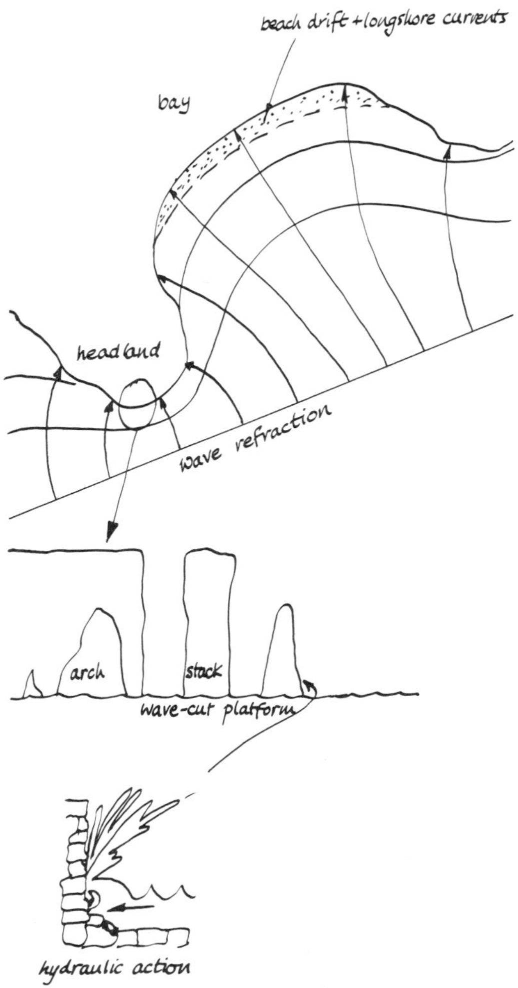

2 *Beaches* occur where there is insufficient energy to remove all the sediment.
3 Bare *wave-cut platforms* are exposed only in high-energy environments.
4 On coasts with beaches, energy is used to move water and coarse sediment up the beach (*swash*). Together with *backwash* this action causes *attrition* of beach sediment and allows it to be carried offshore in suspension.
5 On deep-water coasts, waves reach cliffs producing *standing waves* or *breaking waves*. Erosion is due to *hydraulic action* or *solution*.
6 Features of deep-water coasts include *stacks*, *arches* and *wave-cut platforms*.
7 Waves either *plunge*, *spill* or *surge* depending on beach steepness and wave form.

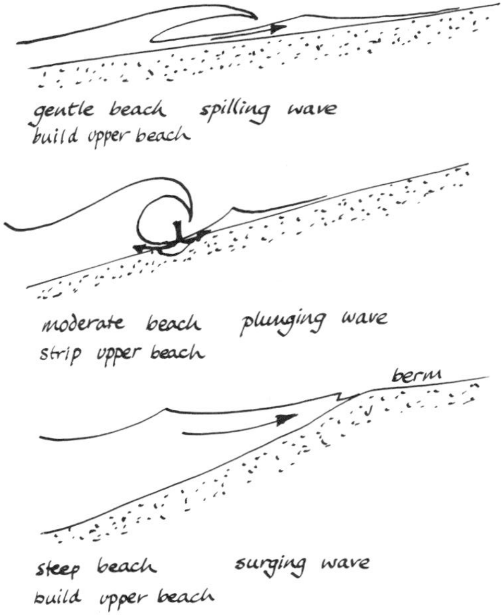

8 Plunging waves comb down the upper beach and deposit sediment below the low water as a *break-point bar*.
9 Surging waves push material up the beach and steepen the beach profile. They cause a swash ridge or *berm*.
10 Spilling waves push material onshore and steepen a beach.
11 Wave action and beach slope are related to the tidal range and to whether the tide is falling or rising.
12 The type of wave is related to seasons: plunging and surging waves are common in winter; spilling waves are more common in summer.
13 Sand dunes behind a beach provide a reservoir of material to replenish beaches after a period of storm-wave attack.
14 Many bays are eroded into a *spiral curve* plan determined by exposure to storm waves: steep beaches with coarse sediment face the storm waves; gentle beaches with fine sediment occur in sheltered areas.

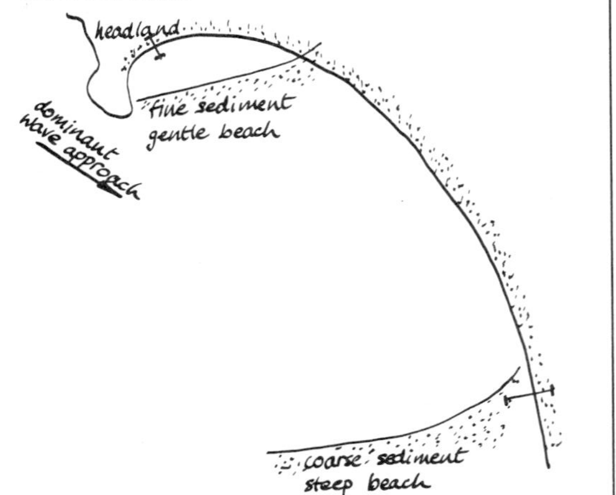

15 Sediment is transported from regions of high energy to those of low energy by *longshore currents* and by swash action on beaches (*beach drift*). Together these create *longshore drift*.
16 Coasts that change direction abruptly (at estuaries or headlands) may cause sediment to build into a *spit*.

sand dunes SINK
river sediment SOURCE
spit SINK
onshore transport SOURCE
bar SINK
cliff erosion SOURCE
longshore drift TRANSFER

17 Spits do not usually extend to form barriers.
18 Most barriers (*bars*) and *barrier islands* are the result of post-glacial sea-level rise and are fossil features.
19 Longshore drift across an estuary produces offshore *sandbanks* (bars).
20 Coastal *deltas* form where rivers deposit sediment faster than the sea can remove it.
21 Estuaries have deposits (equivalent to deltas) in the form of intertidal sandbanks and mudbanks.
22 Action by waves and currents moves sediment in *cells* consisting of *source areas, transfer regions* and *sinks*.

23 Changes in sea level produce *raised beaches* and *marine erosion surfaces* when sea levels fall, and *drowned valleys* when they rise.

Coasts and man

1 Longshore drift and sediment cells can be disrupted by coastal structures such as *jetties* and *piers*.
2 Erosion on the downdrift side of structures may cause beach loss and coastal retreat, and may threaten homes.
3 Groynes and *sea walls* are built to stabilise sediment transfer, but they only move the problem along the shore.
4 The best solution to beach loss is *beach nourishment* or *sediment bypassing*.

Part H Soils

Soils

Properties

1 A *soil* is a mixture of mineral and organic fragments organised into *horizons* by physical, chemical and biological processes.
2 The basic properties of a soil are derived from both mineral and organic materials. They are:

(a) *colour*, mostly due to humic (black) or iron-oxide (orange or red) staining;
(b) *texture*, due to the relative proportions of sand, silt and clay fragments;
(c) *structure*, the natural soil groupings (peds) formed as a result either of electrical forces or of polysaccharide gums;
(d) *acidity*, the proportion of hydrogen ions in the soil water;
(e) *porosity*, the total space not occupied by particles;
(f) *permeability*, the rate at which water can move through a soil.

3 There are three basic and separate soil-forming processes. They are:

(a) the weathering of soil parent-material;
(b) the decay and incorporation of plant and animal materials;
(c) the reorganisation of these materials into soil horizons.

4 Mineral-matter weathering, mainly by water as hydrolysis, produces clay minerals (a precipitate), which have a negative surface charge; releases iron and aluminium oxides (sesquioxides); and puts into solution a range of ions.
5 Cations hold clay particles together; anions play no soil-forming role and are mostly washed away. Both types are essential nutrients for plants.
6 Organic-matter decay, mainly by soil micro-organisms (bacteria and fungi), converts celluloses of plants into soluble organic materials (acids) which afterwards recombine, using cations, to form polymers called *humus*. Humus has a negative surface charge (like clay). Biochemical processes also produce polysaccharides, long filaments which bind soil mineral particles together.
7 The rate of breakdown and incorporation by earthworms into the lower soil is reflected in the *mull*, *moder* and *mor* types of organic matter pattern on the soil surface.

Processes and profiles

1 Soils may either by dominated by *in-situ weathering* or *translocation*.
2 Translocation involves *loss* (by leaching of materials in solution and by washing of solid particles in suspension) and *gain* (by reprecipitation or settling out). Horizons that show signs of loss are called *eluvial*; those that show signs of gain *illuvial*.

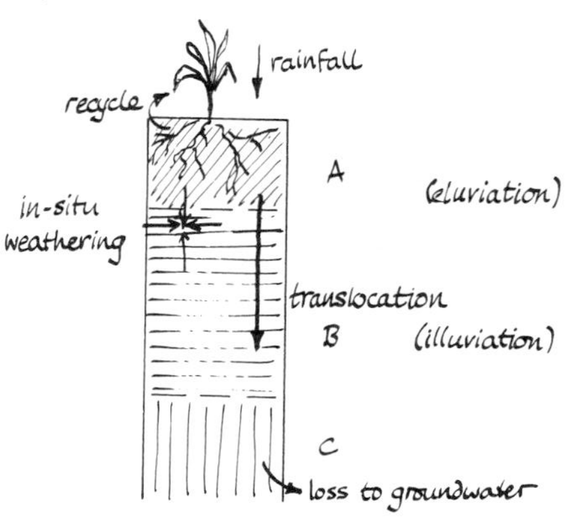

PROCESSES

3 Soils that form by weathering and that remain stable because of extensive cation bridges between clays and because of webs of gums show only a gradual change from the surface to the parent material. They have Ah–Bw–C profiles and are *brown earths*. In these soils water passes through the profile and leaches ions away, but they are replaced quickly by plant decay and rock weathering.

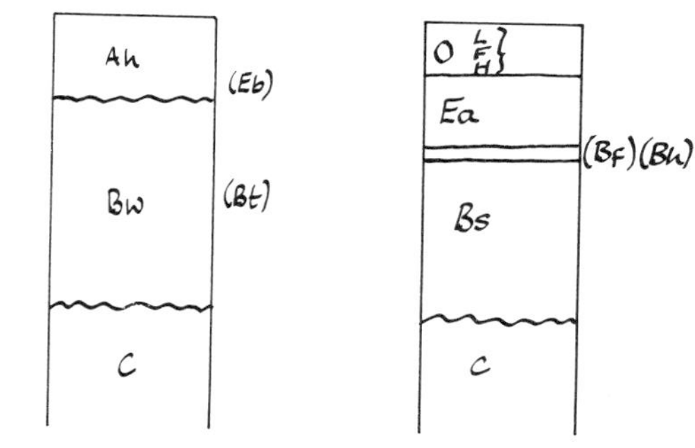

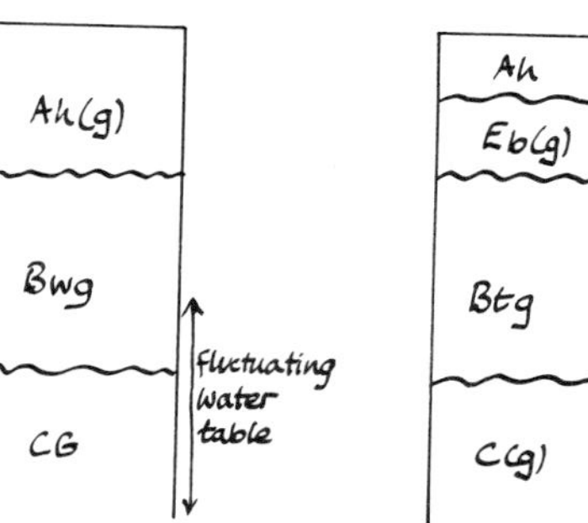

4 Brown earths under woodland may recycle cations less well and some clay eluviation occurs, giving a clay-rich subsoil (Eb–Bt–C profiles).
5 On cation-deficient parent material, and in places of high rainfall and of poorly recycling vegetation, *podzolisation* occurs (yielding Ea–(Bh)(Bfe)Bs–C profiles). Organic acids do not form humus readily and instead combine with iron oxides, making them soluble and able to be leached away. An ash-coloured horizon results.
6 Waterlogging and anaerobic conditions (*gleying*) can occur if the soil is poorly drained due to heavy clay texture (surface-water gley) or to a high water table (ground-water gley). This can affect all soil. Gleying is shown by blue-grey colours.
7 In tropical regions seasonal movements of the water table can concentrate sesquioxides (*ferrallitisation*) and cause the development of a *laterite* horizon.

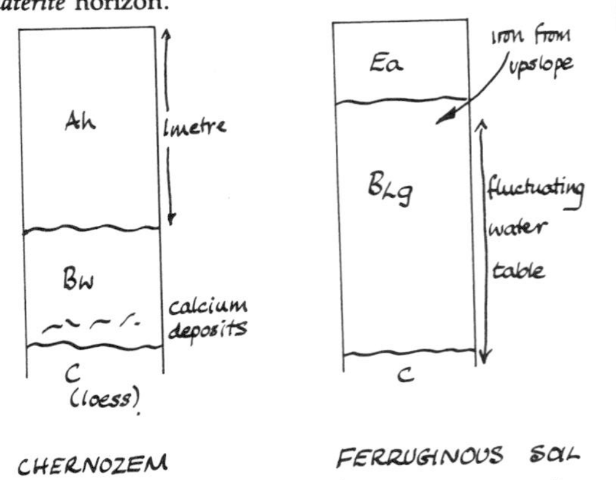

8 In semi-arid regions water percolates through the soils infrequently and calcium rarely becomes leached beyond the subsoil. Roots searching deeply for water provide organic matter as deep as 1 m and decay to give a black Ah. The resulting profile is a *chernozem*. It is rare for water to move upward in a soil, but if (in arid regions) the water table is only just below the surface, water and its contained salts may be drawn to the surface, where the salts will be deposited as crusts (*salinisation*).

Patterns and mapping

1 For ease of description, sequences of soils need to be given shorthand names. The most useful broad groups (*major soil groups*) are those which reflect basic processes: brown earth, podzol, chernozem, gley. Because soils vary gradually from one place to another it is always hard to draw lines separating soils (for soil maps). Major soil groups do not differentiate important soil contrasts of texture, depth and so on. In surveys of small areas the *soil series* is used as a mapping unit.
2 The pattern of world soils is determined by five environmental factors:

(a) *climate*, which sets the broad scene;
(b) *parent material*, which determines soil texture and fertility;
(c) *vegetation and soil organisms*, which determine the efficiency of cation recycling, the rate of tissue decay, and the soil structure (which depends on gums and soil mixing);
(d) *relief*, which affects waterlogging and soil stability;
(e) *time*, which affects how far the soil has developed toward an equilibrium.

Fertility, erosion and conservation

1 A *fertile soil* is one that will allow the unrestrained growth of plants. Loss of fertility, due to increased leaching of cations or overuse by man from cropping, will also lead to soil-structure breakdown and an increased tendency for accelerated erosion.
2 The study of *land capability* seeks to identify the *potential* of a soil under reasonably careful management using detailed knowledge both of soil properties and of the needs of crops. It is of direct use both to farmers and planners in assessing the value of their resource. Capability is divided into seven classes. The best land (classes 1 and 2) should be allocated to the most demanding types of land use if possible. Capability surveys show whether land is worth improving and, if it is, the most suitable type of improvement to make (e.g. drainage).
3 *Accelerated erosion* occurs when an unsuitable type of land use is adopted on fragile soils. Erosion is either by *wind* or *water*, and of these water erosion is the most widespread.
4 *Wind erosion* occurs wherever soils are left exposed to drying winds. Soils with poor structures (due to lack of cation bonding and gums) on flat land are the most vulnerable. Arable fields left bare over winter are commonly affected. Wind erosion removes soil, leaving less fertile subsoil behind; and blowing soil may abrade leaves from seedlings and destroy a growing crop. Wind erosion is best controlled by improving soil structure, but it can be helped by using *windbreaks*, *intercropping*, *minimum cultivation* and *stubble mulching*.
5 *Water erosion* also removes productive topsoil, especially on sloping land. The main agents are *rainsplash* and *sheetwash* but if there is sufficient overland flow it may concentrate and form *gullies*. Water erosion is most troublesome in semi-arid areas with high-intensity rainstorms and low-density vegetation. Conservation involves improving soil structure to prevent loss of material and providing a vegetation cover to reduce raindrop impact. *Terraces*, *contour ploughing* and *grass baulks* all help.

Biogeography

1 Living things perform certain basic functions: (a) they feed; (b) they grow; (c) they reproduce; (d) they move; (e) they respire; (f) they are sensitive to their surroundings; (g) they excrete waste substances.
2 Plants build tissue and grow using organic substances obtained directly from the air, water and soil, and the energy of the sun in the process of *photosynthesis*.
3 The distribution of plants is very closely related to the global variation of the inorganic and solar-energy resources.
4 Animals build tissue and grow using organic substances obtained by consuming plants.
5 Animals are mobile and not as closely tied to one part of their physical environment as are plants.
6 All living things have *evolved* by *natural selection* and *mutation* to fit their present *habitats*.

7 For long periods of time plants are bounded by certain *environmental barriers* beyond which they cannot normally flourish.

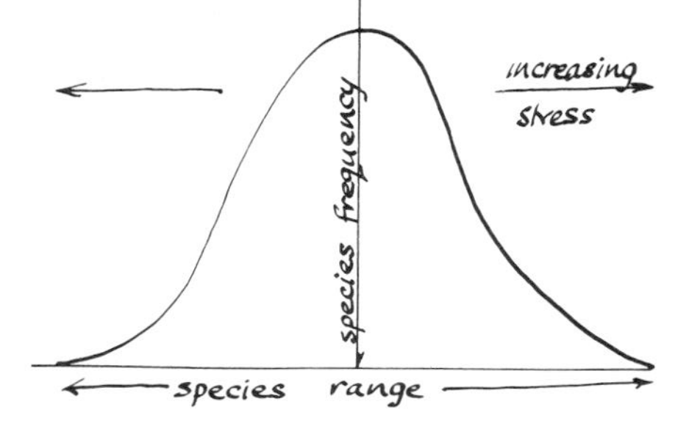

8 The most important limitations on the distribution of a species are imposed by temperature, moisture and light.
9 An *ecosystem* is a more or less closed system in which the limited resources of light, nutrients, moisture and shelter are as fully used as possible.
10 Plants are grouped into *biomes*, major units related to climate. Each biome contains many stable ecosystems.
11 Stable ecosystems have to provide for a continual renewal of life through *food chains* and *food webs*.

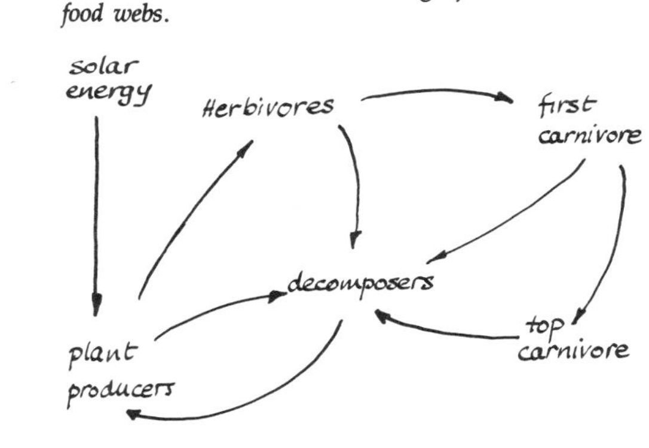

12 Within a food web there is a pyramidal structure of dependent organisms; each *trophic level* converts into usable energy only 10 per cent of the energy in the food it eats.

13 Ecosystem *colonisation* proceeds by:

(a) colonisation;
(b) competition;
(c) successions;
(d) stabilisation.

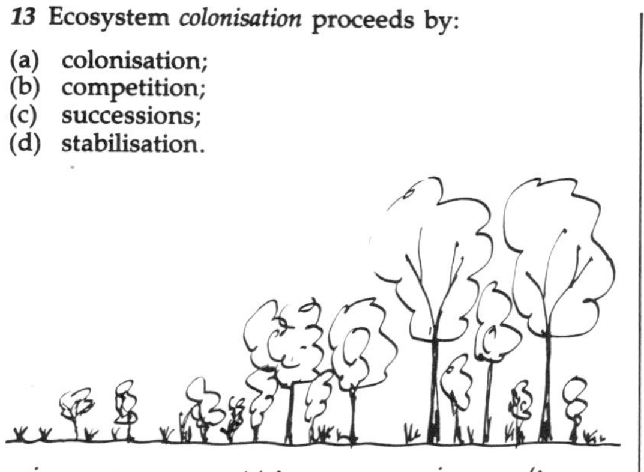

14 An ecologically stable vegetation assemblage is called the *climax* vegetation.
15 Within an ecosystem there are a certain number of *ecological niches* to be filled, although exactly which species are present is unpredictable and varies between continents.

climax
disturbance
invasion by
pioneer
invasion by
inhibitor
succession

COLONISATION ALTERNATIVES

World vegetation

Tropical rainforest

1 The characteristics of rainforest are:

(a) the monotonous green high-level canopy foliage;
(b) the lack of large, brightly coloured flowers;
(c) the presence of lianas and buttress roots;
(d) the apparent scarcity of animal life except insects.

2 Mature rainforests are free from shrubs because the tree canopy shades the ground.
3 Any dead organic tissue is rapidly decomposed.
4 Incorporation of organic matter into the soil is achieved primarily by insects.

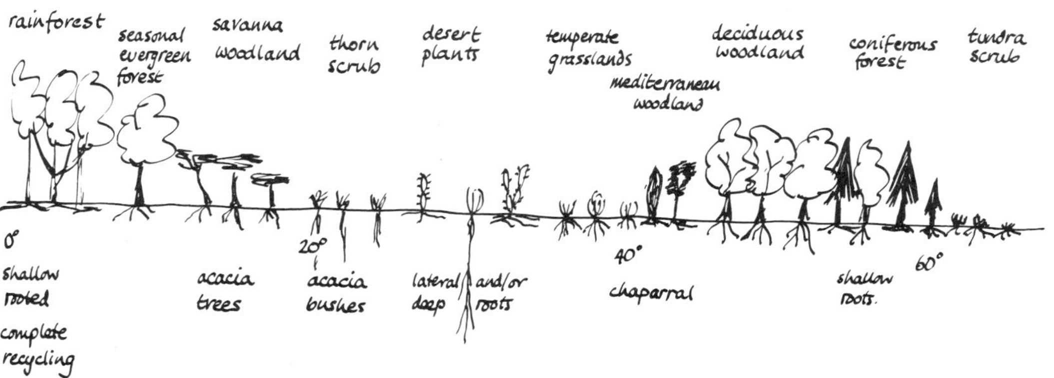

5 There is a great variety of species.
6 Each species has a low population *density*.
7 The lack of animals results from the shortage of edible plants in a rainforest.
8 The overwhelming proportion of nutrients needed by the rainforest are held in the vegetation, not in the soil. Additional supplies come from rainfall.
9 Soils are severely leached and infertile.
10 Nutrients from decaying tissue are taken directly to plant roots by fungi.

Dry-season lands

TROPICAL SAVANNA

1 Low seasonal rainfall and constantly high potential evapotranspiration rates are the main limits to plant growth
2 The main ecosystem is dominated by *deciduous trees* (e.g. *Acacia*) and very *tall grasses* (e.g. *Themeda*).
3 Grasses are shallow-rooted and grow, ripen and set seed in the short period from the onset of the rains to the early part of the dry season.
4 Savanna trees are short, rarely more than 6 m.
5 All plants are especially adapted to resist fire damage.
6 Organic matter is incorporated into the soil by termites.

TEMPERATE GRASSLANDS

1 These cover large areas of the semi-arid mid-latitudes where summers may be as hot as in the Tropics, but where winters are cold, frost is common and winter precipitation falls as snow.
2 There are no trees.
3 Grasses are short and have more biomass below the surface than above, in the form of extensive roots.
4 Grasses are bunch or tussock varieties.
5 There are two grass varieties: those with shallow roots that grow quickly in spring; and those with deep roots that grow slowly into autumn.
6 This ecosystem has been maintained as grassland by its natural fauna.

Temperate deciduous forests

1 These areas are affected by seasonal temperature changes with winter montly temperatures below 6 °C which cause a brake on growth.
2 Vegetation adapts to the seasonal climate by adopting a *deciduous habit*.
3 Vegetation has broad, thin leaves which enable it to gather all available energy.
4 Propagation is by wind pollination – seeds are hard-cased.
5 There are far fewer species in temperate ecosystems than in the Tropics.
6 These are *transitional zones* where soil, slope and drainage strongly affect the species.
7 *Understorey plants* receive little energy and have adapted either to tolerate shade, or to flower in spring and set seed early.
8 Nutrients are largely stored in the soil, not in the plant.

Boreal forest

1 Coniferous forests have little species variety.
2 Trees are shallow-rooted and capture nutrients using fungi.
3 The vegetation is adapted to low temperature, low insolation and high snowfall – for example, it has evergreen waxy needles.

Tundra

1 The tundra vegetation mosaic depends on the microclimate and on soils.
2 The biomass is a tenth of that of the boreal forest, with up to 80 per cent below the surface.
3 Severe waterlogging is a major feature.
4 Nutrient cycling is achieved by rodents.

Deserts

1 Adaptations create the most diverse of all biome forms.
2 Plants survive:

(a) by forming seeds, as part of an otherwise rapid life cycle;
(b) by having very deep roots;
(c) by retaining water;
(d) by keeping most biomass below the surface.

Farming

1 *Environmental factors* (such as sunlight, temperature, effective precipitation, topography and soil) determine whether it is feasible to introduce a certain crop in a given area. The actual choice of crops, however, is also related to the economic and cultural environment of the society doing the farming. Density of population, distance to market, level of technology and cultural heritage can play as large a role as natural forces.
2 *Cereals* are grown over half the world because they are easy to grow, harvest, store and transport. They also have a high food content.
3 Half of the world is climatically unsuited for cereal cultivation; *livestock* farming then dominates.

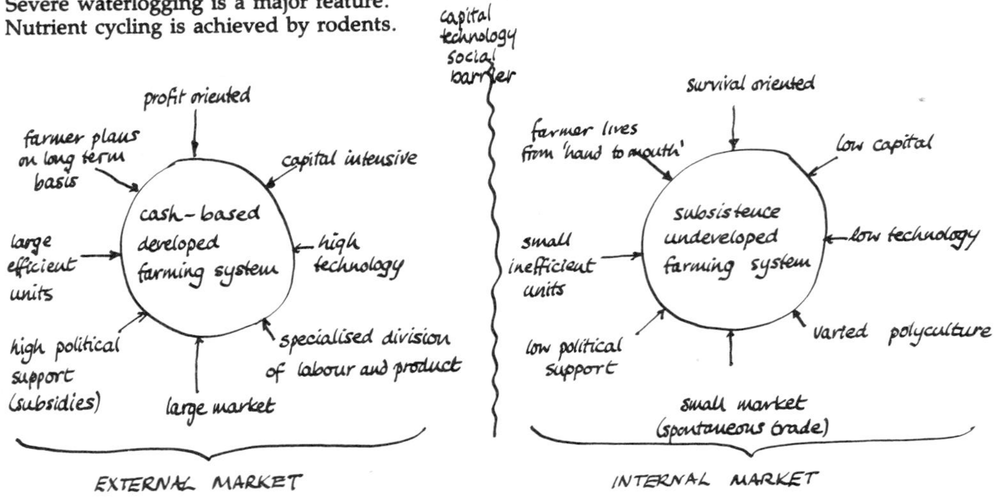

4 Farming systems can be classified into:

(a) shifting cultivation;
(b) fallow;
(c) ley and dairy;
(d) permanent cultivation;
(e) arable irrigation;
(f) perennial crops;
(g) grazing.

5 Each system may be stable at a high level of soil fertility and output (e.g. wet rice); or stable at a low level (e.g. upland cultivation); or unstable, with fertility maintained artificially (e.g. western Europe) or improving or degrading (e.g. dry farming in Africa).

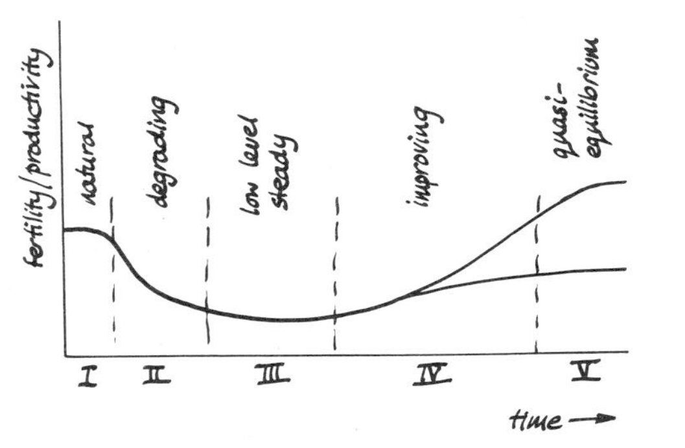

6 Commercial permanent *cultivation* of perennial crops in the Tropics and sub-Tropics is often by plantation on large estates. Labour is often substituted for the machines used in temperate lands.

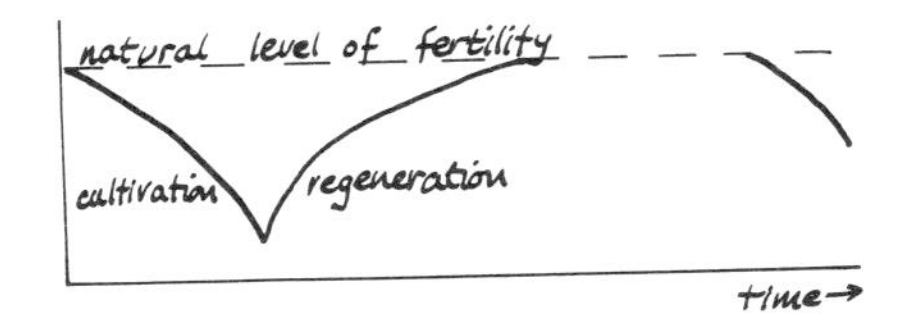

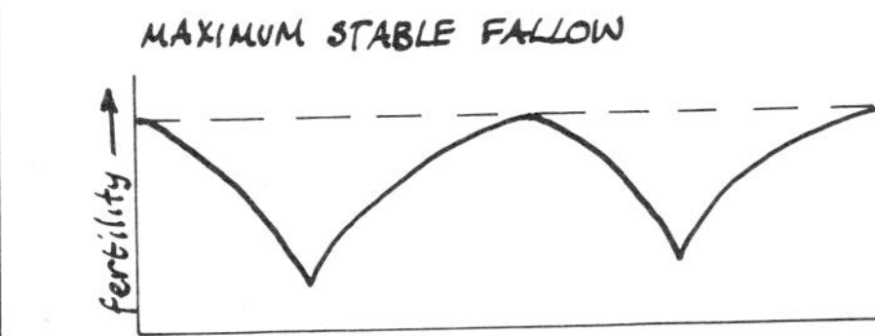

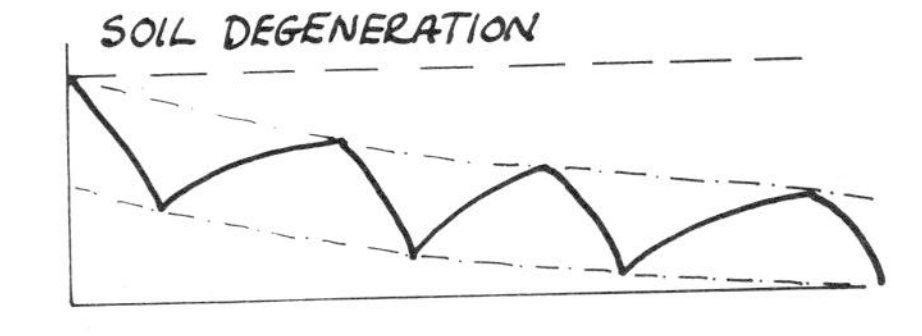

7 Grazing systems vary from *nomadism* through *transhumance* to forms of *ranching* and *settled grazing*.
8 There is a tendency for all farming systems to become more intensive, and for grazing and shifting cultivation to change to permanent cultivation.

9 Farming systems can only change from subsistence level or from a downward spiral to commercial systems if there are adequate inputs of technology and money.

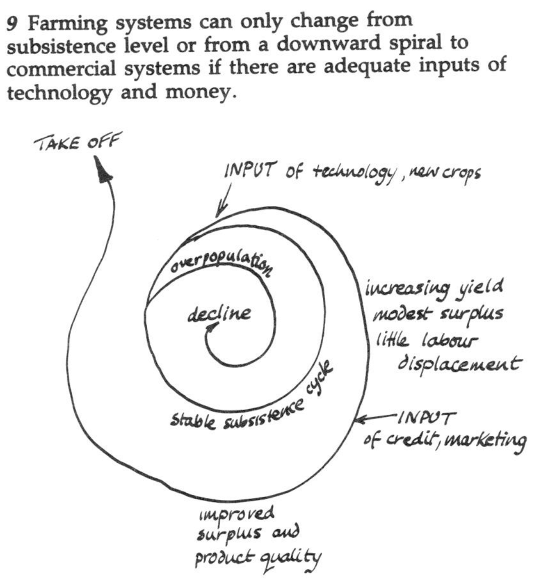

Von Thünen's rural land-use model

1 Land use changes with distance from the *farm*, activities requiring the most attention occurring near the farm.
2 Land use changes with distance from *market* (Von Thünen).

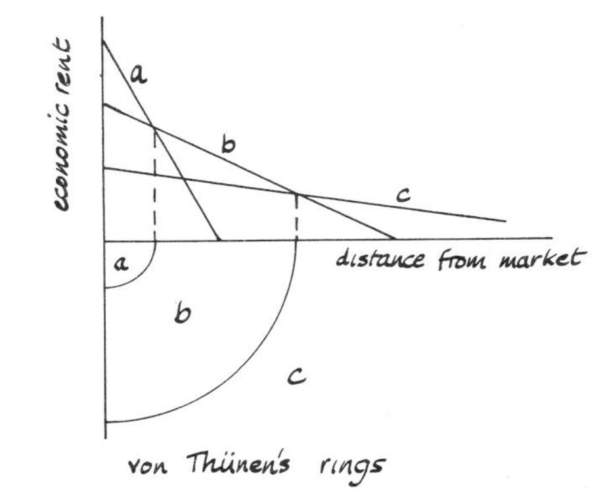

3 In any given field one may expect to find the *type of farming* that yields the highest net return of money per hectare of land.
4 Assuming that the cost of cultivation is constant for all crops, it is only the *cost of transport* that reduces the profit.
5 Costs vary with each type of product, each declining at different rates from the market.
6 Land uses change when costs of transport exceed profit forming rings centred on the local market.

7 On uniform land supplying only one market, variations in transport costs will produce a pattern of land-use rings, each ring enclosing a different type of farming product.
8 The simple ring pattern can be distorted:

(a) by variations in land quality;
(b) by variations in transport access to market;
(c) by farmers choosing different profit levels and so changing their product (those wanting the maximum possible versus those satisfied with less);
(d) by the desire of farmers to diversify in case one product fails;
(e) by the influence of government subsidies.

Part I Population

1 The *distribution* of the world's people is influenced:

(a) by physical factors;
(b) by economic factors;
(c) by cultural factors;
(d) by political factors;
(e) by disasters.

2 Regions of high and low population can also be correlated with the periods for which they have been occupied. Asia and Africa are the source regions for man; America has only recently been colonised.
3 The world can be classed into four *population–resource–technology regions*:

(a) high-technology: either low population–resource ratio (USA type) or high population–resource ratio (British type);
(b) low-technology: either low population–resource ratio (Brazil type) or high population–resource ratio (Egypt type).

The Egypt type offers the lowest prospect for improved living standards.

4 Populations can be described by *age–sex pyramids*. Usually pyramids are:

(a) progressive, when populations are expanding;
(b) stationary;
(c) regressive, when populations are declining.

Type (a) is common in the developing world; types (b) and (c) in the developed world.

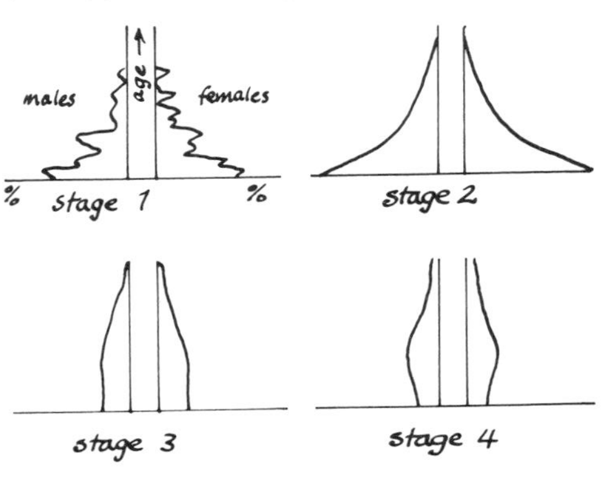

Population growth

1 *Malthus* suggested that populations grow *geometrically* but that food supply grows *arithmetically*. Sooner or later there will be a world food crisis.

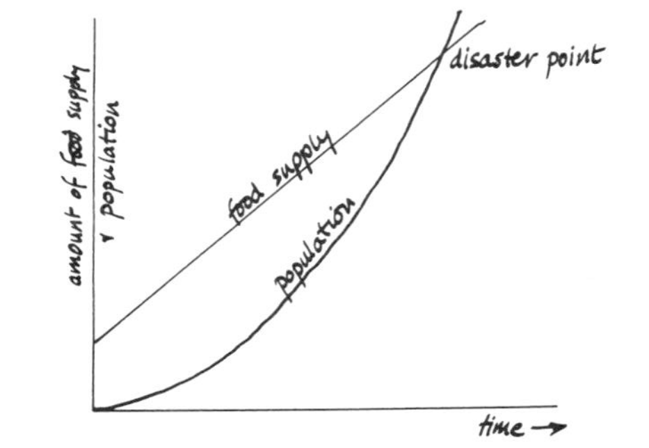

2 Populations rate controlled by the rate of births and deaths. At present the *world population* has a net birth rate of over 2 per cent and will double to 8 billion in 35 years if this rate is maintained.
3 With progress from a rural subsistence to an urban, developed society, populations undergo a *demographic transition*.

4 There are four stages to the demographic transition:

(a) 'primitive' (high birth rate, high death rate, little growth);
(b) 'early expanding' (high birth rate, lowering death rate);
(c) 'late expanding' (lowering birth rate, low death rate);
(d) 'advanced' (low birth and death rates; little growth).

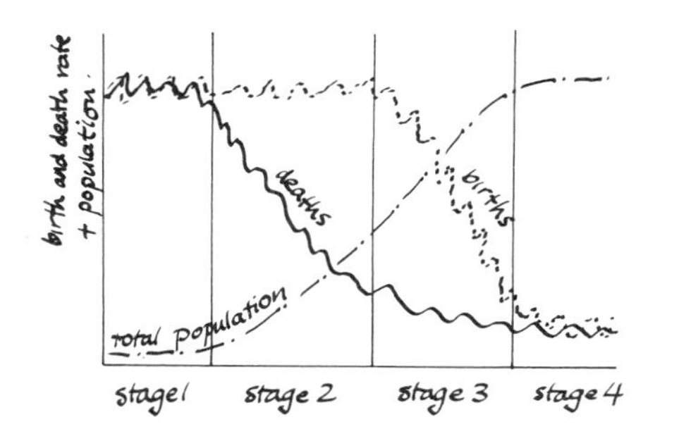

5 The demographic transition accompanied the *Industrial Revolution* in Europe and the growing population could find work. Deaths decreased slowly, the growth rate never exceeded 2 per cent, and the populations were initially low.
6 The demographic transition has begun very suddenly in the developing world. Improved nutrition and medical care have caused death rates to drop rapidly . Populations are growing by up to 4 per cent. There will be little work for the larger population in an age of *mechanisation*.
7 Each stage of the transition has a characteristic *population pyramid*.
8 At present *agricultural revolutions* have kept food supplies in step with population on a world level, although *regional shortages* and *famines* still occur.

Migration

1 *Migration* can be:

(a) by drift;
(b) by compulsory movements;
(c) by voluntary movements.

2 Migration can be due to *push factors* (no land; overpopulated rural or urban areas; unemployment) and *pull factors* (better wages and conditions; improved status).
3 Laws of migration:

(a) most people move short distances;
(b) migration occurs in steps;
(c) long-distance migration is to large urban areas;
(d) urban dwellers migrate more than rural dwellers;
(e) men migrate greater distances than women.

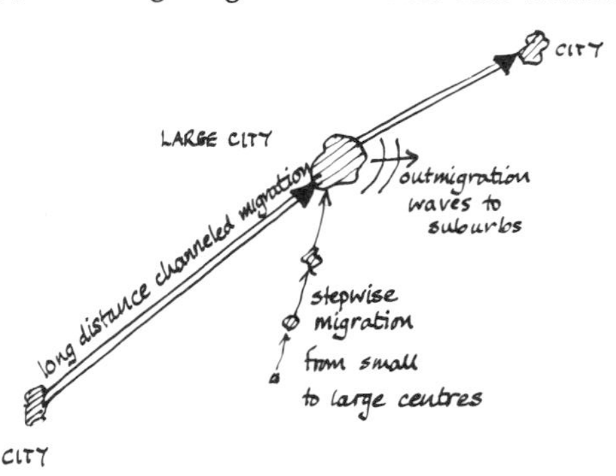

THE PATTERNS OF MIGRATION

4 A nation gains migrants from countries less developed than itself and sends migrants to countries more developed than itself.

5 Vance's *social mercantile model* suggests that migration occurs in two stages:

(a) exploration and reporting back; gathering in coastal social entrepots; no mixing;
(b) new exploration within the new country; development of new interior entrepots; start of diffusion of immigrants.

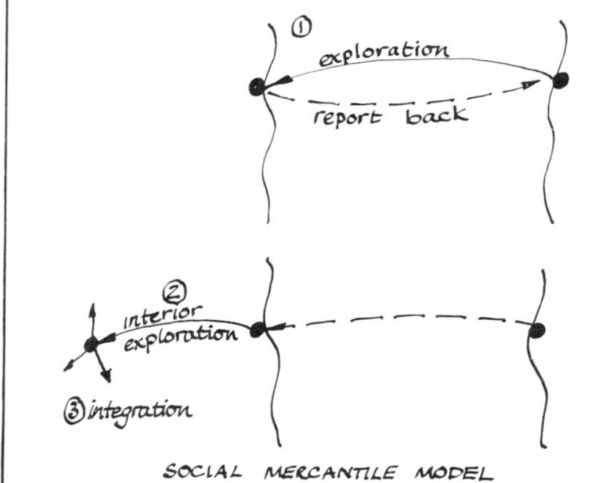

SOCIAL MERCANTILE MODEL

POPULATION STRUCTURES

6 Migration distorts the population structures of exporting and importing countries.
7 *Internal migration* is:

(a) rural–urban: people seeking jobs;
(b) urban–urban: people seeking jobs;
(c) urban–rural (more wealthy in the developed world);
(d) urban–rural and urban–coastal: retired people (in the developed world).

8 Governments have tried to counter *internal* migration because it causes stress in areas of gain (increased housing need, greater demand on social services, congestion, and job competition) and also causes stress in areas of loss (depopulation and loss of rural identity, loss of only part of the population, usually the most active).

Urban activity

1 Urban activity is partly related to the location of natural resources and of sources of energy.
2 Energy resources come mainly from the indirect sources of fossil fuels, of which oil is overwhelmingly the most important.
3 Wood and charcoal still meet 80 per cent of rural needs in the developing world. This has led to deforestation and a fuelwood crisis.

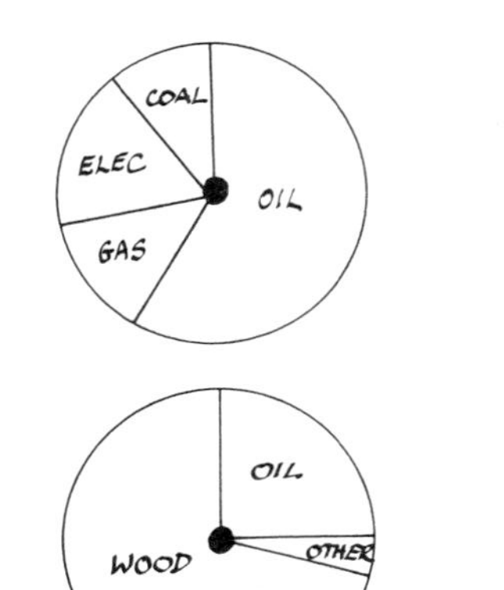

9 Governments have tried to counter *international* migration because:
(a) it increases social stress on areas of gain when jobs are scarce;
(b) it adds to the less-skilled sector;
(c) it removes the most skilled sector.

Part J Urban Activity

4 Communications consist of *networks* of *nodes* joined by *links*.
5 Transport systems have varying cost–distance curves.
6 Routes are designed:

(a) to be the shortest distance between points;
(b) to maximise traffic;
(c) to optimise distance and traffic.

The actual route is often a compromise between the conflicting needs of the builder and of the user.

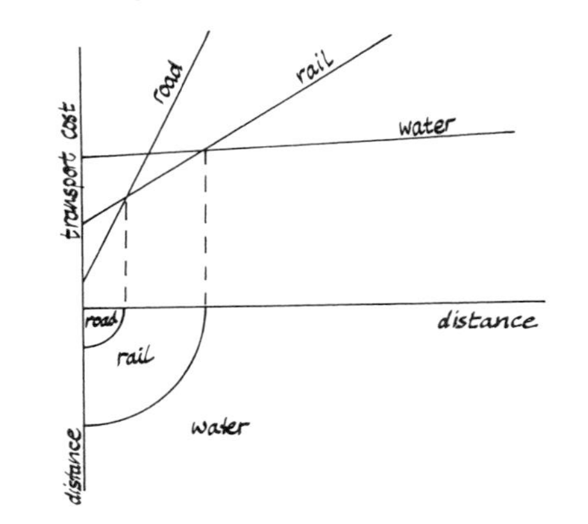

7 Networks can be analysed according to their degree of *connectivity*, using:

(a) the index;
(b) the König number;
(c) the Shimbel-Katz index.

10 In Britain migration has been:
(a) north to south;
(b) urban-core to urban-periphery (suburbs);
(c) rural-periphery to urban;
(d) urban to coast.

8 City shape is affected by transport systems. *City models* include:

(a) the 'walking city';
(b) the 'public-transport city';
(c) the 'rubber city';
(d) the 'motorway city'.

9 Trade develops by the *law of comparative advantage*, using specialisation of products.
10 Trade barriers go against the law of comparative advantage, and *trading blocs* need to be large to be successful.

Industry

1 Industrial activity forms the economic base on which the *living standard* of countries depends.

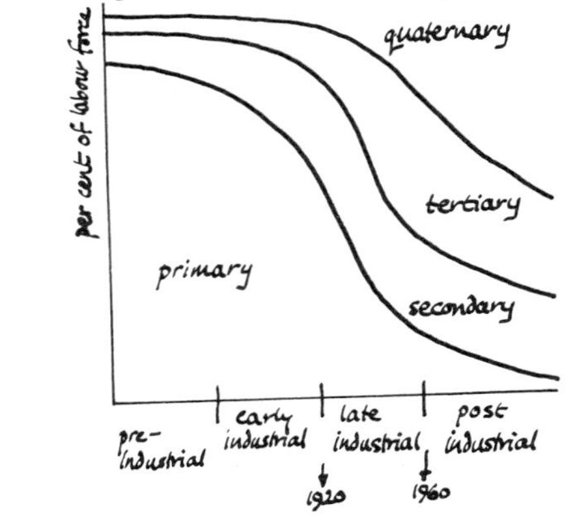

2 There are four *types of industry:*

(a) *primary* (extractive);
(b) *secondary* (manufacturing);
(c) *tertiary* (service);
(d) *quaternary* (office).

3 Developed countries in which (c) and (d) are larger employers than (b) are in a *post-industrial phase.*
4 *Weber* suggested that manufacturing industry would locate at a point where the combined transport costs are at a minimum.

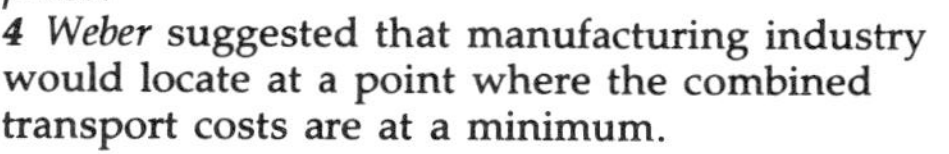

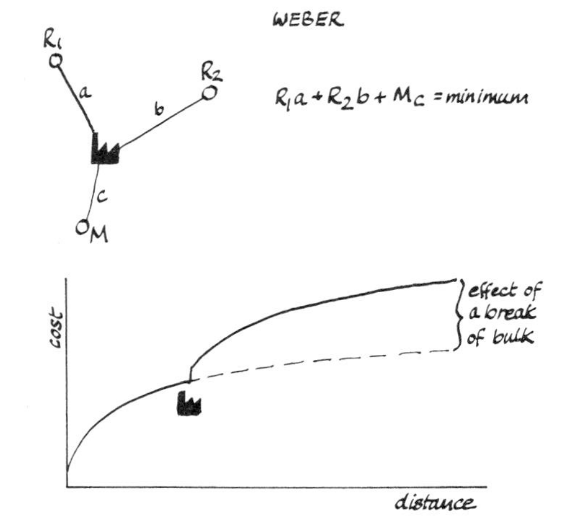

5 Break-of-bulk points add loading and unlcading costs.
6 Short journeys use transport at its highest cost rate.
7 Lösch suggested that industries will locate in areas where they can derive greatest revenue by being central to their markets.
8 Industrial plant often gains *economies of scale* because it allows:

(a) **specialisation by each worker;**
(b) **larger, more economic sizes of machines;**
(c) **integration of various processes.**

9 When plants become too large the average distance to the market is so great that there are transport cost penalties. Therefore there is a certain most economic size in each area of manufacturing.

10 Large free-standing plants gain economies internally.
11 Medium-sized plants tend to swarm together because they are dependent on each other.

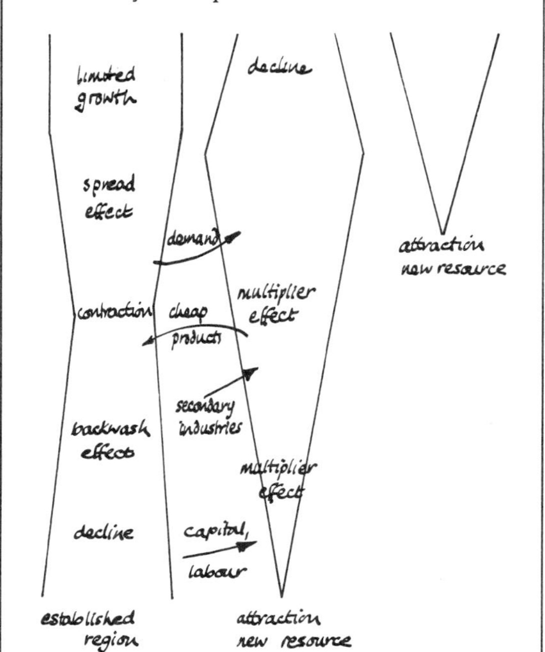

12 *Industrial location* is often due to inertia and locational reasons that were important in the past but that are often not obvious today.

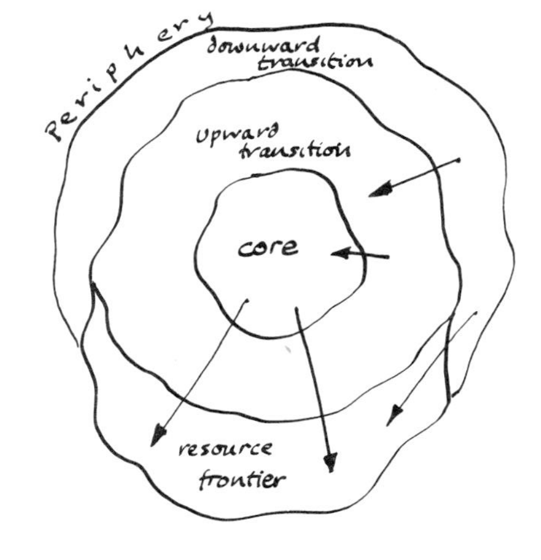

13 *Individual products* go through a sequence of evolutionary change from:

(a) initial innovation near the firm's headquarters;
(b) growth to export and mass production in less costly peripheral areas;
(c) maturity, in which the product is made in very low-cost areas to counter competition.

14 *Multinational companies* are most likely to open and close factories to keep costs down. This may result in local hardship when people are made *redundant.*
15 Industry will respond to social and economic pulls in the future:

(a) to cause less environmental pollution;
(b) to reduce energy costs;
(c) to integrate land uses (social, housing, industrial);
(d) to grow in specialised clusters;
(e) to cater for the social needs of employees;
(f) to meet technological changes.

16 Today not only are *services* the largest sector of industrial activity in the world, they can directly lead to growth.

17 Many countries are moving towards a *post-industrial society*.
18 The cities of the developed world are experiencing *deindustrialisation*.
19 Services can be both dependent and self-supporting; *tourism* can use natural resources as a raw material.
20 Services are located close to their markets.
21 The 'head office' function is less important than in manufacturing; most of the firm's assets are at the selling points and there is widespread dispersal from CBDs.

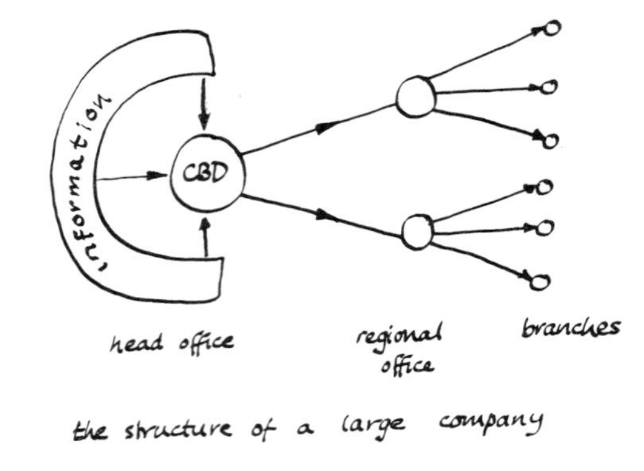

the structure of a large company

Part K Settlement Patterns

Rural settlement

1 A *settlement* is a place where a group of people live more or less permanently.
2 Present-day settlements are nearly all the result of a long heritage of response to environmental, social and economic change.
3 In Britain many settlements date back to the Dark Ages when people were acutely aware of their environment.
4 *Land clearance* around a settlement was often circular to minimise travelling distance. In a homogeneous landscape this eventually produced a uniformly spaced settlement pattern, with territories later being confirmed as parishes.
5 Shared landholding and a system of rotational fallow led to a *nucleated* settlement, centrally placed within the fields.
6 In heterogeneous land, settlements were often located according to frequency of access to different type of land, nearness of a water supply, or availability of a dry site away from floods.
7 The *density of settlement* is largely related to land quality.

8 Settlements have a variety of forms:

(a) *nucleated*;
(b) *open* (green), associated with pastoralism;
(c) *linear*, dicatated by topography or routeways.

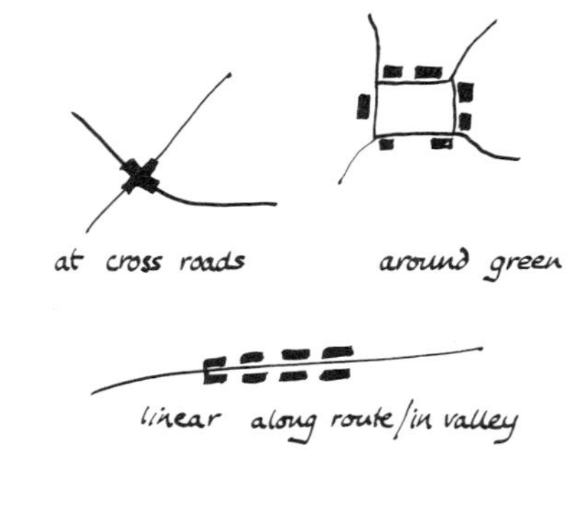

VILLAGE SHAPES

9 Since *enclosures* rural settlements have taken on a new role as farmers have built houses on their farmland. Most rural settlements are now small service centres.

10 Worldwide rural settlement patterns depend on:

(a) the density of population;
(b) the actual area exploited;
(c) the nature of the rural economy.

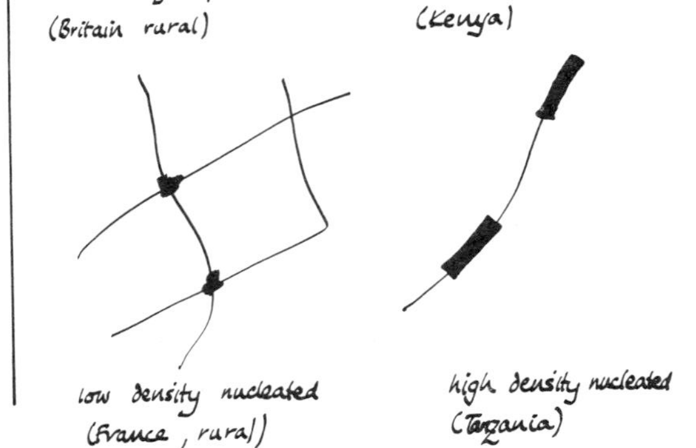

11 There has been a large swing from rural to urban areas due to *push factors*:
(a) absolute decline in numbers required to farm the land;
(b) absolute increase in rural population;
(c) change from rural (cottage) to urban (factory) industry;
and to *pull factors:*
(a) increasing employment in towns;
(b) better working conditions and pay.

Urbanisation

1 Towns grow as people with specialised skills and goods seek a central place to have access to as wide a market (*hinterland*) as possible.
2 Market towns grow in two stages:

(a) randomly spaced with individual variation, unrelated to the overall needs of a successful trading network;
(b) selection by competition to produce organisation and self-regulation.

3 Urban areas have two separate functions:

(a) local trade;
(b) inter-regional trade.

Growth can only be generated by inter-regional trade.

4 Small market towns are still dominated by local trade and have little industry. This is why they have not grown and have changed little over the centuries.
5 The *spacing of trading centres* relates to population density and transport available (compare India and USA).
6 Urban growth was related to industrialisation in the developed world.
7 Urban growth can only follow an agricultural revolution and improvements in transport which allow food to reach the city.

8 Rural settlements were designed to provide homes for those working on the surrounding land; *urban settlements* provide homes for those engaged in manufactures and services.

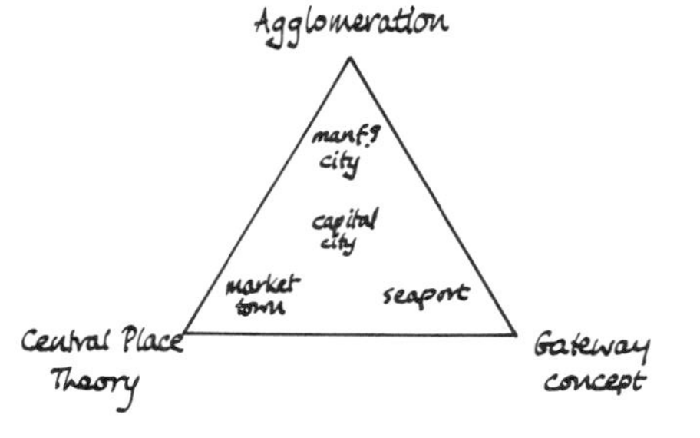

9 Britain's settlement network evolved in stages:

(a) primary colonisation;
(b) complete cover;
(c) gradual emergence of larger settlements;
(d) stabilisation after competition.

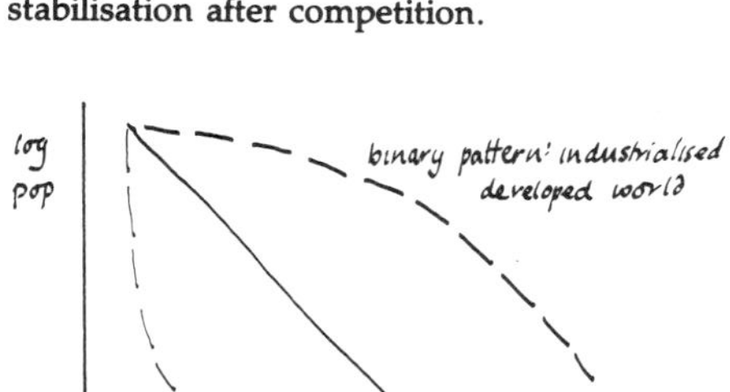

10 The rank–size rule shows a logarithmic relationship between settlement size and rank.

11 Christaller's central-place theory considers a stable society evolving without inter-regional trade. There is a hierarchy of trading settlements nested as hexagonal trade areas ($K=3$: trade; $K=4$: transport; $K=7$: administration).

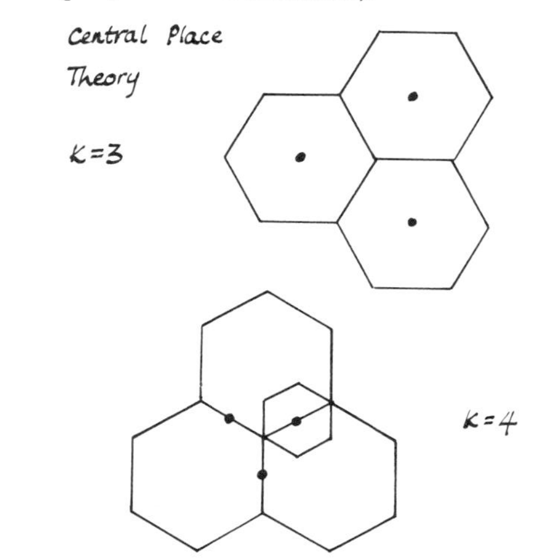

12 Lösch modified the Christaller network by assuming that settlements are influenced not by one economic pull but by many. His pattern:

(a) gives 'city rich' and 'city poor' sectors;
(b) produces a nearly continuous sequence of settlement sizes;
(c) allows settlements of the same size to have different functions.

13 Large urban areas contain more than a trading function.
14 Large urban areas depend on inter-regional trade and are not dependent on the local hinterland.
15 Inter-regional trade causes the growth of urban areas that will not fit into central-place theory.

16 Urban areas growing on inter-regional trade vary worldwide according to Vance's mercantile model.

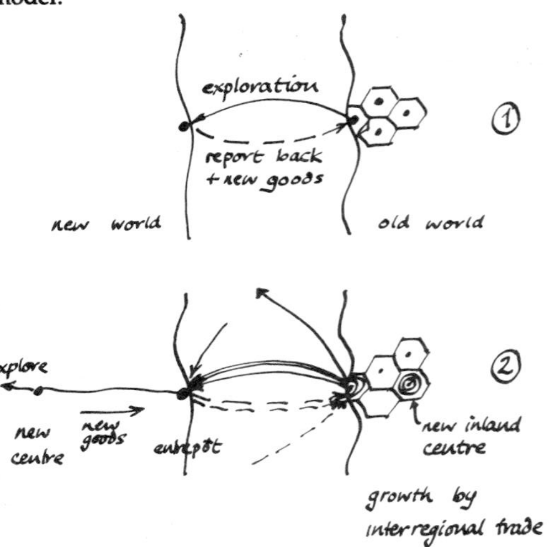

17 Agglomeration causes a *multiplier effect* on population and services.
18 A conurbation is a group of urban areas with interlocked *urban fields*.
19 In developing-world countries few urban centres are needed and a *primate city* develops.
20 Change from developing to developed society results in the city pattern of a country changing from primate to rank–size rule to binary.

The shape and structure of cities

1 Pre-industrial cities were small; government and religious activities dominated the central core. Trade was only a minor activity. The rich lived in the centre, the poor in outer zones.
2 Pre-industrial cities only survive as long as most people use the land as a resource.
3 Industrial cities developed by:

(a) the rise of trade;
(b) concentration of political power;
(c) the growing idea of orderliness.

4 Growth based on agriculture is slow; growth based on manufactures can be rapid.
5 The location of a shop, office or factory within the urban structure is often of vital importance in ensuring their success. The pattern of resulting competition is described by *bid-rent analysis*.

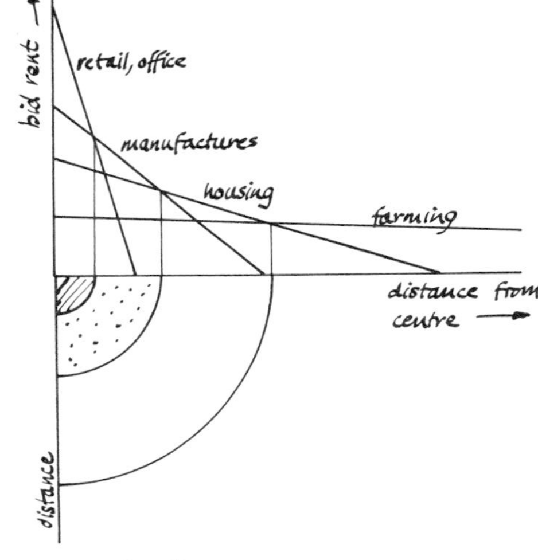

6 Bid-rent assumes that at any location the land is used by the function that will derive most benefit.
7 The bid-rent sequence is: retail, wholesale, manufacturing, housing.
8 Bid-rent works not just from the centre but also from suburban route-centre nodes on ring roads, and the like.
9 Bid-rent analysis assumes that there is no government interference (planning) and that land is of uniform quality.
10 A city is in a state of dynamic equilibrium representing a balance between *centrifugal forces* and *centripetal forces*.
11 Centrifugal forces include:

(a) lack of space to expand;
(b) lack of a suitably shaped site;
(c) high bid-rents;
(d) congestion;
(e) government planning;
(f) the desire for a more spacious living environment.

12 Centripetal forces include;
(a) central-area prestige;
(b) high accessibility;
(c) advantages from clustering;
(d) the city environment's nearness to entertainments.

13 There are three basic city-structure models, derived from observation:

(a) Burgess's concentric-ring model;
(b) Hoyt's sector model;
(c) Harris and Ullman's multiple-nuclei model.

14 The sector model best fits towns in which there are contrasts in land quality.

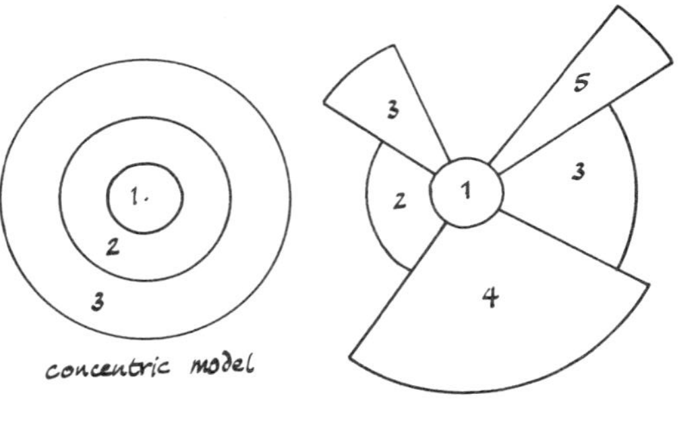

15 The multiple-nuclei model is suited to the complex structure of large urban areas with more than one well-defined centre.
16 Very large urban areas tend to break up into *cells* of common function.
17 Offices cluster in the CBD for personal contact.

18 Large cities offer accessibility both to sources of new information and to branches scattered throughout the country.
19 *Industry* does not need a central location but does need an adequate communications network which allows raw materials to be brought in and goods to be efficiently despatched.
20 Central areas are expensive and congested.
21 Peripheral *trading* or *industrial estates* offer many advantages.
22 Industry is rapidly leaving city centres and moving to the periphery.
23 The population of a city has everyday domestic needs that will support a *local shopping area;* more specialised needs that will support a *neighbourhood shopping centre;* and very specialised needs that will support *city-centre shops.* This gives a hierarchical pattern.
24 Shopping-area quality and shape vary with the mobility and purchasing power of the neighbourhood.
25 *Suburbs* grew by outward leapfrogging as communications improve first because of railways, then because of roads.
26 Planning has produced areas zoned according to their functions.
27 Planned urban renewal of the old industrial-housing areas led to *the crisis of the inner city.*
28 The *green belt* and *new towns* were an attempt to limit city growth and to create new, self-contained towns housing the city overspill population.

Developing-world urbanisation

1 Developing-world cities are remarkable because of their rate of urbanisation.
2 They are poorly served by transport and do not develop zoned structures as do developed-world cities.
3 Developing-world cities mostly grow by *marginal accretion* of *spontaneous settlements* which gradually stabilise.
4 Such structure as exists has the 'Western' CBD-core mixed together with a crowded inner city of poor people; then relatively affluent inner suburbs; finally an outer area of high-density *shanty towns.*
5 The élite often concentrate in a *spine* near the better communications. This is also the zone that contains industry.

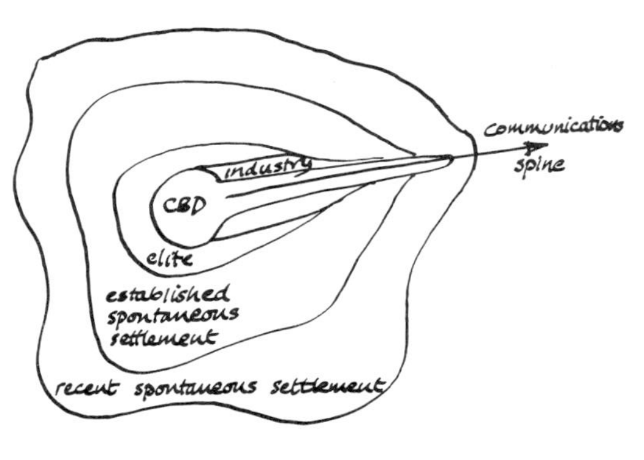

Third world city model

6 CBD, industry and high-quality housing often expand in the same direction.
7 There is no outward growth in developing cities because there is no cash-based economy to provide the money for people to move from one standard of housing to another.

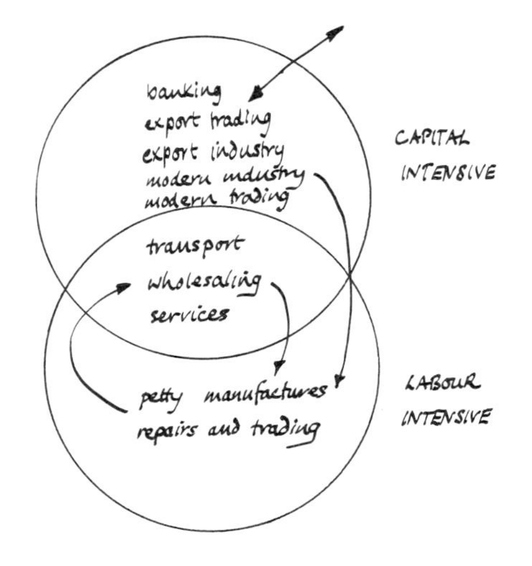

Dual circuit economics of the third world

Further Reading

Physical geography

General

Bradshaw, M. 1978. *The Earth's changing surface*. Sevenoaks: Hodder & Stoughton.
Hilton, K. 1979. *Process and pattern in physical geography*. Slough: University Tutorial Press.
Strahler, A. and A. 1979. *Elements of physical geography*. Chichester: Wiley.
Whittow, J. W. 1984. *Penguin dictionary of physical geography*. London: Penguin.

Geomorphology

Finlayson, B. and I. Statham 1980. *Hillslope analysis*. London: Butterworth.
Small, R. 1978. *The study of landforms*, 2nd edn. Cambridge: Cambridge University Press.
Sparkes, B. 1972. *Geomorphology*. London: Longman.
Thomas, M. 1974. *Tropical geomorphology*. London: Macmillan.
Weyman, D. 1981. *Tectonic processes*. London: George Allen & Unwin.
Weyman, D. and V. 1977. *Landscape processes*. London: George Allen & Unwin.
Young, A. 1972. *Slopes*. London: Longman.

Hydrology and meteorology

Chandler, T. 1972. *Modern meteorology and climatology*. Walton-on-Thames: Nelson.
Hanwell, J. 1980. *Atmospheric processes*. London: George Allen & Unwin.
Knapp, B. J. 1979. *Elements of geographical hydrology*. London: George Allen & Unwin.
Miller, A. and M. Parry 1975. *Everyday meteorology*. London: Hutchinson.
Riley, D. and H. Spolton 1974. *World weather and climate*, 2nd edn. Cambridge: Cambridge University Press.
Smith, D. I. and P. Stopp 1978. *The river basin*. Cambridge: Cambridge University Press.

Soils and vegetation

Bridges, M. 1978. *World soils*, 2nd edn. Cambridge: Cambridge University Press.
Knapp, B. J. 1981. *Soil processes*. London: George Allen & Unwin.
Hudson, N. 1971. *Soil conservation*. Cambridge: Cambridge University Press.
Simmons, I. 1982. *Biogeographical processes*. London: George Allen & Unwin.
Simmons, I. 1979. *Biogeography: natural and cultural*. London: Edward Arnold.

Man and the environment

Drew, D. 1983. *Man–environment processes*. London: George Allen & Unwin.
Goudie, A. 1981. *The human impact*. Oxford: Basil Blackwell.
Slater, F. (ed.) 1985. *People and environment*. London: Collins.
Tivy, J. and G. O'Hare 1981. *Human impact on the ecosystem*. Edinburgh: Oliver & Boyd.
Watson, J. 1983. *Geology and man*. London: George Allen & Unwin.

Human geography

General

Briggs, K. 1982. *Human geography*. Sevenoaks: Hodder & Stoughton.
Brunn, S. and J. Williams 1983. *Cities of the World*. New York: Harper & Row.
Carter, H. 1981. *The study of urban geography*. London: Edward Arnold.
Chapman, K. 1979. *People, pattern and processes*. London: Edward Arnold.

Haggett, P., A. Cliff and A. Frey 1977. *Locational analysis in human geography*. London: Edward Arnold.
Johnson, J. H. 1972. *Urban geography*, 2nd edn. Oxford: Pergamon.
Lloyd, E. and P. Dicken *Location in space*, 2nd edn. New York: Harper & Row.
Whynne-Hammond, C. 1985. *Elements of human geography*, 2nd edn. London: George Allen & Unwin.

Third World

Adejuwon, J. O. 1979. *An introduction to the geography of the Tropics*. Walton-on-Thames: Nelson.
Barke, M. and G. O'Hare 1984. *The Third World*. Edinburgh: Oliver & Boyd
Bromley, R. and R. Bromley 1982. *South American development*. Cambridge: Cambridge University Press.
Chisholm, M. 1982. *Modern world development*. London: Hutchinson.
Mountjoy, A. 1982. *Industrialization and developing countries*. London: Hutchinson.
Roberts, B. 1978. *Cities of peasants*. London: Edward Arnold.
Turner, B. 1983. *Africa*. Harlow: Longman.
World Bank. *Annual reports*. Oxford: Oxford University Press.

Industrial location

Bale, J. 1981. *The location of manufacturing industry*. Plymouth: Macdonald & Evans.
Daniels, P. 1982. *Service industries: growth and location*. Cambridge: Cambridge University Press.
Day, M., I. Meyer and S. Day 1983. *Industrial location*. Walton-on-Thames: Nelson.
Goddard, J. 1975. *Office location in urban and regional development*. London: Oxford University Press.
Hoare, A. G. 1983. *The location of industry in Britain*. Cambridge: Cambridge University Press.
Jarret, H. 1977. *A geography of manufacturing*. Plymouth: Macdonald & Evans.
Sant, M. 1975. *Regional disparities*. London: Macmillan.

Agriculture

Bayliss-Smith, T. *The ecology of agricultural systems*. Cambridge: Cambridge University Press.
Grove, A. T. and F. M. G. Klein 1979. *Rural Africa*. Cambridge: Cambridge University Press.
Newbury, P. A. K. 1980. *A geography of agriculture*. Plymouth: Macdonald & Evans.
Ruthenberg, H. 1978. *Farming systems in the Tropics*. Oxford: Oxford University Press.

Population

Clarke, J. 1982. *Population geography*, 2nd edn. Oxford: Pergamon.
Hornby, W. F. and M. Jones 1980. *An introduction to population geography*. Cambridge: Cambridge University Press.

Transport

Robinson, R. 1977. *Ways to move*. Cambridge: Cambridge University Press.

INDEX

ACKNOWLEDGEMENTS

The compilation of this book involved numerous individuals and institutions who generously gave of their time. In particular, I would like to thank the following for commenting on the typescript in its various stages and providing material: Don Agerskow, Mike Hart, Harold Carter, Alan Gilbert, Ken Sealey, Bryan Roberts, Roland Goldring, Maurice Parry, Stanley Schumm, Jo Anderson, Dick Chorley, Chris Wilson, Bruce Atkinson, Bruce Webb, Denys Brunsden, Ian Douglas, David Sugden, Julian Orford, Bernard Smith, Bob Evans, Ian Simmons, Richard Munton, Philip Ogden, Peter Dicken and Brian Robson. The diagrams owe much to: the artistic efforts of Duncan McCrae who helped to formulate my two-dimensional ideas into something that made sense in three dimensions; Harry Walkland, who processed many of the photographs; and Chris Howitt, who produced the illustration lettering and photographic prints. The whole project would never have reached completion without the patient help and encouragement of Roger Jones, Geoff Palmer, and others of the publishing staff.

I would like to thank the following individuals and organisations who have given permission for the reproduction of copyright material:

Hachette Guides Bleus (1.3); Figures 1.7, 1.9, 1.10, 1.11, 1.12, 2.2, 2.3, 2.4, 30.6 and 30.7 reproduced from *Scientific American* by permission of W. H. Freeman; D. Weyman (1.8, 1.13, 1.15); Wiley (1.14, 1.15); J. Watson (1.16); USGS (2.1, 2.4, 2.5, 2.6, 2.7, 2.9, 2.13, 11.3, 11.6, 12.9, 14.9b, 15.1, 15.5, 18.5, 21.1, 33.13, 38.12, 43.2, 44.4, Part A title); Royal Geographical Society (2.11, 2.12).

European Space Agency (3.3, 3.6, 5.6); M. Parry (3.13); A. N. Strahler and Wiley (3.14, 5.21); A. Miller, M. Parry and Hutchinson (4.1); Methuen (4.3, 5.20); Nelson and T. Chandler (4.4); Hutchinson (4.7, 35.1, 36.3, 36.4, 36.5, 36.6); Edward Arnold (5.5, 14.19, 16.6, 30.1, 30.2, 30.3, 32.3, 34.4, 39.14, 44.13, 45.2, 45.3, 45.21); R. Barry, R. Chorley and Methuen (5.13, 5.19); Electronics Laboratory, University of Dundee (5.15, 5.22, Part B title); J. Gale (6.1); World Meteorological Organisation (6.1, 6.3, 6.5); Joint Matriculation Board (6.4); W. Kellog (6.7); Friends of the Earth (6.8b).

S. M. Burn (7.8d); N. Salisbury (8.14); Figures 8.4, 21.20, 22.4, 24.6 and 24.15b reproduced by permission of the Director, British Geological Survey: Figure 8.6 reproduced from the *Geological Society of America Bulletin* (1962) by permission of the Society, R. J. Chorley and M. A. Morgan; D. Brunsden and the Geographical Association (8.8); S. A. Schumm and R. Parker (8.9); S. A. Schumm and the Institute of British Geographers (8.11, 16.3); S. A. Schumm (8.12, 9.1d,e); W. H. Freeman (9.2, 9.11); E. A. Keller (9.7, 31.5); United States Corps of Engineers (9.9c, 10.3); C. Harris (9.12); A. Schumm and Water Resources Publications (Table 9.1); The Editor, *The Denver Post* (11.3b); G. Hollis and the Institute of British Geographers (11.5); Institute of British Geographers (11.5b, 12.14, 13.7, 34.5, 45.13–18).

The Norwegian Geotechnical Institute (13.5b,c); University of Lille (13.10); Gebrüder Borntraeger (13.12, 25.3, 25.4, 25.6); D. Walling and the Institute of British Geographers (13.13); Institute of Civil Engineers (14.9a, 14.10); Cambridge University Collection (14.11, 21.13b, 21.15, 21.18b, 21.22, 24.15a, 24.23, 24.25a, 29.3, 29.4); D. Brunsden and the Royal Society (14.12, 14.20); Cambridge University Press (14.17); Geological Society of London (15.3, 18.3, 23.13, 24.21); D. Brunsden and the Geological Society of London (15.4, 15.7, 18.2).

ARDEA (17.1a); M. Thomas and Macmillan (17.3); I. Douglas (17.4); Macmillan (17.5); A. Schick (17.7); Bernard Smith (17.9); Batsford (17.10a); R. U. Cooke (17.10c); R. Jones (17.11); CBS College Publishing (17.13); Zoological Society of London (17.15).

Popperfoto (19.1); Oxford University Press (19.3, 20.4, 20.10, 20.13, 20.16, 32.9, 33.6, 33.7, 33.26, 33.27, 33.28, 45.11); T. Péwé (19.2, 20.1, 20.9d, Part F title); R. J. E. Brown and T. Péwé (20.2); S. Rudberg (20.7, 20.18); Geodaetisk Institute, Denmark (21.5); A. Dreimanis, U. Vagners and the Arctic Institute of North America (21.9); G. Boulton (21.11); T. Péwé and the Alaska Division of Geological and Geophysical Surveys (22.1, 22.2); British Petroleum (22.3).

Andrew Besley (23.1); J. Orford (23.7, 23.8); Aerofilms (24.5, 33.21); W. Bascom and the American Geophysical Union (24.8); US Army Corps of Engineers (24.10); J. R. L. Allen (24.16, 24.17a).

B. Whalley (27.5); J. Anderson and J. Stout (27.8); I. Fenwick (28.4); A. Rapp (29.5, 29.7); University of Reading (29.8); Elsevier (31.1); Greater Manchester Council (31.2); Blackie (31.3, 31.4); E. Keller (31.5); Crown copyright (31.6b); Countryside Commission (31.6a); Institute of British Geographers (31.7, 34.5, 45.13, 45.14, 45.16, 45.17, 45.18); E. P. Odum and CBS College Publishing (32.1); V. Collinson (32.16); H. Sergeant (33.2, 37.2); The Editor, *Geographical Review* (33.4, 33.5); Grant Heilman (33.19a); United States Department of the Interior (33.19c); Association of American Geographers (33.20); Mark Edwards/Earthscan (33.23, 33.24, 38.9, 39.1, 46.17); M. Chisholm and Hutchinson (34.1).

A. A. Miller and Methuen (35.2); D. McCrae (35.3, 42.2); Figure 35.4 reproduced from M. Hansen and O. Dunea, *The study of population*, by permission of the University of Chicago Press (35.4); W. Zelinsky (35.5); World Bank (36.1, 36.9); Office of Population Surveys (36.7, 37.9, 37.10, Table 37.2); J. Vance (37.3, 37.4, 39.9, 44.16); P. Jones (37.5); Heinemann (37.6, 37.12).

J. Watson (38.1, 38.13); Marcos Satili/Earthscan (38.4); Longman (38.5, 46.1, 46.3, 46.16); Swedish Hydroelectricity Board (38.11); W. Bunge (39.6, 39.7); Professor Tamsma (39.12b); Dennis Wompra Studios (40.8); J. Bale and Oliver & Boyd (40.10); Urban Land Institute, Washington (40.19); MFI (40.20); School of Geography, University of Manchester (41.1, 41.3, Table 40.2); P. Davis and Wiley (41.2); M. Sant and Macmillan (42.7); Figures 42.12, 43.15, 43.16, 43.17, 43.18, 43.19, 43.20 and 44.3 reproduced by permission of the

Copyright Controller, HM Ordnance Survey, Crown copyright reserved; Geographical Association (42.14); Government of Kuwait (42.16).

Eurostat (43.1); B. K. Roberts and GeoBooks (43.4, 43.15); Government of Cameroon (43.5); Figures 43.6 and 44.5 reproduced by kind permission of the Director of Surveys, Kenya; The Editor, *Geografiska Annaller* (43.9); Topografische Dienst, Delft (43.11, 43.12); Institut Géographique National (43.11); Figure 43.13 reproduced with permission A44/84 of Geodaetisk Institut, Denmark; Taunton and District Civic Society (44.1); D. Preston and GeoBooks (44.7); J. Johnson and Pergamon Press (44.8, 44.9, 44.12, 44.17); Geographical Association (44.15); Clyde Surveys (45.1); Department of Geography, University of Chicago (45.4); The Editor, *The Economist* (45.5); J. B. Goddard and Oxford University Press (45.12); University of Cambridge Local Examinations Syndicate (45.19); Wimpey Homes (45.23b); Harlow Development Corporation (45.24); D. Dwyer and Longman (46.1, 46.3); Sanjay Acharya/Earthscan (46.2b); Edwin Huffman/World Bank (46.4a); UNICEF/Dutt (46.4b); Cambridge University Press (46.5, Table 40.2); Harper & Row (46.10, 46.15, 46.18, 46.19); Felix Tisnes/Earthscan (46.20).

All remaining photographs were supplied by Earthscope

The following kindly gave permission for the reproduction of boxed material:

The Editor, *The Guardian*, and Peter J. Smith (Ch. 2); J. Andrews (Ch. 12); The Editor, *Daily Telegraph* (Ch. 15); The Editor, *The Sunday Times* and the Editor, *The Farmer's Gazette* (Ch. 31); The Editor, *The London Standard* (Ch. 39); The Editor, *The Guardian* (Ch. 41); The Editor, *The Economist* (Ch. 45).